RESERVOIR ENGINEERING

HANDBOOK

Second Edition

RESERVOIR
ENGINEERING
HANDBOOK

Second Edition   **Tarek Ahmed**

Gulf Professional Publishing
Boston • London • Auckland • Johannesbourg • Melbourne • New Delhi

Gulf Professional Publishing is an imprint of Butterworth-Heinemann.

Copyright © 2001 by Butterworth-Heinemann

 A member of the Reed Elsevier group

Previously copyrighted © 2000 by Gulf Publishing Company, Houston, Texas

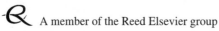
Library of Congress Cataloging-in-Publication Data
Ahmed, Tared H., 1946-
   Reservoir engineering handbook / Tarek Ahmed.
      p.cm.
   Includes bibliographical references and index.
   ISBN 0-88415-770-9 (alk. paper)
   1. Oil reservoir engineering. 2. Oil fields. 3. Gas reservoirs. I. Title.

TN871 .A337 2000
622'.3382--dc21

                                                              99-005377

British Library Cataloguing-in-Publication Data
A catalogue record for this book is available from the British Library.

The publisher offers special discounts on bulk orders of this book. For information, please contact:

Manager of Special Sales
Butterworth-Heinemann
225 Wildwood Avenue
Woburn, MA 01801–2041
Tel: 781-904-2500
Fax: 781-904-2620

For information on all Butterworth-Heinemann publications available, contact our World Wide Web home page at: http://www.bh.com

10 9 8 7 6 5 4 3 2 1

Printed in the United States

To my gorgeous wife Shanna,

And my beautiful children

Jennifer

Justin

Brittany

Carsen

CONTENTS

Viscosity, 108; Methods of Calculating Viscosity of the Dead Oil, 109;
Methods of Calculating the Saturated Oil Viscosity, 111;
Methods of Calculating the Viscosity of the Undersaturated Oil, 112;
Surface/Interfacial Tension, 115; Properties of Reservoir Water, 118;
Water Formation Volume Factor, 118; Water Viscosity, 119;
Gas Solubility in Water, 119; Water Isothermal Compressibility, 120;
Problems, 120; References, 126.

ACKNOWLEDGMENTS

Much of the material on which this book is based was drawn from the publications of the Society of Petroleum Engineers. Tribute is due to the SPE and the petroleum engineers, scientists, and authors who have made numerous and significant contributions to the field of reservoir engineering. I would like to express my appreciation to a large number of my colleagues within the petroleum industry and academia who offered suggestions and critiques on the first edition; special thanks go to Dr. Wenxia Zhang with TotalFinaElf E&P USA, Inc, for her suggestions and encouragements. I am also indebted to my students at Montana Tech of the University of Montana, whose enthusiasm has made teaching a pleasure; I think! Special thanks to my colleagues and friends: Dr. Gil Cady, Professor John Evans; and Dr. Margaret Ziaja for making valuable suggestions for the improvement of this book. I would like to acknowledge and express my appreciation to Gary Kolstad, Vice President and General Manager with Schlumberger, and Darrell McKenna, Vice President with Schlumberger; for their continued support.

I would like to thank the editorial staff of Butterworth-Heinemann and Gulf Professional Publishing for their concise and thorough work. I greatly appreciate the assistance that Karen Forster has given me during my work on the second edition.

PREFACE TO THE SECOND EDITION

I have attempted to construct the chapters following a sequence that I have used for several years in teaching three undergraduate courses in reservoir engineering. Two new chapters have been included in this second edition; Chapter 14 and 15. Chapter 14 reviews principles of waterflooding with emphasis on the design of a waterflooding project. Chapter 15 is intended to introduce and document the practical applications of equations of state in the area of vapor-liquid phase equilibria. A comprehensive review of different equations of state is presented with an emphasis on the Peng-Robinson equation of state.

PREFACE TO THE FIRST EDITION

This book explains the fundamentals of reservoir engineering and their practical application in conducting a comprehensive field study. Chapter 1 reviews fundamentals of reservoir fluid behavior with an emphasis on the classification of reservoir and reservoir fluids. Chapter 2 documents reservoir-fluid properties, while Chapter 3 presents a comprehensive treatment and description of the routine and specialized PVT laboratory tests. The fundamentals of rock properties are discussed in Chapter 4 and numerous methodologies for generating those properties are reviewed. Chapter 5 focuses on presenting the concept of relative permeability and its applications in fluid flow calculations.

The fundamental mathematical expressions that are used to describe the reservoir fluid flow behavior in porous media are discussed in Chapter 6, while Chapters 7 and 8 describe the principle of oil and gas well performance calculations, respectively. Chapter 9 provides the theoretical analysis of coning and outlines many of the practical solutions for calculating water and gas coning behavior. Various water influx calculation models are shown in Chapter 10, along with detailed descriptions of the computational steps involved in applying these models. The objective of Chapter 11 is to introduce the basic principle of oil recovery mechanisms and to present the generalized form of the material balance equation. Chapters 12 and 13 focus on illustrating the practical applications of the material balance equation in oil and gas reservoirs.

FUNDAMENTALS OF RESERVOIR FLUID BEHAVIOR

Naturally occurring hydrocarbon systems found in petroleum reservoirs are mixtures of organic compounds which exhibit multiphase behavior over wide ranges of pressures and temperatures. These hydrocarbon accumulations may occur in the gaseous state, the liquid state, the solid state, or in various combinations of gas, liquid, and solid.

These differences in phase behavior, coupled with the physical properties of reservoir rock that determine the relative ease with which gas and liquid are transmitted or retained, result in many diverse types of hydrocarbon reservoirs with complex behaviors. Frequently, petroleum engineers have the task to study the behavior and characteristics of a petroleum reservoir and to determine the course of future development and production that would maximize the profit.

The objective of this chapter is to review the basic principles of reservoir fluid phase behavior and illustrate the use of phase diagrams in classifying types of reservoirs and the native hydrocarbon systems.

CLASSIFICATION OF RESERVOIRS AND RESERVOIR FLUIDS

Petroleum reservoirs are broadly classified as oil or gas reservoirs. These broad classifications are further subdivided depending on:

- The composition of the reservoir hydrocarbon mixture
- Initial reservoir pressure and temperature
- Pressure and temperature of the surface production

The conditions under which these phases exist are a matter of considerable practical importance. The experimental or the mathematical determinations of these conditions are conveniently expressed in different types of diagrams commonly called *phase diagrams.* One such diagram is called the *pressure-temperature diagram.*

Pressure-Temperature Diagram

Figure 1-1 shows a typical pressure-temperature diagram of a multicomponent system with a specific overall composition. Although a different hydrocarbon system would have a different phase diagram, the general configuration is similar.

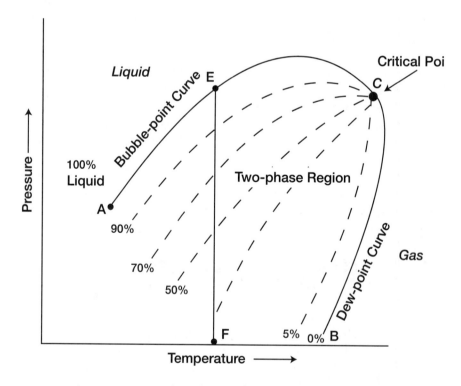

Figure 1-1. Typical p-T diagram for a multicomponent system.

These multicomponent pressure-temperature diagrams are essentially used to:

- Classify reservoirs
- Classify the naturally occurring hydrocarbon systems
- Describe the phase behavior of the reservoir fluid

To fully understand the significance of the pressure-temperature diagrams, it is necessary to identify and define the following key points on these diagrams:

- **Cricondentherm (T_{ct})**—The Cricondentherm is defined as the maximum temperature above which liquid cannot be formed regardless of pressure (point E). The corresponding pressure is termed the Cricondentherm pressure p_{ct}.
- **Cricondenbar (p_{cb})**—The Cricondenbar is the maximum pressure above which no gas can be formed regardless of temperature (point D). The corresponding temperature is called the Cricondenbar temperature T_{cb}.
- **Critical point**—The critical point for a multicomponent mixture is referred to as the state of pressure and temperature at which all intensive properties of the gas and liquid phases are equal (point C). At the critical point, the corresponding pressure and temperature are called the critical pressure p_c and critical temperature T_c of the mixture.
- **Phase envelope (two-phase region)**—The region enclosed by the bubble-point curve and the dew-point curve (line BCA), wherein gas and liquid coexist in equilibrium, is identified as the phase envelope of the hydrocarbon system.
- **Quality lines**—The dashed lines within the phase diagram are called quality lines. They describe the pressure and temperature conditions for equal volumes of liquids. Note that the quality lines converge at the critical point (point C).
- **Bubble-point curve**—The bubble-point curve (line BC) is defined as the line separating the liquid-phase region from the two-phase region.
- **Dew-point curve**—The dew-point curve (line AC) is defined as the line separating the vapor-phase region from the two-phase region.

In general, reservoirs are conveniently classified on the basis of the location of the point representing the initial reservoir pressure p_i and temperature T with respect to the pressure-temperature diagram of the reservoir fluid. Accordingly, reservoirs can be classified into basically two types. These are:

• **Oil reservoirs**—If the reservoir temperature T is less than the critical temperature T_c of the reservoir fluid, the reservoir is classified as an oil reservoir.

• **Gas reservoirs**—If the reservoir temperature is greater than the critical temperature of the hydrocarbon fluid, the reservoir is considered a gas reservoir.

Oil Reservoirs

Depending upon initial reservoir pressure p_i, oil reservoirs can be sub-classified into the following categories:

1. **Undersaturated oil reservoir.** If the initial reservoir pressure p_i (as represented by point 1 on Figure 1-1), is greater than the bubble-point pressure p_b of the reservoir fluid, the reservoir is labeled an undersaturated oil reservoir.

2. **Saturated oil reservoir.** When the initial reservoir pressure is equal to the bubble-point pressure of the reservoir fluid, as shown on Figure 1-1 by point 2, the reservoir is called a saturated oil reservoir.

3. **Gas-cap reservoir.** If the initial reservoir pressure is below the bubble-point pressure of the reservoir fluid, as indicated by point 3 on Figure 1-1, the reservoir is termed a gas-cap or two-phase reservoir, in which the gas or vapor phase is underlain by an oil phase. The appropriate quality line gives the ratio of the gas-cap volume to reservoir oil volume.

Crude oils cover a wide range in physical properties and chemical compositions, and it is often important to be able to group them into broad categories of related oils. In general, crude oils are commonly classified into the following types:

• Ordinary black oil
• Low-shrinkage crude oil
• High-shrinkage (volatile) crude oil
• Near-critical crude oil

The above classifications are essentially based upon the properties exhibited by the crude oil, including physical properties, composition, gas-oil ratio, appearance, and pressure-temperature phase diagrams.

1. **Ordinary black oil.** A typical pressure-temperature phase diagram for ordinary black oil is shown in Figure 1-2. It should be noted that quality lines which are approximately equally spaced characterize this

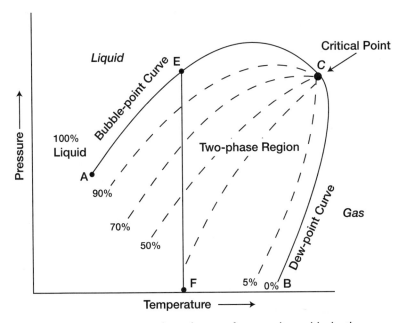

Figure 1-2. A typical p-T diagram for an ordinary black oil.

black oil phase diagram. Following the pressure reduction path as indicated by the vertical line EF on Figure 1-2, the liquid shrinkage curve, as shown in Figure 1-3, is prepared by plotting the liquid volume percent as a function of pressure. The liquid shrinkage curve approximates a straight line except at very low pressures. When produced, ordinary black oils usually yield gas-oil ratios between 200–700 scf/STB and oil gravities of 15 to 40 API. The stock tank oil is usually brown to dark green in color.

2. **Low-shrinkage oil.** A typical pressure-temperature phase diagram for low-shrinkage oil is shown in Figure 1-4. The diagram is characterized by quality lines that are closely spaced near the dew-point curve. The liquid-shrinkage curve, as given in Figure 1-5, shows the shrinkage characteristics of this category of crude oils. The other associated properties of this type of crude oil are:

• Oil formation volume factor less than 1.2 bbl/STB
• Gas-oil ratio less than 200 scf/STB
• Oil gravity less than 35° API
• Black or deeply colored

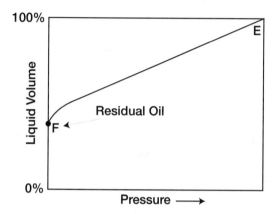

Figure 1-3. Liquid-shrinkage curve for black oil.

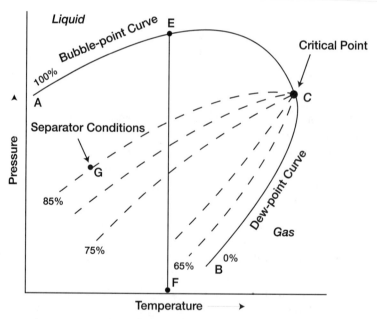

Figure 1-4. A typical phase diagram for a low-shrinkage oil.

- Substantial liquid recovery at separator conditions as indicated by point G on the 85% quality line of Figure 1-4.
3. **Volatile crude oil.** The phase diagram for a volatile (high-shrinkage) crude oil is given in Figure 1-6. Note that the quality lines are close

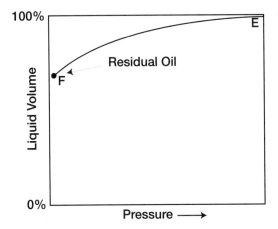

Figure 1-5. Oil-shrinkage curve for low-shrinkage oil.

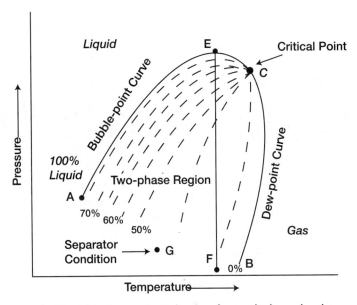

Figure 1-6. A typical p-T diagram for a volatile crude oil.

together near the bubble-point and are more widely spaced at lower pressures. This type of crude oil is commonly characterized by a high liquid shrinkage immediately below the bubble-point as shown in Figure 1-7. The other characteristic properties of this oil include:

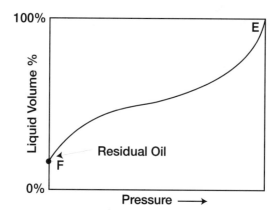

Figure 1-7. A typical liquid-shrinkage curve for a volatile crude oil.

- Oil formation volume factor less than 2 bbl/STB
- Gas-oil ratios between 2,000–3,200 scf/STB
- Oil gravities between 45–55° API
- Lower liquid recovery of separator conditions as indicated by point G on Figure 1-6
- Greenish to orange in color

Another characteristic of volatile oil reservoirs is that the API gravity of the stock-tank liquid will increase in the later life of the reservoirs.

4. **Near-critical crude oil.** If the reservoir temperature T is near the critical temperature T_c of the hydrocarbon system, as shown in Figure 1-8, the hydrocarbon mixture is identified as a near-critical crude oil. Because all the quality lines converge at the critical point, an isothermal pressure drop (as shown by the vertical line EF in Figure 1-8) may shrink the crude oil from 100% of the hydrocarbon pore volume at the bubble-point to 55% or less at a pressure 10 to 50 psi below the bubble-point. The shrinkage characteristic behavior of the near-critical crude oil is shown in Figure 1-9. The near-critical crude oil is characterized by a high GOR in excess of 3,000 scf/STB with an oil formation volume factor of 2.0 bbl/STB or higher. The compositions of near-critical oils are usually characterized by 12.5 to 20 mol% heptanes-plus, 35% or more of ethane through hexanes, and the remainder methane.

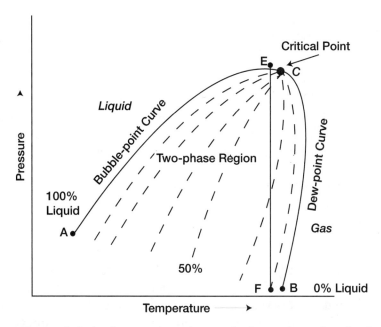

Figure 1-8. A schematic phase diagram for the near-critical crude oil.

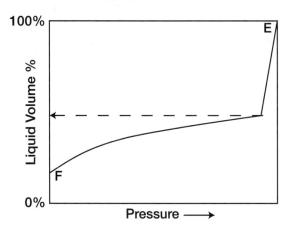

Figure 1-9. A typical liquid-shrinkage curve for the near-critical crude oil.

Figure 1-10 compares the characteristic shape of the liquid-shrinkage curve for each crude oil type.

Gas Reservoirs

In general, if the reservoir temperature is above the critical temperature of the hydrocarbon system, the reservoir is classified as a natural gas reservoir. On the basis of their phase diagrams and the prevailing reservoir conditions, natural gases can be classified into four categories:

• Retrograde gas-condensate
• Near-critical gas-condensate
• Wet gas
• Dry gas

Retrograde gas-condensate reservoir. If the reservoir temperature T lies between the critical temperature T_c and cricondentherm T_{ct} of the reservoir fluid, the reservoir is classified as a retrograde gas-condensate reservoir. This category of gas reservoir is a unique type of hydrocarbon accumulation in that the special thermodynamic behavior of the reservoir fluid is the controlling factor in the development and the depletion process of the reservoir. When the pressure is decreased on these mix-

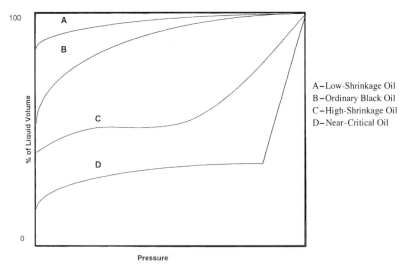

A–Low-Shrinkage Oil
B–Ordinary Black Oil
C–High-Shrinkage Oil
D–Near-Critical Oil

Figure 1-10. Liquid shrinkage for crude oil systems.

tures, instead of expanding (if a gas) or vaporizing (if a liquid) as might be expected, they vaporize instead of condensing.

Consider that the initial condition of a retrograde gas reservoir is represented by point 1 on the pressure-temperature phase diagram of Figure 1-11. Because the reservoir pressure is above the upper dew-point pressure, the hydrocarbon system exists as a single phase (i.e., vapor phase) in the reservoir. As the reservoir pressure declines isothermally during production from the initial pressure (point 1) to the upper dew-point pressure (point 2), the attraction between the molecules of the light and heavy components causes them to move further apart further apart. As this occurs, attraction between the heavy component molecules becomes more effective; thus, liquid begins to condense.

This retrograde condensation process continues with decreasing pressure until the liquid dropout reaches its maximum at point 3. Further reduction in pressure permits the heavy molecules to commence the normal vaporization process. This is the process whereby fewer gas molecules strike the liquid surface and causes more molecules to leave than

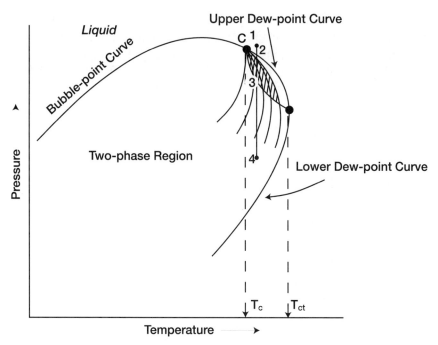

Figure 1-11. A typical phase diagram of a retrograde system.

enter the liquid phase. The vaporization process continues until the reservoir pressure reaches the lower dew-point pressure. This means that all the liquid that formed must vaporize because the system is essentially all vapors at the lower dew point.

Figure 1-12 shows a typical liquid shrinkage volume curve for a condensate system. The curve is commonly called the **liquid dropout curve.** In most gas-condensate reservoirs, the condensed liquid volume seldom exceeds more than 15%–19% of the pore volume. This liquid saturation is not large enough to allow any liquid flow. It should be recognized, however, that around the wellbore where the pressure drop is high, enough liquid dropout might accumulate to give two-phase flow of gas and retrograde liquid.

The associated physical characteristics of this category are:

- Gas-oil ratios between 8,000 to 70,000 scf/STB. Generally, the gas-oil ratio for a condensate system increases with time due to the liquid dropout and the loss of heavy components in the liquid.
- Condensate gravity above 50° API
- Stock-tank liquid is usually water-white or slightly colored.

There is a fairly sharp dividing line between oils and condensates from a compositional standpoint. Reservoir fluids that contain heptanes and are heavier in concentrations of more than 12.5 mol% are almost always in the liquid phase in the reservoir. Oils have been observed with hep-

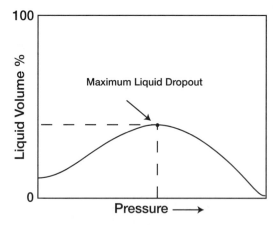

Figure 1-12. A typical liquid dropout curve.

tanes and heavier concentrations as low as 10% and condensates as high as 15.5%. These cases are rare, however, and usually have very high tank liquid gravities.

Near-critical gas-condensate reservoir. If the reservoir temperature is near the critical temperature, as shown in Figure 1-13, the hydrocarbon mixture is classified as a near-critical gas-condensate. The volumetric behavior of this category of natural gas is described through the isothermal pressure declines as shown by the vertical line 1-3 in Figure 1-13 and also by the corresponding liquid dropout curve of Figure 1-14. Because all the quality lines converge at the critical point, a rapid liquid buildup will immediately occur below the dew point (Figure 1-14) as the pressure is reduced to point 2.

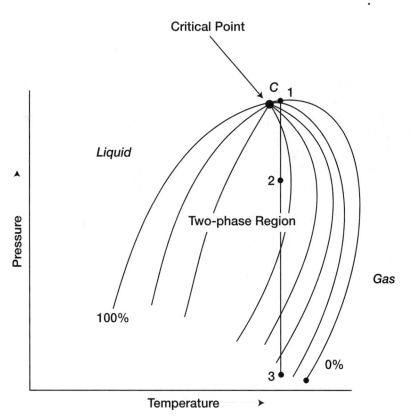

Figure 1-13. A typical phase diagram for a near-critical gas condensate reservoir.

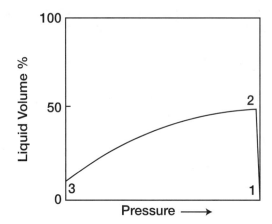

Figure 1-14. Liquid-shrinkage curve for a near-critical gas-condensate system.

This behavior can be justified by the fact that several quality lines are crossed very rapidly by the isothermal reduction in pressure. At the point where the liquid ceases to build up and begins to shrink again, the reservoir goes from the retrograde region to a normal vaporization region.

Wet-gas reservoir. A typical phase diagram of a wet gas is shown in Figure 1-15, where reservoir temperature is above the cricondentherm of the hydrocarbon mixture. Because the reservoir temperature exceeds the cricondentherm of the hydrocarbon system, the reservoir fluid will always remain in the vapor phase region as the reservoir is depleted isothermally, along the vertical line A-B.

As the produced gas flows to the surface, however, the pressure and temperature of the gas will decline. If the gas enters the two-phase region, a liquid phase will condense out of the gas and be produced from the surface separators. This is caused by a sufficient decrease in the kinetic energy of heavy molecules with temperature drop and their subsequent change to liquid through the attractive forces between molecules.

Wet-gas reservoirs are characterized by the following properties:

• Gas oil ratios between 60,000 to 100,000 scf/STB
• Stock-tank oil gravity above 60° API
• Liquid is water-white in color
• Separator conditions, i.e., separator pressure and temperature, lie within the two-phase region

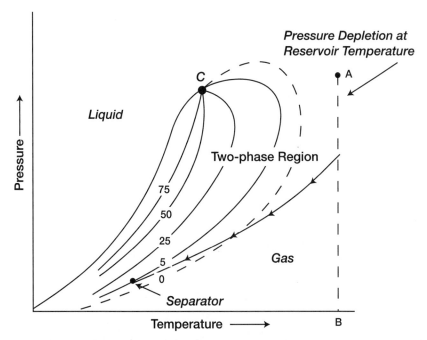

Figure 1-15. Phase diagram for a wet gas. (*After Clark, N.J.* Elements of Petroleum Reservoirs, *SPE, 1969.*)

Dry-gas reservoir. The hydrocarbon mixture exists as a gas both in the reservoir and in the surface facilities. The only liquid associated with the gas from a dry-gas reservoir is water. A phase diagram of a dry-gas reservoir is given in Figure 1-16. Usually a system having a gas-oil ratio greater than 100,000 scf/STB is considered to be a dry gas.

Kinetic energy of the mixture is so high and attraction between molecules so small that none of them coalesce to a liquid at stock-tank conditions of temperature and pressure.

It should be pointed out that the classification of hydrocarbon fluids might be also characterized by the initial composition of the system. McCain (1994) suggested that the heavy components in the hydrocarbon mixtures have the strongest effect on fluid characteristics. The ternary diagram, as shown in Figure 1-17, with equilateral triangles can be conveniently used to roughly define the compositional boundaries that separate different types of hydrocarbon systems.

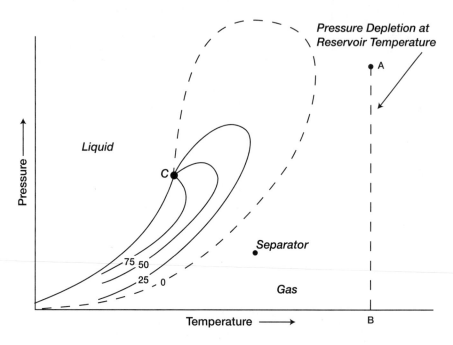

Figure 1-16. Phase diagram for a dry gas. (*After Clark, N.J.* Elements of Petroleum Reservoirs, *SPE, 1969.*)

From the foregoing discussion, it can be observed that hydrocarbon mixtures may exist in either the gaseous or liquid state, depending on the reservoir and operating conditions to which they are subjected. The qualitative concepts presented may be of aid in developing quantitative analyses. Empirical equations of state are commonly used as a quantitative tool in describing and classifying the hydrocarbon system. These equations of state require:

• Detailed compositional analyses of the hydrocarbon system
• Complete descriptions of the physical and critical properties of the mixture individual components

Many characteristic properties of these individual components (in other words, pure substances) have been measured and compiled over the years. These properties provide vital information for calculating the thermodynamic properties of pure components, as well as their mixtures. The most important of these properties are:

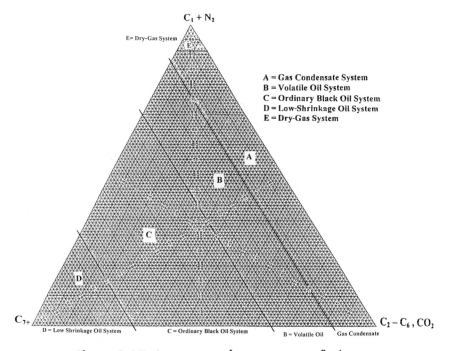

Figure 1-17. Compositions of various reservoir fluid types.

- Critical pressure, p_c
- Critical temperature, T_c
- Critical volume, V_c
- Critical compressibility factor, z_c
- Acentric factor, T
- Molecular weight, M

Table 1-2 documents the above-listed properties for a number of hydrocarbon and nonhydrocarbon components.

Katz and Firoozabadi (1978) presented a generalized set of physical properties for the petroleum fractions C_6 through C_{45}. The tabulated properties include the average boiling point, specific gravity, and molecular weight. The authors' proposed a set of tabulated properties that were generated by analyzing the physical properties of 26 condensates and crude oil systems. These generalized properties are given in Table 1-1.

(text continued on page 24)

Table 1-1
Generalized Physical Properties

Group	T_b (°R)	γ	K	M	T_c (°R)	P_c (psia)	ω	V_c (ft³/lb)	Group
C_6	607	0.690	12.27	84	923	483	0.250	0.06395	C_6
C_7	658	0.727	11.96	96	985	453	0.280	0.06289	C_7
C_8	702	0.749	11.87	107	1,036	419	0.312	0.06264	C_8
C_9	748	0.768	11.82	121	1,085	383	0.348	0.06258	C_9
C_{10}	791	0.782	11.83	134	1,128	351	0.385	0.06273	C_{10}
C_{11}	829	0.793	11.85	147	1,166	325	0.419	0.06291	C_{11}
C_{12}	867	0.804	11.86	161	1,203	302	0.454	0.06306	C_{12}
C_{13}	901	0.815	11.85	175	1,236	286	0.484	0.06311	C_{13}
C_{14}	936	0.826	11.84	190	1,270	270	0.516	0.06316	C_{14}
C_{15}	971	0.836	11.84	206	1,304	255	0.550	0.06325	C_{15}
C_{16}	1,002	0.843	11.87	222	1,332	241	0.582	0.06342	C_{16}
C_{17}	1,032	0.851	11.87	237	1,360	230	0.613	0.06350	C_{17}
C_{18}	1,055	0.856	11.89	251	1,380	222	0.638	0.06362	C_{18}
C_{19}	1,077	0.861	11.91	263	1,400	214	0.662	0.06372	C_{19}
C_{20}	1,101	0.866	11.92	275	1,421	207	0.690	0.06384	C_{20}
C_{21}	1,124	0.871	11.94	291	1,442	200	0.717	0.06394	C_{21}
C_{22}	1,146	0.876	11.95	300	1,461	193	0.743	0.06402	C_{22}
C_{23}	1,167	0.881	11.95	312	1,480	188	0.768	0.06408	C_{23}
C_{24}	1,187	0.885	11.96	324	1,497	182	0.793	0.06417	C_{24}

C_{25}	1,207	0.888	11.99	337	1,515	177	0.819	0.06431
C_{26}	1,226	0.892	12.00	349	1,531	173	0.844	0.06438
C_{27}	1,244	0.896	12.00	360	1,547	169	0.868	0.06443
C_{28}	1,262	0.899	12.02	372	1,562	165	0.894	0.06454
C_{29}	1,277	0.902	12.03	382	1,574	161	0.915	0.06459
C_{30}	1,294	0.905	12.04	394	1,589	158	0.941	0.06468
C_{31}	1,310	0.909	12.04	404	1,603	143	0.897	0.06469
C_{32}	1,326	0.912	12.05	415	1,616	138	0.909	0.06475
C_{33}	1,341	0.915	12.05	426	1,629	134	0.921	0.06480
C_{34}	1,355	0.917	12.07	437	1,640	130	0.932	0.06489
C_{35}	1,368	0.920	12.07	445	1,651	127	0.942	0.06490
C_{36}	1,382	0.922	12.08	456	1,662	124	0.954	0.06499
C_{37}	1,394	0.925	12.08	464	1,673	121	0.964	0.06499
C_{38}	1,407	0.927	12.09	475	1,683	118	0.975	0.06506
C_{39}	1,419	0.929	12.10	484	1,693	115	0.985	0.06511
C_{40}	1,432	0.931	12.11	495	1,703	112	0.997	0.06517
C_{41}	1,442	0.933	12.11	502	1,712	110	1.006	0.06520
C_{42}	1,453	0.934	12.13	512	1,720	108	1.016	0.06529
C_{43}	1,464	0.936	12.13	521	1,729	105	1.026	0.06532
C_{44}	1,477	0.938	12.14	531	1,739	103	1.038	0.06538
C_{45}	1,487	0.940	12.14	539	1,747	101	1.048	0.06540

Table 1-2
Physical Properties for Pure Components

Physical Constants

Number	Compound	See Note No. -> Formula	A. Molar mass (molecular weight)	B. Boiling point, °F 14.696 psia	Vapor pressure, psia 100°F	C. Freezing point, °F 14.696 psia	D. Refractive index, n_D 60°F	Critical constants Pressure, psia	Temperature, °F	Volume, ft³/lbm	Number
1	Methane	CH₄	16.043	-258.73	(5000)*	-296.44*	1.00042*	666.4	-116.67	0.0988	1
2	Ethane	C₂H₆	30.070	-127.49	(800)*	-297.04*	1.20971*	706.5	89.92	0.0783	2
3	Propane	C₃H₈	44.097	-43.75	188.64	-305.73*	1.29480*	616.0	206.06	0.0727	3
4	Isobutane	C₄H₁₀	58.123	10.78	72.581	-255.28	1.3245*	527.9	274.46	0.0714	4
5	n-Butane	C₄H₁₀	58.123	31.08	51.706	-217.05	1.33588*	550.6	305.62	0.0703	5
6	Isopentane	C₅H₁₂	72.150	82.12	20.445	-255.82	1.35631	490.4	369.10	0.0679	6
7	n-Pentane	C₅H₁₂	72.150	96.92	15.574	-201.51	1.35992	488.6	385.8	0.0675	7
8	Neopentane	C₅H₁₂	72.150	49.10	36.69	2.17	1.342*	464.0	321.13	0.0673	8
9	n-Hexane	C₆H₁₄	86.177	155.72	4.9597	-139.58	1.37708	436.9	453.6	0.0688	9
10	2-Methylpentane	C₆H₁₄	86.177	140.47	6.769	-244.62	1.37387	436.6	435.83	0.0682	10
11	3-Methylpentane	C₆H₁₄	86.177	145.89	6.103		1.37888	453.1	448.4	0.0682	11
12	Neohexane	C₆H₁₄	86.177	121.52	9.859	-147.72	1.37126	446.8	420.13	0.0667	12
13	2,3-Dimethylbutane	C₆H₁₄	86.177	136.36	7.406	-199.38	1.37730	453.5	440.29	0.0665	13
14	n-Heptane	C₇H₁₆	100.204	209.16	1.620	-131.05	1.38989	396.8	512.7	0.0691	14
15	2-Methylhexane	C₇H₁₆	100.204	194.09	2.272	-180.89	1.38714	396.5	495.00	0.0673	15
16	3-Methylhexane	C₇H₁₆	100.204	197.33	2.131		1.39091	408.1	503.80	0.0646	16
17	3-Ethylpentane	C₇H₁₆	100.204	200.25	2.013	-181.48	1.39566	419.3	513.39	0.0665	17
18	2,2-Dimethylpentane	C₇H₁₆	100.204	174.54	3.494	-190.86	1.38446	402.2	477.23	0.0665	18
19	2,4-Dimethylpentane	C₇H₁₆	100.204	176.89	3.293	-182.63	1.38379	396.9	475.95	0.0668	19
20	3,3-Dimethylpentane	C₇H₁₆	100.204	186.91	2.774	-210.01	1.38564	427.2	505.87	0.0662	20
21	Triptane	C₇H₁₆	100.204	177.58	3.375	-12.81	1.39168	428.4	496.44	0.0636	21
22	n-Octane	C₈H₁₈	114.231	258.21	0.53694	-70.18	1.39956	360.7	564.22	0.0690	22
23	Diisobutyl	C₈H₁₈	114.231	228.39	1.102	-132.11	1.39461	360.6	530.44	0.0676	23
24	Isooctane	C₈H₁₈	114.231	210.63	1.709	-161.27	1.38624	372.4	519.46	0.0656	24

No.	Compound	Formula	Mol. wt.								No.
25	n-Nonane	C9H20	128.258	303.47	0.17953	-64.28	1.40746	331.8	610.68	0.0684	25
26	n-Decane	C10H22	142.285	345.48	0.06088	-21.36	1.41385	305.2	652.0	0.0679	26
27	Cyclopentane	C5H10	70.134	120.65	9.915	-136.91	1.40896	653.8	461.2	0.0594	27
28	Methylcyclopentane	C6H12	84.161	161.25	4.503	-224.40	1.41210	548.9	499.35	0.0607	28
29	Cyclohexane	C6H12	84.161	177.29	3.266	43.77	1.42862	590.8	536.6	0.0586	29
30	Methylcyclohexane	C7H14	98.188	213.68	1.609	-195.87	1.42538	503.5	570.27	0.0600	30
31	Ethene(Ethylene)	C2H4	28.054	-154.73	(1400)*	-272.47*	(1.228)*	731.0	48.54	0.0746	31
32	Propene(Propylene)	C3H6	42.081	-53.84	227.7	-301.45*	1.3130*	668.6	197.17	0.0689	32
33	1-Butene(Butylene)	C4H8	56.108	20.79	62.10	-301.63*	1.3494*	583.5	295.48	0.0685	33
34	cis-2-Butene	C4H8	56.108	38.69	45.95	-218.06	1.3665*	612.1	324.37	0.0668	34
35	trans-2-Butene	C4H8	56.108	33.58	49.87	-157.96	1.3563*	587.4	311.86	0.0679	35
36	Isobutene	C4H8	56.108	19.59	63.02	-220.65	1.3512*	580.2	292.55	0.0682	36
37	1-Pentene	C5H10	70.134	85.93	19.12	-265.39	1.37426	511.8	376.93	0.0676	37
38	1,2-Butadiene	C4H6	54.092	51.53	36.53	-213.16		(653.)*	(340.)*	(0.065)*	38
39	1,3-Butadiene	C4H6	54.092	24.06	59.46	-164.02	1.3975*	627.5	305.	(0.065)*	39
40	Isoprene	C5H8	68.119	93.31	16.68	-230.73	1.42498	(558.)*	(412.)*	(0.065)*	40
41	Acetylene	C2H2	26.038	-120.49*		-114.5*		890.4	95.34	0.0695	41
42	Benzene	C6H6	78.114	176.18	3.225	41.95	1.50396	710.4	552.22	0.0531	42
43	Toluene	C7H8	92.141	231.13	1.033	-139.00	1.49942	595.5	605.57	0.0550	43
44	Ethylbenzene	C8H10	106.167	277.16	0.3716	-138.966	1.49826	523.0	651.29	0.0565	44
45	o-Xylene	C8H10	106.167	291.97	0.2643	-13.59	1.50767	541.6	674.92	0.0557	45
46	m-Xylene	C8H10	106.167	282.41	0.3265	-54.18	1.49951	512.9	651.02	0.0567	46
47	p-Xylene	C8H10	106.167	281.07	0.3424	55.83	1.49810	509.2	649.54	0.0570	47
48	Styrene	C8H8	104.152	293.25	0.2582	-23.10	1.54937	587.8	(703.)*	0.0534	48
49	Isopropylbenzene	C9H12	120.194	306.34	0.1884	-140.814	1.49372	465.4	676.3	0.0572	49
50	Methyl alcohol	CH4O	32.042	148.44	4.629	-143.79	1.33034	1174.	463.08	0.0590	50
51	Ethyl alcohol	C2H6O	46.069	172.90	2.312	-173.4	1.36346	890.1	465.39	0.0581	51
52	Carbon monoxide	CO	28.010	-312.68		-337.00*	1.00036*	507.5	-220.43	0.0532	52
53	Carbon dioxide	CO2	44.010	-109.257*		-69.83*	1.00048*	1071.	87.91	0.0344	53
54	Hydrogen sulfide	H2S	34.08	-76.497	394.59	-121.88*	1.00060*	1300.	212.45	0.0461	54
55	Sulfur dioxide	SO2	64.06	14.11	85.46	-103.86*	1.00062*	1143.	315.8	0.0305	55
56	Ammonia	NH3	17.0305	-27.99	211.9	-107.88*	1.00036*	1646.	270.2	0.0681	56
57	Air	N2+O2	28.9625	-317.8			1.00028*	546.9	-221.31	0.0517	57
58	Hydrogen	H2	2.0159	-422.955*		-435.26*	1.00013*	188.1	-399.9	0.5165	58
59	Oxygen	O2	31.9988	-297.332*		-361.820*	1.00027*	731.4	-181.43	0.0367	59
60	Nitrogen	N2	28.0134	-320.451		-346.00*	1.00028*	493.1	-232.51	0.0510	60
61	Chlorine	Cl2	70.906	-29.13	157.3	-149.73*	1.3878*	1157.	290.75	0.0280	61
62	Water	H2O	18.0153	212.000*	0.9501	32.00	1.33335	3198.8	705.16	0.04975	62
63	Helium	He	4.0026	-452.09			1.00003*	32.99	-450.31	0.2300	63
64	Hydrogen chloride	HCl	36.461	-121.27	906.71	-173.52*	1.00042*	1205.	124.27	0.0356	64

(table continued on next page)

Table 1-2 (continued)

Physical Constants

*See the Table of Notes and References.

Number	E. Density of liquid 14.696 psia, 60°F — Relative density (specific gravity) 60°F/60°F	E. lbm/gal	E. gal/lb·mole	F. Temperature coefficient of density, 1/°F	G. Acentric factor, ω	H. Compressibility factor of real gas, Z 14.696 psia, 60°F	I. Relative density (specific gravity) Air = 1	I. ft³ gas/lbm	I. ft³ gas/gal liquid	J. Cp, Ideal gas	J. Cp, Liquid	Number
1	(0.3)•	(2.5)•	(6.4172)•		0.0104	0.9980	0.5539	23.654	(59.135)•	0.52669	0.97225	1
2	0.35619•	2.9696•	10.126•	-0.00162•	0.0979	0.9919	1.0382	12.620	37.476•	0.40782	0.61996	2
3	0.50699•	4.2268•	10.433•	-0.00119•	0.1522	0.9825	1.5226	8.6059	36.375•	0.38852	0.57066	3
4	0.56287•	4.6927•	12.386•	-0.00106•	0.1852	0.9711	2.0068	6.5291	30.639•	0.38669	0.57272	4
5	0.58401•	4.8690•	11.937•		0.1995	0.9667	2.0068	6.5291	31.790•	0.39499		5
6	0.62470	5.2082	13.853	-0.00090	0.2280		2.4912	5.2596	27.393	0.38440	0.53331	6
7	0.63112	5.2617	13.712	-0.00086•	0.2514		2.4912	5.2596	27.674	0.38825	0.54363	7
8	0.59666•	4.9744•	14.504•	-0.00106•	0.1963	0.9582	2.4912	5.2596	26.163•	0.39038	0.55021	8
9	0.66383	5.5344	15.571	-0.00075	0.2994		2.9755	4.4035	24.371	0.38628	0.53327	9
10	0.65785	5.4846	15.713	-0.00076	0.2780		2.9755	4.4035	24.152	0.38526	0.52732	10
11	0.66901	5.5776	15.451	-0.00076	0.2732		2.9755	4.4035	24.561	0.37902	0.51876	11
12	0.65385	5.4512	15.809	-0.00076	0.2326		2.9755	4.4035	24.005	0.38231	0.51367	12
13	0.66631	5.5551	15.513	-0.00076	0.2469		2.9755	4.4035	24.462	0.37762	0.51308	13
14	0.68820	5.7376	17.464	-0.00068	0.3494		3.4598	3.7872	21.729	0.38447	0.52802	14
15	0.68310	5.6951	17.595	-0.00070	0.3298		3.4598	3.7872	21.568	0.38041	0.52199	15
16	0.69165	5.7664	17.377	-0.00070	0.3232		3.4598	3.7872	21.838	0.37882	0.51019	16
17	0.70276	5.8590	17.103	-0.00069	0.3105		3.4598	3.7872	22.189	0.38646	0.51410	17
18	0.67829	5.6550	17.720	-0.00070	0.2871		3.4598	3.7872	21.416	0.38594	0.51678	18
19	0.67733	5.6470	17.745	-0.00073	0.3026		3.4598	3.7872	21.386	0.39414	0.52440	19
20	0.69772	5.8170	17.226	-0.00067	0.2674		3.4598	3.7872	22.030	0.38306	0.50138	20
21	0.69457	5.7907	17.304	-0.00068	0.2503		3.4598	3.7872	21.930	0.37724	0.49920	21
22	0.70696	5.8940	19.381	-0.00064	0.3977		3.9441	3.3220	19.580	0.38331	0.52406	22
23	0.69793	5.8187	19.632	-0.00067	0.3564		3.9441	3.3220	19.330	0.37571	0.51130	23
24	0.69624	5.8046	19.679	-0.00065	0.3035		3.9441	3.3220	19.283	0.38222	0.48951	24
25	0.72187	6.0183	21.311	-0.00061	0.4445		4.4284	2.9588	17.807	0.38246	0.52244	25
26	0.73421	6.1212	23.245	-0.00057	0.4898		4.9127	2.6671	16.326	0.38179	0.52103	26

No	(1)	(2)	(3)	(4)	(5)	(6)	(7)	(8)	(9)	(10)	(11)
27	0.42182	0.27199	33.856	5.4110	2.4215	——	0.1950	-0.00073	11.209	6.2570	0.75050
28	0.44126	0.30100	28.325	4.5090	2.9059	——	0.2302	-0.00069	13.397	6.2819	0.75349
29	0.43584	0.28817	29.452	4.5090	2.9059	——	0.2096	-0.00065	12.885	6.5319	0.78347
30	0.44012	0.31700	24.940	3.8649	3.3902	——	0.2358	-0.00062	15.216	6.4529	0.77400
31	0.57116	0.35697	39.167*	13.527	0.9686	0.9936	0.0865	-0.00173*	9.6889*	4.3432*	0.52095*
32	0.54533	0.35714	35.894*	9.0179	1.4529	0.9844	0.1356	-0.00112*	11.197*	5.0112*	0.60107*
33	0.52980	0.35446	35.366*	6.7636	1.9373	0.9699	0.1941	-0.00105*	10.731*	5.2288*	0.62717*
34	0.54215	0.33754	34.395*	6.7636	1.9373	0.9665	0.2029	-0.00106*	11.033*	5.0853*	0.60996*
35	0.54839	0.35574	33.856*	6.7636	1.9373	0.9667	0.2128	-0.00117*	11.209*	5.0056*	0.60040*
36	0.51782	0.37690	29.129	6.7636	1.9373	0.9700	0.1999	-0.00089	13.028	5.3834	0.64571
37	0.54029	0.36351	38.485*	5.4110	2.4215	——	0.2333	-0.00101*	9.8605*	5.4857*	0.65799*
38	0.53447	0.34347	36.687*	7.0156	1.8677	——	0.2540	-0.00110*	10.344*	5.2293*	0.62723*
39	0.51933	0.34120		7.0156	1.8677	(0.969)	0.2007				
40		0.35072	31.869	5.5710	2.3520	(0.965)	0.1568	-0.00082	11.908	5.7205	0.68615
41		0.39754	35.824	14.574	0.8990	0.9930	0.1949	-0.00067	(7.473)	(3.4842)	(0.41796)
42	0.40989	0.24296	29.937	4.8581	2.6971	——	0.2093	-0.00059	10.593	7.3740	0.88448
43	0.40095	0.26370	26.976	4.1184	3.1814	——	0.2633	-0.00056	12.676	7.2691	0.87190
44	0.41139	0.27792	26.363	3.5744	3.6657	——	0.3027	-0.00054	14.609	7.2673	0.87168
45	0.41620	0.28964	25.889	3.5744	3.6657	——	0.3942	-0.00053	14.594	7.3756	0.88467
46	0.40545	0.27427	25.800	3.5744	3.6657	——	0.3257	-0.00056	14.658	7.2429	0.86875
47	0.40255	0.27471	27.675	3.6435	3.5961	——	0.3216	-0.00053	14.708	7.2181	0.86578
48	0.41220	0.27110	22.804	3.1573	4.1500	——	(0.2412)	-0.00055	13.712	7.5958	0.91108
49	0.42053	0.29170				——	0.3260		16.641	7.2228	0.86634
50	0.59187	0.32316	78.622	11.843	1.1063		0.5649	-0.00066	4.8267	6.6385	0.79626
51	0.56610	0.33222	54.527	8.2372	1.5906		0.6438	-0.00058	6.9595	6.6196	0.79399
52		0.24847	89.163*	13.548	0.9671	0.9959	0.0484		4.2561*	6.5812*	0.78939*
53		0.19911	58.807*	8.6229	1.5196	0.9943	0.2667		6.4532*	6.8199*	0.81802*
54	0.50418	0.23827	74.401*	11.135	1.1767	0.9846	0.0948	-0.00583*	5.1005*	6.6817*	0.80144*
55	0.32460	0.14804	69.012*	5.9238	2.2118	0.9802	0.2548	-0.00157*	5.4987*	11.650*	1.3974*
56	1.1209	0.49677	114.87*	22.283	0.5880	0.9877	0.2557		3.3037*	5.1550*	0.61832*
57		0.23988	95.557*	13.103	0.0000	1.0000			3.9713*	7.2930*	0.87476*
58		3.4038	111.54*	188.25	0.06960	1.0006	-0.2202		3.4022*	0.59252*	0.071070*
59		0.21892	112.93*	11.859	0.1048	0.9992	0.0216		3.3605*	9.5221*	1.1421*
60		0.24828	91.413*	13.546	0.9672	0.9997	0.0372		4.1513*	6.7481*	0.80940*
61		0.11377	63.554*	5.3519	2.4482	(0.9875)	0.0878		5.9710*	11.875*	1.4244*
62	0.99974	0.44457	175.62	21.065	0.62202	1.0006	0.3443	-0.00009	2.1609	8.33712*	1.00000
63		1.2404	98.891*	94.814	0.1382	0.9923	0.		3.8376*	1.0430*	0.12510*
64		0.19086	73.869*	10.408	1.2589		0.1259	-0.00300*	5.1373*	7.0973*	0.85129*

(text continued from page 17)

Ahmed (1985) correlated Katz-Firoozabadi-tabulated physical properties with the number of carbon atoms of the fraction by using a regression model. The generalized equation has the following form:

$$\theta = a_1 + a_2 n + a_3 n^2 + a_4 n^3 + (a_5/n) \tag{1-1}$$

where θ = any physical property
    n = number of carbon atoms, i.e., 6. 7. . . . ., 45
    a_1–a_5 = coefficients of the equation and are given in Table 1-3

Table 1-3
Coefficients of Equation 1-1

θ	a_1	a_2	a_3	a_4	a_5
M	−131.11375	24.96156	−0.34079022	2.4941184×10^{-3}	468.32575
T_c, °R	915.53747	41.421337	−0.7586859	5.8675351×10^{-3}	-1.3028779×10^{3}
P_c, psia	275.56275	−12.522269	0.29926384	$-2.8452129 \times 10^{-3}$	1.7117226×10^{-3}
T_b, °R	434.38878	50.125279	−0.9097293	7.0280657×10^{-3}	−601.85651
T	−0.50862704	8.700211×10^{-2}	$-1.8484814 \times 10^{-3}$	1.4663890×10^{-5}	1.8518106
γ	0.86714949	3.4143408×10^{-3}	-2.839627×10^{-5}	2.4943308×10^{-8}	−1.1627984
V_c, ft³/lb	5.223458×10^{-2}	$7.87091369 \times 10^{-4}$	$-1.9324432 \times 10^{-5}$	1.7547264×10^{-7}	4.4017952×10^{-2}

Undefined Petroleum Fractions

Nearly all naturally occurring hydrocarbon systems contain a quantity of heavy fractions that are not well defined and are not mixtures of discretely identified components. These heavy fractions are often lumped together and identified as the plus fraction, e.g., C_{7+} fraction.

A proper description of the physical properties of the plus fractions and other undefined petroleum fractions in hydrocarbon mixtures is essential in performing reliable phase behavior calculations and compositional modeling studies. Frequently, a distillation analysis or a chromatographic analysis is available for this undefined fraction. Other physical properties, such as molecular weight and specific gravity, may also be measured for the entire fraction or for various cuts of it.

To use any of the thermodynamic property-prediction models, e.g., equation of state, to predict the phase and volumetric behavior of complex hydrocarbon mixtures, one must be able to provide the acentric factor, along with the critical temperature and critical pressure, for both the defined and undefined (heavy) fractions in the mixture. The problem of how to adequately characterize these undefined plus fractions in terms of their critical properties and acentric factors has been long recognized in the petroleum industry. Whitson (1984) presented an excellent documentation on the influence of various heptanes-plus (C_{7+}) characterization schemes on predicting the volumetric behavior of hydrocarbon mixtures by equations-of-state.

Riazi and Daubert (1987) developed a simple two-parameter equation for predicting the physical properties of pure compounds and undefined hydrocarbon mixtures. The proposed generalized empirical equation is based on the use of the molecular weight M and specific gravity γ of the undefined petroleum fraction as the correlating parameters. Their mathematical expression has the following form:

$$\theta = a\,(M)^b\,\gamma^c\,EXP\,[d\,(M) + e\,\gamma + f\,(M)\,\gamma] \tag{1-2}$$

where θ = any physical property
  a–f = constants for each property as given in Table 1-4
  γ = specific gravity of the fraction
  M = molecular weight
  T_c = critical temperature, °R
  P_c = critical pressure, psia (Table 1-4)
  T_b = boiling point temperature, °R
  V_c = critical volume, ft³/lb

Table 1-4
Correlation Constants for Equation 1-2

θ	a	b	c	d	e	f
T_c, °R	544.4	0.2998	1.0555	-1.3478×10^{-4}	-0.61641	0.0
P_c, psia	4.5203×10^4	-0.8063	1.6015	-1.8078×10^{-3}	-0.3084	0.0
V_c ft³/lb	1.206×10^{-2}	0.20378	-1.3036	-2.657×10^{-3}	0.5287	2.6012×10^{-3}
T_b, °R	6.77857	0.401673	-1.58262	3.77409×10^{-3}	2.984036	-4.25288×10^{-3}

Edmister (1958) proposed a correlation for estimating the acentric factor T of pure fluids and petroleum fractions. The equation, widely used in the petroleum industry, requires boiling point, critical temperature, and critical pressure. The proposed expression is given by the following relationship:

$$\omega = \frac{3 \,[\log (p_c / 14.70)]}{7 \,[(T_c / T_b - 1)]} - 1 \qquad\qquad (1\text{-}3)$$

where T = acentric factor
$\quad\quad\quad p_c$ = critical pressure, psia
$\quad\quad\quad T_c$ = critical temperature, °R
$\quad\quad\quad T_b$ = normal boiling point, °R

If the acentric factor is available from another correlation, the Edmister equation can be rearranged to solve for any of the three other properties (providing the other two are known).

The critical compressibility factor is another property that is often used in thermodynamic-property prediction models. It is defined as the component compressibility factor calculated at its critical point. This property can be conveniently computed by the real gas equation-of-state at the critical point, or

$$z_c = \frac{p_c \, V_c \, M}{R \, T_c} \qquad\qquad (1\text{-}4)$$

where R = universal gas constant, 10.73 psia-ft^3/lb-mol. °R
$\quad\quad\quad V_c$ = critical volume, ft^3/lb
$\quad\quad\quad M$ = molecular weight

The accuracy of Equation 1-4 depends on the accuracy of the values of p_c, T_c, and V_c used in evaluating the critical compressibility factor. Table 1-5 presents a summary of the critical compressibility estimation methods.

Table 1-5
Critical Compressibility Estimation Methods

Method	Year	z_c	Equation No.
Haugen	1959	$z_c = 1/(1.28\,\omega + 3.41)$	1-5
Reid, Prausnitz, and Sherwood	1977	$z_c = 0.291 - 0.080\,\omega$	1-6
Salerno, et al.	1985	$z_c = 0.291 - 0.080\,\omega - 0.016\,\omega^2$	1-7
Nath	1985	$z_c = 0.2918 - 0.0928$	1-8

Example 1-1

Estimate the critical properties and the acentric factor of the heptanes-plus fraction, i.e., C_{7+}, with a measured molecular weight of 150 and specific gravity of 0.78.

Solution

Step 1. Use Equation 1-2 to estimate T_c, p_c, V_c, and T_b:

- $T_c = 544.2\ (150)^{.2998}\ (.78)^{1.0555}\ \exp[-1.3478 \times 10^{-4}\ (150) - 0.61641\ (.78) + 0] = 1139.4\ °R$
- $p_c = 4.5203 \times 10^4\ (150)^{-.8063}\ (.78)^{1.6015}\ \exp[-1.8078 \times 10^{-3}\ (150) - 0.3084\ (.78) + 0] = 320.3\ \text{psia}$
- $V_c = 1.206 \times 10^{-2}\ (150)^{.20378}\ (.78)^{-1.3036}\ \exp[-2.657 \times 10^{-3}\ (150) + 0.5287\ (.78) = 2.6012 \times 10^{-3}\ (150)\ (.78)] = .06035\ \text{ft}^3/\text{lb}$
- $T_b = 6.77857\ (150)^{.401673}\ (.78)^{-1.58262}\ \exp[3.77409 \times 10^{-3}\ (150) + 2.984036\ (0.78) - 4.25288 \times 10^{-3}\ (150)\ (0.78)] = 825.26\ °R$

Step 2. Use Edmister's Equation (Equation 1-3) to estimate the acentric factor:

$$\omega = \frac{3[\log(320.3/14.7)]}{7[1139.4/825.26 - 1]} - 1 = 0.5067$$

PROBLEMS

1. The following is a list of the compositional analysis of different hydrocarbon systems. The compositions are expressed in the terms of mol%.

Component	System #1	System #2	System #3	System #4
C_1	68.00	25.07	60.00	12.15
C_2	9.68	11.67	8.15	3.10
C_3	5.34	9.36	4.85	2.51
C_4	3.48	6.00	3.12	2.61
C_5	1.78	3.98	1.41	2.78
C_6	1.73	3.26	2.47	4.85
C_{7+}	9.99	40.66	20.00	72.00

Classify these hydrocarbon systems.

2. If a petroleum fraction has a measured molecular weight of 190 and a specific gravity of 0.8762, characterize this fraction by calculating the boiling point, critical temperature, critical pressure, and critical volume of the fraction. Use the Riazi and Daubert correlation.
3. Calculate the acentric factor and critical compressibility factor of the component in the above problem.

REFERENCES

1. Ahmed, T., "Composition Modeling of Tyler and Mission Canyon Formation Oils with CO_2 and Lean Gases," final report submitted to the Montana's on a New Track for Science (MONTS) program (Montana National Science Foundation Grant Program), 1985.

2. Edmister, W. C., "Applied Hydrocarbon Thermodynamic, Part 4: Compressibility Factors and Equations of State," *Petroleum Refiner,* April 1958, Vol. 37, pp. 173–179.

3. Haugen, O. A., Watson, K. M., and Ragatz R. A., *Chemical Process Principles*, 2nd ed. New York: Wiley, 1959, p. 577.

4. Katz, D. L. and Firoozabadi, A., "Predicting Phase Behavior of Condensate/Crude-oil Systems Using Methane Interaction Coefficients," *JPT,* Nov. 1978, pp. 1649–1655.

5. McCain, W. D., "Heavy Components Control Reservoir Fluid Behavior," *JPT,* September 1994, pp. 746–750.

6. Nath, J., "Acentric Factor and Critical Volumes for Normal Fluids," *Ind. Eng. Chem. Fundam.*, 1985, Vol. 21, No. 3, pp. 325–326.

7. Reid, R., Prausnitz, J. M., and Sherwood, T., *The Properties of Gases and Liquids,* 3rd ed., pp. 21. McGraw-Hill, 1977.

8. Riazi, M. R. and Daubert, T. E., "Characterization Parameters for Petroleum Fractions," *Ind. Eng. Chem. Res.*, 1987, Vol. 26, No. 24, pp. 755–759.

9. Salerno, S., et al., "Prediction of Vapor Pressures and Saturated Vol.," *Fluid Phase Equilibria,* June 10, 1985, Vol. 27, pp. 15–34.

C H A P T E R   2

RESERVOIR-FLUID PROPERTIES

To understand and predict the volumetric behavior of oil and gas reservoirs as a function of pressure, knowledge of the physical properties of reservoir fluids must be gained. These fluid properties are usually determined by laboratory experiments performed on samples of actual reservoir fluids. In the absence of experimentally measured properties, it is necessary for the petroleum engineer to determine the properties from empirically derived correlations. The objective of this chapter is to present several of the well-established physical property correlations for the following reservoir fluids:

• Natural gases
• Crude oil systems
• Reservoir water systems

PROPERTIES OF NATURAL GASES

A gas is defined as a homogeneous fluid of low viscosity and density that has no definite volume but expands to completely fill the vessel in which it is placed. Generally, the natural gas is a mixture of hydrocarbon and nonhydrocarbon gases. The hydrocarbon gases that are normally found in a natural gas are methanes, ethanes, propanes, butanes, pentanes, and small amounts of hexanes and heavier. The nonhydrocarbon gases (i.e., impurities) include carbon dioxide, hydrogen sulfide, and nitrogen.

Knowledge of pressure-volume-temperature (PVT) relationships and other physical and chemical properties of gases is essential for solving problems in natural gas reservoir engineering. These properties include:

- Apparent molecular weight, M_a
- Specific gravity, γ_g
- Compressibility factor, z
- Density, ρ_g
- Specific volume, v
- Isothermal gas compressibility coefficient, c_g
- Gas formation volume factor, B_g
- Gas expansion factor, E_g
- Viscosity, μ_g

The above gas properties may be obtained from direct laboratory measurements or by prediction from generalized mathematical expressions. This section reviews laws that describe the volumetric behavior of gases in terms of pressure and temperature and also documents the mathematical correlations that are widely used in determining the physical properties of natural gases.

BEHAVIOR OF IDEAL GASES

The kinetic theory of gases postulates that gases are composed of a very large number of particles called molecules. For an ideal gas, the volume of these molecules is insignificant compared with the total volume occupied by the gas. It is also assumed that these molecules have no attractive or repulsive forces between them, and that all collisions of molecules are perfectly elastic.

Based on the above kinetic theory of gases, a mathematical equation called equation-of-state can be derived to express the relationship existing between pressure p, volume V, and temperature T for a given quantity of moles of gas n. This relationship for perfect gases is called the **ideal gas law** and is expressed mathematically by the following equation:

$$pV = nRT \tag{2-1}$$

where p = absolute pressure, psia
$\qquad$ V = volume, ft^3
$\qquad$ T = absolute temperature, °R

n = number of moles of gas, lb-mole
R = the universal gas constant which, for the above units, has the value 10.730 psia ft^3/lb-mole °R

The number of pound-moles of gas, i.e., n, is defined as the weight of the gas m divided by the molecular weight M, or:

$$n = \frac{m}{M} \tag{2-2}$$

Combining Equation 2-1 with 2-2 gives:

$$pV = \left(\frac{m}{M}\right) RT \tag{2-3}$$

where m = weight of gas, lb
M = molecular weight, lb/lb-mol

Since the density is defined as the mass per unit volume of the substance, Equation 2-3 can be rearranged to estimate the gas density at any pressure and temperature:

$$\rho_g = \frac{m}{V} = \frac{pM}{RT} \tag{2-4}$$

where ρ_g = density of the gas, lb/ft^3

It should be pointed out that lb refers to lbs mass in any of the subsequent discussions of density in this text.

Example 2-1

Three pounds of n-butane are placed in a vessel at 120°F and 60 psia. Calculate the volume of the gas assuming an ideal gas behavior.

Solution

Step 1. Determine the molecular weight of n-butane from Table 1-1 to give:

M = 58.123

Step 2. Solve Equation 2-3 for the volume of gas:

$$V = \left(\frac{m}{M}\right)\frac{RT}{p}$$

$$V = \left(\frac{3}{58.123}\right)\frac{(10.73)(120 + 460)}{60} = 5.35 \text{ ft}^3$$

Example 2-2

Using the data given in the above example, calculate the density n-butane.

Solution

Solve for the density by applying Equation 2-4:

$$\rho_g = \frac{(60)(58.123)}{(10.73)(580)} = 0.56 \text{ lb/ft}^3$$

Petroleum engineers are usually interested in the behavior of mixtures and rarely deal with pure component gases. Because natural gas is a mixture of hydrocarbon components, the overall physical and chemical properties can be determined from the physical properties of the individual components in the mixture by using appropriate mixing rules.

The basic properties of gases are commonly expressed in terms of the apparent molecular weight, standard volume, density, specific volume, and specific gravity. These properties are defined as follows:

Apparent Molecular Weight

One of the main gas properties that is frequently of interest to engineers is the apparent molecular weight. If y_i represents the mole fraction of the *ith* component in a gas mixture, the apparent molecular weight is defined mathematically by the following equation:

$$M_a = \sum_{i=1} y_i M_i \qquad (2-5)$$

where M_a = apparent molecular weight of a gas mixture
  M_i = molecular weight of the *ith* component in the mixture
  y_i = mole fraction of component i in the mixture

Standard Volume

In many natural gas engineering calculations, it is convenient to measure the volume occupied by 1 lb-mole of gas at a reference pressure and temperature. These reference conditions are usually 14.7 psia and 60°F, and are commonly referred to as standard conditions. The standard volume is then defined as the volume of gas occupied by 1 lb-mol of gas at standard conditions. Applying the above conditions to Equation 2-1 and solving for the volume, i.e., the standard volume, gives:

$$V_{sc} = \frac{(1)\, RT_{sc}}{p_{sc}} = \frac{(1)\,(10.73)\,(520)}{14.7}$$

or

$$V_{sc} = 379.4 \text{ scf/lb-mol} \tag{2-6}$$

where V_{sc} = standard volume, scf/lb-mol
$\quad$ scf = standard cubic feet
$\quad T_{sc}$ = standard temperature, °R
$\quad p_{sc}$ = standard pressure, psia

Density

The density of an ideal gas mixture is calculated by simply replacing the molecular weight of the pure component in Equation 2-4 with the apparent molecular weight of the gas mixture to give:

$$\rho_g = \frac{pM_a}{RT} \tag{2-7}$$

where ρ_g = density of the gas mixture, lb/ft^3
$\quad M_a$ = apparent molecular weight

Specific Volume

The specific volume is defined as the volume occupied by a unit mass of the gas. For an ideal gas, this property can be calculated by applying Equation 2-3:

$$v = \frac{V}{m} = \frac{RT}{p\,M_a} = \frac{1}{\rho_g} \tag{2-8}$$

where v = specific volume, ft³/lb
ρ_g = gas density, lb/ft³

Specific Gravity

The specific gravity is defined as the ratio of the gas density to that of the air. Both densities are measured or expressed at the same pressure and temperature. Commonly, the standard pressure p_{sc} and standard temperature T_{sc} are used in defining the gas specific gravity:

$$\gamma_g = \frac{\rho_g}{\rho_{air}} \tag{2-9}$$

Assuming that the behavior of both the gas mixture and the air is described by the ideal gas equation, the specific gravity can then be expressed as:

$$\gamma_g = \frac{\dfrac{p_{sc}\, M_a}{RT_{sc}}}{\dfrac{p_{sc}\, M_{air}}{RT_{sc}}}$$

or

$$\gamma_g = \frac{M_a}{M_{air}} = \frac{M_a}{28.96} \tag{2-10}$$

where γ_g = gas specific gravity
ρ_{air} = density of the air
M_{air} = apparent molecular weight of the air = 28.96
M_a = apparent molecular weight of the gas
p_{sc} = standard pressure, psia
T_{sc} = standard temperature, °R

Example 2-3

A gas well is producing gas with a specific gravity of 0.65 at a rate of 1.1 MMscf/day. The average reservoir pressure and temperature are 1,500 psi and 150°F. Calculate:

a. Apparent molecular weight of the gas
b. Gas density at reservoir conditions
c. Flow rate in lb/day

Solution

a. From Equation 2-10, solve for the apparent molecular weight:

$$M_a = 28.96 \, \gamma_g$$

$$M_a = (28.96)(0.65) = 18.82$$

b. Apply Equation 2-7 to determine gas density:

$$\rho_g = \frac{(1500)(18.82)}{(10.73)(610)} = 4.31 \, \text{lb/ft}^3$$

c. *Step 1.* Because 1 lb-mol of any gas occupies 379.4 scf at standard conditions, then the daily number of moles that the gas well is producing can be calculated from:

$$n = \frac{(1.1)(10)^6}{379.4} = 2899 \, \text{lb-mol}$$

Step 2. Determine the daily mass m of the gas produced from Equation 2-2:

$$m = (n)(M_a)$$

$$m = (2899)(18.82) = 54559 \, \text{lb/day}$$

Example 2-4

A gas well is producing a natural gas with the following composition:

Component	y_i
CO_2	0.05
C_1	0.90
C_2	0.03
C_3	0.02

Assuming an ideal gas behavior, calculate:

a. Apparent molecular weight
b. Specific gravity
c. Gas density at 2000 psia and 150°F
d. Specific volume at 2000 psia and 150°F

Solution

Component	y_i	M_i	$y_i \cdot M_i$
CO_2	0.05	44.01	2.200
C_1	0.90	16.04	14.436
C_2	0.03	30.07	0.902
C_3	0.02	44.11	0.882
			$M_a = 18.42$

a. Apply Equation 2-5 to calculate the apparent molecular weight:

$$M_a = 18.42$$

b. Calculate the specific gravity by using Equation 2-10:

$$\gamma_g = 18.42 / 28.96 = 0.636$$

c. Solve for the density by applying Equation 2-7:

$$\rho_g = \frac{(2000)(18.42)}{(10.73)(610)} = 5.628 \text{ lb/ft}^3$$

d. Determine the specific volume from Equation 2-8:

$$v = \frac{1}{5.628} = 0.178 \text{ ft}^3/\text{lb}$$

BEHAVIOR OF REAL GASES

In dealing with gases at a very low pressure, the ideal gas relationship is a convenient and generally satisfactory tool. At higher pressures, the use of the ideal gas equation-of-state may lead to errors as great as 500%, as compared to errors of 2–3% at atmospheric pressure.

Basically, the magnitude of deviations of real gases from the conditions of the ideal gas law increases with increasing pressure and temperature and varies widely with the composition of the gas. Real gases behave differently than ideal gases. The reason for this is that the perfect gas law was derived under the assumption that the volume of molecules is insignificant and that no molecular attraction or repulsion exists between them. This is not the case for real gases.

Numerous equations-of-state have been developed in the attempt to correlate the pressure-volume-temperature variables for real gases with experimental data. In order to express a more exact relationship between the variables p, V, and T, a correction factor called the *gas compressibility factor, gas deviation factor,* or simply the *z-factor,* must be introduced into Equation 2-1 to account for the departure of gases from ideality. The equation has the following form:

$$pV = znRT \tag{2-11}$$

where the gas compressibility factor z is a dimensionless quantity and is defined as the ratio of the actual volume of n-moles of gas at T and p to the ideal volume of the same number of moles at the same T and p:

$$z = \frac{V_{actual}}{V_{ideal}} = \frac{V}{(nRT)/p}$$

Studies of the gas compressibility factors for natural gases of various compositions have shown that compressibility factors can be generalized with sufficient accuracies for most engineering purposes when they are expressed in terms of the following two dimensionless properties:

• Pseudo-reduced pressure
• Pseudo-reduced temperature

These dimensionless terms are defined by the following expressions:

$$p_{pr} = \frac{p}{p_{pc}} \tag{2-12}$$

$$T_{pr} = \frac{T}{T_{pc}} \tag{2-13}$$

where   p = system pressure, psia

p_{pr} = pseudo-reduced pressure, dimensionless

T = system temperature, °R

T_{pr} = pseudo-reduced temperature, dimensionless

p_{pc}, T_{pc} = pseudo-critical pressure and temperature, respectively, and defined by the following relationships:

$$P_{pc} = \sum_{i=1} y_i \, P_{ci} \qquad\qquad (2\text{-}14)$$

$$T_{pc} = \sum_{i=1} y_i \, T_{ci} \qquad\qquad (2\text{-}15)$$

It should be pointed out that these pseudo-critical properties, i.e., p_{pc} and T_{pc}, do not represent the actual critical properties of the gas mixture. These pseudo properties are used as correlating parameters in generating gas properties.

Based on the concept of pseudo-reduced properties, Standing and Katz (1942) presented a generalized gas compressibility factor chart as shown in Figure 2-1. The chart represents compressibility factors of sweet natural gas as a function of p_{pr} and T_{pr}. This chart is generally reliable for natural gas with minor amount of nonhydrocarbons. It is one of the most widely accepted correlations in the oil and gas industry.

Example 2-5

A gas reservoir has the following gas composition: the initial reservoir pressure and temperature are 3000 psia and 180°F, respectively.

Component	y_i
CO_2	0.02
N_2	0.01
C_1	0.85
C_2	0.04
C_3	0.03
i - C_4	0.03
n - C_4	0.02

Calculate the gas compressibility factor under initial reservoir conditions.

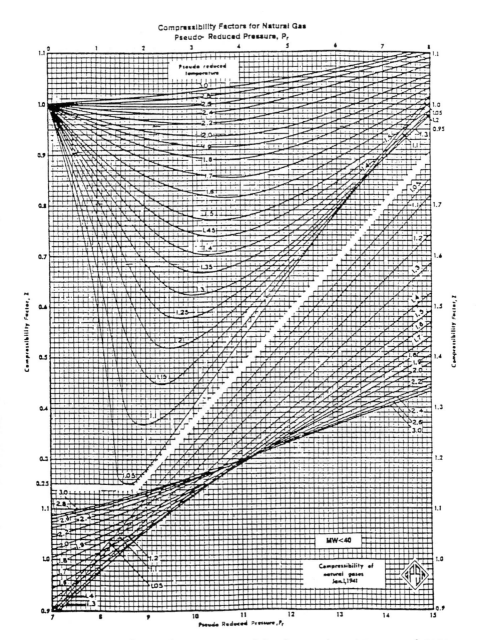

Figure 2-1. Standing and Katz compressibility factors chart. (*Courtesy of GPSA and* GPA Engineering Data Book, *EO Edition, 1987.*)

Solution

Component	y_i	$T_{ci}, °R$	$y_i T_{ci}$	p_{ci}	$y_i p_{ci}$
CO_2	0.02	547.91	10.96	1071	21.42
N_2	0.01	227.49	2.27	493.1	4.93
C_1	0.85	343.33	291.83	666.4	566.44
C_2	0.04	549.92	22.00	706.5	28.26
C_3	0.03	666.06	19.98	616.4	18.48
$i - C_4$	0.03	734.46	22.03	527.9	15.84
$n - C_4$	0.02	765.62	15.31	550.6	11.01
			$T_{pc} = 383.38$		$p_{pc} = 666.38$

Step 1. Determine the pseudo-critical pressure from Equation 2-14:

$$p_{pc} = 666.18$$

Step 2. Calculate the pseudo-critical temperature from Equation 2-15:

$$T_{pc} = 383.38$$

Step 3. Calculate the pseudo-reduced pressure and temperature by applying Equations 2-12 and 2-13, respectively:

$$p_{pr} = \frac{3000}{666.38} = 4.50$$

$$T_{pr} = \frac{640}{383.38} = 1.67$$

Step 4. Determine the z-factor from Figure 2-1, to give:

$$z = 0.85$$

Equation 2-11 can be written in terms of the apparent molecular weight M_a and the weight of the gas m:

$$pV = z \left(\frac{m}{M_a} \right) RT$$

Solving the above relationship for the gas specific volume and density, give:

$$v = \frac{V}{m} = \frac{zRT}{pM_a} \qquad (2\text{-}16)$$

$$\rho_g = \frac{1}{v} = \frac{pM_a}{zRT} \qquad (2\text{-}17)$$

where v = specific volume, ft^3/lb
ρ_g = density, lb/ft^3

Example 2-6

Using the data in Example 2-5 and assuming real gas behavior, calculate the density of the gas phase under initial reservoir conditions. Compare the results with that of ideal gas behavior.

Solution

Component	y_i	M_i	$y_i \cdot M_i$	$T_{ci}, °R$	$y_i T_{ci}$	P_{ci}	$y_i P_{ci}$
CO_2	0.02	44.01	0.88	547.91	10.96	1071	21.42
N_2	0.01	28.01	0.28	227.49	2.27	493.1	4.93
C_1	0.85	16.04	13.63	343.33	291.83	666.4	566.44
C_2	0.04	30.1	1.20	549.92	22.00	706.5	28.26
C_3	0.03	44.1	1.32	666.06	19.98	616.40	18.48
i - C_4	0.03	58.1	1.74	734.46	22.03	527.9	15.84
n - C_4	0.02	58.1	1.16	765.62	15.31	550.6	11.01
			$M_a = 20.23$		$T_{pc} = 383.38$		$P_{pc} = 666.38$

Step 1. Calculate the apparent molecular weight from Equation 2-5:

$M_a = 20.23$

Step 2. Determine the pseudo-critical pressure from Equation 2-14:

$p_{pc} = 666.18$

Step 3. Calculate the pseudo-critical temperature from Equation 2-15:

$T_{pc} = 383.38$

Step 4. Calculate the pseudo-reduced pressure and temperature by applying Equations 2-12 and 2-13, respectively:

$$p_{pr} = \frac{3000}{666.38} = 4.50$$

$$T_{pr} = \frac{640}{383.38} = 1.67$$

Step 5. Determine the z-factor from Figure 2-1:

$$z = 0.85$$

Step 6. Calculate the density from Equation 2-17:

$$\rho_g = \frac{(3000)(20.23)}{(0.85)(10.73)(640)} = 10.4 \text{ lb} / \text{ft}^3$$

Step 7. Calculate the density of the gas assuming an ideal gas behavior from Equation 2-7:

$$\rho_g = \frac{(3000)(20.23)}{(10.73)(640)} = 8.84 \text{ lb} / \text{ft}^3$$

The results of the above example show that the ideal gas equation estimated the gas density with an absolute error of 15% when compared with the density value as predicted with the real gas equation.

In cases where the composition of a natural gas is not available, the pseudo-critical properties, i.e., p_{pc} and T_{pc}, can be predicted solely from the specific gravity of the gas. Brown et al. (1948) presented a graphical method for a convenient approximation of the pseudo-critical pressure and pseudo-critical temperature of gases when only the specific gravity of the gas is available. The correlation is presented in Figure 2-2. Standing (1977) expressed this graphical correlation in the following mathematical forms:

Case 1: Natural Gas Systems

$$T_{pc} = 168 + 325 \gamma_g - 12.5 \gamma_g^2 \qquad (2\text{-}18)$$

$$p_{pc} = 677 + 15.0 \gamma_g - 37.5 \gamma_g^2 \qquad (2\text{-}19)$$

Case 2: Gas-Condensate Systems

$$T_{pc} = 187 + 330 \gamma_g - 71.5 \gamma_g^2 \qquad (2\text{-}20)$$

Pseudo-critical Properties of Natural Gases

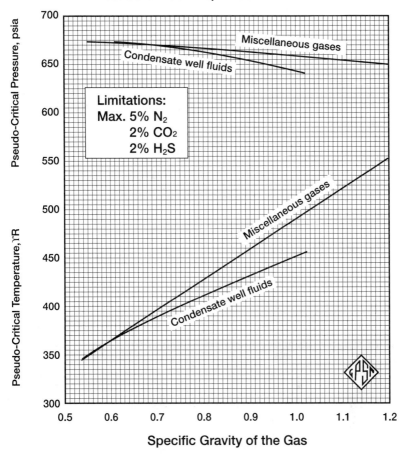

Figure 2-2. Pseudo-critical properties of natural gases. (*Courtesy of GPSA and GPA Engineering Data Book, 10th Edition, 1987.*)

$$p_{pc} = 706 - 51.7 \, \gamma_g - 11.1 \, \gamma_g^2 \tag{2-21}$$

where T_{pc} = pseudo-critical temperature, °R
p_{pc} = pseudo-critical pressure, psia
γ_g = specific gravity of the gas mixture

Example 2-7

Rework Example 2-5 by calculating the pseudo-critical properties from Equations 2-18 and 2-19.

Solution

Step 1. Calculate the specific gravity of the gas:

$$\gamma_g = \frac{M_a}{28.96} = \frac{20.23}{28.96} = 0.699$$

Step 2. Solve for the pseudo-critical properties by applying Equations 2-18 and 2-19:

$$T_{pc} = 168 + 325\,(0.699) - 12.5\,(0.699)^2 = 389.1°R$$

$$p_{pc} = 677 + 15\,(0.699) - 37.5\,(0.699)^2 = 669.2 \text{ psia}$$

Step 3. Calculate p_{pr} and T_{pr}.

$$p_{pr} = \frac{3000}{669.2} = 4.48$$

$$T_{pr} = \frac{640}{389.1} = 1.64$$

Step 4. Determine the gas compressibility factor from Figure 2-1:

$$z = 0.824$$

Step 5. Calculate the density from Equation 2-17:

$$\rho_g = \frac{(3000)\,(20.23)}{(0.845)\,(10.73)\,(640)} = 10.46 \text{ lb/ft}^3$$

EFFECT OF NONHYDROCARBON COMPONENTS ON THE Z-FACTOR

Natural gases frequently contain materials other than hydrocarbon components, such as nitrogen, carbon dioxide, and hydrogen sulfide. Hydrocarbon gases are classified as sweet or sour depending on the hydrogen sulfide content. Both sweet and sour gases may contain nitrogen, carbon dioxide, or both. A hydrocarbon gas is termed a sour gas if it contains one grain of H_2S per 100 cubic feet.

The common occurrence of small percentages of nitrogen and carbon dioxide is, in part, considered in the correlations previously cited. Concentrations of up to 5 percent of these nonhydrocarbon components will not seriously affect accuracy. Errors in compressibility factor calculations as large as 10 percent may occur in higher concentrations of nonhydrocarbon components in gas mixtures.

Nonhydrocarbon Adjustment Methods

There are two methods that were developed to adjust the pseudo-critical properties of the gases to account for the presence of the nonhydrocarbon components. These two methods are the:

• Wichert-Aziz correction method
• Carr-Kobayashi-Burrows correction method

The Wichert-Aziz Correction Method

Natural gases that contain H_2S and or CO_2 frequently exhibit different compressibility-factors behavior than do sweet gases. Wichert and Aziz (1972) developed a simple, easy-to-use calculation procedure to account for these differences. This method permits the use of the Standing-Katz chart, i.e., Figure 2-1, by using a pseudo-critical temperature adjustment factor, which is a function of the concentration of CO_2 and H_2S in the sour gas. This correction factor is then used to adjust the pseudo-critical temperature and pressure according to the following expressions:

$$T'_{pc} = T_{pc} - \varepsilon \qquad (2\text{-}22)$$

$$p'_{pc} = \frac{p_{pc} T'_{pc}}{T_{pc} + B(1 - B)\varepsilon} \qquad (2\text{-}23)$$

where T_{pc} = pseudo-critical temperature, °R
$\quad$ p_{pc} = pseudo-critical pressure, psia
$\quad$ T'_{pc} = corrected pseudo-critical temperature, °R
$\quad$ p'_{pc} = corrected pseudo-critical pressure, psia
$\quad$ B = mole fraction of H_2S in the gas mixture
$\quad$ ε = pseudo-critical temperature adjustment factor and is defined mathematically by the following expression

$$\varepsilon = 120 [A^{0.9} - A^{1.6}] + 15 (B^{0.5} - B^{4.0}) \qquad (2\text{-}24)$$

where the coefficient A is the sum of the mole fraction H_2S and CO_2 in the gas mixture, or:

$$A = y_{H_2S} + y_{CO_2}$$

The computational steps of incorporating the adjustment factor ε into the z-factor calculations are summarized below:

Step 1. Calculate the pseudo-critical properties of the whole gas mixture by applying Equations 2-18 and 2-19 or Equations 2-20 and 2-21.

Step 2. Calculate the adjustment factor ε from Equation 2-24.

Step 3. Adjust the calculated p_{pc} and T_{pc} (as computed in Step 1) by applying Equations 2-22 and 2-23.

Step 4. Calculate the pseudo-reduced properties, i.e., p_{pr} and T_{pr}, from Equations 2-11 and 2-12.

Step 5. Read the compressibility factor from Figure 2-1.

Example 2-8

A sour natural gas has a specific gravity of 0.7. The compositional analysis of the gas shows that it contains 5 percent CO_2 and 10 percent H_2S. Calculate the density of the gas at 3500 psia and 160°F.

Solution

Step 1. Calculate the uncorrected pseudo-critical properties of the gas from Equations 2-18 and 2-19:

$$T_{pc} = 168 + 325\,(0.7) - 12.5\,(0.7)^2 = 389.38°R$$

$$p_{pc} = 677 + 15\,(0.7) - 37.5\,(0.7)^2 = 669.1 \text{ psia}$$

Step 2. Calculate the pseudo-critical temperature adjustment factor from Equation 2-24:

$$\varepsilon = 120\,(0.15^{0.9} - 0.15^{1.6}) + 15\,(0.1^{0.5} - 0.1^4) = 20.735$$

Step 3. Calculate the corrected pseudo-critical temperature by applying Equation 2-22:

$$T'_{pc} = 389.38 - 20.735 = 368.64$$

Step 4. Adjust the pseudo-critical pressure p_{pc} by applying Equation 2-23:

$$p'_{pc} = \frac{(669.1)(368.64)}{389.38 + 0.1(1-0.1)(20.635)}$$

Step 5. Calculate p_{pr} and T_{pr}:

$$p_{pr} = \frac{3500}{630.44} = 5.55$$

$$T_{pr} = \frac{160 + 460}{368.64} = 1.68$$

Step 6. Determine the z-factor from Figure 2-1:

$$z = 0.89$$

Step 7. Calculate the apparent molecular weight of the gas from Equation 2-10:

$$M_a = (28.96)(0.7) = 20.27$$

Step 8. Solve for gas density:

$$\rho_g = \frac{(3500)(20.27)}{(0.89)(10.73)(620)} = 11.98 \text{ lb/ft}^3$$

The Carr-Kobayashi-Burrows Correction Method

Carr, Kobayashi, and Burrows (1954) proposed a simplified procedure to adjust the pseudo-critical properties of natural gases when nonhydrocarbon components are present. The method can be used when the composition of the natural gas is not available. The proposed procedure is summarized in the following steps:

Step 1. Knowing the specific gravity of the natural gas, calculate the pseudo-critical temperature and pressure by applying Equations 2-18 and 2-19.

Step 2. Adjust the estimated pseudo-critical properties by using the following two expressions:

$$T'_{pc} = T_{pc} - 80 \, y_{CO_2} + 130 \, y_{H_2S} - 250 \, y_{N_2} \qquad (2\text{-}25)$$

$$p'_{pc} = p_{pc} + 440 \, y_{CO_2} + 600 \, y_{H_2S} - 170 \, y_{N_2} \qquad (2\text{-}26)$$

where T'_{pc} = the adjusted pseudo-critical temperature, °R
   T_{pc} = the unadjusted pseudo-critical temperature, °R
   y_{CO_2} = mole fraction of CO_2
   y_{N_2} = mole fraction of H_2S in the gas mixture
   = mole fraction of Nitrogen
   p'_{pc} = the adjusted pseudo-critical pressure, psia
   p_{pc} = the unadjusted pseudo-critical pressure, psia

Step 3. Use the adjusted pseudo-critical temperature and pressure to calculate the pseudo-reduced properties.

Step 4. Calculate the z-factor from Figure 2-1.

Example 2-9

Using the data in Example 2-8, calculate the density by employing the above correction procedure.

Solution

Step 1. Determine the corrected pseudo-critical properties from Equations 2-25 and 2-26:

$$T'_{pc} = 389.38 - 80 \, (0.05) + 130 \, (0.10) - 250 \, (0) = 398.38°R$$

$$p'_{pc} = 669.1 + 440 \, (0.05) + 600 \, (0.10) - 170 \, (0) = 751.1 \text{ psia}$$

Step 2. Calculate p_{pr} and T_{pr}:

$$p_{pr} = \frac{3500}{751.1} = 4.56$$

$$T_{pr} = \frac{620}{398.38} = 1.56$$

Step 3. Determine the gas compressibility factor from Figure 2-1:

$z = 0.820$

Step 4. Calculate the gas density:

$$\rho_g = \frac{(3500)\,(20.27)}{(0.82)\,(10.73)\,(620)} = 13.0 \text{ lb/ft}^3$$

CORRECTION FOR HIGH-MOLECULAR WEIGHT GASES

It should be noted that the Standing and Katz compressibility factor chart (Figure 2-1) was prepared from data on binary mixtures of methane with propane, ethane, and butane, and on natural gases, thus covering a wide range in composition of hydrocarbon mixtures containing methane. No mixtures having molecular weights in excess of 40 were included in preparing this plot.

Sutton (1985) evaluated the accuracy of the Standing-Katz compressibility factor chart using laboratory-measured gas compositions and z-factors, and found that the chart provides satisfactory accuracy for engineering calculations. However, Kay's mixing rules, i.e., Equations 2-13 and 2-14 (or comparable gravity relationships for calculating pseudo-critical pressure and temperature), result in unsatisfactory z-factors for high molecular weight reservoir gases. The author observed that large deviations occur to gases with high heptanes-plus concentrations. He pointed out that Kay's mixing rules should not be used to determine the pseudo-critical pressure and temperature for reservoir gases with specific gravities greater than about 0.75.

Sutton proposed that this deviation can be minimized by utilizing the mixing rules developed by Stewart et al. (1959), together with newly introduced empirical adjustment factors (F_J, E_J, and E_K) that are related to the presence of the heptane-plus fraction in the gas mixture. The proposed approach is outlined in the following steps:

Step 1. Calculate the parameters J and K from the following relationships:

$$J = \frac{1}{3}\left[\sum_i y_i\,(T_{ci}/p_{ci})\right] + \frac{2}{3}\left[\sum_i y_i\,(T_{ci}/p_{ci})^{0.5}\right]^2 \qquad (2\text{-}27)$$

$$K = \sum_i [y_i T_{ci} / \sqrt{p_{ci}}] \qquad (2-28)$$

where J = Stewart-Burkhardt-Voo correlating parameter, °R/psia
       K = Stewart-Burkhardt-Voo correlating parameter, °R/psia
       y_i = mole fraction of component i in the gas mixture.

Step 2. Calculate the adjustment parameters F_J, E_J, and E_K from the following expressions:

$$F_J = \frac{1}{3}[y(T_c/p_c)]_{C_{7+}} + \frac{2}{3}[y(T_c/p_c)^{0.5}]^2_{C_{7+}} \qquad (2-29)$$

$$E_J = 0.6081\, F_J + 1.1325\, F_J^2 - 14.004\, F_J\, y_{C_{7+}}$$
$$+ 64.434\, F_J\, y^2_{C_{7+}} \qquad (2-30)$$

$$E_K = [T_c/M\, p_c]_{C7+}\, [0.3129\, y_{C_{7+}} - 4.8156\, (y_{C_{7+}})^2$$
$$+ 27.3751\, (y_{C_{7+}})^3] \qquad (2-31)$$

where $y_{C_{7+}}$ = mole fraction of the heptanes-plus component
        $(T_c)_{C_{7+}}$ = critical temperature of the C_{7+}
        $(p_c)_{C_{7+}}$ = critical pressure of the C_{7+}

Step 3. Adjust the parameters J and K by applying the adjustment factors E_J and E_K, according to the relationships:

$$J' = J - E_J \qquad (2-32)$$

$$K' = K - E_K \qquad (2-33)$$

where   J, K = calculated from Equations 2-27 and 2-28
        E_J, E_K = calculated from Equations 2-30 and 2-31

Step 4. Calculate the adjusted pseudo-critical temperature and pressure from the expressions:

$$T'_{pc} = \frac{(K')^2}{J'} \qquad (2-34)$$

$$p'_{pc} = \frac{T'_{pc}}{J'} \qquad (2-35)$$

Step 5. Having calculated the adjusted T_{pc} and p_{pc}, the regular procedure of calculating the compressibility factor from the Standing and Katz chart is followed.

Sutton's proposed mixing rules for calculating the pseudo-critical properties of high-molecular-weight reservoir gases, i.e., $\gamma_g > 0.75$, should significantly improve the accuracy of the calculated z-factor.

Example 2-10

A hydrocarbon gas system has the following composition:

Component	y
C_1	0.83
C_2	0.06
C_3	0.03
$n\text{-}C_4$	0.02
$n\text{-}C_5$	0.02
C_6	0.01
C_{7+}	0.03

The heptanes-plus fraction is characterized by a molecular weight and specific gravity of 161 and 0.81, respectively.

a. Using Sutton's methodology, calculate the density of the gas 2000 psi and 150°F.
b. Recalculate the gas density without adjusting the pseudo-critical properties.

Solution

Part A.

Step 1. Calculate the critical properties of the heptanes-plus fraction by the Riazi-Daubert correlation (Chapter 1, Equation 1-2):

$$(T_c)_{C_{7+}} = 544.2 \; 161^{0.2998} 0.81^{1.0555}$$
$$\exp^{[-1.3478(10)^{-4}(150)-0.61641(0.81)]} = 1189°R$$

$$(p_c)_{C_{7+}} = 4.5203(10)^4 \; 161^{-.8063} \; 0.81^{1.6015}$$
$$\exp^{[-1.8078(10)^{-3}(150)-0.3084(0.81)]} = 318.4 \; psia$$

Step 2. Construct the following table:

Component	y_i	M_i	T_{ci}	p_{ci}	$y_i M_i$	$y_i(T_{ci}/p_{ci})$	$y_i z \overline{(T_c/p_c)_i}$	$y_i[T_c/z\overline{p_c}]_i$
C_1	0.83	16.0	343.33	666.4	13.31	.427	.596	11.039
C_2	0.06	30.1	549.92	706.5	1.81	.047	.053	1.241
C_3	0.03	44.1	666.06	616.4	1.32	.032	.031	.805
n-C_4	0.02	58.1	765.62	550.6	1.16	.028	.024	.653
n-C_5	0.02	72.2	845.60	488.6	1.45	.035	.026	.765
C_6	0.01	84.0	923.00	483.0	0.84	.019	.014	.420
C_{7+}	0.03	161.	1189.0	318.4	4.83	.112	.058	1.999
Total					27.72	0.700	0.802	16.972

Step 3. Calculate the parameters J and K from Equations 2-27 and 2-28:

$$J = (1/3) [0.700] + (2/3) [0.802]^2 = 0.662$$

$$K = 16.922$$

Step 4. Determine the adjustment factors F_J, E_J and E_K by applying Equations 2-29 through 2-31:

$$F_J = \frac{1}{3}[0.112] + \frac{2}{3}[0.058]^2 = 0.0396$$

$$E_J = 0.6081 (0.04) + 1.1325 (0.04)^2 - 14.004 (0.04) (0.03)$$
$$+ 64.434 (0.04) 0.3^2 = 0.012$$

$$E_K = 66.634 [0.3129 (0.03) - 4.8156 (0.03)^2$$
$$+ 27.3751 (0.03)^3] = 0.386$$

Step 5. Calculate the parameters J' and K' from Equations 2-32 and 2-33:

$$J' = 0.662 - 0.012 = 0.650$$

$$K' = 16.922 - 0.386 = 16.536$$

Step 6. Determine the adjusted pseudo-critical properties from Equations 2-33 and 2-36:

$$T'_{pc} = \frac{(16.536)^2}{0.65} = 420.7$$

$$p'_{pc} = \frac{420.7}{0.65} = 647.2$$

Step 7. Calculate the pseudo-reduced properties of the gas by applying Equations 2-11 and 2-12, to give:

$$p_{pr} = \frac{2000}{647.2} = 3.09$$

$$T_{pr} = \frac{610}{420.7} = 1.45$$

Step 8. Calculate the z-factor from Figure 2-1, to give:

$$z = 0.745$$

Step 9. From Equation 2-16, calculate the density of the gas:

$$\rho_g = \frac{(2000)(24.73)}{(10.73)(610)(.745)} = 10.14 \, \text{lb/ft}^3$$

Part B.

Step 1. Calculate the specific gravity of the gas:

$$\gamma_g = \frac{M_a}{28.96} = \frac{24.73}{28.96} = 0.854$$

Step 2. Solve for the pseudo-critical properties by applying Equations 2-18 and 2-19:

$$T_{pc} = 168 + 325 \,(0.854) - 12.5 \,(0.854)^2 = 436.4°R$$

$$p_{pc} = 677 + 15 \,(0.854) - 37.5 \,(0.854)^2 = 662.5 \, \text{psia}$$

Step 3. Calculate p_{pr} and T_{pr}:

$$p_{pr} = \frac{2000}{662.5} = 3.02$$

$$T_{pr} = \frac{610}{436.4} = 1.40$$

Step 4. Calculate the z-factor from Figure 2-1, to give:

$$z = 0.710$$

Step 5. From Equation 2-16, calculate the density of the gas:

$$\rho_g = \frac{(2000)\,(24.73)}{(10.73)\,(610)\,(.710)} = 10.64 \text{lb}/\text{ft}^3$$

DIRECT CALCULATION OF COMPRESSIBILITY FACTORS

After four decades of existence, the Standing-Katz z-factor chart is still widely used as a practical source of natural gas compressibility factors. As a result, there has been an apparent need for a simple mathematical description of that chart. Several empirical correlations for calculating z-factors have been developed over the years. The following three empirical correlations are described below:

• Hall-Yarborough
• Dranchuk-Abu-Kassem
• Dranchuk-Purvis-Robinson

The Hall-Yarborough Method

Hall and Yarborough (1973) presented an equation-of-state that accurately represents the Standing and Katz z-factor chart. The proposed expression is based on the Starling-Carnahan equation-of-state. The coefficients of the correlation were determined by fitting them to data taken from the Standing and Katz z-factor chart. Hall and Yarborough proposed the following mathematical form:

$$z = \left[\frac{0.06125\, p_{pr} t}{Y}\right] \exp[-1.2(1-t)^2] \qquad (2\text{-}36)$$

where p_{pr} = pseudo-reduced pressure
$\quad\quad\ \ t$ = reciprocal of the pseudo-reduced temperature, i.e., T_{pc}/T
$\quad\quad\ \ Y$ = the reduced density that can be obtained as the solution of the following equation:

$$F(Y) = X1 + \frac{Y + Y^2 + Y^3 + Y^4}{(1-Y)^3} - (X2) Y^2 + (X3) Y^{X4} = 0 \qquad (2\text{-}37)$$

where $X1 = -0.06125 \, p_{pr} \, t \, \exp[-1.2 \, (1-t)^2]$
$X2 = (14.76 \, t - 9.76 \, t^2 + 4.58 \, t^3)$
$X3 = (90.7 \, t - 242.2 \, t^2 + 42.4 \, t^3)$
$X4 = (2.18 + 2.82 \, t)$

Equation 2-37 is a nonlinear equation and can be conveniently solved for the reduced density Y by using the Newton-Raphson iteration technique. The computational procedure of solving Equation 2-37 at any specified pseudo-reduced pressure p_{pr} and temperature T_{pr} is summarized in the following steps:

Step 1. Make an initial guess of the unknown parameter, Y^k, where k is an iteration counter. An appropriate initial guess of Y is given by the following relationship:

$$Y^k = 0.0125 \, p_{pr} \, t \, \exp[-1.2 \, (1-t)^2]$$

Step 2. Substitute this initial value in Equation 2-37 and evaluate the nonlinear function. Unless the correct value of Y has been initially selected, Equation 2-37 will have a nonzero value of F(Y):

Step 3. A new improved estimate of Y, i.e., Y^{k+1}, is calculated from the following expression:

$$Y^{k+1} = Y^k - \frac{f(Y^k)}{f'(Y^k)} \qquad (2\text{-}38)$$

where $f'(Y^k)$ is obtained by evaluating the derivative of Equation 2-37 at Y^k, or:

$$f'(Y) = \frac{1 + 4Y + 4Y^2 - 4Y^3 + Y^4}{(1-Y)^4} - 2 \, (X2) \, Y$$
$$+ (X3) \, (X4) \, Y^{(X4-1)} \qquad (2\text{-}39)$$

Step 4. Steps 2–3 are repeated n times, until the error, i.e., abs($Y^k - Y^{k+1}$), becomes smaller than a preset tolerance, e.g., 10^{-12}:

Step 5. The correct value of Y is then used to evaluate Equation 2-36 for the compressibility factor.

Hall and Yarborough pointed out that the method is not recommended for application if the pseudo-reduced temperature is less than one.

The Dranchuk-Abu-Kassem Method

Dranchuk and Abu-Kassem (1975) derived an analytical expression for calculating the reduced gas density that can be used to estimate the gas compressibility factor. The reduced gas density ρ_r is defined as the ratio of the gas density at a specified pressure and temperature to that of the gas at its critical pressure or temperature, or:

$$\rho_r = \frac{\rho}{\rho_c} = \frac{p\,M_a/[zRT]}{p_c\,M_a/[z_c RT_c]} = \frac{p/[zT]}{p_c/[z_c T_c]}$$

The critical gas compressibility factor z_c is approximately 0.27 which leads to the following simplified expression for the reduced gas density:

$$\rho_r = \frac{0.27\,p_{pr}}{z\,T_{pr}} \qquad (2\text{-}40)$$

The authors proposed the following eleven-constant equation-of-state for calculating the reduced gas density:

$$f(\rho_r) = (R_1)\rho_r - \frac{R_2}{\rho_r} + (R_3)\rho_r^2 - (R_4)\rho_r^5$$

$$+\,(R_5)(1 + A_{11}\,\rho_r^2)\exp[-A_{11}\,\rho_r^2] + 1 = 0 \qquad (2\text{-}41)$$

With the coefficients R_1 through R_5 as defined by the following relations:

$$R_1 = \left[A_1 + \frac{A_2}{T_{pr}} + \frac{A_3}{T_{pr}^3} + \frac{A_r}{T_{pr}^4} + \frac{A_t}{T_{pr}^5} \right]$$

$$R_2 = \left[\frac{0.27\,p_{pr}}{T_{pr}} \right]$$

$$R_3 = \left[A_6 + \frac{A_7}{T_{pr}} + \frac{A_8}{T_{pr}^2} \right]$$

$$R_4 = A_9 \left[\frac{A_7}{T_{pr}} + \frac{A_8}{T_{pr}^2} \right]$$

$$R_5 = \left[\frac{A_{10}}{T_{pr}^3} \right]$$

(2-42)

The constants A_1 through A_{11} were determined by fitting the equation, using nonlinear regression models, to 1,500 data points from the Standing and Katz z-factor chart. The coefficients have the following values:

$A_1 = 0.3265$ $A_2 = -1.0700$ $A_3 = -0.5339$ $A_4 = 0.01569$
$A_5 = -0.05165$ $A_6 = 0.5475$ $A_7 = -0.7361$ $A_8 = 0.1844$
$A_9 = 0.1056$ $A_{10} = 0.6134$ $A_{11} = 0.7210$

Equation 2-41 can be solved for the reduced gas density ρ_r by applying the Newton-Raphson iteration technique as summarized in the following steps:

Step 1. Make an initial guess of the unknown parameter, ρ_r^k, where k is an iteration counter. An appropriate initial guess of ρ_r^k is given by the following relationship:

$$\rho_r = \frac{0.27\, p_{pr}}{T_{pr}}$$

Step 2. Substitute this initial value in Equation 2-41 and evaluate the nonlinear function. Unless the correct value of ρ_r^k has been initially selected, Equation 2-41 will have a nonzero value for the function $f(\rho_r^k)$.

Step 3. A new improved estimate of ρ_r, i.e., ρ_r^{k+1}, is calculated from the following expression:

$$\rho_r^{k+1} = \rho_r^k - \frac{f(\rho_r^k)}{f'(\rho_r^k)}$$

where

$$f'(\rho_r) = (R_1) + \frac{R_2}{\rho_r^2} + 2(R_3)\rho_r - 5(R_4)\rho_r^4 + 2(R_5)\rho_r$$
$$\exp[-A_{11}\rho_r^2][(1 + 2A_{11}\rho_r^3) - A_{11}\rho_r^2(1 + A_{11}\rho_r^2)]$$

Step 4. Steps 2–3 are repeated n times, until the error, i.e., abs($\rho_r^k - \rho_r^{k+1}$), becomes smaller than a preset tolerance, e.g., 10^{-12}.

Step 5. The correct value of ρ_r is then used to evaluate Equation 2-40 for the compressibility factor, i.e.,:

$$z = \frac{0.27 \, p_{pr}}{\rho_r T_{pr}}$$

The proposed correlation was reported to duplicate compressibility factors from the Standing and Katz chart with an average absolute error of 0.585 percent and is applicable over the ranges:

$$0.2'' \, p_{pr} < 30$$

$$1.0 < T_{pr}'' \, 3.0$$

The Dranchuk-Purvis-Robinson Method

Dranchuk, Purvis, and Robinson (1974) developed a correlation based on the Benedict-Webb-Rubin type of equation-of-state. Fitting the equation to 1,500 data points from the Standing and Katz z-factor chart optimized the eight coefficients of the proposed equations. The equation has the following form:

$$1 + T_1\,\rho_r + T_2\,\rho_r^2 + T_3\,\rho_r^5 + [T_4\rho_r^2\,(1 + A_8\,\rho_r^2)$$
$$\exp(-A_8\,\rho_r^2)] - \frac{T_5}{\rho_r} = 0 \qquad (2\text{-}43)$$

with

$$T_1 = \left[A_1 + \frac{A_2}{T_{pr}} + \frac{A_3}{T_{pr}^3}\right]$$

$$T_2 = \left[A_4 + \frac{A_5}{T_{pr}} \right]$$

$$T_3 = [A_5 A_6 / T_{pr}]$$

$$T_4 = [A_7 / T_{pr}^3]$$

$$T_5 = [0.27 \, p_{pr} / T_{pr}]$$

where ρ_r is defined by Equation 2-41 and the coefficients A_1 through A_8 have the following values:

$A_1 = 0.31506237$ $A_5 = -0.61232032$
$A_2 = -1.0467099$ $A_6 = -0.10488813$
$A_3 = -0.57832720$ $A_7 = 0.68157001$
$A_4 = 0.53530771$ $A_8 = 0.68446549$

The solution procedure of Equation 2-43 is similar to that of Dranchuk and Abu-Kassem.

The method is valid within the following ranges of pseudo-reduced temperature and pressure:

$$1.05 \, '' \, T_{pr} < 3.0$$

$$0.2 \, '' \, p_{pr} \, '' \, 3.0$$

COMPRESSIBILITY OF NATURAL GASES

Knowledge of the variability of fluid compressibility with pressure and temperature is essential in performing many reservoir engineering calculations. For a liquid phase, the compressibility is small and usually assumed to be constant. For a gas phase, the compressibility is neither small nor constant.

By definition, the isothermal gas compressibility is the change in volume per unit volume for a unit change in pressure or, in equation form:

$$c_g = -\frac{1}{V} \left(\frac{\partial V}{\partial p} \right)_T \qquad (2 \text{-} 44)$$

where c_g = isothermal gas compressibility, 1/psi.

From the real gas equation-of-state:

$$V = \frac{nRTz}{p}$$

Differentiating the above equation with respect to pressure at constant temperature T gives:

$$\left(\frac{\partial V}{\partial p}\right)_T = nRT\left[\frac{1}{p}\left(\frac{\partial z}{\partial p}\right) - \frac{z}{p^2}\right]$$

Substituting into Equation 2-44 produces the following generalized relationship:

$$c_g = \frac{1}{p} - \frac{1}{z}\left(\frac{\partial z}{\partial p}\right)_T \qquad (2\text{-}45)$$

For an ideal gas, $z = 1$ and $(\partial z/\partial p)_T = 0$, therefore:

$$c_g = \frac{1}{p} \qquad (2\text{-}46)$$

It should be pointed out that Equation 2-46 is useful in determining the expected order of magnitude of the isothermal gas compressibility.

Equation 2-45 can be conveniently expressed in terms of the pseudo-reduced pressure and temperature by simply replacing p with ($p_{pc}\, p_{pr}$), or:

$$c_g = \frac{1}{p_{pr}\, p_{pc}} - \frac{1}{z}\left[\frac{\partial z}{\partial (p_{pr}\, p_{pc})}\right]_{T_{pr}}$$

Multiplying the above equation by p_{pc} yields:

$$c_g\, p_{pc} = c_{pr} = \frac{1}{p_{pr}} - \frac{1}{z}\left[\frac{\partial z}{\partial p_{pr}}\right]_{T_{pr}} \qquad (2\text{-}47)$$

The term c_{pr} is called the isothermal pseudo-reduced compressibility and is defined by the relationship

$$c_{pr} = c_g p_{pc} \qquad (2\text{-}48)$$

where c_{pr} = isothermal pseudo-reduced compressibility
c_g = isothermal gas compressibility, psi^{-1}
p_{pc} = pseudo-reduced pressure, psi

Values of $(\partial z / \partial p_{pr})_{T_{pr}}$ can be calculated from the slope of the T_{pr} isotherm on the Standing and Katz z-factor chart.

Example 2-10

A hydrocarbon gas mixture has a specific gravity of 0.72. Calculate the isothermal gas compressibility coefficient at 2000 psia and 140°F by assuming:

a. An ideal gas behavior
b. A real gas behavior

Solution

a. Assuming an ideal gas behavior, determine c_g by applying Equation 2-45:

$$c_g = \frac{1}{2000} = 500 \times 10^{-6} \ psi^{-1}$$

b. Assuming a real gas behavior

Step 1. Calculate T_{pc} and p_{pc} by applying Equations 2-17 and 2-18

$$T_{pc} = 168 + 325 \ (0.72) - 12.5 \ (0.72)^2 = 395.5 \ °R$$

$$P_{Pc} = 677 + 15 \ (0.72) - 37.5 \ (0.72)^2 = 668.4 \ psia$$

Step 2. Compute p_{pr} and T_{pr} from Equations 2-11 and 2-12.

$$p_{pr} = \frac{2000}{668.4} = 2.99$$

$$T_{pr} = \frac{600}{395.5} = 1.52$$

Step 3. Determine the z-factor from Figure 2-1:

$$z = 0.78$$

Step 4. Calculate the slope $[\partial z/\partial p_{pr}]_{T_{pr} = 1.52}$:

$$\left[\frac{\partial z}{\partial p_{pr}}\right]_{T_{pr}} = -0.022$$

Step 5. Solve for c_{pr} by applying Equation 2-47:

$$c_{pr} = \frac{1}{2.99} - \frac{1}{0.78} [-0.022] = 0.3627$$

Step 6. Calculate c_g from Equation 2-48:

$$c_g = \frac{0.327}{668.4} = 543 \times 10^{-6} \text{ psi}^{-1}$$

Trube (1957) presented graphs from which the isothermal compressibility of natural gases may be obtained. The graphs, as shown in Figures 2-3 and 2-4 give the isothermal pseudo-reduced compressibility as a function of pseudo-reduced pressure and temperature.

Example 2-11

Using Trube's generalized charts, rework Example 2-10.

Solution

Step 1. From Figure 2-3, find c_{pr}:

$$c_{pr} = 0.36$$

Step 2. Solve for c_g by applying Equation 2-49:

$$c_g = \frac{0.36}{668.4} = 539 \times 10^{-6} \text{ psi}^{-1}$$

Matter, Brar, and Aziz (1975) presented an analytical technique for calculating the isothermal gas compressibility. The authors expressed c_{pr}

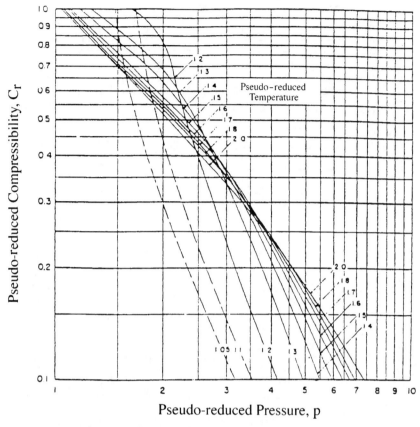

Figure 2-3. Trube's pseudo-reduced compressibility for natural gases. (*Permission to publish by the Society of Petroleum Engineers of AIME. Copyright SPE-AIME.*)

as a function of $\partial p / \partial \rho_r$ rather than $\partial p / \partial p_{pr}$.

Equation 2-41 is differentiated with respect to p_{pr} to give:

$$\left[\frac{\partial z}{\partial p_{pr}} \right] = \frac{0.27}{z \, T_{pr}} \left[\frac{(\partial z / \partial \rho_r)_{T_{pr}}}{1 + \frac{\rho_r}{z} \, (\partial z / \partial \rho_r)_{T_{pr}}} \right] \qquad (2\text{-}49)$$

Equation 2-49 may be substituted into Equation 2-47 to express the pseudo-reduced compressibility as:

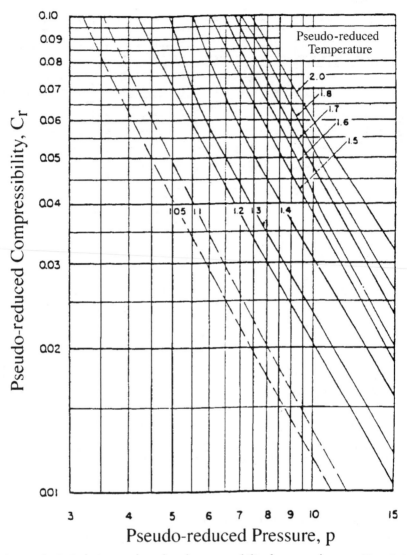

Figure 2-4. Trube's pseudo-reduced compressibility for natural gases. (*Permission to publish by the Society of Petroleum Engineers of AIME. Copyright SPE-AIME.*)

$$c_{pr} = \frac{1}{p_{pr}} - \frac{0.27}{z^2\,T_{pr}}\left[\frac{(\partial z/\partial \rho_r)_{T_{pr}}}{1 + \dfrac{\rho_r}{z}\,(\partial z/\partial \rho_r)_{T_{pr}}}\right] \qquad (2\text{-}50)$$

where ρ_r = pseudo-reduced gas density.

The partial derivative appearing in Equation 2-50 is obtained from Equation 2-43 to give:

$$\left[\frac{\partial z}{\partial \rho_r} \right]_{Tpr} = T_1 + 2\,T_2\,\rho_r + 5\,T_3\,\rho_r^4 + 2\,T_4\,\rho_r\,(1 + A_8\,\rho_r^2 - A_8^2\,\rho_r^4)$$
$$\times \exp\left(-A_8\,\rho_r^2\right) \qquad (2\text{-}51)$$

where the coefficients T_1 through T_4 and A_1 through A_8 are defined previously by Equation 2-43.

GAS FORMATION VOLUME FACTOR

The gas formation volume factor is used to relate the volume of gas, as measured at reservoir conditions, to the volume of the gas as measured at standard conditions, i.e., 60°F and 14.7 psia. This gas property is then defined as the actual volume occupied by a certain amount of gas at a specified pressure and temperature, divided by the volume occupied by the same amount of gas at standard conditions. In an equation form, the relationship is expressed as

$$B_g = \frac{V_{p,T}}{V_{sc}} \qquad (2\text{-}52)$$

where B_g = gas formation volume factor, ft³/scf
$\quad V_{p,T}$ = volume of gas at pressure p and temperature, T, ft³
$\quad V_{sc}$ = volume of gas at standard conditions, scf

Applying the real gas equation-of-state, i.e., Equation 2-11, and substituting for the volume V, gives:

$$B_g = \frac{\dfrac{zn\,RT}{p}}{\dfrac{z_{sc}\,n\,R\,T_{sc}}{p_{sc}}} = \frac{p_{sc}}{T_{sc}}\,\frac{zT}{p}$$

where z_{sc} = z-factor at standard conditions = 1.0
$\quad p_{sc}, T_{sc}$ = standard pressure and temperature

Assuming that the standard conditions are represented by $p_{sc} = 14.7$ psia and $T_{sc} = 520$, the above expression can be reduced to the following relationship:

$$B_g = 0.02827 \frac{zT}{p} \qquad (2-53)$$

where B_g = gas formation volume factor, ft^3/scf
$\quad\quad$ z = gas compressibility factor
$\quad\quad$ T = temperature, °R

In other field units, the gas formation volume factor can be expressed in bbl/scf, to give:

$$B_g = 0.005035 \frac{zT}{p} \qquad (2-54)$$

The reciprocal of the gas formation volume factor is called the gas expansion factor and is designated by the symbol E_g, or:

$$E_g = 35.37 \frac{p}{zT}, \text{scf/ft}^3 \qquad (2-55)$$

In other units:

$$E_g = 198.6 \frac{p}{zT}, \text{scf/bbl} \qquad (2-56)$$

Example 2-12

A gas well is producing at a rate of 15,000 ft^3/day from a gas reservoir at an average pressure of 2,000 psia and a temperature of 120°F. The specific gravity is 0.72. Calculate the gas flow rate in scf/day.

Solution

Step 1. Calculate the pseudo-critical properties from Equations 2-17 and 2-18, to give:

$\quad\quad T_{Pc} = 395.5$ °R $\quad\quad p_{pc} = 668.4$ psia

Step 2. Calculate the p_{pr} and T_{pr}:

$$p_{pr} = \frac{2000}{668.4} = 2.99$$

$$T_{pr} = \frac{600}{395.5} = 1.52$$

Step 3. Determine the z-factor from Figure 2-1:

$$z = 0.78$$

Step 4. Calculate the gas expansion factor from Equation 2-55:

$$E_g = 35.37 \frac{2000}{(0.78)\,(600)} = 151.15 \text{ scf/ft}^3$$

Step 5. Calculate the gas flow rate in scf/day by multiplying the gas flow rate (in ft³/day) by the gas expansion factor E_g as expressed in scf/ft³:

Gas flow rate = (151.15) (15,000) = 2.267 MMscf/day

GAS VISCOSITY

The viscosity of a fluid is a measure of the internal fluid friction (resistance) to flow. If the friction between layers of the fluid is small, i.e., low viscosity, an applied shearing force will result in a large velocity gradient. As the viscosity increases, each fluid layer exerts a larger frictional drag on the adjacent layers and velocity gradient decreases.

The viscosity of a fluid is generally defined as the ratio of the shear force per unit area to the local velocity gradient. Viscosities are expressed in terms of poises, centipoise, or micropoises. One poise equals a viscosity of 1 dyne-sec/cm² and can be converted to other field units by the following relationships:

1 poise = 100 centipoises
$\qquad$ = 1×10^6 micropoises
$\qquad$ = 6.72×10^{-2} lb mass/ft-sec
$\qquad$ = 2.09×10^{-3} lb-sec/ft²

The gas viscosity is not commonly measured in the laboratory because it can be estimated precisely from empirical correlations. Like all intensive properties, viscosity of a natural gas is completely described by the following function:

$$\mu_g = (p, T, y_i)$$

where μ_g = the viscosity of the gas phase. The above relationship simply states that the viscosity is a function of pressure, temperature, and composition. Many of the widely used gas viscosity correlations may be viewed as modifications of that expression.

METHODS OF CALCULATING THE VISCOSITY OF NATURAL GASES

Two popular methods that are commonly used in the petroleum industry are the:

• Carr-Kobayashi-Burrows Correlation Method
• Lee-Gonzalez-Eakin Method

The Carr-Kobayashi-Burrows Correlation Method

Carr, Kobayashi, and Burrows (1954) developed graphical correlations for estimating the viscosity of natural gas as a function of temperature, pressure, and gas gravity. The computational procedure of applying the proposed correlations is summarized in the following steps:

Step 1. Calculate the pseudo-critical pressure, pseudo-critical temperature, and apparent molecular weight from the specific gravity or the composition of the natural gas. Corrections to these pseudo-critical properties for the presence of the nonhydrocarbon gases (CO_2, N_2, and H_2S) should be made if they are present in concentrations greater than 5 mole percent.

Step 2. Obtain the viscosity of the natural gas at one atmosphere and the temperature of interest from Figure 2-5. This viscosity, as denoted by μ_1, must be corrected for the presence of nonhydrocarbon components by using the inserts of Figure 2-5. The nonhydrocarbon fractions tend to increase the viscosity of the gas phase. The effect of nonhydrocarbon components on the viscosity of the nat-

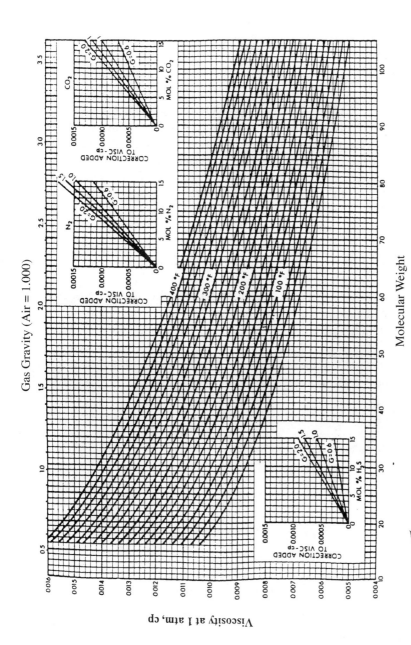

Figure 2-5. Carr's atmospheric gas viscosity correlation. (*Permission to publish by the Society of Petroleum Engineers of AIME. Copyright SPE-AIME.*)

ural gas can be expressed mathematically by the following relationships:

$$\mu_1 = (\mu_1)_{\text{uncorrected}} + (\Delta\mu)_{N_2} + (\Delta\mu)_{CO_2} + (\Delta\mu)_{H_2S} \qquad (2\text{-}57)$$

where
$$\mu_1 = \text{``corrected'' gas viscosity at one atmospheric pressure and reservoir temperature, cp}$$
$$(\Delta\mu)_{N_2} = \text{viscosity corrections due to the presence of } N_2$$
$$(\Delta\mu)_{CO_2} = \text{viscosity corrections due to the presence of } CO_2$$
$$(\Delta\mu)_{H_2S} = \text{viscosity corrections due to the presence of } H_2S$$
$$(\mu_1)_{\text{uncorrected}} = \text{uncorrected gas viscosity, cp}$$

Step 3. Calculate the pseudo-reduced pressure and temperature.

Step 4. From the pseudo-reduced temperature and pressure, obtain the viscosity ratio (μ_g/μ_1) from Figure 2-6. The term μ_g represents the viscosity of the gas at the required conditions.

Step 5. The gas viscosity, μ_g, at the pressure and temperature of interest is calculated by multiplying the viscosity at one atmosphere and system temperature, μ_1, by the viscosity ratio.

The following examples illustrate the use of the proposed graphical correlations:

Example 2-13

Using the data given in Example 2-12, calculate the viscosity of the gas.

Solution

Step 1. Calculate the apparent molecular weight of the gas:

$$M_a = (0.72)(28.96) = 20.85$$

Step 2. Determine the viscosity of the gas at 1 atm and 140°F from Figure 2-5:

$$\mu_1 = 0.0113$$

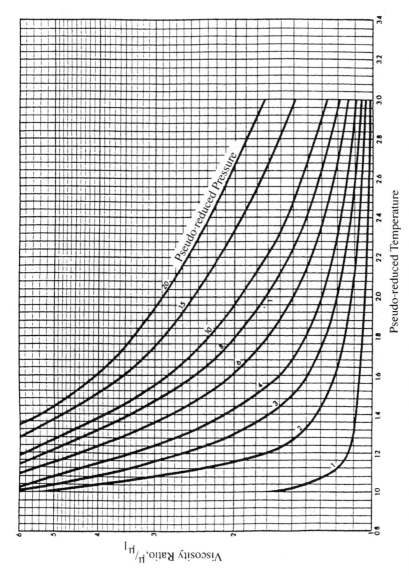

Figure 2-6. Carr's viscosity ratio correlation. (Permission to publish by the Society of Petroleum Engineers of AIME. Copyright SPE-AIME.)

Step 3. Calculate p_{pr} and T_{pr}:

$$p_{pr} = 2.99$$

$$T_{pr} = 1.52$$

Step 4. Determine the viscosity rates from Figure 2-6:

$$\frac{\mu_g}{\mu_1} = 1.5$$

Step 5. Solve for the viscosity of the natural gas:

$$\mu_g = \frac{\mu_g}{\mu_1} (\mu_1) = (1.5)(0.0113) = 0.01695 \text{ cp}$$

Standing (1977) proposed a convenient mathematical expression for calculating the viscosity of the natural gas at atmospheric pressure and reservoir temperature, μ_1. Standing also presented equations for describing the effects of N_2, CO_2, and H_2S on μ_1. The proposed relationships are:

$$\mu_1 = (\mu_1)_{uncorrected} + (\Delta\mu)_{CO_2} + (\Delta\mu)_{H_2S} + (\Delta\mu)_{N_2} \qquad (2\text{-}58)$$

where:

$$(\mu_1)_{uncorrected} = [1.709 \, (10^{-5} - 2.062 \ 10^{-6} \ \gamma_g] \, (T - 460)$$
$$+ 8.118 \, (10^{-3}) - 6.15 \, (10^{-3}) \ \log \, (\gamma_g) \qquad (2\text{-}59)$$

$$(\Delta\mu)_{CO_2} = y_{CO_2} \left[(9.08 \times 10^{-3}) \log \gamma_g + (6.24 \times 10^{-3}) \right]$$

$$(\Delta\mu)_{N_2} = y_{N_2} [8.48 \, (10^{-3}) \log(\gamma_g) + 9.59 \, (10^{-3})] \qquad (2\text{-}60)$$

$$(\Delta\mu)_{H_2S} = y_{H_2S} [8.49 \, (10^{-3}) \log \, (\gamma_g) + 3.73 \, (10^{-3})] \qquad (2\text{-}61)$$

where μ_1 = viscosity of the gas at atmospheric pressure and reservoir temperature, cp

         T = reservoir temperature, °R

         γ_g = gas gravity

$y_{N_2}, y_{CO_2}, y_{H_2S}$ = mole fraction of N_2, CO_2, and H_2S, respectively

Dempsey (1965) expressed the viscosity ratio μ_g/μ_1 by the following relationship:

$$\ln\left[T_{pr}\left(\frac{\mu_g}{\mu_1}\right)\right] = a_0 + a_1\ P_{pr} + a_2\ p_{pr}^2 + a_3\ p_{pr}^3 + T_{pr}\ (a_4 + a_5\ p_{pr}$$

$$+ a_6\ p_{pr}^2 + a_7\ p_{pr}^3) + T_{pr}^2\ (a_8 + a_9\ P_{pr} + a_{10}\ p_{pr}^2 + a_{11}\ p_{pr}^3)$$

$$+ T_{pr}^3\ (a_{12} + a_{13}\ P_{pr} + a_{14}\ p_{pr}^2 + a_{15}\ p_{pr}^3) \qquad (2\text{-}62)$$

where T_{pr} = pseudo-reduced temperature of the gas mixture, °R

$\qquad p_{pr}$ = pseudo-reduced pressure of the gas mixture, psia

$a_0 \ldots a_{17}$ = coefficients of the equations are given below:

$a_0 = -2.46211820$ $a_8 = -7.93385648\ (10^{-1})$

$a_1 = 2.970547414$ $a_9 = 1.39643306$

$a_2 = -2.86264054\ (10^{-1})$ $a_{10} = -1.49144925\ (10^{-1})$

$a_3 = 8.05420522\ (10^{-3})$ $a_{11} = 4.41015512\ (10^{-3})$

$a_4 = 2.80860949$ $a_{12} = 8.39387178\ (10^{-2})$

$a_5 = -3.49803305$ $a_{13} = -1.86408848\ (10^{-1})$

$a_6 = 3.60373020\ (10^{-1})$ $a_{14} = 2.03367881\ (10^{-2})$

$a_7 = -1.044324\ (10^{-2})$ $a_{15} = -6.09579263\ (10^{-4})$

The Lee-Gonzalez-Eakin Method

Lee, Gonzalez, and Eakin (1966) presented a semi-empirical relationship for calculating the viscosity of natural gases. The authors expressed the gas viscosity in terms of the reservoir temperature, gas density, and the molecular weight of the gas. Their proposed equation is given by:

$$\mu_g = 10^{-4}\ K \exp\left[X\left(\frac{\rho_g}{62.4}\right)^Y\right] \qquad (2\text{-}63)$$

where

$$K = \frac{(9.4 + 0.02\ M_a)\ T^{1.5}}{209 + 19\ M_a + T} \qquad (2\text{-}64)$$

$$X = 3.5 + \frac{986}{T} + 0.01\ M_a \qquad (2\text{-}65)$$

$$Y = 2.4 - 0.2\ X \qquad (2\text{-}66)$$

ρ_g = gas density at reservoir pressure and temperature, lb/ft^3
T = reservoir temperature, °R
M_a = apparent molecular weight of the gas mixture

The proposed correlation can predict viscosity values with a standard deviation of 2.7% and a maximum deviation of 8.99%. The correlation is less accurate for gases with higher specific gravities. The authors pointed out that the method cannot be used for sour gases.

Example 2-14

Rework Example 2-13 and calculate the gas viscosity by using the Lee-Gonzalez-Eakin method.

Step 1. Calculate the gas density from Equation 2-16:

$$\rho_g = \frac{(2000)(20.85)}{(10.73)(600)(0.78)} = 8.3 \text{ lb/ft}^3$$

Step 2. Solve for the parameters K, X, and Y by using Equations 2-64, 2-65, and 2-66, respectively:

$$K = \frac{[9.4 + 0.02(20.85)](600)^{1.5}}{209 + 19(20.85) + 600} = 119.72$$

$$X = 3.5 + \frac{986}{600} + 0.01(20.85) = 5.35$$

$$Y = 2.4 - 0.2(5.35) = 1.33$$

Step 3. Calculate the viscosity from Equation 2-63:

$$\mu_g = 10^{-4}(119.72)\exp\left[5.35\left(\frac{8.3}{62.4}\right)^{1.33}\right] = 0.0173 \text{ cp}$$

PROPERTIES OF CRUDE OIL SYSTEMS

Petroleum (an equivalent term is crude oil) is a complex mixture consisting predominantly of hydrocarbons and containing sulfur, nitrogen, oxygen, and helium as minor constituents. The physical and chemical properties of crude oils vary considerably and are dependent on the concentration of the various types of hydrocarbons and minor constituents present.

An accurate description of physical properties of crude oils is of a considerable importance in the fields of both applied and theoretical science and especially in the solution of petroleum reservoir engineering problems. Physical properties of primary interest in petroleum engineering studies include:

• Fluid gravity
• Specific gravity of the solution gas
• Gas solubility
• Bubble-point pressure
• Oil formation volume factor
• Isothermal compressibility coefficient of undersaturated crude oils
• Oil density
• Total formation volume factor
• Crude oil viscosity
• Surface tension

Data on most of these fluid properties are usually determined by laboratory experiments performed on samples of actual reservoir fluids. In the absence of experimentally measured properties of crude oils, it is necessary for the petroleum engineer to determine the properties from empirically derived correlations.

Crude Oil Gravity

The crude oil density is defined as the mass of a unit volume of the crude at a specified pressure and temperature. It is usually expressed in pounds per cubic foot. The specific gravity of a crude oil is defined as the ratio of the density of the oil to that of water. Both densities are measured at 60°F and atmospheric pressure:

$$\gamma_o = \frac{\rho_o}{\rho_w} \qquad (2\text{-}67)$$

where γ_o = specific gravity of the oil
ρ_o = density of the crude oil, lb/ft^3
ρ_w = density of the water, lb/ft^3

It should be pointed out that the liquid specific gravity is dimensionless, but traditionally is given the units 60°/60° to emphasize the fact that both densities are measured at standard conditions. The density of the water is approximately 62.4 lb/ft³, or:

$$\gamma_o = \frac{\rho_o}{62.4}, 60°/60°$$

Although the density and specific gravity are used extensively in the petroleum industry, the API gravity is the preferred gravity scale. This gravity scale is precisely related to the specific gravity by the following expression:

$$°API = \frac{141.5}{\gamma_o} - 131.5 \qquad (2\text{-}68)$$

The API gravities of crude oils usually range from 47° API for the lighter crude oils to 10° API for the heavier asphaltic crude oils.

Example 2-15

Calculate the specific gravity and the API gravity of a crude oil system with a measured density of 53 lb/ft³ at standard conditions.

Solution

Step 1. Calculate the specific gravity from Equation 2-67:

$$\gamma_o = \frac{53}{62.4} = 0.849$$

Step 2. Solve for the API gravity:

$$API = \frac{141.5}{0.849} - 131.5 = 35.2° API$$

Specific Gravity of the Solution Gas

The specific gravity of the solution gas γ_g is described by the weighted average of the specific gravities of the separated gas from each separator.

This weighted-average approach is based on the separator gas-oil ratio, or:

$$\gamma_g = \frac{\sum\limits_{i=1}^{n} (R_{sep})_i \, (\gamma_{sep})_i + R_{st} \, \gamma_{st}}{\sum\limits_{i=1}^{n} (R_{sep})_i + R_{st}} \qquad (2\text{-}69)$$

where   n = number of separators
   R_{sep} = separator gas-oil ratio, scf/STB
   γ_{sep} = separator gas gravity
   R_{st} = gas-oil ratio from the stock tank, scf/ STB
   γ_{st} = gas gravity from the stock tank

Example 2-16

Separator tests were conducted on a crude oil sample. Results of the test in terms of the separator gas-oil ration and specific gravity of the separated gas are given below:

Separator #	Pressure psig	Temperature °F	Gas-Oil Ratio scf/STB	Gas Specific Gravity
Primary	660	150	724	0.743
Intermediate	75	110	202	0.956
Stock tank	0	60	58	1.296

Calculate the specific gravity of the separated gas.

Solution

Estimate the specific gravity of the solution by using Equation 2-69:

$$\gamma_g = \frac{(724)\,(0.743) + (202)\,(0.956) + (58)\,(1.296)}{724 + 202 + 58} = 0.819$$

Gas Solubility

The gas solubility R_s is defined as the number of standard cubic feet of gas which will dissolve in one stock-tank barrel of crude oil at certain pressure and temperature. The solubility of a natural gas in a crude oil is a strong function of the pressure, temperature, API gravity, and gas gravity.

For a particular gas and crude oil to exist at a constant temperature, the solubility increases with pressure until the saturation pressure is reached. At the saturation pressure (bubble-point pressure) all the available gases are dissolved in the oil and the gas solubility reaches its maximum value. Rather than measuring the amount of gas that will dissolve in a given stock-tank crude oil as the pressure is increased, it is customary to determine the amount of gas that will come out of a sample of reservoir crude oil as pressure decreases.

A typical gas solubility curve, as a function of pressure for an under-saturated crude oil, is shown in Figure 2-7. As the pressure is reduced from the initial reservoir pressure p_i, to the bubble-point pressure p_b, no gas evolves from the oil and consequently the gas solubility remains constant at its maximum value of R_{sb}. Below the bubble-point pressure, the solution gas is liberated and the value of R_s decreases with pressure. The following five empirical correlations for estimating the gas solubility are given below:

• Standing's correlation
• The Vasquez-Beggs correlation
• Glaso's correlation
• Marhoun's correlation
• The Petrosky-Farshad correlation

Standing's Correlation

Standing (1947) proposed a graphical correlation for determining the gas solubility as a function of pressure, gas specific gravity, API gravity, and system temperature. The correlation was developed from a total of 105 experimentally determined data points on 22 hydrocarbon mixtures from California crude oils and natural gases. The proposed correlation has an average error of 4.8%. Standing (1981) expressed his proposed graphical correlation in the following more convenient mathematical form:

$$R_s = \gamma_g \left[\left(\frac{p}{18.2} + 1.4 \right) 10^x \right]^{1.2048} \qquad (2\text{-}70)$$

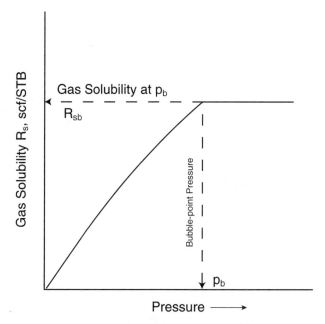

Figure 2-7. Gas-solubility pressure diagram.

with

$$x = 0.0125 \, API - 0.00091(T - 460)$$

where T = temperature, °R
        p = system pressure, psia
        γ_g = solution gas specific gravity

It should be noted that Standing's equation is valid for applications at and below the bubble-point pressure of the crude oil.

Example 2-17

The following experimental PVT data on six different crude oil systems are available. Results are based on two-stage surface separation.

Oil #	T	p_b	R_s	B_o	ρ_o	c_o at $p > p_b$	p_{sep}	T_{sep}	API	γ_g
1	250	2377	751	1.528	38.13	22.14×10^{-6} at 2689	150	60	47.1	0.851
2	220	2620	768	1.474	40.95	18.75×10^{-6} at 2810	100	75	40.7	0.855
3	260	2051	693	1.529	37.37	22.69×10^{-6} at 2526	100	72	48.6	0.911
4	237	2884	968	1.619	38.92	21.51×10^{-6} at 2942	60	120	40.5	0.898
5	218	3045	943	1.570	37.70	24.16×10^{-6} at 3273	200	60	44.2	0.781
6	180	4239	807	1.385	46.79	11.45×10^{-6} at 4370	85	173	27.3	0.848

where   T = reservoir temperature, °F

p_b = bubble-point pressure, psig

B_o = oil formation volume factor, bbl/STB

p_{sep} = separator pressure, psig

T_{sep} = separator temperature, °F

c_o = isothermal compressibility coefficient of the oil at a specified pressure, psi^{-1}

Using Standing's correlation, estimate the gas solubility at the bubble-point pressure and compare with the experimental value in terms of the **absolute average error** (AAE).

Solution

Apply Equation 2-70 to determine the gas solubility. Results of the calculations are given in the following tabulated form:

Oil #	X	10^X	Predicted R_S Equation 2-70	Measured R_S	% Error
1	0.361	2.297	838	751	11.6
2	0.309	2.035	817	768	6.3
3	0.371	2.349	774	693	11.7
4	0.312	2.049	969	968	0.108
5	0.322	2.097	1012	943	7.3
6	0.177	1.505	998	807	23.7

AAE = 10.1%

The Vasquez-Beggs Correlation

Vasquez and Beggs (1980) presented an improved empirical correlation for estimating R_s. The correlation was obtained by regression analysis using 5,008 measured gas solubility data points. Based on oil gravity, the measured data were divided into two groups. This division was made at a value of oil gravity of 30°API. The proposed equation has the following form:

$$R_s = C_1 \, \gamma_{gs} \, p^{C_2} \, \exp\left[C_3\left(\frac{API}{T}\right)\right]$$

(2-71)

Values for the coefficients are as follows:

Coefficient	API " 30	API > 30
C_1	0.0362	0.0178
C_2	1.0937	1.1870
C_3	25.7240	23.931

Realizing that the value of the specific gravity of the gas depends on the conditions under which it is separated from the oil, Vasquez and Beggs proposed that the value of the gas specific gravity as obtained from a separator pressure of 100 psig be used in the above equation. This reference pressure was chosen because it represents the average field separator conditions. The authors proposed the following relationship for adjustment of the gas gravity γ_g to the reference separator pressure:

$$\gamma_{gs} = \gamma_g \left[1 + 5.912\,(10^{-5})\,(API)\,(T_{sep} - 460)\log\left(\frac{p_{sep}}{114.7}\right)\right]$$

(2-72)

where γ_{gs} = gas gravity at the reference separator pressure
   γ_g = gas gravity at the actual separator conditions of p_{sep} and T_{sep}
   p_{sep} = actual separator pressure, psia
   T_{sep} = actual separator temperature, °R

The gas gravity used to develop all the correlations reported by the authors was that which would result from a two-stage separation. The first-stage pressure was chosen as 100 psig and the second stage was the stock tank. If the separator conditions are unknown, the unadjusted gas gravity may be used in Equation 2-71.

An independent evaluation of the above correlation by Sutton and Farashad (1984) shows that the correlation is capable of predicting gas solubilities with an average absolute error of 12.7%.

Example 2-18

Using the PVT of the six crude oil systems of Example 2-17, solve for the gas solubility.

Solution

Oil #	γ_{gs} From Equation 2-72	Predicted R_S Equation 2-71	Measured R_S	% Error
1	0.8731	779	751	3.76
2	0.855	733	768	−4.58
3	0.911	702	693	1.36
4	0.850	820	968	15.2
5	0.814	947	943	0.43
6	0.834	841	807	4.30

AAE = 4.9%

Glaso's Correlation

Glaso (1980) proposed a correlation for estimating the gas solubility as a function of the API gravity, pressure, temperature, and gas specific gravity. The correlation was developed from studying 45 North Sea crude oil samples. Glaso reported an average error of 1.28% with a standard deviation of 6.98%. The proposed relationship has the following form:

$$R_s = \gamma_g \left[\left(\frac{API^{0.989}}{(T - 460)^{0.172}} \right) (p_b^*) \right]^{1.2255} \qquad (2\text{-}73)$$

where p_b^* is a correlating number and is defined by the following expression:

$$p_b^* = 10^x$$

with

$$x = 2.8869 - [14.1811 - 3.3093 \log (p)]^{0.5}$$

Example 2-19

Rework Example 2-17 and solve for the gas solubility by using Glaso's correlation.

Solution

Oil #	x	p_b^*	Predicted R_s Equation 2-73	Measured R_s	% Error
1	1.155	14.286	737	751	−1.84
2	1.196	15.687	714	768	−6.92
3	1.095	12.450	686	693	−0.90
4	1.237	17.243	843	968	−12.92
5	1.260	18.210	868	943	−7.95
6	1.413	25.883	842	807	4.34

$$AAE = 5.8\%$$

Marhoun's Correlation

Marhoun (1988) developed an expression for estimating the saturation pressure of the Middle Eastern crude oil systems. The correlation originates from 160 experimental saturation pressure data. The proposed correlation can be rearranged and solved for the gas solubility to give:

$$R_s = \left[a \, \gamma_g^b \, \gamma_o^c \, T^d \, p \right]^e \tag{2-74}$$

where γ_g = gas specific gravity
   γ_o = stock-tank oil gravity
   T = temperature, °R
   a–e = coefficients of the above equation having these values:
   a = 185.843208
   b = 1.877840
   c = −3.1437
   d = −1.32657
   e = 1.398441

Example 2-20

Resolve Example 2-17 by using Marhoun's correlation.

Solution

Oil #	Predicted R_S Equation 2-74	Measured R_S	% Error
1	740	751	−1.43
2	792	768	3.09
3	729	693	5.21
4	1041	968	7.55
5	845	943	−10.37
6	1186	807	47.03

$$\text{AAE} = 12.4\%$$

The Petrosky-Farshad Correlation

Petrosky and Farshad (1993) used a nonlinear multiple regression software to develop a gas solubility correlation. The authors constructed a PVT database from 81 laboratory analyses from the Gulf of Mexico crude oil system. Petrosky and Farshad proposed the following expression:

$$R_s = \left[\left(\frac{p}{112.727} + 12.340 \right) \gamma_g^{0.8439} 10^x \right]^{1.73184} \tag{2-75}$$

with

$$x = 7.916 \, (10^{-4}) \, (API)^{1.5410} - 4.561(10^{-5}) \, (T - 460)^{1.3911}$$

where p = pressure, psia
       T = temperature, °R

Example 2-21

Test the predictive capability of the Petrosky and Farshad equation by resolving Example 2-17.

Solution

Oil #	x	Predicted R_S Equation 2-75	Measured R_S	% Error
1	0.2008	772	751	2.86
2	0.1566	726	768	-5.46
3	0.2101	758	693	9.32
4	0.1579	875	968	-9.57
5	0.1900	865	943	-8.28
6	0.0667	900	807	11.57

AAE = 7.84%

The gas solubility can also be calculated rigorously from the experimental measured PVT data at the specified pressure and temperature. The following expression relates the gas solubility R_s to oil density, specific gravity of the oil, gas gravity, and the oil formation volume factor:

$$R_s = \frac{B_o\, \rho_o - 62.4\, \gamma_o}{0.0136\, \gamma_g} \tag{2-76}$$

where ρ_o = oil density, lb/ft^3
  B_o = oil formation volume factor, bbl/STB
  γ_o = specific gravity of the stock-tank oil
  γ_g = specific gravity of the solution gas

McCain (1991) pointed out that the weight average of separator and stock-tank gas specific gravities should be used for γ_g. The error in calculating R_s by using the above equation will depend only on the accuracy of the available PVT data.

Example 2-22

Using the data of Example 2-17, estimate R_s by applying Equation 2-76.

Solution

Oil #	Predicted R_S Equation 2-76	Measured R_S	% Error
1	762	751	1.53
2	781	768	1.73
3	655	693	−5.51
4	956	968	−1.23
5	841	943	−10.79
6	798	807	−1.13

AAE = 3.65%

Bubble-Point Pressure

The bubble-point pressure p_b of a hydrocarbon system is defined as the highest pressure at which a bubble of gas is first liberated from the oil. This important property can be measured experimentally for a crude oil system by conducting a constant-composition expansion test.

In the absence of the experimentally measured bubble-point pressure, it is necessary for the engineer to make an estimate of this crude oil property from the readily available measured producing parameters. Several graphical and mathematical correlations for determining p_b have been proposed during the last four decades. These correlations are essentially based on the assumption that the bubble-point pressure is a strong function of gas solubility R_s, gas gravity γ_g, oil gravity API, and temperature T, or:

$$p_b = f (R_S, \gamma_g, API, T)$$

Several ways of combining the above parameters in a graphical form or a mathematical expression are proposed by numerous authors, including:

- Standing
- Vasquez and Beggs
- Glaso
- Marhoun
- Petrosky and Farshad

The empirical correlations for estimating the bubble-point pressure proposed by the above-listed authors are given below.

Standing's Correlation

Based on 105 experimentally measured bubble-point pressures on 22 hydrocarbon systems from California oil fields, Standing (1947) proposed a graphical correlation for determining the bubble-point pressure of crude oil systems. The correlating parameters in the proposed correlation are the gas solubility R_s, gas gravity γ_g, oil API gravity, and the system temperature. The reported average error is 4.8%.

In a mathematical form, Standing (1981) expressed the graphical correlation by the following expression:

$$p_b = 18.2\ [(R_s/\gamma_g)^{0.83}\ (10)^a - 1.4] \tag{2-77}$$

with

$$a = 0.00091\ (T - 460) - 0.0125\ (API) \tag{2-78}$$

where p_b = bubble-point pressure, psia
T = system temperature, °R

Standing's correlation should be used with caution if nonhydrocarbon components are known to be present in the system.

Example 2-23

The experimental data given in Example 2-17 are repeated here for convenience.

Oil #	T	P_b	R_s	B_o	ρ_o	c_o at $p > p_b$	p_{sep}	T_{sep}	API	γ_g
1	250	2377	751	1.528	38.13	22.14×10^{-6} at 2689	150	60	47.1	0.851
2	220	2620	768	1.474	40.95	18.75×10^{-6} at 2810	100	75	40.7	0.855
3	260	2051	693	1.529	37.37	22.69×10^{-6} at 2526	100	72	48.6	0.911
4	237	2884	968	1.619	38.92	21.51×10^{-6} at 2942	60	120	40.5	0.898
5	218	3065	943	1.570	37.70	24.16×10^{-6} at 3273	200	60	44.2	0.781
6	180	4239	807	1.385	46.79	11.65×10^{-6} at 4370	85	173	27.3	0.848

Predict the bubble-point pressure by using Standing's correlation.

Solution

Oil #	Coeff. a Equation 2-78	Predicted p_b Equation 2-77	Measured p_b	% Error
1	−0.3613	2181	2392	−8.8
2	−0.3086	2503	2635	−5.0
3	−0.3709	1883	2066	−8.8
4	−0.3115	2896	2899	−0.1
5	−0.3541	2884	3060	−5.7
6	−0.1775	3561	4254	−16.3

AAE = 7.4%

McCain (1991) suggested that by replacing the specific gravity of the gas in Equation 2-77 with that of the separator gas, i.e., excluding the gas from the stock tank would improve the accuracy of the equation.

Example 2-24

Using the data of Example 2-23 and given the following separator gas gravities, estimate the bubble-point pressure by applying Standing's correlation.

Oil #	Separator Gas Gravity
1	0.755
2	0.786
3	0.801
4	0.888
5	0.705
6	0.813

Solution

Oil #	Predicted p_b	Measured p_b	% Error
1	2411	2392	0.83
2	2686	2635	1.93
3	2098	2066	1.53
4	2923	2899	0.84
5	3143	3060	2.70
6	3689	4254	−13.27

AAE = 3.5%

The Vasquez-Beggs Correlation

Vasquez and Beggs' gas solubility correlation as presented by Equation 2-71 can be solved for the bubble-point pressure p_b to give:

$$p_b = \left[(C_1 R_s / \gamma_{gs}) (10)^a \right]^{C_2} \qquad (2\text{-}79)$$

with

$$a = -C_3 \, API/T$$

The gas specific gravity γ_{gs} at the reference separator pressure is defined by Equation 2-72. The coefficients C_1, C_2, and C_3 have the following values:

Coefficient	API ″ 30	API > 30
C_1	27.624	56.18
C_2	0.914328	0.84246
C_3	11.172	10.393

Example 2-25

Rework Example 2-23 by applying Equation 2-79.

Solution

Oil #	γ_{gs} Equation 2-72	a	Predicted p_b	Measured p_b	% Error
1	0.873	−0.689	2319	2392	−3.07
2	0.855	−0.622	2741	2635	4.03
3	0.911	−0.702	2043	2066	−1.14
4	0.850	−0.625	3331	2899	14.91
5	0.814	−0.678	3049	3060	−0.36
6	0.834	−0.477	4093	4254	−3.78

AAE= 4.5%

Glaso's Correlation

Glaso (1980) used 45 oil samples, mostly from the North Sea hydrocarbon system, to develop an accurate correlation for bubble-point pressure prediction. Glaso proposed the following expression:

$$\log(p_b) = 1.7669 + 1.7447 \log(p_b^*) - 0.30218 \, [\log(p_b^*)]^2 \qquad (2\text{-}80)$$

where p_b^* is a correlating number and defined by the following equation:

$$p_b^* = (R_s/\gamma_g)^a (t)^b (API)^c \qquad (2\text{-}81)$$

where R_s = gas solubility, scf/STB
$\quad$ t = system temperature, °F
$\quad \gamma_g$ = average specific gravity of the total surface gases
a, b, c = coefficients of the above equation having the following values:

$\quad$ a = 0.816
$\quad$ b = 0.172
$\quad$ c = −0.989

For volatile oils, Glaso recommends that the temperature exponent b of Equation 2-81 be slightly changed, to the value of 0.130.

Example 2-26

Resolve Example 2-23 by using Glaso's correlation.

Solution

Oil #	p_b^* Equation 2-81	p_b Equation 2-80	Measured p_b	% Error
1	14.51	2431	2392	1.62
2	16.63	2797	2635	6.14
3	12.54	2083	2066	0.82
4	19.30	3240	2899	11.75
5	19.48	3269	3060	6.83
6	25.00	4125	4254	−3.04

AAE = 5.03%

Marhoun's Correlation

Marhoun (1988) used 160 experimentally determined bubble-point pressures from the PVT analysis of 69 Middle Eastern hydrocarbon mixtures to develop a correlation for estimating p_b. The author correlated the bubble-point pressure with the gas solubility R_s, temperature T, and specific gravity of the oil and the gas. Marhoun proposed the following expression:

$$p_b = a R_s^b \gamma_g^c \gamma_o^d T^e \qquad (2\text{-}82)$$

where   T = temperature, °R
  γ_o = stock-tank oil specific gravity
  γ_g = gas specific gravity
  a–e = coefficients of the correlation having the following values:

  a = 5.38088 × 10⁻³   b = 0.715082
  c = –1.87784   d = 3.1437
  e = 1.32657

The reported average absolute relative error for the correlation is 3.66% when compared with the experimental data used to develop the correlation.

Example 2-27

Using Equation 2-82, rework Example 2-23

Solution

Oil #	Predicted p_b	Measured p_b	% Error
1	2417	2392	1.03
2	2578	2635	–2.16
3	1992	2066	–3.57
4	2752	2899	–5.07
5	3309	3060	8.14
6	3229	4254	–24.09

AAE = 7.3%

The Petrosky-Farshad Correlation

The Petrosky and Farshad gas solubility equation, i.e., Equation 2-75, can be solved for the bubble-point pressure to give:

$$p_b = \left[\frac{112.727 R_s^{0.577421}}{\gamma_g^{0.8439} (10)^x} \right] - 1391.051 \qquad (2\text{-}83)$$

where the correlating parameter x is previously defined by Equation 2-75.

The authors concluded that the correlation predicts measured bubble-point pressures with an average absolute error of 3.28%.

Example 2-28

Use the Petrosky and Farshad correlation to predict the bubble-point pressure data given in Example 2-23.

Solution

Oil #	X	Predicted p_b	Measured p_b	% Error
1	0.2008	2331	2392	−2.55
2	0.1566	2768	2635	5.04
3	0.2101	1893	2066	−8.39
4	0.1579	3156	2899	8.86
5	0.1900	3288	3060	7.44
6	0.0667	3908	4254	−8.13

AAE = 6.74%

Oil Formation Volume Factor

The oil formation volume factor, B_o, is defined as the ratio of the volume of oil (plus the gas in solution) at the prevailing reservoir temperature and pressure to the volume of oil at standard conditions. B_o is always greater than or equal to unity. The oil formation volume factor can be expressed mathematically as:

$$B_o = \frac{(V_o)_{p,T}}{(V_o)_{sc}} \qquad (2\text{-}84)$$

where B_o = oil formation volume factor, bbl/STB

$(V_o)p,T$ = volume of oil under reservoir pressure p and temperature T, bbl

$(V_o)_{sc}$ = volume of oil is measured under standard conditions, STB

A typical oil formation factor curve, as a function of pressure for an undersaturated crude oil ($p_i > p_b$), is shown in Figure 2-8. As the pressure is reduced below the initial reservoir pressure p_i, the oil volume increases due to the oil expansion. This behavior results in an increase in the oil formation volume factor and will continue until the bubble-point pressure is reached. At p_b, the oil reaches its maximum expansion and consequently attains a maximum value of B_{ob} for the oil formation volume factor. As the pressure is reduced below p_b, volume of the oil and B_o are decreased as the solution gas is liberated. When the pressure is reduced to atmospheric pressure and the temperature to 60°F, the value of B_o is equal to one.

Most of the published empirical B_o correlations utilize the following generalized relationship:

$$B_o = f(R_s, \gamma_g, \gamma_o, T)$$

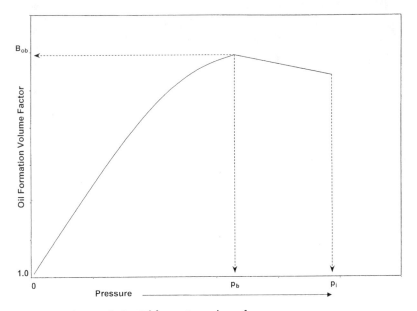

Figure 2-8. Oil formation volume factor versus pressure.

Six different methods of predicting the oil formation volume factor are presented below:

- Standing's correlation
- The Vasquez-Beggs correlation
- Glaso's correlation
- Marhoun's correlation
- The Petrosky-Farshad correlation
- Other correlations

It should be noted that all the correlations could be used for any pressure equal to or below the bubble-point pressure.

Standing's Correlation

Standing (1947) presented a graphical correlation for estimating the oil formation volume factor with the gas solubility, gas gravity, oil gravity, and reservoir temperature as the correlating parameters. This graphical correlation originated from examining a total of 105 experimental data points on 22 different California hydrocarbon systems. An average error of 1.2% was reported for the correlation.

Standing (1981) showed that the oil formation volume factor can be expressed more conveniently in a mathematical form by the following equation:

$$\beta_o = 0.9759 + 0.000120 \left[R_s \left(\frac{\gamma_g}{\gamma_o} \right)^{0.5} + 1.25(T - 460) \right]^{1.2} \qquad (2\text{-}85)$$

where T = temperature, °R
γ_o = specific gravity of the stock-tank oil
γ_g = specific gravity of the solution gas

The Vasquez-Beggs Correlation

Vasquez and Beggs (1980) developed a relationship for determining B_o as a function of R_s, γ_o, γ_g, and T. The proposed correlation was based on 6,000 measurements of B_o at various pressures. Using the regression analysis technique, Vasquez and Beggs found the following equation to be the best form to reproduce the measured data:

$$B_o = 1.0 + C_1 R_s + (T - 520) \left(\frac{API}{\gamma_{gs}} \right) [C_2 + C_3 R_s] \qquad (2-86)$$

where   R = gas solubility, scf/STB
   T = temperature, °R
   γ_{gs} = gas specific gravity as defined by Equation 2-72

Values for the coefficients C_1, C_2 and C_3 are given below:

Coefficient	API ″ 30	API > 30
C_1	4.677×10^{-4}	4.670×10^{-4}
C_2	1.751×10^{-5}	1.100×10^{-5}
C_3	-1.811×10^{-8}	1.337×10^{-9}

Vasquez and Beggs reported an average error of 4.7% for the proposed correlation.

Glaso's Correlation

Glaso (1980) proposed the following expressions for calculating the oil formation volume factor:

$$B_o = 1.0 + 10^A \qquad (2-87)$$

where

$$A = -6.58511 + 2.91329 \log B_{ob}^* - 0.27683 (\log B_{ob}^*)^2 \qquad (2-88)$$

B_{ob}^* is a *correlating number* and is defined by the following equation:

$$B_{ob}^* = R_s \left(\frac{\gamma_g}{\gamma_o} \right)^{0.526} + 0.968(T - 460) \qquad (2-89)$$

where   T = temperature, °R
   γ_o = specific gravity of the stock-tank oil

The above correlations were originated from studying PVT data on 45 oil samples. The average error of the correlation was reported at −0.43% with a standard deviation of 2.18%.

Sutton and Farshad (1984) concluded that Glaso's correlation offers the best accuracy when compared with the Standing and Vasquez-Beggs correlations. In general, Glaso's correlation underpredicts formation volume factor. Standing's expression tends to overpredict oil formation volume factors greater than 1.2 bbl/STB. The Vasquez-Beggs correlation typically overpredicts the oil formation volume factor.

Marhoun's Correlation

Marhoun (1988) developed a correlation for determining the oil formation volume factor as a function of the gas solubility, stock-tank oil gravity, gas gravity, and temperature. The empirical equation was developed by use of the nonlinear multiple regression analysis on 160 experimental data points. The experimental data were obtained from 69 Middle Eastern oil reserves. The author proposed the following expression:

$$B_o = 0.497069 + 0.862963 \times 10^{-3}\, T + 0.182594 \times 10^{-2}\, F \\ + 0.318099 \times 10^{-5}\, F^2 \tag{2-90}$$

with the correlating parameter F as defined by the following equation:

$$F = R_s^a\, \gamma_g^b\, \gamma_o^c \tag{2-91}$$

The coefficients a, b and c have the following values:

a = 0.742390
b = 0.323294
c = −1.202040

where T is the system temperature in °R.

The Petrosky-Farshad Correlation

Petrosky and Farshad (1993) proposed a new expression for estimating B_o. The proposed relationship is similar to the equation developed by Standing; however, the equation introduces three additional fitting parameters in order to increase the accuracy of the correlation.

The authors used a nonlinear regression model to match experimental crude oil from the Gulf of Mexico hydrocarbon system. Their correlation has the following form:

$$B_o = 1.0113 + 7.2046 \, (10^{-5})$$

$$\left[R_s^{0.3738} \left(\frac{\gamma_g^{0.2914}}{\gamma_o^{0.6265}} \right) + 0.24626 \, (T - 460)^{0.5371} \right]^{3.0936} \tag{2-92}$$

where T = temperature, °R
  γ_o = specific gravity of the stock-tank oil

Material Balance Equation

Following the definition of B_o as expressed mathematically by Equation 2-84, it can be shown that:

$$B_o = \frac{62.4 \, \gamma_o + 0.0136 \, R_s \, \gamma_g}{\rho_o} \tag{2-93}$$

where ρ_o = density of the oil at the specified pressure and temperature, lb/ft^3.

The error in calculating B_o by using Equation 2-93 will depend only on the accuracy of the input variables (R_s, γ_g, and γ_o) and the method of calculating ρ_o.

Example 2-29

The following experimental PVT data on six different crude oil systems are available. Results are based on two-stage surface separation.

Oil #	T	P$_b$	R$_s$	B$_o$	ρ_o	c$_o$ at p > p$_b$	P$_{sep}$	T$_{sep}$	API	γ_g
1	250	2377	751	1.528	38.13	22.14×10^{-6} at 2689	150	60	47.1	0.851
2	220	2620	768	1.474	40.95	18.75×10^{-6} at 2810	100	75	40.7	0.855
3	260	2051	693	1.529	37.37	22.69×10^{-6} at 2526	100	72	48.6	0.911
4	237	2884	968	1.619	38.92	21.51×10^{-6} at 2942	60	120	40.5	0.898
5	218	3065	943	1.570	37.70	24.16×10^{-6} at 3273	200	60	44.2	0.781
6	180	4239	807	1.385	46.79	11.65×10^{-6} at 4370	85	173	27.3	0.848

Calculate the oil formation volume factor at the bubble-point pressure by using the six different correlations. Compare the results with the experimental values and calculate the absolute average error (AAE).

Solution

Crude Oil	Exp. B_o	Method 1	Method 2	Method 3	Method 4	Method 5	Method 6
1	1.528	1.506	1.474	1.473	1.516	1.552	1.525
2	1.474	1.487	1.450	1.459	1.477	1.508	1.470
3	1.529	1.495	1.451	1.461	1.511	1.556	1.542
4	1.619	1.618	1.542	1.589	1.575	1.632	1.623
5	1.570	1.571	1.546	1.541	1.554	1.584	1.599
6	1.385	1.461	1.389	1.438	1.414	1.433	1.387
%AAE	—	1.7	2.8	2.8	1.3	1.8	0.6

where Method 1 = Standing's correlation
      Method 2 = Vasquez-Beggs correlation
      Method 3 = Glaso's correlation
      Method 4 = Marhoun's correlation
      Method 5 = Petrosky-Farshad correlation
      Method 6 = Material balance equation

Isothermal Compressibility Coefficient of Crude Oil

Isothermal compressibility coefficients are required in solving many reservoir engineering problems, including transient fluid flow problems, and they are also required in the determination of the physical properties of the undersaturated crude oil.

By definition, the isothermal compressibility of a substance is defined mathematically by the following expression:

$$c = -\frac{1}{V}\left(\frac{\partial V}{\partial p}\right)_T$$

For a crude oil system, the isothermal compressibility coefficient of the oil phase c_o is defined for *pressures above the bubble-point* by one of the following equivalent expressions:

$$c_o = -(1/V)(\partial V/\partial p)_T \qquad (2\text{-}94)$$

$$c_o = -(1/B_o)(\partial B_o/\partial p)_T \tag{2-95}$$

$$c_o = (1/\rho_o)(\partial \rho_o/\partial p)_T \tag{2-96}$$

where c_o = isothermal compressibility, psi^{-1}
ρ_o = oil density lb/ft^3
B_o = oil formation volume factor, bbl/STB

At *pressures below the bubble-point pressure,* the oil compressibility is defined as:

$$c_o = \frac{-1}{B_o}\frac{\partial B_o}{\partial p} + \frac{B_g}{B_o}\frac{\partial R_s}{\partial p} \tag{2-97}$$

where B_g = gas formation volume factor, bbl/scf

There are several correlations that are developed to estimate the oil compressibility at pressures *above* the bubble-point pressure, i.e., undersaturated crude oil system. Three of these correlations are presented below:

• The Vasquez-Beggs correlation
• The Petrosky-Farshad correlation
• McCain's correlation

The Vasquez-Beggs Correlation

From a total of 4,036 experimental data points used in a linear regression model, Vasquez and Beggs (1980) correlated the isothermal oil compressibility coefficients with R_s, T, °API, γ_g, and p. They proposed the following expression:

$$c_o = \frac{-1,433 + 5R_{sb} + 17.2(T-460) - 1,180\,\gamma_{gs} + 12.61°\text{API}}{10^5 p} \tag{2-98}$$

where   T = temperature, °R
p = pressure above the bubble-point pressure, psia
R_{sb} = gas solubility at the bubble-point pressure
γ_{gs} = corrected gas gravity as defined by Equation 2-72

The Petrosky-Farshad Correlation

Petrosky and Farshad (1993) proposed a relationship for determining the oil compressibility for undersaturated hydrocarbon systems. The equation has the following form:

$$c_o = 1.705 \times 10^{-7} \, R_{sb}^{0.69357} \, \gamma_g^{0.1885} \, API^{0.3272}$$
$$(T-460)^{0.6729} \, p^{-0.5906} \qquad\qquad (2\text{-}99)$$

where T = temperature, °R
 R_{sb} = gas solubility at the bubble-point pressure, scf/STB

Example 2-30

Using the experimental data given in Example 2-29, estimate the undersaturated oil compressibility coefficient by using the Vasquez-Beggs and the Petrosky-Farshad correlations. Calculate the AAE.

Solution

Oil #	Pressure	Measured c_o 10^{-6} psi	Vasquez-Beggs 10^{-6} psi	Petrosky-Farshad 10^{-6} psi
1	2689	22.14	22.88	22.24
2	2810	18.75	20.16	19.27
3	2526	22.60	23.78	22.92
4	2942	21.51	22.31	21.78
5	3273	24.16	20.16	20.39
6	4370	11.45	11.54	11.77
AAE			6.18%	4.05%

Below the bubble-point pressure, McCain and coauthors (1988) correlated the oil compressibility with pressure ρ, the oil API gravity, gas solubility at the bubble-point R_{sb}, and the temperature T in °R. Their proposed relationship has the following form:

$$c_o = \exp(A) \qquad\qquad (2\text{-}100)$$

where the correlating parameter A is given by the following expression:

$$A = -7.633 - 1.497 \ln(p) + 1.115 \ln(T) + 0.533 \ln(API)$$
$$+ 0.184 \ln(R_{sp}) \qquad\qquad (2\text{-}101)$$

The authors suggested that the accuracy of the Equation 2-100 can be substantially improved if the bubble-point pressure is known. They improved correlating parameter A by including the bubble-point pressure p_b as one of the parameters in the above equation, to give:

$$A = -7.573 - 1.45 \ln (p) - 0.383 \ln (P_b) + 1.402 \ln (T)$$
$$+ 0.256 \ln (API) + 0.449 \ln (R_{sb}) \qquad (2\text{-}102)$$

Analytically, Standing's correlations for R_s (Equation 2-70) and β_o (Equation 2-85) can be differentiated with respect to the pressure p to give:

$$\frac{\partial R_s}{\partial p} = \frac{R_s}{0.83\,p + 21.75} \qquad (2\text{-}103)$$

$$\frac{\partial B_o}{\partial p} = \left[\frac{0.000144\,R_s}{0.83\,p+21.75}\right]\left(\frac{\gamma_g}{\gamma_o}\right)^{0.5}$$
$$\times \left[R_s\left(\frac{\gamma_g}{\gamma_o}\right)^{0.5} + 1.25\,(T-460)\right]^{0.12} \qquad (2\text{-}104)$$

The above two expressions can be substituted into Equation 2-97 to give the following relationship:

$$c_o = \frac{-R_s}{B_o(0.83\,p+21.75)}$$
$$\times \left\{0.00014\sqrt{\frac{\gamma_g}{\gamma_o}}\left[R_s\sqrt{\frac{\gamma_g}{\gamma_o}} + 1.25\,(T-460)\right]^{0.12} - B_g\right\} \qquad (2\text{-}105)$$

where p = pressure, psia
T = temperature, °R
B_g = gas formation volume factor at pressure p, bbl/scf
R_s = gas solubility at pressure p, scf/STB
B_o = oil formation volume factor at p, bbl/STB
γ_o = specific gravity of the stock-tank oil
γ_g = specific gravity of the solution gas

Example 2-31

A crude oil system exists at 1650 psi and a temperature of 250°F. The system has the following PVT properties:

$API = 47.1$ $p_b = 2377$ $\gamma_g = 0.851$ $\gamma_{gs} = 0.873$
$R_{sb} = 751$ scf/STB $B_{ob} = 1.528$ bbl/STB

The laboratory measured oil PVT data at 1650 psig are listed below:

$B_o = 1.393$ bbl/STB $R_s = 515$ scf/STB
$B_g = 0.001936$ bbl/scf $c_o = 324.8 \times 15^{-6}$ psi^{-1}

Estimate the oil compressibility by using:

a. McCain's correlation
b. Equation 2-105

Solution

McCain's Correlation:
• Calculate the correlating parameter A by applying Equation 2-102

$$A = -7.573 - 1.45 \ln (1665) - 0.383 \ln (2392) + 1.402 \ln (710)$$
$$+ 0.256 \ln (47.1) + 0.449 \ln (451) = -8.1445$$

• Solve for c_o by using Equation 2-100

$$c_o = \exp(-8.1445) = 290.3 \times 10^{-6} \text{ psi}^{-1}$$

Oil Compressibility Using Equation 2-105

$$c_o = \frac{-515}{1.393[0.83(1665)+21.75]}$$

$$\times \left\{ 0.00014 \sqrt{\frac{0.851}{0.792}} \left[515 \sqrt{\frac{.851}{.792}} + 1.25(250) \right]^{0.12} - 0.001936 \right\}$$

$$c_o = 424 \times 10^{-6} \text{ psi}^{-1}$$

It should be pointed out that when it is necessary to establish PVT relationships for the hydrocarbon system through correlations or by extrapolation, care should be exercised to see that the PVT functions are consistent.

This consistency is assured if the increase in oil volume with increasing pressure is less than the decrease in volume associated with the gas going into solution. Since the oil compressibility coefficient c_o as expressed by Equation 2-97 must be positive, that leads to the following consistency criteria:

$$\frac{\partial B_o}{\partial p} < B_g \frac{\partial R_s}{\partial p} \tag{2-106}$$

This consistency can easily be checked in the tabular form of PVT data. The PVT consistency errors most frequently occur at higher pressures where the gas formation volume factor, B_g, assumes relatively small values.

Oil Formation Volume Factor for Undersaturated Oils

With increasing pressures above the bubble-point pressure, the oil formation volume factor decreases due to the compression of the oil, as illustrated schematically in Figure 2-9.

To account for the effects of oil compression on B_o, the oil formation volume factor at the bubble-point pressure is first calculated by using any of the methods previously described. The calculated B_o is then adjusted to account for the effect if increasing the pressure above the bubble-point pressure. This adjustment step is accomplished by using the isothermal compressibility coefficient as described below.

The isothermal compressibility coefficient (as expressed mathematically by Equation 2-94) can be equivalently written in terms of the oil formation volume factor:

$$c_o = \frac{-1}{B_o} \frac{\partial B_o}{\partial p}$$

The above relationship can be rearranged and integrated to produce

$$\int_{p_b}^{p} -c_o \, dp = \int_{B_{ob}}^{B_o} \frac{1}{B_o} dB_o \tag{2-107}$$

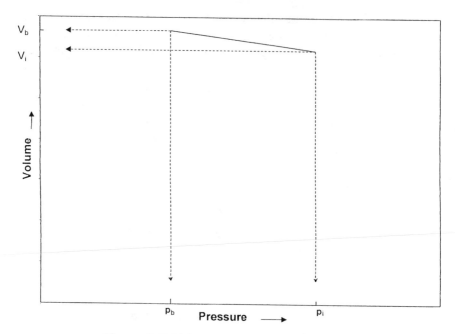

Figure 2-9. Volume versus pressure relationship.

Evaluating c_o at the arithmetic average pressure and concluding the integration procedure to give:

$$B_o = B_{ob} \exp [-c_o (p - p_b)] \tag{2-108}$$

where B_o = oil formation volume factor at the pressure of interest, bbl/STB

B_{ob} = oil formation volume factor at the bubble-point pressure, bbl/STB

p = pressure of interest, psia

p_b = bubble-point pressure, psia

Replacing with the Vasquez-Beggs' c_o expression, i.e., Equation 2-98, and integrating the resulting equation gives:

$$B_o = B_{ob} \exp \left[-A \ln \left(\frac{p}{p_b} \right) \right] \tag{2-109}$$

where

$$A = 10^{-5} \, [-1,433 + 5 \, R_{sb} + 17.2(T - 460) - 1,180 \, \gamma_{gs} + 12.61 \, API]$$

Replacing c_o in Equation 2-107 with the Petrosky-Farshad expression (i.e., Equation 2-99) and integrating gives:

$$B_o = B_{ob} \, \exp \, [-A \, (p^{0.4094} - p_b^{0.4094})] \qquad (2 - 110)$$

with the correlating parameter A as defined by:

$$A = 4.1646 \, (10^{-7}) \, R_{sb}^{0.69357} \, \gamma_g^{0.1885} \, (API)^{0.3272} \, (T - 460)^{0.6729} \quad (2\text{-}111)$$

where T = temperature, °R
  p = pressure, psia
  R_{sb} = gas solubility at the bubble-point pressure

Example 2-32

Using the PVT data given in Example 2-31, calculate the oil formation volume factor at 5000 psig by using:

a. Equation 2-109
b. Equation 2-110

The experimental measured B_o is 1.457 bbl/STB.

Solution

Using Equation 2-109:

• Calculate the parameter A:

$$A = 10^{-5} \, [-1433 + 5 \, (751) + 17.2 \, (250) - 1180 \, (0.873)$$
$$+ \, 12.61 \, (47.1)] = 0.061858$$

• Apply Equation 2-109:

$$B_o = 1.528 \exp\left[-0.061858 \ln\left(\frac{5015}{2392}\right)\right] = 1.459 \, bbl/STB$$

Using Equation 2-110:
- Calculate the correlating parameter A from Equation 2-111:

$$A = 4.1646 \times 10^{-7} (751)^{0.69357} (0.851)^{0.1885} (47.1)^{0.3272}$$
$$\times (250)^{0.6729} = 0.005778$$

- Solve for B_o by applying Equation 2-110:

$$B_o = 1.528 \exp [-0.005778 (5015^{.4094} - 2392^{.4096})] = 1.453 \text{ bbl/STB}$$

Crude Oil Density

The crude oil density is defined as the mass of a unit volume of the crude at a specified pressure and temperature. It is usually expressed in pounds per cubic foot. Several empirical correlations for calculating the density of liquids of unknown compositional analysis have been proposed. The correlations employ limited PVT data such as gas gravity, oil gravity, and gas solubility as correlating parameters to estimate liquid density at the prevailing reservoir pressure and temperature.

Equation 2-93 may be used to calculate the density of the oil at pressure below or equal to the bubble-point pressure. Solving Equation 2-93 for the oil density gives:

$$\rho_o = \frac{62.4\gamma_o + 0.0136 R_s \gamma_g}{B_o} \qquad (2\text{-}112)$$

where γ_o = specific gravity of the stock-tank oil
R_s = gas solubility, scf/STB
ρ_o = oil density, lb/ft^3

Standing (1981) proposed an empirical correlation for estimating the oil formation volume factor as a function of the gas solubility R_s, the specific gravity of stock-tank oil γ_o, the specific gravity of solution gas γ_g, and the system temperature T. By coupling the mathematical definition of the oil formation volume factor (as discussed in a later section) with Standing's correlation, the density of a crude oil at a specified pressure and temperature can be calculated from the following expression:

$$\rho_o = \frac{62.4\,\gamma_o + 0.0136R_s\,\gamma_g}{\left[0.972 + 0.000147\left[R_s\left(\frac{\gamma_g}{\gamma_o}\right)^{.5} + 1.25(T-460)\right]\right]^{1.175}} \qquad (2\text{-}113)$$

where   T = system temperature, $^\circ$R

        γ_o = specific gravity of the stock-tank oil

Example 2-33

Using the experimental PVT data given in Example 2-29 for the six different crude oil systems, calculate the oil density by using Equations 2-112 and 2-113. Compare the results with the experimental values and calculate the absolute average error (AAE).

Solution

Crude Oil	Measured Oil Density	Equation 2-112	Equation 2-113
1	38.13	38.04	38.31
2	40.95	40.85	40.18
3	37.37	37.68	38.26
4	42.25	41.52	40.39
5	37.70	38.39	38.08
6	46.79	46.86	44.11
AAE		0.84%	2.65%

Density of the oil at pressures above the bubble-point pressure can be calculated with:

$$\rho_o = \rho_{ob}\,\exp\left[c_o\,(p - p_b)\right] \qquad (2\text{-}114)$$

where ρ_o = density of the oil at pressure p, lb/ft^3

       ρ_{ob} = density of the oil at the bubble-point pressure, lb/ft^3

       c_o = isothermal compressibility coefficient at average pressure, psi^{-1}

Vasquez-Beggs' oil compressibility correlation and the Petrosky-Farshad c_o expression can be incorporated in Equation 2-114 to give:

For the Vasquez-Beggs c_o equation:

$$\rho_o = \rho_{ob} \, \exp\left[A \ln\left(\frac{p}{p_b} \right) \right] \tag{2-115}$$

where

$$A = 10^{-5} \, [-1{,}433 + 5 \, R_{sb} + 17.2 \, (T - 460) - 1{,}180 \, \gamma_{gs} + 12.61 \, °API]$$

For the Petrosky-Farshad c_o expression:

$$\rho_o = \rho_{ob} \, \exp[A \, (p^{0.4094} - p_b^{0.4094})] \tag{2-116}$$

with the correlating parameter A as given by Equation 2-111

Crude Oil Viscosity

Crude oil viscosity is an important physical property that controls and influences the flow of oil through porous media and pipes. The viscosity, in general, is defined as the internal resistance of the fluid to flow.

The oil viscosity is a strong function of the temperature, pressure, oil gravity, gas gravity, and gas solubility. Whenever possible, oil viscosity should be determined by laboratory measurements at reservoir temperature and pressure. The viscosity is usually reported in standard PVT analyses. If such laboratory data are not available, engineers may refer to published correlations, which usually vary in complexity and accuracy depending upon the available data on the crude oil.

According to the pressure, the viscosity of crude oils can be classified into three categories:

- **Dead-Oil Viscosity**
  The dead-oil viscosity is defined as the viscosity of crude oil at atmospheric pressure (no gas in solution) and system temperature.
- **Saturated-Oil Viscosity**
  The saturated (bubble-point)-oil viscosity is defined as the viscosity of the crude oil at the bubble-point pressure and reservoir temperature.

• Undersaturated-Oil Viscosity

The undersaturated-oil viscosity is defined as the viscosity of the crude oil at a pressure above the bubble-point and reservoir temperature.

Estimation of the oil viscosity at pressures equal to or below the bubble-point pressure is a two-step procedure:

Step 1. Calculate the viscosity of the oil without dissolved gas (dead oil), μ_{od}, at the reservoir temperature.

Step 2. Adjust the dead-oil viscosity to account for the effect of the gas solubility at the pressure of interest.

At pressures greater than the bubble-point pressure of the crude oil, another adjustment step, i.e. Step 3, should be made to the bubble-point oil viscosity, μ_{ob}, to account for the compression and the degree of undersaturation in the reservoir. A brief description of several correlations that are widely used in estimating the oil viscosity in the above three steps is given below.

METHODS OF CALCULATING VISCOSITY OF THE DEAD OIL

Several empirical methods are proposed to estimate the viscosity of the dead oil, including:

• Beal's correlation
• The Beggs-Robinson correlation
• Glaso's correlation

These three methods are presented below.

Beal's Correlation

From a total of 753 values for dead-oil viscosity at and above 100°F, Beal (1946) developed a graphical correlation for determining the viscosity of the dead oil as a function of temperature and the API gravity of the crude. Standing (1981) expressed the proposed graphical correlation in a mathematical relationship as follows:

$$\mu_{od} = \left(0.32 + \frac{1.8(10^7)}{API^{4.53}}\right)\left(\frac{360}{T - 260}\right)^a \qquad (2\text{-}117)$$

with

$$a = 10^{(0.43 + 8.33/API)}$$

where μ_{od} = viscosity of the dead oil as measured at 14.7 psia and
          reservoir temperature, cp
       T = temperature, °R

The Beggs-Robinson Correlation

Beggs and Robinson (1975) developed an empirical correlation for determining the viscosity of the dead oil. The correlation originated from analyzing 460 dead-oil viscosity measurements. The proposed relationship is expressed mathematically as follows:

$$\mu_{od} = 10^x - 1 \qquad (2\text{-}118)$$

where $x = Y\,(T - 460)^{-1.163}$
       $Y = 10^Z$
       $Z = 3.0324 - 0.02023\,°API$

An average error of −0.64% with a standard deviation of 13.53% was reported for the correlation when tested against the data used for its development. Sutton and Farshad (1980) reported an error of 114.3% when the correlation was tested against 93 cases from the literature.

Glaso's Correlation

Glaso (1980) proposed a generalized mathematical relationship for computing the dead-oil viscosity. The relationship was developed from experimental measurements on 26 crude oil samples. The correlation has the following form:

$$\mu_{od} = [3.141\,(10^{10})]\,(T - 460)^{-3.444}[\log\,(API)]^a \qquad (2\text{-}119)$$

where the coefficient a is given by:

$$a = 10.313\,[\log(T - 460)] - 36.447$$

The above expression can be used within the range of 50–300°F for the system temperature and 20–48° for the API gravity of the crude.

Sutton and Farshad (1986) concluded that Glaso's correlation showed the best accuracy of the three previous correlations.

METHODS OF CALCULATING THE SATURATED OIL VISCOSITY

Several empirical methods are proposed to estimate the viscosity of the saturated oil, including:

• The Chew-Connally correlation
• The Beggs-Robinson correlation

These two correlations are presented below.

The Chew-Connally Correlation

Chew and Connally (1959) presented a graphical correlation to adjust the dead-oil viscosity according to the gas solubility at saturation pressure. The correlation was developed from 457 crude oil samples. Standing (1977) expressed the correlation in a mathematical form as follows:

$$\mu_{ob} = (10)^a \, (\mu_{od})^b \tag{2-120}$$

with $\quad a = R_s \, [2.2(10^{-7}) \, R_s - 7.4(10^{-4})]$

$$b = \frac{0.68}{10^c} + \frac{0.25}{10^d} + \frac{0.062}{10^e}$$

$$c = 8.62(10^{-5})R_s$$
$$d = 1.1(10^{-3})R_s$$
$$e = 3.74(10^{-3})R_s$$

where μ_{ob} = viscosity of the oil at the bubble-point pressure, cp
$\quad \mu_{od}$ = viscosity of the dead oil at 14.7 psia and reservoir temperature, cp

The experimental data used by Chew and Connally to develop their correlation encompassed the following ranges of values for the independent variables:

Pressure, psia: 132–5,645
Temperature, °F: 72–292
Gas solubility, scf/STB: 51–3,544
Dead oil viscosity, cp: 0.377–50

The Beggs-Robinson Correlation

From 2,073 saturated oil viscosity measurements, Beggs and Robinson (1975) proposed an empirical correlation for estimating the saturated-oil viscosity. The proposed mathematical expression has the following form:

$$\mu_{ob} = a(\mu_{od})^b \tag{2-121}$$

where $a = 10.715(R_s + 100)^{-0.515}$
$\quad b = 5.44(R_s + 150)^{-0.338}$

The reported accuracy of the correlation is −1.83% with a standard deviation of 27.25%.

The ranges of the data used to develop Beggs and Robinson's equation are:

Pressure, psia: 132–5,265
Temperature, °F: 70–295
API gravity: 16–58
Gas solubility, scf/STB: 20–2,070

METHODS OF CALCULATING THE VISCOSITY OF THE UNDERSATURATED OIL

Oil viscosity at pressures above the bubble point is estimated by first calculating the oil viscosity at its bubble-point pressure and adjusting the bubble-point viscosity to higher pressures. Vasquez and Beggs proposed a simple mathematical expression for estimating the viscosity of the oil above the bubble-point pressure. This method is discussed below.

The Vasquez-Beggs Correlation

From a total of 3,593 data points, Vasquez and Beggs (1980) proposed the following expression for estimating the viscosity of undersaturated crude oil:

$$\mu_o = \mu_{ob}\left(\frac{p}{p_b}\right)^m \qquad (2\text{-}123)$$

where

$$m = 2.6\, p^{1.187}\, 10^a$$

with

$$a = -3.9(10^{-5})\, p - 5$$

The data used in developing the above correlation have the following ranges:

Pressure, psia: 141–9,151
Gas solubility, scf/STB: 9.3–2,199
Viscosity, cp: 0.117–148
Gas gravity: 0.511–1.351
API gravity: 15.3–59.5

The average error of the viscosity correlation is reported as −7.54%.

Example 2-34

In addition to the experimental PVT data given in Example 2-29, the following viscosity data are available:

Oil #	Dead Oil μ_{od} @ T	Saturated Oil μ_{ob}, cp	Undersaturated Oil μ_o @ p
1	0.765 @ 250°F	0.224	0.281 @ 5000 psi
2	1.286 @ 220°F	0.373	0.450 @ 5000 psi
3	0.686 @ 260°F	0.221	0.292 @ 5000 psi
4	1.014 @ 237°F	0.377	0.414 @ 6000 psi
5	1.009 @ 218°F	0.305	0.394 @ 6000 psi
6	4.166 @ 180°F	0.950	1.008 @ 5000 psi

Using all the oil viscosity correlations discussed in this chapter, please calculate μ_{od}, μ_{ob}, and the viscosity of the undersaturated oil.

Solution

Dead-oil viscosity

Oil #	Measured μ_{od}	Beal's	Beggs-Robinson	Glaso's
1	0.765	0.322	0.568	0.417
2	0.286	0.638	1.020	0.775
3	0.686	0.275	0.493	0.363
4	1.014	0.545	0.917	0.714
5	1.009	0.512	0.829	0.598
6	4.166	4.425	4.246	4.536
AAE		44.9%	17.32%	35.26%

Saturated-oil viscosity

Oil #	Measured μ_{ob}	Chew-Connally	Beggs-Robinson
1	0.224	0.313*	0.287*
2	0.373	0.426	0.377
3	0.221	0.308	0.279
4	0.377	0.311	0.297
5	0.305	0.316	0.300
6	0.950	0.842	0.689
AAE		21%	17%

*Using the measured μ_{od}.

Undersaturated-oil viscosity

Oil #	Measured μ_o	Beal's	Vasquez-Beggs
1	0.281	0.273*	0.303*
2	0.45	0.437	0.485
3	0.292	0.275	0.318
4	0.414	0.434	0.472
5	0.396	0.373	0.417
6	1.008	0.945	1.016
AAE		3.8%	7.5%

Using the measured μ_{ob}.

Surface/Interfacial Tension

The surface tension is defined as the force exerted on the boundary layer between a liquid phase and a vapor phase per unit length. This force is caused by differences between the molecular forces in the vapor phase and those in the liquid phase, and also by the imbalance of these forces at the interface. The surface can be measured in the laboratory and is unusually expressed in dynes per centimeter. The surface tension is an important property in reservoir engineering calculations and designing enhanced oil recovery projects.

Sugden (1924) suggested a relationship that correlates the surface tension of a pure liquid in equilibrium with its own vapor. The correlating parameters of the proposed relationship are molecular weight M of the pure component, the densities of both phases, and a newly introduced temperature independent parameter P_{ch}. The relationship is expressed mathematically in the following form:

$$\sigma = \left[\frac{P_{ch} (\rho_L - \rho_v)}{M} \right]^4 \qquad (2\text{-}124)$$

where σ is the surface tension and P_{ch} is a temperature independent parameter and is called the *parachor.*

The parachor is a dimensionless constant characteristic of a pure compound and is calculated by imposing experimentally measured surface tension and density data on Equation 2-124 and solving for P_{ch}. The Parachor values for a selected number of pure compounds are given in Table 2-1 as reported by Weinaug and Katz (1943).

Table 2-1
Parachor for Pure Substances

Component	Parachor	Component	Parachor
CO_2	78.0	$n\text{-}C_4$	189.9
N_2	41.0	$i\text{-}C_5$	225.0
C_1	77.0	$n\text{-}C_5$	231.5
C_2	108.0	$n\text{-}C_6$	271.0
C_3	150.3	$n\text{-}C_7$	312.5
$i\text{-}C_4$	181.5	$n\text{-}C_8$	351.5

Fanchi (1985) correlated the parachor with molecular weight with a simple linear equation. This linear is only valid for components heavier than methane. Fanchi's linear equation has the following form:

$$(P_{ch})_i = 69.9 + 2.3 \, M_i \tag{2-125}$$

where M_i = molecular weight of component i
$(P_{ch})_i$ = parachor of component i

For a complex hydrocarbon mixture, Katz et al. (1943) employed the Sugden correlation for mixtures by introducing the compositions of the two phases into Equation 2-124. The modified expression has the following form:

$$\sigma^{1/4} = \sum_{i=1}^{n} \left[(P_{ch})_i \, (Ax_i - By_i) \right] \tag{2-126}$$

with the parameters A and B as defined by:

$$A = \frac{\rho_o}{62.4 M_o}$$

$$B = \frac{\rho_g}{62.4 M_g}$$

where ρ_o = density of the oil phase, lb/ft^3
M_o = apparent molecular weight of the oil phase
ρ_g = density of the gas phase, lb/ft^3
M_g = apparent molecular weight of the gas phase
x_i = mole fraction of component i in the oil phase
y_i = mole fraction of component i in the gas phase
n = total number of components in the system

Example 2-35

The composition of a crude oil and the associated equilibrium gas is given below. The reservoir pressure and temperature are 4,000 psia and 160°F, respectively.

Component	x_i	y_i
C_1	0.45	0.77
C_2	0.05	0.08
C_3	0.05	0.06
$n\text{-}C_4$	0.03	0.04
$n\text{-}C_5$	0.01	0.02
C_6	0.01	0.02
C_{7+}	0.40	0.01

The following additional PVT data are available:

Oil density = 46.23 lb/ft^3
Gas density = 18.21 lb/ft^3
Molecular weight of C_{7+} = 215

Calculate the surface tension.

Solution

Step 1. Calculate the apparent molecular weight of the liquid and gas phase:

$$M_o = 100.253 \quad M_g = 24.99$$

Step 2. Calculate the coefficients A and B:

$$A = \frac{46.23}{(62.4)(100.253)} = 0.00739$$

$$B = \frac{18.21}{(62.6)(24.99)} = 0.01168$$

Step 3. Calculate the parachor of C_{7+} from Equation 2-125:

$$(P_{ch})_{C_{7+}} = 69.9 + (2.3)(215) = 564.4$$

Step 4. Construct the following working table:

Component	P_{ch}	Ax_i	By_i	$P_{ch}(Ax_i - By_i)$
C_1	77	0.00333	0.0090	−0.4361
C_2	108	0.00037	0.00093	−0.0605
C_3	150.3	0.00037	0.00070	−0.0497
$n\text{-}C_4$	189.9	0.00022	0.00047	−0.0475
$n\text{-}C_5$	231.5	0.00007	0.00023	−0.0370
C_6	271.0	0.000074	0.00023	−0.0423
C_{7+}	564.4	0.00296	0.000117	1.6046
				0.9315

Step 5. $\sigma = (0.9315)^4 = 0.753$ dynes/cm

PROPERTIES OF RESERVOIR WATER

Water Formation Volume Factor

The water formation volume factor can be calculated by the following mathematical expression:*

$$B_w = A_1 + A_2\, p + A_3\, p^2 \tag{2-127}$$

where the coefficients $A_1 - A_3$ are given by the following expression:

$$A_i = a_1 + a_2(T - 460) + a_3(T - 460)^2$$

with $a_1 - a_3$ given for gas-free and gas-saturated water:

Gas-Free Water

A_i	a_1	a_2	a_3
A_1	0.9947	$5.8(10^{-6})$	$1.02(10^{-6})$
A_2	$-4.228(10^{-6})$	$1.8376(10^{-8})$	$-6.77(10^{-11})$
A_3	$1.3(10^{-10})$	$-1.3855(10^{-12})$	$4.285(10^{-15})$

Gas-Saturated Water

A_i	a_1	a_2	a_3
A_1	0.9911	$6.35(10^{-5})$	$8.5(10^{-7})$
A_2	$-1.093(10^{-6})$	$-3.497(10^{-9})$	$4.57(10^{-12})$
A_3	$-5.0(10^{-11})$	$6.429(10^{-13})$	$-1.43(10^{-15})$

*Hewlett-Packard H.P. 41C Petroleum Fluids PAC manual, 1982.

The temperature T in Equation 2-127 is in °R.

Water Viscosity

Meehan (1980) proposed a water viscosity correlation that accounts for both the effects of pressure and salinity:

$$\mu_w = \mu_{wD} \, [1 + 3.5 \times 10^{-2} \, p^2 \, (T - 40)] \qquad (2\text{-}128)$$

with

$$\mu_{wD} = A + B/T$$
$$A = 4.518 \times 10^{-2} + 9.313 \times 10^{-7} Y - 3.93 \times 10^{-12} Y^2$$
$$B = 70.634 + 9.576 \times 10^{-10} Y^2$$

where μ_w = brine viscosity at p and T, cp
  μ_{wD} = brine viscosity at p = 14.7, T, cp
  p = pressure of interest, psia
  T = temperature of interest, T °F
  Y = water salinity, ppm

Brill and Beggs (1978) presented a simpler equation, which considers only temperature effects:

$$\mu_w = \exp \, (1.003 - 1.479 \times 10^{-2} T + 1.982 \times 10^{-5} T^2) \qquad (2\text{-}129)$$

where T is in °F and μ_w is in cp.

Gas Solubility in Water

The following correlation can be used to determine the gas solubility in water:

$$R_{sw} = A + B \, p + C \, p^2 \qquad (2\text{-}130)$$

where $A = 2.12 + 3.45 \, (10^{-3}) \, T - 3.59 \, (10^{-5}) \, T^2$
   $B = 0.0107 - 5.26 \, (10^{-5}) \, T + 1.48 \, (10^{-7}) \, T^2$
   $C = 8.75 \, (10^{-7}) + 3.9 \, (10^{-9}) \, T - 1.02 \, (10^{-11}) \, T^2$

The temperature T in above equations is expressed in °F.

Water Isothermal Compressibility

Brill and Beggs (1978) proposed the following equation for estimating water isothermal compressibility, ignoring the corrections for dissolved gas and solids:

$$C_w = (C_1 + C_2 T + C_3 T^2) \times 10^{-6} \tag{2-131}$$

where $C_1 = 3.8546 - 0.000134\ p$
$\qquad C_2 = -0.01052 + 4.77 \times 10^{-7}\ p$
$\qquad C_3 = 3.9267 \times 10^{-5} - 8.8 \times 10^{-10} p$
$\qquad T = °F$
$\qquad p = psia$
$\qquad C_w = psi^{-1}$

PROBLEMS

1. Assuming an ideal gas behavior, calculate the density of n-butane at 220°F and 50 psia.
2. Show that:

$$y_i = \frac{(w_i / M_i)}{\sum_i (w_i / M_i)}$$

3. Given the following gas:

Component	Weight Fraction
C_1	0.65
C_2	0.15
C_3	0.10
n-C_4	0.06
n-C_5	0.04

Calculate:

a. Mole fraction of the gas
b. Apparent molecular weight
c. Specific gravity
d. Specific volume at 300 psia and 120°F by assuming an ideal gas behavior

4. An ideal gas mixture has a density of 1.92 lb/ft^3 at 500 psia and 100°F. Calculate the apparent molecular weight of the gas mixture.

5. Using the gas composition as given in Problem 3, and assuming real gas behavior, calculate:

   a. Gas density at 2,000 psia and 150°F

   b. Specific volume at 2,000 psia and 150°F

   c. Gas formation volume factor in scf/ft^3

6. A natural gas with a specific gravity of 0.75 has a gas formation volume factor of 0.00529 ft^3/scf at the prevailing reservoir pressure and temperature. Calculate the density of the gas.

7. A natural gas has the following composition:

Component	y_i
C_1	0.75
C_2	0.10
C_3	0.05
i-C_4	0.04
n-C_4	0.03
i-C_5	0.02
n-C_5	0.01

Reservoir conditions are 3,500 psia and 200°F. Calculate:

   a. Isothermal gas compressibility coefficient

   b. Gas viscosity by using the

      1. Carr-Kobayashi-Burrows method

      2. Lee-Gonzales-Eakin method

8. Given the following gas composition:

Component	y_i
CO_2	0.06
N_2	0.03
C_1	0.75
C_2	0.07
C_3	0.04
n-C_4	0.03
n-C_5	0.02

If the reservoir pressure and temperature are 2,500 psia and 175°F, respectively, calculate:

   a. Gas density by accounting for the presence of nonhydrocarbon
      components by using the

   1. Wichert-Aziz method
   2. Carr-Kobayashi-Burrows method

   b. Isothermal gas compressibility coefficient
   c. Gas viscosity by using the

   1. Carr-Kobayashi-Burrows method
   2. Lee-Gonzales-Eakin method

9. A crude oil system exists at its bubble-point pressure of 1,708.7 psia
   and a temperature of 131°F. Given the following data:

   API = 40°

   Average specific gravity of separator gas = 0.85
   Separator pressure = 100 psig

   a. Calculate R_{sb} by using

   1. Standing's correlation
   2. The Vasquez-Beggs method
   3. Glaso's correlation
   4. Marhoun's equation
   5. The Petrosky-Farshad correlation

   b. Calculate B_{ob} by applying methods listed in Part a.

10. Estimate the bubble-point pressure of a crude oil system with the fol-
    lowing limited PVT data:

    API = 35°   T = 160°F R_{sb} = 700 scf/STB γ_g = 0.75

    Use the six different methods listed in Problem 9.

11. A crude oil system exists at an initial reservoir pressure of 4500 psi
    and 85°F. The bubble-point pressure is estimated at 2109 psi. The oil
    properties at the bubble-point pressure are as follows:

    B_{ob} = 1.406 bbl/STB R_{sb} = 692 scf/STB
    γ_g = 0.876   API = 41.9°

    Calculate:

    a. Oil density at the bubble-point pressure
    b. Oil density at 4,500 psi
    c. B_o at 4500 psi

12. A high-pressure cell has a volume of 0.33 ft^3 and contains gas at
    2,500 psia and 130°F, at which conditions its z-factor is 0.75. When
    43.6 scf of the gas are bled from the cell, the pressure dropped to

1,000 psia, the temperature remaining at 130°F. What is the gas deviation factor at 1,000 psia and 130°F?

13. A hydrocarbon gas mixture with a specific gravity of 0.7 has a density of 9 lb/ft^3 at the prevailing reservoir pressure and temperature. Calculate the gas formation volume factor in bbl/scf.

14. A gas reservoir exists at a 150°F. The gas has the following composition:

Component	Mole%
C_1	89
C_2	7
C_3	4

The gas expansion factor E_g was calculated as 204.648 scf/ft^3 at the existing reservoir pressure and temperature. Calculate the viscosity of the gas.

15. A 20 cu ft tank at a pressure of 2500 psia and 212°F contains ethane gas. How many pounds of ethane are in the tank?

16. The PVT data as shown below were obtained on a crude oil sample taken from the Nameless Field. The initial reservoir pressure was 3600 psia at 160°F. The average specific gravity of the solution gas is 0.65. The reservoir contains 250 mm bbl of oil initially in place. The oil has a bubble-point pressure of 2500 psi.

a. Calculate the two-phase oil formation volume factor at:

1. 3200 psia
2. 2800 psia
3. 1800 psia

b. What is the initial volume of dissolved gas in the reservoir?
c. Oil compressibility coefficient at 3200 psia.

Pressure, psia	Solution gas, scf/STB at 1407 psia and 60°F	Formation Volume Factor, bbl/STB
3600		1.310
3200		1.317
2800		1.325
2500	567	1.333
2400	554	1.310
1800	436	1.263
1200	337	1.210
600	223	1.140
200	143	1.070

17. The following PVT data were obtained from the analysis of a bottom-hole sample.

p psia	Relative Volume V/V_{sat}
3000	1.0000
2927	1.0063
2703	1.0286
2199	1.1043
1610	1.2786
1206	1.5243
999	1.7399

a. Plot the Y-function versus pressure on rectangular coordinate paper, see Equation 3-3

b. Determine the constants in the equation

$$Y = mp + b$$

by using method of least squares.

c. Recalculate relative oil volume from the equation, see Equation 3-5

18. A 295-cc crude oil sample was placed in a PVT at an *initial* pressure of 3500 psi. The cell temperature was held at a constant temperature of 220°F. A differential liberation test was then performed on the crude oil sample with the recorded measurements as given below:

p, psi	T, °F	Total Volume, cc	Vol. of Liquids, cc	Vol. of Liberated Gas, scf	Specific Gravity of Liberated Gas
3500	220	290	290	0	—
3300	220	294	294	0	—
*3000	220	300	300	0	—
2000	220	323.2	286.4	0.1627	0.823
1000	220	375.2	271.5	0.1840	0.823
14.7	60	—	179.53	0.5488	0.823

*Bubble-point pressure

Using the bore-recorded measurements and assuming an oil gravity of 40° API, calculate the following PVT properties:

a. Oil formation volume factor at 3500 psi.

b. Gas solubility at 3500 psi.

c. Oil viscosity at 3500 psi.

d. Isothermal compressibility coefficient at 3300 psi.

e. Oil density at 1000 psi.

19. Experiments were made on a bottom-hole crude oil sample taken from the North Grieve Field to determine the gas solubility and oil formation volume factor as a function of pressure. The initial reservoir pressure was recorded as 3600 psia and reservoir temperature was 130°F. The following data were obtained from the measurements:

Pressure psia	R_S scf/STB	B_o bbl/STB
3600	567	1.310
3200	567	1.317
2800	567	1.325
2500	567	1.333
2400	554	1.310
1800	436	1.263
1200	337	1.210
600	223	1.140
200	143	1.070

At the end of the experiments, the API gravity of the oil was measured as 40°. If the average specific gravity of the solution gas is 0.7, calculate:

a. Total formation volume factor at 3200 psia

b. Oil viscosity at 3200 psia

c. Isothermal compressibility coefficient at 1800 psia

20. You are producing a 35°API crude oil from a reservoir at 5000 psia and 140°F. The bubble-point pressure of the reservoir liquids is 4000 psia at 140°F. Gas with a gravity of 0.7 is produced with the oil at a rate of 900 scf/STB. Calculate:

a. Density of the oil at 5000 psia and 140°F

b. Total formation volume factor at 5000 psia and 140°F

21. An undersaturated-oil reservoir exists at an initial reservoir pressure 3112 psia and a reservoir temperature of 125°F. The bubble point of the oil is 1725 psia. The crude oil has the following pressure versus oil formation volume factor relationship:

Pressure psia	B_o bbl/STB
3112	1.4235
2800	1.4290
2400	1.4370
2000	1.4446
1725	1.4509
1700	1.4468
1600	1.4303
1500	1.4139
1400	1.3978

The API gravity of the crude oil and the specific gravity of the solution gas are 40° and 0.65, respectively. Calculate the density of the crude oil at 3112 psia and 125°F.

22. A PVT cell contains 320 cc of oil and its bubble-point pressure of 2500 psia and 200°F. When the pressure was reduced to 2000 psia, the volume increased to 335.2 cc. The gas was bled off and found to occupy a volume of 0.145 scf. The volume of the oil was recorded as 303 cc. The pressure was reduced to 14.7 psia and the temperature to 60°F while 0.58 scf of gas was evolved leaving 230 cc of oil with a gravity of 42°API. Calculate:

a. Gas compressibility factor at 2000 psia
b. Gas solubility at 2000 psia

REFERENCES

1. Ahmed, T., "Compositional Modeling of Tyler and Mission Canyon Formation Oils with CO_2 and Lean Gases," final report submitted to Montana's on a New Track for Science (MONTS) (Montana National Science Foundation Grant Program), 1985–1988.

2. Baker, O. and Swerdloff, W., "Calculations of Surface Tension-3: Calculations of Surface Tension Parachor Values," *OGJ*, December 5, 1955, Vol. 43, p. 141.

3. Beal, C., "The Viscosity of Air, Water, Natural Gas, Crude Oils and its Associated Gases at Oil Field Temperatures and Pressures," *Trans. AIME*, 1946, Vol. 165, pp. 94–112.

4. Beggs, H. D. and Robinson, J. R., "Estimating the Viscosity of Crude Oil Systems," *JPT,* September 1975, pp. 1140–1141.

5. Brill, J. and Beggs, H., *Two-Phase Flow in Pipes.* Tulsa, OK: The University of Tulsa, 1978.

6. Brown, et al., "Natural Gasoline and the Volatile Hydrocarbons," Tulsa: NGAA, 1948.

7. Carr, N., Kobayashi, R., and Burrows, D., "Viscosity of Hydrocarbon Gases Under Pressure," *Trans. AIME,* 1954, Vol. 201, pp. 270–275.

8. Chew, J., and Connally, Jr., C. A., "A Viscosity Correlation for Gas-Saturated Crude Oils," *Trans. AIME,* 1959, Vol. 216, pp. 23–25.

9. Dean, D. E., and Stiel, L. I., "The Viscosity of Non-polar Gas Mixtures at Moderate and High Pressure," *AIChE Jour.,* 1958, Vol. 4, pp. 430–436.

10. Dempsey, J. R., "Computer Routine Treats Gas Viscosity as a Variable," *Oil and Gas Journal,* Aug. 16, 1965, pp. 141–143.

11. Dranchuk, P. M., and Abu-Kassem, J. H., "Calculation of Z-factors for Natural Gases Using Equations-of-State," *JCPT,* July–Sept., 1975, pp. 34–36.

12. Dranchuk, P. M., Purvis, R. A., and Robinson, D. B., "Computer Calculations of Natural Gas Compressibility Factors Using the Standing and Katz Correlation," *Inst. Of Petroleum Technical Series,* No. IP 74-008, 1974.

13. Fanchi, J. R., "Calculation of Parachors for Composition Simulation," *JPT,* November 1985, pp. 2049–2050.

14. Glaso, O., "Generalized Pressure-Volume-Temperature Correlations," *JPT,* May 1980, pp. 785–795.

15. Hall, K. R., and Yarborough, L., "A New Equation-of-State for Z-factor Calculations," *Oil and Gas Journal,* June 18, 1973, pp. 82–92.

16. Hankinson, R. W., Thomas, L. K., and Phillips, K. A., "Predict Natural Gas Properties," *Hydrocarbon Processing,* April 1969, pp. 106–108.

17. Kay, W. B., "Density of Hydrocarbon Gases and Vapor," *Industrial and Engineering Chemistry,* 1936, Vol. 28, pp. 1014–1019.

18. Lee, A. L., Gonzalez, M. H., and Eakin, B. E., "The Viscosity of Natural Gases," *JTP,* August 1966, pp. 997–1000.

19. Marhoun, M. A., "PVT Correlation for Middle East Crude Oils," *JPT,* May 1988, pp. 650–665.

20. Mattar, L. G., Brar, S., and Aziz, K., "Compressibility of Natural Gases," *Journal of Canadian Petroleum Technology,* October–November 1975, pp. 77–80.

21. Meehan, D. N., "A Correlation for Water Compressibility," *Petroleum Engineer,* November 1980, pp. 125–126.

22. Papay, J., "A Termelestechnologiai Parameterek Valtozasa a Gazlelepk Muvelese Soran," *OGIL MUSZ,* Tud, Kuzl., Budapest, 1968, pp. 267–273.

23. Petrosky, G. E., and Farshad, F., "Pressure-Volume-Temperature Correlations for Gulf of Mexico Crude Oils," SPE Paper 26644, presented at the 68th Annual Technical Conference of the SPE in Houston, Texas, 3–6 October, 1993.

24. Standing, M. B., *Volumetric and Phase Behavior of Oil Field Hydrocarbon Systems,* pp. 125–126. Dallas: Society of Petroleum Engineers, 1977.

25. Standing, M. B. and Katz, D. L., "Density of Natural Gases," *Trans. AIME,* 1942, Vol. 146, pp. 140–149.

26. Stewart, W. F., Burkhard, S. F., and Voo, D., "Prediction of Pseudo-Critical Parameters for Mixtures," paper presented at the AIChE Meeting, Kansas City, MO, 1959.

27. Sugden, S., "The Variation of Surface Tension, VI. The Variation of Surface Tension with Temperature and Some Related Functions," *J. Chem. Soc.,* 1924, Vol. 125, pp. 32–39.

28. Sutton, R. P., "Compressibility Factors for High-Molecular-Weight Reservoir Gases," SPE Paper 14265, presented at the 60th Annual Technical Conference and Exhibition of the Society of Petroleum Engineers, Las Vegas, Sept. 22–25, 1985.

29. Sutton, R. P., and Farshad, F. F., "Evaluation of Empirically Derived PVT Properties for Gulf of Mexico Crude Oils," SPE Paper 13172, presented at the 59th Annual Technical Conference, Houston, Texas, 1984.

30. Takacs, G., "Comparisons Made for Computer Z-Factor Calculation," *Oil and Gas Journal,* Dec. 20, 1976, pp. 64–66.

31. Trube, A. S., "Compressibility of Undersaturated Hydrocarbon Reservoir Fluids," *Trans. AIME,* 1957, Vol. 210, pp. 341–344.

32. Trube, A. S., "Compressibility of Natural Gases," *Trans AIME,* 1957, Vol. 210, pp. 355–357.

33. Vasquez, M., and Beggs, D., "Correlations for Fluid Physical Properties Prediction," *JPT,* June 1980, pp. 968–970.

34. Weinaug, C., and Katz, D. L., "Surface Tension of Methane-Propane Mixtures," *Ind. Eng. Chem.*, 1943, Vol. 25, pp. 35–43.

35. Wichert, E., and Aziz, K., "Calculation of Z's for Sour Gases," *Hydrocarbon Processing*, 1972, Vol. 51, No. 5, pp. 119–122.

36. Yarborough, L., and Hall, K. R., "How to Solve Equation-of-State for Z-factors," *Oil and Gas Journal*, Feb 18, 1974, pp. 86–88.

LABORATORY ANALYSIS OF RESERVOIR FLUIDS

Accurate laboratory studies of PVT and phase-equilibria behavior of reservoir fluids are necessary for characterizing these fluids and evaluating their volumetric performance at various pressure levels. There are many laboratory analyses that can be made on a reservoir fluid sample. The amount of data desired determines the number of tests performed in the laboratory. In general, there are three types of laboratory tests used to measure hydrocarbon reservoir samples:

1. **Primary tests**

   These are simple, routine field (on-site) tests involving the measurements of the specific gravity and the gas-oil ratio of the produced hydrocarbon fluids.

2. **Routine laboratory tests**

   These are several laboratory tests that are routinely conducted to characterize the reservoir hydrocarbon fluid. They include:
   • Compositional analysis of the system
   • Constant-composition expansion
   • Differential liberation
   • Separator tests
   • Constant-volume depletion

3. **Special laboratory PVT tests**

   These types of tests are performed for very specific applications. If a reservoir is to be depleted under miscible gas injection or a gas cycling scheme, the following tests may be performed:
   • Slim-tube test
   • Swelling test

The objective of this chapter is to review the PVT laboratory tests and to illustrate the proper use of the information contained in PVT reports.

COMPOSITION OF THE RESERVOIR FLUID

It is desirable to obtain a fluid sample as early in the life of a field as possible so that the sample will closely approximate the original reservoir fluid. Collection of a fluid sample early in the life of a field reduces the chances of free gas existing in the oil zone of the reservoir.

Most of the parameters measured in a reservoir fluid study can be calculated with some degree of accuracy from the composition. It is the most complete description of reservoir fluid that can be made. In the past, reservoir fluid compositions were usually measured to include separation of the component methane through hexane, with the heptanes and heavier components grouped as a single component reported with the average molecular weight and density.

With the development of more sophisticated equations-of-state to calculate fluid properties, it was learned that a more complete description of the heavy components was necessary. It is recommended that compositional analyses of the reservoir fluid should include a separation of components through C_{10} as a minimum. The more sophisticated research laboratories now use equations-of-state that require compositions through C_{30} or higher.

Table 3-1 shows a chromatographic "fingerprint" compositional analysis of the Big Butte crude oil system. The table includes the mole fraction, weight fraction, density, and molecular weight of the individual component.

CONSTANT-COMPOSITION EXPANSION TESTS

Constant-composition expansion experiments are performed on gas condensates or crude oil to simulate the pressure-volume relations of these hydrocarbon systems. The test is conducted for the purposes of determining:

• Saturation pressure (bubble-point or dew-point pressure)
• Isothermal compressibility coefficients of the single-phase fluid in excess of saturation pressure
• Compressibility factors of the gas phase
• Total hydrocarbon volume as a function of pressure

(*text continued on page 134*)

Table 3-1
Hydrocarbon Analysis of Reservoir Fluid Sample

Component Name	Mol %	Wt %	Liquid Density (gm/cc)	MW
				Composition of Reservoir Fluid Sample (by Flash, Extended-Capillary Chromatography)
Hydrogen Sulfide	0.00	0.00	0.8006	34.08
Carbon Dioxide	0.25	0.11	0.8172	44.01
Nitrogen	0.88	0.25	0.8086	28.013
Methane	23.94	3.82	0.2997	16.043
Ethane	11.67	3.49	0.3562	30.07
Propane	9.36	4.11	0.5070	44.097
iso-Butane	1.39	0.81	0.5629	58.123
n-Butane	4.61	2.66	0.5840	58.123
iso-Pentane	1.50	1.07	0.6244	72.15
n-Pentane	2.48	1.78	0.6311	72.15
Hexanes	3.26	2.73	0.6850	84
Heptanes	5.83	5.57	0.7220	96
Octanes	5.52	5.88	0.7450	107
Nonanes	3.74	4.50	0.7640	121
Decanes	3.38	4.50	0.7780	134

Total Sample Properties

Molecular Weight 100.55
Equivalent Liquid Density, gm/scc 0.7204

			Density	MW
Undecanes	2.57	3.76	0.7890	147
Dodecanes	2.02	3.23	0.8000	161
Tridecanes	2.02	3.52	0.8110	175
Tetradecanes	1.65	3.12	0.8220	190
Pentadecanes	1.48	3.03	0.8320	206
Hexadecanes	1.16	2.57	0.8390	222
Heptadecanes	1.06	2.50	0.8470	237
Octadecanes	0.93	2.31	0.8520	251
Nonadecanes	0.88	2.31	0.8570	263
Eicosanes	0.77	2.11	0.8620	275
Heneicosanes	0.68	1.96	0.8670	291
Docosanes	0.60	1.83	0.8720	305
Tricosanes	0.55	1.74	0.8770	318
Tetracosanes	0.48	1.57	0.8810	331
Pentacosanes	0.47	1.60	0.8850	345
Hexacosanes	0.41	1.46	0.8890	359
Heptacosanes	0.36	1.33	0.8930	374
Octacosanes	0.37	1.41	0.8960	388
Nonacosanes	0.34	1.34	0.8990	402
Triacontanes plus	3.39	16.02	1.0440	474
Totals	100.00	100.00		

Plus Fractions	Mol %	Wt %	Density	MW
Heptanes plus	40.66	79.17	0.8494	196
Undecanes plus	22.19	58.72	0.8907	266
Pentadecanes plus	13.93	45.09	0.9204	326
Eicosanes plus	8.42	32.37	0.9540	387
Pentacosanes plus	5.34	23.16	0.9916	437
Triacontanes plus	3.39	16.02	1.0440	474

(text continued from page 131)

The experimental procedure, as shown schematically in Figure 3-1 involves placing a hydrocarbon fluid sample (oil or gas) in a visual PVT cell at reservoir temperature and at a pressure in excess of the initial reservoir pressure (Figure 3-1, Section A). The pressure is reduced in steps at constant temperature by removing mercury from the cell, and the change in the *total* hydrocarbon volume V_t is measured for each pressure increment.

The saturation pressure (bubble-point or dew-point pressure) and the corresponding volume are observed and recorded and used as a reference volume V_{sat} (Figure 3-1, Section C). The volume of the hydrocarbon system as a function of the cell pressure is reported as the ratio of the reference volume. This volume is termed the relative volume and is expressed mathematically by the following equation:

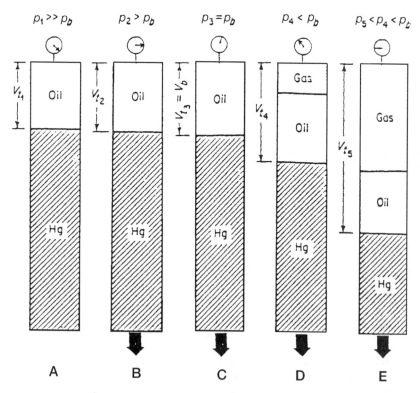

Figure 3-1. Constant-composition expansion test.

$$V_{rel} = \frac{V_t}{V_{sat}} \qquad (3\text{-}1)$$

where V_{rel} = relative volume
V_t = total hydrocarbon volume
V_{sat} = volume at the saturation pressure

The relative volume is equal to **one** at the saturation pressure. This test is commonly called pressure-volume relations, flash liberation, flash vaporization, or flash expansion.

It should be noted that no hydrocarbon material is removed from the cell, thus, the composition of the total hydrocarbon mixture in the cell remains fixed at the original composition.

Table 3-2 shows the results of the flash liberation test (the constant composition expansion test) for the Big Butte crude oil system. The bubble-point pressure of the hydrocarbon system is 1930 psi at 247°F. In addition to the reported values of the relative volume, the table includes the measured values of the oil density at and above the saturation pressure.

The density of the oil at the saturation pressure is 0.6484 gm/cc and is determined from direct weight-volume measurements on the sample in the PVT cell. Above the bubble-point pressure, the density of the oil can be calculated by using the recorded relative volume:

$$\rho = \rho_{sat} / V_{rel} \qquad (3\text{-}2)$$

where ρ = density at any pressure above the saturation pressure
ρ_{sat} = density at the saturation pressure
V_{rel} = relative volume at the pressure of interest

Example 3-1

Given the experimental data in Table 3-2, verify the oil density values at 4000 and 6500 psi.

Solution

Using Equation 3-2 gives:

• At 4000 psi

Table 3-2
Constant Composition Expansion Data

Pressure-Volume Relations
(at 247°F)

Pressure psig	Relative Volume (A)	Y-Function (B)	Density gm/cc
6500	0.9371		0.6919
6000	0.9422		0.6882
5500	0.9475		0.6843
5000	0.9532		0.6803
4500	0.9592		0.6760
4000	0.9657		0.6714
3500	0.9728		0.6665
3000	0.9805		0.6613
2500	0.9890		0.6556
2400	0.9909		0.6544
2300	0.9927		0.6531
2200	0.9947		0.6519
2100	0.9966		0.6506
2000	0.9987		0.6493
b≫1936	1.0000		0.6484
1930	1.0014		
1928	1.0018		
1923	1.0030		
1918	1.0042		
1911	1.0058		
1878	1.0139		
1808	1.0324		
1709	1.0625	2.108	
1600	1.1018	2.044	
1467	1.1611	1.965	
1313	1.2504	1.874	
1161	1.3694	1.784	
1035	1.5020	1.710	
782	1.9283	1.560	
600	2.4960	1.453	
437	3.4464	1.356	

(A) Relative volume: V/V_{sat} or volume at indicated pressure per volume at saturation pressure.

(B) Where Y-function $\dfrac{(p_{sat} - p)}{(p_{abs}) \bullet (V/V_{sat} - 1)}$

$$\rho_o = \frac{0.6484}{0.9657} = 0.6714\,\text{gm/cc}$$

- At 6500 psi

$$\rho_o = \frac{0.6484}{0.9371} = 0.6919$$

The relative volume data frequently require smoothing to correct for laboratory inaccuracies in measuring the total hydrocarbon volume just below the saturation pressure and also at lower pressures. A dimensionless compressibility function, commonly called the **Y-function,** is used to smooth the values of the relative volume. The function in its mathematical form is only defined below the saturation pressure and is given by the following expression:

$$Y = \frac{p_{sat} - p}{p(V_{rel} - 1)} \qquad (3\text{-}3)$$

where p_{sat} = saturation pressure, psia
$\qquad\quad$ p = pressure, psia
$\qquad\ V_{rel}$ = relative volume at pressure p

Column 3 in Table 3-2 lists the computed values of the Y-function as calculated by using Equation 3-3. To smooth the relative volume data below the saturation pressure, the Y-function is plotted as a function of pressure on a Cartesian scale. When plotted, the Y-function forms a straight line or has only a small curvature. Figure 3-2 shows the Y-function versus pressure for the Big Butte crude oil system. The figure illustrates the erratic behavior of the data near the bubble-point pressure.

The following steps summarize the simple procedure of smoothing and correcting the relative volume data:

Step 1. Calculate the Y-function for all pressures below the saturation pressure by using Equation 3-3.

Step 2. Plot the Y-function versus pressure on a Cartesian scale.

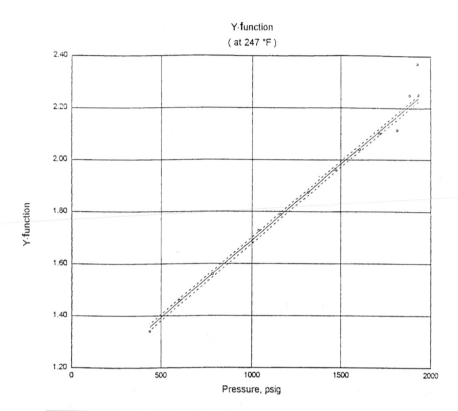

Figure 3-2. Y-function versus pressure.

Step 3. Determine the coefficients of the best straight fit of the data, or:

$$Y = a + bp \qquad (3\text{-}4)$$

where a and b are the intercept and slope of the lines, respectively.

Step 4. Recalculate the relative volume at all pressure below the saturation pressure from the following expression:

$$V_{rel} = 1 + \frac{p_{sat} - p}{p(a + bp)} \qquad (3\text{-}5)$$

Example 3-2

The best straight fit of the Y-function as a function of pressure for the Big Butte oil system is given by:

where $Y = a + bp$
        $a = 1.0981$
        $b = 0.000591$

Smooth the recorded relative volume data of Table 3-2.

Solution

Pressure	Measured V_{rel}	Smoothed V_{rel} Equation 3-5
1936	—	—
1930	—	1.0014
1928	—	1.0018
1923	—	1.0030
1918	—	1.0042
1911	—	1.0058
1878	—	1.0139
1808	—	1.0324
1709	1.0625	1.0630
1600	1.1018	1.1028
1467	1.1611	1.1626
1313	1.2504	1.2532
1161	1.3696	1.3741
1035	1.5020	1.5091
782	1.9283	1.9458
600	2.4960	2.5328
437	3.4464	3.5290

The oil compressibility coefficient c_o above the bubble-point pressure is also obtained from the relative volume data as listed in Table 3-3 for the Big Butte oil system.

Table 3-3
Undersaturated Compressibility Data

Volumetric Data
(at 247°F)

Saturation Pressure (P_{sat}) .... 1936 psig
Density at P_{sat} .... 0.6484 gm/cc
Thermal Exp @ 6500 psig .... 1.10401 V at 247°F/V at 60°F

Average Single-Phase Compressibilities

Pressure Range psig			Single-Phase Compressibility v/v/psi
6500	to	6000	10.73 E-6
6000	to	5500	11.31 E-6
5500	to	5000	11.96 E-6
5000	to	4500	12.70 E-6
4500	to	4000	13.57 E-6
4000	to	3500	14.61 E-6
3500	to	3000	15.86 E-6
3000	to	2500	17.43 E-6
2500	to	2000	19.47 E-6
2000	to	1936	20.79 E-6

The oil compressibility is defined by Equations 2-94 through 2-96 and equivalently can be written in terms of the relative volume, as:

$$c_o = \frac{-1}{V_{rel}} \frac{\partial V_{rel}}{\partial p} \qquad (3\text{-}6)$$

Commonly, the relative volume data above the bubble-point pressure is plotted as a function of pressure as shown in Figure 3-3. To evaluate c_o at any pressure p, it is only necessary to graphically differentiate the curve by drawing a tangent line and determining the slope of the line, i.e., $\partial V_{rel}/\partial p$.

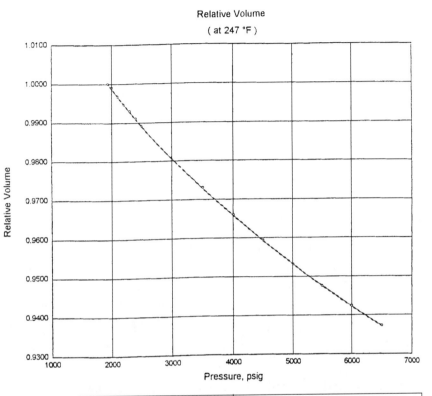

Relative Volume
(at 247 °F)

Relative Volume Expression:		LEGEND	
$y = a + b (Xd)^i + c (Xd)^j + d \log(Xd)^k$			
where:			o — Laboratory Data
a= 1.11371e+ 00	i= 0.500	- - - - - Confidence Limits	
b= -1.48699e- 01	j= 0.750	———— Analytical Expression	
c= 3.49924e- 02	k= 1.000		
d= 1.73284e- 02		Saturation Pressure: 1936 psig	
Note: Xd (dimensionless 'X') = Pi / Psat, psig		Current Reservoir Pressure: 2900 psig	
Confidence level:	99 %	**Pressure-Volume Relations**	
Confidence interval:	+/- 0.00015	Figure A-1	
'r squared':	.999928		

Figure 3-3. Relative volume data above the bubble-point pressure.

Example 3-3

Using Figure 3-3, evaluate c_o at 3000 psi.

Solution

• Draw a tangent line to the curve and determine the slope.

$$\partial V_{rel}/\partial p = -14.92 \times 10^{-6}$$

• Apply Equation 3-6 to give

$$c_o = \left(\frac{-1}{0.98}\right)\left(-14.92 \times 10^{-6}\right) = 15.23 \times 10^{-6} \ psi^{-1}$$

It should be noted that Table 3-3 lists the compressibility coefficient at several ranges of pressure, e.g. 6500–6000. These values are determined by calculating the changes in the relative volume at the indicated pressure interval and evaluating the relative volume at the lower pressure, or

$$c_o = \frac{-1}{\left[V_{rel}\right]_2} \ \frac{\left(V_{rel}\right)_1 - \left(V_{rel}\right)_2}{p_1 - p_2} \tag{3-7}$$

where the subscripts 1 and 2 represent the corresponding values at the higher and lower pressure range, respectively.

Example 3-4

Using the measured relative volume data in Table 3-2 for the Big Butte crude oil system, calculate the average oil compressibility in the pressure range of 2500 to 2000 psi.

Solution

Apply Equation 3-7 to give

$$c_o = \frac{-1}{0.9987} \ \frac{0.9890 - 0.9987}{2500 - 2000} = 19.43 \times 10^{-6} \ psi^{-1}$$

DIFFERENTIAL LIBERATION (VAPORIZATION) TEST

In the differential liberation process, the solution gas that is liberated from an oil sample during a decline in pressure is continuously removed from contact with the oil, and before establishing equilibrium with the liquid phase. This type of liberation is characterized by a varying composition of the total hydrocarbon system.

The experimental data obtained from the test include:

• Amount of gas in solution as a function of pressure
• The shrinkage in the oil volume as a function of pressure
• Properties of the evolved gas including the composition of the liberated gas, the gas compressibility factor, and the gas specific gravity
• Density of the remaining oil as a function of pressure

The differential liberation test is considered to better describe the separation process taking place in the reservoir and is also considered to simulate the flowing behavior of hydrocarbon systems at conditions above the critical gas saturation. As the saturation of the liberated gas reaches the critical gas saturation, the liberated gas begins to flow, leaving behind the oil that originally contained it. This is attributed to the fact that gases have, in general, higher mobility than oils. Consequently, this behavior follows the differential liberation sequence.

The test is carried out on reservoir oil samples and involves charging a visual PVT cell with a liquid sample at the bubble-point pressure and at reservoir temperature. As shown schematically in Figure 3-4, the pressure is reduced in steps, usually 10 to 15 pressure levels, and all the liberated gas is removed and its volume is measured at standard conditions. The volume of oil remaining V_L is also measured at each pressure level. It should be noted that the remaining oil is subjected to continual compositional changes as it becomes progressively richer in the heavier components.

The above procedure is continued to atmospheric pressure where the volume of the residual (remaining) oil is measured and converted to a volume at 60°F, V_{sc}. The differential oil formation volume factors B_{od} (commonly called the relative oil volume factors) at all the various pressure levels are calculated by dividing the recorded oil volumes V_L by the volume of residual oil V_{sc}, or:

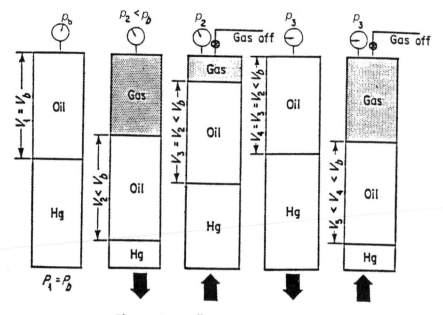

Figure 3-4. Differential vaporization test.

$$B_{od} = \frac{V_L}{V_{sc}} \qquad\qquad (3\text{-}8)$$

The differential solution gas-oil ratio R_{sd} is also calculated by dividing the volume of gas in solution by the residual oil volume.

Table 3-4 shows the results of a differential liberation test for the Big Butte crude. The test indicates that the differential gas-oil ratio and differential relative oil volume at the bubble-point pressure are 933 scf/STB and 1.730 bbl/STB, respectively. The symbols R_{sdb} and B_{odb} are used to represent these two values, i.e.:

$$R_{sdb} = 933 \text{ scf/STB} \quad \text{and} \quad B_{odb} = 1.730 \text{ bbl/STB}$$

Column C of Table 3-4 shows the relative total volume B_{td} from differential liberation as calculated from the following expression:

$$B_{td} = B_{od} + (R_{sdb} - R_{sd}) B_g \qquad\qquad (3\text{-}9)$$

where B_{td} = relative total volume, bbl/STB
  B_g = gas formation volume factor, bbl/scf

Table 3-4
Differential Liberation Data

Differential Vaporization
(at 247°F)

Pressure psig	Solution Gas/Oil Ratio R_{sd} (A)	Relative Oil Volume B_{od} (B)	Relative Total Volume B_{td} (C)	Oil Density gm/cc	Deviation Factor Z	Gas Formation Volume Factor (D)	Incremental Gas Gravity (Air = 1.000)
b≫1936	933	1.730	1.730	0.6484			
1700	841	1.679	1.846	0.6577	0.864	0.01009	0.885
1500	766	1.639	1.982	0.6650	0.869	0.01149	0.894
1300	693	1.600	2.171	0.6720	0.876	0.01334	0.901
1100	622	1.563	2.444	0.6790	0.885	0.01591	0.909
900	551	1.525	2.862	0.6863	0.898	0.01965	0.927
700	479	1.486	3.557	0.6944	0.913	0.02559	0.966
500	400	1.440	4.881	0.7039	0.932	0.03626	1.051
300	309	1.382	8.138	0.7161	0.955	0.06075	1.230
185	242	1.335	13.302	0.7256	0.970	0.09727	1.423
120	195	1.298	20.439	0.7328	0.979	0.14562	1.593
0	0	1.099		0.7745			2.375
	@ 60°F = 1.000						

Gravity of residual oil = 34.6°API at 60°F
Density of residual oil = 0.8511 gm/cc at 60°F
(A) Cubic feet of gas at 14.73 psia and 60°F per barrel of residual oil at 60°F.
(B) Barrels of oil at indicated pressure and temperature per barrel of residual oil at 60°F.
(C) Barrels of oil plus liberated gas at indicated pressure and temperature per barrel of residual oil at 60°F.
(D) Cubic feet of gas at indicated pressure and temperature per cubic feet at 14.73 psia and 60°F.

The gas deviation z-factor listed in column 6 of Table 3-4 represents the z-factor of the liberated (removed) solution gas at the specific pressure and these values are calculated from the recorded gas volume measurements as follows:

$$z = \left(\frac{V p}{T}\right)\left(\frac{T_{sc}}{V_{sc} \, p_{sc}}\right) \qquad (3\text{-}10)$$

where   V = volume of the liberated gas in the PVT cell at p and T
V_{sc} = volume of the removed gas at standard column 7 of Table 3-4 contains the gas formation volume factor B_g as expressed by the following equation:

$$B_g = \left(\frac{p_{sc}}{T_{sc}}\right)\frac{zT}{p} \qquad\qquad (3\text{-}11)$$

where B_g = gas formation volume factor, ft^3/scf
   T = temperature, °R
   p = cell pressure, psia
   T_{sc} = standard temperature, °R
   p_{sc} = standard pressure, psia

Moses (1986) pointed out that reporting the experimental data in relation to the residual oil volume at 60°F (as shown graphically in Figures 3-5 and 3-6) gives the relative oil volume B_{od} and that the differential gas-oil ratio R_{sc} curves the appearance of the oil formation volume factor B_o and the solution gas solubility R_s curves, leading to their misuse in reservoir calculations.

It should be pointed out that the differential liberation test represents the behavior of the oil in the reservoir as the pressure declines. We must find a way of bringing this oil to the surface through separators and into the stock tank. This process is a flash or separator process.

SEPARATOR TESTS

Separator tests are conducted to determine the changes in the volumetric behavior of the reservoir fluid as the fluid passes through the separator (or separators) and then into the stock tank. The resulting volumetric behavior is influenced to a large extent by the operating conditions, i.e., pressures and temperatures, of the surface separation facilities. The primary objective of conducting separator tests, therefore, is to provide the essential laboratory information necessary for determining the optimum surface separation conditions, which in turn will maximize the stock-tank oil production. In addition, the results of the test, when appropriately combined with the differential liberation test data, provide a means of obtaining the PVT parameters (B_o, R_s, and B_t) required for petroleum engineering calculations. These separator tests are performed only on the original oil at the bubble point.

Relative Oil Volume
(at 247 °F)

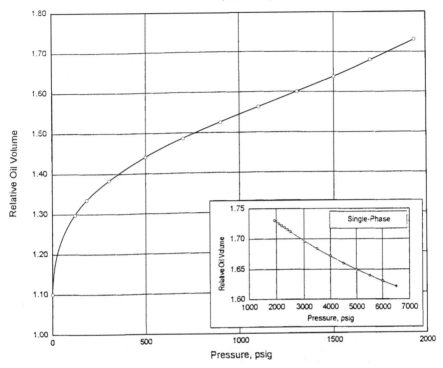

Pressure, psig

Relative Oil Volume Expression:		LEGEND	
$y = a + b\,(Xi)^{\wedge}i + c\,(Xi)^{\wedge}j + d\,(Xi)^{\wedge}k$			
where:			
a= 1.09883e+ 00	i= 1.075	○ Laboratory Data	
b= -1.08945e- 04	j= 0.449	------ Confidence Limits	
c= 2.52865e- 02	k= 2.000	——— Analytical Expression	
d= 6.59813e- 08			
Note: XI (incremental 'X') = pressure, psig		Saturation Pressure: 1936 psig	
Confidence level:	99 %	**Differential Vaporization**	
Confidence interval:	+/- 0.00028	Figure B-1	
'r squared':	.999997		

Figure 3-5. Relative volume versus pressure.

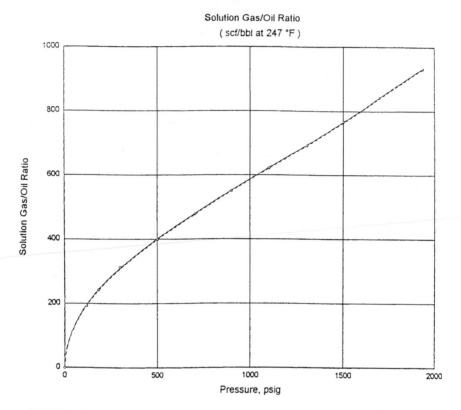

Solution Gas/Oil Ratio
(scf/bbl at 247 °F)

Solution Gas/Oil Ratio Expression:		LEGEND
$y = a + b\,(Xi)^{\wedge}i + c\,(Xi)^{\wedge}j + d\,(Xi)^{\wedge}k$		
where:		○ Laboratory Data
a= -2.13685e- 01 i= 0.515		- - - - - Confidence Limits
b= 1.69108e+ 01 j= 1.482		—— Analytical Expression
c= -5.05326e- 03 k= 1.906		
d= 2.58392e- 04		Saturation Pressure: 1936 psig
Note: Xi (incremental 'X') = pressure, psig		
Confidence level: 99 %		**Differential Vaporization**
Confidence interval: +/- 1.47 scf/bbl		
'r squared': .999966		**Figure B-2**

Figure 3-6. Solution gas-oil ratio versus pressure.

The test involves placing a hydrocarbon sample at its saturation pressure and reservoir temperature in a PVT cell. The volume of the sample is measured as V_{sat}. The hydrocarbon sample is then displaced and flashed through a laboratory multistage separator system—commonly one to three stages. The pressure and temperature of these stages are set to represent the desired or actual surface separation facilities. The gas liberated from each stage is removed and its specific gravity and volume at standard conditions are measured. The volume of the remaining oil in the last stage (representing the stock-tank condition) is measured and recorded as $(V_o)_{st}$. These experimental, measured data can then be used to determine the oil formation volume factor and gas solubility at the bubble-point pressure as follows:

$$B_{ofb} = \frac{V_{sat}}{(V_o)_{st}} \tag{3-12}$$

$$R_{sfb} = \frac{(V_g)_{sc}}{(V_o)_{st}} \tag{3-13}$$

where B_{ofb} = bubble-point oil formation volume factor, as measured by flash liberation, bbl of the bubble-point oil/STB

R_{sfb} = bubble-point solution gas-oil ratio as measured by flash liberation, scf/STB

$(V_g)_{sc}$ = total volume of gas removed from separators, scf

The above laboratory procedure is repeated at a series of different separator pressures and at a fixed temperature. It is usually recommended that four of these tests be used to determine the optimum separator pressure, which is usually considered the separator pressure that results in minimum oil formation volume factor. At the same pressure, the stock-tank oil gravity will be a maximum and the total evolved gas, i.e., the separator gas and the stock-tank gas will be at a minimum.

A typical example of a set of separator tests for a two-stage separation system, as reported by Moses (1986), is shown in Table 3-5. By examining the laboratory results reported in Table 3-5, it should be noted that

Table 3-5
Separator Tests
(Permission to publish by the Society of Petroleum Engineers of AIME. Copyright SPE-AIME.)

Separator Pressure (psig)	Temperature (°F)	GOR, R_{stb}*	Stock-Tank Oil Gravity (°API at 60°F)	FVF, B_{otb}**
50	75	737		
to 0	75	_41_	40.5	1.481
		778		
100	75	676		
to 0	75	_92_	40.7	1.474
		768		
200	75	602		
to 0	75	178	40.4	1.483
		780		
300	75	549		
to 0	75	246	40.1	1.495
		795		

*GOR in cubic feet of gas at 14.65 psia and 60°F per barrel of stock-tank oil at 60°F.
**FVF is barrels of saturated oil at 2.620 psig and 220°F per barrel of stock-tank oil at 60°F.

the optimum separator pressure is 100 psia, considered to be the separator pressure that results in the minimum oil formation volume factor. It is important to notice that the oil formation volume factor varies from 1.474 bbl/STB to 1.495 bbl/STB while the gas solubility ranges from 768 scf/STB to 795 scf/STB.

Table 3-5 indicates that the values of the crude oil PVT data are dependent on the method of surface separation. Table 3-6 presents the results of performing a separator test on the Big Butte crude oil. The differential liberation data, as expressed in Table 3-4, show that the solution gas-oil ratio at the bubble point is 933 scf/STB as compared with the measured value of 646 scf/STB from the separator test. This significant difference is attributed to the fact that the processes of obtaining residual oil and stock-tank oil from bubble-point oil are different.

The differential liberation is considered as a multiple series of flashes at the elevated reservoir temperatures. The separator test is generally a one- or two-stage flash at low pressure and low temperature. The quantity of gas released will be different and the quantity of final liquid will be different. Again, it should be pointed out that oil formation volume fac-

Table 3-6
Separator Tests Data

Separator Flash Analysis

Flash Conditions psig °F	Gas/Oil Ratio (scf/bbl) (A)	Gas/Oil Ratio (scf/STbbl) (B)	Stock Tank Oil Gravity at 60°F (°API)	Formation Volume Factor Bofb (C)	Separator Volume Factor (D)	Specific Gravity of Flashed Gas (Air = 1.000)	Oil Phase Density
1936 247							0.6484
28 130	593	632			1.066	1.132*	0.7823
0 80	13	13	38.8	1.527	1.010	**	0.8220
		Rsfb = 646					

Collected and analyzed in the laboratory by gas chromatography.
**Insufficient quantity for measurement.*
(A) Cubic feet of gas at 14.73 psia and 60°F per barrel of oil at indicated pressure and temperature.
(B) Cubic feet of gas at 14.73 psia and 60°F per barrel of stock-tank oil at 60°F.
(C) Barrels of saturated oil at 1936 psig and 247°F per barrel of stock-tank oil at 60°F.
(D) Barrels of oil at indicated pressure and temperature per barrel of stock-tank oil at 60°F.

tor, as expressed by Equation 3-12, is defined as "the volume of oil at reservoir pressure and temperature divided by the resulting stock-tank oil volume after it passes through the surface separators."

Adjustment of Differential Liberation Data to Separator Conditions

To perform material balance calculations, the oil formation volume factor B_o and gas solubility R_s as a function of the reservoir pressure must be available. The ideal method of obtaining these data is to place a large crude oil sample in a PVT cell at its bubble-point pressure and reservoir temperature. At some pressure a few hundred psi below the bubble-point pressure, a small portion of the oil is removed and flashed at temperatures and pressures equal to those in the surface separators and stock tank. The liberated gas volume and stock-tank oil volume are measured to obtain B_o and R_s. This process is repeated at several progressively lower reservoir pressures until complete curves of B_o and R_s versus pressure have been obtained. This procedure is occasionally conducted in the laboratory. This experimental methodology was originally proposed by Dodson (1953) and is called the Dodson Method.

Amyx et al. (1960) and Dake (1978) proposed a procedure for constructing the oil formation volume factor and gas solubility curves by using the differential liberation data (as shown in Table 3-4) in conjunction with the experimental separator flash data (as shown in Table 3-6) for a given set of separator conditions. The method is summarized in the following steps:

Step 1. Calculate the differential shrinkage factors at various pressures by dividing each relative oil volume factors B_{od} by the relative oil volume factor at the bubble-point B_{odb}, or:

$$S_{od} = \frac{B_{od}}{B_{odb}} \tag{3-14}$$

where B_{od} = differential relative oil volume factor at pressure p, bbl/STB

B_{odb} = differential relative oil volume factor at the bubble-point pressure p_b, psia, bbl/STB

S_{od} = differential oil shrinkage factor, bbl/bbl of bubble-point oil

The differential oil shrinkage factor has a value of one at the bubble-point and a value less than one at subsequent pressures below p_b.

Step 2. Adjust the relative volume data by multiplying the separator (flash) formation volume factor at the bubble-point B_{ofb} (as defined by Equation 3-12) by the differential oil shrinkage factor S_{od} (as defined by Equation 3-14) at various reservoir pressures. Mathematically, this relationship is expressed as follows:

$$B_o = B_{ofb}\, S_{od} \tag{3-15}$$

where B_o = oil formation volume factor, bbl/STB

B_{ofb} = bubble-point oil formation volume factor, bbl of the bubble-point oil/STB (as obtained from the separator test)

S_{od} = differential oil shrinkage factor, bbl/bbl of bubble-point oil

Step 3. Calculate the oil formation volume factor at pressures above the bubble-point pressure by multiplying the relative oil volume data V_{rel}, as generated from the constant-composition expansion test, by B_{ofb}, or:

$$B_o = (V_{rel})\,(B_{ofb}) \tag{3-16}$$

where B_o = oil formation volume factor above the bubble-point
        pressure, bbl/STB
V_{rel} = relative oil volume, bbl/bbl

Step 4. Adjust the differential gas solubility data R_{sd} to give the required
gas solubility factor R_s

$$R_s = R_{sfb} - (R_{sdb} - R_{sd}) \frac{B_{ofb}}{B_{odb}} \qquad (3-17)$$

where R_s = gas solubility, scf/STB
      R_{sfb} = bubble-point solution gas-oil ratio from the separator
             test, scf/STB
      R_{sdb} = solution gas-oil at the bubble-point pressure as
             measured by the differential liberation test, scf/STB
      R_{sd} = solution gas-oil ratio at various pressure levels as
             measured by the differential liberation test, scf/STB

These adjustments will typically produce lower formation volume
factors and gas solubilities than the differential liberation data.

Step 5. Obtain the two-phase (total) formation volume factor B_t by multiplying values of the relative oil volume V_{rel} below the bubble-
point pressure by B_{ofb}, or:

$$B_t = (B_{ofb}) (V_{rel}) \qquad (3-18)$$

where B = two-phase formation volume factor, bbl/STB
      V_{rel} = relative oil volume below the p_b, bbl/bbl

Similar values for B_t can be obtained from the differential liberation test by multiplying the relative total volume B_{td} (see Table 3-4,
Column C) by B_{ofb}, or

$$B_t = (B_{td}) (B_{ofb})/B_{odb} \qquad (3-19)$$

It should be pointed out that Equations 3-16 and 3-17 usually produce values less than one for B_o and negative values for R_s at low
pressures. The calculated curves of B_o and R_s versus pressures
must be manually drawn to $B_o = 1.0$ and $Rs = 0$ at atmospheric
pressure.

Example 3-5

The constant-composition expansion test, differential liberation test, and separator test for the Big Butte crude oil system are given in Tables 3-2, 3-4, and 3-6, respectively. Calculate:

• Oil formation volume factor at 4000 and 1100 psi
• Gas solubility at 1100 psi
• The two-phase formation volume factor at 1300 psi

Solution

Step 1. Determine B_{odb}, R_{sdb}, B_{ofb}, and R_{sfb} from Tables 3-4 and 3-6

$$B_{odb} = 1.730 \text{ bbl/STB} \qquad R_{sdb} = 933 \text{ scf/STB}$$

$$B_{ofb} = 1.527 \text{ bbl/STB} \qquad R_{sfb} = 646 \text{ scf/STB}$$

Step 2. Calculate B_o at 4000 by applying Equation 3-16

$$B_o = (0.9657)\,(1.57) = 1.4746 \text{ bbl/STB}$$

Step 3. Calculate B_o at 1100 psi by applying Equations 3-14 and 3-15.

$$S_{od} = \frac{1.563}{1.730} = 0.9035$$

$$B_o = (0.9035)\,(1.527) = 1.379 \text{ bbl/STB}$$

Step 4. Calculate the gas solubility at 1100 psi by using Equation 3-17.

$$R_s = 646 - (933 - 622)\left(\frac{1.527}{1.730}\right) = 371 \text{ scf/STB}$$

Step 5. From the pressure-volume relations (i.e., constant-composition data) of Table 3-2 the relative volume at 1300 PSI in 1.2579 bbl/bbl. Using Equation 3-18, calculate B_t to give:

$$B_t = (1.527)\,(1.2579) = 1.921 \text{ bbl/STB}$$

Applying Equation 3-19 gives:

$$B_t = (2.171)\,(1.527)/1.73 = 1.916 \text{ bbl/STB}$$

Table 3-7 presents a complete documentation of the adjusted differential vaporization data for the Big Butte crude oil system. Figures 3-7 and 3-8 compare graphically the adjusted values of R_s

Table 3-7
Adjusted Differential Liberation Data

Differential Vaporization
Adjusted to Separator Conditions*

Pressure psig	Solution Gas/Oil Ratio R_s (A)	Formation Volume Factor B_o (B)	Gas Formation Volume Factor (C)	Oil Density gm/cc	Oil/Gas Viscosity Ratio
6500	646	1.431		0.6919	
6000	646	1.439		0.6882	
5500	646	1.447		0.6843	
5000	646	1.456		0.6803	
4500	646	1.465		0.6760	
4000	646	1.475		0.6714	
3500	646	1.486		0.6665	
3000	646	1.497		0.6613	
2500	646	1.510		0.6556	
2400	646	1.513		0.6544	
2300	646	1.516		0.6531	
2200	646	1.519		0.6519	
2100	646	1.522		0.6506	
2000	646	1.525		0.6493	
b≫1936	646	1.527		0.6484	
1700	564	1.482	0.01009	0.6577	19.0
1500	498	1.446	0.01149	0.6650	21.3
1300	434	1.412	0.01334	0.6720	23.8
1100	371	1.379	0.01591	0.6790	26.6
900	309	1.346	0.01965	0.6863	29.8
700	244	1.311	0.02559	0.6944	33.7
500	175	1.271	0.03626	0.7039	38.6
300	95	1.220	0.06075	0.7161	46.0
185	36	1.178	0.09727	0.7256	52.8
120		1.146	0.14562	0.7328	58.4
0				0.7745	

*Separator Conditions

Fist Stage	28 psig at 130°F
Stock Tank	0 psig at 80°F

(A) Cubic feet of gas at 14.73 psia and 60°F per barrel of stock-tank oil at 60°F.
(B) Barrel of oil at indicated pressure and temperature per barrel of stock-tank oil at 60°F.
(C) Cubic feet of gas at indicated pressure and temperature per cubic feet at 14.73 psia and 60°F.

and B_o with those of the unadjusted PVT data. It should be noted that no adjustments are needed for the gas formation volume factor, oil density, or viscosity data.

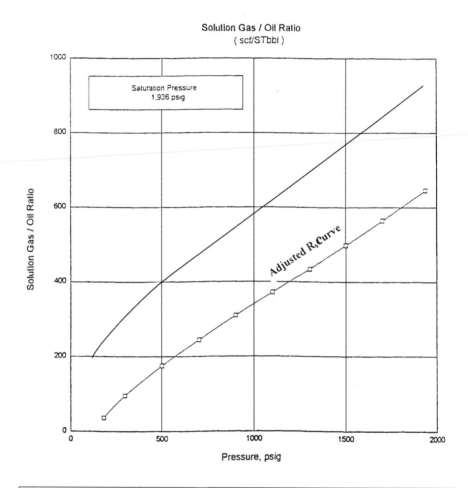

Solution Gas / Oil Ratio
(scf/STbbl)

Saturation Pressure
1,936 psig

Adjusted R_s Curve

Solution Gas / Oil Ratio

Pressure, psig

LEGEND	DV Adjusted to Separator
—— Differential Vaporization ⊡ 28 psig at 130 °F	Figure D-1

Figure 3-7. Adjusted gas solubility versus pressure.

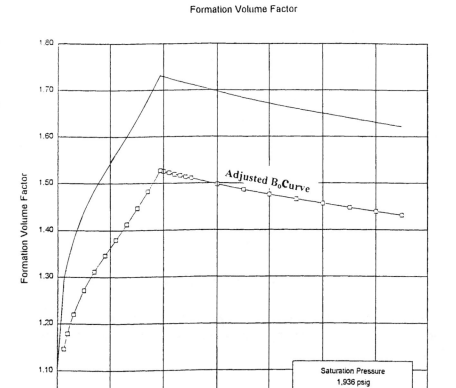

Formation Volume Factor

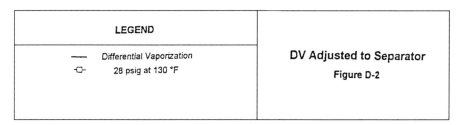

Figure 3-8. Adjusted oil formation volume factor versus pressure.

EXTRAPOLATION OF RESERVOIR FLUID DATA

In partially depleted reservoirs or in fields that originally existed at the bubble-point pressure, it is difficult to obtain a fluid sample, which usually represents the original oil in the reservoir at the time of discovery. Also, in collecting fluid samples from oil wells, the possibility exists of obtaining samples with a saturation pressure that might be lower than or higher than the actual saturation pressure of the reservoir. In these cases, it is necessary to correct or adjust the laboratory PVT measured data to reflect the actual saturation pressure. The proposed correction procedure for adjusting the following laboratory test data is described in the subsequent sections:

- Constant-composition expansion (CCE) test
- Differential expansion (DE) test
- Oil viscosity test
- Separator tests

Correcting Constant-Composition Expansion Data

The correction procedure, summarized in the following steps, is based on calculating the Y-function value for each point below the "old" saturation pressure.

Step 1. Calculate the Y-function, as expressed by Equation 3-3, for each point by using the old saturation pressure.

Step 2. Plot the values of the Y-function versus pressure on a Cartesian scale and draw the best straight line. Points in the neighborhood of the saturation pressure may be erratic and need not be used.

Step 3. Calculate the coefficients a and b of the straight-line equation, i.e.:

$$Y = a + bp$$

Step 4. Recalculate the relative volume V_{rel} values by applying Equation 3-5 and using the "new" saturation pressure, or:

$$V_{rel} = 1 + \frac{p_{sat}^{new} - p}{p(a + bp)} \tag{3-20}$$

To determine points above the "new" saturation pressure, apply the following steps:

Step 1. Plot the "old" relative volume values above the "old" saturation pressure versus pressure on a regular scale and draw the best straight line through these points.

Step 2. Calculate the slope of the Line S. It should be noted that the slope is negative, i.e., $S < 0$.

Step 3. Draw a straight line that passes through the point ($V_{rel} = 1$, p_{sat}^{new}) and parallel to the line of Step 1.

Step 4. Relative volume data above the new saturation pressure are read from the straight line or determined from the following expression at any pressure p:

$$V_{rel} = 1 - S\ (p_{sat}^{new} - p) \tag{6-21}$$

where S = slope of the line
p = pressure

Example 3-6

The pressure-volume relations of the Big Butte crude oil system is given in Table 3-2. The test indicates that the oil has a bubble-point pressure of 1930 psig at 247°F. The Y-function for the oil system is expressed by the following linear equation:

Y = 1.0981 + 0.000591p

Above the bubble-point pressure, the relative volume data versus pressure exhibit a straight-line relationship with a slope of −0.0000138.

The surface production data of the field suggest that the actual bubble-point pressure is approximately 2500 psig. Reconstruct the pressure-volume data using the new reported saturation pressure.

Solution

Using Equations 3-30 and 3-31, gives:

Pressure psig	Old V_{rel}	New V_{rel}	Comments
6500	0.9371	0.9448	Equation 3-21
6000	0.9422	0.9517	
5000	0.9532	0.9655	
4000	0.9657	0.9793	
3000	0.9805	0.9931	
$p_b^{new} = 2500$	0.9890	1.0000	
2000	0.9987	1.1096	Equation 3-20
$p_b^{old} = 1936$	1.0000	1.1299	
1911	1.0058	1.1384	
1808	1.0324	1.1767	
1600	1.1018	1.1018	
600	2.4960	2.4960	
437	3.4404	3.4404	

Correcting Differential Liberation Data

Relative oil volume B_{od} versus pressure:

The laboratory measured B_{od} data must be corrected to account for the new bubble-point pressure p_b^{new}. The proposed procedure is summarized in the following steps:

Step 1. Plot the B_{od} data versus gauge pressure on a regular scale.

Step 2. Draw the best straight line through the *middle pressure range* of 30%–90% p_b.

Step 3. Extend the straight line to the new bubble-point pressure, as shown schematically in Figure 3-9.

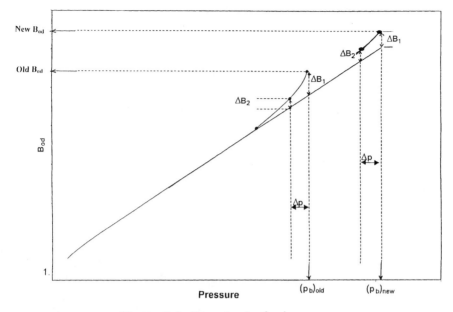

Figure 3-9. Correcting B_{od} for the new p_b.

Step 4. Transfer any curvature at the end of the original curve, i.e., ΔB_{ol} at p_b^{old}, to the new bubble-point pressure by placing ΔB_{ol} above or below the straight line at p_b^{new}.

Step 5. Select any differential pressure Δp below the p_b^{old} and transfer the corresponding curvature to the pressure ($p_b^{new} - \Delta p$).

Step 6. Repeat the above process and draw a curve that connects the generated B_{od} points with original curve at the point of intersection with the straight line. Below this point no change is needed.

Solution gas-oil ratio:

The correction procedure for the isolation gas-oil ratio R_{sd} data is identical to that of the relative oil volume data.

Correcting Oil Viscosity Data

The oil viscosity data can be extrapolated to a new higher bubble-point pressure by applying the following steps:

Step 1. Defining the *fluidity* as the reciprocal of the oil viscosity, i.e., $1/\mu_o$, calculate the fluidity for each point *below* the original saturation pressure.

Step 2. Plot fluidity versus pressure on a Cartesian scale (see Figure 3-10).

Step 3. Draw the best straight line through the points and extend it to the new saturation pressure p_b^{old}.

Step 4. New oil viscosity values above p_b^{old} are read from the straight line.

To obtain the oil viscosity for pressures above the new bubble-point pressure p_b^{new}, follow these steps:

Step 1. Plot the viscosity values for all points *above* the old saturation pressure on a Cartesian coordinate as shown schematically in Figure 3-11, and draw the best straight line through them, as Line A.

Step 2. Through the point on the extended viscosity curve at p_b^{new}, draw a straight line (Line B) parallel to A.

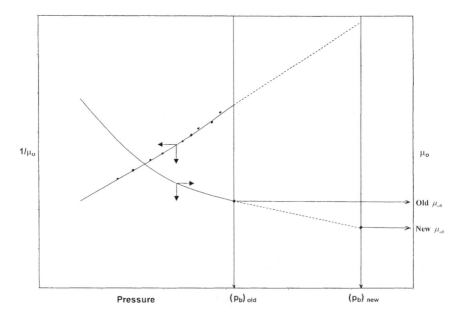

Figure 3-10. Extrapolating μ_o to new p_b.

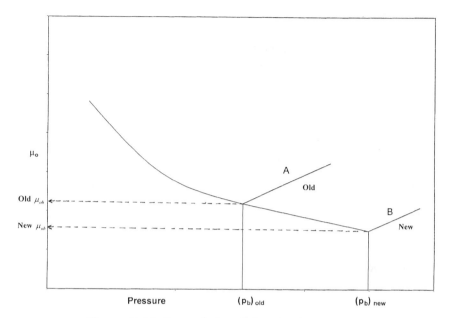

Figure 3-11. Extrapolating oil viscosity above new p_b.

Step 3. Viscosities above the new saturation pressure are then read from Line A.

Correcting the Separator Tests Data

Stock-tank gas-oil ratio and gravity:

No corrections are needed for the stock-tank gas-oil ratio and the stock-tank API gravity.

Separator gas-oil ratio:

The *total* gas-oil ratio R_{sfb} is changed in the same proportion as the differential ratio was changed, or

$$R_{sfb}^{new} = R_{sfb}^{old} \left(R_{sdb}^{new} / R_{sdb}^{old} \right) \tag{3-22}$$

The *separator* gas-oil ratio is then the difference between the new (corrected) gas solubility R_{sfb}^{new} and the unchanged stock-tank gas-oil ratio.

Formation volume factor:

The separator oil formation volume factor B_{ofb} is adjusted in the same proportion as the differential liberation values:

$$B_{ofb}^{new} = B_{ofb}^{old} \left(B_{odb}^{new} / B_{odb}^{old} \right) \tag{3-23}$$

Example 3-7

Results of the differential liberation and the separator tests on the Big Butte crude oil system are given in Tables 3-4 and 3-6, respectively. New field and production data indicate that the bubble-point pressure is better described by a value of 2500 psi as compared with the laboratory reported value of 1936 psi. The correction procedure for B_{od} and R_{sd} as described previously was applied, to give the following values at the new bubble point:

$$B_{odb}^{new} = 2.013 \, bbl/STB \qquad R_{sbd}^{new} = 1134 \, scf/STB$$

Using the separator test data as given in Table 3-6, calculate the gas solubility and the oil formation volume factor at the new bubble-point pressure.

Solution

• Gas solubility: from Equation 3-22

$$R_{sb} = 646 \left(\frac{1134}{933} \right) = 785 \, scf/STB$$

Separator GOR = 785 − 13 = 772 scf/STB

• Oil formation volume factor

Applying Equation 3-23, gives

$$B_{ob} = 1.527 \left(\frac{2.013}{1.730} \right) = 1.777 \, bbl/STB$$

LABORATORY ANALYSIS OF GAS CONDENSATE SYSTEMS

In the laboratory, a standard analysis of a gas-condensate sample consists of:

• Recombination and analysis of separator samples
• Measuring the pressure-volume relationship, i.e., constant-composition expansion test
• Constant-volume depletion test (CVD)

Recombination of Separator Samples

Obtaining a representative sample of the reservoir fluid is considerably more difficult for a gas-condensate fluid than for a conventional black-oil reservoir. The principal reason for this difficulty is that liquid may condense from the reservoir fluid during the sampling process, and if representative proportions of both liquid and gas are not recovered then an erroneous composition will be calculated.

Because of the possibility of erroneous compositions and also because of the limited volumes obtainable, subsurface sampling is seldom used in gas-condensate reservoirs. Instead, surface sampling techniques are used, and samples are obtained only after long stabilized flow periods. During this stabilized flow period, volumes of liquid and gas produced in the surface separation facilities are accurately measured, and the fluid samples are then recombined in these proportions.

The hydrocarbon composition of separator samples is also determined by chromatography or low-temperature fractional distillation or a combination of both. Table 3-7 shows the hydrocarbon analyses of the separator liquid and gas samples taken from the Nameless Field. The gas and liquid samples are recombined in the proper ratio to obtain the well stream composition as given in Table 3-8. The laboratory data indicates that the overall well-stream system contains 63.71 mol% Methane and 10.75 mol% Heptanes-plus.

Frequently, the surface gas is processed to remove and liquefy all hydrocarbon components that are heavier than methane, i.e., ethane, propanes, etc. These liquids are called *plant products*. These quantities of

Table 3-8
Hydrocarbon Analyses of Separator Products
and Calculated Wellstream

Component	Separator mol %	Separator Gas mol %	GPM	Well Stream mol %	GPM
Hydrogen Sulfide	Nil	Nil		Nil	
Carbon Dioxide	0.29	1.17		0.92	
Nitrogen	0.13	0.38		0.31	
Methane	18.02	81.46		63.71	
Ethane	12.08	11.46		11.63	
Propane	11.40	3.86	1.083	5.97	1.675
iso-Butane	3.05	0.49	0.163	1.21	0.404
n-Butane	5.83	0.71	0.228	2.14	0.688
iso-Pentane	3.07	0.18	0.067	0.99	0.369
Pentane	2.44	0.12	0.044	0.77	0.284
Hexanes	5.50	0.09	0.037	1.60	0.666
Heptanes-plus	38.19	0.08	0.037	10.75	7.944
	100.00	100.00	1.659	100.00	12.030

Properties of Heptanes-plus
   API gravity @ 60°F   43.4
   Specific gravity @
     60/60°F   0.8091   0.809
     Molecular weight   185   103   185

Calculated separator gas gravity (air = 1.000) = 0.687
Calculated gross heating value for separator gas = 1209 BTU
  per cubic foot of dry gas @ 15.025 psia and 60°F.

Primary separator gas collected @ 745 psig and 74°F.
Primary separator liquid collected @ 745 psig and 74°F.

Primary separator gas/separator liquid ratio	2413 scf/bbl @ 60°F
Primary separator liquid/stock-tank liquid ratio	1.360 bbl @ 60°F
Primary separator gas/wellstream ratio	720.13 Mscf/MMscf
Stock-tank liquid/wellstream ratio	219.4 bbl/MMscf

liquid products are expressed in gallons of liquid per thousand standard cubic feet of gas processed, i.e., gal/Mscf, or GPM. McCain (1990) derived the following expression for calculating the anticipated GPM for each component in the gas phase:

$$GPM_i = 11.173 \left(\frac{p_{sc}}{T_{sc}} \right) \left(\frac{y_i\, M_i}{\gamma_{oi}} \right) \qquad (3\text{-}24)$$

where p_{sc} = standard pressure, psia

T_{sc} = standard temperature, °R

y_i = mole fraction of component i in the gas phase

M_i = molecular weight of component i

γ_{oi} = specific gravity of component i as a liquid at standard conditions (Chapter 1, Table 1-1, Column E)

McCain pointed out that the complete recovery of these products is not feasible. He proposed that, as a rule of thumb, 5 to 25% of ethane, 80 to 90% of the propane, 95% or more of the butanes, and 100% of the heavier components can be recovered from a simple surface facility.

Example 3-8

Table 3-8 shows the wellstream compositional analysis of the Nameless Field. Using Equation 3-24, calculate the maximum available liquid products assuming 100% plant efficiency.

Solution

• Using the standard conditions as given in Table 3-8, gives:

$$GPM = 11.173 \left(\frac{15.025}{520} \right) \left(\frac{y_i \, M_i}{\gamma_{oi}} \right) = 0.3228 \left(\frac{y_i \, M_i}{\gamma_{oi}} \right)$$

• Construct the following working table:

Component	y_i	M_i	γ_{oi}	GPM_i
CO_2	0.0092			
N_2	0.0031			
C_1	0.6371			
C_2	0.1163	30.070	0.35619	1.069
C_3	0.0597	44.097	0.50699	1.676
i-C_4	0.0121	58.123	0.56287	0.403
n-C_4	0.0214	58.123	0.58401	0.688
i-C_5	0.0099	72.150	0.63112	0.284
n-C5	0.0077	72.150	0.63112	0.284
C_6	0.0160	86.177	0.66383	0.670
C_7^+	0.1075	185.00	0.809	7.936

15.20 GPM

Constant-Composition Test

This test involves measuring the pressure-volume relations of the reservoir fluid at reservoir temperature with a visual cell. This usual PVT cell allows the visual observation of the condensation process that results from changing the pressures. The experimental test procedure is similar to that conducted on crude oil systems. The CCE test is designed to provide the dew-point pressure p_d at reservoir temperature and the total relative volume V_{rel} of the reservoir fluid (relative to the dew-point volume) as a function of pressure. The relative volume is equal to one at p_d. The gas compressibility factor at pressures greater than or equal to the saturation pressure is also reported. It is only necessary to experimentally measure the z-factor at one pressure p_1 and determine the gas deviation factor at the other pressure p from:

$$z = z_1 \left(\frac{p}{p_1} \right) \frac{V_{rel}}{(V_{rel})_1} \tag{3-25}$$

where   z = gas deviation factor at p
V_{rel} = relative volume at pressure p
$(V_{rel})_1$ = relative volume at pressure p_1

If the gas compressibility factor is measured at the dew-point pressure, then:

$$z = z_d \left(\frac{p}{p_d} \right) (V_{rel}) \tag{3-26}$$

where z_d = gas compressibility factor at the dew-point pressure p_d
p_d = dew-point pressure, psia
p = pressure, psia

Table 3-9 shows the dew-point determination and the pressure-volume relations of the Nameless Field. The dew-point pressure of the system is reported as 4968 psi at 262°F. The measured gas compressibility factor at the dew point is 1.043.

Example 3-9

Using Equation 3-26 and the data in Table 3-9, calculate the gas deviation factor at 6000 and 8100 psi.

Table 3-9
Pressure-Volume Relations of Reservoir Fluid at 262°F
(Constant-Composition Expansion)

Pressure psig	Relative Volume	Deviation Factor Z
8100	0.8733	1.484
7800	0.8806	1.441
7500	0.8880	1.397
7000	0.9036	1.327
6500	0.9195	1.254
6000	0.9397	1.184
5511	0.9641	1.116
5309	0.9764	1.089
5100	0.9909	1.061
5000	0.9979	1.048
4968 Dew-point Pressure	1.0000	1.048
4905	1.0057	
4800	1.0155	
4600	1.0369	
4309	1.0725	
4000	1.1177	
3600	1.1938	
3200	1.2970	
2830	1.4268	
2400	1.6423	
2010	1.9312	
1600	2.4041	
1230	3.1377	
1000	3.8780	
861	4.5249	
770	5.0719	

Gas Expansion Factor = 1.2854 Mscf/bbl.

Solution

• At 6000 psi

$$z = 1.043 \left(\frac{8100+15.025}{4968+15.025} \right) (0.9397) = 1.183$$

• At 8100 psi

$$z = 1.043 \left(\frac{8100 + 15.025}{4968 + 15.025} \right) (0.8733) = 1.483$$

Constant-Volume Depletion (CVD) Test

Constant-volume depletion (CVD) experiments are performed on gas condensates and volatile oils to simulate reservoir depletion performance and compositional variation. The test provides a variety of useful and important information that is used in reservoir engineering calculations.

The laboratory procedure of the test is shown schematically in Figure 3-12 and is summarized in the following steps:

Step 1. A measured amount of a representative sample of the original reservoir fluid with a known overall composition of z_i is charged to a visual PVT cell at the dew-point pressure p_d ("a" in Figure 3-12). The temperature of the PVT cell is maintained at the reservoir temperature T throughout the experiment. The initial volume V_i of the saturated fluid is used as a reference volume.

Step 2. The initial gas compressibility factor is calculated from the real gas equation

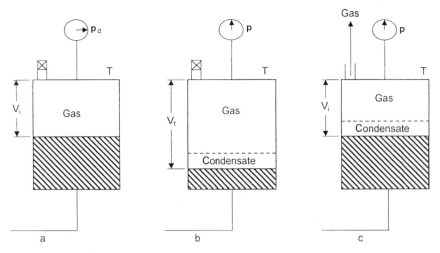

Figure 3-12. A schematic illustration of the constant-volume depletion test.

$$z_d = \frac{p_d V_i}{n_i RT} \tag{3-27}$$

where p_d = dew-point pressure, psia
V_i = initial gas volume, ft^3
n_i = initial number of moles of the gas = m/M_a
R = gas constant, 10.73
T = temperature, °R
z_d = compressibility factor at dew-point pressure

Step 3. The cell pressure is reduced from the saturation pressure to a pre-determined level P. This can be achieved by withdrawing mercury from the cell, as illustrated in column b of Figure 3-12. During the process, a second phase (retrograde liquid) is formed. The fluid in the cell is brought to equilibrium and the gas volume V_g and volume of the retrograde liquid V_L are visually measured. This retrograde volume is reported as a percent of the initial volume V_i which basically represents the retrograde liquid saturation S_L:

$$S_L = \left(\frac{V_L}{V_i}\right) 100$$

Step 4. Mercury is reinjected into the PVT cell at constant pressure P while an equivalent volume of gas is simultaneously removed. When the initial volume V_i is reached, mercury injection is ceased, as illustrated in column c of Figure 3-12. This step simulates a reservoir producing only gas, with retrograde liquid remaining immobile in the reservoir.

Step 5. The removed gas is charged to analytical equipment where its composition y_i is determined, and its volume is measured at standard conditions and recorded as $(V_{gp})_{sc}$. The corresponding moles of gas produced can be calculated from the expression

$$n_p = \frac{p_{sc}\left(V_{gp}\right)_{sc}}{R\,T_{sc}} \tag{3-28}$$

where n_p = moles of gas produced
$(V_{gp})_{sc}$ = volume of gas produced measured at standard
conditions, scf
T_{sc} = standard temperature, °R
p_{sc} = standard pressure, psia
$R = 10.73$

Step 6. The gas compressibility factor at cell pressure and temperature is
calculated from the real gas equation-of-state as follows:

$$z = \frac{p\left(V_g\right)}{n_p\, R\, T} \tag{3-29}$$

Another property, the two-phase compressibility factor, is also calculated. The two-phase compressibility factor represents the total compressibility of all the remaining fluid (gas and retrograde liquid) in the
cell and is computed from the real gas law as

$$z_{\text{two-phase}} = \frac{p\, V_i}{(n_i - n_p)\, RT} \tag{3-30}$$

where $(n_i - n_p)$ = the remaining moles of fluid in the cell
n_i = initial moles in the cell
n_p = cumulative moles of gas removed

The two-phase z-factor is a significant property because it is used
when the p/z versus cumulative-gas produced plot is constructed for
evaluating gas-condensate production.

Equation 3-30 can be expressed in a more convenient form by replacing moles of gas, i.e., n_i and n_p, with their equivalent gas volumes, or:

$$z_{\text{two-phase}} = \left(\frac{z_d}{P_d}\right)\left[\frac{p}{1-(G_p/\text{GIIP})}\right] \tag{3-31}$$

where z_d = gas deviation factor at the dew-point pressure
P_d = dew-point pressure, psia
P = reservoir pressure, psia
GIIP = initial gas in place, scf
G_p = cumulative gas produced at pressure p, scf

Step 7. The volume of gas produced as a percentage of gas initially in place is calculated by dividing the cumulative volume of the produced gas by the gas initially in place, both at standard conditions

$$\% G_p = \left[\frac{\sum (V_{gp})_{sc}}{GIIP} \right] 100 \tag{3-32}$$

or

$$\% G_p = \left[\frac{\sum n_p}{(n_i)_{original}} \right] 100$$

The above experimental procedure is repeated several times until a minimum test pressure is reached, after which the quantity and composition of the gas and retrograde liquid remaining in the cell are determined.

The test procedure can also be conducted on a volatile oil sample. In this case, the PVT cell initially contains liquid, instead of gas, at its bubble-point pressure.

The results of the pressure-depletion study for the Nameless Field are illustrated in Tables 3-10 and 3-11. Note that the composition listed in the 4968 psi pressure column in Table 3-10 is the composition of the reservoir fluid at the dew point and exists in the reservoir in the gaseous state. Table 3-10 and Figure 3-13 show the changing composition of the wellstream during depletion. Notice the progressive reduction of C_{7+} below the dew point and increase in the Methane fraction, i.e., C_1.

The concentrations of *intermediates,* i.e., C_2–C_6, are also seen to decrease (they condense) as pressure drops down to about 2,000 psi, then increase as they revaporize at the lower pressures. The final column shows the composition of the liquid remaining in the cell (or reservoir) at the abandonment pressure of 700 psi; the predominance of C_{7+} components in the liquid is apparent.

The z-factor of the equilibrium gas and the two-phase z are presented. (Note: if a (p/z) versus G_p analysis is to be done, the two-phase compressibility factors are the appropriate values to use.)

(*text continued on page 176*)

Table 3-10
Depletion Study at 262°F

Hydrocarbon Analyses of Produced Wellstream-Mol Percent

Component	Reservoir Pressure—psig							
	4968	4300	3500	2800	2000	1300	700	700*
Carbon dioxide	0.92	0.97	0.99	1.01	1.02	1.03	1.03	0.30
Nitrogen	0.31	0.34	0.37	0.39	0.39	0.37	0.31	0.02
Methane	63.71	69.14	71.96	73.24	73.44	72.48	69.74	12.09
Ethane	11.63	11.82	11.87	11.92	12.25	12.67	13.37	5.86
Propane	5.97	5.77	5.59	5.54	5.65	5.98	6.80	5.61
iso-Butane	1.21	1.14	1.07	1.04	1.04	1.13	1.32	1.61
n-Butane	2.14	1.99	1.86	1.79	1.76	1.88	2.24	3.34
iso-Pentane	0.99	0.88	0.79	0.73	0.72	0.77	0.92	2.17
n-Pentane	0.77	0.68	0.59	0.54	0.53	0.56	0.68	1.88
Hexanes	1.60	1.34	1.12	0.98	0.90	0.91	1.07	5.34
Heptanes plus	10.75	5.93	3.79	2.82	2.30	2.22	2.52	61.78
	100.00	100.00	100.00	100.00	100.00	100.00	100.00	100.00

Molecular weight of heptanes-plus	185	143	133	125	118	114	112	203
Specific gravity of heptanes-plus	0.809	0.777	0.768	0.760	0.753	0.749	0.747	0.819
Deviation Factor-Z								
Equilibrium gas	1.043	0.927	0.874	0.862	0.879	0.908	0.946	
Two-phase	1.043	0.972	0.897	0.845	0.788	0.720	0.603	
Wellstream produced—								
Cumulative percent of initial	0.000	7.021	17.957	30.268	46.422	61.745	75.172	
GPM from smooth compositions								
Propane-plus	12.030	7.303	5.623	4.855	4.502	4.624	5.329	
Butanes-plus	10.354	5.683	4.054	3.301	2.916	2.946	3.421	
Pentanes-plus	9.263	4.664	3.100	2.378	2.004	1.965	2.261	

*Equilibrium liquid phase, representing 13.323 percent of original well stream.

Table 3-11
Retrograde Condensation During Gas Depletion at 262°F

Pressure psig	Retrograde Liquid Volume Percent of Hydrocarbon Pore Space
4968 Dew-point Pressure	0.0
4905	19.3
4800	25.0
4600	29.9
4300 First Depletion Level	33.1
3500	34.4
2800	34.1
2000	32.5
1300	30.2
700	27.3
0	21.8

(text continued from page 173)

The row in the table, "Wellstream Produced, % of initial GPM from smooth compositions," gives the fraction of the total moles (of scf) in the cell (or reservoir) that has been produced. This is *total* recovery of wellstream and has not been separated here into surface gas and oil recoveries.

In addition to the composition of the produced wellstream at the final depletion pressure, the composition of the retrograde liquid is also measured. The composition of the liquid is reported in the last column of Table 3-10 at 700 psi. These data are included as a control composition in the event the study is used for compositional material-balance purposes.

The volume of the retrograde liquid, i.e., liquid dropout, measured during the course of the depletion study is shown in Table 3-11. The data are reshown as a percent of hydrocarbon pore space. The measurements indicate that the maximum liquid dropout of 34.4% occurs at 3500 psi. The liquid dropout can be expressed as a percent of the pore volume, i.e., saturation, by adjusting the reported values to account for the presence of the initial water saturation, or

$$(S_o = (LDO) (1 - S_{wi}) \tag{3-33}$$

where S_o = retrograde liquid (oil) saturation, %
   LDO = liquid dropout, %
   S_{wi} = initial water saturation, fraction

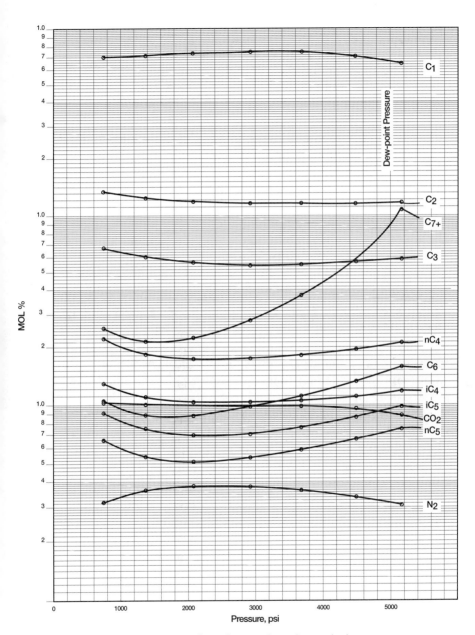

Figure 3-13. Hydrocarbon analysis during depletion.

Example 3-10

Using the experimental data of the Nameless gas-condensate field given in Table 3-10, calculate the two-phase compressibility factor at 2000 psi by applying Equation 3-31.

Solution

The laboratory report indicates that the base (standard) pressure is 15.025 psia. Applying Equation 3-31 gives:

$$z_{2-\text{phase}} = \left[\frac{1.043}{4968+15.025}\right]\left[\frac{2000+15.025}{1-0.46422}\right] = 0.787$$

PROBLEMS

Table 3-12 shows the experimental results performed on a crude oil sample taken from the Mtech field. The results include the CCE, DE, and separator tests.

• Select the optimum separator conditions and generate B_o, R_s, and B_t values for the crude oil system. Plot your results and compare with the unadjusted values.
• Assume that new field indicates that the bubble-point pressure is better described by a value of 2500 psi. Adjust the PVT to reflect for the new bubble-point pressure.

(text continued on page 182)

Table 3-12
Pressure-Volume Relations of Reservoir Fluid at 260°F
(Constant-Composition Expansion)

Pressure psig	Relative Volume
5000	0.9460
4500	0.9530
4000	0.9607
3500	0.9691
3000	0.9785
2500	0.9890
2300	0.9938
2200	0.9962
2100	0.9987
<u>2051</u>	1.0000
2047	1.0010
2041	1.0025
2024	1.0069
2002	1.0127
1933	1.0320
1843	1.0602
1742	1.0966
1612	1.1524
1467	1.2299
1297	1.3431
1102	1.5325
862	1.8992
653	2.4711
482	3.4050

Table 3-12 (Continued)
Differential Vaporization at 260°F

Pressure psig	Solution Gas/Oil Ratio (1)	Relative Oil Volume (2)	Relative Total Volume (3)	Oil Density gm/cc	Deviation Factor Z	Gas Formation Volume Factor (4)	Incremental Gas Gravity
2051	1004	1.808	1.808	0.5989			
1900	930	1.764	1.887	0.6063	0.880	0.00937	0.843
1700	838	1.708	2.017	0.6165	0.884	0.01052	0.840
1500	757	1.660	2.185	0.6253	0.887	0.01194	0.844
1300	678	1.612	2.413	0.6348	0.892	0.01384	0.857
1100	601	1.566	2.743	0.6440	0.899	0.01644	0.876
900	529	1.521	3.229	0.6536	0.906	0.02019	0.901
700	456	1.476	4.029	0.6635	0.917	0.02616	0.948
500	379	1.424	5.537	0.6755	0.933	0.03695	0.018
300	291	1.362	9.214	0.6896	0.955	0.06183	1.188
170	223	1.309	16.246	0.7020	0.974	0.10738	1.373
0	0	1.110		0.7298			2.230

at 60°F = 1.000

Gravity of Residual Oil = 43.1° API at 60°F

(1) Cubic feet of gas at 14.73 psia and 60°F per barrel of residual oil at 60°F.
(2) Barrels of oil at indicated pressure and temperature per barrel of residual oil at 60°F.
(3) Barrels of oil plus liberated gas at indicated pressure and temperature per barrel of residual oil at 60°F.
(4) Cubic feet of gas at indicated pressure and temperature per cubic foot at 14.73 psia and 60°F.

Table 3-12 (Continued)
Separator Tests of Reservoir Fluid Sample

Separator Pressure PSI Gauge	Separator Temperature °F	Gas/Oil Ratio (1)	Gas/Oil Ratio (2)	Stock Tank Gravity °API @ 60°F	Formation Volume Factor (3)	Separator Volume Factor (4)	Specific Gravity of Flashed Gas
200 to 0	71	431	490			1.138	0.739*
	71	222	223	48.2	1.549	1.006	1.367
100 to 0	72	522	566			1.083	0.801*
	72	126	127	48.6	1.529	1.006	1.402
50 to 0	71	607	632			1.041	0.869*
	71	54	54	48.6	1.532	1.006	1.398
25 to 0	70	669	682			1.020	0.923*
	70	25	25	48.4	1.558	1.006	1.340

*Collected and analyzed in the laboratory

(1) Gas/oil ratio in cubic feet of gas @ 60°F and 14.75 psi absolute per barrel of oil @ indicated pressure and temperature.

(2) Gas/oil ratio in cubic feet of gas @ 60°F and 14.73 psi absolute per barrel of stock-tank oil @ 60°F.

(3) Formation volume factor in barrels of saturated oil @ 2051 psi gauge and 260°F per barrel of stock-tank oil @ 60°F.

(4) Separator volume factor in barrels of oil @ indicated pressure and temperature per barrel of stock-tank oil @ 60°F.

(*text continued from page 178*)

REFERENCES

1. Amyx, J. M., Bass, D. M., and Whiting, R., *Petroleum Reservoir Engineering-Physical Properties.* New York: McGraw-Hill Book Company, 1960.

2. Dake, L. P., *Fundamentals of Reservoir Engineering.* Amsterdam: Elsevier Scientific Publishing Company, 1978.

3. Dodson, L. P., "Application of Laboratory PVT Data to Reservoir Engineering Problems," *JPT,* December 1953, pp. 287–298.

4. McCain, W., *The Properties of Petroleum Fluids.* Tulsa, OK: PennWell Publishing Company, 1990.

5. Moses, P., "Engineering Application of Phase Behavior of Crude Oil and Condensate Systems," *JPT,* July 1986, pp. 715–723.

FUNDAMENTALS OF ROCK PROPERTIES

The material of which a petroleum reservoir rock may be composed can range from very loose and unconsolidated sand to a very hard and dense sandstone, limestone, or dolomite. The grains may be bonded together with a number of materials, the most common of which are silica, calcite, or clay. Knowledge of the physical properties of the rock and the existing interaction between the hydrocarbon system and the formation is essential in understanding and evaluating the performance of a given reservoir.

Rock properties are determined by performing laboratory analyses on cores from the reservoir to be evaluated. The cores are removed from the reservoir environment, with subsequent changes in the core bulk volume, pore volume, reservoir fluid saturations, and, sometimes, formation wettability. The effect of these changes on rock properties may range from negligible to substantial, depending on characteristics of the formation and property of interest, and should be evaluated in the testing program.

There are basically two main categories of core analysis tests that are performed on core samples regarding physical properties of reservoir rocks. These are:

Routine core analysis tests

• Porosity
• Permeability
• Saturation

Special tests

- Overburden pressure
- Capillary pressure
- Relative permeability
- Wettability
- Surface and interfacial tension

The above rock property data are essential for reservoir engineering calculations as they directly affect both the quantity and the distribution of hydrocarbons and, when combined with fluid properties, control the flow of the existing phases (i.e., gas, oil, and water) within the reservoir.

POROSITY

The porosity of a rock is a measure of the storage capacity (pore volume) that is capable of holding fluids. Quantitatively, the porosity is the ratio of the pore volume to the total volume (bulk volume). This important rock property is determined mathematically by the following generalized relationship:

$$\phi = \frac{\text{pore volume}}{\text{bulk volume}}$$

where ϕ = porosity

As the sediments were deposited and the rocks were being formed during past geological times, some void spaces that developed became isolated from the other void spaces by excessive cementation. Thus, many of the void spaces are interconnected while some of the pore spaces are completely isolated. This leads to two distinct types of porosity, namely:

- Absolute porosity
- Effective porosity

Absolute porosity

The absolute porosity is defined as the ratio of the total pore space in the rock to that of the bulk volume. A rock may have considerable absolute porosity and yet have no conductivity to fluid for lack of pore interconnection. The absolute porosity is generally expressed mathematically by the following relationships:

$$\phi_a = \frac{\text{total pore volume}}{\text{bulk volume}} \qquad (4-1)$$

or

$$\phi_a = \frac{\text{bulk volume} - \text{grain volume}}{\text{bulk volume}} \qquad (4-2)$$

where ϕ_a = absolute porosity.

Effective porosity

The effective porosity is the percentage of *interconnected* pore space with respect to the bulk volume, or

$$\phi = \frac{\text{interconnected pore volume}}{\text{bulk volume}} \qquad (4-3)$$

where ϕ = effective porosity.

The effective porosity is the value that is used in all reservoir engineering calculations because it represents the interconnected pore space that contains the recoverable hydrocarbon fluids.

Porosity may be classified according to the mode of origin as original induced.

The *original* porosity is that developed in the deposition of the material, while *induced* porosity is that developed by some geologic process subsequent to deposition of the rock. The intergranular porosity of sandstones and the intercrystalline and oolitic porosity of some limestones typify original porosity. Induced porosity is typified by fracture development as found in shales and limestones and by the slugs or solution cavities commonly found in limestones. Rocks having original porosity are more uniform in their characteristics than those rocks in which a large part of the porosity is included. For direct quantitative measurement of porosity, reliance must be placed on formation samples obtained by coring.

Since effective porosity is the porosity value of interest to the petroleum engineer, particular attention should be paid to the methods used to determine porosity. For example, if the porosity of a rock sample was determined by saturating the rock sample 100 percent with a fluid of known density and then determining, by weighing, the increased weight due to the saturating fluid, this would yield an effective porosity measurement because the saturating fluid could enter only the interconnected pore spaces. On the other hand, if the rock sample were crushed with a mortar and pestle to determine the actual volume of the solids in the core sample, then an absolute porosity measurement would result because the identity of any isolated pores would be lost in the crushing process.

One important application of the effective porosity is its use in determining the original hydrocarbon volume in place. Consider a reservoir with an areal extent of A acres and an average thickness of h feet. The total bulk volume of the reservoir can be determined from the following expressions:

$$\text{Bulk volume} = 43,560 \, Ah, \quad ft^3 \tag{4-4}$$

or

$$\text{Bulk volume} = 7,758 \, Ah, \quad bbl \tag{4-5}$$

where A = areal extent, acres
       h = average thickness

The reservoir pore volume PV can then be determined by combining Equations 4-4 and 4-5 with 4-3. Expressing the reservoir pore volume in cubic feet gives:

$$PV = 43,560 \, Ah\phi, \quad ft^3 \tag{4-6}$$

Expressing the reservoir pore volume in barrels gives:

$$PV = 7,758 \, Ah\phi, \quad bbl \tag{4-7}$$

Example 4-1

An oil reservoir exists at its bubble-point pressure of 3000 psia and temperature of 160°F. The oil has an API gravity of 42° and gas-oil ratio of 600 scf/STB. The specific gravity of the solution gas is 0.65. The following additional data are also available:

- Reservoir area = 640 acres
- Average thickness = 10 ft
- Connate water saturation = 0.25
- Effective porosity = 15%

Calculate the initial oil in place in STB.

Solution

Step 1. Determine the specific gravity of the stock-tank oil from Equation 2-68.

$$\gamma_o = \frac{141.5}{42 + 131.5} = 0.8156$$

Step 2. Calculate the initial oil formation volume factor by applying Standing's equation, i.e., Equation 2-85, to give:

$$\beta_o = 0.9759 + 0.00012 \left[600 \left(\frac{0.65}{0.8156} \right)^{0.5} + 1.25(160) \right]^{1.2}$$
$$= 1.306 \, \text{bbl/STB}$$

Step 3. Calculate the pore volume from Equation 4-7.

Pore volume = 7758 (640) (10) (0.15) = 7,447,680 bbl

Step 4. Calculate the initial oil in place.

Initial oil in place = 12,412,800 (1 − 0.25)/1.306 = 4,276,998 STB

The reservoir rock may generally show large variations in porosity vertically but does not show very great variations in porosity parallel to the bedding planes. In this case, the arithmetic average porosity or the thickness-weighted average porosity is used to describe the average reservoir porosity. A change in sedimentation or depositional conditions, however, can cause the porosity in one portion of the reservoir to be greatly different from that in another area. In such cases, the areal-weighted average or the volume-weighted average porosity is used to characterize the average rock porosity. These averaging techniques are expressed mathematically in the following forms:

Arithmetic average	$\phi = \Sigma\phi_i/n$	(4-8)
Thickness-weighted average	$\phi = \Sigma\phi_i h_i/\Sigma h_i$	(4-9)
Areal-weighted average	$\phi = \Sigma\phi_i A_i/\Sigma A_i$	(4-10)
Volumetric-weighted average	$\phi = \Sigma\phi_i A_i h_i/\Sigma A_i h_i$	(4-11)

where   n = total number of core samples
      h_i = thickness of core sample i or reservoir area i
      ϕ_i = porosity of core sample i or reservoir area i
      A_i = reservoir area i

Example 4-2

Calculate the arithmetic average and thickness-weighted average from the following measurements:

Sample	Thickness, ft	Porosity, %
1	1.0	10
2	1.5	12
3	1.0	11
4	2.0	13
5	2.1	14
6	1.1	10

Solution

• Arithmetic average

$$\phi = \frac{10 + 12 + 11 + 13 + 14 + 10}{6} = 11.67\%$$

• Thickness-weighted average

$$\phi = \frac{(1)\,(10) + (1.5)\,(12) + (1)\,(11) + (2)\,(13) + (2.1)\,(14) + (1.1)\,(10)}{1 + 1.5 + 1 + 2 + 2.1 + 1.1}$$
$$= 12.11\%$$

SATURATION

Saturation is defined as that fraction, or percent, of the pore volume occupied by a particular fluid (oil, gas, or water). This property is expressed mathematically by the following relationship:

$$\text{fluid saturation} = \frac{\text{total volume of the fluid}}{\text{pore volume}}$$

Applying the above mathematical concept of saturation to each reservoir fluid gives

$$S_o = \frac{\text{volume of oil}}{\text{pore volume}} \tag{4-12}$$

$$S_g = \frac{\text{volume of gas}}{\text{pore volume}} \tag{4-13}$$

$$S_w = \frac{\text{volume of water}}{\text{pore volume}} \tag{4-14}$$

where S_o = oil saturation
S_g = gas saturation
S_w = water saturation

Thus, all saturation values are based on *pore volume* and not on the gross reservoir volume.

The saturation of each individual phase ranges between zero to 100 percent. By definition, the sum of the saturations is 100%, therefore

$$S_g + S_o + S_w = 1.0 \tag{4-15}$$

The fluids in most reservoirs are believed to have reached a state of equilibrium and, therefore, will have become separated according to their density, i.e., oil overlain by gas and underlain by water. In addition to the bottom (or edge) water, there will be connate water distributed throughout the oil and gas zones. The water in these zones will have been reduced to some irreducible minimum. The forces retaining the water in the oil and gas zones are referred to as *capillary forces* because they are important only in pore spaces of capillary size.

Connate (interstitial) water saturation S_{wc} is important primarily because it reduces the amount of space available between oil and gas. It is generally not uniformly distributed throughout the reservoir but varies with permeability, lithology, and height above the free water table.

Another particular phase saturation of interest is called the *critical saturation* and it is associated with each reservoir fluid. The definition and the significance of the critical saturation for each phase is described below.

Critical oil saturation, S_{oc}

For the oil phase to flow, the saturation of the oil must exceed a certain value which is termed critical oil saturation. At this particular saturation, the oil remains in the pores and, for all practical purposes, will not flow.

Residual oil saturation, S_{or}

During the displacing process of the crude oil system from the porous media by water or gas injection (or encroachment) there will be some remaining oil left that is quantitatively characterized by a saturation value that is larger than the *critical oil saturation.* This saturation value is called the *residual oil saturation, S_{or}.* The term residual saturation is usually associated with the nonwetting phase when it is being displaced by a wetting phase.

Movable oil saturation, S_{om}

Movable oil saturation S_{om} is another saturation of interest and is defined as the fraction of pore volume occupied by movable oil as expressed by the following equation:

$$S_{om} = 1 - S_{wc} - S_{oc}$$

where S_{wc} = connate water saturation
S_{oc} = critical oil saturation

Critical gas saturation, S_{gc}

As the reservoir pressure declines below the bubble-point pressure, gas evolves from the oil phase and consequently the saturation of the gas increases as the reservoir pressure declines. The gas phase remains immobile until its saturation exceeds a certain saturation, called *critical gas saturation,* above which gas begins to move.

Critical water saturation, S_{wc}

The critical water saturation, connate water saturation, and irreducible water saturation are extensively used interchangeably to define the maximum water saturation at which the water phase will remain immobile.

Average Saturation

Proper averaging of saturation data requires that the saturation values be weighted by both the interval *thickness* h_i and interval *porosity* ϕ. The average saturation of each reservoir fluid is calculated from the following equations:

$$S_o = \frac{\sum\limits_{i=1}^{n} \phi_i \, h_i \, S_{oi}}{\sum\limits_{i=1}^{n} \phi_i \, h_i} \tag{4-16}$$

$$S_w = \frac{\sum\limits_{i=1}^{n} \phi_i \, h_i \, S_{wi}}{\sum\limits_{i=1}^{n} \phi_i \, h_i} \tag{4-17}$$

$$S_g = \frac{\sum\limits_{i=1}^{n} \phi_i \, h_i \, S_{gi}}{\sum\limits_{i=1}^{n} \phi_i \, h_i} \tag{4-18}$$

where the subscript i refers to any individual measurement and h_i represents the depth interval to which ϕ_i, S_{oi}, S_{gi}, and S_{wi} apply.

Example 4-3

Calculate average oil and connate water saturation from the following measurements:

Sample	h_i, ft	ϕ, %	S_o, %	S_{wc}, %
1	1.0	10	75	25
2	1.5	12	77	23
3	1.0	11	79	21
4	2.0	13	74	26
5	2.1	14	78	22
6	1.1	10	75	25

Solution

Construct the following table and calculate the average saturation for the oil and water phase:

Sample	h_i, ft	ϕ	ϕh	S_o	$S_o\phi h$	S_{wc}	$S_{wc}\phi h$
1	1.0	.10	.100	.75	.0750	.25	.0250
2	1.5	.12	.180	.77	.1386	.23	.0414
3	1.0	.11	.110	.79	.0869	.21	.0231
4	2.0	.13	.260	.74	.1924	.26	.0676
5	2.1	.14	.294	.78	.2293	.22	.0647
6	1.1	.10	.110	.75	.0825	.25	.0275
			1.054		0.8047		0.2493

Calculate average oil saturation by applying Equation 4-16:

$$S_o = \frac{.8047}{1.054} = 0.7635$$

Calculate average water saturation by applying Equation 4-17:

$$S_w = \frac{0.2493}{1.054} = 0.2365$$

WETTABILITY

Wettability is defined as the tendency of one fluid to spread on or adhere to a solid surface in the presence of other immiscible fluids. The concept of wettability is illustrated in Figure 4-1. Small *drops* of three liquids—mercury, oil, and water—are placed on a clean glass plate. The three droplets are then observed from one side as illustrated in Figure 4-1. It is noted that the mercury retains a spherical shape, the oil droplet develops an approximately hemispherical shape, but the water tends to spread over the glass surface.

The tendency of a liquid to spread over the surface of a solid is an indication of the *wetting* characteristics of the liquid for the solid. This spreading tendency can be expressed more conveniently by measuring the angle of contact at the *liquid-solid* surface. This angle, which is always measured through the liquid to the solid, is called the contact angle θ.

The contact angle θ has achieved significance as a measure of wettability. As shown in Figure 4-1, as the contact angle decreases, the wetting characteristics of the liquid increase. Complete wettability would be evidenced by a zero contact angle, and complete nonwetting would be evidenced by a contact angle of 180°. There have been various definitions of *intermediate* wettability but, in much of the published literature, contact angles of 60° to 90° will tend to repel the liquid.

The wettability of reservoir rocks to the fluids is important in that the distribution of the fluids in the porous media is a function of wettability. Because of the attractive forces, the wetting phase tends to occupy the smaller pores of the rock and the nonwetting phase occupies the more open channels.

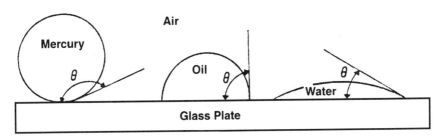

Figure 4-1. Illustration of wettability.

SURFACE AND INTERFACIAL TENSION

In dealing with multiphase systems, it is necessary to consider the effect of the forces at the interface when two immiscible fluids are in contact. When these two fluids are liquid and gas, the term *surface tension* is used to describe the forces acting on the interface. When the interface is between two liquids, the acting forces are called *interfacial tension.*

Surfaces of liquids are usually blanketed with what acts as a thin film. Although this apparent film possesses little strength, it nevertheless acts like a thin membrane and resists being broken. This is believed to be caused by attraction between molecules within a given system. All molecules are attracted one to the other in proportion to the product of their masses and inversely as the squares of the distance between them.

Consider the two immiscible fluids, air (or gas) and water (or oil) as shown schematically in Figure 4-2. A liquid molecule, which is remote from the interface, is surrounded by other liquid molecules, thus having a resulting net attractive force on the molecule of zero. A molecule at the interface, however, has a force acting on it from the air (gas) molecules lying immediately above the interface and from liquid molecules lying below the interface.

Resulting forces are unbalanced and give rise to surface tension. The unbalanced attraction force between the molecules creates a membrane-like surface with a measurable tension, i.e., surface tension. As a matter

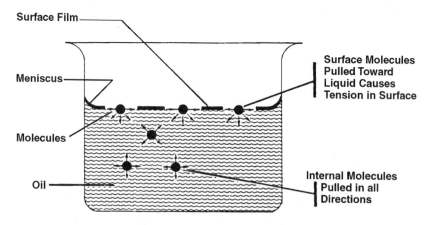

Figure 4-2. Illustration of surface tension. (*After Clark, N. J.,* Elements of Petroleum Reservoirs, *SPE, 1969.*)

of fact, if carefully placed, a needle will float on the surface of the liquid, supported by the thin membrane even though it is considerably more dense than the liquid.

The surface or interfacial tension has the units of force per unit of length, e.g., dynes/cm, and is usually denoted by the symbol σ.

If a glass capillary tube is placed in a large open vessel containing water, the combination of surface tension and wettability of tube to water will cause water to rise in the tube above the water level in the container outside the tube as shown in Figure 4-3.

The water will rise in the tube until the total force acting to pull the liquid upward is balanced by the weight of the column of liquid being

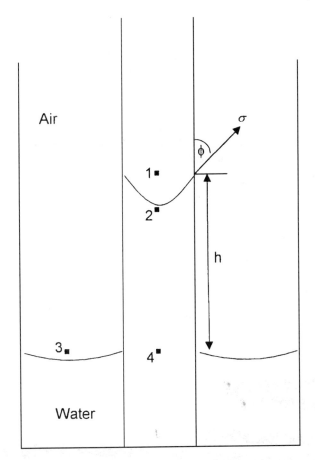

Figure 4-3. Pressure relations in capillary tubes.

supported in the tube. Assuming the radius of the capillary tube is r, the total upward force F_{up}, which holds the liquid up, is equal to the force per unit length of surface times the total length of surface, or

$$F_{up} = (2\pi r)\,(\sigma_{gw})\,(\cos\theta) \qquad\qquad (4\text{-}19)$$

where σ_{gw} = surface tension between air (gas) and water (oil), dynes/cm
        θ = contact angle
        r = radius, cm

The upward force is counteracted by the weight of the water, which is equivalent to a downward force of mass times acceleration, or

$$F_{down} = \pi r^2 h\,(\rho_w - \rho_{air})\,g \qquad\qquad (4\text{-}20)$$

where h = height to which the liquid is held, cm
       g = acceleration due to gravity, cm/sec^2
      ρ_w = density of water, gm/cm^3
     ρ_{air} = density of gas, gm/cm^3

Because the density of air is negligible in comparison with the density of water, Equation 4-20 is reduced to:

$$F_{down} = \pi r^2 \rho_w g \qquad\qquad (4\text{-}21)$$

Equating Equation 4-19 with 4-21 and solving for the surface tension gives:

$$\sigma_{gw} = \frac{r\,h\,\rho_w\,g}{2\cos\theta} \qquad\qquad (4\text{-}22)$$

The generality of Equations 4-19 through 4-22 will not be lost by applying them to the behavior of two liquids, i.e., water and oil. Because the density of oil is not negligible, Equation 4-22 becomes

$$\sigma_{ow} = \frac{r\,h\,g\,(\rho_w - \rho_o)}{2\cos\theta} \qquad\qquad (4\text{-}23)$$

where ρ_o = density of oil, gm/cm^3
     σ_{ow} = interfacial tension between the oil and the water, dynes/cm

CAPILLARY PRESSURE

The capillary forces in a petroleum reservoir are the result of the combined effect of the surface and interfacial tensions of the rock and fluids, the pore size and geometry, and the wetting characteristics of the system. Any curved surface between two immiscible fluids has the tendency to contract into the smallest possible area per unit volume. This is true whether the fluids are oil and water, water and gas (even air), or oil and gas. When two immiscible fluids are in contact, a discontinuity in pressure exists between the two fluids, which depends upon the curvature of the interface separating the fluids. We call this pressure difference the *capillary pressure* and it is referred to by p_c.

The displacement of one fluid by another in the pores of a porous medium is either aided or opposed by the surface forces of capillary pressure. As a consequence, in order to maintain a porous medium partially saturated with nonwetting fluid and while the medium is also exposed to wetting fluid, it is necessary to maintain the pressure of the nonwetting fluid at a value greater than that in the wetting fluid.

Denoting the pressure in the wetting fluid by p_w and that in the nonwetting fluid by p_{nw}, the capillary pressure can be expressed as:

Capillary pressure = (pressure of the nonwetting phase) − (pressure of the wetting phase)

$$p_c = p_{nw} - p_w \qquad (4\text{-}24)$$

That is, the pressure excess in the nonwetting fluid is the capillary pressure, and this quantity is a function of saturation. This is the defining equation for capillary pressure in a porous medium.

There are three types of capillary pressure:

- Water-oil capillary pressure (denoted as P_{cwo})
- Gas-oil capillary pressure (denoted as P_{cgo})
- Gas-water capillary pressure (denoted as P_{cgw})

Applying the mathematical definition of the capillary pressure as expressed by Equation 4-24, the three types of the capillary pressure can be written as:

$$p_{cwo} = p_o - p_w$$

$$p_{cgo} = p_g - p_o$$

$$p_{cgw} = p_g - p_w$$

where p_g, p_o, and p_w represent the pressure of gas, oil, and water, respectively.

If all the three phases are continuous, then:

$$p_{cgw} = p_{cgo} + p_{cwo}$$

Referring to Figure 4-3, the pressure difference across the interface between Points 1 and 2 is essentially the capillary pressure, i.e.:

$$p_c = p_1 - p_2 \qquad (4\text{-}25)$$

The pressure of the water phase at Point 2 is equal to the pressure at point 4 minus the head of the water, or:

$$p_2 = p_4 - gh\rho_w \qquad (4\text{-}26)$$

The pressure just above the interface at Point 1 represents the pressure of the air and is given by:

$$p_1 = p_3 - gh\rho_{air} \qquad (4\text{-}27)$$

It should be noted that the pressure at Point 4 within the capillary tube is the same as that at Point 3 outside the tube. Subtracting Equation 4-26 from 4-27 gives:

$$p_c = gh\,(\rho_w - \rho_{air}) = gh\Delta\rho \qquad (4\text{-}28)$$

where $\Delta\rho$ is the density difference between the wetting and nonwetting phase. The density of the air (gas) is negligible in comparison with the water density.

In practical units, Equation 4-28 can be expressed as:

$$p_c = \left(\frac{h}{144}\right)\Delta\rho$$

where p_c = capillary pressure, psi
$\qquad$ h = capillary rise, ft
$\qquad$ $\Delta\rho$ = density difference, lb/ft³

In the case of an oil-water system, Equation 4-28 can be written as:

$$p_c = gh \, (\rho_w - \rho_o) = gh\Delta\rho \tag{4-29}$$

and in practical units

$$p_c = \left(\frac{h}{144}\right)(\rho_w - \rho_o)$$

The capillary pressure equation can be expressed in terms of the surface and interfacial tension by combining Equations 4-28 and 4-29 with Equations 4-22 and 4-23 to give:

• **Gas-liquid system**

$$P_c = \frac{2 \, \sigma_{gw} \, (\cos \theta)}{r} \tag{4-30}$$

and

$$h = \frac{2 \, \sigma_{gw} \, (\cos \theta)}{r \, g \, (\rho_w - \rho_{gas})} \tag{4-31}$$

where ρ_w = water density, gm/cm^3
σ_{gw} = gas-water surface tension, dynes/cm
r = capillary radius, cm
θ = contact angle
h = capillary rise, cm
g = acceleration due to gravity, cm/sec^2
p_c = capillary pressure, dynes/cm^2

• **Oil-water system**

$$P_c = \frac{2 \, \sigma_{ow} \, (\cos \theta)}{r} \tag{4-32}$$

and

$$h = \frac{2 \, \sigma_{wo} \, (\cos \theta)}{r \, g \, (\rho_w - \rho_o)} \tag{4-33}$$

where σ_{wo} is the water-oil interfacial tension.

Example 4-4

Calculate the pressure difference, i.e., capillary pressure, and capillary rise in an oil-water system from the following data:

$$\theta = 30° \qquad \rho_w = 1.0 \text{ gm/cm}^3 \qquad \rho_o = 0.75 \text{ gm/cm}^3$$
$$r = 10^{-4} \text{ cm} \qquad \sigma_{ow} = 25 \text{ dynes/cm}$$

Solution

Step 1. Apply Equation 4-32 to give

$$p_c = \frac{(2)\,(25)\,(\cos\ 30°)}{0.0001} = 4.33 \times 10^5 \text{ dynes/cm}^2$$

Since 1 dyne/cm^2 = 1.45 × 10^{B5} psi, then

$$p_c = 6.28 \text{ psi}$$

This result indicates that the oil-phase pressure is 6.28 psi higher than the water-phase pressure.

Step 2. Calculate the capillary rise by applying Equation 4-33.

$$h = \frac{(2)\,(25)\,(\cos\ 30°)}{(0.0001)\,(980.7)\,(1.0 - 0.75)} = 1766 \text{ cm} = 75.9 \text{ ft}$$

Capillary Pressure of Reservoir Rocks

The interfacial phenomena described above for a single capillary tube also exist when bundles of interconnected capillaries of varying sizes exist in a porous medium. The capillary pressure that exists within a porous medium between two immiscible phases is a function of the interfacial tensions and the average size of the capillaries which, in turn, controls the curvature of the interface. In addition, the curvature is also a function of the saturation distribution of the fluids involved.

Laboratory experiments have been developed to simulate the displacing forces in a reservoir in order to determine the magnitude of the capillary forces in a reservoir and, thereby, determine the fluid saturation distributions and connate water saturation. One such experiment is called the *restored capillary pressure technique* which was developed primarily to determine the magnitude of the connate water saturation. A diagrammatic sketch of this equipment is shown in Figure 4-4.

Briefly, this procedure consists of saturating a core 100% with the reservoir water and then placing the core on a porous membrane which is saturated 100% with water and is permeable to the water only, under the pressure drops imposed during the experiment. Air is then admitted into the core chamber and the pressure is increased until a small amount of water is displaced through the porous, semi-permeable membrane into the graduated cylinder. Pressure is held constant until no more water is displaced, which may require several days or even several weeks, after

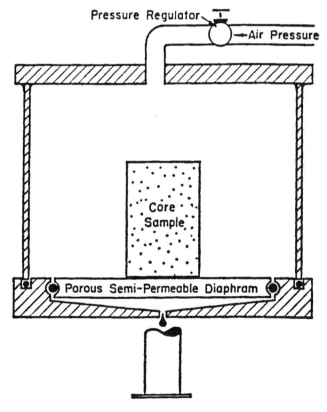

Figure 4-4. Capillary pressure equipment. (*After Cole, F., 1969.*)

which the core is removed from the apparatus and the water saturation determined by weighing. The core is then replaced in the apparatus, the pressure is increased, and the procedure is repeated until the water saturation is reduced to a minimum.

The data from such an experiment are shown in Figure 4-5. Since the pressure required to displace the wetting phase from the core is exactly equal to the capillary forces holding the remaining water within the core after equilibrium has been reached, the pressure data can be plotted as capillary pressure data. Two important phenomena can be observed in Figure 4-5. First, there is a finite capillary pressure at 100% water saturation that is necessary to force the nonwetting phase into a capillary filled with the wetting phase. This minimum capillary pressure is known as the *displacement pressure*, p_d.

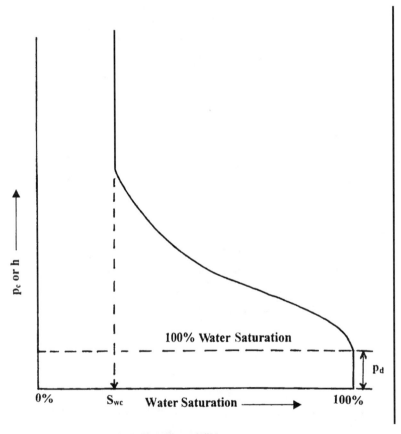

Figure 4-5. Capillary pressure curve.

If the largest capillary opening is considered as circular with a radius of r, the pressure needed for forcing the nonwetting fluid out of the core is:

$$P_c = \frac{2\,\sigma\,(\cos\theta)}{r}$$

This is the minimum pressure that is required to displace the wetting phase from the largest capillary pore because any capillary of smaller radius will require a higher pressure.

As the wetting phase is displaced, the second phenomenon of any immiscible displacement process is encountered, that is, the reaching of some finite minimum irreducible saturation. This irreducible water saturation is referred to as connate water.

It is possible from the capillary pressure curve to calculate the average size of the pores making up a stated fraction of the total pore space. Let p_c be the average capillary pressure for the 10% between saturation of 40 and 50%. The average capillary radius is obtained from

$$r = \frac{2\,\sigma\,(\cos\theta)}{p_c}$$

The above equation may be solved for r providing that the interfacial tension σ, and the angle of contact θ may be evaluated.

Figure 4-6 is an example of typical oil-water capillary pressure curves. In this case, capillary pressure is plotted versus water saturation for four rock samples with permeabilities increasing from k_1 to k_4. It can be seen that, for decreases in permeability, there are corresponding increases in capillary pressure at a constant value of water saturation. This is a reflection of the influence of pore size since the smaller diameter pores will invariably have the lower permeabilities. Also, as would be expected the capillary pressure for any sample increases with decreasing water saturation, another indication of the effect of the radius of curvature of the water-oil interface.

Capillary Hysteresis

It is generally agreed that the pore spaces of reservoir rocks were originally filled with water, after which oil moved into the reservoir, displacing some of the water and reducing the water to some residual saturation. When discovered, the reservoir pore spaces are filled with a connate-water saturation and an oil saturation. All laboratory experiments are

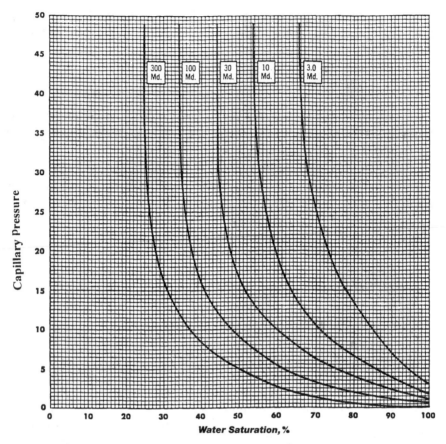

Figure 4-6. Variation of capillary pressure with permeability.

designed to duplicate the saturation history of the reservoir. The process of generating the capillary pressure curve by displacing the wetting phase, i.e., water, with the nonwetting phase (such as with gas or oil), is called the *drainage process.*

This drainage process establishes the fluid saturations which are found when the reservoir is discovered. The other principal flow process of interest involves reversing the drainage process by displacing the nonwetting phase (such as with oil) with the wetting phase, (e.g., water). This displacing process is termed the *imbibition process* and the resulting curve is termed the *capillary pressure imbibition curve.* The process of *saturating* and *desaturating* a core with the nonwetting phase is called *capillary hysteresis.* Figure 4-7 shows typical drainage and imbibition

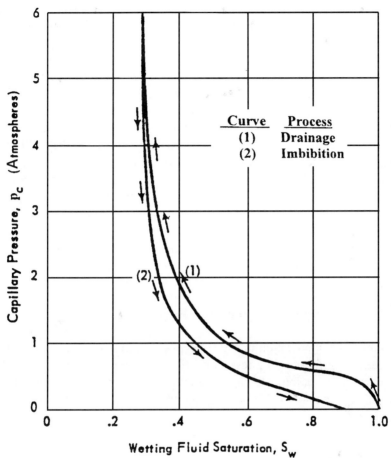

Figure 4-7. Capillary pressure hysteresis.

capillary pressure curves. The two capillary pressure-saturation curves are not the same.

This difference in the saturating and desaturating of the capillary-pressure curves is closely related to the fact that the advancing and receding contact angles of fluid interfaces on solids are different. Frequently, in natural crude oil-brine systems, the contact angle or wettability may change with time. Thus, if a rock sample that has been thoroughly cleaned with volatile solvents is exposed to crude oil for a period of time, it will behave as though it were oil wet. But if it is exposed to brine after cleaning, it will appear water wet. At the present time, one of the greatest unsolved problems in the petroleum industry is that of wettability of reservoir rock

Another mechanism that has been proposed by McCardell (1955) to account for capillary hysteresis is called the *ink-bottle effect.* This phenomenon can be easily observed in a capillary tube having variations in radius along its length. Consider a capillary tube of axial symmetry having roughly sinusoidal variations in radius. When such a tube has its lower end immersed in water, the water will rise in the tube until the hydrostatic fluid head in the tube becomes equal to the capillary pressure. If then the tube is lifted to a higher level in the water, some water will drain out, establishing a new equilibrium level in the tube.

When the meniscus is advancing and it approaches a constriction, it *jumps* through the neck, whereas when receding, it halts without passing through the neck. This phenomenon explains why a given capillary pressure corresponds to a higher saturation on the drainage curve than on the imbibition curve.

Initial Saturation Distribution in a Reservoir

An important application of the concept of capillary pressures pertains to the fluid distribution in a reservoir prior to its exploitation. The capillary pressure-saturation data can be converted into height-saturation data by arranging Equation 4-29 and solving for the height h above the free-water level.

$$h = \frac{144 \, p_c}{\Delta \rho} \tag{4-34}$$

where p_c = capillary pressure, psia
$\Delta \rho$ = density difference between the wetting and nonwetting phase, lb/ft^3
H = height above the free-water level, ft

Figure 4-8 shows a plot of the water saturation distribution as a function of distance from the free-water level in an oil-water system.

It is essential at this point to introduce and define four important concepts:

• Transition zone
• Water-oil contact (WOC)
• Gas-oil contact (GOC)
• Free water level (FWL)

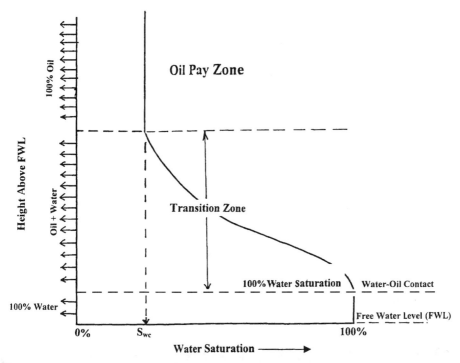

Figure 4-8. Water saturation profile.

Figure 4-9 illustrates an idealized gas, oil, and water distribution in a reservoir. The figure indicates that the saturations are gradually charging from 100% water in the water zone to irreducible water saturation some vertical distance above the water zone. This vertical area is referred to as the *transition zone,* which must exist in any reservoir where there is a bottom water table. The transition zone is then defined as the vertical thickness over which the water saturation ranges from 100% saturation to irreducible water saturation S_{wc}. The important concept to be gained from Figure 4-9 is that there is no abrupt change from 100% water to maximum oil saturation. The creation of the oil-water transition zone is one of the major effects of capillary forces in a petroleum reservoir.

Similarly, the total liquid saturation (i.e. oil and water) is smoothly changing from 100% in the oil zone to the connate water saturation in the gas cap zone. A similar transition exists between the oil and gas zone. Figure 4-8 serves as a definition of what is meant by gas-oil and water-oil contacts. The WOC is defined as the "uppermost depth in the reservoir where a 100% water saturation exists." The GOC is defined as the "mini-

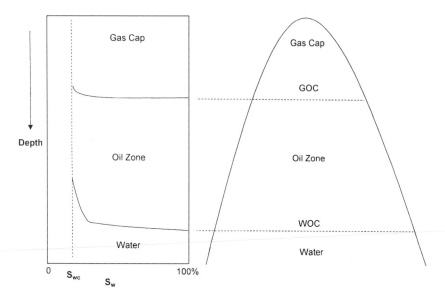

Figure 4-9. Initial saturation profile in a combination-drive reservoir.

mum depth at which a 100% liquid, i.e., oil + water, saturation exists in the reservoir."

Section A of Figure 4-10 shows a schematic illustration of a core that is represented by five different pore sizes and completely saturated with water, i.e., wetting phase. Assume that we subject the core to oil (the nonwetting phase) with increasing pressure until some water is displaced from the core, i.e., displacement pressure p_d. This water displacement will occur from the largest pore size. The oil pressure will have to increase to displace the water in the second largest pore. This sequential process is shown in sections B and C of Figure 4-10.

It should be noted that there is a difference between the free water level (FWL) and the depth at which 100% water saturation exists. From a reservoir engineering standpoint, the free water level is defined by *zero capillary pressure*. Obviously, if the largest pore is so large that there is no capillary rise in this size pore, then the free water level and 100% water saturation level, i.e., WOC, will be the same. This concept can be expressed mathematically by the following relationship:

$$FWL = WOC + \frac{144\, p_d}{\Delta \rho} \qquad (4\text{-}35)$$

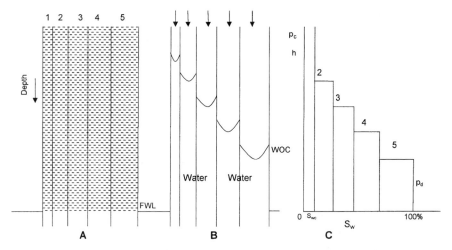

Figure 4-10. Relationship between saturation profile and pore-size distribution.

where p_d = displacement pressure, psi
       $\Delta\rho$ = density difference, lb/ft^3
       FWL = free water level, ft
       WOC = water-oil contact, ft

Example 4-5

The reservoir capillary pressure-saturation data of the Big Butte Oil reservoir is shown graphically in Figure 4-11. Geophysical log interpretations and core analysis establish the WOC at 5023 ft. The following additional data are available:

• Oil density = 43.5 lb/ft^3
• Water density = 64.1 lb/ft^3
• Interfacial tension = 50 dynes/cm

Calculate:

• Connate water saturation (S_{wc})
• Depth to FWL
• Thickness of the transition zone
• Depth to reach 50% water saturation

Solution

a. From Figure 4-11, connate-water saturation is 20%.
b. Applying Equation 4-35 with a displacement pressure of 1.5 psi gives

$$\text{FWL} = 5023 + \frac{(144)\,(1.5)}{(64.1 - 43.5)} = 5033.5 \text{ ft}$$

c. Thickness of transition zone $= \dfrac{144\,(6.0 - 1.5)}{(64.1 - 43.5)} = 31.5$ ft

d. P_c at 50% water saturation = 3.5 psia
   Equivalent height above the FWL = (144) (3.5)/(64.1 − 432.5) = 24.5 ft
   Depth to 50% water saturation = 5033.5 − 24.5 = 5009 ft

The above example indicates that only oil will flow in the interval between the top of the pay zone and depth of 4991.5 ft. In the transition zone, i.e., the interval from 4991.5 ft to the WOC, oil production would be accompanied by simultaneous water production.

It should be pointed out that the thickness of the transition zone may range from few feet to several hundred feet in some reservoirs. Recalling the capillary rise equation, i.e., height above FWL,

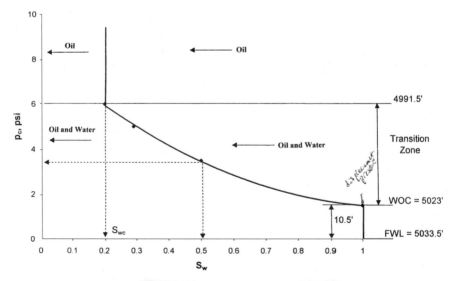

Figure 4-11. Capillary pressure saturation data.

$$h = \frac{2\,\sigma\,(\cos\phi)}{r\,g\,\Delta\rho}$$

The above relationship suggests that the height above FWL increases with decreasing the density difference $\Delta\rho$.

From a practical standpoint, this means that in a gas reservoir having a gas-water contact, the thickness of the transition zone will be a minimum since $\Delta\rho$ will be large. Also, if all other factors remain unchanged, a low API gravity oil reservoir with an oil-water contact will have a longer transition zone than a high API gravity oil reservoir. Cole (1969) illustrated this concept graphically in Figure 4-12.

The above expression also shows that as the radius of the pore r increases the volume of h decreases. Therefore, a reservoir rock system with small pore sizes will have a longer transition zone than a reservoir rock system comprised of large pore sizes.

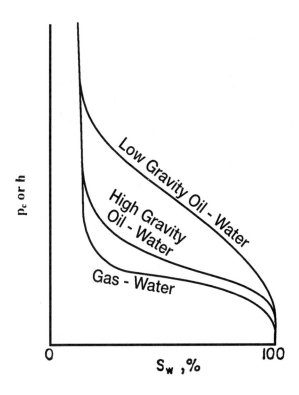

Figure 4-12. Variation of transition zone with fluid gravity. (*After Cole, F., 1969.*)

The reservoir pore size can often be related approximately to permeability, and where this applies, it can be stated that high permeability reservoirs will have shorter transition zones than low permeability reservoirs as shown graphically in Figure 4-13. As shown by Cole (Figure 4-14), a tilted water-oil contact could be caused by a change in permeability across the reservoir. It should be emphasized that the factor responsible for this change in the location of the water-oil contact is actually a change in the size of the pores in the reservoir rock system.

The previous discussion of capillary forces in reservoir rocks has assumed that the reservoir pore sizes, i.e., permeabilities, are essentially uniform. Cole (1969) discussed the effect of reservoir non-uniformity on the distribution of the fluid saturation through the formation. Figure 4-15 shows a hypothetical reservoir rock system that is comprised of seven layers. In addition, the seven layers are characterized by only two different pore sizes, i.e., permeabilities, and corresponding capillary pressure

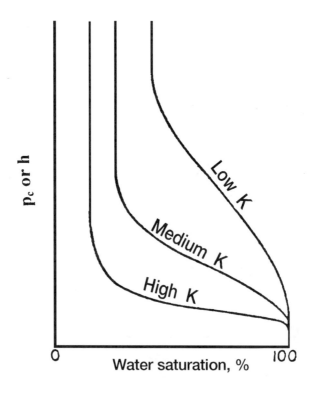

Figure 4-13. Variation of transition zone with permeability.

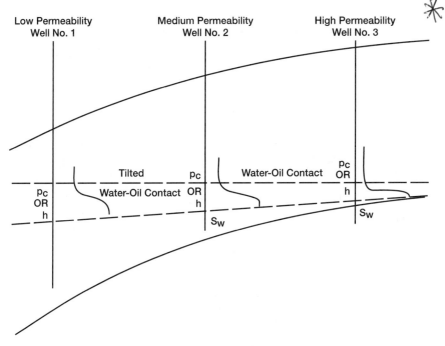

Figure 4-14. Tilted WOC. (*After Cole, F., 1969.*)

curves as shown in section A of Figure 4-15. The resulting capillary pressure curve for the layered reservoir would resemble that shown in section B of Figure 4-15. If a well were drilled at the point shown in section B of Figure 4-15, Layers 1 and 3 would not produce water, while Layer 2, which is above Layer 3, would produce water since it is located in the transition zone.

Example 4-6

A four-layer oil reservoir is characterized by a set of reservoir capillary pressure-saturation curves as shown in Figure 4-16. The following additional data are also available.

Layer	Depth, ft	Permeability, md
1	4000–4010	80
2	4010–4020	190
3	4020–4035	70
4	4035–4060	100

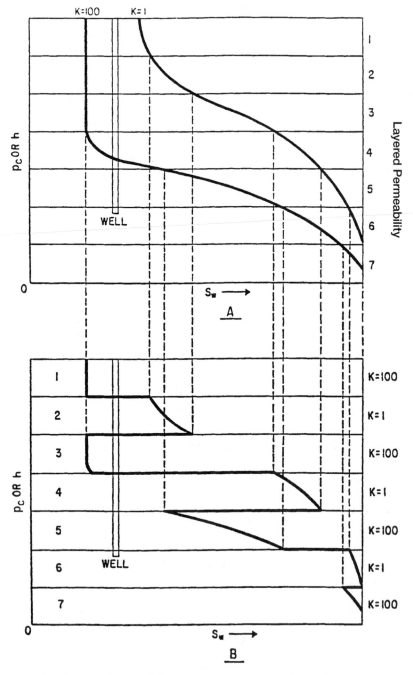

Figure 4-15. Effect of permeability on water saturation profile. (*After Cole, F., 1969.*)

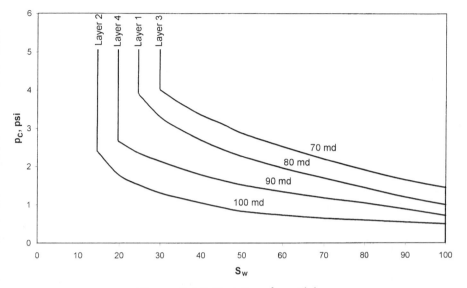

Figure 4-16. Variation of p_c with k.

WOC = 4060 ft
Water density = 65.2 lb/ft^3
Oil density = 55.2 lb/ft^3

Calculate and plot water saturation versus depth for this reservoir.

Solution

Step 1. Establish the FWL by determining the displacement pressure p_d for the bottom layer, i.e., Layer 4, and apply Equation 4-37:

- $p_d = 0.75$ psi

$$FWL = 4060 + \frac{(144)\,(0.75)}{(65.2 - 55.2)} = 4070.8 \text{ ft}$$

Step 2. The top of the bottom layer is located at a depth of 4035 ft which is 35.8 ft above the FWL. Using that height h of 35.8 ft, calculate the capillary pressure at the top of the bottom layer.

- $p_c = \left(\dfrac{h}{144}\right) \Delta \rho = \left(\dfrac{35.8}{144}\right)(65.2 - 55.2) = 2.486$ psi

- From the capillary pressure-saturation curve designated for Layer 4, read the water saturation that corresponds to a p_c of 2.486 to give $S_w = 0.23$.
- Assume different values of water saturations and convert the corresponding capillary pressures into height above the FWL by applying Equation 4-34.

$$h = \frac{144\, p_c}{\rho_w - \rho_o}$$

S_w	p_c, psi	h, ft	Depth = FWL – h
0.23	2.486	35.8	4035
0.25	2.350	33.84	4037
0.30	2.150	30.96	4040
0.40	1.800	25.92	4045
0.50	1.530	22.03	4049
0.60	1.340	19.30	4052
0.70	1.200	17.28	4054
0.80	1.050	15.12	4056
0.90	0.900	12.96	4058

Step 3. The top of Layer 3 is located at a distance of 50.8 ft from the FWL (i.e., h = 4070.8 – 4020 = 50.8 ft). Calculate the capillary pressure at the top of the third layer:

- $$p_c = \left(\frac{50.8}{144}\right)(65.2 - 55.2) = 3.53 \text{ psi}$$

- The corresponding water saturation as read from the curve designated for Layer 3 is 0.370.
- Construct the following table for Layer 3.

S_w	p_c, psi	h, ft	Depth = FWL – h
0.37	3.53	50.8	4020
0.40	3.35	48.2	4023
0.50	2.75	39.6	4031
0.60	2.50	36.0	4035

Step 4. • Distance from the FWL to the top of Layer 2 is:

> h = 4070.8 − 4010 = 60.8 ft

$$• \; p_c = \left(\frac{60.8}{144}\right)(65.2 - 55.2) = 4.22 \text{ psi}$$

• S_w at p_c of 4.22 psi is 0.15.
• Distance from the FWL to the bottom of the layer is 50.8 ft that corresponds to a p_c of 3.53 psi and S_w of 0.15. This indicates that the second layer has a uniform water saturation of 15%.

Step 5. For Layer 1, distance from the FWL to the top of the layer:

• h = 4070.8 − 4000 = 70.8 ft

$$• \; p_c = \left(\frac{70.8}{144}\right)(10) = 4.92 \text{ psi}$$

• S_w at the top of Layer 1 = 0.25
• The capillary pressure at the bottom of the layer is 3.53 psi with a corresponding water saturation of 0.27.

Step 6. Figure 4-17 documents the calculated results graphically. The figure indicates that Layer 2 will produce 100% oil while all remaining layers produce oil and water simultaneously.

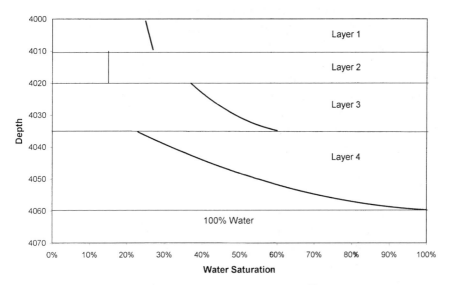

Figure 4-17. Water saturation profile.

Leverett J-Function

Capillary pressure data are obtained on small core samples that represent an extremely small part of the reservoir and, therefore, it is necessary to combine all capillary data to classify a particular reservoir. The fact that the capillary pressure-saturation curves of nearly all naturally porous materials have many features in common has led to attempts to devise some general equation describing all such curves. Leverett (1941) approached the problem from the standpoint of dimensional analysis.

Realizing that capillary pressure should depend on the porosity, interfacial tension, and mean pore radius, Leverett defined the dimensionless function of saturation, which he called the J-function, as

$$J\left(S_w\right) = 0.21645 \frac{p_c}{\sigma} \sqrt{\frac{k}{\phi}} \qquad (4\text{-}36)$$

where $J(S_w)$ = Leverett J-function
p_c = capillary pressure, psi
σ = interfacial tension, dynes/cm
k = permeability, md
ϕ = fractional porosity

In doing so, Leverett interpreted the ratio of permeability, k, to porosity, ϕ, as being proportional to the square of a mean pore radius.

The J-function was originally proposed as a means of converting all capillary-pressure data to a universal curve. There are significant differences in correlation of the J-function with water saturation from formation to formation, so that no universal curve can be obtained. For the same formation, however, this dimensionless capillary-pressure function serves quite well in many cases to remove discrepancies in the p_c versus S_w curves and reduce them to a common curve. This is shown for various unconsolidated sands in Figure 4-18.

Example 4-7

A laboratory capillary pressure test was conducted on a core sample taken from the Nameless Field. The core has a porosity and permeability of 16% and 80 md, respectively. The capillary pressure-saturation data are given as follows:

S_w	p_c, psi
1.0	0.50
0.8	0.60
0.6	0.75
0.4	1.05
0.2	1.75

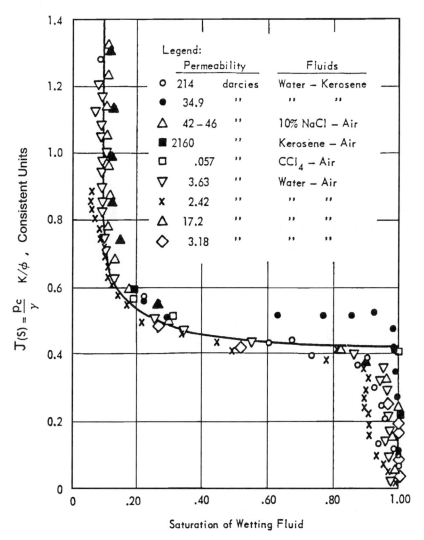

Figure 4-18. The Leverett J-function for unconsolidated sands. (*After Leverett, 1941.*)

The interfacial tension is measured at 50 dynes/cm. Further reservoir engineering analysis indicated that the reservoir is better described at a porosity value of 19% and an absolute permeability of 120 md. Generate the capillary pressure data for the reservoir.

Solution

Step 1. Calculate the J-function using the measured capillary pressure data.

$$J\,(S_w) = 0.21645\,(p_c/50)\,\sqrt{80/0.16} = 0.096799\,p_c$$

S_w	p_c, psi	$J\,(S_w) = 0.096799\,(p_c)$
1.0	0.50	0.048
0.8	0.60	0.058
0.6	0.75	0.073
0.4	1.05	0.102
0.2	1.75	0.169

Step 2. Using the new porosity and permeability values, solve Equation 4-36 for the capillary pressure p_c.

$$p_c = J\,(S_w)\,\sigma \Big/ \left[0.21645\,\sqrt{\dfrac{k}{\phi}}\,\right]$$

$$p_c = J\,(S_w)\,50 \Big/ \left[0.21645\,\sqrt{\dfrac{120}{0.19}}\,\right]$$

$$p_c = 9.192\,J\,(S_w)$$

Step 3. Reconstruct the capillary pressure-saturation table.

S_w	$J(S_w)$	$p_c = 9.192\,J(S_w)$
1.0	0.048	0.441
0.8	0.058	0.533
0.6	0.073	0.671
0.4	0.102	0.938
0.2	0.169	1.553

Converting Laboratory Capillary Pressure Data

For experimental convenience, it is common in the laboratory determination of capillary pressure to use air-mercury or air-brine systems, rather than the actual water-oil system characteristic of the reservoir. Since the laboratory fluid system does not have the same surface tension as the reservoir system, it becomes necessary to convert laboratory capillary pressure to reservoir capillary pressure. By assuming that the Leverett J-function is a property of rock and does not change from the laboratory to the reservoir, we can calculate reservoir capillary pressure as show below.

$$(p_c)_{res} = (p_c)_{lab} \frac{\sigma_{res}}{\sigma_{lab}}$$

Even after the laboratory capillary pressure has been corrected for surface tension, it may be necessary to make further corrections for permeability and porosity. The reason for this is that the core sample that was used in performing the laboratory capillary pressure test may not be representative of the average reservoir permeability and porosity. If we assume that the J-function will be invariant for a given rock type over a range of porosity and permeability values, then the reservoir capillary pressure can be expressed as

$$(p_c)_{res} = (p_c)_{lab} \frac{\sigma_{res}}{\sigma_{lab}} \sqrt{(\phi_{res} \, k_{core}) / (\phi_{core} \, k_{res})} \qquad (4\text{-}37)$$

where $(p_c)_{res}$ = reservoir capillary pressure
σ_{res} = reservoir surface or interfacial tension
k_{res} = reservoir permeability
ϕ_{res} = reservoir porosity
$(p_c)_{lab}$ = laboratory measured capillary pressure
ϕ_{core} = core porosity
k_{core} = core permeability

PERMEABILITY

Permeability is a property of the porous medium that measures the capacity and ability of the formation to transmit fluids. The rock permeability, k, is a very important rock property because it controls the directional movement and the flow rate of the reservoir fluids in the formation. This rock characterization was first defined mathematically by Henry Darcy in 1856. In fact, the equation that defines permeability in terms of measurable quantities is called **Darcy's Law.**

Darcy developed a fluid flow equation that has since become one of the standard mathematical tools of the petroleum engineer. If a horizontal linear flow of an incompressible fluid is established through a core sample of length L and a cross-section of area A, then the governing fluid flow equation is defined as

$$v = -\frac{k}{\mu}\frac{dp}{dL} \qquad\qquad (4\text{-}38)$$

where v = apparent fluid flowing velocity, cm/sec
      k = proportionality constant, or permeability, Darcys
      μ = viscosity of the flowing fluid, cp
   dp/dL = pressure drop per unit length, atm/cm

The velocity, v, in Equation 4-38 is not the actual velocity of the flowing fluid but is the apparent velocity determined by dividing the flow rate by the cross-sectional area across which fluid is flowing. Substituting the relationship, q/A, in place of v in Equation 4-38 and solving for q results in

$$q = -\frac{kA}{\mu}\frac{dp}{dL} \qquad\qquad (4\text{-}39)$$

where q = flow rate through the porous medium, cm^3/sec
      A = cross-sectional area across which flow occurs, cm^2

With a flow rate of one cubic centimeter per second across a cross-sectional area of one square centimeter with a fluid of one centipoise viscosity and a pressure gradient at one atmosphere per centimeter of length, it is obvious that k is unity. For the units described above, k has been arbitrarily assigned a unit called *Darcy* in honor of the man responsible for the development of the theory of flow through porous media. Thus, when all other parts of Equation 4-39 have values of unity, k has a value of one Darcy.

One Darcy is a relatively high permeability as the permeabilities of most reservoir rocks are less than one Darcy. In order to avoid the use of fractions in describing permeabilities, the term *millidarcy* is used. As the term indicates, one millidarcy, i.e., 1 md, is equal to one-thousandth of one Darcy or,

1 Darcy = 1000 md

The negative sign in Equation 4-39 is necessary as the pressure increases in one direction while the length increases in the opposite direction.

Equation 4-39 can be integrated when the geometry of the system through which fluid flows is known. For the simple linear system shown in Figure 4-19, the integration is performed as follows:

$$q \int_0^L dL = -\frac{kA}{\mu} \int_{p_1}^{p_2} dp$$

Integrating the above expression yields:

$$qL = -\frac{kA}{\mu}(p_2 - p_1)$$

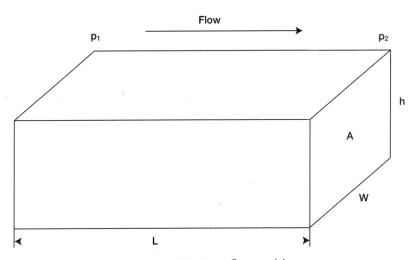

Figure 4-19. Linear flow model.

It should be pointed out that the volumetric flow rate, q, is constant for liquids because the density does not change significantly with pressure.

Since p_1 is greater than p_2, the pressure terms can be rearranged, which will eliminate the negative term in the equation. The resulting equation is:

$$q = \frac{kA\,(p_1 - p_2)}{\mu L} \qquad\qquad (4\text{-}40)$$

Equation 4-40 is the conventional linear flow equation used in fluid flow calculations.

Standard laboratory analysis procedures will generally provide reliable data on permeability of core samples. If the rock is not homogeneous, the whole core analysis technique will probably yield more accurate results than the analysis of core plugs (small pieces cut from the core). Procedures that have been used for improving the accuracy of the permeability determination include cutting the core with an oil-base mud, employing a pressure-core barrel, and conducting the permeability tests with reservoir oil.

Permeability is reduced by overburden pressure, and this factor should be considered in estimating permeability of the reservoir rock in deep wells because permeability is an isotropic property of porous rock in some defined regions of the system, that is, it is directional. Routine core analysis is generally concerned with plug samples drilled parallel to bedding planes and, hence, parallel to direction of flow in the reservoir. These yield horizontal permeabilities (k_h).

The measured permeability on plugs that are drilled perpendicular to bedding planes are referred to as vertical permeability (k_v). Figure 4-20 shows a schematic illustration of the concept of the core plug and the associated permeability.

As shown in Figure 4-20, there are several factors that must be considered as possible sources of error in determining reservoir permeability. These factors are:

1. Core sample may not be representative of the reservoir rock because of reservoir heterogeneity.
2. Core recovery may be incomplete.
3. Permeability of the core may be altered when it is cut, or when it is cleaned and dried in preparation for analysis. This problem is likely to occur when the rock contains reactive clays.
4. Sampling process may be biased. There is a temptation to select the best parts of the core for analysis.

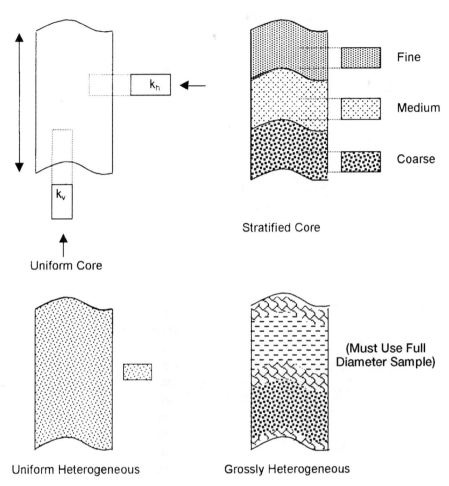

Figure 4-20. Representative samples of porous media.

Permeability is measured by passing a fluid of known viscosity μ through a core plug of measured dimensions (A and L) and then measuring flow rate q and pressure drop Δp. Solving Equation 4-40 for the permeability, gives:

$$k = \frac{q \, \mu \, L}{A \, \Delta p}$$

where L = length of core, cm
       A = cross-sectional area, cm^2

The following conditions must exist during the measurement of permeability:

• Laminar (viscous) flow
• No reaction between fluid and rock
• Only single phase present at 100% pore space saturation

This measured permeability at 100% saturation of a single phase is called the *absolute permeability* of the rock.

Example 4-8

A brine is used to measure the absolute permeability of a core plug. The rock sample is 4 cm long and 3 cm^2 in cross section. The brine has a viscosity of 1.0 cp and is flowing a constant rate of 0.5 cm^3/sec under a 2.0 atm pressure differential. Calculate the absolute permeability.

Solution

Applying Darcy's equation, i.e., Equation 4-40, gives:

$$0.5 = \frac{(k)\ (3)\ (2)}{(1)\ (4)}$$

k = 0.333 Darcys

Example 4-9

Rework the above example assuming that an oil of 2.0 cp is used to measure the permeability. Under the same differential pressure, the flow rate is 0.25 cm^3/sec.

Solution

Applying Darcy's equation yields:

$$0.25 = \frac{(k)\,(3)\,(2)}{(2)\,(4)}$$

$$k = 0.333 \text{ Darcys}$$

Dry gas is usually used (air, N_2, He) in permeability determination because of its convenience, availability, and to minimize fluid-rock reaction.

The measurement of the permeability should be restricted to the low (laminar/viscous) flow rate region, where the pressure remains proportional to flow rate within the experimental error. For high flow rates, Darcy's equation as expressed by Equation 4-40 is inappropriate to describe the relationship of flow rate and pressure drop.

In using dry gas in measuring the permeability, the gas volumetric flow rate q varies with pressure because the gas is a highly compressible fluid. Therefore, the value of q at the average pressure in the core must be used in Equation 4-40. Assuming the used gases follow the ideal gas behavior (at low pressures), the following relationships apply:

$$p_1 V_1 = p_2 V_2 = p_m V_m$$

In terms of the flow rate q, the above equation can be equivalently expressed as:

$$p_1 q_1 = p_2 q_2 = p_m q_m \tag{4-41}$$

with the mean pressure p_m expressed as:

$$p_m = \frac{p_1 + p_2}{2}$$

where p_1, p_2, p_m = inlet, outlet, and mean pressures, respectively, atm
V_1, V_2, V_m = inlet, outlet, and mean gas volume, respectively, cm^3
q_1, q_2, q_m = inlet, outlet, and mean gas flow rate, respectively, cm^3/sec

The gas flow rate is usually measured at base (atmospheric) pressure p_b and, therefore, the term Q_{gsc} is introduced into Equation 4-41 to produce:

$$Q_{gsc}\, p_b = q_m\, p_m$$

where Q_{gsc} = gas flow rate at standard conditions, cm³/sec
p_b = base pressure (atmospheric pressure), atm

Substituting Darcy's Law in the above expression gives

$$Q_{gsc}\, p_b = \frac{k\, A\, (p_1 - p_2)}{\mu_g\, L} \left(\frac{p_1 + p_2}{2} \right)$$

or

$$Q_{gsc} = \frac{k\, A\, (p_1^2 - p_2^2)}{2\, \mu_g\, L\, p_b} \qquad\qquad (4\text{-}42)$$

where k = absolute permeability, Darcys
μ_g = gas viscosity, cp
p_b = base pressure, atm
p_1 = inlet (upstream) pressure, atm
p_2 = outlet (downstream) pressure, atm
L = length of the core, cm
A = cross-sectional area, cm²
Q_{gsc} = gas flow rate at standard conditions, cm³/sec

The Klinkenberg Effect

Klinkenberg (1941) discovered that permeability measurements made with air as the flowing fluid showed different results from permeability measurements made with a liquid as the flowing fluid. The permeability of a core sample measured by flowing air is always greater than the permeability obtained when a liquid is the flowing fluid. Klinkenberg postulated, on the basis of his laboratory experiments, that liquids had a zero velocity at the sand grain surface, while gases exhibited some finite velocity at the sand grain surface. In other words, the gases exhibited *slippage* at the sand grain surface. This slippage resulted in a higher flow rate for the gas at a given pressure differential. Klinkenberg also found that for a given porous medium as the mean pressure increased the calculated permeability decreased.

Mean pressure is defined as upstream flowing plus downstream flowing pressure divided by two, $[p_m = (p_1 + p_2)/2]$. If a plot of measured permeability versus $1/p_m$ were extrapolated to the point where $1/p_m = 0$, in other words, where p_m = infinity, this permeability would be approximately equal to the liquid permeability. A graph of this nature is shown in Figure 4-21. The absolute permeability is determined by extrapolation as shown in Figure 4-21.

The magnitude of the Klinkenberg effect varies with the core permeability and the type of the gas used in the experiment as shown in Figures 4-22 and 4-23. The resulting straight-line relationship can be expressed as

$$k_g = k_L + c \left[\frac{1}{p_m} \right] \tag{4-43}$$

where k_g = measured gas permeability

p_m = mean pressure

k_L = equivalent liquid permeability, i.e., absolute permeability, k

c = slope of the line

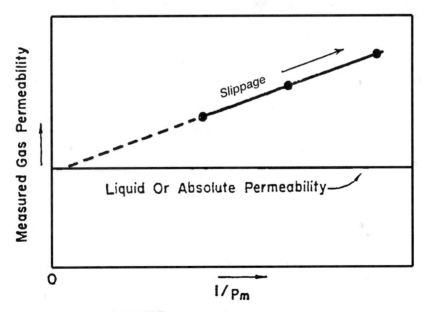

Figure 4-21. The Klinkenberg effect in gas permeability measurements.

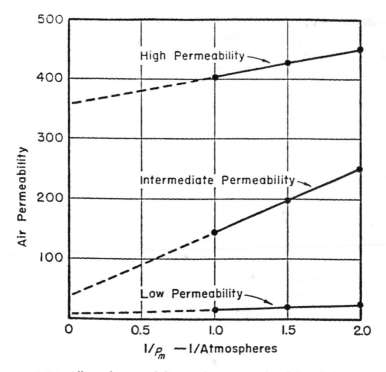

Figure 4-22. Effect of permeability on the magnitude of the Klinkenberg effect. (*After Cole, F., 1969.*)

Klinkenberg suggested that the slope Ac ≅ is a function of the following factors:

- Absolute permeability k, i.e., permeability of medium to a single phase completely filling the pores of the medium k_L.
- Type of the gas used in measuring the permeability, e.g., air.
- Average radius of the rock capillaries.

Klinkenberg expressed the slope c by the following relationship:

$$c = bk_L \qquad (4\text{-}44)$$

where k_L = equivalent liquid permeability, i.e., absolute permeability, k
     b = constant which depends on the size of the pore openings and is inversely proportional to radius of capillaries.

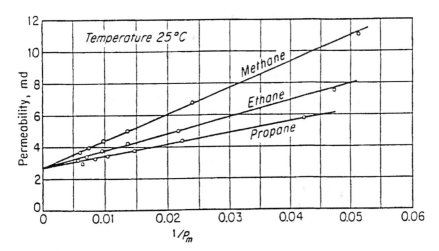

Figure 4-23. Effect of gas pressure on measured permeability for various gases. (After Calhoun, J., 1976.)

Combining Equation 4-44 with 4-43 gives:

$$k_g = k_L + (b\ k_L)\left[\frac{1}{p_m}\right] \qquad (4\text{-}45)$$

where k_g is the gas permeability as measured at the average pressure p_m.

Jones (1972) studied the gas slip phenomena for a group of cores for which porosity, liquid permeability k_L (absolute permeability), and air permeability were determined. He correlated the parameter b with the liquid permeability by the following expression:

$$b = 6.9\ k_L^{-0.36} \qquad (4\text{-}46)$$

The usual measurement of permeability is made with air at mean pressure just above atmospheric pressure (1 atm). To evaluate the slip phenomenon and the Klinkenberg effect, it is necessary to at least measure the gas permeability at two mean-pressure levels. In the absence of such data, Equations 4-45 and 4-46 can be combined and arranged to give:

$$6.9\ k_L^{0.64} + p_m\ k_L - p_m\ k_g = 0 \qquad (4\text{-}47)$$

where p_m = mean pressure, psi
$\quad k_g$ = air permeability at p_m, psi
$\quad k_L$ = absolute permeability (k), md

Equation 4-47 can be used to calculate the absolute permeability when only one gas permeability measurement (k_g) of a core sample is made at p_m. This nonlinear equation can be solved iteratively by using the Newton-Raphson iterative methods. The proposed solution method can be conveniently written as

$$k_{i+1} = k_i - \frac{f(k_i)}{f'(k_i)}$$

where k_i = initial guess of the absolute permeability, md
$\quad k_{i+1}$ = new permeability value to be used for the next iteration
$\quad i$ = iteration level
$\quad f(k_i)$ = Equation 4-47 as evaluated by using the assumed value of k_i.
$\quad f'(k_i)$ = first-derivative of Equation 4-47 as evaluated at k_i

The first derivative of Equation 4-47 with respect to k_i is:

$$f'(k_i) = 4.416\, k_i^{-0.36} + p_m \tag{4-48}$$

The iterative procedure is repeated until convergence is achieved when $f(k_i)$ approaches zero or when no changes in the calculated values of k_i are observed.

Example 4-10

The permeability of a core plug is measured by air. Only one measurement is made at a mean pressure of 2.152 psi. The air permeability is 46.6 md. Estimate the absolute permeability of the core sample. Compare the result with the actual absolute permeability of 23.66 md.

Solution

Step 1. Substitute the given values of p_m and k_g into Equations 4-47 and 4-48, to give:

$$f(k_i) = 6.9\, k_i^{0.64} + 2.152\, k_i - (2.152)(46.6) \quad f'(k_i) = 4.416\, k_i^{-0.36}$$
$$+ 2.152$$

Step 2. Assume $k_i = 30$ and apply the Newton-Raphson method to find the required solution as shown below.

i	k_i	$f(k_i)$	$f'(k_i)$	k_{i+1}
1	30.000	25.12	3.45	22.719
2	22.719	−0.466	3.29	22.861
3	22.861	0.414	3.29	22.848

After three iterations, the Newton-Raphson method converges to an absolute value for the permeability of 22.848 md.

Equation 4-39 can be expanded to describe flow in any porous medium where the geometry of the system is not too complex to integrate. For example, the flow into a well bore is not linear, but is more often radial. Figure 4-24 illustrates the type of flow that is typical of that occurring in the vicinity of a producing well. For a radial flow, Darcy's equation in a differential form can be written as:

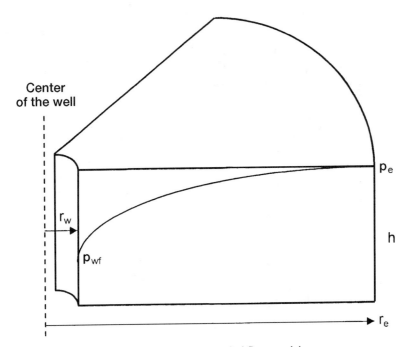

Figure 4-24. Radial flow model.

$$q = \frac{k\,A}{\mu} \frac{dp}{dr}$$

Integrating Darcy's equation gives:

$$q \int_{r_w}^{r_e} dr = \frac{kA}{\mu} \int_{P_{wf}}^{P_e} dp$$

The term dL has been replaced by dr as the length term has now become a radius term. The minus sign is no longer required for the radial system shown in Figure 4-24 as the radius increases in the same direction as the pressure. In other words, as the radius increases going away from the well bore, the pressure also increases. At any point in the reservoir, the cross-sectional area across which flow occurs will be the surface area of a cylinder, which is $2\pi rh$. Since the cross-sectional area is related to r, then A must be included within the integral sign as follows:

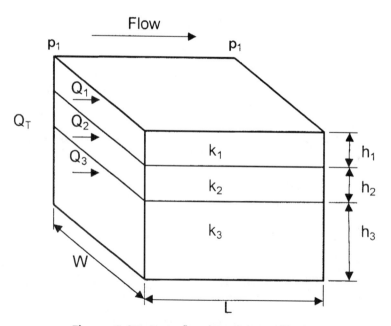

Figure 4-25. Linear flow through layered beds.

$$q \int_{r_w}^{r_e} \frac{dr}{2 \pi rh} = \frac{k}{\mu} \int_{p_{wf}}^{p_e} dp$$

rearranging

$$\frac{q}{2 \pi h} \int_{r_w}^{r_e} \frac{dr}{r} = \frac{k}{\mu} \int_{p_{wf}}^{p_e} dp$$

and integrating

$$\frac{q}{2 \pi h} (\ln r_e - \ln r_w) = \frac{k}{\mu} (p_e - p_{wf})$$

Solving for the flow rate, q, results in:

$$q = \frac{2 \pi kh (p_e - p_{wf})}{\mu \ln (r_e/r_w)} \qquad (4\text{-}49)$$

The above equation assumes that the reservoir is homogeneous and is completely saturated with a single liquid phase (appropriate modifications will be discussed in later sections to account for the presence of other fluids), where:

q = flow rate, reservoir cm^3/sec
k = absolute permeability, Darcy
h = thickness, cm
r_e = drainage radius, cm
r_w = well bore radius, cm
p_e = pressure at drainage radius, atm
p_{wf} = bottom-hole flowing pressure
μ = viscosity, cp

Averaging Absolute Permeabilities

The most difficult reservoir properties to determine usually are the level and distribution of the absolute permeability throughout the reservoir. They are more variable than porosity and more difficult to measure. Yet an adequate knowledge of permeability distribution is critical to the prediction of reservoir depletion by any recovery process. It is rare to encounter a homogeneous reservoir in actual practice. In many cases, the reservoir contains distinct layers, blocks, or concentric rings of varying permeabilities. Also, because smaller-scale heterogeneities always exist, core permeabilities must be averaged to represent the flow characteristics of the entire reservoir or individual reservoir layers (units). The proper way of averaging the permeability data depends on how permeabilities were distributed as the rock was deposited.

There are three simple permeability-averaging techniques that are commonly used to determine an appropriate average permeability to represent an equivalent homogeneous system. These are:

• Weighted-average permeability
• Harmonic-average permeability
• Geometric-average permeability

Weighted-Average Permeability

This averaging method is used to determine the average permeability of layered-parallel beds with different permeabilities. Consider the case where the flow system is comprised of three parallel layers that are separated from one another by thin impermeable barriers, i.e., no cross flow, as shown in Figure 4-25. All the layers have the same width w with a cross-sectional area of A.

The flow from each layer can be calculated by applying Darcy's equation in a linear form as expressed by Equation 4-40, to give:

Layer 1

$$q_1 = \frac{k_1 \, w \, h_1 \, \Delta p}{\mu \, L}$$

Layer 2

$$q_2 = \frac{k_2 \, w \, h_2 \, \Delta p}{\mu \, L}$$

Layer 3

$$q_3 = \frac{k_3 \, w \, h_3 \, \Delta p}{\mu \, L}$$

The total flow rate from the entire system is expressed as

$$q_t = \frac{k_{avg}\, w\, h_t\, \Delta p}{\mu L}$$

where q_t = total flow rate
k_{avg} = average permeability for the entire model
w = width of the formation
$\Delta p = p_1 \, B \, p_2$
h_t = total thickness

The total flow rate q_t is equal to the sum of the flow rates through each layer or:

$$q_t = q_1 + q_2 + q_3$$

Combining the above expressions gives:

$$\frac{k_{avg}\, w\, h_t\, \Delta p}{\mu L} = \frac{k_1\, w\, h_1\, \Delta p}{\mu L} + \frac{k_2\, w\, h_2\, \Delta p}{\mu L} + \frac{k_3\, w\, h_3\, \Delta p}{\mu L}$$

or

$$k_{avg}\, h_t = k_1\, h_1 + k_2\, h_2 + k_3\, h_3$$

$$k_{avg} = \frac{k_1\, h_1 + k_2\, h_2 + k_3\, h_3}{h_t}$$

The average absolute permeability for a parallel-layered system can be expressed in the following form:

$$k_{avg} = \frac{\displaystyle\sum_{j=1}^{n} k_j h_j}{\displaystyle\sum_{j=1}^{n} h_j} \qquad (4\text{-}50)$$

Equation 4-50 is commonly used to determine the average permeability of a reservoir from core analysis data.

Figure 4-26 shows a similar layered system with variable layers width. Assuming no cross-flow between the layers, the average permeability can be approximated in a manner similar to the above derivation to give:

$$k_{avg} = \frac{\sum\limits_{j=1}^{n} k_j A_j}{\sum\limits_{j=1}^{n} A_j} \qquad (4\text{-}51)$$

with

$$A_j = h_j w_j$$

where A_j = cross-sectional area of layer j
w_j = width of layer j

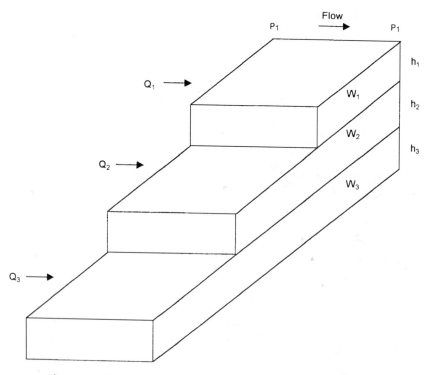

Figure 4-26. Linear flow through layered beds with variable area.

Example 4-11

Given the following permeability data from a core analysis report, calculate the average permeability of the reservoir.

Depth, ft	Permeability, md
3998-02	200
4002-04	130
4004-06	170
4006-08	180
4008-10	140

Solution

h_i, ft	k_i	$h_i k_i$
4	200	800
2	130	260
2	170	340
2	180	360
2	140	280
$h_t = 12$		• $h_i k_i = 2040$

$$k_{avg} = \frac{2040}{12} = 170 \text{ md}$$

Harmonic-Average Permeability

Permeability variations can occur laterally in a reservoir as well as in the vicinity of a well bore. Consider Figure 4-27 which shows an illustration of fluid flow through a series combination of beds with different permeabilities.

For a steady-state flow, the flow rate is constant and the total pressure drop Δp is equal to the sum of the pressure drops across each bed, or

$$\Delta p = \Delta p_1 + \Delta p_2 + \Delta p_3$$

Substituting for the pressure drop by applying Darcy's equation, i.e., Equation 4-40, gives:

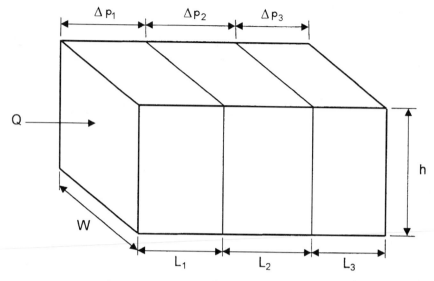

Figure 4-27. Linear flow through series beds.

$$\frac{q\mu L}{A\, k_{avg}} = \frac{q\mu\, L_1}{A\, k_1} + \frac{q\mu\, L_2}{A\, k_2} + \frac{q\mu\, L_3}{A\, k_3}$$

Canceling the identical terms and simplifying gives:

$$k_{avg} = \frac{L}{(L/k)_1 + (L/k)_2 + (L/k)_3}$$

The above equation can be expressed in a more generalized form to give:

$$k_{avg} = \frac{\displaystyle\sum_{i=1}^{n} L_i}{\displaystyle\sum_{i=1}^{n} (L/k)_i} \qquad (4\text{-}52)$$

where L_i = length of each bed
k_i = absolute permeability of each bed

In the radial system shown in Figure 4-28, the above averaging methodology can be applied to produce the following generalized expression:

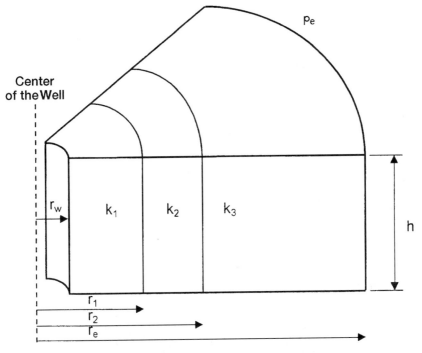

Figure 4-28. Flow through series beds.

$$k_{avg} = \frac{Ln \, (r_e/r_w)}{\sum\limits_{j=1}^{n} \left[\dfrac{Ln \, (r_j/r_{j-1})}{k_j} \right]} \tag{4-53}$$

The relationship in Equation 4-53 can be used as a basis for estimating a number of useful quantities in production work. For example, the effects of mud invasion, acidizing, or well shooting can be estimated from it.

Example 4-12

A hydrocarbon reservoir is characterized by five distinct formation segments that are connected in series. Each segment has the same formation thickness. The length and permeability of each section of the five-bed reservoir are given below:

Length, ft	Permeability, md
150	80
200	50
300	30
500	20
200	10

Calculate the average permeability of the reservoir by assuming:

a. Linear flow system
b. Radial flow system

Solution

For a linear system:

L_i, ft	k_i	L_i/k_i
150	80	1.8750
200	50	4.0000
300	30	10.000
500	20	25.000
200	10	20.000
1350		$\Sigma\, L_i/k_i = 60.875$

Using Equation 4-52 gives:

$$k_{avg} = \frac{1350}{60.875} = 22.18 \text{ md}$$

For a radial system:

The solution of the radial system can be conveniently expressed in the following tabulated form. The solution is based on Equation 4-53 and assuming a wellbore radius of 0.25 ft:

Segment	r_i, ft	$\ln(r_i/r_{iB1})$	k_i	$[\ln(r_i/r_{iB1})]/k_i$
well bore	0.25	—	—	—
1	150	6.397	80	0.080
2	350	0.847	50	0.017
3	650	0.619	30	10.021
4	1150	0.571	20	0.029
5	1350	0.160	10	0.016
				0.163

From Equation 4-53,

$$k_{avg} = \frac{\ln(1350/0.25)}{0.163} = 52.72 \text{ md}$$

Geometric-Average Permeability

Warren and Price (1961) illustrated experimentally that the most probable behavior of a heterogeneous formation approaches that of a uniform system having a permeability that is equal to the geometric average. The geometric average is defined mathematically by the following relationship:

$$k_{avg} = \exp\left[\frac{\sum_{i=1}^{n}(h_i \ln(k_i))}{\sum_{i=1}^{n} h_i}\right] \tag{4-54}$$

where k_i = permeability of core sample i
h_i = thickness of core sample i
n = total number of samples

If the thicknesses (h_i) of all core samples are the same, Equation 4-57 can be simplified as follows:

$$k_{avg} = (k_1 \, k_2 \, k_3 \dots k_n)^{\frac{1}{n}} \tag{4-55}$$

Example 4-13

Given the following core data, calculate the geometric average permeability:

Sample	h_i, ft	k_i, md
1	1.0	10
2	1.0	30
3	0.5	100
4	1.5	40
5	2.0	80
6	1.5	70
7	1.0	15
8	1.0	50
9	1.5	35
10	0.5	20

Solution

Sample	h_i, ft	k_i, md	$h_i \cdot Ln\,(k_i)$
1	1.0	10	2.303
2	1.0	30	3.401
3	0.5	100	2.303
4	1.5	40	5.533
5	2.0	80	8.764
6	1.5	70	6.373
7	1.0	15	2.708
8	1.0	50	3.912
9	1.5	35	5.333
10	0.5	20	1.498
	11.5		42.128

$$k_{avg} = \exp\left[\frac{42.128}{11.5}\right] = 39 \text{ md}$$

Absolute Permeability Correlations

The determination of connate water by capillary-pressure measurements has allowed the evaluation of connate-water values on samples of varying permeability and within a given reservoir to a wider extent and to a greater accuracy than was possible beforehand. These measurements have accumulated to the point where it is possible to correlate connate-water content with the permeability of the sample in a given reservoir and to a certain extent between reservoirs.

Calhoun (1976) suggested that in an ideal pore configuration of uniform structure, the irreducible connate water would be independent of permeability, lower permeabilities being obtained merely by a scaled reduction in particle size. In an actual porous system formed by deposition of graded particles or by some other natural means, the connate water might be expected to increase as permeability decreases. This conclusion results from the thought that lower permeabilities result from increasing non-uniformity of pore structure by a gradation of particles rather than by a scaled reduction of particles. In this sense, connate-water content is a function of permeability only insofar as permeability is dependent upon the variation of pore structure. Thus, for unconsolidated sands formed of uniform particles of one size, the connate-water content would be independent of permeability.

Calhoun (1976) pointed out that any correlation found between various reservoir properties would be anticipated to apply only within the rather narrow limits of a single reservoir or perhaps of a given formation. Beyond these bounds, a general correspondence between permeability and pore structure would not be known. It would be anticipated, however, that for formations of similar characteristics, a similar dependence of permeability on pore structure and, consequently, similar correlation of connate water and permeability would be found.

It has been generally considered for many years that connate water reached higher values in lower permeabilities. This observation amounted to nothing more than a trend. The data from capillary pressure measurements have indicated that the relationship is semi-logarithmic, although it is not yet certain from published data that this is the exact relationship. No generalizations are apparent from this amount of data, although it can now be quite generally stated that within a given reservoir the connate water (if an irreducible value) will increase proportionally to the decrease in the logarithm of the permeability. It is apparent, moreover, that one cannot state the value of connate water expected in any new formation unless one knows something of its pore makeup.

Experience indicates a general relationship between reservoir porosity (ϕ) and irreducible water saturation (S_{wc}) provided the rock type and/or the grain size does not vary over the zone of interest. This relationship is defined by the equation

$$C = (S_{wi})\,(\phi)$$

where C is a constant for a particular rock type and/or grain size.

Several investigators suggest that the constant C that describes the rock type can be correlated with the absolute permeability of the rock. Two commonly used empirical methods are the Timur equation and the Morris-Biggs equation.

The Timur Equation

Timur (1968) proposed the following expression for estimating the permeability from connate water saturation and porosity:

$$k = 8.58102 \, \frac{\phi^{4.4}}{S_{wc}^2} \tag{4-56}$$

The Morris-Biggs Equation

Morris and Biggs (1967) presented the following two expressions for estimating the permeability if oil and gas reservoirs:
For an oil reservoir:

$$k = 62.5 \left(\frac{\phi^3}{S_{wc}} \right)^2 \qquad (4 - 57)$$

For a gas reservoir:

$$k = 2.5 \left(\frac{\phi^3}{S_{wc}} \right)^2 \qquad (4 - 58)$$

where   k = absolute permeability, Darcy
ϕ = porosity, fraction
S_{wc} = connate-water saturation, fraction

Example 4-14

Estimate the absolute permeability of an oil zone with a connate-water saturation and average porosity of 25% and 19%, respectively.

Solution

Applying the Timur equation:

$$k = 8.58102 \frac{(0.19)^{4.4}}{(0.25)^2} = 0.0921 \text{ Darcy}$$

From the Morris and Biggs correlation:

$$k = 62.5 \left[\frac{(.29)^3}{0.25} \right]^2 = 0.047 \text{ Darcy}$$

In the previous discussion of Darcy's Law and absolute permeability measurements, it was assumed that the entire porous medium is fully saturated with a single phase, i.e., 100% saturation. In hydrocarbon reservoir, however, the rocks are usually saturated with two or more fluids.

Therefore, the concept of absolute permeability must be modified to describe the fluid flowing behavior when more than one fluid is present in the reservoir. If a core sample is partially saturated with a fluid (other than the test fluid) and both saturations are maintained constant throughout the flow, the measured permeability to the test fluid will be reduced below the permeability which could be measured if the core were 100 percent saturated with the test fluid.

As the saturation of a particular phase decreases, the permeability to that phase also decreases. The measured permeability is referred to as the *effective permeability* and is a relative measure of the conductance of the porous medium for one fluid when the medium is saturated with more than one fluid. This implies that the effective permeability is an associated property with each reservoir fluid, i.e., gas, oil, and water. These effective permeabilities for the three reservoir fluids are represented by:

k_g = effective gas permeability
k_o = effective oil permeability
k_w = effective water permeability

One of the phenomena of multiphase effective permeabilities is that the sum of the effective permeabilities is always less than or equal to the absolute permeability, i.e.,

$$k_g + k_o + k_w \leq k$$

The effective permeability is used mathematically in Darcy's Law in place of the absolute permeability. For example, the expression for flow through the linear system under a partial saturation of oil is written

$$q_o = \frac{k_o \, A \, (p_1 - p_2)}{\mu_o \, L} \qquad (4\text{-}59)$$

where q_o = oil flow rate, cc/sec
μ_o = oil viscosity, cm
k_o = oil effective permeability, Darcys

Effective permeabilities are normally measured directly in the laboratory on small core samples. Owing to the many possible combinations of saturation for a single medium, however, laboratory data are usually summarized and reported as relative permeability. Relative permeability is defined as the ratio of the effective permeability to a given fluid at a definite saturation to the permeability at 100% saturation. The terminology most widely used is simply k_g/k, k_o/k, k_w/k, meaning the relative permeability to gas, oil, and water, respectively. Since k is a constant for a given porous material, the relative permeability varies with the fluid saturation in the same fashion as does the effective permeability. The relative permeability to a fluid will vary from a value of zero at some low saturation of that fluid to a value of 1.0 at 100% saturation of that fluid. Thus, the relative permeability can be expressed symbolically as

$$k_{rg} = \frac{k_g}{k}$$

$$k_{ro} = \frac{k_o}{k}$$

$$k_{rw} = \frac{k_w}{k}$$

which are relative permeabilities to gas, oil, and water, respectively. A comprehensive treatment of the relative permeability is presented in Chapter 5.

ROCK COMPRESSIBILITY

A reservoir thousands of feet underground is subjected to an overburden pressure caused by the weight of the overlying formations. Overburden pressures vary from area to area depending on factors such as depth, nature of the structure, consolidation of the formation, and possibly the geologic age and history of the rocks. Depth of the formation is the most important consideration, and a typical value of overburden pressure is approximately one psi per foot of depth.

The weight of the overburden simply applies a compressive force to the reservoir. The pressure in the rock pore spaces does not normally approach the overburden pressure. A typical pore pressure, commonly referred to as the reservoir pressure, is approximately 0.5 psi per foot of depth, assuming that the reservoir is sufficiently consolidated so the overburden pressure is not transmitted to the fluids in the pore spaces.

The pressure difference between overburden and internal pore pressure is referred to as the *effective overburden* pressure. During pressure depletion operations, the internal pore pressure decreases and, therefore, the effective overburden pressure increases. This increase causes the following effects:

• The bulk volume of the reservoir rock is reduced.
• Sand grains within the pore spaces expand.

These two volume changes tend to reduce the pore space and, therefore, the porosity of the rock. Often these data exhibit relationships with both porosity and the effective overburden pressure. Compressibility typically decreases with increasing porosity and effective overburden pressure.

Geertsma (1957) points out that there are three different types of compressibility that must be distinguished in rocks:

- **Rock-matrix compressibility, c_r**

  Is defined as the fractional change in volume of the solid rock material (grains) with a unit change in pressure. Mathematically, the rock compressibility coefficient is given by

$$c_r = -\frac{1}{V_r}\left(\frac{\partial V_r}{\partial p}\right)_T \tag{4-60}$$

  where c_r = rock-matrix compressibility, psi^{-1}
  V_r = volume of solids

  The subscript T indicates that the derivative is taken at constant temperature.

- **Rock-bulk compressibility, c_B**

  Is defined as the fractional change in volume of the bulk volume of the rock with a unit change in pressure. The rock-bulk compressibility is defined mathematically by:

$$c_B = -\frac{1}{V_B}\left(\frac{\partial V_B}{\partial p}\right)_T \tag{4-61}$$

  where c_B = rock-bulk compressibility coefficient, psi^{-1}
  V_B = bulk volume

- **Pore compressibility, c_p**

  The pore compressibility coefficient is defined as the fractional change in pore volume of the rock with a unit change in pressure and given by the following relationship:

$$c_p = \frac{-1}{V_p}\left(\frac{\partial V_p}{\partial p}\right)_T \tag{4-62}$$

where p = pore pressure, psi
  c_p = pore compressibility coefficient, psi^{-1}
  V_p = pore volume

Equation 4-62 can be expressed in terms of the porosity ϕ by noting that ϕ increases with the increase in the pore pressure; or:

$$c_p = \frac{1}{\phi} \frac{\partial \phi}{\partial p}$$

For most petroleum reservoirs, the rock and bulk compressibility are considered small in comparison with the pore compressibility c_p. The *formation compressibility* c_f is the term commonly used to describe the total compressibility of the formation and is set equal to c_p, i.e.:

$$c_f = c_p = \frac{1}{\phi} \frac{\partial \phi}{\partial p} \tag{4-63}$$

Typical values for the formation compressibility range from 3×10^{-6} to 25×10^{-6} psi^{-1}. Equation 4-62 can be rewritten as:

$$c_f = \frac{1}{V_p} \frac{\Delta V_p}{\Delta p}$$

or

$$\Delta V_p = c_f V_P \Delta p \tag{4-64}$$

where ΔV_p and Δp are the change in the pore volume and pore pressure, respectively.

Geertsma (1957) suggested that the bulk compressibility c_B is related to the pore compressibility c_p by the following expression.

$$c_B \cong c_p \, \phi \tag{4-65}$$

Geertsma has stated that in a reservoir only the vertical component of hydraulic stress is constant and that the stress components in the horizontal plane are characterized by the boundary condition that there is no bulk deformation in those directions. For those boundary conditions, he developed the following approximation for sandstones:

$$c_p \text{ (reservoir)} = 1/2 \, c_p \text{ (laboratory)}$$

Example 4-15

Calculate the reduction in the pore volume of a reservoir due to a pressure drop of 10 psi. The reservoir original pore volume is one million barrels with an estimated formation compressibility of 10×10^{-6} psi^{-1}

Solution

Applying Equation 4-64 gives

$$\Delta V_p = (10 \times 10^{-6})(1 \times 10^6)(10) = 100 \text{ bbl}$$

Although the above value is small, it becomes an important factor in undersaturated reservoirs when calculations are made to determine initial oil-in-place and aquifer contents.

The reduction in the pore volume due to pressure decline can also be expressed in terms of the changes in the reservoir porosity. Equation 4-63 can be rearranged, to give:

$$c_f \, \partial p = \left(\frac{1}{\phi}\right) \partial \phi$$

Integrating the above relation gives:

$$c_f \int_{p_o}^{p} \partial p = \int_{\phi_o}^{\phi} \frac{\partial \phi}{\phi}$$

$$c_f (p - p_o) = \ln\left(\frac{\phi}{\phi_o}\right)$$

or:

$$\phi = \phi_o e^{c_f(p - p_o)} \tag{4-66}$$

where p_o = original pressure, psi
ϕ_o = original porosity
p = current pressure, psi
ϕ = porosity at pressure p

Noting that the e^x expansion series is expressed as:

$$e^x = 1 + x + \frac{x^2}{2!} + \frac{x^3}{3!} + \dots$$

Using the expansion series and truncating the series after the first two terms, gives:

$$\phi = \phi_o [1 + c_f (p - p_o)] \tag{4-67}$$

Example 4-16

Given the following data:

- $c_f = 10 \times 10^{-6}$
- original pressure = 5000 psi
- original porosity = 18%
- current pressure = 4500 psi

Calculate the porosity at 4500 psi.

Solution

$$\phi = 0.18 [1 + (10 \times 10^{-6})(4500 - 5000)] = 0.179$$

It should be pointed out that the total reservoir compressibility c_t is extensively used in the transient flow equation and the material balance equation as defined by the following expression:

$$c_t = S_o c_o + S_w c_s + S_g c_g + c_f \tag{4-68}$$

where S_o, S_w, S_g = oil, water, and gas saturation
c_o = oil compressibility, psi^{-1}
c_w = water compressibility, psi^{-1}
c_g = gas compressibility, psi^{-1}
c_t = total reservoir compressibility

For undersaturated oil reservoirs, the reservoir pressure is above the bubble-point pressure, i.e., no initial gas cap, which reduces Equation 4-68 to:

$$c_t = S_o c_o + S_w c_w + c_f$$

In general, the formation compressibility c_f is the same order of magnitude as the compressibility of the oil and water and, therefore, cannot be regulated.

Several authors have attempted to correlate the pore compressibility with various parameters including the formation porosity. Hall (1953) correlated the pore compressibility with porosity as given by the following relationship:

$$c_f = (1.782/\phi^{0.438})\,10^{-6} \tag{4-69}$$

where c_f = formation compressibility, psi^{-1}
ϕ = porosity, fraction

Newman (1973) used 79 samples for consolidated sandstones and limestones to develop a correlation between the formation compressibility and porosity. The proposed generalized hyperbolic form of the equation is:

$$c_f = \frac{a}{[1 + cb\phi]}$$

where

For consolidated sandstones

$a = 97.32 \times 10^{-6}$
$b = 0.699993$
$c = 79.8181$

For limestones

$a = 0.8535$
$b = 1.075$
$c = 2.202 \times 10^{6}$

Example 4-17

Estimate the compressibility coefficient of a sandstone formation that is characterized by a porosity of 0.2, using:

a. Hall's correlation
b. Newman's correlation

Solution

a. Hall's correlations:

$$c_f = (1.782/0.2^{0.438})\ 10^{-6} = 3.606 \times 10^{-6}\ psi^{-1}$$

b. Newman's correlation:

$$c_f = \frac{97.32 \times 10^{-6}}{[1 + (0.699993)(79.8181)(0.2)]^{1/0.699993}} = 2.74 \times 10^{-6}\ psi^{-1}$$

NET PAY THICKNESS

A fundamental prerequisite to reservoir performance prediction is a satisfactory knowledge of the volume of oil originally in place. The reservoir is necessarily confined to certain geologic and fluid boundaries, i.e., GOC, WOC, and GWC, so accuracy is imperative. Within the confines of such boundaries, oil is contained in what is commonly referred to as *Gross Pay*. *Net Pay* is that part of the reservoir thickness which contributes to oil recovery and is defined by imposing the following criteria:

• Lower limit of porosity
• Lower limit of permeability
• Upper limit of water saturation

All available measurements performed on reservoir samples and in wells, such as core analysis and well logs, are extensively used in evaluating the reservoir net thickness.

The choice of lower limits of porosity and permeability will depend upon such individual characteristics as

• Total reservoir volume
• Total range of permeability values
• Total range of porosity values
• Distribution of the permeability and porosity values

RESERVOIR HETEROGENEITY

It has been proposed that most reservoirs are laid down in a body of water by a long-term process, spanning a variety of depositional environments, in both time and space. As a result of subsequent physical and chemical reorganization, such as compaction, solution, dolomitization and cementation, the reservoir characteristics are further changed. Thus the heterogeneity of reservoirs is, for the most part, dependent upon the depositional environments and subsequent events.

The main geologic characteristic of all the physical rock properties that have a bearing on reservoir behavior when producing oil and gas is the extreme variability in such properties within the reservoir itself, both laterally and vertically, and within short distances. It is important to recognize that there are no homogeneous reservoirs, only varying degrees of heterogeneity.

The reservoir heterogeneity is then defined as a variation in reservoir properties as a function of space. Ideally, if the reservoir is homogeneous, measuring a reservoir property at any location will allow us to fully describe the reservoir. The task of reservoir description is very simple for homogeneous reservoirs. On the other hand, if the reservoir is heterogeneous, the reservoir properties vary as a function of a spatial location. These properties may include permeability, porosity, thickness, saturation, faults and fractures, rock facies and rock characteristics. For a proper reservoir description, we need to predict the variation in these reservoir properties as a function of spatial locations. There are essentially two types of heterogeneity:

• Vertical heterogeneity
• Areal heterogeneity

Geostatistical methods are used extensively in the petroleum industry to quantitatively describe the two types of the reservoir heterogeneity. It is obvious that the reservoir may be nonuniform in all intensive properties such as permeability, porosity, wettability, and connate water saturation. We will discuss heterogeneity of the reservoir in terms of permeability.

Vertical Heterogeneity

One of the first problems encountered by the reservoir engineer in predicting or interpreting fluid displacement behavior during secondary recovery and enhanced oil recovery processes is that of organizing and using the large amount of data available from core analysis. Permeabilities pose particular problems in organization because they usually vary by more than an order of magnitude between different strata. The engineer must be able then to:

• Describe the degree of the vertical heterogeneity in mathematical terms, and
• Describe and define the proper permeability stratification of the pay zone. This task is commonly called the *zoning or layering problem*.

It is appropriate to be able to describe the degree of heterogeneity within a particular system in quantitative terms. The *degree of homogeneity* of a reservoir property is a number that characterizes the departure from uniformity or constancy of that particular measured property through the thickness of reservoir. A formation is said to have a uniformity coefficient of zero in a specified property when that property is constant throughout the formation thickness. A completely heterogeneous formation has a uniformity coefficient of unity. Between the two extremes, formations have uniformity coefficients comprised between zero and one. The following are the two most widely used descriptors of the vertical heterogeneity of the formation:

• Dykstra-Parsons permeability variation V
• Lorenz coefficient L

The Dykstra-Parsons Permeability Variation

Dykstra and Parsons (1950) introduced the concept of the permeability variation coefficient V which is a statistical measure of non-uniformity of a set of data. It is generally applied to the property of permeability but can be extended to treat other rock properties. It is generally recognized that the permeability data are log-normally distributed. That is, the geologic processes that create permeability in reservoir rocks appear to leave permeabilities distributed around the geometric mean. Dykstra and Parsons recognized this feature and introduced the permeability variation that characterizes a particular distribution. The required computational steps for determining the coefficient V are summarized below:

Step 1. Arrange the core samples in decreasing permeability sequence, i.e., descending order.

Step 2. For each sample, calculate the percentage of thickness with permeability greater than this sample.

Step 3. Using a log-probability graph paper, plot permeability values on the log scale and the % of thickness on the probability scale. This special graph paper is shown in Figure 4-29.

Step 4. Draw the best straight line through the points.

Step 5. Read the corresponding permeability values at 84.1% and 50% of thickness. These two values are designated as $k_{84.1}$ and k_{50}.

Step 6. The Dykstra-Parsons permeability variation is defined by the following expression:

$$V = \frac{k_{50} - k_{84.1}}{k_{50}} \qquad (4-70)$$

Example 4-18

The following conventional core analysis data are available from three wells:

Well #1			Well #2			Well #3		
Depth ft	k md	φ %	Dept ft	k md	φ %	Dept ft	k md	φ %
5389–5391	166	17.4	5397–5398.5	72	15.7	5401–5403	28	14.0
–5393	435	18.0	–539.95	100	15.6	–5405	40	13.7
–5395	147	16.7	–5402	49	15.2	–5407	20	12.2
–5397	196	17.4	–5404.5	90	15.4	–5409	32	13.6
–5399	254	19.2	–5407	91	16.1	–5411	35	14.2
–5401	105	16.8	–5409	44	14.1	–5413	27	12.6
–5403	158	16.8	–5411	62	15.6	–5415	27	12.3
–5405	153	15.9	–5413	49	14.9	–5417	9	10.6
–5406	128	17.6	–5415	49	14.8	–5419	30	14.1
–5409	172	17.2	–5417	83	15.2			

Calculate the Dykstra-Parsons permeability variation.

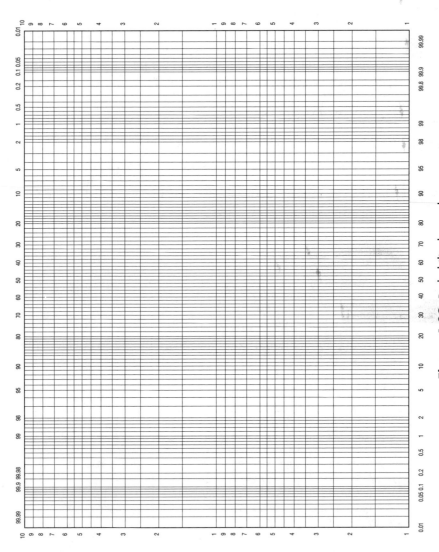

Figure 4-29. Probability-log scale.

Solution

Step 1. Arrange the entire permeability data in a descending order and calculate % of thickness with greater permeability as shown below:

k md	h ft	h with greater k	% of h with greater k
435	2	0	0
254	2	2	3.6
196	2	4	7.1
172	3	6	10.7
166	2	9	16.1
158	2	11	19.6
153	2	13	23.2
147	2	15	26.8
128	1	17	30.4
105	2	18	32.1
100	1	20	35.7
91	2.5	21	37.5
90	2.5	23.5	42.0
83	2	26	46.4
72	1.5	28	50
62	2	29.5	52.7
49	6.5	31.5	56.3
44	2	38	67.9
40	2	40	71.4
35	2	42	75.0
32	2	44	78.6
30	2	46	82.1
28	2	48	85.7
27	2	50	89.3
20	2	52	92.9
9	2	54	96.4

Total = 56′

Step 2. Plot the permeability versus % of thickness with greater k on a log-probability scale as shown in Figure 4-30 and read

$k_{50} = 68$ md

$k_{84.1} = 29.5$

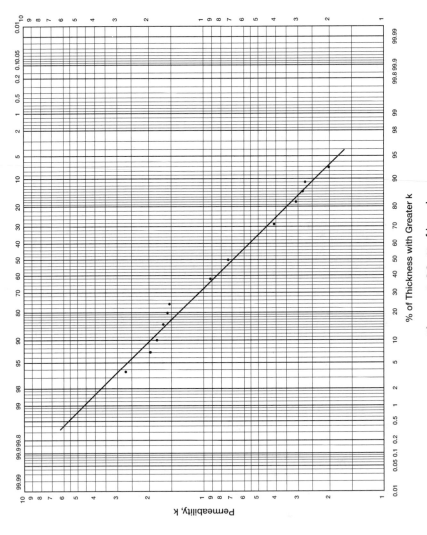

Figure 4-30. % of h vs. k.

Step 3. Calculate V by applying Equation 4-70.

$$V = \frac{68 - 29.5}{68} = 0.57$$

It should be noted that if all the permeabilities are equal, the numerator or Equation 4-70 would be zero, and the V would also be zero. This would be the case for a completely homogeneous system. The Dykstra-Parsons method is commonly referred to as a *Permeability Ordering Technique.*

In water flooding calculations, it is frequently desired to divide the reservoir into layers that have equal thickness and different permeability. The log-probability scale can be used in this case to assign the permeability scale into equal percent increments and to read the corresponding permeability at the midpoint of each interval.

Example 4-19

Using the data given in Example 4-18, determine the average layer permeability for a 10-layered system, assuming a uniform porosity.

Solution

Using the Dykstra-Parsons's log-probability plot as shown in Figure 4-30, determine the permeability for the 10-layered system as follows:

Layer	% Probability	k, md
1	5	265
2	15	160
3	25	120
4	35	94
5	45	76
6	55	60
7	65	49
8	75	39
9	85	29
10	95	18

Although permeability and porosity are not related in a strict technical sense, they should correlate in rock of similar lithology and pore size distribution. In many cases, the logarithm of permeability versus porosity plots is frequently made and the best straight line is drawn through the points.

Lorenz Coefficient L

Schmalz and Rahme (1950) introduced a single parameter that describes the degree of heterogeneity within a pay zone section. The term is called *Lorenz coefficient* and varies between zero, for a completely homogeneous system, to one for a completely heterogeneous system.

The following steps summarize the methodology of calculating Lorenz coefficient:

Step 1. Arrange all the available permeability values in a descending order.

Step 2. Calculate the cumulative permeability capacity Σkh and cumulative volume capacity $\Sigma \phi h$.

Step 3. Normalize both cumulative capacities such that each cumulative capacity ranges from 0 to 1.

Step 4. Plot the normalized cumulative permeability capacity versus the normalized cumulative volume capacity on a Cartesian scale.

Figure 4-31 shows an illustration of the flow capacity distribution. A completely uniform system would have all permeabilities equal, and a plot of the normalized Σkh versus $\Sigma \phi h$ would be a straight line. Figure 4-31 indicates that as the degree of contrast between high and low values of permeability increases the plot exhibits greater concavity toward the upper left corner. This would indicate more heterogeneity, i.e., the severity of deviation from a straight line is an indication of the degree of heterogeneity. The plot can be used to describe the reservoir heterogeneity quantitatively by calculating the Lorenz coefficient. The coefficient is defined by the following expression:

$$L = \frac{\text{Area above the straight line}}{\text{Area below the straight line}} \tag{4-71}$$

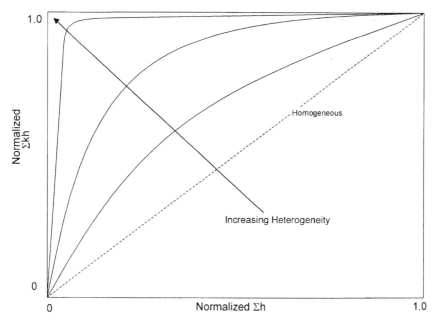

Figure 4-31. Normalized flow capacity.

where the Lorenz coefficient L can vary between 0 and 1.

0 = completely homogeneous
1 = completely heterogeneous

Figure 4-32 shows the relation of the permeability variation V and Lorenz coefficient L for log-normal permeability distributions as proposed by Warren and Price (1961). This relationship can be expressed mathematically by the following two expressions:

Lorenz coefficient in terms of permeability variation:

$$L = 0.0116356 + 0.339794V + 1.066405V^2 - 0.3852407V^3 \qquad (4\text{-}72)$$

Permeability variation in terms of Lorenz coefficient:

$$V = -5.05971(10^{-4}) + 1.747525L - 1.468855\,L^2 + 0.701023\,L^3 \quad (4\text{-}73)$$

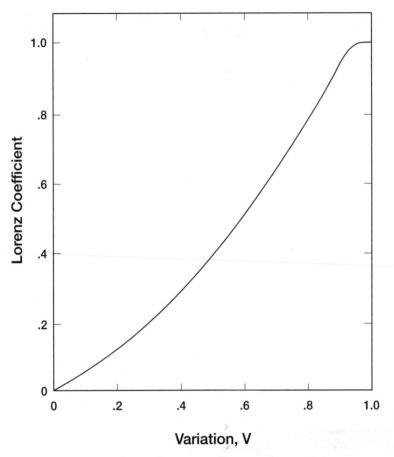

Figure 4-32. Correlation of Lorenz coefficient and permeability variation.

The above two expressions are applicable between $0 < L < 1$ and $0 < V < 1$.

Example 4-20

Using the data given in Example 4-18, calculate the Lorenz coefficient assuming a uniform porosity.

Solution

Step 1. Tabulate the permeability data in a descending order and calculate the normalized Σkh and Σh as shown below:

k, md	h, ft	kh	Σkh	Σkh/5646.5	Σh	Σh/56
435	2	870	870	0.154	2	0.036
254	2	508	1378	0.244	4	0.071
196	2	392	1770	0.313	6	0.107
172	3	516	2286	0.405	9	0.161
166	2	332	2618	0.464	11	0.196
158	2	316	2934	0.520	13	0.232
153	2	306	3240	0.574	15	0.268
147	2	294	3534	0.626	17	0.304
128	1	128	3662	0.649	18	0.321
105	2	210	3872	0.686	20	0.357
100	1	100	3972	0.703	21	0.375
91	2.5	227.5	4199.5	0.744	23.5	0.420
90	2.5	225	4424.5	0.784	26	0.464
83	2	166	4590.5	0.813	28	0.50
72	1.5	108	4698.5	0.832	29.5	0.527
62	2	124	4822.5	0.854	31.5	0.563
49	6.5	294	5116.5	0.906	38.0	0.679
44	2	88	5204.5	0.922	40.0	0.714
40	2	80	5284.5	0.936	42	0.750
35	2	70	5354.4	0.948	44	0.786
32	2	64	5418.5	0.960	46	0.821
30	2	60	5478.5	0.970	48	0.857
28	2	56	5534.5	0.980	50	0.893
27	2	54	5588.5	0.990	52	0.929
20	2	40	5628.5	0.997	54	0.964
9	2	18	5646.5	1.000	56	1.000

Step 2. Plot the normalized capacities on a Cartesian scale as shown in Figure 4-33.

Step 3. Calculate the Lorenz coefficient by dividing the area above the straight line (area A) by the area under the straight line (area B) to give:

$$L = 0.42$$

A plot of the cumulative permeability capacity Σkh versus Σh (without normalization) is commonly constructed, as shown in Figure 4-34, and used to assign average permeability values for a selected number of reservoir layers. If the intervals of the thickness are chosen, as shown in Figure 4-34, then the average values of permeability for each thickness interval (layer) can be calculated by dividing the incremental (kh) by the incremental thickness.

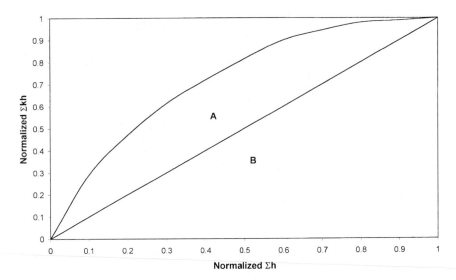

Figure 4-33. Normalized flow capacity for Example 4-20.

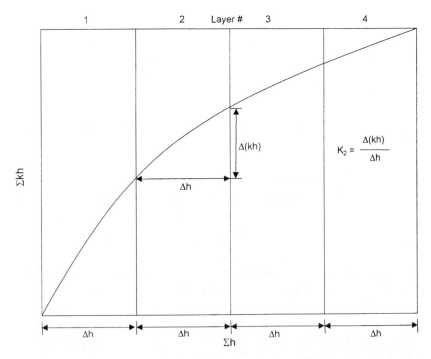

Figure 4-34. Cumulative permeability capacity vs. cumulative thickness.

It should be noted that it is not necessary that equal thickness sections be chosen. They may be selected at irregular increments as desired. There are also some advantages of selecting layer properties so that each layer has the same permeability thickness product.

Example 4-21

Using the data given in Example 4-18, calculate the average permeability for a 10-layered system reservoir. Compare the results with those of the Dykstra-Parsons method.

Solution

Step 1. Using the calculated values of Σkh and Σh of Example 4-20, plot Σkh versus Σh on a Cartesian coordinate as shown in Figure 4-35.

Step 2. Divide the x-axis into 10 equal segments*, each with 5.6 ft.

Step 3. Calculate the average permeability $\bar{k}$ for each interval, to give:

Layer	$\bar{k}$	$\bar{k}$ from Dykstra-Parsons, Example 4-19
1	289	265
2	196.4	160
3	142.9	120
4	107.1	94
5	83.9	76
6	67.9	60
7	44.6	49
8	35.7	39
9	32.1	29
10	17.2	18

The permeability sequencing (ordering) methods of zonation do not consider the physical location of the rocks with the vertical column. All the data are considered to be a statistical sampling, which will describe the statistical distribution of permeability, porosity, and thickness within the reservoir. All the values of equal permeability are presumed to be in communication with each other.

---

*It should be noted that the 56 feet *do not* equal the reservoir net thickness. It essentially represents the cumulative thickness of the core samples.

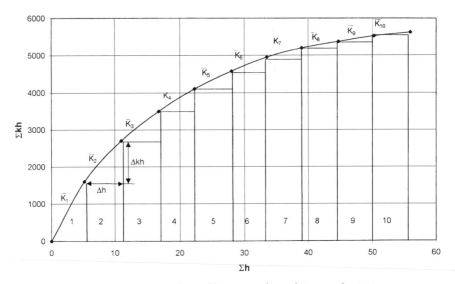

Figure 4-35. Cumulative kh vs. cumulative h (Example 4-21).

Miller and Lents (1947) suggested that the fluid movement in the reservoir remains in the same relative vertical position, i.e., remains in the same elevation, and that the permeability in this elevation (layer) is better described by the *geometric mean average permeability.* This method is called the *positional method.* Thus, to describe the layering system, or a reservoir using the positional approach, it is necessary to calculate the geometric mean average permeability (Equations 4-54 and 4-55) for each elevation and treat each of these as an individual layer.

AREAL HETEROGENEITY

Since the early stages of oil production, engineers have recognized that most reservoirs vary in permeability and other rock properties in the lateral direction. To understand and predict the behavior of an underground reservoir, one must have as accurate and detailed knowledge as possible of the subsurface. Indeed, water and gas displacement is conditioned by the storage geometry (structural shape, thickness of strata) and the local values of the physical parameters (variable from one point to another) characteristic of the porous rock. Hence, prediction accuracy is closely related to the detail in which the reservoir is described.

Johnson and co-workers (1966) devised a well testing procedure, called *pulse testing,* to generate rock properties data between wells. In this procedure, a series of producing rate changes or pluses is made at one well with the response being measured at adjacent wells. The technique provides a measure of the formation flow capacity (kh) and storage capacity (ϕh). The most difficult reservoir properties to define usually are the level and distribution of permeability. They are more variable than porosity and more difficult to measure. Yet an adequate knowledge of permeability distribution is critical to the prediction of reservoir depletion by any recovery process.

A variety of geostatistical estimation techniques has been developed in an attempt to describe accurately the spatial distribution of rock properties. The concept of spatial continuity suggests that data points close to one another are more likely to be similar than are data points farther apart from one another. One of the best geostatistical tools to represent this continuity is a visual map showing a data set value with regard to its location. Automatic or computer contouring and girding is used to prepare these maps. These methods involve interpolating between known data points, such as elevation or permeability, and extrapolating beyond these known data values. These rock properties are commonly called regionalized variables. These variables usually have the following contradictory characteristics:

• A random characteristic showing erratic behavior from point to point
• A structural characteristic reflecting the connections among data points

For example, net thickness values from a limited number of wells in a field may show randomness or erratic behavior. They also can display a connecting or smoothing behavior as more wells are drilled or spaced close together.

To study regionalized variables, a proper formulation must take this double aspect of randomness and structure into account. In geostatistics, a variogram is used to describe the randomness and spatial correlations of the regionalized variables.

There are several conventional interpolation and extrapolation methods that can be applied to values of a regionalized variable at different locations. Most of these methods use the following generalized expression:

$$Z^*(x) = \sum_{i=1}^{n} \lambda_i \, Z(x_i) \qquad (4\text{-}74)$$

with

$$\sum_{i-1}^{n} \lambda_i = 1 \qquad\qquad (4\text{-}75)$$

where $Z^*(x)$ = estimate of the regionalized variable at location x
  $Z(x_i)$ = measured value of the regionalized variable at position x_i
  λ_i = weight factor
  n = number of nearby data points

The difference between the commonly used interpolation and extrapolation methods is in the mathematical algorithm employed to compute the weighting factors λ_i. Compared to other interpolation methods, the geostatistical originality stems from the intuition that the accuracy of the estimation at a given point (and the λ_i) depends on two factors, the first one being of geometrical nature, the second related to the statistical spatial characteristics of the considered phenomenon.

The first factor is the geometry of the problem that is the relative positions of the measured points to the one to be estimated. When a point is well surrounded by experimental points, it can be estimated with more accuracy than one located in an isolated area. This fact is taken into account by classical interpolation methods (polynomial, multiple regression, least-squares) but these appear to be inapplicable as soon as the studied phenomenon shows irregular variations or measurement errors.

Three simple conventional interpolation and/or extrapolation methods are briefly discussed below:

• **The Polygon Method**
  This technique is essentially based on assigning the nearest measured value of the regionalized variable to the designated location. This implies that all the weighting factors, i.e., λ_i, in Equation 4-72 are set equal to zero except the corresponding λ_i for the nearest point is set equal to one.

• **The Inverse Distance Method**
  With *inverse distance,* data points are weighted during interpolation such that the influences of one data point relative to another declines with distance from the desired location.

The inverse distance method assigns a weight factor λ_i to each measured regionalized variable by the inverse distance between the measured value and the point being estimated, or

$$\lambda_i = \left(\frac{1}{d_i}\right) \bigg/ \sum_{i=1}^{n}\left(\frac{1}{d_i}\right) \tag{4-76}$$

where d_i = distance between the measured value and location of interest
n = number of nearby points

• The Inverse Distance Squared Method

The method assigns a weight to each measured regionalized variable by the inverse distance squared of the sample to the point being estimated, i.e.,

$$\lambda_i = \left(\frac{1}{d_i}\right)^2 \bigg/ \sum_{i=1}^{n}\left(\frac{1}{d_i}\right)^2 \tag{4-77}$$

While this method accounts for all nearby wells with recorded rock properties, it gives proportionately more weight to near wells than the previous method.

Example 4-22

Figure 4-36 shows a schematic illustration of the locations of four wells and distances between the wells and point x. The average permeability in each well location is given below:

Well #	Permeability, md
1	73
2	110
3	200
4	140

Estimate the permeability at location x by the polygon and the two inverse distance methods.

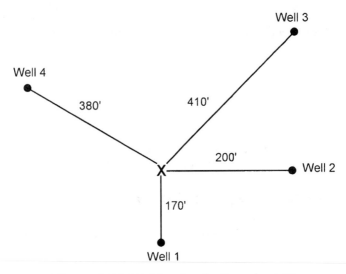

Figure 4-36. Well locations for Examples 4-22.

Solution

The Polygon Method

The nearest well location to point x is Well #1 with a distance of 170 ft. The recorded average permeability at this well is 73 md; therefore, the permeability in location x is

$$k = (1)(73) + (0)(110) + (0)(200) + (0)(140) = 73 \text{ md}$$

The Inverse Distance Method

Step 1. Calculate the weighting factors by applying Equation 4-76

Distance d_i ft	$1/d_i$	$\lambda_i = \left(\dfrac{1}{d_i} \right) \Big/ 0.0159$	k, md
170	0.0059	0.3711	73
200	0.0050	0.3145	110
410	0.0024	0.1509	200
380	0.0026	0.1635	140
	Sum = 0.0159		

Step 2. Estimate the permeability at location x by applying Equation 4-74

$$k = (0.3711)(73) + (0.3145)(110) + (0.1509)(200) + (0.1635)$$
$$(140) = 114.8 \text{ md}$$

The Inverse Distance Squared

Step 1. Apply Equation 4-77 to determine the weighting factors.

d_i ft	$\left(\dfrac{1}{d_i}\right)^2$	$\lambda_i = \left(\dfrac{1}{d_i}\right)^2 / 0.000073$	k, md
170	0.000035	0.4795	73
200	0.000025	0.3425	110
410	0.000006	0.0822	200
380	0.000007	0.958	140

Sum = 0.000073

Step 2. Estimate the permeability in location x by using Equation 4-72

$$k = (0.4795)(73) + (0.3425)(110) + (0.0822)(200) + (0.0958)$$
$$(140) = 102.5 \text{ md}$$

PROBLEMS

1. Given:

$p_i = 3500$ $p_b = 3500$ $T = 160°F$
$A = 1000$ acres $h = 25$ ft $S_{wi} = 30\%$
$\phi = 12\%$ $API = 45°$ $R_{sb} = 750$ scf/STB
$\gamma_g = 0.7$

Calculate:

a. Initial oil in place as expressed in STB
b. Volume of gas originally dissolved in the oil

2. The following measurements on pay zone are available:

Sample	Thickness, ft	ϕ, %	S_{oi}, %
1	2	12	75
2	3	16	74
3	1	10	73
4	4	14	76
5	2	15	75
6	2	15	72

Calculate:

a. Average porosity
b. Average oil and water saturations (assuming no gas).

3. The capillary pressure data for a water-oil system are given below:

S_w	P_c
0.25	35
0.30	16
0.40	8.5
0.50	5
1.0	0

The core sample used in generalizing the capillary pressure data was taken from a layer that is characterized by an absolute permeability of 300 md and a porosity of 17%. Generate the capillary pressure data for a different layer that is characterized by a porosity and permeability of 15%, 200 md, respectively. The interfacial tension is measured at 35 dynes/cm.

4. A five-layer oil reservoir is characterized by a set of capillary pressure-saturation curves as shown in Figure 4-6. The following additional data are also available:

Layer	Depth, ft	Permeability
1	6000–6016	10
2	6016–6025	300
3	6025–6040	100
4	6040–6055	30
5	6055–6070	3

- WOC = 6070 ft
- Water density = 65 lb/ft^3
- Oil density = 32 lb/ft^3

Calculate and plot the water and oil saturation profiles for this reservoir.

5. Assuming a steady-state laminar flow, calculate the permeability from the following measurement made on core sample by using air.

flow rate = 2 cm^3/sec   T = 65°F
upstream pressure = 2 atm   downstream pressure = 1 atm
A = 2 cm^2 L = 3 cm   viscosity = 0.018 cp

6. Calculate average permeability from the following core analysis data.

Depth, ft	k, md
4000–4002	50
4002–4005	20
4005–4006	70
4006–4008	100
4008–4010	85

7. Calculate the average permeability of a formation that consists of four beds in series, assuming:

a. Linear system
b. Radial system with $r_w = 0.3$ and $r_e = 1450$ ft.

Bed	Length of bed Linear or radial	k, md
1	400	70
2	250	400
3	300	100
4	500	60

8. Estimate the absolute permeability of a formation that is characterized by an average porosity and connate water saturation of 15% and 20% md, respectively.

9. Given:

Depth, ft	k, md
4100–4101	295
4101–4102	262
4102–4103	88
4103–4104	87
4104–4105	168
4105–4106	71
4106–4107	62
4107–4108	187
4108–4109	369
4109–4110	77
4110–4111	127
4111–4112	161
4112–4113	50
4113–4114	58
4114–4115	109
4115–4116	228
4116–4117	282
4117–4118	776
4118–4119	87
4119–4120	47
4120–4121	16
4121–4122	35
4122–4123	47
4123–4124	54
4124–4125	273
4125–4126	454
4126–4127	308
4127–4128	159
4128–4129	178

Calculate:

a. Average permeability
b. Permeability variation
c. Lorenz coefficient
d. Assuming four-layer reservoir system with equal length, calculate the permeability for each layer.

10. Three layers of 4, 6, and 10 feet thick respectively, are conducting fluid in parallel flow.

The depth to the top of the first layer is recorded as 5,012 feet. Core analysis report shows the following permeability data for each layer.

Layer #1		Layer #2		Layer #3	
Depth ft	Permeability md	Depth ft	Permeability md	Depth ft	Permeability md
5012–5013	485	5016–5017	210	5022–5023	100
5013–5014	50	5017–5018	205	5023–5024	95
5014–5015	395	5018–5019	60	5024–5025	20
5015–5016	110	5019–5020	203	5025–5026	96
		5020–5021	105	5026–5027	98
		5021–5022	195	5027–5028	30
				5028–5029	89
				5029–5030	86
				5030–5031	90
				5031–5032	10

Calculate the *average* permeability of the entire pay zone (i.e., 5012–5032′).

11. A well has a radius of 0.25 ft and a drainage radius of 660 ft. The sand that penetrates is 15 ft thick and has an absolute permeability of 50 md. The sand contains crude oil with the following PVT properties.

Pressure psia	B_o bbl/STB	μ_o cp
3500	1.827	1.123
3250	1.842	1.114
3000	1.858	1.105
2746*	1.866	1.100
2598	1.821	1.196
2400	1.771	1.337
2200	1.725	1.497
600	1.599	2.100

*Bubble point

The reservoir pressure (i.e., p_e) and the bubble-point pressure are 3500 and 2746 psia, respectively. If the bottom-hole flowing pressure is 2500 psia, calculate the oil-flow rate.

12. Test runs on three core samples from three wells in the mythical field yielded the following three sets of values for water saturation (S_w),

porosity (ϕ), and permeability (k). It is believed that these three properties can be used to determine the recovery fraction (RF).

	Core 1	Core 2	Core 3
ϕ	.185	.157	.484
S_w	0.476	.527	.637
k	.614	.138	.799
Recovery factor	.283	.212	.141

The recovery factor can be expressed by the following equation:

$$RF = a_o \, \phi + a_1 \, S_w + a_2 \, k$$

where a_o, a_1, and a_2 are constants

Calculate RF if:

$S_w = .75$, $\phi = .20$, and $k = .85$

REFERENCES

1. Calhoun, J. R., *Fundamentals of Reservoir Engineering.* University of Oklahoma Press, 1976.

2. Cole, Frank, *Reservoir Engineering Manual.* Houston: Gulf Publishing Company, 1969.

3. Dykstra, H., and Parsons, R. L., "The Prediction of Oil Recovery by Water Flood," *In Secondary Recovery of Oil in the United States,* 2nd ed., pp. 160–174. API, 1950.

4. Geertsma, J., "The Effect of Fluid Pressure Decline on Volumetric Changes of Porous Rocks," *Trans. AIME,* 1957, pp. 210, 331–340.

5. Hall, H. N., "Compressibility of Reservoir Rocks," *Trans. AIME,* 1953, p. 309.

6. Hustad, O., and Holt, H., "Gravity Stable Displacement of Oil by Gas after WaterFlooding," SPE Paper 24116, SPE/DOE Symposium on EOR, Tulsa, OK, April 22–24, 1972.

7. Johnson C. R., Careenkorn, R. A. and Woods, E. G., "Pulse Testing: A New Method for Describing Reservoir Flow Properties Between Wells," *JPT,* Dec. 1966, pp. 1599–1604

8. Jones, S. C., "A Rapid Accurate Unsteady State Klinkenberg Parameter," *SPEJ,* 1972, Vol. 12, No. 5, pp. 383–397.

9. Klinkenberg, L. J., "The Permeability of Porous Media to Liquids and Gases," API *Drilling and Production Practice,* 1941, p. 200.

10. Leverett, M. C., "Capillary Behavior in Porous Solids," *Trans. AIME,* 1941.

11. McCardell, W. M., "A Review of the Physical Basis for the Use of the J-Function," Eighth Oil Recovery Conference, Texas Petroleum Research Committee, 1955.

12. Miller, M. G., and Lents, M. R., "Performance of Bodcaw Reservoir, Cotton Valley Field Cycling Project: New Methods of Predicting Gas-Condensate Reservoir," *SPEJ,* Sept. 1966, pp. 239.

13. Morris, R. L., and Biggs, W. P., "Using Log-Derived Values of Water Saturation and Porosity." *SPWLA,* Paper X, 1967.

14. Newman, G. H., "Pore-Volume Compressibility," *JPT,* Feb. 1973, pp. 129–134

15. Schmalz, J. P., and Rahme, H. D., "The Variation of Waterflood Performance With Variation in Permeability Profile," *Prod. Monthly,* 1950, Vol. 15, No. 9, pp. 9–12.

16. Timur, A., "An Investigation of Permeability, Porosity, and Residual Water Saturation Relationships," *AIME,* June 1968.

17. Warren, J. E. and Price, H. S., "Flow in Heterogeneous Porous Media," *SPEJ,* Sept. 1961, pp. 153–169.

RELATIVE PERMEABILITY CONCEPTS

Numerous laboratory studies have concluded that the effective permeability of any reservoir fluid is a function of the reservoir fluid saturation and the wetting characteristics of the formation. It becomes necessary, therefore, to specify the fluid saturation when stating the effective permeability of any particular fluid in a given porous medium. Just as k is the accepted universal symbol for the absolute permeability, k_o, k_g, and k_w are the accepted symbols for the effective permeability to oil, gas, and water, respectively. The saturations, i.e., S_o, S_g, and S_w, must be specified to completely define the conditions at which a given effective permeability exists.

Effective permeabilities are normally measured directly in the laboratory on small core plugs. Owing to many possible combinations of saturation for a single medium, however, laboratory data are usually summarized and reported as relative permeability.

The absolute permeability is a property of the porous medium and is a measure of the capacity of the medium to transmit fluids. When two or more fluids flow at the same time, the relative permeability of each phase at a specific saturation is the ratio of the effective permeability of the phase to the absolute permeability, or:

$$k_{ro} = \frac{k_o}{k}$$

$$k_{rg} = \frac{k_g}{k}$$

$$k_{rw} = \frac{k_w}{k}$$

where k_{ro} = relative permeability to oil
  k_{rg} = relative permeability to gas
  k_{rw} = relative permeability to water
  k = absolute permeability
  k_o = effective permeability to oil for a given oil saturation
  k_g = effective permeability to gas for a given gas saturation
  k_w = effective permeability to water at some given water saturation

For example, if the absolute permeability k of a rock is 200 md and the effective permeability k_o of the rock at an oil saturation of 80 percent is 60 md, the relative permeability k_{ro} is 0.30 at $S_o = 0.80$.

Since the effective permeabilities may range from zero to k, the relative permeabilities may have any value between zero and one, or:

$$0 \leq k_{rw}, k_{ro}, k_{rg} \leq 1.0$$

It should be pointed out that when three phases are present the *sum of the relative permeabilities* ($k_{ro} + k_{rg} + k_{rw}$) *is both variable and always less than or equal to unity.* An appreciation of this observation and of its physical causes is a prerequisite to a more detailed discussion of two- and three-phase relative permeability relationships.

It has become a common practice to refer to the relative permeability curve for the nonwetting phase as k_{nw} and the relative permeability for the wetting phase as k_w.

TWO-PHASE RELATIVE PERMEABILITY

When a wetting and a nonwetting phase flow together in a reservoir rock, each phase follows separate and distinct paths. The distribution of the two phases according to their wetting characteristics results in characteristic wetting and nonwetting phase relative permeabilities. Since the wetting phase occupies the smaller pore openings at small saturations, and these pore openings do not contribute materially to flow, it follows that the presence of a small wetting phase saturation will affect the nonwetting phase permeability only to a limited extent. Since the nonwetting phase occupies the central or larger pore openings which contribute materially to fluid flow through the reservoir, however, a small nonwetting phase saturation will drastically reduce the wetting phase permeability.

Figure 5-1 presents a typical set of relative permeability curves for a water-oil system with the water being considered the wetting phase. Figure 5-1 shows the following four distinct and significant points:

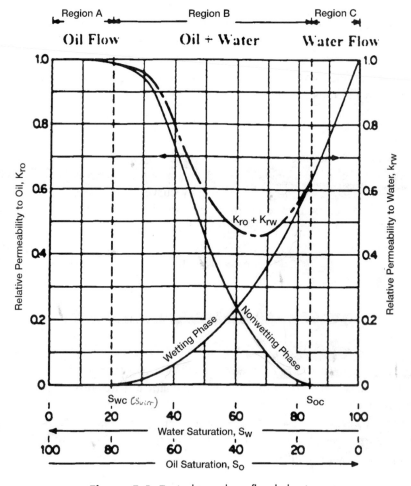

Figure 5-1. Typical two-phase flow behavior.

✱ Water-oil, rel. perm. Curve

- **Point 1**

Point 1 on the wetting phase relative permeability shows that a small saturation of the nonwetting phase will drastically reduce the relative permeability of the wetting phase. The reason for this is that the nonwetting phase occupies the larger pore spaces, and it is in these large pore spaces that flow occurs with the least difficulty.

- **Point 2**

Point 2 on the nonwetting phase relative permeability curve shows that the nonwetting phase begins to flow at the relatively low saturation of the nonwetting phase. The saturation of the oil at this point is called *critical oil saturation* S_{oc}.

- **Point 3**

Point 3 on the wetting phase relative permeability curve shows that the wetting phase will cease to flow at a relatively large saturation. This is because the wetting phase preferentially occupies the smaller pore spaces, where capillary forces are the greatest. The saturation of the water at this point is referred to as the *irreducible water saturation* S_{wir} or *connate water saturation* S_{wi}—both terms are used interchangeably.

- **Point 4**

Point 4 on the nonwetting phase relative permeability curve shows that, at the lower saturations of the wetting phase, changes in the wetting phase saturation have only a small effect on the magnitude of the nonwetting phase relative permeability curve. The reason for the phenomenon at Point 4 is that at the low saturations the wetting phase fluid occupies the small pore spaces which do not contribute materially to flow, and therefore changing the saturation in these small pore spaces has a relatively small effect on the flow of the nonwetting phase.

This process could have been visualized in reverse just as well. It should be noted that this example portrays oil as nonwetting and water as wetting. The curve shapes shown are typical for wetting and nonwetting phases and may be mentally reversed to visualize the behavior of an oil-wet system. Note also that the total permeability to both phases, $k_{rw} + k_{ro}$, is less than 1, in regions B and C.

The above discussion may be also applied to gas-oil relative permeability data, as can be seen for a typical set of data in Figure 5-2. Note that this might be termed gas-liquid relative permeability since it is plotted versus the liquid saturation. This is typical of gas-oil relative permeability data in the presence of connate water. Since the connate (irreducible) water normally occupies the smallest pores in the presence of oil

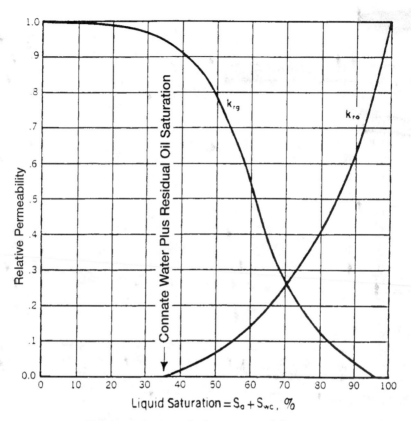

Figure 5-2. Gas-oil relative permeability curves.

and gas, it appears to make little difference whether water or oil that would also be immobile in these small pores occupies these pores. Consequently, in applying the gas-oil relative permeability data to a reservoir, the total liquid saturation is normally used as a basis for evaluating the relative permeability to the gas and oil.

Note that the relative permeability curve representing oil changes completely from the shape of the relative permeability curve for oil in the water-oil system. In the water-oil system, as noted previously, oil is normally the nonwetting phase, whereas in the presence of gas the oil is the wetting phase. Consequently, in the presence of water only, the oil relative permeability curve takes on an S shape whereas in the presence of gas the oil relative-permeability curve takes on the shape of the wetting phase, or is concave upward. Note further that the critical gas saturation S_{gc} is generally very small.

Another important phenomenon associated with fluid flow through porous media is the concept of residual saturations. As when one immiscible fluid is displacing another, it is impossible to reduce the saturation of the displaced fluid to zero. At some small saturation, which is presumed to be the saturation at which the displaced phase ceases to be continuous, flow of the displaced phase will cease. This saturation is often referred to as the *residual* saturation. This is an important concept as it determines the maximum recovery from the reservoir. Conversely, a fluid must develop a certain minimum saturation before the phase will begin to flow. This is evident from an examination of the relative permeability curves shown in Figure 5-1. The saturation at which a fluid will just begin to flow is called the *critical* saturation.

Theoretically, the critical saturation and the residual saturation should be exactly equal for any fluid; however, they are not identical. **Critical saturation is measured in the direction of increasing saturation, while irreducible saturation is measured in the direction of reducing saturation.** Thus, the saturation histories of the two measurements are different.

As was discussed for capillary-pressure data, there is also a saturation history effect for relative permeability. The effect of saturation history on relative permeability is illustrated in Figure 5-3. If the rock sample is initially saturated with the wetting phase (e.g., water) and relative-permeability data are obtained by decreasing the wetting-phase saturation while flowing nonwetting fluid (e.g., oil) in the core, the process is classified as *drainage* or *desaturation.*

If the data are obtained by increasing the saturation of the wetting phase, the process is termed *imbibition* or *resaturation.* The nomenclature is consistent with that used in connection with capillary pressure. This difference in permeability when changing the saturation history is called *hysteresis.* Since relative permeability measurements are subject to hysteresis, it is important to duplicate, in the laboratory, the saturation history of the reservoir.

Drainage Process

It is generally agreed that the pore spaces of reservoir rocks were originally filled with water, after which oil moved into the reservoir, displacing some of the water, and reducing the water to some residual saturation. When discovered, the reservoir pore spaces are filled with a connate water saturation and an oil saturation. If gas is the displacing agent, then gas moves into the reservoir, displacing the oil.

This same history must be duplicated in the laboratory to eliminate the effects of hysteresis. The laboratory procedure is to first saturate the core with water, then displace the water to a residual, or connate, water saturation with oil after which the oil in the core is displaced by gas. This flow process is called the gas drive, or drainage, depletion process. In the gas drive depletion process, the nonwetting phase fluid is continuously increased, and the wetting phase fluid is continuously decreased.

Imbibition Process

The *imbibition process* is performed in the laboratory by first saturating the core with the water (wetting phase), then displacing the water to its irreducible (connate) saturation by injection oil. This "drainage" procedure is designed to establish the original fluid saturations that are found when the reservoir is discovered. The wetting phase (water) is reintroduced into the core and the water (wetting phase) is continuously increased. This is the imbibition process and is intended to produce the relative permeability data needed for water drive or water flooding calculations.

Figure 5-3 schematically illustrates the difference in the drainage and imbibition processes of measuring relative permeability. It is noted that the imbibition technique causes the nonwetting phase (oil) to lose its mobility at higher values of water saturation than does the drainage process. The two processes have similar effects on the wetting phase (water) curve. The drainage method causes the wetting phase to lose its mobility at higher values of wetting-phase saturation than does the imbibition method.

Two-phase Relative Permeability Correlations

In many cases, relative permeability data on actual samples from the reservoir under study may not be available, in which case it is necessary to obtain the desired relative permeability data in some other manner. Field relative permeability data can usually be calculated, and the procedure will be discussed more fully in Chapter 6. The field data are unavailable for future production, however, and some substitute must be devised. Several methods have been developed for calculating relative permeability relationships. Various parameters have been used to calculate the relative permeability relationships, including:

• Residual and initial saturations
• Capillary pressure data

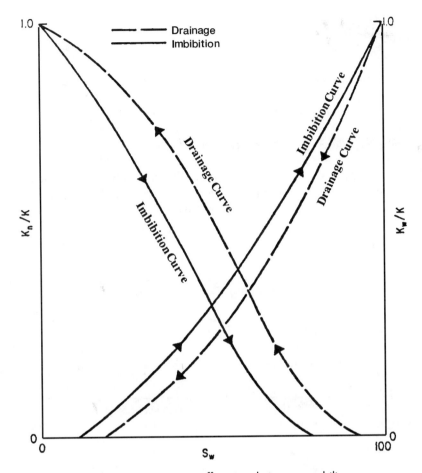

Figure 5-3. Hysteresis effects in relative permeability.

In addition, most of the proposed correlations use the effective phase saturation as a correlating parameter. The effective phase saturation is defined by the following set of relationships:

$$S_o^* = \frac{S_o}{1 - S_{wc}} \qquad (5\text{-}1)$$

$$S_w^* = \frac{S_w - S_{wc}}{1 - S_{wc}} \qquad (5\text{-}2)$$

$$S_g^* = \frac{S_g}{1 - S_{wc}} \tag{5-3}$$

where S_o^*, S_w^*, S_g^* = effective oil, water, and gas saturation, respectively
  S_o, S_w, S_g = oil, water and gas saturation, respectively
    S_{wc} = connate (irreducible) water saturation

1. Wyllie and Gardner Correlation

Wyllie and Gardner (1958) observed that, in some rocks, the relationship between the reciprocal capillary pressure squared $(1/P_c^2)$ and the effective water saturation S_w^* is linear over a wide range of saturation. Honapour et al. (1988) conveniently tabulated Wyllie and Gardner correlations as shown below:

| Drainage Oil-Water Relative Permeabilities | | | |
Type of formation	k_{ro}	k_{rw}	Equation
Unconsolidated sand, well sorted	$(1 - S_w^*)$	$(S_w^*)^3$	(5-4)
Unconsolidated sand, poorly sorted	$(1 - S_w^*)^2 (1 - S_w^{*\,1.5})$	$(S_w^*)^{3.5}$	(5-5)
Cemented sandstone, oolitic limestone	$(1 - S_w^*)^2 (1 - S_w^{*\,2})$	$(S_w^*)^4$	(5-6)

| Drainage Gas-Oil Relative Permeabilities | | | |
Type of formation	k_{ro}	k_{rg}	Equation
Unconsolidated sand, well sorted	$(S_o^*)^3$	$(1 - S_o^*)^3$	(5-7)
Unconsolidated sand, poorly sorted	$(S_o^*)^{3.5}$	$(1 - S_o^*)^2 (1 - S_o^{*\,1.5})$	(5-8)
Cemented sandstone, oolitic limestone, rocks with vugular porosity	$(S_o^*)^4$	$(1 - S_o^*)^2 (1 - S_o^{*\,2})$	(5-9)

Wyllie and Gardner have also suggested the following two expressions that can be used when one relative permeability is available:

• Oil-water system

$$k_{rw} = (S_w^*)^2 - k_{ro} \left[\frac{S_w^*}{1 - S_w^*} \right] \tag{5-10}$$

• Gas-oil system

$$k_{ro} = (S_o^*) - k_{rg} \left[\frac{S_o^*}{1 - S_o^*} \right] \qquad (5\text{-}11)$$

2. Torcaso and Wyllie Correlation

Torcaso and Wyllie (1958) developed a simple expression to determine the relative permeability of the oil phase in a gas-oil system. The expression permits the calculation of k_{ro} from the measurements of k_{rg}. The equation has the following form:

$$k_{ro} = k_{rg} \left[\frac{(S_o^*)^4}{(1 - S_o^*)^2 (1 - (S_o^*)^2)} \right] \qquad (5\text{-}12)$$

The above expression is very useful since k_{rg} measurements are easily made and k_{ro} measurements are usually made with difficulty.

3. Pirson's Correlation

From petrophysical considerations, Pirson (1958) derived generalized relationships for determining the wetting and nonwetting phase relative permeability for both imbibition and drainage processes. The generalized expressions are applied for water-wet rocks.

For the water (wetting) phase

$$k_{rw} = \sqrt{S_w^*} \; S_w^3 \qquad (5\text{-}13)$$

The above expression is valid for both the imbibition and drainage processes.

For the nonwetting phase

• Imbibition

$$(k_r)_{nonwetting} = \left[1 - \left(\frac{S_w - S_{wc}}{1 - S_{wc} - S_{nw}} \right) \right]^2 \qquad (5\text{-}14)$$

• **Drainage**

$$(k_r)_{nonwetting} = (1 - S_w^*)\left[1 - (S_w^*)^{0.25}\sqrt{S_w}\right]^{0.5} \qquad (5\text{-}15)$$

where S_{nw} = saturation of the nonwetting phase
S_w = water saturation
S_w^* = effective water saturation as defined by Equation 5-2

Example 5-1

Generate the drainage relative permeability data for an unconsolidated well-sorted sand by using the Wyllie and Gardner method. Assume the following critical saturation values:

$$S_{oc} = 0.3, \quad S_{wc} = 0.25, \quad S_{gc} = 0.05$$

Solution

Generate the oil-water relative permeability data by applying Equation 5-4 in conjunction with Equation 5-2, to give:

S_w	$S_w^* = \dfrac{S_w - S_{wc}}{1 - S_{wc}}$	$k_{ro} = (1 - S_w^*)^3$	$K_{rw} = (S_w^*)^3$
0.25	0.0000	1.000	0.0000
0.30	0.0667	0.813	0.0003
0.35	0.1333	0.651	0.0024
0.40	0.2000	0.512	0.0080
0.45	0.2667	0.394	0.0190
0.50	0.3333	0.296	0.0370
0.60	0.4667	0.152	0.1017
0.70	0.6000	0.064	0.2160

Apply Equation 5-7 in conjunction with Equation 5-1 to generate relative permeability data for the gas-oil system.

S_g	$S_o = 1 - S_g - S_{wc}$	$S_o^* = \dfrac{S_o}{1-S_{wc}}$	$k_{ro} = (S_o^*)^3$	$k_{rg} = (1 - S_o^*)^3$
0.05	0.70	0.933	0.813	—
0.10	0.65	0.867	0.651	0.002
0.20	0.55	0.733	0.394	0.019
0.30	0.45	0.600	0.216	0.064
0.40	0.35	0.467	0.102	0.152
0.50	0.25	0.333	0.037	0.296
0.60	0.15	0.200	0.008	0.512
0.70	0.05	0.067	0.000	0.813

Example 5-2

Resolve Example 5-1 by using Pirson's correlation for the water-oil system.

Solution

S_w	$S_w^* = \dfrac{S_w - S_{wc}}{1 - S_{wc}}$	$k_{rw} = \sqrt{S_w^*}\, S_w^3$	$k_{ro} = (1 - S_w^*)\left[1 - (S_w^*)^{0.25}\sqrt{S_w}\right]^{0.5}$
0.25	0.0000	0.000	1.000
0.30	0.0667	0.007	0.793
0.35	0.1333	0.016	0.695
0.40	0.2000	0.029	0.608
0.45	0.2667	0.047	0.528
0.50	0.3333	0.072	0.454
0.60	0.4667	0.148	0.320
0.70	0.6000	0.266	0.205

4. Corey's Method

Corey (1954) proposed a simple mathematical expression for generating the relative permeability data of the gas-oil system. The approximation is good for drainage processes, i.e., gas-displacing oil.

$$k_{ro} = (1 - S_g^*)^4 \qquad (5\text{-}16)$$

$$k_{rg} = (S_g^*)(2 - S_g^*) \qquad (5\text{-}17)$$

where the effective gas saturation S_g^* is defined in Equation 5-3.

Example 5-3

Use Corey's approximation to generate the gas-oil relative permeability for a formation with a connate water saturation of 0.25.

Solution

S_g	$S_g^* = \dfrac{S_g}{1 - S_{wc}}$	$k_{ro} = (1 - S_g^*)^4$	$k_{rg} = (S_g^*)^3 (2 - S_g^*)$
0.05	0.0667	0.759	0.001
0.10	0.1333	0.564	0.004
0.20	0.2667	0.289	0.033
0.30	0.4000	0.130	0.102
0.40	0.5333	0.047	0.222
0.50	0.6667	0.012	0.395
0.60	0.8000	0.002	0.614
0.70	0.9333	0.000	0.867

5. Relative Permeability from Capillary Pressure Data

Rose and Bruce (1949) showed that capillary pressure p_c is a measure of the fundamental characteristics of the formation and could also be used to predict the relative permeabilities. Based on the concepts of tortuosity, Wyllie and Gardner (1958) developed the following mathematical expression for determining the drainage water-oil relative permeability from capillary pressure data:

$$k_{rw} = \left(\frac{S_w - S_{wc}}{1 - S_{wc}} \right)^2 \frac{\displaystyle\int_{S_{wc}}^{S_w} dS_w / p_c^2}{\displaystyle\int_{S_{wc}}^{1} dS_w / p_c^2} \qquad (5\text{-}18)$$

$$k_{ro} = \left(\frac{1 - S_w}{1 - S_{wc}} \right)^2 \frac{\displaystyle\int_{S_w}^{1} dS_w / p_c^2}{\displaystyle\int_{S_{wc}}^{1} dS_w / (p_c)^2} \qquad (5\text{-}19)$$

Wyllie and Gardner also presented two expressions for generating the oil and gas relative permeabilities in the presence of the connate water saturation. The authors considered the connate water as part of the rock matrix to give:

$$k_{ro} = \left(\frac{S_o - S_{or}}{1 - S_{or}}\right)^2 \frac{\int_0^{S_o} dS_o/p_c^2}{\int_0^1 dS_o/p_c^2} \qquad (5\text{-}20)$$

$$k_{rg} = \left(1 - \frac{S_o - S_{or}}{S_g - S_{gc}}\right)^2 \frac{\int_{S_o}^1 dS_o/p_c^2}{\int_0^1 dS_o/p_c^2} \qquad (5\text{-}21)$$

where S_{gc} = critical gas saturation
S_{wc} = connate water saturation
S_{or} = residual oil saturation

Example 5-4

The laboratory capillary pressure curve for a water-oil system between the connate water saturation and a water saturation of 100% is represented by the following linear equation:

$$P_c = 22 - 20\,S_w$$

The connate water saturation is 30%. Using Wyllie and Gardner methods, generate the relative permeability data for the oil-water system.

Solution

Step 1. Integrate the capillary pressure equation, to give:

$$I = \int_a^b \frac{dS_w}{(22 - 20\,S_w)^2} = \left[\frac{1}{440 - 400b}\right] - \left[\frac{1}{440 - 400a}\right]$$

Step 2. Evaluate the above integral at the following limits:

$$\bullet \int_{0.3}^{1} \frac{dS_w}{(22-20\,S_w)^2} = \left[\frac{1}{440-400(1)} - \frac{1}{440-400(0.3)} \right] = 0.02188$$

$$\bullet \int_{.3}^{S_w} \frac{dS_w}{(22-20\,S_w)^2} = \left[\frac{1}{440-400\,S_w} - 0.00313 \right]$$

$$\bullet \int_{S_w}^{1} \frac{dS_w}{(22-20\,S_w)^2} = \left[0.025 - \frac{1}{440-400\,S_w} \right]$$

Step 3. Construct the following working table:

S_w	k_{rw} Equation 5-18	k_{ro} Equation 5-19
0.3	0.0000	1.0000
0.4	0.0004	0.7195
0.5	0.0039	0.4858
0.6	0.0157	0.2985
0.7	0.0466	0.1574

6. Relative Permeability from Analytical Equations

Analytical representations for individual-phase relative permeabilities are commonly used in numerical simulators. The most frequently used functional forms for expressing the relative-permeability and capillary-pressure data are given below:

Oil-Water Systems:

$$k_{ro} = (k_{ro})_{S_{wc}} \left[\frac{1-S_w-S_{orw}}{1-S_{wc}-S_{orw}} \right]^{n_o} \qquad (5-22)$$

$$k_{rw} = (k_{rw})_{S_{orw}} \left[\frac{S_w - S_{wc}}{1 - S_{wc} - S_{orw}} \right]^{n_w} \tag{5-23}$$

$$p_{cwo} = (p_c)_{S_{wc}} \left(\frac{1 - S_w - S_{orw}}{1 - S_{wc} - S_{orw}} \right)^{n_p} \tag{5-24}$$

Gas-Oil Systems:

$$k_{ro} = (k_{ro})_{S_{gc}} \left[\frac{1 - S_g - S_{lc}}{1 - S_{gc} - S_{lc}} \right]^{n_{go}} \tag{5-25}$$

$$k_{rg} = (k_{rg})_{S_{wc}} \left[\frac{S_g - S_{gc}}{1 - S_{lc} - S_{gc}} \right]^{n_g} \tag{5-26}$$

$$p_{cgo} = (p_c)_{S_{lc}} \left[\frac{S_g - S_{gc}}{1 - S_{lc} - S_{gc}} \right]^{n_{pg}} \tag{5-27}$$

with

$$S_{lc} = S_{wc} + S_{org}$$

where S_{lc} = total critical liquid saturation
$(k_{ro})_{S_{wc}}$ = oil relative permeability at connate water saturation
$(k_{ro})_{S_{gc}}$ = oil relative permeability at critical gas saturation
S_{orw} = residual oil saturation in the water-oil system
S_{org} = residual oil saturation in the gas-oil system
S_{gc} = critical gas saturation
$(k_{rw})_{S_{orw}}$ = water relative permeability at the residual oil saturation
n_o, n_w, n_g, n_{go} = exponents on relative permeability curves
p_{cwo} = capillary pressure of water-oil systems

$(p_c)_{S_{wc}}$ = capillary pressure at connate water saturation

n_p = exponent of the capillary pressure curve for the oil-water system

p_{cgo} = capillary pressure of gas-oil system

n_{p_g} = exponent of the capillary pressure curve in gas-oil system

$(p_c)_{S_{lc}}$ = capillary pressure at critical liquid saturation.

The exponents and coefficients of Equations 5-22 through 5-26 are usually determined by the least-squares method to match the experimental or field relative permeability and capillary pressure data.

Figures 5-4 and 5-5 schematically illustrate the key critical saturations and the corresponding relative permeability values that are used in Equations 5-22 through 5-27.

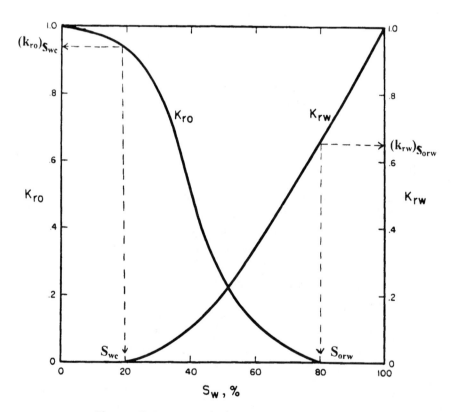

Figure 5-4. Water-oil relative permeability curves.

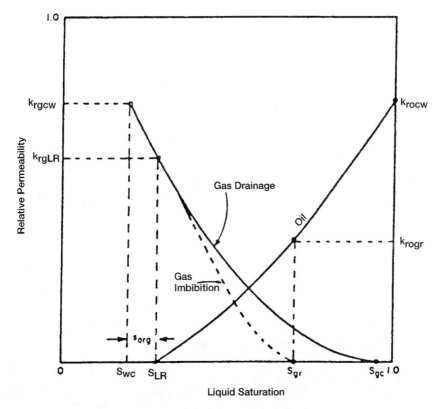

Figure 5-5. Gas-oil relative permeability curves.

Example 5-5

Using the analytical expressions of Equations 5-22–5-27, generate the relative permeability and capillary pressure data. The following information on the water-oil and gas-oil systems is available:

$$S_{wc} = 0.25 \qquad S_{orw} = 0.35 \qquad S_{gc} = 0.05 \qquad S_{org} = .23$$
$$(k_{ro})_{S_{wc}} = 0.85 \qquad (k_{rw})_{S_{orw}} = 0.4 \qquad (P_c)_{S_{wc}} = 20 \text{ psi}$$
$$(k_{ro})_{S_{gc}} = 0.60 \qquad (k_{rg})_{S_{wc}} = 0.95$$
$$n_o = 0.9 \qquad n_w = 1.5 \qquad n_p = 0.71$$
$$n_{go} = 1.2 \qquad n_g = 0.6 \qquad (p_c)_{S_{lc}} = 30 \text{ psi}$$
$$n_{pg} = 0.51$$

Solution

Step 1. Calculate residual liquid saturation S_{lc}.

$$S_{lc} = S_{wc} + S_{org}$$
$$= 0.25 + 0.23 = 0.48$$

Step 2. Generate relative permeability and capillary pressure data for oil-water system by applying Equations 5-22 through 5-24.

S_w	k_{ro} Equations 5-22	k_{rw} Equation 5-23	P_c Equation 5-24
0.25	0.850	0.000	20.00
0.30	0.754	0.018	18.19
0.40	0.557	0.092	14.33
0.50	0.352	0.198	9.97
0.60	0.131	0.327	4.57
0.65	0.000	0.400	0.00

Step 3. Apply Equations 5-25 through 5-27 to determine the relative permeability and capillary data for the gas-oil system.

S_g	k_{ro} Equation 5-25	k_{rg} Equation 5-26	P_c Equation 5-27
0.05	0.600	0.000	0.000
0.10	0.524	0.248	9.56
0.20	0.378	0.479	16.76
0.30	0.241	0.650	21.74
0.40	0.117	0.796	25.81
0.52	0.000	0.95	30.00

RELATIVE PERMEABILITY RATIO

Another useful relationship that derives from the relative permeability concept is the relative (or effective) permeability ratio. This quantity lends itself more readily to analysis and to the correlation of flow performances than does relative permeability itself. The relative permeability ratio expresses the ability of a reservoir to permit flow of one fluid as related to its ability to permit flow of another fluid under the same circumstances. The two most useful permeability ratios are k_{rg}/k_{ro} the relative permeability to gas with respect to that to oil and k_{rw}/k_{ro} the relative permeability to water with respect to that to oil, it being understood that both quantities in the ratio are determined simultaneously on a given system. The relative permeability ratio may vary in magnitude from zero to infinity.

In describing two-phase flow mathematically, it is always the relative permeability ratio (e.g., k_{rg}/k_{ro} or k_{ro}/k_{rw}) that is used in the flow equations. Because the wide range of the relative permeability ratio values, the permeability ratio is usually plotted on the log scale of semilog paper as a function of the saturation. Like many relative permeability ratio curves, the central or the main portion of the curve is quite linear.

Figure 5-6 shows a plot of k_{rg}/k_{ro} versus gas saturation. It has become common usage to express the central straight-line portion of the relationship in the following analytical form:

$$\frac{k_{rg}}{k_{ro}} = a\, e^{bS_g} \tag{5-28}$$

The constants a and b may be determined by selecting the coordinate of two different points on the straight-line portion of the curve and substituting in Equation 5-28. The resulting two equations can be solved simultaneously for the constants a and b. To find the coefficients of Equation 5-28 for the straight-line portion of Figure 5-6, select the following two points:

Point 1: at $S_g = 0.2$, the relative permeability ratio $k_{rg}/k_{ro} = 0.07$
Point 2: at $S_g = 0.4$, the relative permeability ratio $k_{rg}/k_{ro} = 0.70$

Imposing the above points on Equation 5-28, gives:

$0.07 = a\, e^{0.2b}$
$0.70 = a\, e^{0.4b}$

Solving simultaneously gives:

• The intercept $a = 0.0070$
• The slope $b = 11.513$

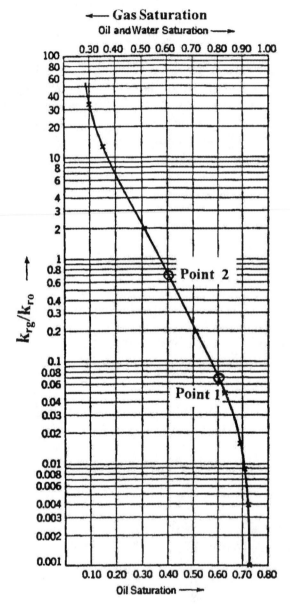

Figure 5-6. k_{rg}/k_{ro} as a function of saturation.

or

$$\frac{k_{rg}}{k_{ro}} = 0.0070 \ e^{11.513 S_g}$$

In a similar manner, Figure 5-7 shows a semilog plot of k_{ro}/k_{rw} versus water saturation.

The middle straight-line portion of the curve is expressed by a relationship similar to that of Equation 5-28

$$\frac{k_{ro}}{k_{rw}} = a \ e^{b S_w} \tag{5-29}$$

where the slope b has a negative value.

DYNAMIC PSEUDO-RELATIVE PERMEABILITIES

For a multilayered reservoir with each layer as described by a set of relative permeability curves, it is possible to treat the reservoir by a single layer

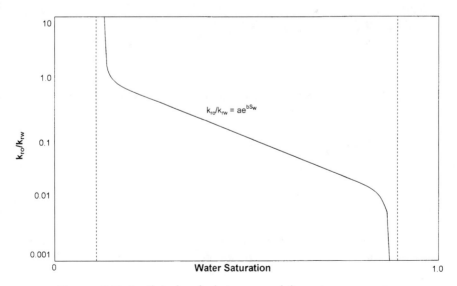

Figure 5-7. Semilog plot of relative permeability ratio vs. saturation.

that is characterized by a weighted-average porosity, absolute permeability, and a set of dynamic pseudo-relative permeability curves. These averaging properties are calculated by applying the following set of relationships:

Average Porosity

$$\phi_{avg} = \frac{\sum\limits_{i=1}^{N} \phi_i\, h_i}{\sum h_i} \tag{5-30}$$

Average Absolute Permeability

$$k_{avg} = \frac{\sum\limits_{i=1}^{N} k_i\, h_i}{\sum h_i} \tag{5-31}$$

Average Relative Permeability for the Wetting Phase

$$\overline{k}_{rw} = \frac{\sum\limits_{i=1}^{N} (k\,h)_i\, (k_{rw})_i}{\sum\limits_{i=1}^{N} (k\,h)_i} \tag{5-32}$$

Average Relative Permeability for the Nonwetting Phase

$$\overline{k}_{rnw} = \frac{\sum\limits_{i=1}^{N} (k\,h)_i\, (k_{rnw})_i}{\sum\limits_{i=1}^{N} (k\,h)_i} \tag{5-33}$$

The corresponding average saturations should be determined by using Equations 4-16 through 4-18. These equations are given below for convenience:

Average Oil Saturation

$$\overline{S}_o = \frac{\sum\limits_{i=1}^{n} \phi_i \, h_i \, S_{o_i}}{\sum\limits_{i=1}^{n} \phi_i \, h_i}$$

Average Water Saturation

$$\overline{S}_w = \frac{\sum\limits_{i=1}^{n} \phi_i \, h_i \, S_{w_i}}{\sum\limits_{i=1}^{n} \phi_i \, h_i}$$

Average Gas Saturation

$$\overline{S}_g = \frac{\sum\limits_{i=1}^{n} \phi_i \, h_i \, S_{g_i}}{\sum\limits_{i=1}^{n} \phi_i \, h_i}$$

where   n = total number of layers
  h_i = thickness of layer i
  k_i = absolute permeability of layer i
  $\overline{k}_{rw}$ = average relative permeability of the wetting phase
  $\overline{k}_{rnw}$ = average relative permeability of the nonwetting phase

In Equations 5-22 and 5-23, the subscripts *w* and n_w represent wetting and *nonwetting*, respectively. The resulting dynamic pseudo-relative permeability curves are then used in a single-layer model. The objective of the single-layer model is to produce results similar to those from the multilayered, cross-sectional model.

NORMALIZATION AND AVERAGING RELATIVE PERMEABILITY DATA

Results of relative permeability tests performed on several core samples of a reservoir rock often vary. Therefore, it is necessary to average the relative permeability data obtained on individual rock samples. Prior to usage for oil recovery prediction, the relative permeability curves should first be normalized to remove the effect of different initial water and critical oil saturations. The relative permeability can then be de-normalized and assigned to different regions of the reservoir based on the existing critical fluid saturation for each reservoir region.

The most generally used method adjusts all data to reflect assigned end values, determines an average adjusted curve and finally constructs an average curve to reflect reservoir conditions. These procedures are commonly described as normalizing and de-normalizing the relative permeability data.

To perform the normalization procedure, it is helpful to set up the calculation steps for each core sample i in a tabulated form as shown below:

Relative Permeability Data for Core Sample i					
(1)	(2)	(3)	(4)	(5)	(6)
S_w	k_{ro}	k_{rw}	$S^*_w = \dfrac{S_w - S_{wc}}{1 - S_{wc} - S_{oc}}$	$k^*_{ro} = \dfrac{k_{ro}}{(k_{ro})_{S_{wc}}}$	$k^*_{rw} = \dfrac{k_{rw}}{(k_{rw})_{S_{oc}}}$

The following normalization methodology describes the necessary steps for a water-oil system as outlined in the above table.

Step 1. Select several values of S_w starting at S_{wc} (column 1), and list the corresponding values of k_{ro} and k_{rw} in columns 2 and 3.

Step 2. Calculate the normalized water saturation S^*_w for each set of relative permeability curves and list the calculated values in column 4 by using the following expression:

$$S^*_w = \frac{S_w - S_{wc}}{1 - S_{wc} - S_{oc}} \qquad (5\text{-}34)$$

where S_{oc} = critical oil saturation
S_{wc} = connate water saturation
S_w^* = normalized water saturation

Step 3. Calculate the normalized relative permeability for the oil phase at different water saturation by using the relation (column 5):

$$k_{ro}^* = \frac{k_{ro}}{(k_{ro})_{Swc}} \qquad (5\text{-}35)$$

where k_{ro} = relative permeability of oil at different S_w
$(k_{ro})_{Swc}$ = relative permeability of oil at connate water saturation
k_{ro}^* = normalized relative permeability of oil

Step 4. Normalize the relative permeability of the water phase by applying the following expression and document results of the calculation in column 6

$$k_{rw}^* = \frac{k_{rw}}{(k_{rw})_{Soc}} \qquad (5\text{-}36)$$

where $(k_{rw})_{Soc}$ is the relative permeability of water at the critical oil saturation.

Step 5. Using regular Cartesian coordinate, plot the normalized k_{ro}^* and k_{rw}^* versus S_w^* for *all* core samples on the same graph.

Step 6. Determine the average normalized relative permeability values for oil and water as a function of the normalized water saturation by select arbitrary values of S_w^* and calculate the average of k_{ro}^* and k_{rw}^* by applying the following relationships:

$$(k_{ro}^*)_{avg} = \frac{\sum\limits_{i=1}^{n} (h \, k \, k_{ro}^*)_i}{\sum\limits_{i=1}^{n} (h \, k)_i} \qquad (5\text{-}37)$$

$$(k_{rw}^*)_{avg} = \frac{\sum\limits_{i=1}^{n} (h\,k\,k_{rw}^*)_i}{\sum\limits_{i=1}^{n} (h\,k)_i} \qquad (5\text{-}38)$$

where n = total number of core samples
h_i = thickness of sample i
k_i = absolute permeability of sample i

Step 7. The last step in this methodology involves de-normalizing the average curve to reflect actual reservoir and conditions of S_{wc} and S_{oc}. These parameters are the most critical part of the methodology and, therefore, a major effort should be spent in determining representative values. The S_{wc} and S_{oc} are usually determined by averaging the core data, log analysis, or correlations, versus graphs, such as: $(k_{ro})_{S_{wc}}$ vs. S_{wc}, $(k_{rw})_{S_{oc}}$ vs. S_{oc}, and S_{oc} vs. S_{wc} which should be constructed to determine if a significant correlation exists. Often, plots of S_{wc} and S_{or} versus log $z\,k/\phi$ may demonstrate a reliable correlation to determine end-point saturations as shown schematically in Figure 5-8. When representative end values have been estimated, it is again convenient to perform the de-normalization calculations in a tabular form as illustrated below:

(1)	(2)	(3)	(4)	(5)	(6)
S_w^*	$(k_{ro}^*)_{avg}$	$(k_{rw}^*)_{avg}$	$S_w = S_w^*\,(1 - S_{wc} - S_{oc}) + S_{wc}$	$k_{ro} = (k_{ro}^*)_{avg}\,(\bar{k}_{ro})_{S_{wc}}$	$k_{rw} = (k_{rw}^*)_{avg}\,(\bar{k}_{rw})_{S_{oc}}$

Where $(k_{ro})_{S_{wc}}$ and $(k_{ro})_{S_{oc}}$ are the average relative permeability of oil and water at connate water and critical oil, respectively, and given by:

$$(\bar{k}_{ro})_{S_{wc}} = \frac{\sum\limits_{i=1}^{n} \left[h\,k\,(k_{ro})_{S_{wc}} \right]_i}{\sum\limits_{i=1}^{n} (h\,k)_i} \qquad (5\text{-}39)$$

$$(\bar{k}_{rw})_{S_{oc}} = \frac{\sum\limits_{i=1}^{n} \left[h\,k\,(k_{rw})_{S_{oc}} \right]_i}{\sum\limits_{i=1}^{n} (h\,k)_i} \qquad (5\text{-}40)$$

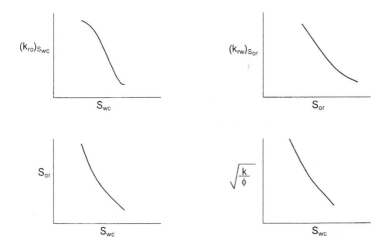

Figure 5-8. Critical saturation relationships.

Example 5-6

Relative permeability measurements are made on three core samples. The measured data are summarized below:

	Core Sample #1		Core Sample #2		Core Sample #3	
	$h = 1$ft		$h = 1$ ft		$h = 1$ ft	
	$k = 100$ md		$k = 80$ md		$k = 150$ md	
	$S_{oc} = 0.35$		$S_{oc} = 0.28$		$S_{oc} = 0.35$	
	$S_{wc} = 0.25$		$S_{wc} = 0.30$		$S_{wc} = 0.20$	
S_w	k_{ro}	k_{rw}	k_{ro}	k_{rw}	k_{ro}	k_{rw}
0.20	—	—	—	—	1.000*	0.000
0.25	0.850*	0.000	—	—	0.872	0.008
0.30	0.754	0.018	0.800	0	0.839	0.027
0.40	0.557	0.092	0.593	0.077	0.663	0.088
0.50	0.352	0.198	0.393	0.191	0.463	0.176
0.60	0.131	0.327	0.202	0.323	0.215	0.286
0.65	0.000	0.400*	0.111	0.394	0.000	0.350*
0.72	—	—	0.000	0.500*	—	—

Values at critical saturations

It is believed that a connate water saturation of 0.27 and a critical oil saturation of 30% better describe the formation. Generate the oil and water relative permeability data using the new critical saturations.

Solution

Step 1. Calculate the normalized water saturation for each core sample by using Equation 5-34.

S_w^*	Core Sample #1 S_w^*	Core Sample #2 S_w^*	Core Sample #3 S_w^*
0.20	—	—	0.000
0.25	0.000	—	0.111
0.30	0.125	0.000	0.222
0.40	0.375	0.238	0.444
0.50	0.625	0.476	0.667
0.60	0.875	0.714	0.889
0.65	1.000	0.833	1.000
0.72	—	1.000	—

Step 2. Determine relative permeability values at critical saturation for each core sample.

	Core 1	Core 2	Core 3
$(k_{ro})S_{wc}$	0.850	0.800	1.000
$(k_{rw})S_{or}$	0.400	0.500	0.35

Step 3. Calculate $(\overline{k}_{ro})_{S_{wc}}$ and $(\overline{k}_{rw})_{S_{or}}$ by applying Equations 5-39 and 5-40 to give:

$$(\overline{k}_{ro})_{S_{wc}} = 0.906$$

$$(\overline{k}_{rw})_{S_{oc}} = 0.402$$

Step 4. Calculate the normalized k_{ro}^* and k_{rw}^* for all core samples:

S_w	Core 1 S_w^*	k_{ro}^*	k_{rw}^*	Core 2 S_w^*	k_{ro}^*	k_{rw}^*	Core 3 S_w^*	k_{ro}^*	k_{rw}^*
0.20	—	—	—	—	—	—	0.000	1.000	0
0.25	0.000	1.000	0	—	—	—	0.111	0.872	0.023
0.30	0.125	0.887	0.045	0.000	1.000	0	0.222	0.839	0.077
0.40	0.375	0.655	0.230	0.238	0.741	0.154	0.444	0.663	0.251
0.50	0.625	0.414	0.495	0.476	0.491	0.382	0.667	0.463	0.503
0.60	0.875	0.154	0.818	0.714	0.252	0.646	0.889	0.215	0.817
0.65	1.000	0.000	1.000	0.833	0.139	0.788	1.000	0.000	1.000
0.72	—	—	—	1.000	0.000	1.000	—	—	—

Step 5. Plot the normalized values of k_{ro}^* and k_{rw}^* versus S_w^* for each core on a regular graph paper as shown in Figure 5-9.

Step 6. Select arbitrary values of S_w^* and calculate the average k_{ro}^* and k_{rw}^* by applying Equations 5-37 and 5-38.

S_w^*	k_{ro}^*			$(k_{ro}^*)_{Avg}$	k_{rw}^*			$(k_{rw}^*)_{avg}$
	Core 1	Core 2	Core 3		Core 1	Core 2	Core 3	
0.1	0.91	0.88	0.93	0.912	0.035	0.075	0.020	0.038
0.2	0.81	0.78	0.85	0.821	0.100	0.148	0.066	0.096
0.3	0.72	0.67	0.78	0.735	0.170	0.230	0.134	0.168
0.4	0.63	0.51	0.70	0.633	0.255	0.315	0.215	0.251
0.5	0.54	0.46	0.61	0.552	0.360	0.405	0.310	0.348
0.6	0.44	0.37	0.52	0.459	0.415	0.515	0.420	0.442
0.7	0.33	0.27	0.42	0.356	0.585	0.650	0.550	0.585
0.8	0.23	0.17	0.32	0.256	0.700	0.745	0.680	0.702
0.9	0.12	0.07	0.18	0.135	0.840	0.870	0.825	0.833

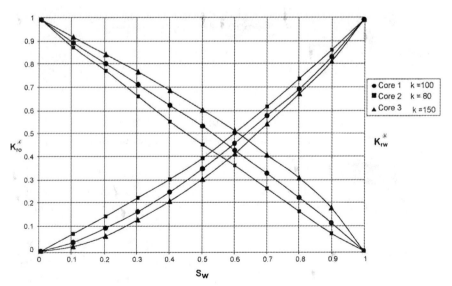

Figure 5-9. Averaging relative permeability data.

Step 7. Using the desired formation S_{oc} and S_{wc} (i.e., $S_{oc} = 0.30$, $S_{wc} = 0.27$), de-normalize the data to generate the required relative permeability data as shown below:

S^*_w	$(k^*_{ro})_{avg}$	$(k^*_{rw})_{avg}$	$S_w = S^*_w (1 - S_{wc} - S_{oc})$ + S_{wc}	$k_{ro} = 0.906$ ✻ $(k^*_{ro})_{avg}$	$k_{rw} = 0.402$ ✻ $(k^*_{rw})_{avg}$
0.1	0.912	0.038	0.313	0.826	0.015
0.2	0.821	0.096	0.356	0.744	0.039
0.3	0.735	0.168	0.399	0.666	0.068
0.4	0.633	0.251	0.442	0.573	0.101
0.5	0.552	0.368	0.485	0.473	0.140
0.6	0.459	0.442	0.528	0.416	0.178
0.7	0.356	0.585	0.571	0.323	0.235
0.8	0.256	0.702	0.614	0.232	0.282
0.9	0.135	0.833	0.657	0.122	0.335

It should be noted that the proposed normalization procedure for water-oil systems as outlined above could be extended to other systems, i.e., gas-oil or gas-water.

THREE-PHASE RELATIVE PERMEABILITY

The relative permeability to a fluid is defined as the ratio of effective permeability at a given saturation of that fluid to the absolute permeability at 100% saturation. Each porous system has unique relative permeability characteristics, which must be measured experimentally. Direct experimental determination of three-phase relative permeability properties is extremely difficult and involves rather complex techniques to determine the fluid saturation distribution along the length of the core. For this reason, the more easily measured two-phase relative permeability characteristics are experimentally determined.

In a three-phase system of this type it is found that the relative permeability to water depends only upon the water saturation. Since the water can flow only through the smallest interconnect pores that are present in the rock and able to accommodate its volume, it is hardly surprising that the flow of water does not depend upon the nature of the fluids occupying the other pores. Similarly, the gas relative permeability depends only upon the gas saturation. This fluid, like water, is restricted to a particular range of pore sizes and its flow is not influenced by the nature of the fluid or fluids that fill the remaining pores.

The pores available for flow of oil are those that, in size, are larger than pores passing only water, and smaller than pores passing only gas. The number of pores occupied by oil depends upon the particular size distribution of the pores in the rock in which the three phases coexist and upon the oil saturation itself.

In general, the relative permeability of each phase, i.e., water, gas, and oil, in a three-phase system is essentially related to the existing saturation by the following functions:

$$k_{rw} = f(S_w) \tag{5-41}$$

$$k_{rg} = f(S_g) \tag{5-42}$$

$$k_{ro} = f(S_w, S_g) \tag{5-43}$$

Function 5-43 is rarely known and, therefore, several practical approaches are proposed and based on estimating the three-phase relative permeability from two sets of two-phase data:

Set 1: Oil-Water System

$$k_{row} = f(S_w)$$

$$k_{rw} = f(S_w)$$

Set 2: Oil-Gas System

$$k_{rog} = f(S_g)$$

$$k_{rg} = f(S_g)$$

where k_{row} and k_{rog} are defined as the relative permeability to oil in the water-oil two-phase system and similarly k_{rog} is the relative permeability of oil in the gas-oil system. The symbol k_{ro} is reserved for the oil relative permeability in the three-phase system.

The triangular graph paper is commonly used to illustrate the changes in the relative permeability values when three phases are flowing simultaneously, as illustrated in Figures 5-10 and 5-11. The relative permeability data are plotted as lines of constant percentage relative permeability (oil, water, and gas isoperms). Figures 5-10 and 5-11 show that the relative permeability data, expressed as isoperms, are dependent on the saturation values for all three phases in the rock.

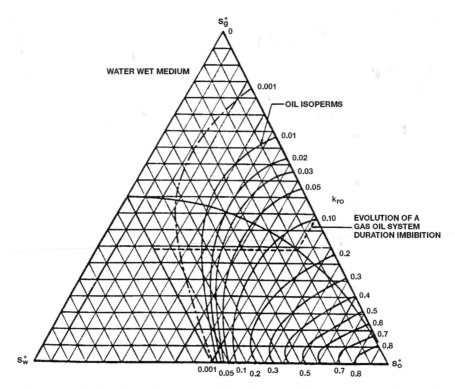

Figure 5-10. Three-plate relative permeability imbibition. (*After Honarpour et al., 1988.*)

Three-Phase Relative Permeability Correlations

Honarpour, Keoderitz, and Harvey (1988) provided a comprehensive treatment of the two- and three-phase relative permeabilities. The authors listed numerous correlations for estimating relative permeabilities. The simplest approach to predict the relative permeability to the oil phase in a three-phase system is defined as:

$$k_{ro} = k_{row}k_{rog} \tag{5-44}$$

There are several practical and more accurate correlations that have developed over the years, including:

- Wyllie's Correlations
- Stone's Model I
- Stone's Model II
- The Hustad-Holt Correlation

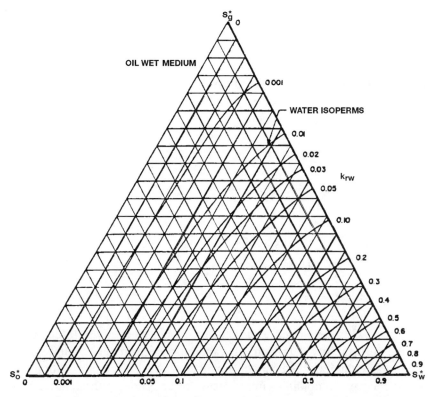

Figure 5-11. Three-phase drainage. (*After Honarpour et al., 1988.*)

Wyllie's Correlations

Wyllie (1961) proposed the following equations for three-phase relative permeabilities in a water-wet system:

In a cemented sandstone, Vugular rock, or oolitic limestone:

$$k_{rg} = \frac{S_g^2 \left[(1 - S_{wc})^2 - (S_w + S_o - S_{wc})^2 \right]}{(1 - S_{wc})^4} \qquad (5\text{-}45)$$

$$k_{ro} = \frac{S_o^3 (2S_w + S_o - 2S_{wc})}{(1 - S_{wc})^4} \qquad (5\text{-}46)$$

$$k_{rw} = \left(\frac{S_w - S_{wc}}{1 - S_{wc}} \right)^4 \tag{5-47}$$

In unconsolidated, well-sorted sand:

$$k_{rw} = \left(\frac{S_w - S_{wc}}{1 - S_{wi}} \right)^3 \tag{5-48}$$

$$k_{ro} = \frac{(S_o)^3}{(1 - S_{wc})^3} \tag{5-49}$$

$$k_{rg} = \frac{(S_o)^3 (2 S_w + S_o - 2 S_{wc})^4}{(1 - S_{wi})^4} \tag{5-50}$$

Stone's Model I

Stone (1970) developed a probability model to estimate three-phase relative permeability data from the laboratory-measured two-phase data. The model combines the channel flow theory in porous media with probability concepts to obtain a simple result for determining the relative permeability to oil in the presence of water and gas flow. The model accounts for hysteresis effects when water and gas saturations are changing in the same direction of the two sets of data.

The use of the channel flow theory implies that water-relative permeability and water-oil capillary pressure in the three-phase system are functions of water saturation alone, irrespective of the relative saturations of oil and gas. Moreover, they are the same function in the three-phase system as in the two-phase water-oil system. Similarly, the gas-phase relative permeability and gas-oil capillary pressure are the same functions of gas saturation in the three-phase system as in the two-phase gas-oil system.

Stone suggested that a nonzero residual oil saturation, called *minimum oil saturation,* S_{om} exists when oil is displaced simultaneously by water and gas. It should be noted that this minimum oil saturation S_{om} is different than the critical oil saturation in the oil-water system (i.e., S_{orw}) and the residual oil saturation in the gas-oil system, i.e., S_{org}. Stone introduced the following normalized saturations:

$$S_o^* = \frac{S_o - S_{om}}{(1 - S_{wc} - S_{om})}, \text{ for } S_o \geq S_{om} \tag{5-51}$$

$$S_w^* = \frac{S_w - S_{wc}}{(1 - S_{wc} - S_{om})}, \text{ for } S_w \geq S_{wc} \tag{5-52}$$

$$S_g^* = \frac{S_g}{(1 - S_{wc} - S_{om})} \tag{5-53}$$

The oil-relative permeability in a three-phase system is then defined as:

$$k_{ro} = S_o^* \beta_w \beta_g \tag{5-54}$$

The two multipliers β_w and β_g are determined from:

$$\beta_w = \frac{k_{row}}{1 - S_w^*} \tag{5-55}$$

$$\beta_g = \frac{k_{rog}}{1 - S_g^*} \tag{5-56}$$

where S_{om} = minimum oil saturation
k_{row} = oil relative permeability as determined from the oil-water two-phase relative permeability at S_w
k_{rog} = oil relative permeability as determined from the gas-oil two-phase relative permeability at S_g

The difficulty in using Stone's first model is selecting the minimum oil saturation S_{om}. Fayers and Mathews (1984) suggested an expression for determining S_{om}.

$$S_{om} = \alpha \, S_{orw} + (1 - \alpha) \, S_{org} \tag{5-57}$$

with

$$\alpha = 1 - \frac{S_g}{1 - S_{wc} - S_{org}} \tag{5-58}$$

where S_{orw} = residual oil saturation in the oil-water relative
   permeability system
   S_{org} = residual oil saturation in the gas-oil relative permeability
   system

Aziz and Sattari (1979) pointed out that Stone's correlation could give k_{ro} values greater than unity. The authors suggested the following normalized form of Stone's model:

$$k_{ro} = \frac{S_o^*}{(1-S_w^*)\,(1-S_g^*)} \left(\frac{k_{row}\,k_{rog}}{(k_{ro})_{S_{wc}}} \right) \quad (5\text{-}59)$$

where $(k_{ro})_{S_{wc}}$ is the value of the relative permeability of the oil at the connate water saturation as determined from the oil-water relative permeability system. It should be noted that it is usually assumed that k_{rg} and krog curves are measured in the presence of connate water.

Stone's Model II

It was the difficulties in choosing S_{om} that led to the development of Stone's Model II. Stone (1973) proposed the following normalized expression:

$$k_{ro} = (k_{ro})_{S_{wc}} \left[\left(\frac{k_{row}}{(k_{ro})_{S_{wc}}} + k_{rw} \right) \right.$$
$$\left. \left(\frac{k_{rog}}{(k_{ro})_{S_{wc}}} + k_{rg} \right) - \left(k_{rw} + k_{rg} \right) \right] \quad (5\text{-}60)$$

This model gives a reasonable approximation to the three-phase relative permeability.

The Hustad-Holt Correlation

Hustad and Holt (1992) modified Stone's Model I by introducing an exponent term n to the normalized saturations to give:

$$k_{ro} = \left[\frac{k_{row} \ k_{rog}}{(k_{ro})_{S_{wc}}} \right] (\beta)^n \tag{5-61}$$

where

$$\beta = \frac{S_o^*}{(1 - S_w^*) \ (1 - S_g^*)} \tag{5-62}$$

$$S_o^* = \frac{S_o - S_{om}}{1 - S_{wc} - S_{om} - S_{gc}} \tag{5-63}$$

$$S_g^* = \frac{S_g - S_{gc}}{1 - S_{wc} - S_{om} - S_{gc}} \tag{5-64}$$

$$S_w^* = \frac{S_w - S_{wc}}{1 - S_{wc} - S_{om} - S_{gc}} \tag{5-65}$$

The β term may be interpreted as a variable that varies between zero and one for low- and high-oil saturations, respectively. If the exponent n is one, the correlation is identical to Stone's first model. Increasing n above unity causes the oil isoperms at low oil saturations to spread from one another. n values below unity have the opposite effect.

Example 5-5

Two-phase relative permeability tests were conducted on core sample to generate the permeability data for oil-water and oil-gas systems. The following information is obtained from the test:

$S_{gc} = 0.10$ $S_{wc} = 0.15$
$S_{orw} = 0.15$ $S_{org} = 0.05$
$(k_{ro})S_{wc} = 0.88$

At the existing saturation values of $S_o = 40\%$, $S_w = 30\%$, and $S_g = 30\%$ the two-phase relative permeabilities are listed below:

$k_{row} = 0.403$
$k_{rw} = 0.030$

$$k_{rg} = 0.035$$
$$k_{rog} = 0.175$$

Estimate the three-phase relative permeability at the existing saturations by using:

a. Stone's Model I
b. Stone's Model II

Solution

a. Stone's Model I

Step 1. Calculate S_{om} by applying Equations 5-58 and 5-57, to give:

$$\alpha = 1 - \frac{0.3}{1 - 0.15 - 0.05} = 0.625$$

$$S_{om} = (0.625)(0.15) + (1 - 0.625)(0.05) = 0.1125$$

Step 2. Calculate the normalized saturations by applying Equations 5-51 through 5-53.

$$S_o^* = \frac{0.4 - 0.1125}{1 - 0.15 - 0.1125} = 0.3898$$

$$S_w^* = \frac{0.30 - 0.15}{1 - 0.15 - 0.1125} = 0.2034$$

$$S_g^* = \frac{0.3}{1 - 0.15 - 0.1125} = 0.4068$$

Step 3. Estimate k_{ro} by using Equation 5-59.

$$k_{ro} = \frac{0.3898}{(1 - 0.2034)(1 - 0.4068)} \left[\frac{(0.406)(0.175)}{0.88} \right] = 0.067$$

b. Stone's Model II

Apply Equation 5-60 to give:

$$k_{ro} = 0.88 \left[\left(\frac{0.406}{0.88} + 0.03 \right) \left(\frac{0.175}{0.88} + 0.035 \right) - (0.03 + 0.035) \right] = 0.044$$

PROBLEMS

1. Given:

   • $S_{wc} = 0.30$ $S_{gc} = 0.06$ $S_{oc} = 0.35$
   • unconsolidated-well sorted sand

   Generate the drainage relative permeability data by using:

   a. The Wyllie-Gardner correlation
   b. Pirson's correlation
   c. Corey's method

2. The capillary pressure data for an oil-water system are given below:

S_w	p_c, psi
0.25	35
0.30	16
0.40	8.5
0.50	5
1.00	0

   a. Generate the relative permeability data for this system.
   b. Using the relative permeability ratio concept, plot k_{ro}/k_{rw} versus S_w on a semi-log scale and determine the coefficients of the following expression:

   $$k_{ro}/k_{rw} = ae^{bS_w}$$

3. Using the relative permeability data of Example 5-6, generate the relative permeability values for a layer in the reservoir that is characterized by the following critical saturations:

   $S_{oc} = 0.25$ $S_{wc} = 0.25$ $h = 1$

4. Prepare a k_{rg}/k_{ro} versus S_g plot for the following laboratory data:

k_{rg}/k_{ro}	S_g
1.9	0.50
0.109	0.30

Find the coefficients of the following relationship:

$$k_{rg}/k_{ro} = ae^{b \, S_g}$$

REFERENCES

1. Aziz, K., and Sattari, A., *Petroleum Reservoir Simulation.* London: Applied Science Publishers Ltd., 1979.

2. Corey, A. T., and Rathjens, C. H., "Effect of Stratification on Relative Permeability," *Trans. AIME,* 1956, pp. 207, 358.

3. Corey, A. T., "The Interrelation Between Gas and Oil Relative Permeabilities," *Prod. Mon.,* 1954, pp. 19, 38, 1954.

4. Fayers, F., and Matthews, J. D., "Evaluation of Normalized Stone's Method for Estimating Three-Phase Relative Permeabilities," *SPEJ,* April 1984, pp. 224–239.

5. Honarpour, M. M., Koederitz, L. F., and Harvey, A. H., "Empirical Equations for Estimating Two-Phase Relative Permeability in Consolidated Rock," *Trans. AIME,* 1982, pp. 273, 290.

6. Honarpour, M. M., Koederitz, L. F., and Harvey, A. H., *Relative Permeability of Petroleum Reservoirs.* CRC Press, Inc., 1988.

7. Hustad, O. S., and Holt, T., "Gravity Stable Displacement of Oil by Hydrocarbon Gas after Waterflooding," SPE Paper 24116, EOR Symposium, Tulsa, OK, 1992.

8. Pirson, S. J. (ed.), *Oil Reservoir Engineering.* New York: McGraw-Hill, 1958.

9. Rose, W. D., and Bruce, W. A., "Evaluation of Capillary Character in Petroleum Reservoir Rock," *Trans. AIME,* 1949, pp. 127, 186.

10. Stone, H. L., "Estimation of Three-Phase Relative Permeability and Residual Oil Data, " *J. of Can. Pet. Technol.,* 1973, pp. 12, 53.

11. Stone, H. L., "Estimation of Three-Phase Relative Permeability, *JPT,* 1970, pp. 2, 214.

12. Torcaso, M. A., and Wyllie, M. R. J., "A Comparison of Calculated k_{rg}/k_{ro} Ratios with Field Data," *JPT,* 1958, pp. 6, 57.

13. Wyllie, M.R.J., and Gardner, G.H.F., "The Generalized Kozeny-Carmen Equation—Its Application to Problems of Multi-Phase Flow in Porous Media," *World Oil,* 1958, pp.121, 146.

14. Wyllie, M.R.J., "Interrelationship Between Wetting and Nonwetting Phase Relative Permeability," *Trans. AIME,* 1961, pp. 83, 192.

FUNDAMENTALS OF RESERVOIR FLUID FLOW

Flow in porous media is a very complex phenomenon and as such cannot be described as explicitly as flow through pipes or conduits. It is rather easy to measure the length and diameter of a pipe and compute its flow capacity as a function of pressure; in porous media, however, flow is different in that there are no clear-cut flow paths which lend themselves to measurement.

The analysis of fluid flow in porous media has evolved throughout the years along two fronts—the experimental and the analytical. Physicists, engineers, hydrologists, and the like have examined experimentally the behavior of various fluids as they flow through porous media ranging from sand packs to fused Pyrex glass. On the basis of their analyses, they have attempted to formulate laws and correlations that can then be utilized to make analytical predictions for similar systems.

The main objective of this chapter is to present the mathematical relationships that are designed to describe the flow behavior of the reservoir fluids. The mathematical forms of these relationships will vary depending upon the characteristics of the reservoir. The primary reservoir characteristics that must be considered include:

• Types of fluids in the reservoir
• Flow regimes
• Reservoir geometry
• Number of flowing fluids in the reservoir

TYPES OF FLUIDS

The isothermal compressibility coefficient is essentially the controlling factor in identifying the type of the reservoir fluid. In general, reservoir fluids are classified into three groups:

• Incompressible fluids
• Slightly compressible fluids
• Compressible fluids

As described in Chapter 2, the isothermal compressibility coefficient c is described mathematically by the following two equivalent expressions:

• In terms of fluid volume:

$$c = \frac{-1}{V} \frac{\partial V}{\partial p} \qquad (6\text{-}1)$$

• In terms of fluid density:

$$c = \frac{1}{\rho} \frac{\partial \rho}{\partial p} \qquad (6\text{-}2)$$

where V and ρ are the volume and density of the fluid, respectively.

Incompressible fluids

An incompressible fluid is defined as the fluid whose volume (or density) does not change with pressure, i.e.:

$$\frac{\partial V}{\partial p} = 0$$

$$\frac{\partial \rho}{\partial p} = 0$$

Incompressible fluids do not exist; this behavior, however, may be assumed in some cases to simplify the derivation and the final form of many flow equations.

Slightly compressible fluids

These "slightly" compressible fluids exhibit small changes in volume, or density, with changes in pressure. Knowing the volume V_{ref} of a slightly compressible liquid at a reference (initial) pressure p_{ref}, the changes in the volumetric behavior of this fluid as a function of pressure p can be mathematically described by integrating Equation 6-1 to give:

$$-c \int_{p_{ref}}^{p} dp = \int_{V_{ref}}^{V} \frac{dV}{V}$$

$$e^{c(p_{ref}-p)} = \frac{V}{V_{ref}}$$

$$V = V_{ref}\, e^{c(p_{ref}-p)} \qquad\qquad (6-3)$$

where p = pressure, psia
 V = volume at pressure p, ft^3
 p_{ref} = initial (reference) pressure, psia
 V_{ref} = fluid volume at initial (reference) pressure, psia

The e^x may be represented by a series expansion as:

$$e^x = 1 + x + \frac{x^2}{2!} + \frac{x^2}{3!} + \cdots + \frac{x^n}{n!} \qquad\qquad (6-4)$$

Because the exponent x [which represents the term c $(p_{ref}-p)$] is very small, the e^x term can be approximated by truncating Equation 6-4 to:

$$e^x = 1 + x \qquad\qquad (6-5)$$

Combining Equation 6-5 with Equation 6-3 gives:

$$V = V_{ref}\,[1 + c\,(p_{ref} - p)] \qquad\qquad (6-6)$$

A similar derivation is applied to Equation 6-2 to give:

$$\rho = \rho_{ref}\,[1 - c\,(p_{ref} - p)] \qquad\qquad (6-7)$$

where V = volume at pressure p
   ρ = density at pressure p
   V_{ref} = volume at initial (reference) pressure p_{ref}
   $ρ_{ref}$ = density at initial (reference) pressure p_{ref}

It should be pointed out that crude oil and water systems fit into this category.

Compressible Fluids

These are fluids that experience large changes in volume as a function of pressure. All gases are considered compressible fluids. The truncation of the series expansion, as given by Equation 6-5, is not valid in this category and the complete expansion as given by Equation 6-4 is used. As shown previously in Chapter 2 in Equation 2-45, the isothermal compressibility of any compressible fluid is described by the following expression:

$$c_g = \frac{1}{p} - \frac{1}{z}\left(\frac{\partial z}{\partial p}\right)_T \qquad (6\text{-}8)$$

Figures 6-1 and 6-2 show schematic illustrations of the volume and density changes as a function of pressure for the three types of fluids.

FLOW REGIMES

There are basically three types of flow regimes that must be recognized in order to describe the fluid flow behavior and reservoir pressure distribution as a function of time. There are three flow regimes:

• Steady-state flow
• Unsteady-state flow
• Pseudosteady-state flow

Steady-State Flow

The flow regime is identified as a steady-state flow if the pressure at every location in the reservoir remains constant, i.e., does not change with time. Mathematically, this condition is expressed as:

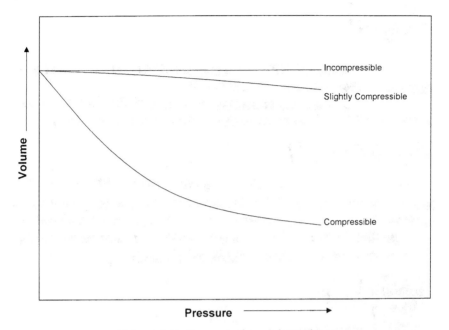

Figure 6-1. Pressure-volume relationship.

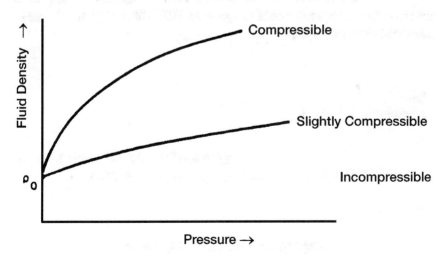

Figure 6-2. Fluid density versus pressure for different fluid types.

$$\left(\frac{\partial p}{\partial t}\right)_i = 0 \qquad (6\text{-}9)$$

The above equation states that the rate of change of pressure p with respect to time t at any location i is zero. In reservoirs, the steady-state flow condition can only occur when the reservoir is completely recharged and supported by strong aquifer or pressure maintenance operations.

Unsteady-State Flow

The unsteady-state flow (frequently called *transient flow*) is defined as the fluid flowing condition at which the rate of change of pressure with respect to time at any position in the reservoir is not zero or constant. This definition suggests that the pressure derivative with respect to time is essentially a function of both position i and time t, thus

$$\left(\frac{\partial p}{\partial t}\right) = f(i, t) \qquad (6\text{-}10)$$

Pseudosteady-State Flow

When the pressure at different locations in the reservoir is declining linearly as a function of time, i.e., at a constant declining rate, the flowing condition is characterized as the pseudosteady-state flow. Mathematically, this definition states that the rate of change of pressure with respect to time at every position is constant, or

$$\left(\frac{\partial p}{\partial t}\right)_i = \text{constant} \qquad (6\text{-}11)$$

It should be pointed out that the pseudosteady-state flow is commonly referred to as semisteady-state flow and quasisteady-state flow.

Figure 6-3 shows a schematic comparison of the pressure declines as a function of time of the three flow regimes.

RESERVOIR GEOMETRY

The shape of a reservoir has a significant effect on its flow behavior. Most reservoirs have irregular boundaries and a rigorous mathematical

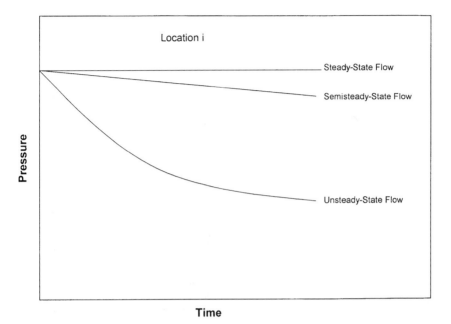

Figure 6-3. Flow regimes.

description of geometry is often possible only with the use of numerical simulators. For many engineering purposes, however, the actual flow geometry may be represented by one of the following flow geometries:

• Radial flow
• Linear flow
• Spherical and hemispherical flow

Radial Flow

In the absence of severe reservoir heterogeneities, flow into or away from a wellbore will follow radial flow lines from a substantial distance from the wellbore. Because fluids move toward the well from all directions and coverage at the wellbore, the term *radial flow* is given to characterize the flow of fluid into the wellbore. Figure 6-4 shows idealized flow lines and iso-potential lines for a radial flow system.

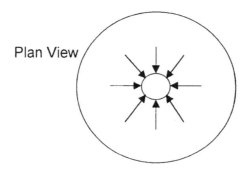

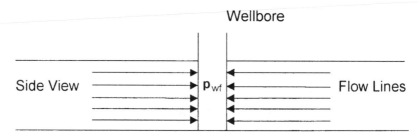

Figure 6-4. Ideal radial flow into a wellbore.

Linear Flow

Linear flow occurs when flow paths are parallel and the fluid flows in a single direction. In addition, the cross sectional area to flow must be constant. Figure 6-5 shows an idealized linear flow system. A common application of linear flow equations is the fluid flow into vertical hydraulic fractures as illustrated in Figure 6-6.

Spherical and Hemispherical Flow

Depending upon the type of wellbore completion configuration, it is possible to have a spherical or hemispherical flow near the wellbore. A well with a limited perforated interval could result in spherical flow in the vicinity of the perforations as illustrated in Figure 6-7. A well that only partially penetrates the pay zone, as shown in Figure 6-8, could result in hemispherical flow. The condition could arise where coning of bottom water is important.

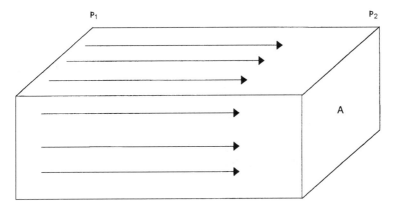

Figure 6-5. Linear flow.

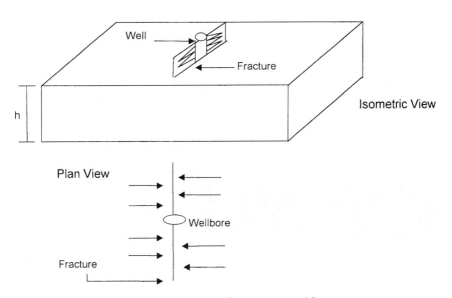

Figure 6-6. Ideal linear flow into vertical fracture.

NUMBER OF FLOWING FLUIDS IN THE RESERVOIR

The mathematical expressions that are used to predict the volumetric performance and pressure behavior of the reservoir vary in forms and complexity depending upon the number of mobile fluids in the reservoir. There are generally three cases of flowing systems:

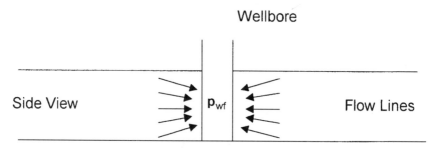

Figure 6-7. Spherical flow due to limited entry.

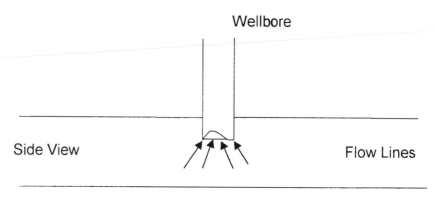

Figure 6-8. Hemispherical flow in a partially penetrating well.

- Single-phase flow (oil, water, or gas)
- Two-phase flow (oil-water, oil-gas, or gas-water)
- Three-phase flow (oil, water, and gas)

The description of fluid flow and subsequent analysis of pressure data becomes more difficult as the number of mobile fluids increases.

FLUID FLOW EQUATIONS

The fluid flow equations that are used to describe the flow behavior in a reservoir can take many forms depending upon the combination of variables presented previously, (i.e., types of flow, types of fluids, etc.). By combining the conservation of mass equation with the transport equation (Darcy's equation) and various equations-of-state, the necessary flow equations can be developed. Since all flow equations to be considered depend on Darcy's Law, it is important to consider this transport relationship first.

Darcy's Law

The fundamental law of fluid motion in porous media is Darcy's Law. The mathematical expression developed by Henry Darcy in 1856 states the velocity of a homogeneous fluid in a porous medium is proportional to the pressure gradient and inversely proportional to the fluid viscosity. For a horizontal linear system, this relationship is:

$$v = \frac{q}{A} = -\frac{k}{\mu}\frac{dp}{dx} \tag{6-12}$$

v is the **apparent** velocity in centimeters per second and is equal to q/A, where q is the volumetric flow rate in cubic centimeters per second and A is total cross-sectional area of the rock in square centimeters. In other words, A includes the area of the rock material as well as the area of the pore channels. The fluid viscosity, μ, is expressed in centipoise units, and the pressure gradient, dp/dx, is in atmospheres per centimeter, taken in the same direction as v and q. The proportionality constant, k, is the *permeability* of the rock expressed in Darcy units.

The negative sign in Equation 6-12 is added because the pressure gradient is negative in the direction of flow as shown in Figure 6-9.

For a horizontal-radial system, the pressure gradient is positive (see Figure 6-10) and Darcy's equation can be expressed in the following generalized radial form:

$$v = \frac{q_r}{A_r} = \frac{k}{\mu}\left(\frac{\partial p}{\partial r}\right)_r \tag{6-13}$$

where q_r = volumetric flow rate at radius r
    A_r = cross-sectional area to flow at radius r
  $(\partial p/\partial r)_r$ = pressure gradient at radius r
    v = apparent velocity at radius r

The cross-sectional area at radius r is essentially the surface area of a cylinder. For a fully penetrated well with a net thickness of h, the cross-sectional area A_r is given by:

$$A_r = 2\,\pi rh$$

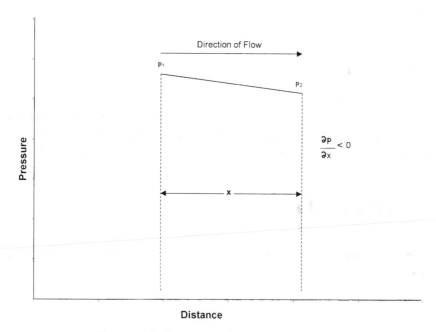

Figure 6-9. Pressure vs. distance in a linear flow.

Darcy's Law applies only when the following conditions exist:

- Laminar (viscous) flow
- Steady-state flow
- Incompressible fluids
- Homogeneous formation

For turbulent flow, which occurs at higher velocities, the pressure gradient increases at a greater rate than does the flow rate and a special modification of Darcy's equation is needed. When turbulent flow exists, the application of Darcy's equation can result in serious errors. Modifications for turbulent flow will be discussed later in this chapter.

STEADY-STATE FLOW

As defined previously, steady-state flow represents the condition that exists when the pressure throughout the reservoir does not change with time. The applications of the steady-state flow to describe the flow behavior of several types of fluid in different reservoir geometries are presented below. These include:

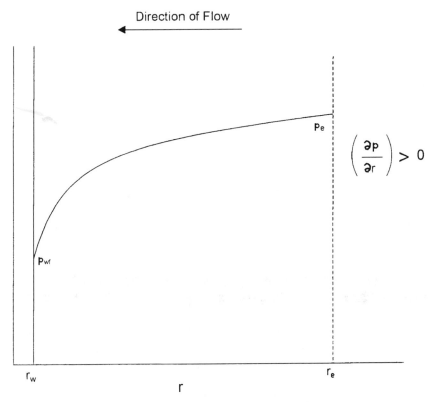

Direction of Flow

$$\left(\frac{\partial p}{\partial r} \right) > 0$$

Figure 6-10. Pressure gradient in radial flow.

- Linear flow of incompressible fluids
- Linear flow of slightly compressible fluids
- Linear flow of compressible fluids
- Radial flow of incompressible fluids
- Radial flow of slightly compressible fluids
- Radial flow of compressible fluids
- Multiphase flow

Linear Flow of Incompressible Fluids

In the linear system, it is assumed the flow occurs through a constant cross-sectional area A, where both ends are entirely open to flow. It is also assumed that no flow crosses the sides, top, or bottom as shown in Figure 6-11.

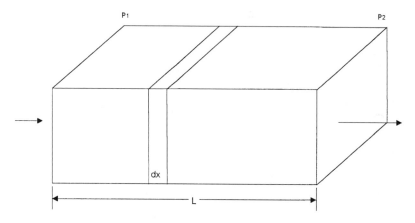

Figure 6-11. Linear flow model.

If an incompressible fluid is flowing across the element dx, then the fluid velocity v and the flow rate q are constants at all points. The flow behavior in this system can be expressed by the differential form of Darcy's equation, i.e., Equation 6-12. Separating the variables of Equation 6-12 and integrating over the length of the linear system gives:

$$\frac{q}{A}\int_{0}^{L} dx = -\frac{k}{\mu}\int_{p1}^{p2} dp$$

or:

$$q = \frac{kA(p_1 - p_2)}{\mu L}$$

It is desirable to express the above relationship in customary field units, or:

$$q = \frac{0.001127\ kA(p_1 - p_2)}{\mu L} \tag{6-14}$$

where q = flow rate, bbl/day
        k = absolute permeability, md
        p = pressure, psia
        μ = viscosity, cp
        L = distance, ft
        A = cross-sectional area, ft^2

Example 6-1

An incompressible fluid flows in a linear porous media with the following properties:

L = 2000 ft	h = 20'	width = 300'
k = 100 md	ϕ = 15%	μ = 2 cp
p_1 = 2000 psi	p_2 = 1990 psi	

Calculate:

a. Flow rate in bbl/day
b. Apparent fluid velocity in ft/day
c. Actual fluid velocity in ft/day

Solution

Calculate the cross-sectional area A:

$$A = (h)(width) = (20)(100) = 6000 \text{ ft}^2$$

a. Calculate the flow rate from Equation 6-14:

$$q = \frac{(0.001127)(100)(6000)(2000 - 1990)}{(2)(2000)} = 1.6905 \text{ bbl/day}$$

b. Calculate the apparent velocity:

$$v = \frac{q}{A} = \frac{(1.6905)(5.615)}{6000} = 0.0016 \text{ ft/day}$$

c. Calculate the actual fluid velocity:

$$v = \frac{q}{\phi A} = \frac{(1.6905)(5.615)}{(0.15)(6000)} = 0.0105 \text{ ft/day}$$

The difference in the pressure (p_1-p_2) in Equation 6-14 is not the only driving force in a tilted reservoir. The gravitational force is the other important driving force that must be accounted for to determine the direction and rate of flow. The fluid gradient force (gravitational force) is always directed *vertically downward* while the force that results from an applied pressure drop may be in any direction. The force causing flow would be then the *vector sum* of these two. In practice, we obtain this result by introducing a new parameter, called fluid potential, which has the same dimensions as pressure, e.g., psi. Its symbol is Φ. The fluid potential at any point in the reservoir is defined as the pressure at that point less the pressure that would be exerted by a fluid head extending to an arbitrarily assigned datum level. Letting Δz_i be the vertical distance from a point i in the reservoir to this datum level.

$$\Phi_i = p_i - \left(\frac{\rho}{144}\right)\Delta z_i \qquad\qquad (6\text{-}15)$$

where ρ is the density in lb/ft^3.

Expressing the fluid density in gm/cc in Equation 6-15 gives:

$$\Phi_i = p_i - 0.433\,\gamma\,\Delta z_i \qquad\qquad (6\text{-}16)$$

where Φ_i = fluid potential at point i, psi
p_i = pressure at point i, psi
Δz_i = vertical distance from point i to the selected datum level
ρ = fluid density, lb/ft^3
γ = fluid density, gm/cm^3

The datum is usually selected at the gas-oil contact, oil-water contact, or at the highest point in formation. In using Equations 6-15 or 6-16 to calculate the fluid potential Φ_i at location i, **the vertical distance Δz_i is assigned as a positive value when the point i is below the datum level and as a negative when it is above the datum level,** i.e.:
If point i is above the datum level:

$$\Phi_i = p_i + \left(\frac{\rho}{144}\right)\Delta z_i$$

and

$$\Phi_i = p_i + 0.433\,\gamma\,\Delta z_i$$

If point i is below the datum level:

$$\Phi_i = p_i - \left(\frac{\rho}{144}\right)\Delta z_i$$

and

$$\Phi_i = p_i - 0.433\,\gamma\,\Delta z_i$$

Applying the above-generalized concept to Darcy's equation (Equation 6-14) gives:

$$q = \frac{0.001127\,kA\,(\Phi_1 - \Phi_2)}{\mu L} \tag{6-17}$$

It should be pointed out that the fluid potential drop $(\Phi_1 - \Phi_2)$ is equal to the pressure drop $(p_1 - p_2)$ only when the flow system is horizontal.

Example 6-2

Assume that the porous media with the properties as given in the previous example is tilted with a dip angle of 5° as shown in Figure 6-12. The incompressible fluid has a density of 42 lb/ft^3. Resolve Example 6-1 using this additional information.

Solution

Step 1. For the purpose of illustrating the concept of fluid potential, select the datum level at half the vertical distance between the two points, i.e., at 87.15 feet, as shown in Figure 6-12.

Step 2. Calculate the fluid potential at Points 1 and 2.

Since Point 1 is below the datum level, then:

$$\Phi_1 = p_1 - \left(\frac{\rho}{144}\right)\Delta z_1 = 2000 - \left(\frac{42}{144}\right)(87.15) = 1974.58\,\text{psi}$$

Since Point 2 is above the datum level, then:

$$\Phi_2 = p_2 + \left(\frac{\rho}{144}\right)\Delta z_2 = 1990 + \left(\frac{42}{144}\right)(87.15) = 2015.42\,\text{psi}$$

Because $\Phi_2 > \Phi_1$, the fluid flows downward from Point 2 to Point 1. The difference in the fluid potential is:

$$\Delta\Phi = 2015.42 - 1974.58 = 40.84\,\text{psi}$$

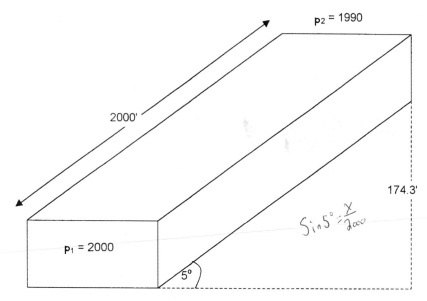

Figure 6-12. Example of a tilted layer.

• Notice, if we select Point 2 for the datum level, then

$$\Phi_1 = 2000 - \left(\frac{42}{144}\right)(174.3) = 1949.16 \, \text{psi}$$

$$\Phi_2 = 1990 + \left(\frac{42}{144}\right)(0) = 1990 \, \text{psi}$$

The above calculations indicate that regardless the position of the datum level, the flow is downward from 2 to 1 with:

$$\Delta\Phi = 1990 - 1949.16 = 40.84 \, \text{psi}$$

Step 3. Calculate the flow rate

$$q = \frac{(0.001127)(100)(6000)(40.84)}{(2)(2000)} = 6.9 \, \text{bbl/day}$$

Step 4. Calculate the velocity:

$$\text{Apparent velocity} = \frac{(6.9)(5.615)}{6000} = 0.0065 \text{ ft/day}$$

$$\text{Actual velocity} = \frac{(6.9)(5.615)}{(0.15)(6000)} = 0.043 \text{ ft/day}$$

Linear Flow of Slightly Compressible Fluids

Equation 6-6 describes the relationship that exists between pressure and volume for slightly compressible fluid, or:

$$V = V_{ref} [1 + c (p_{ref} - p)]$$

The above equation can be modified and written in terms of flow rate as:

$$q = q_{ref} [1 + c (p_{ref} - p)] \tag{6-18}$$

where q_{ref} is the flow rate at some reference pressure p_{ref}. Substituting the above relationship in Darcy's equation gives:

$$\frac{q}{A} = \frac{q_{ref}[1 + c(p_{ref} - p)]}{A} = -0.001127 \frac{k}{\mu} \frac{dp}{dx}$$

Separating the variables and arranging:

$$\frac{q_{ref}}{A} \int_0^L dx = -0.001127 \frac{k}{\mu} \int_{p_1}^{p_2} \left[\frac{dp}{1 + c(p_{ref} - p)} \right]$$

Integrating gives:

$$q_{ref} = \left[\frac{0.001127 \text{ kA}}{\mu c L} \right] \ln \left[\frac{1 + c(p_{ref} - p_2)}{1 + c(p_{ref} - p_1)} \right] \tag{6-19}$$

where q_{ref} = flow rate at a reference pressure p_{ref}, bbl/day
p_1 = upstream pressure, psi

p_2 = downstream pressure, psi
k = permeability, md
μ = viscosity, cp
c = average liquid compressibility, psi^{-1}

Selecting the upstream pressure p_1 as the reference pressure p_{ref} and substituting in Equation 6-19 gives the flow rate at Point 1 as:

$$q_1 = \left[\frac{0.001127\, kA}{\mu cL} \right] \ln\left[1 + c\,(p_1 - p_2) \right] \tag{6-20}$$

Choosing the downstream pressure p_2 as the reference pressure and substituting in Equation 6-19 gives:

$$q_2 = \left[\frac{0.001127\, kA}{\mu cL} \right] \ln\left[\frac{1}{1 + c\,(p_2 - p_1)} \right] \tag{6-21}$$

where q_1 and q_2 are the flow rates at point 1 and 2, respectively.

Example 6-3

Consider the linear system given in Example 6-1 and, assuming a slightly compressible liquid, calculate the flow rate at both ends of the linear system. The liquid has an average compressibility of 21×10^{-5} psi^{-1}.

Solution

• Choosing the upstream pressure as the reference pressure gives:

$$q_1 = \left[\frac{(0.001127)(100)(6000)}{(2)(21 \times 10^{-5})(2000)} \right] \ln[1 + (21 \times 10^{-5})(2000 - 1990)]$$
$$= 1.689\, bbl/day$$

• Choosing the downstream pressure, gives:

$$q_2 = \left[\frac{(0.001127)(100)(6000)}{(2)(21 \times 10^{-5})(2000)} \right] \ln\left[\frac{1}{1 + (21 \times 10^{-5})(1990 - 2000)} \right]$$
$$= 1.692\, bbl/day$$

The above calculations show that q_1 and q_2 are not largely different, which is due to the fact that the liquid is slightly incompressible and its volume is not a strong function of pressure.

Linear Flow of Compressible Fluids (Gases)

For a viscous (laminar) gas flow in a homogeneous-linear system, the real-gas equation-of-state can be applied to calculate the number of gas moles n at pressure p, temperature T, and volume V:

$$n = \frac{pV}{zRT}$$

At standard conditions, the volume occupied by the above n moles is given by:

$$V_{sc} = \frac{n z_{sc} R T_{sc}}{p_{sc}}$$

Combining the above two expressions and assuming $z_{sc} = 1$ gives:

$$\frac{pV}{zT} = \frac{p_{sc} V_{sc}}{T_{sc}}$$

Equivalently, the above relation can be expressed in terms of the flow rate as:

$$\frac{5.615pq}{zT} = \frac{p_{sc} Q_{sc}}{T_{sc}}$$

Rearranging:

$$\left(\frac{p_{sc}}{T_{sc}}\right)\left(\frac{zT}{p}\right)\left(\frac{Q_{sc}}{5.615}\right) = q \qquad (6-22)$$

where   q = gas flow rate at pressure p in bbl/day
   Q_{sc} = gas flow rate at standard conditions, scf/day
   z = gas compressibility factor
   T_{sc}, p_{sc} = standard temperature and pressure in °R and psia, respectively

Replacing the gas flow rate q with that of Darcy's Law, i.e., Equation 6-12, gives:

$$\frac{q}{A} = \left(\frac{p_{sc}}{T_{sc}}\right)\left(\frac{zT}{p}\right)\left(\frac{Q_{sc}}{5.615}\right)\left(\frac{1}{A}\right) = -0.001127\frac{k}{\mu}\frac{dp}{dx}$$

The constant 0.001127 is to convert from Darcy's units to field units. Separating variables and arranging yields:

$$\left[\frac{q_{sc}\,p_{sc}\,T}{0.006328\,k\,T_{sc}\,A}\right]\int_0^L dx = -\int_{p_1}^{p_2}\frac{p}{z\mu_g}dp$$

Assuming constant z and μ_g over the specified pressures, i.e., p_1 and p_2, and integrating gives:

$$Q_{sc} = \frac{0.003164\,T_{sc}\,A\,k\,(p_1^2 - p_2^2)}{p_{sc}\,T\,L\,z\,\mu_g}$$

where Q_{sc} = gas flow rate at standard conditions, scf/day
  k = permeability, md
  T = temperature, °R
  μ_g = gas viscosity, cp
  A = cross-sectional area, ft²
  L = total length of the linear system, ft

Setting p_{sc} =14.7 psi and T_{sc} = 520 °R in the above expression gives:

$$Q_{sc} = \frac{0.111924\,A\,k\,(p_1^2 - p_2^2)}{T\,L\,z\,\mu_g} \tag{6-23}$$

It is essential to notice that those gas properties z and μ_g are a very strong function of pressure, but they have been removed from the integral to simplify the final form of the gas flow equation. The above equation is valid for applications when the pressure < 2000 psi. The gas properties must be evaluated at the average pressure $\bar{p}$ as defined below.

$$\bar{p} = \sqrt{\frac{p_1^2 + p_2^2}{2}} \tag{6-24}$$

Example 6-4

A linear porous media is flowing a 0.72 specific gravity gas at 120°F. The upstream and downstream pressures are 2100 psi and 1894.73 psi, respectively. The cross-sectional area is constant at 4500 ft². The total length is 2500 feet with an absolute permeability of 60 md. Calculate the gas flow rate in scf/day (p_{sc} = 14.7 psia, T_{sc} = 520°R).

Solution

Step 1. Calculate average pressure by using Equation 6-24.

$$\bar{p} = \sqrt{\frac{2100^2 + 1894.73^2}{2}} = 2000\,psi$$

Step 2. Using the specific gravity of the gas, calculate its pseudo-critical properties by applying Equations 2-17 and 2-18.

$$T_{pc} = 395.5°R \qquad p_{pc} = 668.4\ psia$$

Step 3. Calculate the pseudo-reduced pressure and temperature.

$$p_{pr} = \frac{2000}{668.4} = 2.99$$

$$T_{pr} = \frac{600}{395.5} = 1.52$$

Step 4. Determine the z-factor from the Standing-Katz chart (Figure 2-1) to give:

$$z = 0.78$$

Step 5. Solve for the viscosity of the gas by applying the Lee-Gonzalez-Eakin method (Equations 2-63 through 2-66) to give:

$$\mu_g = 0.0173\ cp$$

Step 6. Calculate the gas flow rate by applying Equation 6-23.

$$Q_{sc} = \frac{(0.111924)(4500)(60)(2100^2 - 1894.73^2)}{(600)(0.78)(2500)(0.0173)}$$

(handwritten: (580))

$$= 1,224,242 \text{ scf/day}$$

(handwritten: $1,266,459 \text{ scf/day}$)

Radial Flow of Incompressible Fluids

In a radial flow system, all fluids move toward the producing well from all directions. Before flow can take place, however, a pressure differential must exist. Thus, if a well is to produce oil, which implies a flow of fluids through the formation to the wellbore, the pressure in the formation at the wellbore must be less than the pressure in the formation at some distance from the well.

The pressure in the formation at the wellbore of a producing well is know as the *bottom-hole flowing pressure* (flowing BHP, p_{wf}).

Consider Figure 6-13 which schematically illustrates the radial flow of an incompressible fluid toward a vertical well. The formation is considered to a uniform thickness h and a constant permeability k. Because the fluid is incompressible, the flow rate q must be constant at all radii. Due to the steady-state flowing condition, the pressure profile around the wellbore is maintained constant with time.

Let p_{wf} represent the maintained bottom-hole flowing pressure at the wellbore radius r_w and p_e denote the external pressure at the external or drainage radius. Darcy's equation as described by Equation 6-13 can be used to determine the flow rate at any radius r:

$$v = \frac{q}{A_r} = 0.001127 \frac{k}{\mu} \frac{dp}{dr} \tag{6-25}$$

where   v = apparent fluid velocity, bbl/day-ft^2
        q = flow rate at radius r, bbl/day
        k = permeability, md
        μ = viscosity, cp
    0.001127 = conversion factor to express the equation in field units
        A_r = cross-sectional area at radius r

The minus sign is no longer required for the radial system shown in Figure 6-13 as the radius increases in the same direction as the pressure. In other words, as the radius increases going away from the wellbore the

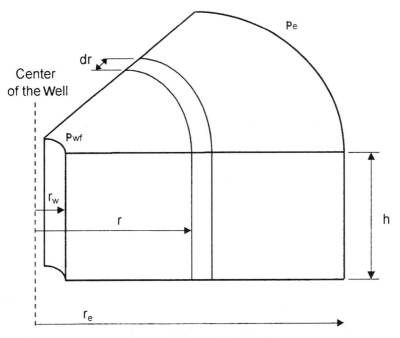

Figure 6-13. Radial flow model.

pressure also increases. At any point in the reservoir the cross-sectional area across which flow occurs will be the surface area of a cylinder, which is $2\pi rh$, or:

$$v = \frac{q}{A_r} = \frac{q}{2\pi rh} = 0.001127 \frac{k}{\mu} \frac{dp}{dr}$$

The flow rate for a crude oil system is customarily expressed in surface units, i.e., stock-tank barrels (STB), rather than reservoir units. Using the symbol Q_o to represent the oil flow as expressed in STB/day, then:

$$q = B_o Q_o$$

where B_o is the oil formation volume factor bbl/STB. The flow rate in Darcy's equation can be expressed in STB/day to give:

$$\frac{Q_o B_o}{2\pi rh} = 0.001127 \frac{k}{\mu_o} \frac{dp}{dr}$$

Integrating the above equation between two radii, r_1 and r_2, when the pressures are p_1 and p_2 yields:

$$\int_{r_1}^{r_2} \left(\frac{Q_o}{2\pi h}\right) \frac{dr}{r} = 0.001127 \int_{P_1}^{P_2} \left(\frac{k}{\mu_o B_o}\right) dp \qquad (6\text{-}26)$$

For incompressible system in a uniform formation, Equation 6-26 can be simplified to:

$$\frac{Q_o}{2\pi h} \int_{r_1}^{r_2} \frac{dr}{r} = \frac{0.001127\,k}{\mu_o\,B_o} \int_{P_1}^{P_2} dp$$

Performing the integration, gives:

$$Q_o = \frac{0.00708\ k\,h\,(p_2 - p_1)}{\mu_o\ B_o\ \ln\,(r_2/r_1)}$$

Frequently the two radii of interest are the wellbore radius r_w and the *external* or *drainage* radius r_e. Then:

$$Q_o = \frac{0.00708\,k\,h\,(p_e - p_w)}{\mu_o\,B_o\,\ln\,(r_e/r_w)} \qquad (6\text{-}27)$$

where Q_o = oil, flow rate, STB/day
p_e = external pressure, psi
p_{wf} = bottom-hole flowing pressure, psi
k = permeability, md
μ_o = oil viscosity, cp
B_o = oil formation volume factor, bbl/STB
h = thickness, ft
r_e = external or drainage radius, ft
r_w = wellbore radius, ft

The external (drainage) radius r_e is usually determined from the well spacing by equating the area of the well spacing with that of a circle, i.e.,

$$\pi\,r_e^2 = 43{,}560\,A$$

or

$$r_e = \sqrt{\frac{43,560\,A}{\pi}} \qquad (6\text{-}28)$$

where A is the well spacing in acres.

In practice, neither the external radius nor the wellbore radius is generally known with precision. Fortunately, they enter the equation as a logarithm, so that the error in the equation will be less than the errors in the radii.

Equation 6-27 can be arranged to solve for the pressure p at any radius r to give:

$$p = p_{wf} + \left[\frac{Q_o\,B_o\,\mu_o}{0.00708\,kh}\right]\ln\!\left(\frac{r}{r_w}\right) \qquad (6\text{-}29)$$

Example 6-5

An oil well in the Nameless Field is producing at a stabilized rate of 600 STB/day at a stabilized bottom-hole flowing pressure of 1800 psi. Analysis of the pressure buildup test data indicates that the pay zone is characterized by a permeability of 120 md and a uniform thickness of 25 ft. The well drains an area of approximately 40 acres. The following additional data is available:

$r_w = 0.25$ ft   A = 40 acres
$B_o = 1.25$ bbl/STB $\mu_o = 2.5$ cp

Calculate the pressure profile (distribution) and list the pressure drop across 1 ft intervals from r_w to 1.25 ft, 4 to 5 ft, 19 to 20 ft, 99 to 100 ft, and 744 to 745 ft.

Solution

Step 1. Rearrange Equation 6-27 and solve for the pressure p at radius r.

$$p = p_{wf} + \left[\frac{\mu_o\,B_o\,Q_o}{0.00708\,k\,h}\right]\ln\,(r/r_w)$$

$$p = 1800 + \left[\frac{(2.5)\,(1.25)\,(600)}{(0.00708)\,(120)\,(25)} \right] \ln \left(\frac{r}{0.25} \right)$$

$$p = 1800 + 88.28 \ln \left(\frac{r}{0.25} \right)$$

Step 2. Calculate the pressure at the designated radii.

r, ft	p, psi	Radius Interval	Pressure drop
0.25	1800		
1.25	1942	0.25–1.25	1942 − 1800 = 142 psi
4	2045		
5	2064	4–5	2064 − 2045 = 19 psi
19	2182		
20	2186	19–20	2186 − 2182 = 4 psi
99	2328		
100	2329	99–100	2329 − 2328 = 1 psi
744	2506.1		
745	2506.2	744–745	2506.2 − 2506.1 = 0.1 psi

Figure 6-14 shows the pressure profile on a function of radius for the calculated data.

Results of the above example reveal that the pressure drop just around the wellbore (i.e., 142 psi) is 7.5 times greater than at the 4–5 ft interval, 36 times greater than at 19–20 ft, and 142 times than that at the 99–100 ft interval. The reason for this large pressure drop around the wellbore is that the fluid is flowing in from a large drainage of 40 acres.

The external pressure pe used in Equation 6-27 cannot be measured readily, but P_e does not deviate substantially from initial reservoir pressure if a strong and active aquifer is present.

Several authors have suggested that the average reservoir pressure p_r, which often is reported in well test results, should be used in performing material balance calculations and flow rate prediction. Craft and Hawkins (1959) showed that the average pressure is located at about 61% of the drainage radius r_e for a steady-state flow condition. Substitute 0.61 r_e in Equation 6-29 to give:

$$p(\text{at } r = 0.61 r_e) = p_r = p_{wf} + \left[\frac{Q_o\, B_o\, \mu_o}{7.08\, k\, h} \right] \ln \left(\frac{0.61 r_e}{r_w} \right)$$

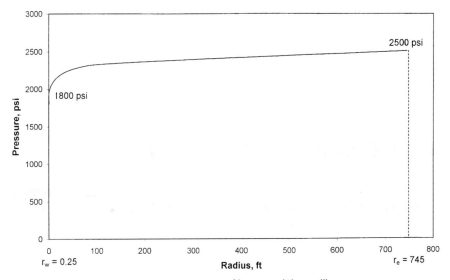

Figure 6-14. Pressure profile around the wellbore.

or in terms of flow rate:

$$Q_o = \frac{0.00708 \, k \, h \, (p_r - p_{wf})}{\mu_o \, B_o \, \ln\left(\dfrac{0.61 r_e}{r_w}\right)} \tag{6-30}$$

But, since $\ln(0.61 r_e / r_w) = \ln\left(\dfrac{r_e}{r_w}\right) - 0.5$, then:

$$Q_o = \frac{0.00708 \, k \, h \, (p_r - p_{wf})}{\mu_o \, B_o \left[\ln\left(\dfrac{r_e}{r_w}\right) - 0.5 \right]} \tag{6-31}$$

Golan and Whitson (1986) suggest a method for approximating drainage area of wells producing from a common reservoir. The authors assume that the volume drained by a single well is proportional to its rate of flow. Assuming constant reservoir properties and a uniform thickness, the approximate drainage area of a single well, A_w, is:

$$A_w = A_T \left(\frac{q_w}{q_T} \right) \tag{6-32}$$

where A_w = drainage area
A_T = total area of the field
q_T = total flow rate of the field
q_w = well flow rate

Radial Flow of Slightly Compressible Fluids

Craft et al. (1990) used Equation 6-18 to express the dependency of the flow rate on pressure for slightly compressible fluids. If this equation is substituted into the radial form of Darcy's Law, the following is obtained:

$$\frac{q}{A_r} = \frac{q_{ref}[1+c(p_{ref}-p)]}{2\pi rh} = 0.001127 \frac{k}{\mu} \frac{dp}{dr}$$

where q_{ref} is the flow rate at some reference pressure p_{ref}.

Separating the variables in the above equation and integrating over the length of the porous medium gives:

$$\frac{q_{ref}\,\mu}{2\pi k h} \int_{r_w}^{r_e} \frac{dr}{r} = 0.001127 \int_{p_{wf}}^{p_e} \frac{dp}{1+c(p_{ref}-p)}$$

or:

$$q_{ref} = \left[\frac{0.00708\,kh}{\mu\,c\ln\left(\frac{r_e}{r_w}\right)} \right] \ln\left[\frac{1+c\,(p_e - p_{ref})}{1+c\,(p_{wf} - p_{ref})} \right]$$

where q_{ref} is oil flow rate at a reference pressure p_{ref}. Choosing the bottom-hole flow pressure p_{wf} as the reference pressure and expressing the flow rate in STB/day gives:

$$Q_o = \left[\frac{0.00708\,kh}{\mu_o\,B_o\,c_o\,\ln\left(\dfrac{r_e}{r_w}\right)} \right] \ln[1+c_o\,(p_e - p_{wf})] \qquad (6\text{-}33)$$

where c_o = isothermal compressibility coefficient, psi^{-1}
$\quad$ Q_o = oil flow rate, STB/day
$\quad$ k = permeability, md

Example 6-6

The following data are available on a well in the Red River Field:

$p_e = 2506$ psi $\qquad$ $p_{wf} = 1800$
$r_e = 745'$ $\qquad$ $r_w = 0.25$
$B_o = 1.25$ $\qquad$ $\mu_o = 2.5$ $\qquad$ $c_o = 25 \times 10^{-6}\,psi^{-1}$
$k = 0.12$ Darcy $\qquad$ $h = 25$ ft.

Assuming a slightly compressible fluid, calculate the oil flow rate. Compare the result with that of incompressible fluid.

Solution

For a slightly compressible fluid, the oil flow rate can be calculated by applying Equation 6-33:

$$Q_o = \left[\frac{(0.00708)(120)(25)}{(2.5)(1.25)(25 \times 10^{-6})\ln(745/0.25)} \right]$$
$$\times \ln\left[1 + (25 \times 10^{-6})(2506 - 1800)\right] = 595\,STB/day$$

Assuming an incompressible fluid, the flow rate can be estimated by applying Darcy's equation, i.e., Equation 6-27:

$$Q_o = \frac{(0.00708)(120)(25)(2506 - 1800)}{(2.5)(1.25)\ln(745/0.25)} = 600\ STB/day$$

Radial Flow of Compressible Gases

The basic differential form of Darcy's Law for a horizontal laminar flow is valid for describing the flow of both gas and liquid systems. For a radial gas flow, the Darcy's equation takes the form:

$$q_{gr} = \frac{0.001127\,(2\pi rh)k}{\mu_g}\frac{dp}{dr} \tag{6-34}$$

where q_{gr} = gas flow rate at radius r, bbl/day
   r = radial distance, ft
   h = zone thickness, ft
   μ_g = gas viscosity, cp
   p = pressure, psi
   0.001127 = conversion constant from Darcy units to field units

The gas flow rate is usually expressed in scf/day. Referring to the gas flow rate at standard condition as Q_g, the gas flow rate q_{gr} under pressure and temperature can be converted to that of standard condition by applying the real gas equation-of-state to both conditions, or

$$\frac{5.615\,q_{gr}\,p}{zRT} = \frac{Q_g\,p_{sc}}{z_{sc}\,R\,T_{sc}}$$

or

$$\left(\frac{p_{sc}}{5.615T_{sc}}\right)\left(\frac{zT}{p}\right)Q_g = q_{gr} \tag{6-35}$$

where p_{sc} = standard pressure, psia
   T_{sc} = standard temperature, °R
   Q_g = gas flow rate, scf/day
   q_{gr} = gas flow rate at radius r, bbl/day
   p = pressure at radius r, psia
   T = reservoir temperature, °R
   z = gas compressibility factor at p and T
   z_{sc} = gas compressibility factor at standard condition $\cong$ 1.0

Combining Equations 6-34 and 6-35 yields:

$$\left(\frac{p_{sc}}{5.615\,T_{sc}}\right)\left(\frac{zT}{p}\right)Q_g = \frac{0.001127\,(2\pi rh)\,k}{\mu_g}\frac{dp}{dr}$$

Assuming that $T_{sc} = 520\ °R$ and $p_{sc} = 14.7$ psia:

$$\left(\frac{T\,Q_g}{kh}\right)\frac{dr}{r} = 0.703\left(\frac{2p}{\mu_g\,z}\right)dp \tag{6-36}$$

Integrating Equation 6-36 from the wellbore conditions (r_w and p_{wf}) to any point in the reservoir (r and p) to give:

$$\int_{r_w}^{r}\left(\frac{T\,Q_g}{kh}\right)\frac{dr}{r} = 0.703\int_{p_{wf}}^{p}\left(\frac{2p}{\mu_g\,z}\right)dp \tag{6-37}$$

Imposing Darcy's Law conditions on Equation 6-37, i.e.:

- **Steady-state flow** which requires that Q_g is constant at all radii
- **Homogeneous formation** which implies that k and h are constant

gives:

$$\left(\frac{TQ_g}{kh}\right)\ln\left(\frac{r}{r_w}\right) = 0.703\int_{p_{wf}}^{p}\left(\frac{2p}{\mu_g\,z}\right)dp$$

The term $\displaystyle\int_{p_{wf}}^{p}\left(\frac{2p}{\mu_g\,z}\right)dp$ can be expanded to give:

$$\int_{p_{wf}}^{p}\left(\frac{2p}{\mu_g\,z}\right)dp = \int_{o}^{p}\left(\frac{2p}{\mu_g\,z}\right)dp - \int_{o}^{p_{wf}}\left(\frac{2p}{\mu_g\,z}\right)dp$$

Combining the above relationships yields:

$$\left(\frac{TQ_g}{kh}\right)\ln\left(\frac{r}{r_w}\right) = 0.703\left[\int_0^p\left(\frac{2p}{\mu_g z}\right)dp - \int_0^{p_{wf}}\left(\frac{2p}{\mu_g z}\right)dp\right] \qquad (6\text{-}38)$$

The integral $\int_0^p 2p/(\mu_g z)\,dp$ is called the *real gas potential* or *real gas pseudopressure* and it is usually represented by m(p) or ψ. Thus

$$m(p) = \psi = \int_0^p\left(\frac{2p}{\mu_g z}\right)dp \qquad (6\text{-}39)$$

Equation 6-38 can be written in terms of the real gas potential to give:

$$\left(\frac{TQ_g}{kh}\right)\ln\frac{r}{r_w} = 0.703\,(\psi - \psi_w)$$

or

$$\psi = \psi_w + \frac{Q_g T}{0.703kh}\ln\frac{r}{r_w} \qquad (6\text{-}40)$$

Equation 6-41 indicates that a graph of ψ vs. ln r/r_w yields a straight line of slope $(Q_g T/0.703kh)$ and intercepts ψ_w (Figure 6-15).
The flow rate is given exactly by

$$Q_g = \frac{0.703kh\,(\psi - \psi_w)}{T\ln\dfrac{r}{r_w}} \qquad (6\text{-}41)$$

In the particular case when $r = r_e$, then:

$$Q_g = \frac{0.703\,kh\,(\psi_e - \psi_w)}{T\left(\ln\dfrac{r_e}{r_w}\right)} \qquad (6\text{-}42)$$

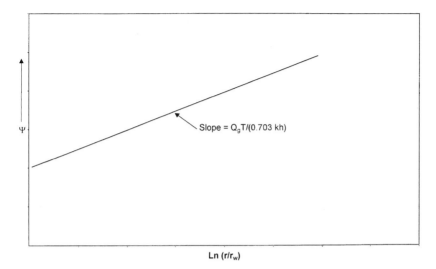

Figure 6-15. Graph of Ψ vs. ln (r/r_w).

where ψ_e = real gas potential as evaluated from 0 to p_e, psi^2/cp
ψ_w = real gas potential as evaluated from 0 to P_{wf}, psi^2/cp
k = permeability, md
h = thickness, ft
r_e = drainage radius, ft
r_w = wellbore radius, ft
Q_g = gas flow rate, scf/day

The gas flow rate is commonly expressed in Mscf/day, or

$$Q_g = \frac{kh(\psi_e - \psi_w)}{1422\,T\left(\ln\dfrac{r_e}{r_w}\right)} \qquad (6\text{-}43)$$

where Q_g = gas flow rate, Mscf/day.

Equation 6-43 can be expressed in terms of the average reservoir pressure p_r instead of the initial reservoir pressure p_e as:

$$Q_g = \frac{kh\,(\psi_r - \psi_w)}{1422\,T\left[\ln\left(\dfrac{r_e}{r_w}\right) - 0.5\right]} \qquad (6\text{-}44)$$

To calculate the integral in Equation 6-43, the values of $2p/\mu_g z$ are calculated for several values of pressure p. Then $(2p/\mu_g z)$ versus p is plotted on a Cartesian scale and the area under the curve is calculated either numerically or graphically, where the area under the curve from p = 0 to any pressure p represents the value of ψ corresponding to p. The following example will illustrate the procedure.

Example 6-7

The following PVT data from a gas well in the Anaconda Gas Field is given below[1]:

p (psi)	μ_g(cp)	z
0	0.0127	1.000
400	0.01286	0.937
800	0.01390	0.882
1200	0.01530	0.832
1600	0.01680	0.794
2000	0.01840	0.770
2400	0.02010	0.763
2800	0.02170	0.775
3200	0.02340	0.797
3600	0.02500	0.827
4000	0.02660	0.860
4400	0.02831	0.896

The well is producing at a stabilized bottom-hole flowing pressure of 3600 psi. The wellbore radius is 0.3 ft. The following additional data is available:

$k = 65$ md $h = 15$ ft $T = 600°R$
$p_e = 4400$ psi $r_e = 1000$ ft

Calculate the gas flow rate in Mscf/day.

Solution

Step 1. Calculate the term $\left(\dfrac{2p}{\mu_g z}\right)$ for each pressure as shown below:

---

[1]Data from "Gas Well Testing, Theory, Practice & Regulations," Donohue and Ertekin, IHRDC Corporation (1982).

p (psi)	μ_g (cp)	z	$\dfrac{2p}{\mu_g z}\left(\dfrac{psia}{cp}\right)$
0	0.0127	1.000	0
400	0.01286	0.937	66391
800	0.01390	0.882	130508
1200	0.01530	0.832	188537
1600	0.01680	0.794	239894
2000	0.01840	0.770	282326
2400	0.02010	0.763	312983
2800	0.02170	0.775	332986
3200	0.02340	0.797	343167
3600	0.02500	0.827	348247
4000	0.02660	0.860	349711
4400	0.02831	0.896	346924

Step 2. Plot the term $\left(\dfrac{2p}{\mu_g z}\right)$ versus pressure as shown in Figure 6-16.

Step 3. Calculate numerically the area under the curve for each value of p. These areas correspond to the real gas potential ψ at each pressure. These ψ values are tabulated below ψ versus p is also plotted in the figure).

p (psi)	$\psi\left(\dfrac{psi^2}{cp}\right)$
400	13.2×10^6
800	52.0×10^6
1200	113.1×10^6
1600	198.0×10^6
2000	304.0×10^6
2400	422.0×10^6
2800	542.4×10^6
3200	678.0×10^6
3600	816.0×10^6
4000	950.0×10^6
4400	1089.0×10^6

Step 4. Calculate the flow rate by applying Equation 6-41.

$$p_w = 816.0 \times 10^6 \qquad p_e = 1089 \times 10^6$$

$$Q_g = \frac{(65)\,(15)\,(1089 - 816)\,10^6}{(1422)\,(600)\,\ln\,(1000/0.25)} = 37,614 \text{ Mscf/day}$$

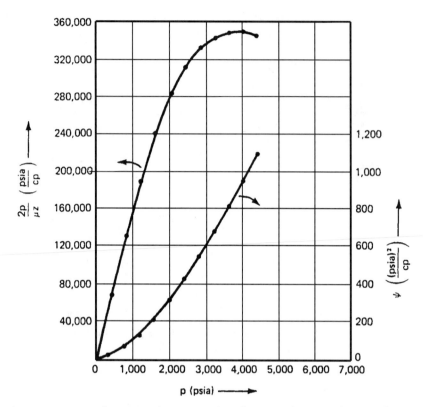

Figure 6-16. Real gas pseudopressure data for Example 6-7 (*After Donohue and Erekin, 1982*).

Approximation of the Gas Flow Rate

The exact gas flow rate as expressed by the different forms of Darcy's Law, i.e., Equations 6-37 through 6-44, can be approximated by removing the term $\dfrac{2}{\mu_g z}$ outside the integral as a constant. It should be pointed out that the $z\mu_g$ is considered constant only under a pressure range of < 2000 psi. Equation 6-43 can be rewritten as:

$$Q_g = \left[\frac{kh}{1422\, T \ln\left(\dfrac{r_e}{r_w}\right)} \right] \int_{P_{wf}}^{P_e} \left(\frac{2p}{\mu_g z} \right) dp$$

Removing the term and integrating gives:

$$Q_g = \frac{kh\,(p_e^2 - p_{wf}^2)}{1422\,T\,(\mu_g\,z)_{avg}\,\ln\left(\dfrac{r_e}{r_w}\right)} \tag{6-45}$$

where Q_g = gas flow rate, Mscf/day
k = permeability, md

The term $(\mu_g\,z)_{avg}$ is evaluated at an average pressure $\bar{p}$ that is defined by the following expression:

$$\bar{p} = \sqrt{\frac{p_{wf}^2 + p_e^2}{2}}$$

The above approximation method is called the *pressure-squared* method and is limited to flow calculations when the reservoir pressure is less that 2000 psi. Other approximation methods are discussed in Chapter 7.

Example 6-8

Using the data given in Example 6-7, re-solve for the gas flow rate by using the pressure-squared method. Compare with the exact method (i.e., real gas potential solution).

Solution

Step 1. Calculate the arithmetic average pressure.

$$\bar{p} = \left[\frac{4400^2 + 3600^2}{2}\right]^{.5} = 4020\,\text{psi}$$

Step 2. Determine gas viscosity and gas compressibility factor at 4020 psi.

$\mu_g = 0.0267$
$z = 0.862$

Step 3. Apply Equation 6-45:

$$Q_g = \frac{(65)\,(15)\,[4400^2 - 3600^2]}{(1422)\,(600)\,(0.0267)\,(0.862)\ln\,(1000/0.25)}$$
$$= 38,314 \text{ Mscf/day}$$

Step 4. Results show that the pressure-squared method approximates the exact solution of 37,614 with an absolute error of 1.86%. This error is due to the limited applicability of the pressure-squared method to a pressure range of <2000 psi.

Horizontal Multiple-Phase Flow

When several fluid phases are flowing simultaneously in a horizontal porous system, the concept of the effective permeability to each phase and the associated physical properties must be used in Darcy's equation. For a radial system, the generalized form of Darcy's equation can be applied to each reservoir as follows:

$$q_o = 0.001127 \left(\frac{2\pi rh}{\mu_o} \right) k_o \frac{dp}{dr}$$

$$q_w = 0.001127 \left(\frac{2\pi rh}{\mu_w} \right) k_w \frac{dp}{dr}$$

$$q_g = 0.001127 \left(\frac{2\pi rh}{\mu_g} \right) k_g \frac{dp}{dr}$$

where k_o, k_w, k_g = effective permeability to oil, water, and gas, md
μ_o, μ_w, μ_g = viscosity to oil, water, and gas, cp
q_o, q_w, q_g = flow rates for oil, water, and gas, bbl/day
k = absolute permeability, md

The effective permeability can be expressed in terms of the relative and absolute permeability, as presented by Equation 5-1 through 5-2, to give:

$$k_o = k_{ro}\, k$$
$$k_w = k_{rw}\, k$$
$$k_g = k_{rg}\, k$$

Using the above concept in Darcy's equation and expressing the flow rate in standard conditions yield:

$$Q_o = 0.00708\,(rhk)\left(\frac{k_{ro}}{\mu_o\,\beta_o}\right)\frac{dp}{dr} \qquad (6\text{-}46)$$

$$Q_w = 0.00708\,(rhk)\left(\frac{k_{rw}}{\mu_w\,\beta_w}\right)\frac{dp}{dr} \qquad (6\text{-}47)$$

$$Q_g = 0.00708\,(rhk)\left(\frac{k_{rg}}{\mu_g\,\beta_g}\right)\frac{dp}{dr} \qquad (6\text{-}48)$$

where Q_o, Q_w = oil and water flow rates, STB/day
B_o, B_w = oil and water formation volume factor, bbl/STB
Q_g = gas flow rate, scf/day
B_g = gas formation volume factor, bbl/scf
k = absolute permeability, md

The gas formation volume factor B_g is previously expressed by Equation 2-54 as:

$$B_g = 0.005035\,\frac{zT}{p}, \text{ bbl/scf}$$

Performing the regular integration approach on Equations 6-46 through 6-48 yields:

• **Oil Phase**

$$Q_o = \frac{0.00708\,(kh)\,(k_{ro})\,(p_e - p_{wf})}{\mu_o\,B_o\,\ln(r_e/r_w)} \qquad (6\text{-}49)$$

• **Water Phase**

$$Q_w = \frac{0.00708\,(kh)\,(k_{rw})\,(p_e - p_{wf})}{\mu_w\,B_w\,\ln(r_e/r_w)} \qquad (6\text{-}50)$$

• **Gas Phase**

In terms of the real gas potential:

$$Q_g = \frac{(kh) k_{rg} (\psi_e - \psi_w)}{1422 \, T \ln (r_e / r_w)} \tag{6-51}$$

In terms of the pressure-squared:

$$Q_g = \frac{(kh) k_{rg} (p_e^2 - p_{wf}^2)}{1422 (\mu_g z)_{avg} \, T \ln (r_e / r_w)} \tag{6-52}$$

where Q_g = gas flow rate, Mscf/day
k = absolute permeability, md
T = temperature, °R

In numerous petroleum engineering calculations, it is convenient to express the flow rate of any phase as a ratio of other flowing phase. Two important flow ratios are the "instantaneous" water-oil ratio (WOR) and "instantaneous" gas-oil ratio (GOR). The generalized form of Darcy's equation can be used to determine both flow ratios.

The water-oil ratio is defined as the ratio of the water flow rate to that of the oil. Both rates are expressed in stock-tank barrels per day, or:

$$WOR = \frac{Q_w}{Q_o}$$

Dividing Equation 6-46 by Equation 6-48 gives:

$$WOR = \left(\frac{k_{rw}}{k_{ro}} \right) \left(\frac{\mu_o B_o}{\mu_w B_w} \right) \tag{6-53}$$

where WOR = water-oil ratio, STB/STB.

The instantaneous GOR, as expressed in scf/STB, is defined as the *total* gas flow rate, i.e., free gas and solution gas, divided by the oil flow rate, or

$$GOR = \frac{Q_o R_s + Q_g}{Q_o}$$

or

$$GOR = R_s + \frac{Q_g}{Q_o} \qquad (6\text{-}54)$$

where GOR = "instantaneous" gas-oil ratio, scf/STB
R_s = gas solubility, scf/STB
Q_g = free gas flow rate, scf/day
Q_o = oil flow rate, STB/day

Substituting Equations 6-46 and 6-48 into Equation 6-54 yields:

$$GOR = R_s + \left(\frac{k_{rg}}{k_{ro}}\right)\left(\frac{\mu_o B_o}{\mu_g B_g}\right) \qquad (6\text{-}55)$$

where B_g is the gas formation volume factor as expressed in bbl/scf.

A complete discussion of the practical applications of the water-oil and gas-oil ratios is given the subsequent chapters.

UNSTEADY-STATE FLOW

Consider Figure 6-17A, which shows a shut-in well that is centered in a homogeneous circular reservoir of radius r_e with a uniform pressure p_i throughout the reservoir. This initial reservoir condition represents the zero producing time. If the well is allowed to flow at a constant flow rate of q, a pressure disturbance will be created at the sand face. The pressure at the wellbore, i.e., p_{wf}, will drop instantaneously as the well is opened. The pressure disturbance will move away from the wellbore at a rate that is determined by:

• Permeability
• Porosity
• Fluid viscosity
• Rock and fluid compressibilities

Section B in Figure 6-17 shows that at time t_1, the pressure disturbance has moved a distance r_1 into the reservoir. Notice that the pressure disturbance radius is continuously increasing with time. This radius is commonly called *radius of investigation* and referred to as r_{inv}. It is also important

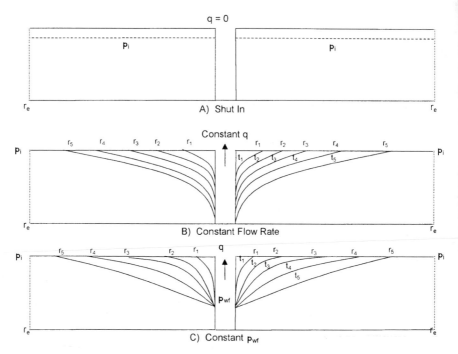

Figure 6-17. Pressure disturbance as a function of time.

to point out that as long as the radius of investigation has not reached the reservoir boundary, i.e., r_e, the reservoir will be acting as if it is *infinite* in size. During this time we say that the reservoir is *infinite acting* because the outer drainage radius r_e can be mathematically *infinite*.

A similar discussion to the above can be used to describe a well that is producing at a constant bottom-hole flowing pressure. Section C in Figure 6-17 schematically illustrates the propagation of the radius of investigation with respect to time. At time t_4, the pressure disturbance reaches the boundary, i.e., $r_{inv} = r_e$. This causes the pressure behavior to change.

Based on the above discussion, the transient (unsteady-state) flow is defined as **that time period during which the boundary has no effect on the pressure behavior in the reservoir and the reservoir will behave as its infinite in size.** Section B in Figure 6-17 shows that the transient flow period occurs during the time interval $0 < t < t_t$ for the constant flow rate scenario and during the time period $0 < t < t_4$ during the constant p_{wf} scenario as depicted by Section C in Figure 6-17.

Basic Transient Flow Equation

Under the steady-state flowing condition, the same quantity of fluid enters the flow system as leaves it. In unsteady-state flow condition, the flow rate into an element of volume of a porous media may not be the same as the flow rate out of that element. Accordingly, the fluid content of the porous medium changes with time. The variables in unsteady-state flow additional to those already used for steady-state flow, therefore, become:

• Time, t
• Porosity, ϕ
• Total compressibility, c_t

The mathematical formulation of the transient-flow equation is based on combining three independent equations and a specifying set of boundary and initial conditions that constitute the unsteady-state equation. These equations and boundary conditions are briefly described below:

a. Continuity Equation
  The continuity equation is essentially a material balance equation that accounts for every pound mass of fluid produced, injected, or remaining in the reservoir.
b. Transport Equation
  The continuity equation is combined with the equation for fluid motion (transport equation) to describe the fluid flow rate "in" and "out" of the reservoir. Basically, the transport equation is Darcy's equation in its generalized differential form.
c. Compressibility Equation
  The fluid compressibility equation (expressed in terms of density or volume) is used in formulating the unsteady-state equation with the objective of describing the changes in the fluid volume as a function of pressure.
d. Initial and Boundary Conditions
  There are two boundary conditions and one initial condition required to complete the formulation and the solution of the transient flow equation. The two boundary conditions are:

• The formation produces at a constant rate into the wellbore.
• There is no flow across the outer boundary and the reservoir behaves as if it were infinite in size, i.e., $r_e = \infty$.

The initial condition simply states the reservoir is at a uniform pressure when production begins, i.e., time = 0.

Consider the flow element shown in Figure 6-18. The element has a width of dr and is located at a distance of r from the center of the well. The porous element has a differential volume of dV. According to the concept of the material-balance equation, the rate of mass flow into an element minus the rate of mass flow out of the element during a differential time Δt must be equal to the mass rate of accumulation during that time interval, or:

$$\begin{bmatrix} \text{mass entering} \\ \text{volume element} \\ \text{during interval} \, \Delta t \end{bmatrix} - \begin{bmatrix} \text{mass leaving} \\ \text{volume element} \\ \text{during interval} \, \Delta t \end{bmatrix}$$

$$= \begin{bmatrix} \text{rate of mass} \\ \text{accumulation} \\ \text{during interval} \, \Delta t \end{bmatrix} \quad (6\text{-}56)$$

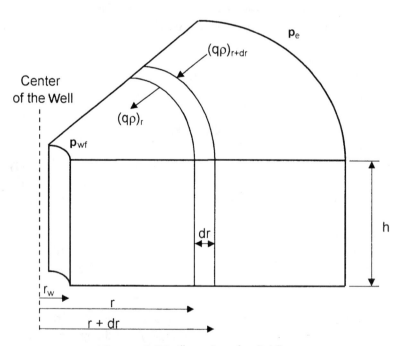

Figure 6-18. Illustration of radial flow.

The individual terms of Equation 6-56 are described below:

Mass entering the volume element during time interval Δt

$$(Mass)_{in} = \Delta t \, [Av\rho]_{r+dr} \qquad (6\text{-}57)$$

where v = velocity of flowing fluid, ft/day
$\qquad \rho$ = fluid density at $(r + dr)$, lb/ft^3
$\qquad A$ = Area at $(r + dr)$
$\qquad \Delta t$ = time interval, days

The area of element at the entering side is:

$$A_{r+dr} = 2\pi(r + dr) \, h \qquad (6\text{-}58)$$

Combining Equation 6-58 with 6-47 gives:

$$[Mass]_{in} = 2\pi \, \Delta t \, (r + dr) \, h \, (v\rho)_{r+dr} \qquad (6\text{-}59)$$

Mass leaving the volume element

Adopting the same approach as that of the leaving mass gives:

$$[Mass]_{out} = 2\pi \, \Delta t \, rh \, (v\rho)_r \qquad (6\text{-}60)$$

Total Accumulation of Mass

The volume of some element with a radius of r is given by:

$$V = \pi \, r^2 \, h$$

Differentiating the above equation with respect to r gives:

$$\frac{dV}{dr} = 2\pi rh$$

or:

$$dV = (2\pi rh)dr \qquad (6\text{-}61)$$

Total mass accumulation during $\Delta t = dV \, [(\phi\rho)_{t + \Delta t} - (\phi\rho)_t]$

Substituting for dV yields:

Total mass accumulation = $(2\pi rh)\ dr\ [(\phi\rho)_{t+\Delta t} - (\phi\rho)_t]$ (6-62)

Replacing terms of Equation 6-56 with those of the calculated relationships gives:

$$2\pi h\ (r+dr)\ \Delta t\ (\phi\rho)_{r+dr} - 2\pi hr\ \Delta t\ (\phi\rho)_r = (2\pi rh)\ dr\ [(\phi\rho)_{t+\Delta t} - (\phi\rho)_t]$$

Dividing the above equation by $(2\pi rh)\ dr$ and simplifying, gives:

$$\frac{1}{(r)dr}[(r+dr)\,(v\rho)_{r+dr} - r\,(v\rho)_r] = \frac{1}{\Delta t}\,[(\phi\rho)_{t+\Delta t} - (\phi\rho)_t]$$

or

$$\frac{1}{r}\frac{\partial}{\partial r}\,[r\,(v\rho)] = \frac{\partial}{\partial t}\,(\phi\rho)$$ (6-63)

where ϕ = porosity
ρ = density, lb/ft^3
v = fluid velocity, ft/day

Equation 6-63 is called the *continuity equation* and it provides the principle of conservation of mass in radial coordinates.

The transport equation must be introduced into the continuity equation to relate the fluid velocity to the pressure gradient within the control volume dV. Darcy's Law is essentially the basic motion equation, which states that the velocity is proportional to the pressure gradient ($\partial p/\partial r$). From Equation 6-25:

$$v = (5.615)\,(0.001127)\,\frac{k}{\mu}\frac{\partial p}{\partial r}$$

$$v = (0.006328)\,\frac{k}{\mu}\frac{\partial p}{\partial r}$$ (6-64)

where k = permeability, md
v = velocity, ft/day

Combining Equation 6-64 with Equation 6-63 results in:

$$\frac{0.006328}{r} \frac{\partial}{\partial r}\left(\frac{k}{\mu}(\rho\,r)\frac{\partial p}{\partial r}\right) = \frac{\partial}{\partial t}(\phi\rho) \tag{6-65}$$

Expanding the right-hand side by taking the indicated derivatives eliminates the porosity from the partial derivative term on the right-hand side:

$$\frac{\partial}{\partial t}(\phi\rho) = \phi\frac{\partial\rho}{\partial t} + \rho\frac{\partial\phi}{\partial t} \tag{6-66}$$

As shown in Chapter 4, porosity is related to the formation compressibility by the following:

$$c_f = \frac{1}{\phi}\frac{\partial\phi}{\partial p} \tag{6-67}$$

Applying the chain rule of differentiation to $\partial\phi/\partial t$,

$$\frac{\partial\phi}{\partial t} = \frac{\partial\phi}{\partial p}\frac{\partial p}{\partial t}$$

Substituting Equation 6-67 into this equation,

$$\frac{\partial\phi}{\partial t} = \phi\,c_f\frac{\partial p}{\partial t}$$

Finally, substituting the above relation into Equation 6-66 and the result into Equation 6-65, gives:

$$\frac{0.006328}{r} \frac{\partial}{\partial r}\left(\frac{k}{\mu}(\rho\,r)\frac{\partial p}{\partial r}\right) = \rho\phi c_f\frac{\partial p}{\partial t} + \phi\frac{\partial\rho}{\partial t} \tag{6-68}$$

Equation 6-68 is the general partial differential equation used to describe the flow of any fluid flowing in a radial direction in porous media. In addition to the initial assumptions, Darcy's equation has been added, which implies that the flow is laminar. Otherwise, the equation is not restricted to any type of fluid and equally valid for gases or liquids. Compressible and slightly compressible fluids, however, must be treated separately in order to develop practical equations that can be used to describe the flow behavior of these two fluids. The treatments of the following systems are discussed below:

• Radial flow of slightly compressible fluids
• Radial flow of compressible fluids

Radial Flow of Slightly Compressible Fluids

To simplify Equation 6-68, assume that the permeability and viscosity are constant over pressure, time, and distance ranges. This leads to:

$$\left[\frac{0.006328 \text{ k}}{\mu \text{ r}}\right] \frac{\partial}{\partial r}\left(r\rho \frac{\partial p}{\partial r}\right) = \rho\phi \text{ c}_f \frac{\partial p}{\partial t} + \phi\frac{\partial\rho}{\partial t} \qquad (6\text{-}69)$$

Expanding the above equation gives:

$$0.006328\left(\frac{k}{\mu}\right)\left[\frac{\rho}{r}\frac{\partial p}{\partial r} + \rho\frac{\partial^2 p}{\partial r^2} + \frac{\partial p}{\partial r}\frac{\partial\rho}{\partial r}\right] = \rho\phi \text{ c}_f\left(\frac{\partial p}{\partial t}\right) + \phi\left(\frac{\partial\rho}{\partial t}\right)$$

Using the chain rule in the above relationship yields:

$$0.006328\left(\frac{k}{\mu}\right)\left[\frac{\rho}{r}\frac{\partial p}{\partial r} + \rho\frac{\partial^2 p}{\partial r^2} + \left(\frac{\partial p}{\partial r}\right)^2\frac{\partial\rho}{\partial p}\right] = \rho\phi \text{ c}_f\left(\frac{\partial p}{\partial t}\right) + \phi\left(\frac{\partial\rho}{\partial t}\right)\left(\frac{\partial\rho}{\partial p}\right)$$

Dividing the above expression by the fluid density ρ gives

$$0.006328\left(\frac{k}{\mu}\right)\left[\frac{1}{r}\frac{\partial p}{\partial r} + \frac{\partial^2 p}{\partial r^2} + \left(\frac{\partial p}{\partial r}\right)^2\left(\frac{1}{\rho}\frac{\partial\rho}{\partial p}\right)\right] = \phi \text{ c}_f\left(\frac{\partial p}{\partial t}\right)$$

$$+ \phi\frac{\partial p}{\partial t}\left(\frac{1}{\rho}\frac{\partial\rho}{\partial p}\right)$$

Recalling that the compressibility of any fluid is related to its density by:

$$c = \frac{1}{\rho}\frac{\partial\rho}{\partial p}$$

Combining the above two equations gives:

$$0.006328\left(\frac{k}{\mu}\right)\left[\frac{\partial^2 p}{\partial r^2}+\frac{1}{r}\frac{\partial p}{\partial r}+c\left[\frac{\partial p}{\partial r}\right]^2\right]=\phi\,c_f\left(\frac{\partial p}{\partial t}\right)+\phi\,c\left(\frac{\partial p}{\partial t}\right)$$

The term $c\left(\dfrac{\partial p}{\partial r}\right)^2$ is considered very small and may be ignored:

$$0.006328\left(\frac{k}{\mu}\right)\left[\frac{\partial^2 p}{\partial r^2}+\frac{1}{r}\frac{\partial p}{\partial r}\right]=\phi(c_f+c)\frac{\partial p}{\partial t} \qquad (6\text{-}70)$$

Define total compressibility, c_t, as:

$$c_t=c+c_f \qquad (6\text{-}71)$$

Combining Equations 6-69 with 6-70 and rearranging gives:

$$\frac{\partial^2 p}{\partial r^2}+\frac{1}{r}\frac{\partial p}{\partial r}=\frac{\phi\mu c_t}{0.006328\,k}\frac{\partial p}{\partial t} \qquad (6\text{-}72)$$

where the time t is expressed in days.

Equation 6-72 is called the *diffusivity equation.* It is one of the most important equations in petroleum engineering. The equation is particularly used in analysis well testing data where the time t is commonly recorded in hours. The equation can be rewritten as:

$$\frac{\partial^2 p}{\partial r^2}+\frac{1}{r}\frac{\partial p}{\partial r}=\frac{\phi\mu c_t}{0.000264\,k}\frac{\partial p}{\partial t} \qquad (6\text{-}73)$$

where k = permeability, md
   r = radial position, ft
   p = pressure, psia
   c_t = total compressibility, psi^{-1}
   t = time, hrs
   ϕ = porosity, fraction
   μ = viscosity, cp

When the reservoir contains more than one fluid, total compressibility should be computed as

$$c_t=c_o S_o+c_w S_w+c_g S_g+c_f \qquad (6\text{-}74)$$

where c_o, c_w and c_g refer to the compressibility of oil, water, and gas, respectively, while S_o, S_w, and S_g refer to the fractional saturation of these fluids. Note that the introduction of c_t into Equation 6-72 does not make Equation 6-72 applicable to multiphase flow; the use of c_t, as defined by Equation 6-73, simply accounts for the compressibility of any *immobile* fluids which may be in the reservoir with the fluid that is flowing.

The term $[0.000264\, k/\phi\mu c_t]$ (Equation 6-73) is called the diffusivity constant and is denoted by the symbol η, or:

$$\eta = \frac{0.000264\, k}{\phi\mu c_t} \qquad (6\text{-}75)$$

The diffusivity equation can then be written in a more convenient form as:

$$\frac{\partial^2 p}{\partial r^2} + \frac{1}{r}\frac{\partial p}{\partial r} = \frac{1}{\eta}\frac{\partial p}{\partial t} \qquad (6\text{-}76)$$

The diffusivity equation as represented by Equation 6-76 is essentially designed to determine the pressure as a function of time t and position r.

Before discussing and presenting the different solutions to the diffusivity equation, it is necessary to summarize the assumptions and limitations used in developing Equation 6-76:

1. Homogeneous and isotropic porous medium
2. Uniform thickness
3. Single phase flow
4. Laminar flow
5. Rock and fluid properties independent of pressure

Notice that for a steady-state flow condition, the pressure at any point in the reservoir is constant and does not change with time, i.e., $\partial p/\partial t = 0$, and therefore Equation 6-76 reduces to:

$$\frac{\partial^2 p}{\partial r^2} + \frac{1}{r}\frac{\partial p}{\partial r} = 0 \qquad (6\text{-}77)$$

Equation 6-77 is called Laplace's equation for steady-state flow.

Example 6-9

Show that the radial form of Darcy's equation is the solution to Equation 6-77.

Solution

Step 1. Start with Darcy's Law as expressed by Equation 6-29

$$p = p_{wf} + \left[\frac{Q_o B_o u_o}{0.00708\,k\,h}\right] \ln\left(\frac{r}{r_w}\right)$$

Step 2. For a steady-state incompressible flow, the term between the two brackets is constant and labeled as C, or:

$$p = p_{wf} + [C]\ln\left(\frac{r}{r_w}\right)$$

Step 3. Evaluate the above expression for the first and second derivative to give:

$$\frac{\partial p}{\partial r} = [C]\left(\frac{1}{r}\right)$$

$$\frac{\partial^2 p}{\partial r^2} = [C]\left(\frac{-1}{r^2}\right)$$

Step 4. Substitute the above two derivatives in Equation 6-77

$$\frac{-1}{r^2}[C] + \left(\frac{1}{r}\right)[C]\left(\frac{-1}{r}\right) = 0$$

Step 5. Results of Step 4 indicate that Darcy's equation satisfies Equation 6-77 and is indeed the solution to Laplace's equation.

To obtain a solution to the diffusivity equation (Equation 6-76), it is necessary to specify an initial condition and impose two boundary conditions. The initial condition simply states that the reservoir is at a uniform pressure p_i when production begins. The two boundary conditions require that the well is producing at a constant production rate and that the reservoir behaves as if it were infinite in size, i.e., $r_e = \infty$.

Based on the boundary conditions imposed on Equation 6-76, there are two generalized solutions to the diffusivity equation:

• Constant-terminal-pressure solution
• Constant-terminal-rate solution

The **constant-terminal-pressure solution** is designed to provide the cumulative flow at any particular time for a reservoir in which the pressure at one boundary of the reservoir is held constant. This technique is frequently used in water influx calculations in gas and oil reservoirs.

The **constant-terminal-rate solution** of the radial diffusivity equation solves for the pressure change throughout the radial system providing that the flow rate is held constant at one terminal end of the radial system, i.e., at the producing well. These are two commonly used forms of the constant-terminal-rate solution:

• The E_i-function solution
• The dimensionless pressure p_D solution

CONSTANT-TERMINAL-PRESSURE SOLUTION

In the constant-rate solution to the radial diffusivity equation, the flow rate is considered to be constant at certain radius (usually wellbore radius) and the pressure profile around that radius is determined as a function of time and position. In the constant-terminal-pressure solution, the pressure is known to be constant at some particular radius and the solution is designed to provide with the cumulative fluid movement across the specified radius (boundary).

The constant-pressure solution is widely used in water influx calculations. A detailed description of the solution and its practical reservoir engineering applications is appropriately discussed in the water influx chapter of the book (Chapter 10).

CONSTANT-TERMINAL-RATE SOLUTION

The constant-terminal-rate solution is an integral part of most transient test analysis techniques, such as with drawdown and pressure buildup analyses. Most of these tests involve producing the well at a **constant flow rate** and recording the flowing pressure as a function of time, i.e.,

$p(r_w,t)$. There are two commonly used forms of the constant-terminal-rate solution:

- The E_i-function solution
- The dimensionless pressure p_D solution

These two popular forms of solution are discussed below.

The E$_i$-Function Solution

Matthews and Russell (1967) proposed a solution to the diffusivity equation that is based on the following assumptions:

- Infinite acting reservoir, i.e., the reservoir is infinite in size.
- The well is producing at a constant flow rate.
- The reservoir is at a uniform pressure, p_i, when production begins.
- The well, with a wellbore radius of r_w, is centered in a cylindrical reservoir of radius r_e.
- No flow across the outer boundary, i.e., at r.

Employing the above conditions, the authors presented their solution in the following form:

$$p(r, t) = p_i + \left[\frac{70.6 \, Q_o \, \mu_o b_o}{kh} \right] E_i \left[\frac{-948 \phi \mu_o c_t r^2}{kt} \right] \qquad (6\text{-}78)$$

where p (r,t) = pressure at radius r from the well after t hours

  t = time, hrs

  k = permeability, md.

  Q_o = flow rate, STB/day

The mathematical function, E_i, is called the **exponential integral** and is defined by:

$$E_i(-x) = -\int_x^\infty \frac{e^{-u} \, du}{u} = \left[\ln x - \frac{x}{1!} + \frac{x^2}{2(2!)} - \frac{x^3}{3(3!)} + \text{etc.} \right] \qquad (6\text{-}79)$$

Craft, Hawkins, and Terry (1991) presented the values of the E_i-function in tabulated and graphical forms as shown in Table 6-1 and Figure 6-19, respectively.

The E_i solution, as expressed by Equation 6-78, is commonly referred to as the **line-source solution.** The exponential integral E_i can be approximated by the following equation when its argument x is less than 0.01:

$$E_i(-x) = \ln(1.781x) \tag{6-80}$$

where the argument x in this case is given by:

$$x = \frac{948\,\phi\mu\,c_t\,r^2}{kt}$$

Equation 6-80 approximates the E_i-function with less than 0.25% error. Another expression that can be used to approximate the E_i-function for the range $0.01 < x < 3.0$ is give by:

$$E_i(-x) = a_1 + a_2\ln(x) + a_3\,[\ln(x)]^2 + a_4\,[\ln(x)]^3 + a_5\,x \\ + a_6\,x^2 + a_7\,x^3 + a_8\,/\,x \tag{6-81}$$

With the coefficients a_1 through a_8 have the following values:

$a_1 = -0.33153973$ $\qquad a_2 = -0.81512322$ $\qquad a_3 = 5.22123384(10^{-2})$
$a_4 = 5.9849819(10^{-3})$ $\quad a_5 = 0.662318450$ $\qquad a_6 = -0.12333524$
$a_7 = 1.0832566(10^{-2})$ $\quad a_8 = 8.6709776(10^{-4})$

The above relationship approximated the E_i-values with an average error of 0.5%.

It should be pointed out that for $x > 10.9$, the $E_i(-x)$ can be considered zero for all practical reservoir engineering calculations.

Example 6-10

An oil well is producing at a constant flow rate of 300 STB/day under unsteady-state flow conditions. The reservoir has the following rock and fluid properties:

$B_o = 1.25$ bbl/STB	$\mu_o = 1.5$ cp	$c_t = 12 \times 10^{-6}$ psi^{-1}
$k_o = 60$ md	$h = 15$ ft	$p_i = 4000$ psi
$\phi = 15\%$	$r_w = 0.25$ ft	

Table 6-1
Values of the $-E_i (-x)$ as a function of x
(After Craft, Hawkins, and Terry, 1991)

x	$-E_i(-x)$	x	$-E_i(-x)$	x	$-E_i(-x)$
0.1	1.82292	4.3	0.00263	8.5	0.00002
0.2	1.22265	4.4	0.00234	8.6	0.00002
0.3	0.90568	4.5	0.00207	8.7	0.00002
0.4	0.70238	4.6	0.00184	8.8	0.00002
0.5	0.55977	4.7	0.00164	8.9	0.00001
0.6	0.45438	4.8	0.00145	9.0	0.00001
0.7	0.37377	4.9	0.00129	9.1	0.00001
0.8	0.31060	5.0	0.00115	9.2	0.00001
0.9	0.26018	5.1	0.00102	9.3	0.00001
1.0	0.21938	5.2	0.00091	9.4	0.00001
1.1	0.18599	5.3	0.00081	9.5	0.00001
1.2	0.15841	5.4	0.00072	9.6	0.00001
1.3	0.13545	5.5	0.00064	9.7	0.00001
1.4	0.11622	5.6	0.00057	9.8	0.00001
1.5	0.10002	5.7	0.00051	9.9	0.00000
1.6	0.08631	5.8	0.00045	10.0	0.00000
1.7	0.07465	5.9	0.00040		
1.8	0.06471	6.0	0.00036		
1.9	0.05620	6.1	0.00032		
2.0	0.04890	6.2	0.00029		
2.1	0.04261	6.3	0.00026		
2.2	0.03719	6.4	0.00023		
2.3	0.03250	6.5	0.00020		
2.4	0.02844	6.6	0.00018		
2.5	0.02491	6.7	0.00016		
2.6	0.02185	6.8	0.00014		
2.7	0.01918	6.9	0.00013		
2.8	0.01686	7.0	0.00012		
2.9	0.01482	7.1	0.00010		
3.0	0.01305	7.2	0.00009		
3.1	0.01149	7.3	0.00008		
3.2	0.01013	7.4	0.00007		
3.3	0.00894	7.5	0.00007		
3.4	0.00789	7.6	0.00006		
3.5	0.00697	7.7	0.00005		
3.6	0.00616	7.8	0.00005		
3.7	0.00545	7.9	0.00004		
3.8	0.00482	8.0	0.00004		
3.9	0.00427	8.1	0.00003		
4.0	0.00378	8.2	0.00003		
4.1	0.00335	8.3	0.00003		
4.2	0.00297	8.4	0.00002		

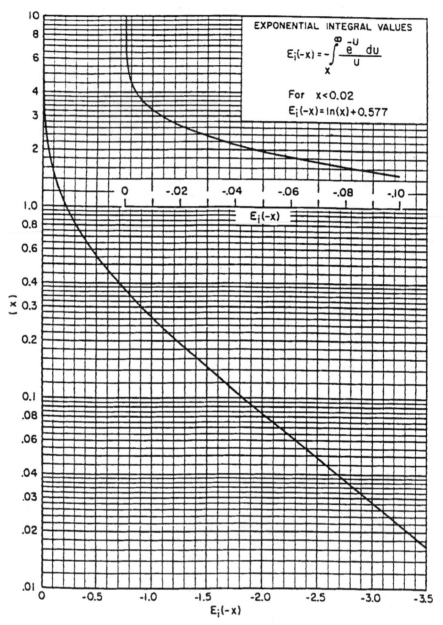

Figure 6-19. The E_i-function. *(After Craft, Hawkins, and Terry, 1991.)*

1. Calculate pressure at radii of 0.25, 5, 10, 50, 100, 500, 1000, 1500, 2000, and 2500 feet, for 1 hour.
   Plot the results as:

   a. Pressure versus logarithm of radius
   b. Pressure versus radius

2. Repeat part 1 for t = 12 hours and 24 hours. Plot the results as pressure versus logarithm of radius.

Solution

Step 1. From Equation 6-78:

$$p(r,t) = 4000 + \left[\frac{70.6\,(300)\,(1.5)\,(1.25)}{(60)\,(15)} \right]$$

$$\times E_i \left[\frac{-948(.15)(1.5)(12 \times 10^{-6})r^2}{(60)\,(t)} \right]$$

$$p(r,t) = 4000 + 44.125\,E_i \left[-42.6(10^{-6})\frac{r^2}{t} \right]$$

Step 2. Perform the required calculations after one hour in the following tabulated form:

Elapsed Time t = 1 hr

r, ft	$x = -42.6(10^{-6})\dfrac{r^2}{1}$	$E_i\,(-x)$	$p(r,1) = 4000 + 44.125\,E_i\,(-x)$
0.25	$-2.6625(10^{-6})$	-12.26*	3459
5	-0.001065	-6.27*	3723
10	-0.00426	-4.88*	3785
50	-0.1065	$-1.76^\dagger$	3922
100	-0.4260	$-0.75^\dagger$	3967
500	-10.65	0	4000
1000	-42.60	0	4000
1500	-95.85	0	4000
2000	-175.40	0	4000
2500	-266.25	0	4000

*As calculated from Equation 6-29
†From Figure 6-19

Step 3. Show results of the calculation graphically as illustrated in Figures 6-20 and 6-21.

Step 4. Repeat the calculation for t = 12 and 24 hrs.

Elapsed Time t = 12 hrs

r, ft	$x = 42.6(10^{-6}) \dfrac{r^2}{12}$	$E_i(-x)$	$p(r,12) = 4000 + 44.125\ E_i(-x)$
0.25	$0.222\ (10^{-6})$	−14.74*	3350
5	$88.75\ (10^{-6})$	−8.75*	3614
10	$355.0\ (10^{-6})$	−7.37*	3675
50	0.0089	−4.14*	3817
100	0.0355	−2.81†	3876
500	0.888	−0.269	3988
1000	3.55	−0.0069	4000
1500	7.99	$-3.77(10^{-5})$	4000
2000	14.62	0	4000
2500	208.3	0	4000

*As calculated from Equation 6-29
†From Figure 6-19

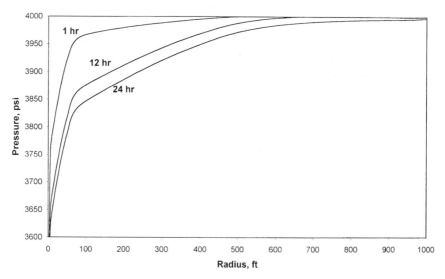

Figure 6-20. Pressure profiles as a function of time.

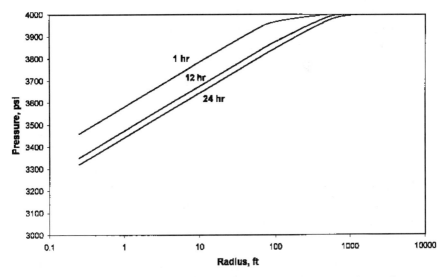

Figure 6-21. Pressure profiles as a function of time on a semi-log scale.

Elapsed Time t = 24 hrs

r, ft	$x = 42.6(10^{-6})\, \dfrac{r^2}{24}$	$E_i\,(-x)$	$p(r,24) = 4000 + 44.125\, E_i\,(-x)$
0.25	$-0.111\,(10^{-6})$	-15.44^*	3319
5	$-44.38\,(10^{-6})$	-9.45^*	3583
10	$-177.5\,(10^{-6})$	-8.06^*	3644
50	-0.0045	-4.83^*	3787
100	-0.0178	$-3.458^\dagger$	3847
500	-0.444	-0.640	3972
1000	-1.775	-0.067	3997
1500	-3.995	-0.0427	3998
2000	-7.310	$8.24\,(10^{-6})$	4000
2500	-104.15	0	4000

*As calculated from Equation 6-29
†From Figure 6-19

Step 5. Results of Step 4 are shown graphically in Figure 6-21.

The above example shows that most of the pressure loss occurs close to the wellbore; accordingly, near-wellbore conditions will exert the greatest influence on flow behavior. Figure 6-21 shows that the pressure profile and the drainage radius are continuously changing with time.

When the parameter x in the E_i-function is less than 0.01, the log approximation as expressed by Equation 6-80 can be used in Equation 6-78 to give:

$$p(r,t) = p_i - \frac{162.6 Q_o B_o \mu_o}{kh} \left[\log\left(\frac{kt}{\phi \mu_o c_t r^2}\right) - 3.23 \right] \qquad (6\text{-}82)$$

For most of the transient flow calculations, engineers are primarily concerned with the behavior of the bottom-hole flowing pressure at the wellbore, i.e., $r = r_w$. Equation 6-82 can be applied at $r = r_w$ to yield:

$$p_{wf} = p_i - \frac{162.6 Q_o B_o \mu_o}{kh} \left[\log\left(\frac{kt}{f \mu_o c_t r_w^2}\right) - 3.23 \right] \qquad (6\text{-}83)$$

where k = permeability, md
        t = time, hr
        c_t = total compressibility, psi^{-1}

It should be noted that Equations 6-82 and 6-83 cannot be used until the flow time t exceeds the limit imposed by the following constraint:

$$t > 9.48 \times 10^4 \, \frac{\phi \mu_o c_t r^2}{k} \qquad (6\text{-}84)$$

where t = time, hr
        k = permeability, md

Example 6-11

Using the data in Example 6-10, estimate the bottom-hole flowing pressure after 10 hours of production.

Solution

Step 1. Equation 6-83 can be used to calculate p_{wf} only if the time exceeds the time limit imposed by Equation 6-84, or:

$$t = 9.48(10^4) \frac{(0.15)(1.5)(12 \times 10^{-6})(0.25)^2}{60} = 0.000267 \text{ hr}$$
$$= 0.153 \text{ sec}$$

For all practical purposes, Equation 6-83 can be used anytime during the transient flow period to estimate the bottom-hole pressure.

Step 2. Since the specified time of 10 hr is greater than 0.000267 hrs, the p_{wf} can be estimated by applying Equation 6-83.

$$p_{wf} = 4000 - \frac{162.6(300)(1.25)(1.5)}{(60)(15)}$$

$$\times \left[\log\left(\frac{(60)(10)}{(0.15)(1.5)(12 \times 10^{-6})(0.25)^2} \right) - 3.23 \right] = 3358 \text{ psi}$$

The second form of solution to the diffusivity equation is called the *dimensionless pressure drop* and is discussed below.

The Dimensionless Pressure Drop (p_D) Solution

Well test analysis often makes use of the concept of the dimensionless variables in solving the unsteady-state flow equation. The importance of dimensionless variables is that they simplify the diffusivity equation and its solution by combining the reservoir parameters (such as permeability, porosity, etc.) and thereby reduce the total number of unknowns.

To introduce the concept of the dimensionless pressure drop solution, consider for example Darcy's equation in a radial form as given previously by Equation 6-27.

$$Q_o = 0.00708 \frac{kh(p_e - p_{wf})}{\mu_o B_o \ln(r_e/r_w)}$$

Rearrange the above equation to give:

$$\frac{p_e - p_{wf}}{\left(\dfrac{Q_o B_o \mu_o}{0.00708 \, kh} \right)} = \ln\left(\frac{r_e}{r_w} \right) \tag{6-85}$$

It is obvious that the right hand side of the above equation has no units (i.e., dimensionless) and, accordingly, the left-hand side must be dimensionless. Since the left-hand side is dimensionless, and ($p_e - p_{wf}$) has the units of psi, it follows that the term $[Q_o B_o \mu_o/(0.00708kh)]$ has units of pressure. In fact, any pressure difference divided by $[Q_o B_o \mu_o/(0.00708kh)]$ is a dimensionless pressure. Therefore, Equation 6-85 can be written in a dimensionless form as:

$$p_D = \ln(r_{eD})$$

where

$$p_D = \frac{p_e - p_{wf}}{\left(\dfrac{Q_o B_o \mu_o}{0.00708 \, k \, h} \right)}$$

This concept can be extended to consider unsteady state equations where the time is a variable. Defining:

$$r_{eD} = \frac{r_e}{r_w}$$

In transient flow analysis, the dimensionless pressure p_D is always a function of dimensionless time that is defined by the following expression:

$$p_D = \frac{p_i - p(r,t)}{\left(\dfrac{Q_o B_o \mu_o}{0.00708 \, k \, h} \right)} \tag{6-86}$$

In transient flow analysis, the dimensionless pressure p_D is always a function of dimensionless time that is defined by the following expression:

$$t_D = \frac{0.000264 \, kt}{\phi \mu c_t \, r_w^2} \tag{6-87}$$

The above expression is only one form of the dimensionless time. Another definition in common usage is t_{DA}, the dimensionless time based on total drainage area.

$$t_{DA} = \frac{0.000264\,kt}{\phi\mu c_t\,A} = t_D\left(\frac{r_w^2}{A}\right) \tag{6-87a}$$

where A = total drainage area = $\pi\,r_e^2$
  r_e = drainage radius, ft
  r_w = wellbore radius, ft

The dimensionless pressure p_D also varies with location in the reservoir as represented by the dimensionless radial distances r_D and r_{eD} that are defined by:

$$r_D = \frac{r}{r_w} \tag{6-88}$$

and

$$r_{eD} = \frac{r_e}{r_w} \tag{6-89}$$

where p_D = dimensionless pressure drop
  r_{eD} = dimensionless external radius
  t_D = dimensionless time
  r_D = dimensionless radius
  t = time, hr
  p(r,t) = pressure at radius r and time t
  k = premeability, md
  μ = viscosity, cp

The above dimensionless groups (i.e., p_D, t_D, and r_D) can be introduced into the diffusivity equation (Equation 6-76) to transform the equation into the following dimensionless form:

$$\frac{\partial^2 p_D}{\partial r_D^2} + \frac{1}{r_D}\frac{\partial p_D}{\partial r_D} = \frac{\partial p_D}{\partial t_D} \tag{6-90}$$

Van Everdingen and Hurst (1949) proposed an analytical solution to the above equation by assuming:

- Perfectly radial reservoir system
- The producing well is in the center and producing at a constant production rate of Q
- Uniform pressure p_i throughout the reservoir before production
- No flow across the external radius r_e

Van Everdingen and Hurst presented the solution to Equation 6-89 in a form of infinite series of exponential terms and Bessel functions. The authors evaluated this series for several values of r_{eD} over a wide range of values for t_D. Chatas (1953) and Lee (1982) conveniently tabulated these solutions for the following two cases:

- Infinite-acting reservoir
- Finite-radial reservoir

Infinite-Acting Reservoir

When a well is put on production at a constant flow rate after a shut-in period, the pressure in the wellbore begins to drop and causes a pressure disturbance to spread in the reservoir. The influence of the reservoir boundaries or the shape of the drainage area does not affect the rate at which the pressure disturbance spreads in the formation. That is why the transient state flow is also called the *infinite acting state*. During the infinite acting period, the declining rate of wellbore pressure and the manner by which the pressure disturbance spreads through the reservoir are determined by reservoir and fluid characteristics such as:

- Porosity, ϕ
- Permeability, k
- Total compressibility, c_t
- Viscosity, μ

For an infinite-acting reservoir, i.e., $r_{eD} = \infty$, the dimensionless pressure drop function p_D is strictly a function of the dimensionless time t_D, or:

$$p_D = f(t_D)$$

Chatas and Lee tabulated the p_D values for the infinite-acting reservoir as shown in Table 6-2. The following mathematical expressions can be used to approximate these tabulated values of p_D:

Table 6-2
p_D vs. t_D—Infinite-Radial System, Constant-Rate at the Inner Boundary (*After Lee, J., Well Testing, SPE Textbook Series.*) (*Permission to publish by the SPE, copyright SPE, 1982*)

t_D	p_D	t_D	p_D	t_D	p_D
0	0	0.15	0.3750	60.0	2.4758
0.0005	0.0250	0.2	0.4241	70.0	2.5501
0.001	0.0352	0.3	0.5024	80.0	2.6147
0.002	0.0495	0.4	0.5645	90.0	2.6718
0.003	0.0603	0.5	0.6167	100.0	2.7233
0.004	0.0694	0.6	0.6622	150.0	2.9212
0.005	0.0774	0.7	0.7024	200.0	3.0636
0.006	0.0845	0.8	0.7387	250.0	3.1726
0.007	0.0911	0.9	0.7716	300.0	3.2630
0.008	0.0971	1.0	0.8019	350.0	3.3394
0.009	0.1028	1.2	0.8672	400.0	3.4057
0.01	0.1081	1.4	0.9160	450.0	3.4641
0.015	0.1312	2.0	1.0195	500.0	3.5164
0.02	0.1503	3.0	1.1665	550.0	3.5643
0.025	0.1669	4.0	1.2750	600.0	3.6076
0.03	0.1818	5.0	1.3625	650.0	3.6476
0.04	0.2077	6.0	1.4362	700.0	3.6842
0.05	0.2301	7.0	1.4997	750.0	3.7184
0.06	0.2500	8.0	1.5557	800.0	3.7505
0.07	0.2680	9.0	1.6057	850.0	3.7805
0.08	0.2845	10.0	1.6509	900.0	3.8088
0.09	0.2999	15.0	1.8294	950.0	3.8355
0.1	0.3144	20.0	1.9601	1,000.0	3.8584
		30.0	2.1470		
		40.0	2.2824		
		50.0	2.3884		

Notes: For $t_D < 0.01$, $p_D \cong 2\sqrt{t_D/x}$.
For $100 < t_D < 0.25\ r_{eD}^2$, $p_D \cong 0.5\ (\ln t_D + 0.80907)$.

• For $t_D < 0.01$:

$$p_D = 2\sqrt{\frac{t_D}{\pi}} \qquad\qquad (6\text{-}91)$$

• For $t_D > 100$:

$$p_D = 0.5[\ln(t_D) + 0.80907] \qquad\qquad (6\text{-}92)$$

• For $0.02 < t_D < 1000$:

$$p_D = a_1 + a_2 \ln (t_D) + a_3 [\ln (t_D)]^2 + a_4 [\ln (t_D)]^3 + a_5 t_D + a_6 (t_D)^2 + a_7 (t_D)^3 + a_8/t_D \qquad (6\text{-}93)$$

where

$a_1 = 0.8085064$	$a_2 = 0.29302022$	$a_3 = 3.5264177(10^{-2})$
$a_4 = -1.4036304(10^{-3})$	$a_5 = -4.7722225(10^{-4})$	$a_6 = 5.1240532(10^{-7})$
$a_7 = -2.3033017(10^{-10})$	$a_8 = -2.6723117(10^{-3})$	

Finite-Radial Reservoir

The arrival of the pressure disturbance at the well drainage boundary marks the end of the transient flow period and the beginning of the semi (pseudo)-steady state. During this flow state, the reservoir boundaries and the shape of the drainage area influence the wellbore pressure response as well as the behavior of the pressure distribution throughout the reservoir. Intuitively, one should not expect the change from the transient to the semi-steady state in this bounded (finite) system to occur instantaneously. There is a short period of time that separates the transient state from the semi-steady state that is called *late-transient state*. Due to its complexity and short duration, the late transient flow is not used in practical well test analysis.

For a finite radial system, the p_D-function is a function of both the dimensionless time and radius, or:

$$p_D = f (t_D, r_{eD})$$

where

$$r_{eD} = \frac{\text{external radius}}{\text{wellbore radius}} = \frac{r_e}{r_w} \qquad (6\text{-}94)$$

Table 6-3 presents p_D as a function of t_D for $1.5 < r_{eD} < 10$. It should be pointed out that Van Everdingen and Hurst principally applied the p_D-function solution to model the performance of water influx into oil reservoirs. Thus, the authors' wellbore radius r_w was in this case the external radius of the reservoir and the r_e was essentially the external boundary radius of the aquifer. Therefore, the range of the r_{eD} values in Table 6-3 are practical for this application.

Table 6-3
p_D vs. t_D—Finite-Radial System, Constant-Rate at the Inner Boundary
(*After Lee, J.,* Well Testing, *SPE Textbook Series.*)
(*Permission to publish by the SPE, copyright SPE, 1982*)

$r_{eD} = 1.5$		$r_{eD} = 2.0$		$r_{eD} = 2.5$		$r_{eD} = 3.0$		$r_{eD} = 3.5$		$r_{eD} = 4.0$	
t_D	p_D	t_D	p_D	t_D	p_D	t_D	p_D	t_D	p_D	t_D	p_D
0.06	0.251	0.22	0.443	0.40	0.565	0.52	0.627	1.0	0.802	1.5	0.927
0.08	0.288	0.24	0.459	0.42	0.576	0.54	0.636	1.1	0.830	1.6	0.948
0.10	0.322	0.26	0.476	0.44	0.587	0.56	0.645	1.2	0.857	1.7	0.968
0.12	0.355	0.28	0.492	0.46	0.598	0.60	0.662	1.3	0.882	1.8	0.988
0.14	0.387	0.30	0.507	0.48	0.608	0.65	0.683	1.4	0.906	1.9	1.007
0.16	0.420	0.32	0.522	0.50	0.618	0.70	0.703	1.5	0.929	2.0	1.025
0.18	0.452	0.34	0.536	0.52	0.628	0.75	0.721	1.6	0.951	2.2	1.059
0.20	0.484	0.36	0.551	0.54	0.638	0.80	0.740	1.7	0.973	2.4	1.092
0.22	0.516	0.38	0.565	0.56	0.647	0.85	0.758	1.8	0.994	2.6	1.123
0.24	0.548	0.40	0.579	0.58	0.657	0.90	0.776	1.9	1.014	2.8	1.154
0.26	0.580	0.42	0.593	0.60	0.666	0.95	0.791	2.0	1.034	3.0	1.184
0.28	0.612	0.44	0.607	0.65	0.688	1.0	0.806	2.25	1.083	3.5	1.255
0.30	0.644	0.46	0.621	0.70	0.710	1.2	0.865	2.50	1.130	4.0	1.324
0.35	0.724	0.48	0.634	0.75	0.731	1.4	0.920	2.75	1.176	4.5	1.392
0.40	0.804	0.50	0.648	0.80	0.752	1.6	0.973	3.0	1.221	5.0	1.460
0.45	0.884	0.60	0.715	0.85	0.772	2.0	1.076	4.0	1.401	5.5	1.527
0.50	0.964	0.70	0.782	0.90	0.792	3.0	1.328	5.0	1.579	6.0	1.594
0.55	1.044	0.80	0.849	0.95	0.812	4.0	1.578	6.0	1.757	6.5	1.660
0.60	1.124	0.90	0.915	1.0	0.832	5.0	1.828			7.0	1.727
0.65	1.204	1.0	0.982	2.0	1.215					8.0	1.861
0.70	1.284	2.0	1.649	3.0	1.506					9.0	1.994
0.75	1.364	3.0	2.316	4.0	1.977					10.0	2.127
0.80	1.444	5.0	3.649	5.0	2.398						

$r_{eD} = 4.5$		$r_{eD} = 5.0$		$r_{eD} = 6.0$		$r_{eD} = 7.0$		$r_{eD} = 8.0$		$r_{eD} = 9.0$		$r_{eD} = 10.0$	
t_D	p_D	t_D	p_D	t_D	p_D	t_D	p_D	t_D	p_D	t_D	p_D	t_D	p_D
2.0	1.023	3.0	1.167	4.0	1.275	6.0	1.436	8.0	1.556	10.0	1.651	12.0	1.732
2.1	1.040	3.1	1.180	4.5	1.322	6.5	1.470	8.5	1.582	10.5	1.673	12.5	1.750
2.2	1.056	3.2	1.192	5.0	1.364	7.0	1.501	9.0	1.607	11.0	1.693	13.0	1.768
2.3	1.702	3.3	1.204	5.5	1.404	7.5	1.531	9.5	1.631	11.5	1.713	13.5	1.784
2.4	1.087	3.4	1.215	6.0	1.441	8.0	1.559	10.0	1.653	12.0	1.732	14.0	1.801
2.5	1.102	3.5	1.227	6.5	1.477	8.5	1.586	10.5	1.675	12.5	1.750	14.5	1.817
2.6	1.116	3.6	1.238	7.0	1.511	9.0	1.613	11.0	1.697	13.0	1.768	15.0	1.832
2.7	1.130	3.7	1.249	7.5	1.544	9.5	1.638	11.5	1.717	13.5	1.786	15.5	1.847
2.8	1.144	3.8	1.259	8.0	1.576	10.0	1.663	12.0	1.737	14.0	1.803	16.0	1.862
2.9	1.158	3.9	1.270	8.5	1.607	11.0	1.711	12.5	1.757	14.5	1.819	17.0	1.890
3.0	1.171	4.0	1.281	9.0	1.638	12.0	1.757	13.0	1.776	15.0	1.835	18.0	1.917

(*table continued on next page*)

Table 6-3 (*continued*)

$r_{eD} = 4.5$		$r_{eD} = 5.0$		$r_{eD} = 6.0$		$r_{eD} = 7.0$		$r_{eD} = 8.0$		$r_{eD} = 9.0$		$r_{eD} = 10.0$	
t_D	p_D	t_D	p_D	t_D	p_D	t_D	p_D	t_D	p_D	t_D	p_D	t_D	p_D
3.2	1.197	4.2	1.301	9.5	1.668	13.0	1.810	13.5	1.795	15.5	1.851	19.0	1.943
3.4	1.222	4.4	1.321	10.0	1.698	14.0	1.845	14.0	1.813	16.0	1.867	20.0	1.968
3.6	1.246	4.6	1.340	11.0	1.757	15.0	1.888	14.5	1.831	17.0	1.897	22.0	2.017
3.8	1.269	4.8	1.360	12.0	1.815	16.0	1.931	15.0	1.849	18.0	1.926	24.0	2.063
4.0	1.292	5.0	1.378	13.0	1.873	17.0	1.974	17.0	1.919	19.0	1.955	26.0	2.108
4.5	1.349	5.5	1.424	14.0	1.931	18.0	2.016	19.0	1.986	20.0	1.983	28.0	2.151
5.0	1.403	6.0	1.469	15.0	1.988	19.0	2.058	21.0	2.051	22.0	2.037	30.0	2.194
5.5	1.457	6.5	1.513	16.0	2.045	20.0	2.100	23.0	2.116	24.0	2.906	32.0	2.236
6.0	1.510	7.0	1.556	17.0	2.103	22.0	2.184	25.0	2.180	26.0	2.142	34.0	2.278
7.0	1.615	7.5	1.598	18.0	2.160	24.0	2.267	30.0	2.340	28.0	2.193	36.0	2.319
8.0	1.719	8.0	1.641	19.0	2.217	26.0	2.351	35.0	2.499	30.0	2.244	38.0	2.360
9.0	1.823	9.0	1.725	20.0	2.274	28.0	2.434	40.0	2.658	34.0	2.345	40.0	2.401
10.0	1.927	10.0	1.808	25.0	2.560	30.0	2.517	45.0	2.817	38.0	2.446	50.0	2.604
11.0	2.031	11.0	1.892	30.0	2.846					40.0	2.496	60.0	2.806
12.0	2.135	12.0	1.975							45.0	2.621	70.0	3.008
13.0	2.239	13.0	2.059							50.0	2.746	80.0	3.210
14.0	2.343	14.0	2.142							60.0	2.996	90.0	3.412
15.0	2.447	15.0	2.225							70.0	3.246	100.0	3.614

Notes: For t_D smaller than values listed in this table for a given r_{eD}, reservoir is infinite acting.
Find p_D in Table 6-2.
For $25 < t_D$ and t_D larger than values in table.

$$p_D \cong \frac{(\frac{1}{2} + 2t_D)}{(r_{eD}^2 - 1)} - \frac{3r_{eD}^4 - 4r_{eD}^4 \ln r_{eD} - 2r_{eD}^2 - 1}{4(r_{eD}^2 - 1)^2}$$

For wells in rebounded reservoirs with
$r_{eD}^2 \gg 1$

$$p_D \cong \frac{2t_D}{r_{eD}^2} + \ln r_{eD} - \frac{3}{4}.$$

Chatas (1953) proposed the following mathematical expression for calculating p_D:

For $25 < t_D$ and $0.25\ r_{eD}^2 < t_D$

$$P_D = \frac{0.5 + 2t_D}{r_{eD}^2 - 1} - \frac{r_{eD}^4 [3 - 4 \ln (r_{eD})] - 2r_{eD}^2 - 1}{4 (r_{eD}^2 - 1)^2} \qquad (6\text{-}95)$$

A special case of Equation 6-95 arises when $r_{eD}^2 \gg 1$, then:

$$P_D = \frac{2t_D}{r_{eD}^2} + \ln (r_{eD}) - 0.75 \qquad (6\text{-}96)$$

The computational procedure of using the p_D-function in determining the bottom-hole flowing pressure changing the transient flow period is summarized in the following steps:

Step 1. Calculate the dimensionless time t_D by applying Equation 6-87.

Step 2. Calculate the dimensionless radius r_{eD} from Equation 6-89.

Step 3. Using the calculated values of t_D and r_{eD}, determine the corresponding pressure function p_D from the appropriate table or equation.

Step 4. Solve for the pressure at the desired radius, i.e., r_w, by applying Equation 6-86, or:

$$p(r_w, t) = p_i - \left(\frac{Q_o \, B_o \, \mu_o}{0.00708 \, k \, h} \right) p_D \qquad (6\text{-}97)$$

Example 6-12

A well is producing at a constant flow rate of 300 STB/day under unsteady-state flow condition. The reservoir has the following rock and fluid properties (see Example 6-10):

$B_o = 1.25$ bbl/STB	$\mu_o = 1.5$ cp	$c_t = 12 \times 10^{-6}$ psi^{-1}
$k = 60$ md	$h = 15$ ft	$p_i = 4000$ psi
$\phi = 15\%$	$r_w = 0.25'$	

Assuming an infinite acting reservoir, i.e., $r_{eD} = \infty$, calculate the bottom-hole flowing pressure after one hour of production by using the dimensionless pressure approach.

Solution

Step 1. Calculate the dimensionless time t_D from Equation 6-87.

$$t_D = \frac{0.000264 \, (60) \, (1)}{(0.15)(1.5)(12 \times 10^{-6})(0.25)^2} = 93,866.67$$

Step 2. Since $t_D > 100$, use Equation 6-92 to calculate the dimensionless pressure drop function:

$$p_D = 0.5 \ [\ln \ (93{,}866.67) + 0.80907] = 6.1294$$

Step 3. Calculate the bottom-hole pressure after 1 hour by applying Equation 6-97:

$$p(0.25, 1) = 4000 - \left[\frac{(300) \ (1.25) \ (1.5)}{0.00708 \ (60) \ (15)} \right] (6.1294) = 3459 \ \text{psi}$$

The above example shows that the solution as given by the p_D-function technique is identical to that of the E_i-function approach. The main difference between the two formulations is that **the p_D-function can be only used to calculate the pressure at radius r when the flow rate Q is constant and known.** In that case, the p_D-function application is essentially restricted to the wellbore radius because the rate is usually known. On the other hand, the E_i-function approach can be used to calculate the pressure at any radius in the reservoir by using the well flow rate Q.

It should be pointed out that, for an infinite-acting reservoir with $t_D > 100$, the p_D-function is related to the E_i-function by the following relation:

$$p_D = 0.5 \left[-E_i \left(\frac{-1}{4 \, t_D} \right) \right] \qquad (6-98)$$

The previous example, i.e., Example 6-12, is not a practical problem, but it is essentially designed to show the physical significance of the p_D solution approach. In transient flow testing, we normally record the bottom-hole flowing pressure as a function of time. Therefore, the dimensionless pressure drop technique can be used to determine one or more of the reservoir properties, e.g., k or kh, as discussed later in this chapter.

Radial Flow of Compressible Fluids

Gas viscosity and density vary significantly with pressure and therefore the assumptions of Equation 6-76 are not satisfied for gas systems, i.e., compressible fluids. In order to develop the proper mathematical function for describing the flow of compressible fluids in the reservoir, the following two addition gas equations must be considered:

• Real density equation

$$\rho = \frac{pM}{zRT}$$

• Gas compressibility equation

$$c_g = \frac{1}{p} - \frac{1}{z}\frac{dz}{dp}$$

Combining the above two basic gas equations with that of Equation 6-68 gives:

$$\frac{1}{r}\frac{\partial}{\partial r}\left(r\frac{p}{\mu z}\frac{\partial p}{\partial r}\right) = \frac{\phi \mu c_t}{0.000264\,k}\frac{p}{\mu z}\frac{\partial p}{\partial t} \qquad (6\text{-}99)$$

where   t = time, hr
        k = permeability, md
        c_t = total isothermal compressibility, psi^{-1}
        ϕ = porosity

Al-Hussainy, Ramey, and Crawford (1966) linearize the above basic flow equation by introducing the real gas potential m(p) to Equation 6-99. Recall the previously defined m(p) equation:

$$m(p) = \int_{0}^{p}\frac{2p}{\mu z}\,dp \qquad (6\text{-}100)$$

Differentiating the above relation with respect to p gives:

$$\frac{\partial m(p)}{\partial p} = \frac{2p}{\mu z} \qquad (6\text{-}101)$$

Obtain the following relationships by applying the chair rule:

$$\frac{\partial m(p)}{\partial r} = \frac{\partial m(p)}{\partial p}\frac{\partial p}{\partial r} \qquad (6\text{-}102)$$

$$\frac{\partial m(p)}{\partial t} = \frac{\partial m(p)}{\partial p}\frac{\partial p}{\partial t} \qquad (6\text{-}103)$$

Substituting Equation 6-101 into Equations 6-102 and 6-103 gives:

$$\frac{\partial p}{\partial r} = \frac{\mu z}{2p}\frac{\partial m(p)}{\partial r} \qquad (6\text{-}104)$$

and

$$\frac{\partial p}{\partial t} = \frac{\mu z}{2p}\frac{\partial m(p)}{\partial t} \qquad (6\text{-}105)$$

Combining Equations 6-104 and 6-105 with 6-99 yields:

$$\frac{\partial^2 m(p)}{\partial r^2} + \frac{1}{r}\frac{\partial m(p)}{\partial r} = \frac{\phi\mu c_t}{0.000264\,k}\frac{\partial m(p)}{\partial t} \qquad (6\text{-}106)$$

Equation 6-106 is the radial diffusivity equation for compressible fluids. This differential equation relates the real gas pseudopressure (real gas potential) to the time t and the radius r. Al-Hussainy, Ramey, and Crawford (1966) pointed out that in gas well testing analysis, the constant-rate solution has more practical applications than that provided by the constant-pressure solution. The authors provided with the exact solution to Equation 6-106 that is commonly referred to as the m(p)-solution method. There are also two other solutions that approximate the exact solution. These two approximation methods are called the pressure-squared method and the *pressure-approximation method.* In general, there are three forms of the mathematical solution to the diffusivity equation:

• The m(p)-Solution Method (Exact Solution)
• The Pressure-Squared Method (p^2-Approximation Method)
• The Pressure Method (p-Approximation Method)

These three methods are presented as follows:

The m(p)-Solution Method (Exact-Solution)

Imposing the constant-rate condition as one of the boundary conditions required to solve Equation 6-106, Al-Hussainy, et al. (1966) proposed the following exact solution to the diffusivity equation:

$$m(p_{wf}) = m(p_i) - 57895.3 \left(\frac{p_{sc}}{T_{sc}} \right) \left(\frac{Q_g T}{kh} \right)$$
$$\left[\log \left(\frac{kt}{\phi \mu_i c_{ti} r_w^2} \right) - 3.23 \right] \tag{6-107}$$

where p_{wf} = bottom-hole flowing pressure, psi
p_e = initial reservoir pressure
Q_g = gas flow rate, Mscf/day
t = time, hr
k = permeability, md
p_{sc} = standard pressure, psi
T_{sc} = standard temperature, °R
T = reservoir temperature
r_w = wellbore radius, ft
h = thickness, ft
μ_i = gas viscosity at the initial pressure, cp
c_{ti} = total compressibility coefficient at p_i, psi^{-1}
ϕ = porosity

When $p_{sc} = 14.7$ psia and $T_{sc} = 520$ °R, Equation 6-107 reduces to:

$$m(p_{wf}) = m(p_i) - \left(\frac{1637 Q_g T}{kh} \right) \left[\log \left(\frac{kt}{\phi \mu_i c_{ti} r_w^2} \right) - 3.23 \right] \tag{6-108}$$

Equation 6-108 can be written equivalently in terms of the dimensionless time t_D as:

$$m(p_{wf}) = m(p_i) - \left(\frac{1637 Q_g T}{kh} \right) \left[\log \left(\frac{4 t_D}{\gamma} \right) \right] \tag{6-109}$$

The dimensionless time is defined previously by Equation 6-86 as:

$$t_D = \frac{0.000264\,kt}{\phi\,\mu_i\,c_{ti}\,r_w^2}$$

The parameter γ is called Euler's constant and given by:

$$\gamma = e^{0.5772} = 1.781 \tag{6-110}$$

The solution to the diffusivity equation as given by Equations 6-108 and 6-109 expresses the bottom-hole real gas pseudopressure as a function of the transient flow time t. The solution as expressed in terms of m(p) is recommended mathematical expression for performing gas-well pressure analysis due to its applicability in all pressure ranges.

The radial gas diffusivity equation can be expressed in a dimensionless form in terms of the dimensionless real gas pseudopressure drop ψ_D. The solution to the dimensionless equation is given by:

$$m\,(p_{wf}) = m\,(p_i) - \left(\frac{1422\,Q_g\,T}{kh}\right)\psi_D \tag{6-111}$$

where Q_g = gas flow rate, Mscf/day
$\quad\quad\;\; k$ = permeability, md

The dimensionless pseudopressure drop ψ_D can be determined as a function of t_D by using the appropriate expression of Equations 6-91 through 6-96. When $t_D > 100$, the ψ_D can be calculated by applying Equation 6-82, or:

$$\psi_D = 0.5\,[\ln\,(t_D) + 0.80907] \tag{6-112}$$

Example 6-13

A gas well with a wellbore radius of 0.3 ft is producing at a constant flow rate of 2000 Mscf/day under transient flow conditions. The initial reservoir pressure (shut-in pressure) is 4400 psi at 140°F. The formation permeability and thickness are 65 md and 15 ft, respectively. The porosity is recorded as 15%. Example 6-7 documents the properties of the gas as well as values of m(p) as a function of pressures. The table is reproduced below for convenience:

p	μ_g (cp)	z	m(p), psi^2/cp
0	0.01270	1.000	0.000
400	0.01286	0.937	13.2×10^6
800	0.01390	0.882	52.0×10^6
1200	0.01530	0.832	113.1×10^6
1600	0.01680	0.794	198.0×10^6
2000	0.01840	0.770	304.0×10^6
2400	0.02010	0.763	422.0×10^6
2800	0.02170	0.775	542.4×10^6
3200	0.02340	0.797	678.0×10^6
3600	0.02500	0.827	816.0×10^6
4000	0.02660	0.860	950.0×10^6
4400	0.02831	0.896	1089.0×10^6

Assuming that the initial total isothermal compressibility is 3×10^{-4} psi^{-1}, calculate, the bottom-hole flowing pressure after 1.5 hours.

Step 1. Calculate the dimensionless time t_D

$$t_D = \frac{(0.000264)(65)(1.5)}{(0.15)(0.02831)(3 \times 10^{-4})(0.3^2)} = 224,498.6$$

Step 2. Solve for m(p_{wf}) by using Equation 6-109

$$m(p_{wf}) = 1089 \times 10^6 - \frac{(1637)(2000)(600)}{(65)(15)} \left[\log \left(\frac{(4)224498.6}{e^{0.5772}} \right) \right]$$
$$= 1077.5 \,(10^6)$$

Step 3. From the given PVT data, interpolate using the value of m(p_{wf}) to give a corresponding p_{wf} of 4367 psi.

An identical solution can be obtained by applying the ψ_D approach as shown below:

Step 1. Calculate ψ_D from Equation 6-112

$$\psi_D = 0.5\,[\ln\,(224498.6) + 0.8090] = 6.565$$

Step 2. Calculate $m(p_{wf})$ by using Equation 6-111

$$m(p_{wf}) = 1089 \times 10^6 - \left(\frac{1422(2000)(600)}{(65)(15)} \right)(6.565) = 1077.5 \times 10^6$$

The Pressure-Squared Approximation Method (p²-method)

The first approximation to the exact solution is to remove the pressure-dependent term (μz) outside the integral that defines $m(p_{wf})$ and $m(p_i)$ to give:

$$m(p_i) - m(p_{wf}) = \frac{2}{\mu z} \int_{p_{wf}}^{p_i} p \, dp \qquad (6-113)$$

or

$$m(p_i) - m(p_{wf}) = \frac{p_i^2 - p_{wf}^2}{\bar{\mu} \, \bar{z}} \qquad (6-114)$$

The bars over μ and z represent the values of the gas viscosity and deviation factor as evaluated at the average pressure $\bar{p}$. This average pressure is given by:

$$\bar{p} = \sqrt{\frac{p_i^2 + p_{wf}^2}{2}} \qquad (6-115)$$

Combining Equation 6-114 with Equation 6-108, 6-109, or 6-111 gives:

$$p_{wf}^2 = p_i^2 - \left(\frac{1637 Q_g \, T \bar{\mu} \, \bar{z}}{kh} \right) \left[\log\left(\frac{kt}{\phi \mu_i \, c_{ti} \, r_w^2} \right) - 3.23 \right] \qquad (6-116)$$

or

$$p_{wf}^2 = p_i^2 - \left(\frac{1637 Q_g \, T \bar{\mu} \, \bar{z}}{kh} \right) \left[\log\left(\frac{4 t_D}{\gamma} \right) \right] \qquad (6-117)$$

or, equivalently:

$$p_{wf}^2 = p_i^2 - \left(\frac{1422 Q_g T \bar{\mu} \bar{z}}{kh}\right) \psi_D \tag{6-118}$$

The above approximation solution forms indicate that the product (μz) is assumed constant at the average pressure $\bar{p}$. This effectively limits the applicability of the p^2-method to reservoir pressures < 2000. It should be pointed out that when the p^2-method is used to determine p_{wf} it is perhaps sufficient to set $\bar{\mu}\,\bar{z} = \mu_i\,z$.

Example 6-14

A gas well is producing at a constant rate of 7454.2 Mscf/day under transient flow conditions. The following data are available:

$k = 50$ md	$h = 10$ ft	$\phi = 20\%$	$p_i = 1600$ psi
$T = 600\ °R$	$r_w = 0.3$ ft	$c_{ti} = 6.25 \times 10^{-4}\ psi^{-1}$	

The gas properties are tabulated below:

p	μ_g , cp	z	m(p) , psi²/cp
0	0.01270	1.000	0.000
400	0.01286	0.937	13.2×10^6
800	0.01390	0.882	52.0×10^6
1200	0.01530	0.832	113.1×10^6
1600	0.01680	0.794	198.0×10^6

Calculate the bottom-hole flowing pressure after 4 hours by using.

a. The m(p)-method
b. The p^2-method

Solution

a. The m(p)-method

Step 1. Calculate t_D

$$t_D = \frac{0.000264\,(50)\,(4)}{(0.2)\,(0.0168)\,(6.25 \times 10^{-4})\,(0.3^2)} = 279,365.1$$

Step 2. Calculate ψ_D:

$$\psi_D = 0.5\,[\text{Ln}(279365.1) + 0.80907] = 6.6746$$

Step 3. Solve for $m(p_{wf})$ by applying Equation 6-111:

$$m(p_{wf}) = (198 \times 10^6) - \left[\frac{1422\,(7454.2)\,(600)}{(50)\,(10)}\right]6.6746 = 113.1 \times 10^6$$

The corresponding value of $p_{wf} = 1200$ psi

b. The p²-method

Step 1. Calculate ψ_D by applying Equation 6-112:

$$\psi_D = 0.5\,[\ln(279365.1) + 0.80907] = 6.6477$$

Step 2. Calculate p_{wf}^2 by applying Equation 6-118:

$$p_{wf}^2 = 1600^2 - \left[\frac{(1422)\,(7454.2)\,(600)\,(0.0168)\,(0.794)}{(50)\,(10)}\right]$$
$$\times 6.6747 = 1,427,491$$

$$p_{wf} = 1195 \text{ psi}$$

Step 3. The absolute average error is 0.4%

The Pressure-Approximation Method

The second method of approximation to the exact solution of the radial flow of gases is to treat the gas as a *pseudoliquid*.

Recalling the gas formation volume factor B_g as expressed in bbl/scf is given by:

$$B_g = \left(\frac{p_{sc}}{5.615 T_{sc}}\right)\left(\frac{zT}{p}\right)$$

Solving the above expression for p/z gives:

$$\frac{p}{z} = \left(\frac{T p_{sc}}{5.615 T_{sc}}\right)\left(\frac{1}{B_g}\right)$$

The difference in the real gas pseudopressure is given by:

$$m(p_i) - (p_{wf}) = \int_{p_{wf}}^{p_i} \frac{2p}{\mu z} dp$$

Combining the above two expressions gives:

$$m(p_i) - m(p_{wf}) = \frac{2T p_{sc}}{5.615 T_{sc}} \int_{p_{wf}}^{p_i} \left(\frac{1}{\mu B_g} \right) dp \qquad (6\text{-}119)$$

Fetkovich (1973) suggested that at high pressures (p > 3000), $1/\mu B_g$ is nearly constant as shown schematically in Figure 6-22. Imposing Fetkovich's condition on Equation 6-119 and integrating gives:

$$m(p_i) - m(p_{wf}) = \frac{2T p_{sc}}{5.615 T_{sc} \bar{\mu} \bar{B}_g} (p_i - p_{wf}) \qquad (6\text{-}120)$$

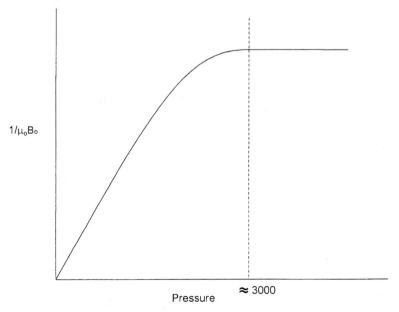

Figure 6-22. $1/\mu_o B_o$ vs. pressure.

Combining Equation 6-120 with Equation 6-108, 6-109, or 6-111 gives:

$$p_{wf} = p_i - \left(\frac{162.5 \times 10^3 \, Q_g \, \bar{\mu} \, \bar{B}_g}{kh}\right)\left[\log\left(\frac{kt}{\phi \bar{\mu} \, \bar{c}_t \, r_w^2}\right) - 3.23\right] \qquad (6\text{-}121)$$

or

$$p_{wf} = p_i - \left(\frac{162.5 \, (10^3) Q_g \, \bar{\mu} \, \bar{B}_g}{kh}\right)\left[\log\left(\frac{4 t_D}{\gamma}\right)\right] \qquad (6\text{-}122)$$

or equivalently in terms of dimensionless pressure drop:

$$p_{wf} = p_i - \left(\frac{141.2 \, (10^3) Q_g \, \bar{\mu} \, \bar{B}_g}{kh}\right) p_D \qquad (6\text{-}123)$$

where Q_g = gas flow rate, Mscf/day
   k = permeability, md
   $\bar{B}_g$ = gas formation volume factor, bbl/scf
   t = time, hr
   p_D = dimensionless pressure drop
   t_D = dimensionless time

It should be noted that the gas properties, i.e., μ, B_g, and c_t, are evaluated at pressure $\bar{p}$ as defined below:

$$\bar{p} = \frac{p_i + p_{wf}}{2} \qquad (6\text{-}124)$$

Again, this method is only limited to applications above 3000 psi. When solving for p_{wf}, it might be sufficient to evaluate the gas properties at p_i.

Example 6-15

Resolve Example 6-13 by using the p-approximation method and compare with the exact solution.

Solution

Step 1. Calculate the dimensionless time t_D.

$$t_D = \frac{(0.000264)(65)(1.5)}{(0.15)(0.02831)(3 \times 10^{-4})(0.3^2)} = 224,498.6$$

Step 2. Calculate B_g at p_i.

$$B_g = 0.00504 \frac{(0.896)(600)}{4400} = 0.0006158 \text{ bbl/scf}$$

Step 3. Calculate the dimensionless pressure p_D by applying Equation 8-92.

$$p_D = 0.5[\ln(224498.6) + 0.80907] = 6.565$$

Step 4. Approximate p_{wf} from Equation 6-123.

$$p_{wf} = 4400 - \left[\frac{141.2 \times 10^3 (2000)(0.02831)(0.0006158)}{(65)(15)} \right] 6.565$$
$$= 4367 \text{ psi}$$

The solution is identical to that of the exact solution.

It should be pointed that Examples 6-10 through 6-15 are designed to illustrate the use of different solution methods. These examples are not practical, however, because in transient flow analysis, the bottom-hole flowing pressure is usually available as a function of time. All the previous methodologies are essentially used to characterize the reservoir by determining the permeability k or the permeability-thickness product (kh).

PSEUDOSTEADY-STATE FLOW

In the unsteady-state flow cases discussed previously, it was assumed that a well is located in a very large reservoir and producing at a constant flow rate. This rate creates a pressure disturbance in the reservoir that travels throughout this infinite-size reservoir. During this transient flow period, reservoir boundaries have no effect on the pressure behavior of the well. Obviously, the time period where this assumption can be imposed is often very short in length. As soon as the pressure disturbance reaches all drainage boundaries, it ends the transient (unsteady-state) flow regime. A different flow regime begins that is called **pseudosteady (semisteady)-state flow.** It is necessary at this point to impose different boundary conditions on the diffusivity equation and drive an appropriate solution to this flow regime.

Consider Figure 6-23, which shows a well in radial system that is producing at a constant rate for a long enough period that eventually affects the entire drainage area. During this semisteady-state flow, the change in pressure with time becomes the same throughout the drainage area. Section B in Figure 6-23 shows that the pressure distributions become paralleled at successive time periods. Mathematically, this important condition can be expressed as:

$$\left(\frac{\partial p}{\partial t}\right)_r = \text{constant}$$

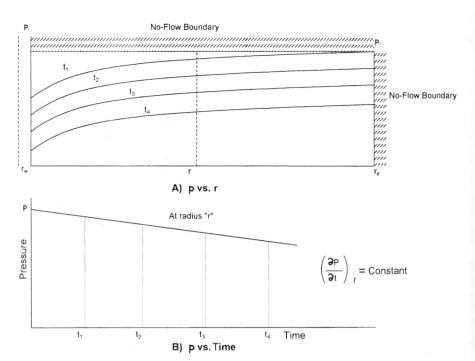

Figure 6-23. Semisteady-state flow regime.

The constant referred to in the above equation can be obtained from a simple material balance using the definition of the compressibility, thus:

$$c = \frac{-1}{V} \frac{dV}{dp}$$

Arranging:

$$cVdp = -dV$$

Differentiating with respect to time t:

$$cV \frac{dp}{dt} = -\frac{dV}{dt} = q$$

or

$$\frac{dp}{dt} = -\frac{q}{cV}$$

Expressing the pressure decline rate dp/dt in the above relation in psi/hr gives:

$$\frac{dp}{dt} = -\frac{q}{24\,cV} = -\frac{Q_o\,B_o}{24\,cV} \qquad (6\text{-}126)$$

where   q = flow rate, bbl/day
   Q_o = flow rate, STB/day
   dp/dt = pressure decline rate, psi/hr
   V = pore volume, bbl

For a radial drainage system, the pore volume is given by:

$$V = \frac{\pi\,r_e^2\,h\phi}{5.615} = \frac{Ah\phi}{5.615} \qquad (6\text{-}127)$$

where A = drainage area, ft^2

Combining Equation 6-127 with Equation 6-126 gives:

$$\frac{dp}{dt} = -\frac{0.23396\,q}{c_t\,\pi\,r_e^2\,h\phi} = \frac{-0.23396\,q}{c_t\,Ah\phi}$$

(6-128)

Examination of the above expression reveals the following important characteristics of the behavior of the pressure decline rate dp/dt during the semisteady-state flow:

• The reservoir pressure declines at a higher rate with an increase in the fluids production rate
• The reservoir pressure declines at a slower rate for reservoirs with higher total compressibility coefficients
• The reservoir pressure declines at a lower rate for reservoirs with larger pore volumes

Example 6-16

An oil well is producing at a constant oil flow rate of 1200 STB/day under a semisteady-state flow regime. Well testing data indicate that the pressure is declining at a constant rate of 4.655 psi/hr. The following additional data is available:

h = 25 ft $\phi = 15\%$ $B_o = 1.3$ bbl/STB
$c_t = 12 \times 10^{-6}$ psi^{-1}

Calculate the well drainage area.

Solution

• $q = Q_o\,B_o$
• $q = (1200)\,(1.3) = 1560$ bb/day
• Apply Equation 6-128 to solve for A.

$$-4.655 = -\frac{0.23396\,(1560)}{(12 \times 10^{-6})(A)(25)(0.15)}$$

A = 1,742,400 ft^2

or

A = 1,742,400 / 43,560 = 40 acres

Matthews, Brons, and Hazebroek (1954) pointed out that once the reservoir is producing under the *semisteady-state condition,* each well will drain from within its own no-flow boundary independently of the other wells. For this condition to prevail, the pressure decline rate dp/dt must be approximately constant throughout the entire reservoir, otherwise flow would occur across the boundaries causing a readjustment in their positions. Because the pressure at every point in the reservoir is changing at the same rate, it leads to the conclusion that the average reservoir pressure is changing at the same rate. This average reservoir pressure is essentially set equal to the volumetric average reservoir pressure $\bar{p}_r$. It is the pressure that is used to perform flow calculations during the semisteady state flowing condition. In the above discussion, $\bar{p}_r$ indicates that, in principal, Equation 6-128 can be used to estimate by replacing the pressure decline rate dp/dt with $(p_i - \bar{p}_r)/t$, or:

$$p_i - \bar{p}_r = \frac{0.23396qt}{c_t\,Ah\phi}$$

or

$$\bar{p}_r = p_i - \frac{0.23396\,q\,t}{c_t Ah\phi} \qquad (6\text{-}129)$$

where t is approximately the elapsed time since the end of the transient flow regime to the time of interest.

It should be noted that when performing material balance calculations, the volumetric average pressure of the entire reservoir is used to calculate the fluid properties. This pressure can be determined from the individual well drainage properties as follows:

$$\bar{p}_r = \frac{\sum_i \bar{p}_{ri} V_i}{\sum_i V_i} \qquad (6\text{-}130)$$

in which V_i = pore volume of the ith drainage volume

$\bar{p}_{ri}$ = volumetric average pressure within the ith drainage
volume.

Figure 6-24 illustrates the concept of the volumetric average pressure. In practice, the V_i's are difficult to determine and, therefore, it is common to use the flow rate q_i in Equation 6-129.

$$\bar{p}_r = \frac{\sum\limits_i (\bar{p}_{ri} q_i)}{\sum\limits_i q_i} \qquad (6\text{-}131)$$

The flow rates are measured on a routing basis throughout the lifetime of the field thus facilitating the calculation of the volumetric average reservoir pressure p_r.

The practical applications of using the pseudosteady-state flow condition to describe the flow behavior of the following two types of fluids are presented below:

• Radial flow of slightly compressible fluids
• Radial flow of compressible fluids

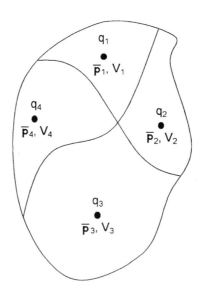

Figure 6-24. Volumetric average reservoir pressure.

Radial Flow of Slightly Compressible Fluids

The diffusivity equation as expressed by Equation 6-73 for the transient flow regime is:

$$\frac{\partial^2 p}{\partial r^2} + \frac{1}{r}\frac{\partial p}{\partial r} = \left(\frac{\phi\mu c_t}{0.000264k}\right)\frac{\partial p}{\partial t}$$

For the semisteady-state flow, the term $(\partial p/\partial t)$ is constant and is expressed by Equation 6-128. Substituting Equation 6-128 into the diffusivity equation gives:

$$\frac{\partial^2 p}{\partial r^2} + \frac{1}{r}\frac{\partial p}{\partial r} = \left(\frac{\phi\mu c_t}{0.000264k}\right)\left(\frac{-0.23396q}{c_t\, Ah\, \phi}\right)$$

or

$$\frac{\partial^2 p}{\partial r^2} + \frac{1}{r}\frac{\partial p}{\partial r} = \frac{-887.22\, q\mu}{Ahk} \qquad (6\text{-}132)$$

Equation 6-132 can be expressed as:

$$\frac{1}{r}\frac{\partial}{\partial r}\left(r\frac{\partial p}{\partial r}\right) = -\frac{887.22\, q\mu}{(\pi r_e^2)hk}$$

Integrating the above equation gives:

$$r\frac{\partial p}{\partial r} = -\frac{887.22\, q\mu}{(\pi r_e^2)\, hk}\left(\frac{r^2}{2}\right) + c_1$$

Where c_1 is the constant of the integration and can be evaluated by imposing the outer no-flow boundary condition [i.e., $(\partial p/\partial r)_{r_e} = 0$] on the above relation to give:

$$c_1 = \frac{141.2\, q\mu}{\pi\, hk}$$

Combining the above two expressions gives:

$$\frac{\partial p}{\partial r} = \frac{141.2\,q\mu}{hk}\left(\frac{1}{r} - \frac{r}{r_e^2}\right)$$

Integrating again:

$$\int_{p_{wf}}^{p_i} dp = \frac{141.2\,q\mu}{hk} \int_{rw}^{re}\left(\frac{1}{r} - \frac{r}{r_e^2}\right)dr$$

Performing the above integration and assuming (r_w^2/r_e^2) is negligible gives:

$$(p_i - p_{wf}) = \frac{141.2\,q\mu}{kh}\left[\ln\left(\frac{r_e}{r_w}\right) - \frac{1}{2}\right] \tag{6-133}$$

A more appropriate form of the above is to solve for the flow rate, to give:

$$Q = \frac{0.00708\,kh\,(p_i - p_{wf})}{\mu B\left[\ln\left(\frac{r_e}{r_w}\right) - 0.5\right]} \tag{6-134}$$

where Q = flow rate, STB/day
       B = formation volume factor, bbl/STB
       k = permeability, md

The volumetric average reservoir pressure $\bar{p}_r$ is commonly used in calculating the liquid flow rate under the semisteady-state flowing condition. Introducing the $\bar{p}_r$ into Equation 6-134 gives:

$$Q = \frac{0.00708\,kh\,(\bar{p}_r - p_{wf})}{\mu B\left[\ln\left(\frac{r_e}{r_w}\right) - 0.75\right]} \tag{6-135}$$

Note that:

$$\ln\left(\frac{0.471 r_e}{r_w}\right) = \ln\left(\frac{r_e}{r_w}\right) - 0.75$$

The above observation suggests that the volumetric average pressure $\bar{p}_r$ occurs at about 47% of the drainage radius during the semisteady-state condition.

It is interesting to notice that the dimensionless pressure p_D solution to the diffusivity equation can be used to derive Equation 6-135. The p_D function for a bounded reservoir was given previously by Equation 6-96 for a bounded system as:

$$p_D = \frac{2 t_D}{r_{eD}^2} + \ln(r_{eD}) - 0.75$$

where the above three dimensionless parameters are given by Equations 6-86 through 6-88 as:

$$p_D = \frac{\dfrac{(p_i - p_{wf})}{Q B \mu}}{0.00708 \, k \, h}$$

$$t_D = \frac{0.000264 \, k \, t}{\phi \, \mu \, c_t \, r_w^2}$$

$$r_{eD} = \frac{r_e}{r_w}$$

Combining the above four relationships gives:

$$p_{wf} = p_i - \frac{Q B \mu}{0.00708 \, k \, h}\left[\frac{0.0005274 \, k \, t}{\phi \, \mu \, c_t \, r_e^2} + \ln\left(\frac{r_e}{r_w}\right) - 0.75\right]$$

Solving Equation 6-130 for the time t gives:

$$t = \frac{c_t A h \phi (p_i - \bar{p}_r)}{0.23396 \, Q \, B} = \frac{c_t (\pi r_e^2) h \phi (p_i - \bar{p}_r)}{0.23396 \, Q \, B}$$

Combining the above two equations and solving for the flow rate Q yields:

$$Q = \frac{0.00708 \, k \, h \, (\bar{p}_r - p_{wf})}{\mu B \left[\ln \left(\frac{r_e}{r_w} \right) - 0.75 \right]}$$

It should be pointed out that the pseudosteady-state flow occurs regardless of the geometry of the reservoir. Irregular geometries also reach this state when they have been produced long enough for the entire drainage area to be affected.

Rather than developing a separate equation for each geometry, Ramey and Cobb (1971) introduced a correction factor that is called the **shape factor,** C_A, which is designed to account for the deviation of the drainage area from the ideal circular form. The shape factor, as listed in Table 6-4, accounts also for the location of the well within the drainage area. Introducing C_A into Equation 6-132 and performing the solution procedure gives the following two solutions:

• In terms of the volumetric average pressure $\bar{p}_r$:

$$p_{wf} = \bar{p}_r - \frac{162.6 \, QB\mu}{kh} \log \left[\frac{4A}{1.781 C_A r_w^2} \right] \qquad (6\text{-}136)$$

• In terms of the initial reservoir pressure p_i:

Recalling Equation 6-129 which shows the changes of the average reservoir pressure as a function of time and initial reservoir pressure p_i:

$$\bar{p}_r = p_i - \frac{0.23396 \, q \, t}{c_t A h \phi}$$

Combining the above equation with Equation 6-136 gives:

$$p_{wf} = \left[p_i - \frac{0.23396\, Q\, Bt}{Ah\, \phi\, c_t} \right] - \frac{162.6\, Q\, B\mu}{kh} \log\left[\frac{4A}{1.781\, C_A\, r_w^2} \right] \quad (6\text{-}137)$$

where k = permeability, md
  A = drainage area, ft²
  C_A = shape factor
  Q = flow rate, STB/day
  t = time, hr
  c_t = total compressibility coefficient, psi⁻¹

Equation 6-136 can be arranged to solve for Q to give:

$$Q = \frac{kh\, (\overline{p}_r - p_{wf})}{162.6\, B\mu\, \log\left[\dfrac{4A}{1.781\, C_A\, r_w^2} \right]} \quad (6\text{-}138)$$

It should be noted that if Equation 6-138 is applied to a circular reservoir of a radius r_e, then:

$$A = \pi\, r_e^2$$

and the shape factor for a circular drainage area as given in Table 6-3 is:

$$C_A = 31.62$$

Substituting in Equation 6-138, it reduces to:

$$p_{wf} = \overline{p}_r - \left(\frac{QB\mu}{0.00708\, kh} \right) \left[\ln\left(\frac{r_e}{r_w} \right) - 0.75 \right]$$

The above equation is identical to that of Equation 6-135.

(text continued on page 416)

Table 6-4
Shape Factors for Various Single-Well Drainage Areas
(*After Earlougher, R.*, Advances in Well Test Analysis, permission to publish by the SPE, copyright SPE, 1977)

In Bounded Reservoirs	C_A	$\ln C_A$	$\frac{1}{2}\ln\left(\dfrac{2.2458}{C_A}\right)$	Exact for $t_{DA} >$	Less than 1% Error For $t_{DA} >$	Use Infinite System Solution with Less Than 1% Error for $t_{DA} <$
(circle)	31.62	3.4538	−1.3224	0.1	0.06	0.10
(hexagon)	31.6	3.4532	−1.3220	0.1	0.06	0.10
(triangle)	27.6	3.3178	−1.2544	0.2	0.07	0.09
(60° parallelogram)	27.1	3.2995	−1.2452	0.2	0.07	0.09
(right triangle, 1/3)	21.9	3.0865	−1.1387	0.4	0.12	0.08
(3,4 triangle)	0.098	−2.3227	+1.5659	0.9	0.60	0.015
(square, centered)	30.8828	3.4302	−1.3106	0.1	0.05	0.09
(square, 4 quadrants)	12.9851	2.5638	−0.8774	0.7	0.25	0.03
(square, offset)	4.5132	1.5070	−0.3490	0.6	0.30	0.025
(square, thirds)	3.3351	1.2045	−0.1977	0.7	0.25	0.01
(2:1 rectangle, centered)	21.8369	3.0836	−1.1373	0.3	0.15	0.025
(2:1 rectangle)	10.8374	2.3830	−0.7870	0.4	0.15	0.025
(2:1 rectangle, offset)	4.5141	1.5072	−0.3491	1.5	0.50	0.06
(2:1 rectangle)	2.0769	0.7309	−0.0391	1.7	0.50	0.02
(2:1 rectangle, thirds)	3.1573	1.1497	−0.1703	0.4	0.15	0.005

In Bounded Reservoirs	C_A	$\ln C_A$	$\frac{1}{2}\ln\left(\dfrac{2.2458}{C_A}\right)$	Exact for $t_{DA} >$	Less than 1% Error For $t_{DA} >$	Use Infinite System Solution with Less than 1% Error for $t_{DA} <$
	0.5813	−0.5425	+0.6758	2.0	0.60	0.02
	0.1109	−2.1991	+1.5041	3.0	0.60	0.005
	5.3790	1.6825	−0.4367	0.8	0.30	0.01
	2.6896	0.9894	−0.0902	0.8	0.30	0.01
	0.2318	−1.4619	+1.1355	4.0	2.00	0.03
	0.1155	−2.1585	+1.4838	4.0	2.00	0.01
	2.3606	0.8589	−0.0249	1.0	0.40	0.025
IN VERTICALLY FRACTURED RESERVOIRS	Use $(x_e/x_f)^2$ in place of A/r_w^2 for fractured systems					
0.1 $=x_f/x_e$	2.6541	0.9761	−0.0835	0.175	0.08	cannot use
0.2	2.0348	0.7104	+0.0493	0.175	0.09	cannot use
0.3	1.9986	0.6924	+0.0583	0.175	0.09	cannot use
0.5	1.6620	0.5080	+0.1505	0.175	0.09	cannot use
0.7	1.3127	0.2721	+0.2685	0.175	0.09	cannot use
1.0	0.7887	−0.2374	+0.5232	0.175	0.09	cannot use
IN WATER-DRIVE RESERVOIRS	19.1	2.95	−1.07	—	—	—
IN RESERVOIRS OF UNKNOWN PRODUCTION CHARACTER	25.0	3.22	−1.20	—	—	—

(*text continued from page 413*)

Example 6-16

An oil well is developed on the center of a 40-acre square-drilling pattern. The well is producing at a constant flow rate of 800 STB/day under a semisteady-state condition. The reservoir has the following properties:

$\phi = 15\%$	$h = 30$ ft	$k = 200$ md
$\mu = 1.5$ cp	$B_o = 1.2$ bbl/STB	$c_t = 25 \times 10^{-6}$ psi-1
$p_i = 4500$ psi	$r_w = 0.25$ ft	$A = 40$ acres

a. Calculate and plot the bottom-hole flowing pressure as a function of time.
b. Based on the plot, calculate the pressure decline rate. What is the decline in the average reservoir pressure from $t = 10$ to $t = 200$ hr ?

Solution

a. p_{wf} calculations:

Step 1. From Table 6-3, determine C_A:

$$C_A = 30.8828$$

Step 2. Convert the area A from acres to ft²:

$$A = (40)\,(43{,}560) = 1{,}742{,}400 \text{ ft}^2$$

Step 3. Apply Equation 6-137:

$$P_{wf} = 4500 - 1.719\,t - 58.536 \log (2{,}027{,}436)$$

or

$$p_{wf} = 4493.69 - 1.719\,t$$

Step 4. Calculate p_{wf} at different assumed times.

t, hr	$P_{wf} = 44369 - 1.719\,t$
10	4476.50
20	4459.31
50	4407.74
100	4321.79
200	4149.89

Step 5. Present the results of Step 4 in a graphical form as shown in Figure 6-25.

b. It is obvious from Figure 6-25 and the above calculation that the bottom-hole flowing pressure is declining at a rate of 1.719 psi/hr, or:

$$\frac{dp}{dt} = -1.719 \text{ psi/hr}$$

The significance of this example is that the rate of pressure decline during the pseudosteady state is the same throughout the drainage area. This means that the *average reservoir pressure,* p_r, is declining at the same rate of 1.719 psi, therefore the change in p_r from 10 to 200 hours is:

$$\Delta \bar{p}_r = (1.719)(200 - 10) = 326.6 \text{ psi}$$

Example 6-17

An oil well is producing under a constant bottom-hole flowing pressure of 1500 psi. The current average reservoir pressure p_r is 3200 psi.

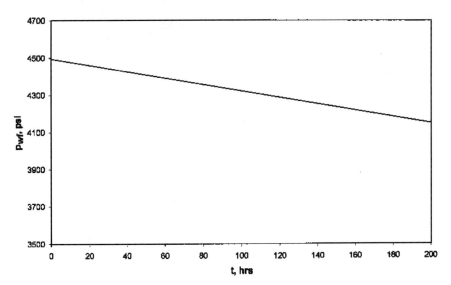

Figure 6-25. Bottom-hole flowing pressure as a function of time.

The well is developed in the center of a 40-acre square drilling pattern. Given the following additional information:

$\phi = 16\%$ $h = 15$ ft $k = 50$ md
$\mu = 26$ cp $B_o = 1.15$ bbl/STB $c_t = 10 \times 10^{-6}$ psi^{-1}
$r_w = 0.25$ ft

calculate the flow rate.

Solution

Because the volumetric average pressure is given, solve for the flow rate by applying Equation 6-138.

$$Q = \frac{(50)(15)(3200-1500)}{(162.6)(1.15)(2.6)\log\left[\frac{(4)(40)(43,560)}{1.781(30.8828)(0.25^2)}\right]}$$

$$= 416 \text{ STB/day}$$

Radial Flow of Compressible Fluids (Gases)

The radial diffusivity equation as expressed by Equation 6-106 was developed to study the performance of compressible fluid under unsteady-state conditions. The equation has the following form:

$$\frac{\partial^2 m(p)}{\partial r^2} + \frac{1}{r}\frac{\partial m(p)}{\partial r} = \frac{\phi\mu c_t}{0.000264\,k}\frac{\partial m(p)}{\partial t}$$

For the semisteady-state flow, the rate of change of the real gas pseudopressure with respect to time is constant, i.e.,

$$\frac{\partial m(p)}{\partial t} = \text{constant}$$

Using the same technique identical to that described previously for liquids gives the following exact solution to the diffusivity equation:

$$Q_g = \frac{kh[m(\bar{p}_r) - m(p_{wf})]}{1422\,T\left[\ln\left(\frac{r_e}{r_w}\right) - 0.75\right]} \tag{6-139}$$

where Q_g = gas flow rate, Mscf/day
$ T$ = temperature, °R
$ k$ = permeability, md

Two approximations to the above solution are widely used. These approximations are:

• Pressure-squared approximation
• Pressure-approximation

Pressure-Squared Approximation Method

As outlined previously, the method provides us with compatible results to that of the exact solution approach when p < 2000. The solution has the following familiar form:

$$Q_g = \frac{kh(\bar{p}_r^2 - p_{wf}^2)}{1422\,T\bar{\mu}\,\bar{z}\left(\ln\dfrac{r_e}{r_w} - 0.75\right)} \tag{6-140}$$

The gas properties $\bar{z}$ and μ are evaluated at:

$$\bar{p} = \sqrt{\frac{(\bar{p}_r)^2 + p_{wf}^2}{2}}$$

Pressure-Approximation Method

This approximation method is applicable at p > 3000 psi and has the following mathematical form:

$$Q_g = \frac{kh(\bar{p}_r - p_{wf})}{1422\,\bar{\mu}\,\bar{B}_g\left(\ln\dfrac{r_e}{r_w} - 0.75\right)} \tag{6-141}$$

with the gas properties evaluated at:

$$\bar{p} = \frac{\bar{p}_r + p_{wf}}{2}$$

where Q_g = gas flow rate, Mscf/day
   k = permeability, md
   $\overline{B}_g$ = gas formation volume factor at average pressure, bbl/scf

The gas formation volume factor is given by the following expression:

$$\overline{B}_g = 0.00504 \frac{\overline{z}\,T}{\overline{p}}$$

In deriving the flow equations, the following two main assumptions were made:

• Uniform permeability throughout the drainage area
• Laminar (viscous) flow

Before using any of the previous mathematical solutions to the flow equations, the solution must be modified to account for the possible deviation from the above two assumptions. Introducing the following two correction factors into the solution of the flow equation can eliminate the above two assumptions:

• Skin factor
• Turbulent flow factor

Skin Factor

It is not unusual for materials such as mud filtrate, cement slurry, or clay particles to enter the formation during drilling, completion or workover operations and reduce the permeability around the wellbore. This effect is commonly referred to as a *wellbore damage* and the region of altered permeability is called the *skin zone*. This zone can extend from a few inches to several feet from the wellbore. Many other wells are stimulated by acidizing or fracturing which in effect increase the permeability near the wellbore. Thus, the permeability near the wellbore is always different from the permeability away from the well where the formation has not been affected by drilling or stimulation. A schematic illustration of the skin zone is shown in Figure 6-26.

Those factors that cause damage to the formation can produce additional localized pressure drop during flow. This additional pressure drop is commonly referred to as Δp_{skin}. On the other hand, well stimulation techniques will normally enhance the properties of the formation and increase the permeability around the wellbore, so that a decrease in pres-

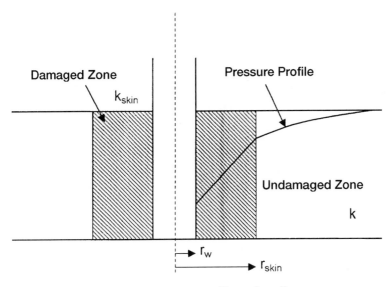

Figure 6-26. Near wellbore skin effect.

sure drop is observed. The resulting effect of altering the permeability around the well bore is called the *skin effect.*

Figure 6-27 compares the differences in the skin zone pressure drop for three possible outcomes:

- First Outcome:
  $\Delta p_{skin} > 0$, indicates an additional pressure drop due to wellbore damage, i.e., $k_{skin} < k$.
- Second Outcome:
  $\Delta p_{skin} < 0$, indicates less pressure drop due to wellbore improvement, i.e., $k_{skin} > k$.
- Third Outcome:
  $\Delta p_{skin} = 0$, indicates no changes in the wellbore condition, i.e., $k_{skin} = k$.

Hawkins (1956) suggested that the permeability in the skin zone, i.e., k_{skin}, is uniform and the pressure drop across the zone can be approximated by Darcy's equation. Hawkins proposed the following approach:

$$\Delta p_{skin} = \begin{bmatrix} \Delta p \text{ in skin zone} \\ \text{due to } k_{skin} \end{bmatrix} - \begin{bmatrix} \Delta p \text{ in the skin zone} \\ \text{due to } k \end{bmatrix}$$

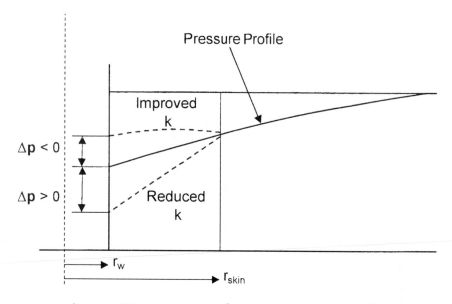

Figure 6-27. Representation of positive and negative skin effects.

Applying Darcy's equation gives:

$$\Delta p_{skin} = \left[\frac{Q_o B_o \mu_o}{0.00708\,h\,k_{skin}}\right] \ln\left(\frac{r_{skin}}{r_w}\right) - \left[\frac{Q_o B_o \mu_o}{0.00708\,h\,k}\right] \ln\left(\frac{r_{skin}}{r_w}\right)$$

or

$$\Delta p_{skin} = \left(\frac{Q_o B_o \mu_o}{0.00708\,kh}\right)\left[\frac{k}{k_{skin}} - 1\right] \ln\left(\frac{r_{skin}}{r_w}\right)$$

where   k = permeability of the formation, md
  k_{skin} = permeability of the skin zone, md

The above expression for determining the additional pressure drop in the skin zone is commonly expressed in the following form:

$$\Delta p_{skin} = \left[\frac{Q_o B_o \mu_o}{0.00708\,kh}\right] s = 141.2\left[\frac{Q_o B_o \mu_o}{kh}\right] s \qquad (6\text{-}142)$$

where s is called the skin factor and defined as:

$$s = \left[\frac{k}{k_{skin}} - 1 \right] \ln \left(\frac{r_{skin}}{r_w} \right) \qquad (6\text{-}143)$$

Equation 6-143 provides some insight into the physical significance of the sign of the skin factor. There are only three possible outcomes in evaluating the skin factor s:

• **Positive Skin Factor, s > 0**
When a damaged zone near the wellbore exists, k_{skin} is less than k and hence s is a positive number. The magnitude of the skin factor increases as k_{skin} decreases and as the depth of the damage r_{skin} increases.
• **Negative Skin Factor, s < 0**
When the permeability around the well k_{skin} is higher than that of the formation k, a negative skin factor exists. This negative factor indicates an improved wellbore condition.
• **Zero Skin Factor, s = 0**
Zero skin factor occurs when no alternation in the permeability around the wellbore is observed, i.e., $k_{skin} = k$.

Equation 6-143 indicates that a negative skin factor will result in a negative value of Δp_{skin}. This implies that a stimulated well will require less pressure drawdown to produce at rate q than an equivalent well with uniform permeability.

The proposed modification of the previous flow equation is based on the concept that the actual total pressure drawdown will increase or decrease by an amount of Δp_{skin}. Assuming that $(\Delta p)_{ideal}$ represents the pressure drawdown for a drainage area with a uniform permeability k, then:

$$(\Delta p)_{actual} = (\Delta p)_{ideal} + (\Delta p)_{skin}$$

or

$$(p_i - p_{wf})_{actual} = (p_i - p_{wf})_{ideal} + \Delta p_{skin} \qquad (6\text{-}144)$$

The above concept as expressed by Equation 6-144 can be applied to all the previous flow regimes to account for the skin zone around the wellbore as follows:

Steady-State Radial Flow

Substituting Equations 6-27 and 6-142 into Equation 6-144 gives:

$$(p_i - p_{wf})_{actual} = \left[\frac{Q_o B_o \mu_o}{0.00708\,kh} \right] \ln\left(\frac{r_e}{r_w}\right) + \left[\frac{Q_o B_o \mu_o}{0.00708\,kh} \right] s$$

or

$$Q_o = \frac{0.00708\,kh\,(p_i - p_{wf})}{\mu_o B_o \left[\ln \dfrac{r_e}{r_w} + s \right]} \tag{6-145}$$

where Q_o = oil flow rate, STB/day
        k = permeability, md
        h = thickness, ft
        s = skin factor
        B_o = oil formation volume factor, bbl/STB
        μ_o = oil viscosity, cp
        p_i = initial reservoir pressure, psi
       p_{wf} = bottom hole flowing pressure, psi

Unsteady-State Radial Flow

• For Slightly Compressible Fluids:

Combining Equations 6-83 and 6-142 with that of Equation 6-144 yields:

$$p_i - p_{wf} = 162.6 \left(\frac{Q_o B_o \mu_o}{kh} \right) \left[\log \frac{kt}{\phi \mu c_t r_w^2} - 3.23 \right]$$
$$+ 141.2 \left(\frac{Q_o B_o \mu_o}{kh} \right) s$$

or

$$p_i - p_{wf} = 162.6 \left(\frac{Q_o B_o \mu_o}{kh} \right) \left[\log \frac{kt}{\phi \mu c_t r_w^2} - 3.23 + 0.87s \right] \tag{6-146}$$

• For Compressible Fluids:

A similar approach to that of the above gives:

$$m(p_{wf}) = m(p_i) - \frac{1637 Q_g T}{kh} \left[\log \frac{kt}{\phi \mu c_{ti} r_w^2} - 3.23 + 0.87s \right] \quad (6\text{-}147)$$

and, in terms of the pressure-squared approach, gives:

$$p_{wf}^2 = p_i^2 - \frac{1037 Q_g T \bar{z} \bar{\mu}}{kh} \left[\log \frac{kt}{\phi \mu_i c_{ti} r_w^2} - 3.23 + 0.87s \right] \quad (6\text{-}148)$$

Pseudosteady-State Flow

• For Slightly Compressible Fluids:

Introducing the skin factor into Equation 6-135 gives:

$$Q_o = \frac{0.00708 \, kh \, (\bar{p}_r - p_{wf})}{\mu_o B_o \left[\ln \left(\frac{r_e}{r_w} \right) - 0.75 + s \right]} \quad (6\text{-}149)$$

• For Compressible Fluids:

$$Q_g = \frac{kh [m (\bar{p}_r) - m(P_{wf})]}{1422 \, T \left[\ln \left(\frac{r_e}{r_w} \right) - 0.75 + s \right]} \quad (6\text{-}150)$$

or, in terms of the pressure-squared approximation, gives:

$$Q_g = \frac{kh (p_r^2 - p_{wf}^2)}{1422 \, T \bar{\mu} \bar{z} \left[\ln \left(\frac{r_e}{r_w} \right) - 0.75 + s \right]} \quad (6\text{-}151)$$

where Q_g = gas flow rate, Mscf/day
$\quad$ k = permeability, md
$\quad$ T = temperature, °R
$\quad$ $(\bar{\mu}_g)$ = gas viscosity at average pressure $\bar{p}$, cp
$\quad$ $\bar{Z}_g$ = gas compressibility factor at average pressure $\bar{p}$

Example 6-18

Calculate the skin factor resulting from the invasion of the drilling fluid to a radius of 2 feet. The permeability of the skin zone is estimated at 20 md as compared with the unaffected formation permeability of 60 md. The wellbore radius is 0.25 ft.

Solution

Apply Equation 6-143 to calculate the skin factor:

$$s = \left[\frac{60}{20} - 1\right] \ln\left(\frac{2}{0.25}\right) = 4.16$$

Matthews and Russell (1967) proposed an alternative treatment to the skin effect by introducing the **effective** or **apparent wellbore radius** r_{wa} that accounts for the pressure drop in the skin. They define r_{wa} by the following equation:

$$r_{wa} = r_w e^{-s} \tag{6-152}$$

All of the ideal radial flow equations can be also modified for the skin by simply replacing wellbore radius r_w with that of the apparent wellbore radius r_{wa}. For example, Equation 6-146 can be equivalently expressed as:

$$p_i - p_{wf} = 162.6 \left(\frac{Q_o B_o \mu_o}{kh}\right)\left[\log \frac{kt}{\phi \mu_o c_t r_{wa}^2} - 3.23\right] \tag{6-153}$$

Turbulent Flow Factor

All of the mathematical formulations presented so far are based on the assumption that laminar flow conditions are observed during flow. During radial flow, the flow velocity increases as the wellbore is approached. This increase in the velocity might cause the development of a turbulent flow around the wellbore. If turbulent flow does exist, it is most likely to occur with gases and causes an additional pressure drop similar to that caused by the skin effect. The term *non-Darcy flow* has been adopted by the industry to describe the additional pressure drop due to the turbulent (non-Darcy) flow.

Referring to the additional real gas pseudopressure drop due to non-Darcy flow as $\Delta\psi$ non-Darcy, the total (actual) drop is given by:

$$(\Delta\psi)_{actual} = (\Delta\psi)_{ideal} + (\Delta\psi)_{skin} + (\Delta\psi)_{non\text{-}Darcy}$$

Wattenburger and Ramey (1968) proposed the following expression for calculating $(\Delta\psi)_{non\text{-}Darcy}$:

$$(\Delta\psi)_{non\text{-}Darcy} = 3.161 \times 10^{-12} \left[\frac{\beta T \gamma_g}{\mu_{gw} h^2 r_w} \right] Q_g^2 \qquad (6\text{-}154)$$

The above equation can be expressed in a more convenient form as:

$$(\Delta\psi)_{non\text{-}Darcy} = FQ_g^2 \qquad (6\text{-}155)$$

where F is called the *non-Darcy flow coefficient* and is given by:

$$F = 3.161 \times 10^{-12} \left[\frac{\beta T \gamma_g}{\mu_{gw} h^2 r_w} \right] \qquad (6\text{-}156)$$

where Q_g = gas flow rate, Mscf/day
μ_{gw} = gas viscosity as evaluated at p_{wf}, cp
γ_g = gas specific gravity
h = thickness, ft
F = non-Darcy flow coefficient, $psi^2/cp/(Mscf/day)^2$
β = turbulence parameter

Jones (1987) proposed a mathematical expression for estimating the turbulence parameter β as:

$$\beta = 1.88 (10^{-10}) (k)^{-1.47} (\phi)^{-0.53} \qquad (6\text{-}157)$$

where   k = permeability, md

ϕ = porosity, fraction

The term $F\,Q_g^2$ can be included in all the compressible gas flow equations in the same way as the skin factor. This non-Darcy term is interpreted as being a *rate-dependent skin*. The modification of the gas flow equations to account for the turbulent flow condition is given below:

Unsteady-State Radial Flow

The gas flow equation for an unsteady-state flow is given by Equation 6-147 and can be modified to include the additional drop in the real gas potential as:

$$m(p_i) - m(p_{wf}) = \left(\frac{1637\,Q_g\,T}{kh}\right)\left[\log\frac{kt}{\phi\,\mu_i\,c_{ti}\,r_w^2} - 3.23 + 0.87s\right]$$
$$+ FQ_g^2 \qquad\qquad (6\text{-}158)$$

Equation 6-158 is commonly written in a more convenient form as:

$$m(p_i) - m(p_{wf}) = \left(\frac{1637\,Q_g\,T}{kh}\right)$$
$$\times\left[\log\frac{kt}{\phi\,\mu_i\,c_{ti}\,r_w^2} - 3.23 + 0.87s + 0.87\,DQ_g\right] \qquad (6\text{-}159)$$

where the term DQ_g is interpreted as the rate dependent skin factor. The coefficient D is called the **inertial** or **turbulent flow factor** and given by:

$$D = \frac{Fkh}{1422T} \qquad\qquad (6\text{-}160)$$

The true skin factor s which reflects the formation damage or stimulation is usually combined with the non-Darcy rate dependent skin and labeled as the **apparent** or **total skin factor**:

$$s' = s + DQ_g \qquad\qquad (6\text{-}161)$$

or

$$m(p_i) - m(p_{wf}) = \left[\frac{1637 Q_g T}{kh} \right]$$

$$\times \left[\log \frac{kt}{\phi \mu_i c_{ti} r_w^2} - 3.23 + 0.87s' \right] \qquad (6\text{-}162)$$

Equation 6-162 can be expressed in the pressure-squared approximation form as:

$$p_i^2 - p_{wf}^2 = \left[\frac{1637 Q_g T \bar{z} \bar{\mu}}{kh} \right] \left[\log \frac{kt}{\phi \mu_i c_{ti} r_w^2} - 3.23 + 0.87s' \right] \qquad (6\text{-}163)$$

where Q_g = gas flow rate, Mscf/day
t = time, hr
k = permeability, md
μ_i = gas viscosity as evaluated at p_i, cp

Semisteady-State Flow

Equations 6-150 and 6-151 can be modified to account for the non-Darcy flow as follows:

$$Q_g = \frac{kh[m(\bar{p}_r) - m(p_{wf})]}{1422 T \left[\ln \left(\frac{r_e}{r_w} \right) - 0.75 + s + DQ_g \right]} \qquad (6\text{-}164)$$

or in terms of the pressure-squared approach:

$$Q_g = \frac{kh(\bar{p}_r^2 - p_{wf}^2)}{1422 T \bar{\mu} \bar{z} \left[\ln \left(\frac{r_e}{r_w} \right) - 0.75 + s + DQ_g \right]} \qquad (6\text{-}165)$$

where the coefficient D is defined as:

$$D = \frac{Fkh}{1422\,T} \tag{6-166}$$

Steady-State Flow

Similar to the above modification procedure, Equations 6-44 and 6-45 can be expressed as:

$$Q_g = \frac{kh[m(p_i) - m(p_{wf})]}{1422\,T\left[\ln\dfrac{r_e}{r_w} - 0.5 + s + DQ_g\right]} \tag{6-167}$$

$$Q_g = \frac{kh(p_e^2 - p_{wf}^2)}{1422\,T\,\bar{\mu}\,\bar{z}\left[\ln\dfrac{r_e}{r_w} - 0.5 + s + DQ_g\right]} \tag{6-168}$$

where D is defined by Equation 6-166.

Example 6-19

A gas well has an estimated wellbore damage radius of 2 feet and an estimated reduced permeability of 30 md. The formation has a permeability and porosity of 55 md and 12%. The well is producing at a rate of 20 MMscf/day with a gas gravity of 0.6. The following additional data is available:

$r_w = 0.25$ $h = 20'$ $T = 140°F$ $\mu_{gw} = 0.013$ cp

Calculate the apparent skin factor.

Solution

Step 1. Calculate skin factor from Equation 6-143

$$s = \left[\frac{55}{30} - 1\right]\ln\left(\frac{2}{0.25}\right) = 1.732$$

Step 2. Calculate the turbulence parameter β by applying Equation 6-155:

$$\beta = 1.88 \ (10)^{-10} \ (55)^{-1.47} \ (0.12)^{-0.53} = 159.904 \times 10^6$$

Step 3. Calculate non-Darcy flow coefficient from Equation 6-156:

$$F = 3.1612 \times 10^{-12} \left[\frac{159.904 \times 10^6 \ (600) \ (0.6)}{(0.013) \ (20)^2 \ (0.25)} \right] = 0.14$$

Step 4. Calculate the coefficient D from Equation 6-160:

$$D = \frac{(0.14) \ (55) \ (20)}{(1422) \ (600)} = 1.805 \times 10^{-4}$$

Step 5. Estimate the apparent skin factor by applying Equation 6-161:

$$s' = 1.732 + (1.805 \times 10^{-4}) \ (20,000) = 5.342$$

PRINCIPLE OF SUPERPOSITION

The solutions to the radial diffusivity equation as presented earlier in this chapter appear to be applicable only for describing the pressure distribution in an infinite reservoir that was caused by a constant production from a single well. Since real reservoir systems usually have several wells that are operating at varying rates, a more generalized approach is needed to study the fluid flow behavior during the unsteady state flow period.

The principle of superposition is a powerful concept that can be applied to remove the restrictions that have been imposed on various forms of solution to the transient flow equation. Mathematically the superposition theorem states that any sum of individual solutions to the diffusivity equation is also a solution to that equation. This concept can be applied to account for the following effects on the transient flow solution:

• Effects of multiple wells
• Effects of rate change
• Effects of the boundary
• Effects of pressure change

Slider (1976) presented an excellent review and discussion of the practical applications of the principle of superposition in solving a wide variety of unsteady-state flow problems.

Effects of Multiple Wells

Frequently, it is desired to account for the effects of more that one well on the pressure at some point in the reservoir. The superposition concept states that the total pressure drop at any point in the reservoir is the sum of the pressure changes at that point caused by flow in each of the wells in the reservoir. In other words, we simply superimpose one effect upon the other.

Consider Figure 6-28 which shows three wells that are producing at different flow rates from an infinite acting reservoir, i.e., unsteady-state flow reservoir. The principle of superposition shows that the total pressure drop observed at any well, e.g., Well 1, is:

$$(\Delta p)_{\text{total drop at well 1}} = (\Delta p)_{\text{drop due to well 1}}$$
$$+ (\Delta p)_{\text{drop due to well 2}}$$
$$+ (\Delta p)_{\text{drop due to well 3}}$$

The pressure drop at well 1 due to its own production is given by the log-approximation to the E_i-function solution presented by Equation 6-146, or:

$$(p_i - p_{wf}) = (\Delta p)_{\text{well1}} = \frac{162.6 Q_{o1} B_o \mu_o}{kh}$$
$$\times \left[\log\left(\frac{kt}{\phi \mu c_t r_w^2}\right) - 3.23 + 0.87s \right]$$

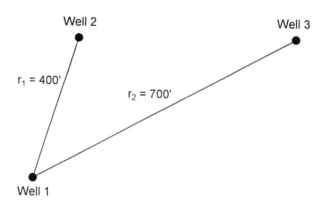

Figure 6-28. Well layout for Example 6-20.

where $t = $ time, hrs.

   $s = $ skin factor

   $k = $ permeability, md

   $Q_{o1} = $ oil flow rate from well 1

The pressure drop at well 1 due to production at Wells 2 and 3 must be written in terms of the E_i-function solution as expressed by Equation 6-78. The log-approximation cannot be used because we are calculating the pressure at a large distance r from the well, i.e., the argument $x > 0.01$, or:

$$(p_i - p_{wf})_{\text{total at well 1}} = \left(\frac{162.6 \, Q_{o1} \, B_o \, \mu_o}{kh} \right)$$

$$\times \left[\log \left(\frac{kt}{\phi \mu c_t \, r_w^2} \right) - 3.23 + 0.87s \right] - \left(\frac{70.6 \, Q_{o2} \, B_o \, \mu_o}{kh} \right)$$

$$\times E_i \left[- \frac{948 \phi \mu c_t \, r_1^2}{kt} \right] - \left(\frac{70.6 \, Q_{o3} \, B_o \, \mu_o}{kh} \right) E_i \left[- \frac{948 \phi \mu c_t \, r_2^2}{kt} \right] \quad (6\text{-}169)$$

where Q_{o1}, Q_{o2}, and Q_{o3} refer to the respective producing rates of Wells 1, 2, and 3.

The above computational approach can be used to calculate the pressure at Wells 2 and 3. Further, it can be extended to include any number of wells flowing under the unsteady-state flow condition. It should also be noted that if the point of interest is an operating well, the skin factor s must be included for that well only.

Example 6-20

Assume that the three wells as shown in Figure 6-28 are producing under a transient flow condition for 15 hours. The following additional data is available:

$Q_{o1} = 100$ STB/day	$h = 20'$
$Q_{o2} = 160$ STB/day	$\phi = 15\%$
$Q_{o3} = 200$ STB/day	$k = 40$ md
$p_i = 4500$ psi	$r_w = 0.25'$
$B_o = 1.20$ bbl/STB	$\mu_o = 2.0$ cp
$c_t = 20 \times 10^{-6}$ psi^{-1}	$r_1 = 400'$
$(s)_{\text{well 1}} = -0.5$	$r_2 = 700'$

If the three wells are producing at a constant flow rate, calculate the sand face flowing pressure at Well 1.

Solution

Step 1. Calculate the pressure drop at Well 1 caused by its own production by using Equation 6-146.

$$(\Delta p)_{well\ 1} = \frac{(162.6)\,(100)\,(1.2)\,(2.0)}{(40)\,(20)}$$

$$\times \left[\log \frac{(40)\,(15)}{(0.15)\,(2)\,(20 \times 10^{-6})\,(0.25)^2} - 3.23 + 0.87\,(0) \right]$$

$$= 270.2\ \text{psi}$$

Step 2. Calculate the pressure drop at Well 1 due to the production from Well 2.

$$(\Delta p)_{due\ to\ well\ 2} = -\frac{(70.6)\,(160)\,(1.2)\,(2)}{(40)\,(20)}$$

$$\times E_i \left[-\frac{(948)\,(0.15)\,(2.0)\,(20 \times 10^{-6})\,(400)^2}{(40)\,(15)} \right]$$

$$= 33.888\,[-E_i\,(-1.5168)]$$

$$= (33.888)\,(0.13) = 4.41\ \text{psi}$$

Step 3. Calculate pressure drop due to production from Well 3.

$$(\Delta p)_{due\ to\ well\ 3} = -\frac{(70.6)\,(200)\,(1.2)\,(2)}{(40)\,(20)}$$

$$\times E_i \left[-\frac{(948)\,(0.15)\,(2.0)\,(20 \times 10^{-6})\,(700)^2}{(40)\,(15)} \right]$$

$$= (42.36)\left[-E_i\,(-4.645) \right]$$

$$= (42.36)\,(1.84 \times 10^{-3}) = 0.08\ \text{psi}$$

Step 4. Calculate total pressure drop at Well 1.

$$(\Delta p)_{\text{total at well 1}} = 270.2 + 4.41 + 0.08 = 274.69 \text{ psi}$$

Step 5. Calculate p_{wf} at Well 1.

$$p_{wf} = 4500 - 274.69 = 4225.31 \text{ psi}$$

Effects of Variable Flow Rates

All of the mathematical expressions presented previously in this chapter require that the wells produce at constant rate during the transient flow periods. Practically all wells produce at varying rates and, therefore, it is important that we be able to predict the pressure behavior when rate changes. For this purpose, the concept of superposition states, "**Every flow rate change in a well will result in a pressure response which is independent of the pressure responses caused by other previous rate changes.**" Accordingly, the total pressure drop that has occurred at any time is the summation of pressure changes causes separately by each *net* flow rate change.

Consider the case of a shut-in well, i.e., $Q = 0$, that was then allowed to produce at a series of constant rates for the different time periods shown in Figure 6-29. To calculate the total pressure drop at the sand face at time t_4, the composite solution is obtained by adding the individual constant-rate solutions at the specified rate-time sequence, or:

$$(\Delta p)_{\text{total}} = (\Delta p)_{\text{due to } (Q_{o1} - 0)} + (\Delta p)_{\text{due to } (Q_{o2} - Q_{o1})} + (\Delta p)_{\text{due to } (Q_{o3} - Q_{o2})}$$
$$+ (\Delta p)_{\text{due to } (Q_{o4} - Q_{o3})}$$

The above expression indicates that there are four contributions to the total pressure drop resulting from the four individual flow rates.

The first contribution results from increasing the rate from 0 to Q_1 and is in effect over the entire time period t_4, thus:

$$(\Delta p)_{q1 - o} = \left[\frac{162.6 \, (Q_1 - 0) \, B\mu}{kh}\right]\left[\log\left(\frac{kt_4}{\phi \mu c_t r_w^2}\right) - 3.23 + 0.87s\right]$$

It is essential to notice the *change* in the rate, i.e., (new rate – old rate), that is used in the above equation. It is the change in the rate that causes the pressure disturbance. Further, it should be noted that the "time" in the equation represents the *total elapsed time* since the change in the rate has been in effect.

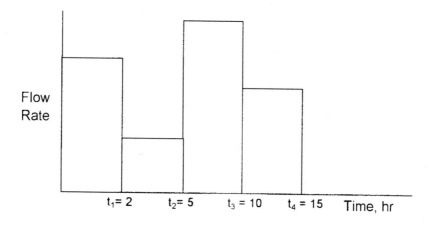

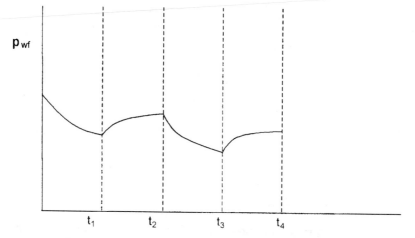

Figure 6-29. Production and pressure history of a well.

Second contribution results from decreasing the rate from Q_1 to Q_2 at t_1, thus:

$$(\Delta p)_{Q_2 - Q_1} = \left[\frac{162.6 \, (Q_2 - Q_1) B \mu}{kh} \right]$$

$$\times \left[\log \left(\frac{k(t_4 - t_1)}{\phi \mu c_t r_w^2} \right) - 3.23 + 0.87s \right]$$

Using the same concept, the contributions from Q_2 to Q_3 and from Q_3 to Q_4 can be computed as:

$$(\Delta p)_{Q3-Q2} = \left[\frac{162.6\,(Q_3 - Q_2)\,B\mu}{kh} \right]$$

$$\times \left[\log\left(\frac{k(t_4 - t_2)}{\phi\mu c_t r_w^2} \right) - 3.23 + 0.87s \right]$$

$$(\Delta p)_{Q4-Q3} = \left[\frac{162.6\,(Q_4 - Q_3)\,B\mu}{kh} \right]$$

$$\times \left[\log\left(\frac{k(t_4 - t_3)}{\phi\mu c_t r_w^2} \right) - 3.23 + 0.87s \right]$$

The above approach can be extended to model a well with several rate changes. Note, however, the above approach is valid only if the well is flowing under the unsteady-state flow condition for the total time elapsed since the well began to flow at its initial rate.

Example 6-21

Figure 6-29 shows the rate history of a well that is producing under transient flow condition for 15 hours. Given the following data:

$p_i = 5000$ psi $h = 20'$
$B_o = 1.1$ bbl/STB $\phi = 15\%$
$\mu_o = 2.5$ cp $r_w = 0.3'$
$c_t = 20 \times 10^{-6}$ psi^{-1} $s = 0$
$k = 40$ md

Calculate the sand face pressure after 15 hours.

Solution

Step 1. Calculate the pressure drop due to the first flow rate for the entire flow period.

$$(\Delta p)_{Q1-0} = \frac{(162.6)\,(100 - 0)\,(1.1)\,(2.5)}{(40)\,(20)}$$

$$\times \left[\log\left[\frac{(40)\,(15)}{(0.15)\,(2.5)\,(20 \times 10^{-6})(0.3)^2} \right] - 3.23 + 0 \right] = 319.6\,\text{psi}$$

Step 2. Calculate the additional pressure change due to the change of the flow rate from 100 to 70 STB/day.

$$(\Delta p)_{Q_2 - Q_1} = \frac{(162.6)(70 - 100)(1.1)(2.5)}{(40)(20)}$$

$$\times \left[\log \left[\frac{(40)(15 - 2)}{(0.15)(2.5)(20 \times 10^{-6})(0.3)^2} \right] - 3.23 \right] = -94.85 \, \text{psi}$$

Step 3. Calculate the additional pressure change due to the change of the flow rate from 70 to 150 STB/day.

$$(\Delta p)_{Q_3 - Q_2} = \frac{(162.6)(150 - 70)(1.1)(2.5)}{(40)(20)}$$

$$\times \left[\log \left[\frac{(40)(15 - 5)}{(0.15)(2.5)(20 \times 10^{-6})(0.3)^2} \right] - 3.23 \right] = 249.18 \, \text{psi}$$

Step 4. Calculate the additional pressure change due to the change of the flow rate from 150 to 85 STB/day.

$$(\Delta p)_{Q_4 - Q_3} = \frac{(162.6)(85 - 150)(1.1)(2.5)}{(40)(20)}$$

$$\times \left[\log \left[\frac{(40)(15 - 10)}{(0.15)(2.5)(20 \times 10^{-6})(0.3)^2} \right] - 3.23 \right] = -190.44 \, \text{psi}$$

Step 5. Calculate the total pressure drop:

$$(\Delta p)_{total} = 319.6 + (-94.85) + 249.18 + (-190.44) = 283.49 \text{ psi}$$

Step 6. Calculate wellbore pressure after 15 hours of transient flow:

$$p_{wf} = 5000 - 283.49 = 4716.51 \text{ psi}$$

Effects of the Reservoir Boundary

The superposition theorem can also be extended to predict the pressure of a well in a bounded reservoir. Consider Figure 6-30 which shows a well that is located a distance r from the non-flow boundary, e.g., sealing fault.

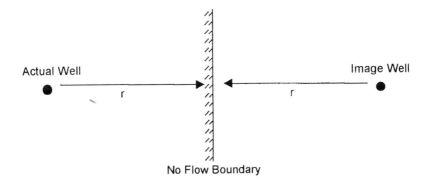

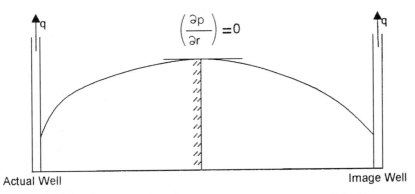

Figure 6-30. Method of images in solving boundary problems.

The no-flow boundary can be represented by the following pressure gradient expression:

$$\left(\frac{\partial p}{\partial r}\right)_{\text{Boundary}} = 0$$

Mathematically, the above boundary condition can be met by placing an *image* well, identical to that of the actual well, on the other side of the fault at exactly distance r. Consequently, the effect of the boundary on the pressure behavior of a well would be the same as the effect from an image well located a distance 2r from the actual well.

In accounting for the boundary effects, the superposition method is frequently called the **method of images.** Thus, for the problem of the system configuration given in Figure 6-30, the problem reduces to one of determining the effect of the image well on the actual well. The total pressure drop at the actual well will be the pressure drop due to its own production plus the additional pressure drop caused by an identical well at a distance of 2r, or:

$$(\Delta p)_{total} = (\Delta p)_{actual\ well} + (\Delta p)_{due\ to\ image\ well}$$

or

$$(\Delta p)_{total} = \frac{162.6\,Q_o\,B_o\mu_o}{kh}\left[\log\left(\frac{kt}{\phi\mu_o\,c_t\,r_w^2}\right) - 3.23 + 0.87s\right]$$

$$-\left(\frac{70.6\,Q_o\,B_o\,\mu_o}{kh}\right)E_i\left(-\frac{948\,\phi\mu_o\,c_t\,(2r)^2}{kt}\right)$$

Notice that this equation assumes the reservoir is infinite except for the indicated boundary. The effect of boundaries is always to cause greater pressure drop than those calculated for infinite reservoirs.

The concept of image wells can be extended to generate the pressure behavior of a well located within a variety of boundary configurations.

Example 6-22

Figure 6-31 shows a well located between two sealing faults at 200 and 100 feet from the two faults. The well is producing under a transient flow condition at a constant flow rate of 200 STB/day.

Given:

$p_i = 500$ psi	$k = 600$ md
$B_o = 1.1$ bbl/STB	$\phi = 17\%$
$\mu_o = 2.0$ cp	$h = 25$ ft
$r_w = 0.3$ ft	$s = 0$
$c_t = 25 \times 10^{-6}$ psi^{-1}	

Calculate the sand face pressure after 10 hours.

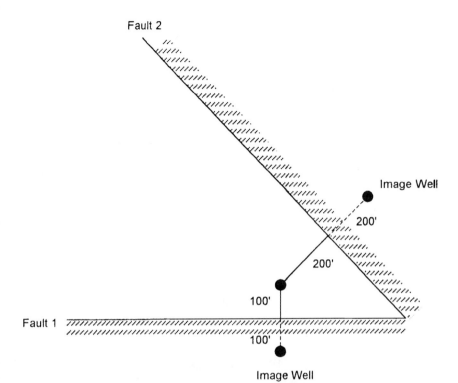

Figure 6-31. Well layout for Example 6-31.

Solution

Step 1. Calculate the pressure drop due to the actual well flow rate.

$$(\Delta p)_{actual} = \frac{(162.6)(200)(1.1)(2.0)}{(60)(25)}$$

$$\times \left[\log \left[\frac{(60)(10)}{(0.17)(2)(0.17)(2)(25 \times 10^{-6})(0.3)^2} \right] - 3.23 + 0 \right]$$

$$= 270.17$$

Step 2. Determine the additional pressure drop due to the first fault (i.e., image well 1):

$$(\Delta p)_{image\,well\,1} = -\frac{(70.6)(200)(1.1)(2.0)}{(60)(25)}$$

$$\times E_i \left[-\frac{(948)(0.17)(2)(25 \times 10^{-6})(2 \times 100)^2}{(6)(10)} \right]$$

$$= 20.71 \left[-E_i\,(-0.537) \right] = 10.64\,psi$$

Step 3. Calculate the effect of the second fault (i.e., image well 2):

$$(\Delta p)_{image\,well\,2} = 20.71 \left[-E_i \left(\frac{-948(0.17)(2)(25 \times 10^{-6})(2 \times 200)^2}{(60)(10)} \right) \right]$$

$$= 20.71[-E_i\,(-2.15)] = 1.0\,psi$$

Step 4. Total pressure drop is:

$$(\Delta p)_{total} = 270.17 + 10.64 + 1.0 = 28.18 \text{ psi}$$

Step 5. $p_{wf} = 5000 - 281.8 = 4718.2$ psi

Accounting for Pressure-Change Effects

Superposition is also used in applying the constant-pressure case. Pressure changes are accounted for in this solution in much the same way that rate changes are accounted for in the constant rate case. The description of the superposition method to account for the pressure-change effect is fully described in the Water Influx section in this book.

TRANSIENT WELL TESTING

Detailed reservoir information is essential to the petroleum engineer in order to analyze the current behavior and future performance of the reservoir. Pressure transient testing is designed to provide the engineer with a quantitative analyze of the reservoir properties. A transient test is essentially conducted by creating a pressure disturbance in the reservoir and recording the pressure response at the wellbore, i.e., bottom-hole flowing pressure p_{wf}, as a function of time. The pressure transient tests most commonly used in the petroleum industry include:

- Pressure drawdown
- Pressure buildup
- Multirate
- Interference
- Pulse
- Drill Stem
- Fall off
- Injectivity
- Step rate

It has long been recognized that the pressure behavior of a reservoir following a rate change directly reflects the geometry and flow properties of the reservoir. Information available from a well test includes:

- Effective permeability
- Formation damage or stimulation
- Flow barriers and fluid contacts
- Volumetric average reservoir pressure
- Drainage pore volume
- Detection, length, capacity of fractures
- Communication between wells

Only the drawdown and buildup tests are briefly described in the following two sections. There are several excellent books that comprehensively address the subject of well testing, notably:

- John Lee, *Well Testing* (1982)
- C. S. Matthews and D. G. Russell, *Pressure Buildup and Flow Test in Wells* (1967)
- Robert Earlougher, *Advances in Well Test Analysis* (1977)
- Canadian Energy Resources Conservation Board, *Theory and Practice of the Testing of Gas Wells* (1975)
- Roland Horn, *Modern Well Test Analysis* (1995)

Drawdown Test

A pressure drawdown test is simply a series of bottom-hole pressure measurements made during a period of flow at constant producing rate.

Usually the well is shut-in prior to the flow test for a period of time sufficient to allow the pressure to equalize throughout the formation, i.e., to reach static pressure. A schematic of the ideal flow rate and pressure history is illustrated by Figure 6-32.

The fundamental objectives of drawdown testing are to obtain the average permeability, k, of the reservoir rock *within the drainage area of the well* and to assess the degree of damage of stimulation induced in the vicinity of the wellbore through drilling and completion practices. Other objectives are to determine the pore volume and to detect reservoir inhomogeneities *within the drainage area of the well.*

During flow at a constant rate of Q_o, the pressure behavior of a well in an infinite-acting reservoir (i.e., during the unsteady-state flow period) is given by Equation 6-146, as:

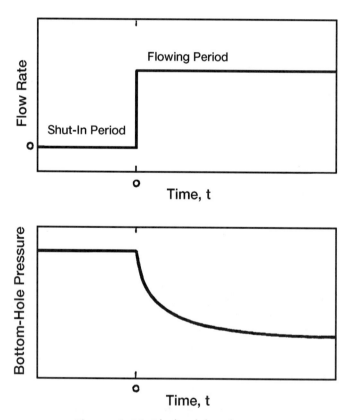

Figure 6-32. Idealized drawdown test.

$$p_{wf} = p_i - \frac{162.6\,Q_o\,B_o\,\mu}{kh}\left[\log\left(\frac{kt}{\phi\mu\,c_t\,r_w^2}\right) - 3.23 + 0.87\,s\right]$$

where k = permeability, md
  t = time, hr
  r_w = wellbore radius
  s = skin factor

The above expression can be written as:

$$p_{wf} = p_i - \frac{162.6\,Q_o\,B_o\,\mu}{kh}$$

$$\times\left[\log(t) + \log\left(\frac{k}{\phi\mu\,c_t\,r_w^2}\right) - 3.23 + 0.87\,s\right] \qquad (6\text{-}170)$$

Equation 6-170 is essentially an equation of a straight line and can be expressed as:

$$p_{wf} = a + m\log(t) \qquad (6\text{-}171)$$

where

$$a = p_i - \frac{162.6\,Q_o\,B_o\,\mu}{kh}\left[\log\left(\frac{k}{\phi\mu\,c_t\,r_w^2}\right) - 3.23 + 0.87\,s\right]$$

The slope m is given by:

$$m = \frac{-162.6\,Q_o\,B_o\,\mu_o}{kh} \qquad (6\text{-}172)$$

Equation 6-171 suggests that a plot of p_{wf} versus time t on semilog graph paper would yield a straight line with a slope m in psi/cycle. This semilog straight-line relationship is illustrated by Figure 6-33.

Equation 6-172 can be also rearranged for the capacity kh of the drainage area of the well. If the thickness is known, then the average permeability is given by:

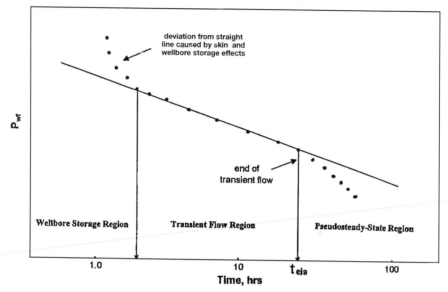

Figure 6-33. Semilog plot of pressure drawdown data.

$$k = \frac{-162.6 Q_o B_o \mu_o}{mh} \qquad (6\text{-}173)$$

where k = average permeability, md
  m = slope, psi/cycle. Notice, slope m is negative

Clearly, kh/μ or k/μ may be also estimated.
The skin effect can be obtained by rearranging Equation 6-170, as:

$$s = 1.151 \left(\frac{p_{wf} - p_i}{m} - \log t - \log \frac{k}{\phi \mu \, c_t \, r_w^2} + 3.23 \right)$$

or, more conveniently, if $p_{wf} = p_{1 \, hr}$ which is found on the extension of the straight line at log t (1 hr), then:

$$s = 1.151 \left(\frac{p_{1 \, hr} - p_i}{m} - \log \frac{k}{\phi \mu \, c_t \, r_w^2} + 3.23 \right) \qquad (6\text{-}174)$$

In Equation 6-174, $p_{1\,hr}$ must be from the semilog straight line. If pressure data measured at 1 hour do not fall on that line, the line *must be extrapolated* to 1 hour and extrapolated value of $p_{1\,hr}$ must be used in Equation 6-174. This procedure is necessary to avoid calculating an incorrect skin by using a wellbore-storage-influenced pressure. Figure 6-33 illustrates the extrapolation to $p_{1\,hr}$.

If the drawdown test is long enough, bottom-hole pressure will deviate from the semilog straight line and make the transition from infinite-acting to pseudosteady state.

Example 6-23[2]

Estimate oil permeability and skin factor from the drawdown data of Figure 6-34.

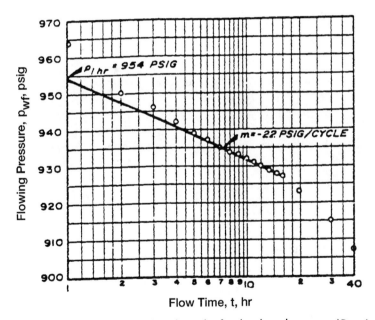

Figure 6-34. Earlougher's semilog data plot for the drawdown test. (*Permission to publish by the SPE, copyright SPE, 1977.*)

---

[2]This example problem and the solution procedure are given by Earlougher, R., "Advances in Well Test Analysis," Monograph Series, SPE, Dallas (1977).

The following reservoir data are available:

$h = 130$ ft $\phi = 20$ percent
$r_w = 0.25$ ft $p_i = 1,154$ psi
$Q_o = 348$ STB/D $m = -22$ psi/cycle
$B_o = 1.14$ bbl/STB
$\mu_o = 3.93$ cp
$c_t = 8.74 \times 10^{-6}$ psi^{-1}

Assuming that the wellbore storage effects are not significant, calculate:

• Permeability
• Skin factor

Solution

Step 1. From Figure 6-34, calculate $p_{1\,hr}$:

$$p_{1\,hr} = 954 \text{ psi}$$

Step 2. Determine the slope of the transient flow line:

$$m = -22 \text{ psi/cycle}$$

Step 3. Calculate the permeability by applying Equation 6-173:

$$k = \frac{-(162.6)\,(348)\,(1.14)\,(3.93)}{(-22)(130)} = 89 \text{ md}$$

Step 4. Solve for the skin factor s by using Equation 6-174:

$$s = 1.151 \left\{ \left(\frac{954 - 1,154}{-22} \right) \right.$$

$$\left. - \log \left[\frac{89}{(0.2)\,(3.93)\,(8.74 \times 10^{-6})(0.25)^2} \right] + 3.2275 \right\} = 4.6$$

Basically, well test analysis deals with the interpretation of the well-bore pressure response to a given change in the flow rate (from zero to a constant value for a drawdown test, or from a constant rate to zero for a buildup test). Unfortunately, the producing rate is controlled at the surface, not at the sand face. Because of the wellbore volume, a constant surface flow rate does not ensure that the entire rate is being produced from the formation. This effect is due to **wellbore storage.** Consider the case of a drawdown test. When the well is first open to flow after a shut-in period, the pressure in the wellbore drops. This drop in the wellbore pressure causes the following two types of wellbore storage:

- Wellbore storage effect caused by **fluid expansion**
- Wellbore storage effect caused by **changing fluid level** in the casing-tubing annulus.

As the bottom hole pressure drops, the wellbore fluid expands and, thus, the initial surface flow rate is not from the formation, but essentially from the fluid that had been stored in the wellbore. This is defined as the wellbore storage due to fluid expansion.

The second type of wellbore storage is due to a changing of the annulus fluid level (falling level during a drawdown test and rising fluid level during a pressure buildup test). When the well is open to flow during a drawdown test, the reduction in pressure causes the fluid level in the annulus to fall. This annulus fluid production joins that from the formation and contributes to the total flow from the well. The falling fluid level is generally able to contribute more fluid than that by expansion.

The above discussion suggests that part of the flow will be contributed by the wellbore instead of the reservoir, i.e.,

$$q = q_f + q_{wb}$$

where q = surface flow rate, bbl/day
q_f = formation flow rate, bbl/day
q_{wb} = flow rate contributed by the wellbore, bbl/day

As production time increases, the wellbore contribution decreases, and the formation rate increases until it eventually equals the surface flow rate. During this period when the formation rate is changed, the measured drawdown pressures will not produce the ideal semilog straight-line behavior that is expected during transient flow. This indicates that the pressure data collected during the duration of the wellbore storage effect cannot be analyzed by using conventional methods.

Each of the above two effects can be quantified in terms of the wellbore storage factor C which is defined as:

$$C = \frac{\Delta V_{wb}}{\Delta p}$$

where   C = wellbore storage volume, bbl/psi
   ΔV_{wb} = change in the volume of fluid in the wellbore, bbl

The above relationship can be applied to mathematically represent the individual effect of wellbore fluid expansion and falling (or rising) fluid level, to give:

• **Wellbore storage effect due to fluid expansion**

$$C = V_{wb}\, c_{wb}$$

where V_{wb} = total wellbore fluid volume, bbl
   c_{wb} = average compressibility of fluid in the wellbore, psi^{-1}

• **Wellbore storage effect due to changing fluid level**

   If A_a is the cross-sectional area of the annulus, and ρ is the average fluid density in the wellbore, the wellbore storage coefficient is given by:

$$C = \frac{144\, A_a}{5.615\, \rho}$$

with:

$$A_a = \frac{\pi \left[(ID_C)^2 - (OD_T)^2 \right]}{4\,(144)}$$

where A_a = annulus cross-sectional area, ft^2
   OD_T = outside diameter of the production tubing, in.
   ID_C = inside diameter of the casing, in.
   ρ = wellbore fluid density, lb/ft^3

This effect is essentially small if a packer is placed near the producing zone. The total storage effect is the sum of both effects. It should be noted during oil well testing that the fluid expansion is generally insignificant due to the small compressibility of liquids. For gas wells, the primary storage effect is due to gas expansion.

To determine the duration of the wellbore storage effect, it is convenient to express the wellbore storage factor in a dimensionless form as:

$$C_D = \frac{5.615\,C}{2\,\pi\,h\phi\,c_t\,r_w^2} = \frac{0.894\,C}{\phi\,h\,c_t\,r_w^2}$$

where C_D = dimensionless wellbore storage factor
$\quad\quad C$ = wellbore storage factor, bbl/psi
$\quad\quad c_t$ = total compressibility coefficient, psi^{-1}
$\quad\quad r_w$ = wellbore radius, ft
$\quad\quad h$ = thickness, ft

Horne (1995) and Earlougher (1977), among other authors, have indicated that the wellbore pressure is directly proportional to the time during the wellbore storage-dominated period of the test and is expressed by:

$$p_D = t_D/C_D$$

where p_D = dimensionless pressure during wellbore storage
$\quad\quad\quad\quad$ domination time
$\quad\quad t_D$ = dimensionless time

Taking the logarithm of both sides of the above relationship, gives:

$$\log(p_D) = \log(t_D) - \log(C_D)$$

The above expression has a characteristic that is diagnostic of wellbore storage effects. It indicates that a plot of p_D versus t_D on a log-log scale will yield as straight line of a unit slope during wellbore storage domination. Since p_D is proportional to Δp and t_D is proportional to time, it is convenient to log $(p_i - p_{wf})$ versus log (t) and observe where the plot has a slope of one cycle in pressure per cycle in time.

The log-log plot is a valuable aid for recognizing wellbore storage effects in transient tests (e.g., drawdown or buildup tests) when early-time pressure recorded data is available. It is recommended that this plot be made a part transient test analysis. As wellbore storage effects become less severe, the formation begins to influence the bottom-hole pressure more and more, and the data points on the log-log plot fall below the unit-slope straight line and signifies the end of wellbore storage effect. At this point, wellbore storage is no longer important and standard semilog data-plotting analysis techniques apply. As a rule of thumb, that time usually occurs about 1 to 1½ cycles in time after the log-log data plot starts deviating significantly from the unit slop. This time may be estimated from:

$$t_D > (60 + 3.5s)\, C_D$$

or approximately:

$$t > \frac{(200,000 + 12,000\ s)C}{(kh\,/\,\mu)}$$

where   t = total time that marks the end of wellbore storage effect and the beginning of the semilog straight line, hr
      k = permeability, md
      s = skin factor
      m = viscosity, cp
      C = wellbore storage coefficient, bbl/psi

Example 6-24

The following data is given for an oil well that is scheduled for a drawdown test:

- Volume of fluid in the wellbore = 180 bbls
- Tubing outside diameter = 2 inches
- Production casing inside diameter = 7.675 inches
- Average oil density in the wellbore = 45 lb/ft^3
- h = 20 ft ϕ = 15% r_w = 0.25 ft
   μ_o = 2 cp   k = 30 md   s = 0
   c_t = 20 × 10^{-6} psi^{-1} c_o = 10 × 10^{-6} psi^{-1}

If this well is placed under a constant production rate, how long will it take for wellbore storage effects to end?

Solution

Step 1. Calculate the cross-sectional area of the annulus A_a:

$$A_a = \frac{\pi \, [(7.675)^2 - (2)^2]}{(4)\,(144)} = 0.2995 \text{ ft}^2$$

Step 2. Calculate the wellbore storage factor caused by fluid expansion:

$$C = V_{wb} \, c_{wb}$$
$$C = (180)\,(10 \times 10^{-6}) = 0.0018 \text{ bbl/psi}$$

Step 3. Determine the wellbore storage factor caused by the falling fluid level:

$$C = \frac{144 \, A_a}{5.615 \, \rho}$$

$$C = \frac{144\,(0.2995)}{(5.615)\,(45)} = 0.1707 \text{ bbl/psi}$$

Step 4. Calculate the total wellbore storage coefficient:

$$C = 0.0018 + 0.1707 = 0.1725 \text{ bbl/psi}$$

The above calculations show that the effect of fluid expansion can generally be neglected in crude oil systems.

Step 5. Determine the time required for wellbore storage influence to end from:

$$t = \frac{(200,000 + 12,000 \text{ s}) \, C \, \mu}{kh}$$

$$t = \frac{(200,000 + 0)\,(0.1725)\,(2)}{(30)\,(20)} = 115 \text{ hrs}$$

The straight line relationship as expressed by Equation 6-171 is only valid during the infinite-acting behavior of the well. Obviously, reservoirs are not infinite in extent, thus the infinite-acting radial flow period cannot last indefinitely. Eventually the effects of the reservoir boundaries will be felt at the well being tested. The time at which the boundary effect is felt is dependent on the following factors:

- Permeability k
- Total compressibility c_t
- Porosity ϕ
- Viscosity μ
- Distance to the boundary
- Shape of the drainage area

Earlougher (1977) suggests the following mathematical expression for estimating the duration of the infinite-acting period.

$$t_{eia} = \left[\frac{\phi \mu c_t A}{0.000264 \, k} \right] (t_{DA})_{eia}$$

where t_{eia} = time to the end of infinite-acting period, hr
  A = well drainage area, ft^2
  c_t = total compressibility, psi^{-1}
  $(t_{DA})_{eia}$ = dimensionless time to the end of the infinite-acting period

Earlougher's expression can be used to predict the end of transient flow in a drainage system of any geometry by obtaining the value of $(t_{DA})_{eia}$ from Table 6-3 as listed under "**Use Infinite System Solution With Less Than 1% Error for t_D <**." For example, for a well centered in a circular reservoir, $(t_{DA})_{eia} = 0.1$, and accordingly:

$$t_{eia} = \frac{380 \, \phi \mu c_t A}{k}$$

Hence, the specific steps involved in a drawdown test analysis are:

1. Plot $(p_i - p_{wf})$ versus t on a log-log scale.
2. Determine the time at which the unit slope line ends.
3. Determine the corresponding time at 1½ log cycle, ahead of the observed time in Step 2. This is the time that marks the end of the wellbore storage effect and the start of the semilog straight line.
4. Estimate the wellbore storage coefficient from:

$$C = \frac{qt}{24\Delta p}$$

where t and Δp are values read from a point on the log-log unit-slope straight line and q is the flow rate in bbl/day.

5. Plot p_{wf} versus t on a semilog scale
6. Determine the start of the straight-line portion as suggested in Step 3 and draw the best line through the points.
7. Calculate the slope of the straight line and determine the permeability k and skin factor s by applying Equations 6-173 and 6-174, respectively.
8. Estimate the time to the end of the infinite-acting (transient flow) period, i.e., t_{eia}, which marks the beginning of the pseudosteady-state flow.
9. Plot all the recorded pressure data after t_{eia} as a function of time on a regular Cartesian scale. These data should form a straight-line relationship.
10. Determine the **slope** of the pseudosteady-state line, i.e., dp/dt (commonly referred to as m′) and use Equation 6-128 to solve for the drainage area "A,"

$$A = \frac{-0.23396\,Q\,B}{c_t\,h\,\phi\,(dp/dt)} = \frac{-0.23396\,Q\,B}{c_t\,h\,\phi\,m'}$$

where m′ = slope of the semisteady-state Cartesian straight-line
$\quad$ Q = fluid flow rate, STB/day
$\quad$ B = formation volume factor, bbl/STB

11. Calculate the shape factor C_A from a expression that has been developed by Earlougher (1977). Earlougher has shown that the reservoir shape factor can be estimated from the following relationship:

$$C_A = 5.456\left(\frac{m}{m'}\right)\exp\left[\frac{2.303\,(p_{1\,hr} - p_{int})}{m}\right]$$

where m = slope of transient semilog straight line, psi/log cycle
$\quad$ m′ = slope of the semisteady-state Cartesian straight-line
$\quad$ p_{1hr} = pressure at t = 1 hr from semilog straight-line, psi
$\quad$ p_{int} = pressure at t = 0 from semisteady-state Cartesian straight-line, psi

12. Use Table 6-4 to determine the drainage configuration of the tested well that has a value of the shape factor C_A closest to that of the calculated one, i.e., Step 11.

Pressure Buildup Test

The use of pressure buildup data has provided the reservoir engineer with one more useful tool in the determination of reservoir behavior. Pressure buildup analysis describes the build up in wellbore pressure with time after a well has been shut in. One of the principal objectives of this analysis is to determine the static reservoir pressure without waiting weeks or months for the pressure in the entire reservoir to stabilize. Because the buildup in wellbore pressure will generally follow some definite trend, it has been possible to extend the pressure buildup analysis to determine:

- Effective reservoir permeability
- Extent of permeability damage around the wellbore
- Presence of faults and to some degree the distance to the faults
- Any interference between producing wells
- Limits of the reservoir where there is not a strong water drive or where the aquifer is no larger than the hydrocarbon reservoir

Certainly all of this information will probably not be available from any given analysis, and the degree of usefulness or any of this information will depend on the experience in the area and the amount of other information available for correlation purposes.

The general formulas used in analyzing pressure buildup data come from a solution of the diffusivity equation. In pressure buildup and drawdown analyses, the following assumptions, with regard to the reservoir, fluid and flow behavior, are usually made:

Reservoir:
- Homogeneous
- Isotropic
- Horizontal of uniform thickness

Fluid:
- Single phase
- Slightly compressible
- Constant μ_o and B_o

Flow:

• Laminar flow
• No gravity effects

Pressure buildup testing requires shutting in a producing well. The most common and the simplest analysis techniques require that the well produce at a constant rate, either from startup or long enough to establish a stabilized pressure distribution, before shut-in. Figure 6-35 schematically shows rate and pressure behavior for an ideal pressure buildup test. In that figure, t_p is the production time and Δt is running shut-in time. The pressure is measured immediately before shut-in and is recorded as a function of time during the shut-in period. The resulting pressure buildup curve is analyzed for reservoir properties and wellbore condition.

Stabilizing the well at a constant rate before testing is an important part of a pressure buildup test. If stabilization is overlooked or is impossible, standard data analysis techniques may provide erroneous information about the formation.

A pressure buildup test is described mathematically by using the principle of superposition. Before the shut-in, the well is allowed to flow at a constant flow rate of Q_o STB/day for t_p days. At the time corresponding to the point of shut-in, i.e., t_p, a second well, superimposed over the location of the first well, is opened to flow at a constant rate equal to $-Q_o$ STB/day for Δt days. The first well is allowed to continue to flow at $+Q_o$ STB/day. When the effects of the two wells are added, the result is that a well has been allowed to flow at rate Q for time t_p and then shut in for time Δt. This simulates the actual test procedure. The time corresponding to the point of shut-in, t_p, can be estimated from the following equation:

$$t_p = \frac{N_p}{Q_o}$$

where N_p = well cumulative oil produced before shut-in, STB
Q_o = stabilized well flow rate before shut-in, STB/day
t_p = total production time

Applying the superposition principle to a shut-in well, the total pressure change, i.e., $(p_i - p_{ws})$, which occurs at the wellbore during the shut-in time Δt, is essentially the sum of the pressure change caused by the constant flow rate Q and that of $-Q$, or:

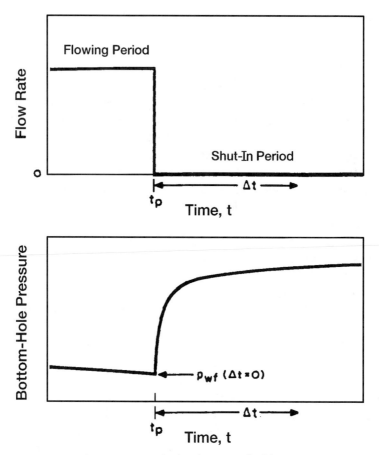

Figure 6-35. Idealized pressure buildup test.

$$p_i - p_{ws} = (p_i - p_{wf})Q_o - 0 + (p_i - p_{wf})0 - Q_o$$

Substituting Equation 6-146 for each of the terms on the right-hand side of the above relationship gives:

$$P_{ws} = p_i - \frac{162.6\,(Q_o - 0)\,\mu\,B_o}{kh}\left[\log\frac{k(t_p + \Delta t)}{\phi\,\mu\,c_t\,r_w^2} - 3.23 + 0.087\,s\right]$$

$$+ \frac{162.6\,(0 - Q_o)\,\mu\,B_o}{kh}\left[\log\frac{k(t_p)}{\phi\,\mu\,c_t\,r_w^2} - 3.23 + 0.087\,s\right] \quad (6\text{-}176)$$

Expanding this equation and canceling terms,

$$p_{ws} = p_i - \frac{162.6 Q_o \mu B}{kh} \left[\log \frac{(t_p + \Delta t)}{\Delta t} \right] \tag{6-177}$$

where p_i = initial reservoir pressure, psi
   p_{ws} = sand-face pressure during pressure buildup, psi
   t_p = flowing time before shut-in, hr
   Δt = shut-in time, hr

The pressure buildup equation, i.e., Equation 6-176 was introduced by Horner (1951) and is commonly referred to as the Horner equation.

Equation 6-177 suggests that a plot of p_{ws} versus $(t_p + \Delta t)/\Delta t$ would produce a straight line relationship with intercept p_i and slope of $-m$, where:

$$m = \frac{162.6 Q_o B_o \mu_o}{kh}$$

or

$$k = \frac{162.6 Q_o B_o \mu_o}{mh} \tag{6-178}$$

This plot, commonly referred to as the Horner plot, is illustrated in Figure 6-36. Note that on the Horner plot, the scale of time ratio increases from left to right. Because of the form of the ratio, however, the shut-in time Δt increases from right to left. It is observed from Equation 6-177 that $p_{ws} = p_i$ when the time ratio is unity. Graphically this means that the initial reservoir pressure, p_i, can be obtained by extrapolating the Horner plot straight line to $(t_p + \Delta t)/\Delta t = 1$.

Earlougher (1977) points out that a result of using the superposition principle is that skin factor, s, does not appear in the general pressure buildup equation, Equation 6-176. As a result, skin factor does not appear in the simplified equation for the Horner plot, Equation 6-177. That means the Horner-plot slope is not affected by the skin factor; however, the skin factor still does affect the shape of the pressure buildup data. In fact, an early-time deviation from the straight line can be caused by skin factor as well as by wellbore storage, as indicated in Figure 6-36. The deviation can be significant for the large negative skins that occur in hydraulically fractured wells. In any case, the skin factor does affect

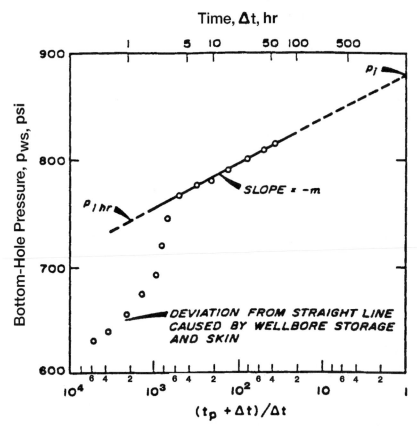

Figure 6-36. Horner plot. (*After Earlougher, R. "Advances in Well Test Analysis."*) (*Permission to publish by the SPE, copyright SPE, 1977.*)

flowing pressure before shut-in, so skin may be estimated from the buildup test data plus the flowing pressure immediately before the buildup test:

$$s = 1.151 \left[\frac{p_{1hr} - p_{wf}(\Delta t = 0)}{m} - \log \frac{k}{\phi \mu c_t r_w^2} + 3.23 \right] \qquad (6\text{-}179)$$

where $p_{wf}(\Delta t = 0)$ = observed flowing bottom-hole pressure immediately before shut-in

$\quad m$ = slope of the Horner plot

$\quad k$ = permeability, md

The value of p_{1hr} must be taken from the Horner straight line. Frequently, pressure data do not fall on the straight line at 1 hour because of wellbore storage effects or large negative skin factors. In that case, the semilog line must be extrapolated to 1 hour and the corresponding pressure is read.

It should be pointed out that when a well is shut in for a pressure buildup test, the well is usually closed at the surface rather than the sandface. Even though the well is shut in, the reservoir fluid continues to flow and accumulates in the wellbore until the well fills sufficiently to transmit the effect of shut-in to the formation. This "after-flow" behavior is caused by the wellbore storage and it has a significant influence on pressure buildup data. During the period of wellbore storage effects, the pressure data points fall below the semilog straight line. The duration of those effects may be estimated by making the log-log data plot described previously. For pressure buildup testing, plot log $[p_{ws} - p_{wf}]$ versus log (Δt). The bottom-hole flow pressure p_{wf} is observed flowing pressure immediately before shut-in. When wellbore storage dominates, that plot will have a unit-slope straight line; as the semilog straight line is approached, the log-log plot bends over to a gently curving line with a low slope.

In all pressure buildup test analyses, the log-log data plot should be made before the straight line is chosen on the semilog data plot. This log-log plot is essential to avoid drawing a semilog straight line through the wellbore storage-dominated data. The beginning of the semilog line can be estimated by observing when the data points on the log-log plot reach the slowly curving low-slope line and adding 1 to 1.5 cycles in time after the end of the unit-slope straight line. Alternatively, the time to the beginning of the semilog straight line can be estimated from:

$$\Delta t > \frac{170,000 \, Ce^{0.14s}}{(kh/\mu)}$$

where Δt = shut-in time, hrs.

C = calculated wellbore storage coefficient, bbl/psi

k = permeability, md

s = skin factor

h = thickness, ft

Example 6-25[3]

Table 6-5 shows pressure buildup data from an oil well with an estimated drainage radius of 2,640 ft.

Table 6-5
Earlougher's Pressure Buildup Data
(Permission to publish by the SPE, copyright SPE, 1977)

Δt (hours)	$t_p + \Delta t$ (hours)	$\dfrac{(t_p + \Delta t)}{\Delta t}$	p_{ws} (psig)
0.0	—	—	2,761
0.10	310.10	3,101	3,057
0.21	310.21	1,477	3,153
0.31	310.31	1,001	3,234
0.52	310.52	597	3,249
0.63	310.63	493	3,256
0.73	310.73	426	3,260
0.84	310.84	370	3,263
0.94	310.94	331	3,266
1.05	311.05	296	3,267
1.15	311.15	271	3,268
1.36	311.36	229	3,271
1.68	311.68	186	3,274
1.99	311.99	157	3,276
2.51	312.51	125	3,280
3.04	313.04	103	3,283
3.46	313.46	90.6	3,286
4.08	314.08	77.0	3,289
5.03	315.03	62.6	3,293
5.97	315.97	52.9	3,297
6.07	316.07	52.1	3,297
7.01	317.01	45.2	3,300
8.06	318.06	39.5	3,303
9.00	319.00	35.4	3,305
10.05	320.05	31.8	3,306
13.09	323.09	24.7	3,310
16.02	326.02	20.4	3,313
20.00	330.00	16.5	3,317
26.07	336.07	12.9	3,320
31.03	341.03	11.0	3,322
34.98	344.98	9.9	3,323
37.54	347.54	9.3	3,323

[3]This example problem and solution procedure are given by Earlougher, R., "Advanced Well Test Analysis," Monograph Series, SPE, Dallas (1977).

Before shut-in, the well had produced at a stabilized rate of 4,900 STB/day for 310 hours. Known reservoir data are:

$$
\begin{aligned}
\text{depth} &= 10{,}476 \text{ ft} \\
r_w &= 0.354 \text{ ft} \\
c_t &= 22.6 \times 10^{-6} \text{ psi}^{-1} \\
Q_o &= 4{,}900 \text{ STB/D} \\
H &= 482 \text{ ft} \\
p_{wf}(\Delta t = 0) &= 2{,}761 \text{ psig} \\
\mu_o &= 0.20 \text{ cp} \\
\phi &= 0.09 \\
B_o &= 1.55 \text{ bbl/STB} \\
\text{casing ID} &= 0.523 \text{ ft} \\
t_p &= 310 \text{ hours}
\end{aligned}
$$

Calculate

• Average permeability k
• Skin factor

Solution

Step 1. Plot p_{ws} versus $(t_p + \Delta t)/\Delta t$ on a semilog scale as shown in Figure 6-37.

Step 2. Identify the correct straight line portion of the curve and determine the slope m to give:

m = 40 psi/cycle

Step 3. Calculate the average permeability by using Equation 6-178 to give:

$$
k = \frac{(162.6)\,(4{,}900)\,(1.55)\,(0.22)}{(40)\,(482)} = 12.8 \text{ md}
$$

Step 4. Determine p_{wf} after 1 hour from the straight-line portion of the curve to give:

$p_{1 \text{ hr}} = 3266$ psi

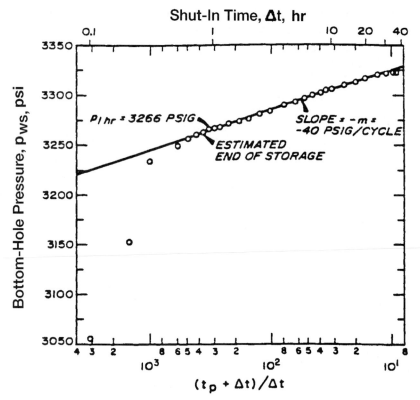

Figure 6-37. Earlougher's semilog data plot for the buildup test. (*Permission to publish by the SPE, copyright SPE, 1977.*)

Step 5. Calculate the skin factor by applying Equation 6-179.

$$s = 1.1513 \left[\frac{3,266 - 2,761}{40} \right.$$

$$\left. - \log \left(\frac{(12.8)\,(12)^2}{(0.09)\,(0.20)\,(22.6 \times 10^{-6})(4.25)^2} \right) + 3.23 \right] = 8.6$$

PROBLEMS

1. An incompressible fluid flows in a linear porous media with the following properties.

$$L = 2500 \text{ ft} \qquad h = 30 \text{ ft} \qquad \text{width} = 500 \text{ ft}$$
$$k = 50 \text{ md} \qquad \phi = 17\% \qquad \mu = 2 \text{ cp}$$
$$\text{inlet pressure} = 2100 \text{ psi} \qquad Q = 4 \text{ bbl/day} \qquad \rho = 45 \text{ lb/ft}^3$$

Calculate and plot the pressure profile throughout the linear system.

2. Assume the reservoir linear system as described in problem 1 is tilted with a dip angle of 7°. Calculate the fluid potential through the linear system.

3. A 0.7 specific gravity gas is flowing in a linear reservoir system at 150°F. The upstream and downstream pressures are 2000 and 1800 psi, respectively. The system has the following properties:

$$L = 2000 \text{ ft} \qquad W = 300 \text{ ft} \qquad h = 15 \text{ ft}$$
$$k = 40 \text{ md} \qquad \phi = 15\%$$

Calculate the gas flow rate.

4. An oil well is producing a crude oil system at 1000 STB/day and 2000 psi of bottom-hole flowing pressure. The pay zone and the producing well have the following characteristics:

$$h = 35 \text{ ft} \qquad r_w = 0.25 \text{ ft} \qquad \text{drainage area} = 40 \text{ acres}$$
$$\text{API} = 45° \qquad \gamma_g = 0.72 \qquad R_s = 700 \text{ scf/STB}$$
$$k = 80 \text{ md} \qquad T = 100°F$$

Assuming steady-state flowing conditions, calculate and plot the pressure profile around the wellbore.

5. Assuming steady-state flow and incompressible fluid, calculate the oil flow rate under the following conditions:

$$p_e = 2500 \text{ psi} \qquad p_{wf} = 2000 \text{ psi} \qquad r_e = 745 \text{ ft}$$
$$r_w = 0.3 \text{ ft} \qquad \mu_o = 2 \text{ cp} \qquad B_o = 1.4 \text{ bbl/STB}$$
$$h = 30 \text{ ft} \qquad k = 60 \text{ md}$$

6. A gas well is flowing under a bottom-hole flowing pressure of 900 psi. The current reservoir pressure is 1300 psi. The following additional data are available:

$T = 140°F$ $\gamma_g = 0.65$ $r_w = 0.3$ ft
$k = 60$ md $h = 40$ ft $r_e = 1000$ ft

Calculate the gas flow rate by using a:
• Real gas pseudo-pressure approach
• Pressure-squared method.

7. An oil well is producing a stabilized flow rate of 500 STB/day under a transient flow condition. Given:

$B_o = 1.1$ bbl/STB $\mu_o = 2$ cp $c_t = 15 \times 10^{-6}$ psi^{-1}
$k_o = 50$ md $h = 20$ ft $\phi = 20\%$
$r_w = 0.3$ ft $p_i = 3500$ psi

Calculate and plot the pressure profile after 1, 5, 10, 15 and 20 hours.

8. An oil well is producing at a constant flow rate of 800 STB/day under a transient flow condition. The following data is available:

$B_o = 1.2$ bbl/STB $\mu_o = 3$ cp $c_t = 15 \times 10^{-6}$ psi^{-1}
$k_o = 100$ md $h = 25$ ft $\phi = 15\%$
$r_w = 0.5$ $p_i = 4000$ psi $r_e = 1000$ ft

Using the E_i-function approach and the p_D-method, calculate the bottom-hole flowing pressure after 1, 2, 3, 5, and 10 hr. Plot the results on a semi-log scale and Cartesian scale.

9. A well is flowing under a drawdown pressure of 350 psi and produces at constant flow rate of 300 STB/day. The net thickness is 25 ft. Given:

$r_e = 660$ ft $r_w = 0.25$ ft $\mu_o = 1.2$ cp $B_o = 1.25$ bbl/STB

Calculate:
• Average permeability
• Capacity of the formation

10. An oil well is producing from the center of 40-acre square drilling pattern. Given:

$\phi = 20\%$ $h = 15$ ft $k = 60$ md

$\mu_o = 1.5$ cp $B_o = 1.4$ bbl/STB $r_w = 0.25$ ft

$\bar{p}_r = 2000$ psi $p_{wf} = 1500$ psi

Calculate the oil flow rate.

11. A shut-in well is located at a distance of 700 ft from one well and 1100 ft from a second well. The first well flows for 5 days at 180 STB/day, at which time the second well begins to flow at 280 STB/day. Calculate the pressure drop in the shut-in well when the second well has been flowing for 7 days. The following additional data is given:

$p_i = 3000$ psi $B_o = 1.3$ bbl/STB $\mu_o = 1.2$ cp $h = 60$ ft

$c_t = 15 \times 10^{-6}$ psi^{-1} $\phi = 15\%$ $k = 45$ md

12. A well is opened to flow at 150 STB/day for 24 hours. The flow rate is then increased to 360 STB/day and lasted for another 24 hours. The well flow rate is then reduced to 310 STB/day for 16 hours. Calculate the pressure drop in a shut-in well 700 ft away from the well given:

$\phi = 15\%$ $h = 20$ ft $k = 100$ md

$\mu_o = 2$ cp $B_o = 1.2$ bbl/STB $r_w = 0.25$ ft

$p_i = 3000$ psi $c_t = 12 \times 10^{-6}$ psi^{-1}

13. A well is flowing under unsteady-state flowing conditions for 5 days at 300 STB/day. The well is located at 350 ft and 420 ft distance from two sealing faults. Given:

$\phi = 17\%$ $c_t = 16 \times 10^{-6}$ psi^{-1} $k = 80$ md

$p_i = 3000$ psi $B_o = 1.3$ bbl/STB $\mu_o = 1.1$ cp

$r_w = 0.25$ ft $h = 25$ ft

Calculate the pressure in the well after 5 days.

14. A drawdown test was conducted on a new well with results as given below:

t, hr	p_{wf}, psi
1.50	2978
3.75	2949
7.50	2927
15.00	2904
37.50	2876
56.25	2863
75.00	2848
112.50	2810
150.00	2790
225.00	2763

Given:

$p_i = 3400$ psi $h = 25$ ft $Q = 300$ STB/day

$c_t = 18 \times 10^{-6}$ psi^{-1} $\mu_o = 1.8$ cp $B_o = 1.1$ bbl/STB

$r_w = 0.25$ ft $\phi = 12\%$

Assuming no wellbore storage, calculate:

• Average permeability
• Skin factor

15. A drawdown test was conducted on a discovery well. The well was flowed at a constant flow rate of 175 STB/day. The fluid and reservoir data are given below:

$S_{wi} = 25\%$ $\phi = 15\%$ $h = 30$ ft $ct = 18 \times 10^{-6}$ psi^{-1}

$r_w = 0.25$ ft $p_i = 4680$ psi $\mu_o = 1.5$ cp $B_o = 1.25$ bbl/STB

The drawdown test data is given below:

t, hr	p_{wf}, psi
0.6	4388
1.2	4367
1.8	4355
2.4	4344
3.6	4334
6.0	4318
8.4	4309
12.0	4300
24.0	4278
36.0	4261
48.0	4258
60.0	4253
72.0	4249
84.0	4244
96.0	4240
108.0	4235
120.0	4230
144.0	4222
180.0	4206

Calculate:

- Drainage radius
- Skin factor
- Oil flow rate at a bottom-hole flowing pressure of 4300 psi, assuming a semisteady-state flowing conditions.

16. A pressure build-up test was conducted on a well that had been producing at 146 STB/day for 53 hours. The reservoir and fluid data are given below.

B_o = 1.29 bbl/STB $\qquad$ μ_o = 0.85 cp $\qquad$ $c_t = 12 \times 10^{-6}$ psi^{-1}
ϕ = 10% $\qquad$ p_{wf} = 1426.9 psig $\qquad$ A = 20 acres

The build-up data is as follows:

Time, hr	p_{ws}, psig
0.167	1451.5
0.333	1476.0
0.500	1498.6
0.667	1520.1
0.833	1541.5
1.000	1561.3
1.167	1581.9
1.333	1599.7
1.500	1617.9
1.667	1635.3
2.000	1665.7
2.333	1691.8
2.667	1715.3
3.000	1736.3
3.333	1754.7
3.667	1770.1
4.000	1783.5
4.500	1800.7
5.000	1812.8
5.500	1822.4
6.000	1830.7
6.500	1837.2
7.000	1841.1
7.500	1844.5
8.000	1846.7
8.500	1849.6
9.000	1850.4
10.000	1852.7
11.000	1853.5
12.000	1854.0
12.667	1854.0
14.620	1855.0

Calculate:

• Average reservoir pressure
• Skin factor
• Formation capacity

REFERENCES

1. Al-Hussainy, R., and Ramey, H. J., Jr., "Application of Real Gas Flow Theory to Well Testing and Deliverability Forecasting," *Jour. of Petroleum Technology,* May 1966; *Theory and Practice of the Testing of Gas Wells,* 3rd ed. Calgary, Canada: Energy Resources Conservation Board, 1975.

2. Al-Hussainy, R., Ramey, H. J., Jr., and Crawford, P. B., "The Flow of Real Gases Through Porous Media," *Trans. AIME,* 1966, pp. 237, 624.

3. Chatas, A. T., "A Practical Treatment of Nonsteady-state Flow Problems in Reservoir Systems," *Pet. Eng.,* Aug 1953, pp. B-44–56.

4. Craft, B., Hawkins, M., and Terry, R., *Applied Petroleum Reservoir Engineering,* 2nd ed. Prentice Hall, 1990.

5. Craft, B., and Hawkins, M., *Applied Petroleum Reservoir Engineering.* Prentice-Hall, 1959.

6. Dake, L. P., *The Practice of Reservoir Engineering.* Amsterdam: Elsevier, 1994.

7. Dake, L., *Fundamentals of Reservoir Engineering.* Amsterdam: Elsevier, 1978.

8. Davis, D. H., "Reduction in Permeability with Overburden Pressure," *Trans. AIME,* 1952, pp. 195, 329.

9. Donohue, D., and Erkekin, T., "Gas Well Testing, Theory and Practice," *IHRDC,* 1982.

10. Earlougher, Robert C., Jr., *Advances in Well Test Analysis,* Monograph Vol. 5, Society of Petroleum Engineers of AIME. Dallas, TX: Millet the Printer, 1977.

11. Fetkovich, M. J., "The Isochronal Testing of Oil Wells," SPE Paper 4529, presented at the SPE Annual meeting, Las Vegas, September 30–October 3, 1973.

12. Golan, M. and Whitson, C., *Well Performance,* 2nd ed. Englewood Cliffs, NJ: Prentice-Hall, 1986.

13. Hawkins, M., "A Note on the Skin Effect," *Trans. AIME,* 1956, pp. 356.

14. Horne, R. *Modern Well Test Analysis.* Palo Alto, CA: Petroway, Inc., 1995.

15. Horner, D. R., "Pressure Build-Up in Wells," *Proc.,* Third World Pet. cong., The Hague (1951), Sec II, 503–523. Also *Reprint Series, No. 9—Pressure Analysis Methods,* pp. 25–43. Dallas: Society of Petroleum Engineers of AIME, 1967.

16. Hurst, W., "Establishment of the Skin Effect and Its Impediment to Fluid Flow into a Wellbore," *Petroleum Engineering,* Oct. 1953, p. 25, B-6.

17. Jones, S. C., "Using the Inertial Coefficient, b, to Characterize Heterogeneity in Reservoir Rock," SPE Paper 16949, presented at the SPE Conference, Dallas, TX, Sept. 27–30, 1987.

18. Joshi, S., *Horizontal Well Technology.* Pennwell Publishing Company, 1991.

19. Lee, J., and Wattenbarger, R., *Gas Reservoir Engineering.* SPE Textbook Series Vol. 5, SPE, 1996.

20. Lee, John W., *Well Testing.* Dallas: Society of Petroleum Engineers Textbook Series, 1982.

21. Matthews, S., Bronz, F., and Hazebroek, P. "A Method for the Determination of Average Pressure in a Bounded Reservoir," *Trans. AIME,* 1954, Vol. 201, pp. 82–191.

22. Matthews, C. S., and Russell, D. G., "Pressure Buildup and Flow Tests in Wells," Monograph Vol. 1, Society of Petroleum Engineers of AIME. Dallas, TX: Millet the Printer, 1967.

23. Ramey, H., and Cobb, W. "A General Pressure Buildup Theory for a Well in a Closed Drainage Area," *JPT,* December 1971, pp. 1493–1505.

24. Russell, D. G., Goodrich, J. H., Perry, G. E., and Bruskotter, J. F., "Methods for Predicting Gas Well Performance," *JPT,* Jan. 1966, pp. 99–108; *Trans. AIME,* p. 237.

25. Slider, H. C., *Practical Petroleum Reservoir Engineering Methods.* Tulsa, OK: Petroleum Publishing Co., 1976.

26. van Everdingen, A. F., "The Skin Effect and Its Influence on the Productive Capacity of a Well," *Trans. AIME,* 1953, pp. 171, 198.

27. van Everdingen, A. F., and Hurst, W., "The Application of the Laplace Transformation to Flow Problems in Reservoirs," *Trans. AIME,* 1949, pp. 186, 305–324.

28. Wattenbarger, Robert A., and Ramey, H. J. Jr., "Gas Well Testing With Turbulence. Damage and Wellbore Storage," *JPT,* 1968, pp. 877–887; *Trans. AIME,* p. 243.

OIL WELL PERFORMANCE

This chapter presents the practical reservoir engineering equations that are designed to predict the performance of vertical and horizontal oil wells. The chapter also describes some of the factors that are governing the flow of fluids from the formation to the wellbore and how these factors may affect the production performance of the well. The analysis of the production performance is essentially based on the following fluid and well characteristics:

• Fluid PVT properties
• Relative permeability data
• Inflow-performance-relationship (IPR)

VERTICAL OIL WELL PERFORMANCE

Productivity Index and IPR

A commonly used measure of the ability of the well to produce is the **Productivity Index.** Defined by the symbol J, the productivity index is the ratio of the total liquid flow rate to the pressure drawdown. For a water-free oil production, the productivity index is given by:

$$J = \frac{Q_o}{\bar{p}_r - p_{wf}} = \frac{Q_o}{\Delta p} \tag{7-1}$$

where Q_o = oil flow rate, STB/day
J = productivity index, STB/day/psi
$\bar{p}_r$ = volumetric average drainage area pressure (static pressure)

p_{wf} = bottom-hole flowing pressure
Δp = drawdown, psi

The productivity index is generally measured during a production test on the well. The well is shut-in until the static reservoir pressure is reached. The well is then allowed to produce at a constant flow rate of Q and a stabilized bottom-hole flow pressure of p_{wf}. Since a stabilized pressure at surface does not necessarily indicate a stabilized p_{wf}, the bottom-hole flowing pressure should be recorded continuously from the time the well is to flow. The productivity index is then calculated from Equation 7-1.

It is important to note that the productivity index is a valid measure of the well productivity potential only if the well is flowing at pseudosteady-state conditions. Therefore, in order to accurately measure the productivity index of a well, it is essential that the well is allowed to flow at a constant flow rate for a sufficient amount of time to reach the pseudosteady-state as illustrated in Figure 7-1. The figure indicates that during the transient flow period, the calculated values of the productivity index will vary depending upon the time at which the measurements of p_{wf} are made.

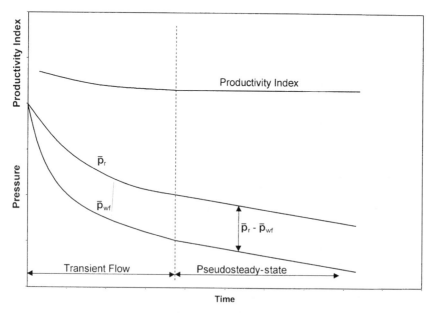

Figure 7-1. Productivity index during flow regimes.

The productivity index can be numerically calculated by recognizing that J must be defined in terms of semisteady-state flow conditions. Recalling Equation 6-149:

$$Q_o = \frac{0.00708\,k_o\,h\,(\bar{p}_r - p_{wf})}{\mu_o\,B_o\left[\ln\left(\frac{r_e}{r_w}\right) - 0.75 + s\right]} \tag{7-2}$$

The above equation is combined with Equation 7-1 to give:

$$J = \frac{0.00708\,k_o\,h}{\mu_o\,B_o\left[\ln\left(\frac{r_e}{r_w}\right) - 0.75 + s\right]} \tag{7-3}$$

where   J = productivity index, STB/day/psi
k_o = effective permeability of the oil, md
s = skin factor
h = thickness, ft

The oil relative permeability concept can be conveniently introduced into Equation 7-3 to give:

$$J = \frac{0.00708\,h\,k}{\left[\ln\left(\frac{r_e}{r_w}\right) - 0.75 + s\right]}\left(\frac{k_{ro}}{\mu_o B_o}\right) \tag{7-4}$$

Since most of the well life is spent in a flow regime that is approximating the pseudosteady-state, the productivity index is a valuable methodology for predicting the future performance of wells. Further, by monitoring the productivity index during the life of a well, it is possible to determine if the well has become damaged due to completion, workover, production, injection operations, or mechanical problems. If a measured J has an unexpected decline, one of the indicated problems should be investigated.

A comparison of productivity indices of different wells in the same reservoir should also indicate some of the wells might have experienced unusual difficulties or damage during completion. Since the productivity indices may vary from well to well because of the variation in thickness of the reservoir, it is helpful to normalize the indices by dividing each by the thickness of the well. This is defined as the **specific productivity index J_s,** or:

$$J_s = \frac{J}{h} = \frac{Q_o}{h(\bar{p}_r - p_{wf})} \tag{7-5}$$

Assuming that the well's productivity index is constant, Equation 7-1 can be rewritten as:

$$Q_o = J(\bar{p}_r - p_{wf}) = J \Delta p \tag{7-6}$$

where Δp = drawdown, psi
J = productivity index

Equation 7-6 indicates that the relationship between Q_o and Δp is a straight line passing through the origin with a slope of J as shown in Figure 7-2.

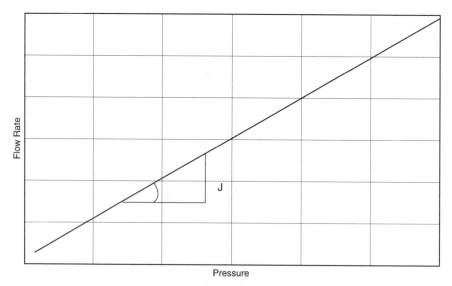

Figure 7-2. Q_o vs. Δp relationship.

Alternatively, Equation 7-1 can be written as:

$$p_{wf} = \bar{p}_r - \left(\frac{1}{J}\right) Q_o \qquad (7\text{-}7)$$

The above expression shows that the plot p_{wf} against Q_o is a straight line with a slope of $(-1/J)$ as shown schematically in Figure 7-3. This graphical representation of the relationship that exists between the oil flow rate and bottom-hole flowing pressure is called the **inflow performance relationship** and referred to as **IPR**.

Several important features of the straight-line IPR can be seen in Figure 7-3:

• When p_{wf} equals average reservoir pressure, the flow rate is zero due to the absence of any pressure drawdown.
• Maximum rate of flow occurs when p_{wf} is zero. This maximum rate is called **absolute open flow** and referred to as **AOF**. Although in practice this may not be a condition at which the well can produce, it is a useful definition that has widespread applications in the petroleum industry

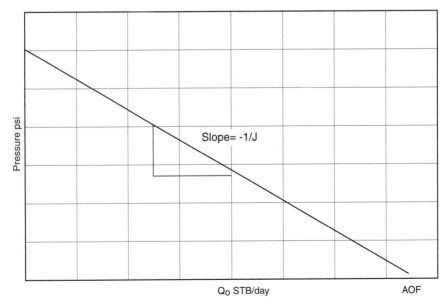

Figure 7-3. IPR.

(e.g., comparing flow potential of different wells in the field). The AOF is then calculated by:

$$AOF = J\,\bar{p}_r$$

• The slope of the straight line equals the reciprocal of the productivity index.

Example 7-1

A productivity test was conducted on a well. The test results indicate that the well is capable of producing at a stabilized flow rate of 110 STB/day and a bottom-hole flowing pressure of 900 psi. After shutting the well for 24 hours, the bottom-hole pressure reached a static value of 1300 psi.
    Calculate:

• Productivity index
• AOF
• Oil flow rate at a bottom-hole flowing pressure of 600 psi
• Wellbore flowing pressure required to produce 250 STB/day

Solution

a. Calculate J from Equation 7-1:

$$J = \frac{110}{1300 - 900} = 0.275 \text{ STB / psi}$$

b. Determine the AOF from:

$$AOF = J\,(\bar{p}_r - 0)$$

$$AOF = 0.275\,(1300 - 0\,) = 375.5 \text{ STB/day}$$

c. Solve for the oil-flow rate by applying Equation 7-1:

$$Q_o = 0.275\,(1300 - 600) = 192.5 \text{ STB/day}$$

d. Solve for p_{wf} by using Equation 7-7:

$$p_{wf} = 1300 - \left(\frac{1}{0.275}\right)250 = 390.9 \text{ psi}$$

Equation 7-6 suggests that the inflow into a well is directly proportional to the pressure drawdown and the constant of proportionality is the productivity index. Muskat and Evinger (1942) and Vogel (1968) observed that when the pressure drops below the bubble-point pressure, the IPR deviates from that of the simple straight-line relationship as shown in Figure 7-4.

Recalling Equation 7-4:

$$J = \left[\frac{0.00708\,h\,k}{\ln\left(\dfrac{r_e}{r_w}\right) - 0.75 + s} \right] \left(\frac{k_{ro}}{\mu_o\,B_o} \right)$$

Treating the term between the two brackets as a constant c, the above equation can be written in the following form:

$$J = c \left(\frac{k_{ro}}{\mu_o\,B_o} \right) \tag{7-8}$$

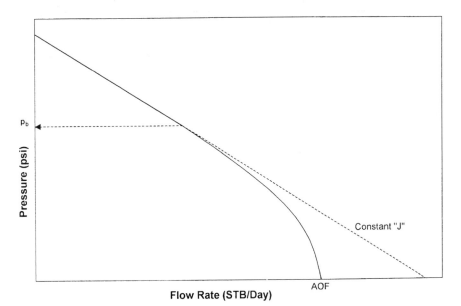

Figure 7-4. IPR below p_b.

With the coefficient c as defined by:

$$c = \frac{0.00708\,k\,h}{\ln\left(\dfrac{r_e}{r_w}\right) - 0.75 + s}$$

Equation 7-8 reveals that the variables affecting the productivity index are essentially those that are pressure dependent, i.e.:

- Oil viscosity μ_o
- Oil formation volume factor B_o
- Relative permeability to oil k_{ro}.

Figure 7-5 schematically illustrates the behavior of those variables as a function of pressure. Figure 7-6 shows the overall effect of changing the pressure on the term $(k_{ro}/\mu_o B_o)$. Above the bubble-point pressure p_b, the relative oil permeability k_{ro} equals unity ($k_{ro} = 1$) and the term $(k_{ro}/\mu_o B_o)$ is almost constant. As the pressure declines below p_b, the gas is released

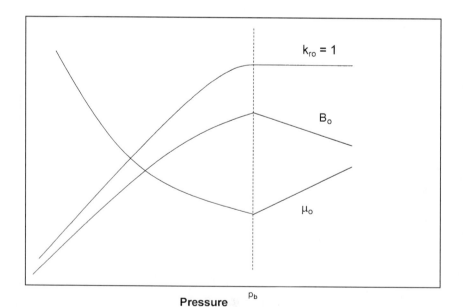

Figure 7-5. Effect of pressure on B_o, μ_o, and k_{ro}.

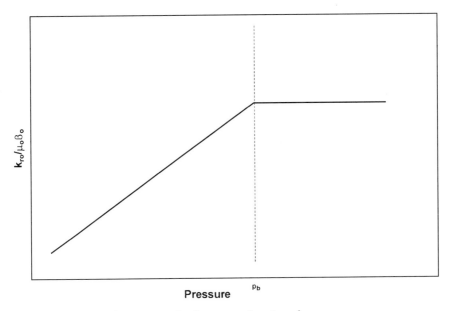

Figure 7-6. $k_{ro}/\mu_o B_o$ as a function of pressure.

from solution which can cause a large decrease in both k_{ro} and $(k_{ro}/\mu_o B_o)$. Figure 7-7 shows qualitatively effect of reservoir depletion on the IPR.

There are several empirical methods that are designed to predict the non-linearity behavior of the IPR for solution gas drive reservoirs. Most of these methods require at least one stabilized flow test in which Q_o and p_{wf} are measured. All the methods include the following two computational steps:

- Using the stabilized flow test data, construct the IPR curve at the current average reservoir pressure $\overline{p}_r$.
- Predict future inflow performance relationships as to the function of average reservoir pressures.

The following empirical methods that are designed to generate the current and future inflow performance relationships:

- Vogel's Method
- Wiggins' Method
- Standing's Method
- Fetkovich's Method
- The Klins-Clark Method

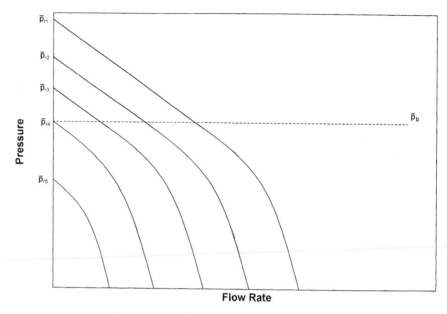

Figure 7-7. Effect of reservoir pressure on IPR.

Vogel's Method

Vogel (1968) used a computer model to generate IPRs for several hypothetical saturated-oil reservoirs that are producing under a wide range of conditions. Vogel normalized the calculated IPRs and expressed the relationships in a dimensionless form. He normalized the IPRs by introducing the following dimensionless parameters:

$$\text{dimensionless pressure} = \frac{p_{wf}}{\bar{p}_r}$$

$$\text{dimensionless pressure} = \frac{Q_o}{(Q_o)_{max}}$$

where $(Q_o)_{max}$ is the flow rate at zero wellbore pressure, i.e., AOF.

Vogel plotted the dimensionless IPR curves for all the reservoir cases and arrived at the following relationship between the above dimensionless parameters:

$$\frac{Q_o}{(Q_o)_{max}} = 1 - 0.2 \left(\frac{p_{wf}}{\bar{p}_r} \right) - 0.8 \left(\frac{p_{wf}}{\bar{p}_r} \right)^2 \qquad (7\text{-}9)$$

where Q_o = oil rate at p_{wf}
  $(Q_o)_{max}$ = maximum oil flow rate at zero wellbore pressure, i.e., AOF
   $\bar{p}_r$ = current average reservoir pressure, psig
   p_{wf} = wellbore pressure, psig

Notice that p_{wf} and $\bar{p}_r$ must be expressed in psig.

Vogel's method can be extended to account for water production by replacing the dimensionless rate with $Q_L/(Q_L)_{max}$ where $Q_L = Q_o + Q_w$. This has proved to be valid for wells producing at water cuts as high as 97%. The method requires the following data:

• Current average reservoir pressure $\bar{p}_r$
• Bubble-point pressure p_b
• Stabilized flow test data that include Q_o at p_{wf}

Vogel's methodology can be used to predict the IPR curve for the following two types of reservoirs:

• Saturated oil reservoirs $\bar{p}_r " p_b$
• Undersaturated oil reservoirs $\bar{p}_r > p_b$

Saturated Oil Reservoirs

When the reservoir pressure equals the bubble-point pressure, the oil reservoir is referred to as a *saturated-oil reservoir.* The computational procedure of applying Vogel's method in a saturated oil reservoir to generate the IPR curve for a well with a stabilized flow data point, i.e., a recorded Q_o value at p_{wf}, is summarized below:

Step 1. Using the stabilized flow data, i.e., Q_o and p_{wf}, calculate:

$(Q_o)_{max}$ from Equation 7-9, or

$$(Q_o)_{max} = Q_o \bigg/ \left[1 - 0.2\left(\frac{p_{wf}}{p_r}\right) - 0.8\left(\frac{p_{wf}}{p_r}\right)^2 \right]$$

Step 2. Construct the IPR curve by assuming various values for p_{wf} and calculating the corresponding Q_o from:

$$Q_o = (Q_o)_{max}\left[1 - 0.2\left(\frac{p_{wf}}{\bar{p}_r}\right) - 0.8\left(\frac{p_{wf}}{\bar{p}_r}\right)^2 \right]$$

Example 7-2

A well is producing from a saturated reservoir with an average reservoir pressure of 2500 psig. Stabilized production test data indicated that the stabilized rate and wellbore pressure are 350 STB/day and 2000 psig, respectively. Calculate:

• Oil flow rate at $p_{wf} = 1850$ psig
• Calculate oil flow rate assuming constant J
• Construct the IPR by using Vogel's method and the constant productivity index approach.

Solution

Part A.

Step 1. Calculate $(Q_o)_{max}$:

$$(Q_o)_{max} = 350 \bigg/ \left[1 - 0.2\left(\frac{2000}{2500}\right) - 0.8\left(\frac{2000}{2500}\right)^2 \right]$$

$$= 1067.1 \, STB/day$$

Step 2. Calculate Q_o at $p_{wf} = 1850$ psig by using Vogel's equation

$$Q_o = (Q_o)_{max} \left[1 - 0.2 \left(\frac{p_{wf}}{\overline{p}_r} \right) - 0.8 \left(\frac{p_{wf}}{\overline{p}_r} \right)^2 \right]$$

$$= 1067.1 \left[1 - 0.2 \left(\frac{1850}{2500} \right) - 0.8 \left(\frac{1850}{2500} \right)^2 \right] = 441.7 \text{ STB/day}$$

Part B.

Calculating oil flow rate by using the constant J approach

Step 1. Apply Equation 7-1 to determine J

$$J = \frac{350}{2500 - 2000} = 0.7 \text{ STB / day / psi}$$

Step 2. Calculate Q_o

$$Q_o = J (\overline{p}_r - p_{wf}) = 0.7 (2500 - 1850) = 455 \text{ STB/day}$$

Part C.

Generating the IPR by using the constant J approach and Vogel's method:

Assume several values for p_{wf} and calculate the corresponding Q_o.

p_{wf}	Vogel's	$Q_o = J (\overline{p}_r - p_{wf})$
2500	0	0
2200	218.2	210
1500	631.7	700
1000	845.1	1050
500	990.3	1400
0	1067.1	1750

Undersaturated Oil Reservoirs

Beggs (1991) pointed out that in applying Vogel's method for undersaturated reservoirs, there are **two possible outcomes to the recorded stabilized flow test data** that must be considered, as shown schematically in Figure 7-8:

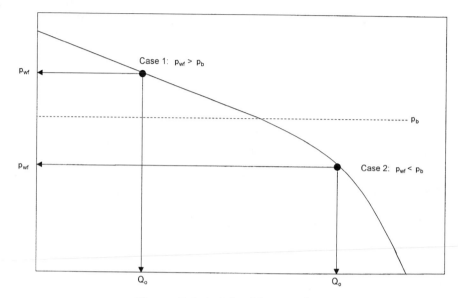

Figure 7-8. Stabilized flow test data.

- The recorded stabilized bottom-hole flowing pressure is greater than or equal to the bubble-point pressure, i.e. $p_{wf} \geq p_b$
- The recorded stabilized bottom-hole flowing pressure is less than the bubble-point pressure $p_{wf} < p_b$

Case 1. The Value of the Recorded Stabilized $p_{wf} \geq p_b$

Beggs outlined the following procedure for determining the IPR when the stabilized bottom-hole pressure is greater than or equal to the bubble-point pressure (Figure 7-8):

Step 1. Using the stabilized test data point (Q_o and p_{wf}) calculate the productivity index J:

$$J = \frac{Q_o}{\overline{p}_r - p_{wf}}$$

Step 2. Calculate the oil flow rate at the bubble-point pressure:

$$Q_{ob} = J\,(\overline{p}_r - P_b) \tag{7-10}$$

where Q_{ob} is the oil flow rate at p_b

Step 3. Generate the IPR values below the bubble-point pressure by assuming different values of $p_{wf} < p_b$ and calculating the correspond oil flow rates by applying the following relationship:

$$Q_o = Q_{ob} + \frac{J\,p_b}{1.8}\left[1 - 0.2\left(\frac{p_{wf}}{p_b}\right) - 0.8\left(\frac{p_{wf}}{p_b}\right)^2\right] \qquad (7\text{-}11)$$

The maximum oil flow rate ($Q_{o\,max}$ or AOF) occurs when the bottom-hole flowing pressure is zero, i.e. $p_{wf} = 0$, which can be determined from the above expression as:

$$Q_{o\,max} = Q_{ob} + \frac{J\,p_b}{1.8}$$

It should be pointed out that when $p_{wf} \geq p_b$, the IPR is linear and is described by:

$$Q_o = J\,(\bar{p}_r - p_{wf}).$$

Example 7-3

An oil well is producing from an undersaturated reservoir that is characterized by a bubble-point pressure of 2130 psig. The current average reservoir pressure is 3000 psig. Available flow test data shows that the well produced 250 STB/day at a stabilized p_{wf} of 2500 psig. Construct the IPR data.

Solution

The problem indicates that the flow test data was recorded above the bubble-point pressure, therefore, the Case 1 procedure for undersaturated reservoirs as outlined previously must be used.

Step 1. Calculate J using the flow test data.

$$J = \frac{250}{3000 - 2500} = 0.5\,\text{STB}/\text{day}/\text{psi}$$

Step 2. Calculate the oil flow rate at the bubble-point pressure by applying Equation 7-10.

$$Q_{ob} = 0.5\,(3000 - 2130) = 435\,\text{STB/day}$$

Step 3. Generate the IPR data by applying the constant J approach for all pressures above p_b and Equation 7-11 for all pressures below p_b.

P_{wf}	Equation #	Q_o
3000	(7-6)	0
2800	(7-6)	100
2600	(7-6)	200
2130	(7-6)	435
1500	(7-11)	709
1000	(7-11)	867
500	(7-11)	973
0	(7-11)	1027

Case 2. The Value of the Recorded Stabilized $p_{wf} < p_b$

When the recorded p_{wf} from the stabilized flow test is below the bubble-point pressure, as shown in Figure 7-8, the following procedure for generating the IPR data is proposed:

Step 1. Using the stabilized well flow test data and combining Equation 7-10 with 7-11, solve for the productivity index J to give:

$$J = \frac{Q_o}{\left(\bar{p}_r - p_b\right) + \dfrac{p_b}{1.8}\left[1 - 0.2\left(\dfrac{p_{wf}}{p_b}\right) - 0.8\left(\dfrac{p_{wf}}{p_b}\right)^2\right]} \qquad (7\text{-}12)$$

Step 2. Calculate Q_{ob} by using Equation 7-10, or:

$$Q_{ob} = J\left(\bar{p}_r - p_b\right)$$

Step 3. Generate the IPR for $p_{wf} \geq p_b$ by assuming several values for p_{wf} above the bubble point pressure and calculating the corresponding Q_o from:

$$Q_o = J\left(\bar{p}_r - p_{wf}\right)$$

Step 4. Use Equation 7-11 to calculate Q_o at various values of p_{wf} below p_b, or:

$$Q_o = Q_{ob} + \frac{J \, p_b}{1.8} \left[1 - 0.2 \left(\frac{p_{wf}}{p_b} \right) - 0.8 \left(\frac{p_{wf}}{p_b} \right)^2 \right]$$

Example 7-4

The well described in Example 7-3 was retested and the following results obtained:

$P_{wf} = 1700$ psig, $Q_o = 630.7$ STB/day

Generate the IPR data using the new test data.

Solution

Notice that the stabilized p_{wf} is less than p_b.

Step 1. Solve for J by applying Equation 7-12.

$$J = \frac{630.7}{(3000 - 2130) + \dfrac{2130}{1.8} \left[1 - \left(\dfrac{1700}{2130} \right) - \left(\dfrac{1700}{2130} \right)^2 \right]}$$

$$= 0.5 \text{ STB} / \text{day} / \text{psi}$$

Step 2. $Q_{ob} = 0.5 \, (3000 - 21300) = 435$ STB/day

Step 3. Generate the IPR data.

Pwf	Equation #	Qo
3000	(7-6)	0
2800	(7-6)	100
2600	(7-6)	200
2130	(7-6)	435
1500	(7-11)	709
1000	(7-11)	867
500	(7-11)	973
0	(7-11)	1027

Quite often it is necessary to predict the well's inflow performance for future times as the reservoir pressure declines. Future well performance calculations require the development of a relationship that can be used to predict future maximum oil flow rates.

There are several methods that are designed to address the problem of how the IPR might shift as the reservoir pressure declines. Some of these prediction methods require the application of the material balance equation to generate future oil saturation data as a function of reservoir pressure. In the absence of such data, there are two simple approximation methods that can be used in conjunction with Vogel's method to predict future IPRs.

First Approximation Method

This method provides a rough approximation of the future maximum oil flow rate $(Q_{omax})_f$ at the specified future average reservoir pressure $(p_r)_f$. This future maximum flow rate $(Q_{omax})_f$ can be used in Vogel's equation to predict the future inflow performance relationships at $(\overline{p}_r)_f$. The following steps summarize the method:

Step 1. Calculate $(Q_{omax})_f$ at $(\overline{p}_r)_f$ from:

$$(Q_{o\,max})_f = (Q_{o\,max})_p \left(\frac{(\overline{p}_r)_f}{(\overline{p}_r)_p} \right) \left[0.2 + 0.8 \left(\frac{(\overline{p}_r)_f}{(\overline{p}_r)_p} \right) \right] \qquad (7\text{-}13)$$

where the subscript f and p represent future and present conditions, respectively.

Step 2. Using the new calculated value of $(Q_{omax})_f$ and $(\overline{p}_r)_f$, generate the IPR by using Equation 7-9.

Second Approximation Method

A simple approximation for estimating future $(Q_{omax})_f$ at $(\overline{p}_r)_f$ is proposed by Fetkovich (1973). The relationship has the following mathematical form:

$$(Q_{o\,max})_f = (Q_{o\,max})_p \left[(\overline{p}_r)_f / (\overline{p}_r)_p \right]^{3.0}$$

Where the subscript f and p represent future and present conditions, respectively. The above equation is intended only to provide a rough estimation of future $(Q_o)_{max}$.

Example 7-5

Using the data given in Example 7-2, predict the IPR where the average reservoir pressure declines from 2500 psig to 2200 psig.

Solution

Example 7-2 shows the following information:

• Present average reservoir pressure $(\bar{p}_r)_p = 2500$ psig
• Present maximum oil rate $(Q_{omax})_p = 1067.1$ STB/day

Step 1. Solve for $(Q_{omax})_f$ by applying Equation 7-13.

$$(Q_{o\,max})_f = 1067.1 \left(\frac{2200}{2500}\right) \left[0.2 + 0.8\left(\frac{2200}{2500}\right)\right] = 849 \text{ STB/day}$$

Step 2. Generate the IPR data by applying Equation 7-9.

p_{wf}	$Q_o = 849 [1 - 0.2 (p_{wf}/2200) - 0.8 (p_{wf}/2200)^2]$
2200	0
1800	255
1500	418
500	776
0	849

It should be pointed out that the main disadvantage of Vogel's methodology lies with its sensitivity to the match point, i.e., the stabilized flow test data point, used to generate the IPR curve for the well.

Wiggins' Method

Wiggins (1993) used four sets of relative permeability and fluid property data as the basic input for a computer model to develop equations to predict inflow performance. The generated relationships are limited by the assumption that the reservoir initially exists at its bubble-point pressure. Wiggins proposed generalized correlations that are suitable for predicting the IPR during three-phase flow. His proposed expressions are similar to that of Vogel's and are expressed as:

$$Q_o = (Q_o)_{max} \left[1 - 0.52 \left(\frac{p_{wf}}{\overline{p}_r} \right) - 0.48 \left(\frac{p_{wf}}{\overline{p}_r} \right)^2 \right] \tag{7-14}$$

$$Q_w = (Q_w)_{max} \left[1 - 0.72 \left(\frac{p_{wf}}{\overline{p}_r} \right) - 0.28 \left(\frac{p_{wf}}{\overline{p}_r} \right)^2 \right] \tag{7-15}$$

where Q_w = water flow rate, STB/day
$(Q_w)_{max}$ = maximum water production rate at $p_{wf} = 0$, STB/day

As in Vogel's method, data from a stabilized flow test on the well must be available in order to determine $(Q_o)_{max}$ and $(Q_w)_{max}$.

Wiggins extended the application of the above relationships to predict future performance by providing with expressions for estimating future maximum flow rates. Wiggins expressed future maximum rates as a function of:

- Current (present) average pressure $(\overline{p}_r)_p$
- Future average pressure $(\overline{p}_r)_f$
- Current maximum oil flow rate $(Q_{omax})_p$
- Current maximum water flow rate $(Q_{wmax})_p$

Wiggins proposed the following relationships:

$$(Q_{omax})_f = (Q_{omax})_p \left\{ 0.15 \left[\frac{(\overline{p}_r)_f}{(\overline{p}_r)_p} \right] + 0.84 \left[\frac{(\overline{p}_r)_f}{(\overline{p}_r)_p} \right]^2 \right\} \tag{7-16}$$

$$(Q_{omax})_f = (Q_{wmax})_p \left\{ 0.59 \left[\frac{(\overline{p}_r)_f}{(\overline{p}_r)_p} \right] + 0.36 \left[\frac{(\overline{p}_r)_f}{(\overline{p}_r)_p} \right]^2 \right\} \tag{7-17}$$

Example 7-6

The information given in Examples 7-2 and 7-5 is repeated here for convenience:

- Current average pressure = 2500 psig
- Stabilized oil flow rate   = 350 STB/day
- Stabilized wellbore pressure = 2000 psig

Generate the current IPR data and predict future IPR when the reservoir pressure declines from 2500 to 2000 psig by using Wiggins' method.

Solution

Step 1. Using the stabilized flow test data, calculate the current maximum oil flow rate by applying Equation 7-14.

$$(Q_{omax})_p = 350 \left/ \left[1 - 0.52 \left(\frac{2000}{2500} \right) - 0.48 \left(\frac{2000}{2500} \right)^2 \right] \right.$$

$$= 1264 \, STB / day$$

Step 2. Generate the current IPR data by using Wiggins' method and compare the results with those of Vogel's.
Results of the two methods are shown graphically in Figure 7-9.

p_{wf}	Wiggins'	Vogel's
2500	0	0
2200	216	218
1500	651	632
1000	904	845
500	1108	990
0	1264	1067

Step 3. Calculate future maximum oil flow rate by using Equation 7-16.

$$(Q_{omax})_f = 1264 \left[0.15 \left(\frac{2200}{2500} \right) + 0.84 \left(\frac{2200}{2500} \right)^2 \right] = 989 \, STB/day$$

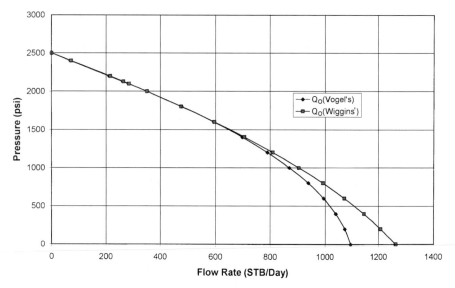

Figure 7-9. IPR curves.

Step 4. Generate future IPR data by using Equation 7-16

p_{wf}	$Q_o = 989 \, [1 - 0.52 \, (p_{wf}/2200) - 0.48 \, (p_{wf}/2200)^2]$
2200	0
1800	250
1500	418
500	848
0	989

Standing's Method

Standing (1970) essentially extended the application of Vogel's to predict future inflow performance relationship of a well as a function of reservoir pressure. He noted that Vogel's equation: Equation 7-9 can be rearranged as:

$$\frac{Q_o}{(Q_o)_{max}} = \left(1 - \frac{p_{wf}}{\overline{p}_r}\right)\left[1 + 0.8\left(\frac{p_{wf}}{\overline{p}_r}\right)\right] \qquad (7\text{-}18)$$

Standing introduced the productivity index J as defined by Equation 7-1 into Equation 7-18 to yield:

$$J = \frac{(Q_o)_{max}}{\bar{p}_r}\left[1 + 0.8\left(\frac{p_{wf}}{\bar{p}_r}\right)\right] \tag{7-19}$$

Standing then defined the present (current) zero drawdown productivity index as:

$$J_p^* = 1.8\left[\frac{(Q_o)_{max}}{\bar{p}_r}\right] \tag{7-20}$$

where J_p^* is Standing's zero-drawdown productivity index. The J_p^* is related to the productivity index J by:

$$\frac{J}{J_p^*} = \frac{1}{1.8}\left[1 + 0.8\left(\frac{p_{wf}}{\bar{p}_r}\right)\right] \tag{7-21}$$

Equation 7-1 permits the calculation of J_p^* from a measured value of J.

To arrive to the final expression for predicting the desired IPR expression, Standing combines Equation 7-20 with Equation 7-18 to eliminate $(Q_o)_{max}$ to give:

$$Q_o = \left[\frac{J_f^*\,(\bar{p}_r)_f}{1.8}\right]\left\{1 - 0.2\left[\frac{p_{wf}}{(\bar{p}_r)_f}\right] - 0.8\left[\frac{p_{wf}}{(\bar{p}_r)_f}\right]^2\right\} \tag{7-22}$$

where the subscript f refers to future condition.

Standing suggested that J_p^* can be estimated from the present value of J_p^* by the following expression:

$$J_f^* = J_p^*\left(\frac{k_{ro}}{\mu_o B_o}\right)_f \Big/ \left(\frac{k_{ro}}{\mu_o B_o}\right)_p \tag{7-23}$$

where the subscript p refers to the present condition.

If the relative permeability data is not available, J_f^* can be roughly estimated from:

$$J_f^* = J_p^* \, [(\bar{p}_r)_f / (\bar{p}_r)_p]^2 \tag{7-24}$$

Standing's methodology for predicting a future IPR is summarized in the following steps:

Step 1. Using the current time condition and the available flow test data, calculate $(Q_o)_{max}$ from Equation 7-9 or Equation 7-18.

Step 2. Calculate J^* at the present condition, i.e., J_p^*, by using Equation 7-20. Notice that other combinations of Equations 7-18 through 7-21 can be used to estimate J_p^*.

Step 3. Using fluid property, saturation and relative permeability data, calculate both $(k_{ro}/\mu_o B_o)_p$ and $(k_{ro}/\mu_o B_o)_f$.

Step 4. Calculate J_f^* by using Equation 7-23. Use Equation 7-24 if the oil relative permeability data is not available.

Step 5. Generate the future IPR by applying Equation 7-22.

Example 7-7

A well is producing from a saturated oil reservoir that exists at its saturation pressure of 4000 psig. The well is flowing at a stabilized rate of 600 STB/day and a p_{wf} of 3200 psig. Material balance calculations provide the following current and future predictions for oil saturation and PVT properties.

	Present	Future
$\bar{p}_r$	4000	3000
μ_o, cp	2.40	2.20
B_o, bbl/STB	1.20	1.15
k_{ro}	1.00	0.66

Generate the future IPR for the well at 3000 psig by using Standing's method.

Solution

Step 1. Calculate the current $(Q_o)_{max}$ from Equation 7-18.

$$(Q_o)_{max} = 600 \bigg/ \left[\left(1 - \frac{3200}{4000}\right)(1 + 0.8)\left(\frac{3200}{4000}\right) \right] = 1829 \, STB/day$$

Step 2. Calculate J_p^* by using Equation 7-20.

$$J_p^* = 1.8 \left[\frac{1829}{4000} \right] = 0.823$$

Step 3. Calculate the following pressure-function:

$$\left(\frac{k_{ro}}{\mu_o B_o} \right)_p = \frac{1}{(2.4)(1.20)} = 0.3472$$

$$\left(\frac{k_{ro}}{\mu_o B_o} \right)_f = \frac{0.66}{(2.2)(1.15)} = 0.2609$$

Step 4. Calculate J_f^* by applying Equation 7-23.

$$J_f^* = 0.823 \left(\frac{0.2609}{0.3472} \right) = 0.618$$

Step 5. Generate the IPR by using Equation 7-22.

P_{wf}	Q_o, STB/day
3000	0
2000	527
1500	721
1000	870
500	973
0	1030

It should be noted that one of the main disadvantages of Standing's methodology is that it requires reliable permeability information; in addition, it also requires material balance calculations to predict oil saturations at future average reservoir pressures.

Fetkovich's Method

Muskat and Evinger (1942) attempted to account for the observed non-linear flow behavior (i.e., IPR) of wells by calculating a theoretical productivity index from the pseudosteady-state flow equation. They expressed Darcy's equation as:

$$Q_o = \frac{0.00708\,kh}{\left[\ln\frac{r_e}{r_w} - 0.75 + s\right]} \int_{P_{wf}}^{\bar{p}_r} f(p)\,dp \qquad (7-25)$$

where the pressure function f(p) is defined by:

$$f(p) = \frac{k_{ro}}{\mu_o B_o} \qquad (7-26)$$

where k_{ro} = oil relative permeability
k = absolute permeability, md
B_o = oil formation volume factor
μ_o = oil viscosity, cp

Fetkovich (1973) suggests that the pressure function f(p) can basically fall into one of the following two regions:

Region 1: Undersaturated Region

The pressure function f(p) falls into this region if $p > p_b$. Since oil relative permeability in this region equals unity (i.e., $k_{ro} = 1$), then:

$$f(p) = \left(\frac{1}{\mu_o B_o}\right)_p \qquad (7-27)$$

Fetkovich observed that the variation in f(p) is only slight and the pressure function is considered constant as shown in Figure 7-10.

Region 2: Saturated Region

In the saturated region where $p < p_b$, Fetkovich shows that the $(k_{ro}/\mu_o B_o)$ changes linearly with pressure and that the straight line passes through the origin. This linear is shown schematically in Figure 7-10 can be expressed mathematically as:

$$f(p) = \left(\frac{1}{\mu_o B_o}\right)_{p_b} \left(\frac{p}{p_b}\right) \qquad (7-28)$$

Where μ_o and B_o are evaluated at the bubble-point pressure. In the application of the straight-line pressure function, there are three cases that must be considered:

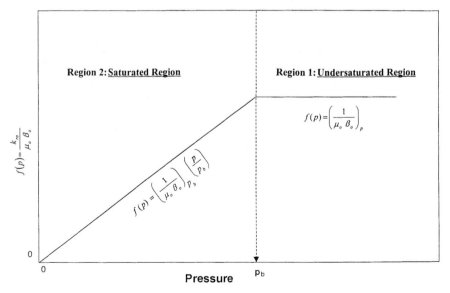

Figure 7-10. Pressure function concept.

- $\bar{p}_r$ and $p_{wf} > p_b$
- $\bar{p}_r$ and $p_{wf} < p_b$
- $\bar{p}_r > p_b$ and $p_{wf} < p_b$

All three cases are presented below.

Case 1: $\bar{p}_r$ and $p_{wf} > p_b$

This is the case of a well producing from an undersaturated oil reservoir where both p_{wf} and $\bar{p}_r$ are greater than the bubble-point pressure. The pressure function f(p) in this case is described by Equation 7-27. Substituting Equation 7-27 into Equation 7-25 gives:

$$Q_o = \frac{0.00708\,kh}{\ln\left(\dfrac{r_e}{r_w}\right) - 0.75 + s} \int_{p_{wf}}^{\bar{p}_r} \left(\frac{1}{\mu_o B_o}\right) dp$$

Since $\left(\dfrac{1}{\mu_o B_o}\right)$ is constant, then:

$$Q_o = \frac{0.00708\,kh}{\mu_o B_o \left[\ln\left(\dfrac{r_e}{r_w}\right) - 0.75 + s\right]} (\bar{p}_r - p_{wf}) \tag{7-29}$$

or

$$Q_o = J\,(\bar{p}_r - p_{wf}) \tag{7-30}$$

The productivity index is defined in terms of the reservoir parameters as:

$$J = \frac{0.00708\,kh}{\mu_o B_o \left[\ln\left(\dfrac{r_e}{r_w}\right) - 0.75 + s\right]} \tag{7-31}$$

where B_o and μ_o are evaluated at $(\bar{p}_r + p_{wf})/2$.

Example 7-8

A well is producing from an undersaturated-oil reservoir that exists at an average reservoir pressure of 3000 psi. The bubble-point pressure is recorded as 1500 psi at 150°F. The following additional data are available:

- stabilized flow rate = 280 STB/day
- stabilized wellbore pressure = 2200 psi
- $h = 20'$ $r_w = 0.3'$ $r_e = 660'$ $s = -0.5$
- $k = 65$ md
- μ_o at 2600 psi = 2.4 cp
- B_o at 2600 psi = 1.4 bbl/STB

Calculate the productivity index by using both the reservoir properties (i.e., Equation 7-31) and flow test data (i.e., Equation 7-30).

Solution

- From Equation 7-31

$$J = \frac{0.00708\,(65)\,(20)}{(2.4)\,(1.4)\left[\ln\left(\dfrac{660}{0.3}\right) - 0.75 - 0.5\right]} = 0.42 \text{ STB/day/psi}$$

- From production data:

$$J = \frac{280}{3000 - 2200} = 0.35 \text{ STB/day/psi}$$

Results show a reasonable match between the two approaches. It should be noted, however, that there are several uncertainties in the values of the parameters used in Equation 7-31 to determine the productivity index. For example, changes in the skin factor k or drainage area would change the calculated value of J.

Case 2: $\bar{p}_r$ and $p_{wf} < p_b$

When the reservoir pressure $\bar{p}_r$ and bottom-hole flowing pressure p_{wf} are both below the bubble-point pressure p_b, the pressure function f(p) is represented by the straight line relationship as expressed by Equation 7-28. Combining Equation 7-28 with Equation 7-25 gives:

$$Q_o = \left[\frac{0.00708 \, kh}{\ln\left(\dfrac{r_e}{r_w}\right) - 0.75 + s} \right] \int_{P_{wf}}^{\bar{p}_r} \frac{1}{(\mu_o B_o)_{pb}} \left(\frac{p}{p_b}\right) dp$$

Since the term $\left[\left(\dfrac{1}{\mu_o B_o}\right)_{pb} \left(\dfrac{1}{p_b}\right)\right]$ is constant, then:

$$Q_o = \left[\frac{0.00708 \, kh}{\ln\left(\dfrac{r_e}{r_w}\right) - 0.75 + s} \right] \frac{1}{(\mu_o B_o)_{pb}} \left(\frac{1}{p_b}\right) \int_{P_{wf}}^{\bar{p}_r} p \, dp$$

Integrating gives:

$$Q_o = \frac{0.00708 \, kh}{(\mu_o B_o)_{pb} \left[\ln\left(\dfrac{r_e}{r_w}\right) - 0.75 + s\right]} \left(\frac{1}{2 p_b}\right) (p_r^{-2} - p_{wf}^2) \qquad (7\text{-}32)$$

Introducing the productivity index into the above equation gives:

$$Q_o = J \left(\frac{1}{2 p_b}\right) (p_r^{-2} - p_{wf}^2) \qquad (7\text{-}33)$$

The term $\left(\dfrac{J}{2 p_b}\right)$ is commonly referred to as the **performance coeffi-cient** C, or:

$$Q_o = C (\bar{p}_r^2 - p_{wf}^2) \qquad (7\text{-}34)$$

To account for the possibility of non-Darcy flow (turbulent flow) in oil wells, Fetkovich introduced the exponent n in Equation 7-35 to yield:

$$Q_o = C(\overline{p}_r^2 - p_{wf}^2)^n \qquad (7\text{-}35)$$

The value of n ranges from 1.000 for a complete laminar flow to 0.5 for highly turbulent flow.

There are two unknowns in Equation 7-35, the performance coefficient C and the exponent n. At least two tests are required to evaluate these two parameters, assuming $\overline{p}_r$ is known:

By taking the log of both sides of Equation 7-35 and solving for log $(p_r^2 - p_{wf}^2)$, the expression can be written as:

$$\log (\overline{p}_r^2 - p_{wf}^2) = \frac{1}{n}\log Q_o - \frac{1}{n}\log C$$

A plot of $\overline{p}_r^2 - p_{wf}^2$ versus Q_o on log-log scales will result in a straight line having a slope of 1/n and an intercept of C at $\overline{p}_r^2 - p_{wf}^2 = 1$. The value of C can also be calculated using any point on the linear plot once n has been determined to give:

$$C = \frac{Q_o}{(\overline{p}_r^2 - p_{wf}^2)^n}$$

Once the values of C and n are determined from test data, Equation 7-35 can be used to generate a complete IPR.

To construct the future IPR when the average reservoir pressure declines to $(\overline{p}_r)_f$, Fetkovich assumes that the performance coefficient C is a linear function of the average reservoir pressure and, therefore, the value of C can be adjusted as:

$$(C)_f = (C)_p \left[(\overline{p}_r)_f/(\overline{p}_r)_p\right] \qquad (7\text{-}36)$$

where the subscripts f and p represent the future and present conditions.

Fetkovich assumes that the value of the exponent n would not change as the reservoir pressure declines. Beggs (1991) presented an excellent and comprehensive discussion of the different methodologies used in constructing the IPR curves for oil and gas wells.

The following example was used by Beggs (1991) to illustrate Fetkovich's method for generating the current and future IPR.

Example 7-9

A four-point stabilized flow test was conducted on a well producing from a saturated reservoir that exists at an average pressure of 3600 psi.

Q_o, STB/day	p_{wf}, psi
263	3170
383	2890
497	2440
640	2150

a. Construct a complete IPR by using Fetkovich's method.
b. Construct the IPR when the reservoir pressure declines to 2000 psi.

Solution

Part A.

Step 1. Construct the following table:

Q_o, STB/day	p_{wf}, psi	$(\bar{p}_r^2 - p_{wf}^2) \times 10^{-6}$, psi^2
263	3170	2.911
383	2890	4.567
497	2440	7.006
640	2150	8.338

Step 2. Plot $(\bar{p}_r^2 - p_{wf}^2)$ verses Q_o on log-log paper as shown in Figure 7-11 and determine the exponent n, or:

$$n = \frac{\log(750) - \log(105)}{\log(10^7) - \log(10^6)} = 0.854$$

Step 3. Solve for the performance coefficient C:

$$C = 0.00079$$

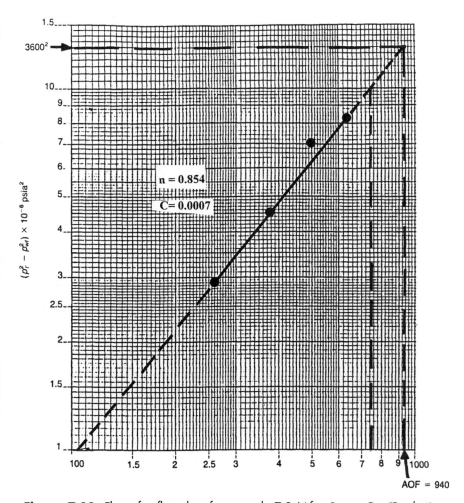

Figure 7-11. Flow-after-flow data for example 7-9 (*After Beggs, D., "Production Optimization Using Nodal Analysis," permission to publish by the OGCI, copyright OGCI, 1991.*)

Step 4. Generate the IPR by assuming various values for p_{wf} and calculating the corresponding flow rate from Equation 7-25:

$$Q_o = 0.00079 \, (3600^2 - p_{wf}^2)^{0.854}$$

P_{wf}	Q_o, STB/day
3600	0
3000	340
2500	503
2000	684
1500	796
1000	875
500	922
0	937

The IPR curve is shown in Figure 7-12. Notice that the AOF, i.e., $(Q_o)_{max}$, is 937 STB/day.

Part B.

Step 1. Calculate future C by applying Equation 7-36

$$(C)_f = 0.00079 \left(\frac{2000}{3600} \right) = 0.000439$$

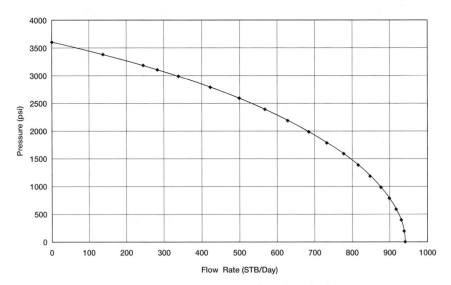

Figure 7-12. IPR using Fetkovich method.

Step 2. Construct the new IPR curve at 2000 psi by using the new calculated C and applying the inflow equation.

$$Q_o = 0.000439 \, (2000^2 - p_{wf}^2)^{0.854}$$

p_{wf}	Q_o
2000	0
1500	94
1000	150
500	181
0	191

Both the present time and future IPRs are plotted in Figure 7-13.

Klins and Clark (1993) developed empirical correlations that correlate the changes in Fetkovich's performance coefficient C and the flow exponent n with the decline in the reservoir pressure. The authors observed the exponent n changes considerably with reservoir pressure. Klins and Clark concluded the "future" values of $(n)_f$ and (C) at pressure $(\bar{p}_r)_f$ are related to the values of n and C at the bubble-point pressure. Denoting C_b

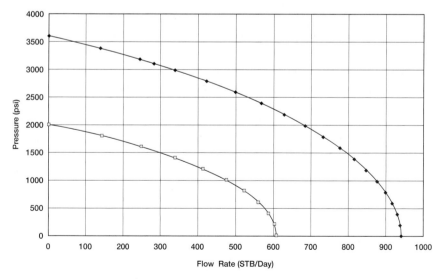

Figure 7-13. Future IPR at 2000 psi.

and n_b as the values of the performance coefficient and the flow exponent at the bubble-point pressure p_b, Klins and Clark introduced the following dimensionless parameters:

- Dimensionless performance coefficient $= C/C_b$
- Dimensionless flow exponent $= n/n_b$
- Dimensionless average reservoir pressure $= \bar{p}_r/p_b$

The authors correlated (C/C_b) and (n/n_b) to the dimensionless pressure by the following two expressions:

$$\left(\frac{n}{n_b}\right) = 1 + 0.0577\left(1 - \frac{\bar{p}_r}{p_b}\right) - 0.2459\left(1 - \frac{\bar{p}_r}{p_b}\right)^2$$
$$+ 0.503\left(1 - \frac{\bar{p}_r}{p_b}\right)^3 \tag{7-37}$$

and

$$\left(\frac{C}{C_b}\right) = 1 - 3.5718\left(1 - \frac{\bar{p}_r}{p_b}\right) + 4.7981\left(1 - \frac{\bar{p}_r}{p_b}\right)^2$$
$$- 2.3066\left(1 - \frac{\bar{p}_r}{p_b}\right)^3 \tag{7-38}$$

where C_b = performance coefficient at the bubble-point pressure
n_b = flow exponent at the bubble-point pressure

The procedure of applying the above relationships in adjusting the coefficients C and n with changing average reservoir pressure is detailed below:

Step 1. Using the available flow-test data in conjunction with Fetkovich's equation, i.e., Equation 7-34, calculate the present (current) values of n and C at the present average pressure $\bar{p}_r$.

Step 2. Using the current values of $\bar{p}_r$, calculate the dimensionless values of (n/n_b) and (C/C_b) by applying Equations 7-37 and 7-38, respectively.

Step 3. Solve for the constants n_b and C_b from:

$$n_b = \frac{n}{n/n_b} \tag{7-39}$$

and

$$C_b = \frac{C}{(C/C_b)} \tag{7-40}$$

It should be pointed out that if the present reservoir pressure equals the bubble-point pressure, the values of n and C as calculated in Step 1 are essentially n_b and C_b.

Step 4. Assume future average reservoir pressure $\bar{p}_r$ and solve for the corresponding future dimensionless parameters (n_f/n_b) and (C_f/C_b) by applying Equations 7-37 and 7-38, respectively.

Step 5. Solve for future values of n_f and C_f from

$$n_f = n_b \, (n/n_b)$$

$$C_f = C_b \, (C_f/C_b)$$

Step 6. Use n_f and C_f in Fetkovich's equation to generate the well's future IPR at the desired average reservoir pressure $(\bar{p}_r)_f$. It should be noted that the maximum oil flow rate $(Q_o)_{max}$ at $(\bar{p}_r)_f$ is given by:

$$(Q_o)_{max} = C_f[(\bar{p}_r)^2]^{n_f} \tag{7-41}$$

Example 7-10

Using the data given in Example 7-9, generate the future IPR data when the reservoir pressure drops to 3200 psi.

Solution

Step 1. Since the reservoir exists at its bubble-point pressure, then:

$$n_b = 0.854 \qquad \text{and } C_b = 0.00079 \qquad \text{at } p_b = 3600 \text{ psi}$$

Step 2. Calculate the future dimensionless parameters at 3200 psi by applying Equations 7-37 and 7-38:

$$\left(\frac{n}{n_b}\right) = 1 + 0.0577\left(1 - \frac{3200}{3600}\right) - 0.2459\left(1 - \frac{3200}{3600}\right)^2$$
$$+ 0.5030\left(1 - \frac{3200}{3600}\right)^3 = 1.0041$$

$$\left(\frac{C}{C_b}\right) = 1 - 3.5718\left(1 - \frac{3200}{3600}\right) + 4.7981\left(1 - \frac{3200}{3600}\right)^2$$
$$- 2.3066\left(1 - \frac{3200}{3600}\right)^3 = 0.6592$$

Step 3. Solve for n_f and C_f:

$n_f = (0.854)(1.0041) = 0.8575$

$C_f = (0.00079)(0.6592) = 0.00052$

Therefore, the flow rate is expressed as:

$Q_o = 0.00052 \, (3200^2 - p_{wf}^2)^{0.8575}$

When the maximum oil flow rate, i.e., AOF, occurs at $p_{wf} = 0$, then:

$(Q_o)_{max} = 0.00052 \, (3200^2 - 0^2)^{0.8575} = 534$ STB/day

Step 4. Construct the following table:

p_{wf}	Q_o
3200	0
2000	349
1500	431
500	523
0	534

Figure 7-14 compares current and future IPRs as calculated in Examples 7-9 and 7-10.

Case 3: $\bar{p}_r > p_b$ and $p_{wf} < p_b$

Figure 7-15 shows a schematic illustration of Case 3 in which it is assumed that $p_{wf} < p_b$ and $\bar{p}_r > p_b$. The integral in Equation 7-25 can be expanded and written as:

$$Q_o = \frac{0.00708\,kh}{\ln\left(\dfrac{r_e}{r_w}\right) - 0.75 + s}\left[\int_{p_{wf}}^{p_b} f(p)\,dp + \int_{p_b}^{\bar{p}_r} f(p)\,dp\right]$$

Substituting Equations 7-27 and 7-18 into the above expression gives:

$$Q_o = \frac{0.00708\,kh}{\ln\left(\dfrac{r_e}{r_w}\right) - 0.75 + s}\left[\int_{p_{wf}}^{p_b} \left(\frac{1}{\mu_o B_o}\right)\left(\frac{p}{p_b}\right)dp + \int_{pb}^{\bar{p}_r} \left(\frac{1}{\mu_o B_o}\right)dp\right]$$

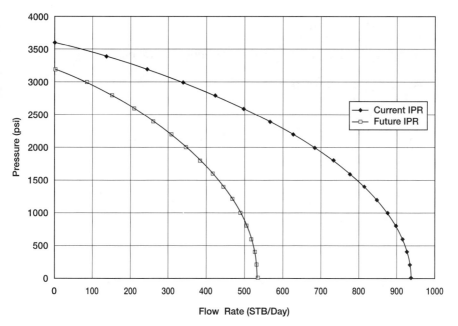

Figure 7-14. IPR.

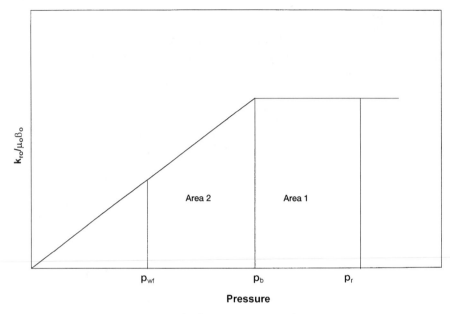

Figure 7-15. $(k_{ro}/\mu_o B_o)$ vs. pressure for Case #3.

where μ_o and B_o are evaluated at the bubble-point pressure p_b.

Arranging the above expression gives:

$$Q_o = \frac{0.00708\,kh}{\mu_o B_o\left[\ln\left(\dfrac{r_e}{r_w}\right) - 0.75 + s\right]}\left[\frac{1}{p_b}\int_{p_{wf}}^{p_b} p\,dp + \int_{p_b}^{\bar{p}_r} dp\right]$$

Integrating and introducing the productivity index J into the above relationship gives:

$$Q_o = J\left[\frac{1}{2 p_b}\,(p_b^2 - p_{wf}^2) + (\bar{p}_r - p_b)\right]$$

or

$$Q_o = J(\bar{p}_r - p_b) + \frac{J}{2\,p_b}(p_b^2 - p_{wf}^2) \qquad (7\text{-}42)$$

Example 7-11

The following reservoir and flow-test data are available on an oil well:

- Pressure data: $\bar{p}_r = 4000$ psi $p_b = 3200$ psi
- Flow test data: $p_{wf} = 3600$ psi $Q_o = 280$ STB/day

Generate the IPR data of the well.

Solution

Step 1. Calculate the productivity index from the flow-test data.

$$J = \frac{280}{4000 - 3600} = 0.7 \text{ STB / day / psi}$$

Step 2. Generate the IPR data by applying Equation 7-30 when the assumed $p_{wf} > p_b$ and using Equation 7-42 when $p_{wf} < p_b$.

p_{wf}	Equation	Q_o
4000	(7-30)	0
3800	(7-30)	140
3600	(7-30)	280
3200	(7-30)	560
3000	(7-42)	696
2600	(7-42)	941
2200	(7-42)	1151
2000	(7-42)	1243
1000	(7-42)	1571
500	(7-42)	1653
0	(7-42)	1680

Results of the calculations are shown graphically in Figure 7-16.

It should be pointed out Fetkovich's method has the advantage over Standing's methodology in that it does not require the tedious material

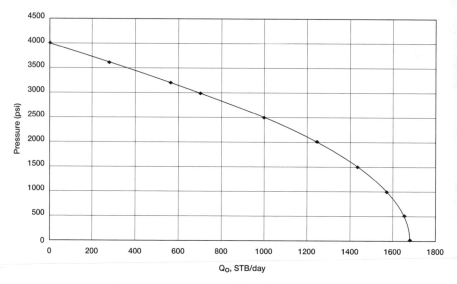

Figure 7-16. IPR using the Fetkovich method.

balance calculations to predict oil saturations at future average reservoir pressures.

The Klins-Clark Method

Klins and Clark (1993) proposed an inflow expression similar in form to that of Vogel's and can be used to estimate future IPR data. To improve the predictive capability of Vogel's equation, the authors introduced a new exponent d to Vogel's expression. The authors proposed the following relationships:

$$\frac{Q_o}{(Q_o)_{max}} = 1 - 0.295 \left(\frac{p_{wf}}{\overline{p}_r} \right) - 0.705 \left(\frac{p_{wf}}{\overline{p}_r} \right)^d \tag{7-43}$$

where

$$d = \left[0.28 + 0.72 \left(\frac{\overline{p}_r}{p_b} \right) \right] (1.24 + 0.001\, p_b) \tag{7-44}$$

The computational steps of the Klins and Clark are summarized below:

Step 1. Knowing the bubble-point pressure and the current reservoir pressure, calculate the exponent d from Equation 7-44.

Step 2. From the available stabilized flow data, i.e., Q_o at p_{wf}, solve Equation 7-43 for $(Q_o)_{max}$.

Step 3. Construct the current IPR by assuming several values of p_{wf} in Equation 7-43 and solving for Q_o.

HORIZONTAL OIL WELL PERFORMANCE

Since 1980, horizontal wells began capturing an ever-increasing share of hydrocarbon production. Horizontal wells offer the following advantages over those of vertical wells:

• Large volume of the reservoir can be drained by each horizontal well.
• Higher productions from thin pay zones.
• Horizontal wells minimize water and gas zoning problems.
• In high permeability reservoirs, where near-wellbore gas velocities are high in vertical wells, horizontal wells can be used to reduce near-wellbore velocities and turbulence.
• In secondary and enhanced oil recovery applications, long horizontal injection wells provide higher injectivity rates.
• The length of the horizontal well can provide contact with multiple fractures and greatly improve productivity.

The actual production mechanism and reservoir flow regimes around the horizontal well are considered more complicated than those for the vertical well, especially if the horizontal section of the well is of a considerable length. Some combination of both linear and radial flow actually exists, and the well may behave in a manner similar to that of a well that has been extensively fractured. Several authors reported that the shape of measured IPRs for horizontal wells is similar to those predicted by the Vogel or Fetkovich methods. The authors pointed out that the productivity gain from drilling 1,500-foot-long horizontal wells is two to four times that of vertical wells.

A horizontal well can be looked upon as a number of vertical wells drilling next to each other and completed in a limited pay zone thickness. Figure 7-17 shows the drainage area of a horizontal well of length L in a reservoir with a pay zone thickness of h. Each end of the horizontal well would drain a half-circular area of radius b, with a rectangular drainage shape of the horizontal well.

Assuming that each end of the horizontal well is represented by a vertical well that drains an area of a half circle with a radius of b, Joshi (1991) proposed the following two methods for calculating the drainage area of a horizontal well.

Method I

Joshi proposed that the drainage area is represented by two half circles of radius b (equivalent to a radius of a vertical well r_{ev}) at each end and a rectangle, of dimensions L(2b), in the center. The drainage area of the horizontal well is given then by:

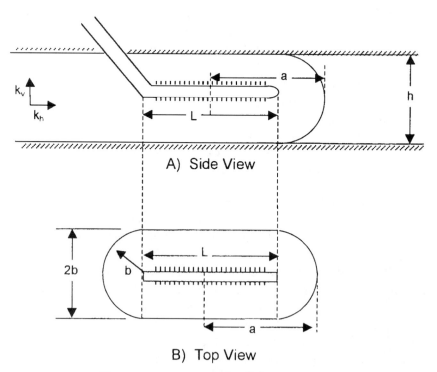

A) Side View

B) Top View

Figure 7-17. Horizontal well drainage area.

$$A = \frac{L(2b) + \pi b^2}{43,560} \tag{7-45}$$

where A = drainage area, acres
   L = length of the horizontal well, ft
   b = half minor axis of an ellipse, ft

Method II

Joshi assumed that the horizontal well drainage area is an ellipse and given by:

$$A = \frac{\pi ab}{43,560} \tag{7-46}$$

with

$$a = \frac{L}{2} + b \tag{7-47}$$

where a is the half major axis of an ellipse.

Joshi noted that the two methods give different values for the drainage area A and suggested assigning the average value for the drainage of the horizontal well. Most of the production rate equations require the value of the drainage radius of the horizontal well, which is given by:

$$r_{eh} = \sqrt{\frac{43,560\,A}{\pi}}$$

where r_{eh} = drainage radius of the horizontal well, ft
   A = drainage area of the horizontal well, acres

Example 7-11

A 480-acre lease is to be developed by using 12 vertical wells. Assuming that each vertical well would effectively drain 40 acres, calculate the possible number of either 1,000- or 2,000-ft-long horizontal wells that will drain the lease effectively.

Solution

Step 1. Calculate the drainage radius of the vertical well:

$$r_{ev} = b = \sqrt{\frac{(40)(43,560)}{\pi}} = 745 \, \text{ft}$$

Step 2. Calculate the drainage area of the 1,000- and 2,000-ft-long horizontal well using Joshi's two methods:

Method I

• For the 1,000-ft horizontal well using Equation 7-45:

$$A = \frac{(1000)(2 \times 745) + \pi (745)^2}{43,560} = 74 \, \text{acres}$$

• For the 2,000-ft horizontal well:

$$A = \frac{(2000)(2 \times 745) + \pi (745)^2}{43,560} = 108 \, \text{acres}$$

Method II

• For the 1,000-ft horizontal well using Equation 7-46:

$$a = \frac{1000}{2} + 745 = 1245'$$

$$A = \frac{\pi (1245)(745)}{43,560} = 67 \, \text{acres}$$

• For the 2,000-ft horizontal well:

$$a = \frac{2000}{2} + 745 = 1745'$$

$$A = \frac{\pi (1745)(745)}{43,560} = 94 \, \text{acres}$$

Step 3. Averaging the values from the two methods:
• Drainage area of 1,000-ft-long well

$$A = \frac{74 + 67}{2} = 71 \text{ acres}$$

• Drainage area of 2,000-ft-long well

$$A = \frac{108 + 94}{2} = 101 \text{ acres}$$

Step 4. Calculate the number of 1,000-ft-long horizontal wells:

$$\text{Total number of } 1,000\text{-ft-long horizontal wells } = \frac{480}{71} = 7 \text{ wells}$$

Step 5. Calculate the number of 2,000-ft-long horizontal wells.

$$\text{Total number of } 2,000\text{-ft-long horizontal wells } = \frac{480}{101} = 5 \text{ wells}$$

From a practical standpoint, inflow performance calculations for horizontal wells are presented here under the following two flowing conditions:

• Steady-state single-phase flow
• Pseudosteady-state two-phase flow

A reference textbook by Joshi (1991) provides an excellent treatment of horizontal well technology and it contains a detailed documentation of recent methodologies of generating inflow performance relationships.

Horizontal Well Productivity under Steady-State Flow

The steady-state analytical solution is the simplest solution to various horizontal well problems. The steady-state solution requires that the pressure at any point in the reservoir does not change with time. The flow rate equation in a steady-state condition is represented by:

$$Q_{oh} = J_h \, \Delta p \tag{7-48}$$

where Q_{oh} = horizontal well flow rate, STB/day

Δp = pressure drop from the drainage boundary to wellbore, psi

J_h = productivity index of the horizontal well, STB/day/psi

The productivity index of the horizontal well J_h can be always obtained by dividing the flow rate Q_{oh} by the pressure drop Δp, or:

$$J_h = \frac{Q_{oh}}{\Delta p}$$

There are several methods that are designed to predict the productivity index from the fluid and reservoir properties. Some of these methods include:

• Borisov's Method
• The Giger-Reiss-Jourdan Method
• Joshi's Method
• The Renard-Dupuy Method

Borisov's Method

Borisov (1984) proposed the following expression for predicting the productivity index of a horizontal well in an isotropic reservoir, i.e., $k_v = k_h$

$$J_h = \frac{0.00708\,h\,k_h}{\mu_o\,B_o\left[\ln\left(\dfrac{4\,r_{eh}}{L}\right)+\left(\dfrac{h}{L}\right)\ln\left(\dfrac{h}{2\pi\,r_w}\right)\right]} \qquad (7\text{-}49)$$

where h = thickness, ft

k_h = horizontal permeability, md

k_v = vertical permeability, md

L = length of the horizontal well, ft

r_{eh} = drainage radius of the horizontal well, ft

r_w = wellbore radius, ft

J_h = productivity index, STB/day/psi

The Giger-Reiss-Jourdan Method

For an isotropic reservoir where the vertical permeability k_v equals the horizontal permeability k_h, Giger et al. (1984) proposed the following expression for determining J_h:

$$J_h = \frac{0.00708\,L\,k_h}{\mu_o\,B_o\left[\left(\dfrac{L}{h}\right)\ln(X)+\ln\left(\dfrac{h}{2\,r_w}\right)\right]} \qquad (7\text{-}50)$$

$$X = \frac{1+\sqrt{1+\left(\dfrac{L}{2\,r_{eh}}\right)^2}}{L/(2\,r_{eh})} \qquad (7\text{-}51)$$

To account for the reservoir anisotropy, the authors proposed the following relationships:

$$J_h = \frac{0.00708\,k_h}{\mu_o\,B_o\left[\left(\dfrac{1}{h}\right)\ln(X)+\left(\dfrac{B^2}{L}\right)\ln\left(\dfrac{h}{2\,r_w}\right)\right]} \qquad (7\text{-}52)$$

With the parameter B as defined by:

$$B = \sqrt{\frac{k_h}{k_v}} \qquad (7\text{-}53)$$

where k_v = vertical permeability, md
  L = Length of the horizontal section, ft

Joshi's Method

Joshi (1991) presented the following expression for estimating the productivity index of a horizontal well in isotropic reservoirs:

$$J_h = \frac{0.00708\,h\,k_h}{\mu_o\,B_o\left[\ln(R)+\left(\dfrac{h}{L}\right)\ln\left(\dfrac{h}{2\,r_w}\right)\right]} \qquad (7\text{-}54)$$

with

$$R = \frac{a+\sqrt{a^2-(L/2)^2}}{(L/2)} \qquad (7\text{-}55)$$

and a is half the major axis of drainage ellipse and given by:

$$a = (L/2)\left[0.5 + \sqrt{0.25 + \left(2\,r_{eh}/L\right)^4}\,\right]^{0.5} \tag{7-56}$$

Joshi accounted for the influence of the reservoir anisotropy by introducing the vertical permeability k_v into Equation 7-54, to give:

$$J_h = \frac{0.00708\,h\,k_h}{\mu_o\,B_o\left[\ln(R) + \left(\dfrac{B^2\,h}{L}\right)\ln\left(\dfrac{h}{2\,r_w}\right)\right]} \tag{7-57}$$

where the parameters B and R are defined by Equations 7-53 and 7-55, respectively.

The Renard-Dupuy Method

For an isotropic reservoir, Renard and Dupuy (1990) proposed the following expression:

$$J_h = \frac{0.00708\,h\,k_h}{\mu_o\,B_o\left[\cosh^{-1}\left(\dfrac{2a}{L}\right) + \left(\dfrac{h}{L}\right)\ln\left(\dfrac{h}{2\,\pi\,r_w}\right)\right]} \tag{7-58}$$

where a is half the major axis of drainage ellipse and given by Equation 7-56.

For anisotropic reservoirs, the authors proposed the following relationship:

$$J_h = \frac{0.00708\,h\,k_h}{\mu_o\,B_o\left[\cosh^{-1}\left(\dfrac{2a}{L}\right) + \left(\dfrac{B\,h}{L}\right)\ln\left(\dfrac{h}{2\,\pi\,r_w'}\right)\right]} \tag{7-59}$$

where

$$r_w' = \frac{(1 + B)\,r_w}{2\,B} \tag{7-60}$$

with the parameter B as defined by Equation 7-53.

Example 7-12

A 2,000-foot-long horizontal well drains an estimated drainage area of 120 acres. The reservoir is characterized by an isotropic with the following properties:

$k_v = k_h = 100$ md $\qquad$ h = 60 ft
$\quad B_o = 1.2$ bbl/STB $\qquad \mu_o = 0.9$ cp
$\quad p_e = 3000$ psi $\qquad p_{wf} = 2500$ psi
$\quad r_w = 0.30$ ft

Assuming a steady-state flow, calculate the flow rate by using:

a. Borisov's Method
b. The Giger-Reiss-Jourdan Method
c. Joshi's Method
d. The Renard-Dupuy Method

Solution

a. Borisov's Method

Step 1. Calculate the drainage radius of the horizontal well:

$$r_{eh} = \sqrt{\frac{(120)(43,560)}{\pi}} = 1290\,\text{ft}$$

Step 2. Calculate J_h by using Equation 7-49:

$$J_h = \frac{(0.00708)(60)(100)}{(0.9)(1.2)\left[\ln\left(\frac{(4)(1290)}{2000}\right) + \left(\frac{60}{2000}\right)\ln\left(\frac{60}{2\pi(0.3)}\right)\right]}$$
$$= 37.4\,\text{STB/day/psi}$$

Step 3. Calculate the flow rate by applying Equation 7-48:

$$Q_{oh} = (37.4)(3000 - 2500) = 18,700\,\text{STB/day}$$

b. The Giger-Reiss-Jourdan Method

Step 1. Calculate the parameter X from Equation 7-51:

$$X = \frac{1 + \sqrt{1 + \left(\dfrac{2000}{(2)(1290)}\right)^2}}{2000/[(2)(1290)]} = 2.105$$

Step 2. Solve for J_h by applying Equation 7-50:

$$J_h = \frac{(1.00708)(2000)(100)}{(0.9)(1.2)\left[\left(\dfrac{2000}{60}\right)\ln(2.105) + \ln\left(\dfrac{60}{2(0.3)}\right)\right]}$$
$$= 44.57 \text{ STB / day}$$

Step 3. Calculate flow rate:

$$Q_{oh} = 44.57 (3000 - 2500) = 22,286 \text{ STB/day}$$

c. Joshi's Method

Step 1. Calculate half major axis of ellipse by using Equation 7-56:

$$a = \left(\frac{2000}{2}\right)\left[0.5 + \sqrt{0.25 + \left[2(1290)/2000\right]^2}\right]^{0.5} = 1372 \text{ ft}$$

Step 2. Calculate the parameter R from Equation 7-55:

$$R = \frac{1372 + \sqrt{(1372)^2 - (2000/2)^2}}{(2000/2)} = 2.311$$

Step 3. Solve for J_h by applying Equation 7-54:

$$J_h = \frac{0.00708\,(60)\,(100)}{(0.9)(1.2)\left[\ln(2.311) + \left(\dfrac{60}{2000}\right)\ln\left(\dfrac{60}{(2)\,(0.3)}\right)\right]}$$

$$= 40.3 \text{ STB/day/psi}$$

Step 4. $Q_{oh} = (40.3)\,(3000 - 2500) = 20{,}154 \text{ STB/day}$

d. The Renard-Dupuy Method

Step 1. Calculate a from Equation 7-56:

$$a = 1372 \text{ ft}$$

Step 2. Apply Equation 7-58 to determine J_h:

$$J_h = \frac{0.00708\,(60)\,(100)}{(0.9)\,(1.2)\left[\cosh^{-1}\left(\dfrac{(2)\,(1372)}{2000}\right) + \left(\dfrac{60}{2000}\right)\ln\left(\dfrac{60}{2\pi\,(0.3)}\right)\right]}$$

$$= 41.77 \text{ STB/day/psi}$$

Step 3. $Q_{oh} = 41.77\,(3000 - 2500) = 20{,}885 \text{ STB/day}$

Example 7-13

Using the data in Example 7-13 and assuming an isotropic reservoir with $k_h = 100$ md and $k_v = 10$ md, calculate flow rate by using:

a. The Giger-Reiss-Jourdan Method
b. Joshi's Method
c. The Renard-Dupuy Method

Solution

a. The Giger-Reiss-Jourdan Method

Step 1. Solve for the permeability ratio B by applying Equation 7-53

$$\beta = \sqrt{\frac{100}{10}} = 3.162$$

Step 2. Calculate the parameter X as shown in Example 7-12 to give:

X = 2.105

Step 3. Determine J_h by using Equation 7-52.

$$J_h = \frac{0.00708\,(100)}{(0.9)\,(1.2)\left[\left(\dfrac{1}{60}\right)\ln(2.105) + \left(\dfrac{3.162^2}{2000}\right)\ln\left(\dfrac{60}{(2)\,(0.3)}\right)\right]}$$

$$= 18.50 \text{ STB/day/psi}$$

Step 4. Calculate Q_{oh}

$$Q_{oh} = (18.50)\,(3000 - 2500) = 9,252 \text{ STB/day}$$

b. Joshi's Method

Step 1. Calculate the permeability ratio β

$$\beta = 3.162$$

Step 2. Calculate the parameters a and R as given in Example 7-12.

A = 1372 ft   R = 2.311

Step 3. Calculate J_h by using Equation 7-54.

$$J_h = \frac{0.00708\,(60)\,(100)}{(0.9)\,(1.2)\left[\ln(2.311) + \left(\dfrac{(3.162)^2\,(60)}{2000}\right)\ln\left(\dfrac{60}{2\,(0.3)}\right)\right]}$$

$$= 17.73 \text{ STB/day/psi}$$

Step 4. $Q_{oh} = (17.73)(3000 - 2500) = 8,863$ STB/day

c. The Renard-Dupuy Method

Step 1. Calculate r'_w from Equation 7-60.

$$r'_w = \frac{(1+3.162)(0.3)}{(2)(3.162)} = 0.1974$$

Step 2. Apply Equation 7-59

$$J_h = \frac{0.00708\,(60)\,(100)}{(0.9)(1.2)\left[\cosh^{-1}\left[\dfrac{(2)(1372)}{2000}\right] + \left[\dfrac{(3.162)^2\,(60)}{2000}\right]\ln\left(\dfrac{60}{(2)\pi\,(0.1974)}\right)\right]}$$

$$= 19.65 \text{ STB/day/psi}$$

Step 3. $Q_{oh} = 19.65\,(3000 - 2500) = 9,825$ STB/day

Horizontal Well Productivity under Semisteady-State Flow

The complex flow regime existing around a horizontal wellbore probably precludes using a method as simple as that of Vogel to construct the IPR of a horizontal well in solution gas drive reservoirs. If at least two stabilized flow tests are available, however, the parameters J and n in the Fetkovich equation (Equation 7-35) could be determined and used to construct the IPR of the horizontal well. In this case, the values of J and n would not only account for effects of turbulence and gas saturation around the wellbore, but also for the effects of nonradial flow regime existing in the reservoir.

Bendakhlia and Aziz (1989) used a reservoir model to generate IPRs for a number of wells and found that a combination of Vogel and Fetkovich equations would fit the generated data if expressed as:

$$\frac{Q_{oh}}{(Q_{oh})_{max}} = \left[1 - V\left(\frac{P_{wf}}{\bar{p}_r}\right) - (1-V)\left(\frac{P_{wf}}{\bar{p}_r}\right)^2\right]^n \qquad (7-61)$$

where $(Q_{oh})_{max}$ = horizontal well maximum flow rate, STB/day
$\qquad$ n = exponent in Fetkovich's equation
$\qquad$ V = variable parameter

In order to apply the equation, at least three stabilized flow tests are required to evaluate the three unknowns $(Q_{oh})_{max}$, V, and n at any given average reservoir pressure $\bar{p}_r$. However, Bendakhlia and Aziz indicated that the parameters V and n are functions of the reservoir pressure or recovery factor and, thus, the use of Equation 7-61 is not convenient in a predictive mode.

Cheng (1990) presented a form of Vogel's equation for horizontal wells that is based on the results from a numerical simulator. The proposed expression has the following form:

$$\frac{Q_{oh}}{(Q_{oh})_{max}} = 1.0 + 0.2055 \left(\frac{p_{wf}}{\bar{p}_r} \right) - 1.1818 \left(\frac{p_{wf}}{\bar{p}_r} \right)^2 \qquad (7-62)$$

Example 7-14

A 1,000-foot-long horizontal well is drilled in a solution gas drive reservoir. The well is producing at a stabilized flow rate of 760 STB/day and wellbore pressure of 1242 psi. The current average reservoir pressure is 2145 psi. Generate the IPR data of this horizontal well by using Cheng's method.

Solution

Step 1. Use the given stabilized flow data to calculate the maximum flow rate of the horizontal well.

$$\frac{760}{(Q_{oh})_{max}} = 1 + 0.2055 \left(\frac{1242}{2145} \right) - 1.1818 \left(\frac{1242}{2145} \right)^2$$

$(Q_{oh})_{max} = 1052$ STB/day

Step 2. Generate the IPR data by applying Equation 7-62.

p_{wf}	Q_{oh}
2145	0
1919	250
1580	536
1016	875
500	1034
0	1052

PROBLEMS

1. An oil well is producing under steady-state flow conditions at 300 STB/day. The bottom-hole flowing pressure is recorded at 2500 psi. Given:

$h = 23$ ft	$k = 50$ md	$\mu_o = 2.3$ cp	$r_w = 0.25$ ft
$B_o = 1.4$ bbl/STB	$r_e = 660$ ft	$s = 0.5$	

   Calculate:

   a. Reservoir pressure
   b. AOF
   c. Productivity index

2. A well is producing from a saturated oil reservoir with an average reservoir pressure of 3000 psig. A stabilized flow test data indicates that the well is capable of producing 400 STB/day at a bottom-hole flowing pressure of 2580 psig.

   a. Oil flow rate at $p_{wf} = 1950$ psig
   b. Construct the IPR curve at the current average pressure.
   c. Construct the IPR curve by assuming a constant J.
   d. Plot the IPR curve when the reservoir pressure is 2700 psig.

3. An oil well is producing from an undersaturated reservoir that is characterized by a bubble-point pressure of 2230 psig. The current average reservoir pressure is 3500 psig. Available flow test data show that the well produced 350 STB/day at a stabilized p_{wf} of 2800 psig. Construct the current IPR data by using:

   a. Vogel's correlation
   b. Wiggins' method
   c. Generate the future IPR curve when the reservoir pressure declines from 3500 psi to 2230 and 2000 psi.

4. A well is producing from a saturated oil reservoir that exists at its saturation pressure of 4500 psig. The well is flowing at a stabilized rate of 800 STB/day and a p_{wf} of 3700 psig. Material balance calculations provide the following current and future predictions for oil saturation and PVT properties.

	Present	Future
$\bar{p}_r$	4500	3300
μ_o, cp	1.45	1.25
B_o, bbl/STB	1.23	1.18
k_{ro}	1.00	0.86

Generate the future IPR for the well at 3300 psig by using Standing's method.

5. A four-point stabilized flow test was conducted on a well producing from a saturated reservoir that exists at an average pressure of 4320 psi.

Q_o, STB/day	p_{wf}, psi
342	3804
498	3468
646	2928
832	2580

a. Construct a complete IPR using Fetkovich's method
b. Construct the IPR when the reservoir pressure declines to 2500 psi

6. The following reservoir and flow-test data are available on an oil well:

- Pressure data: $\bar{p}_r = 3280$ psi $p_b = 2624$ psi
- Flow test data: $p_{wf} = 2952$ psi $Q_o = $ STB/day

Generate the IPR data of the well.

7. A 2,500-foot-long horizontal well drains an estimated drainage area of 120 acres. The reservoir is characterized by an isotropic with the following properties:

$k_v = k_h = 60$ md $h = 70$ ft
$B_o = 1.4$ bbl/STB $\mu_o = 1.9$ cp
$p_e = 3900$ psi $p_{wf} = 3250$ psi
$r_w = 0.30$ ft

Assuming a steady-state flow, calculate the flow rate by using:

a. Borisov's Method
b. The Giger-Reiss-Jourdan Method
c. Joshi's Method
d. The Renard-Dupuy Method

8. A 2,000-foot-long horizontal well is drilled in a solution gas drive reservoir. The well is producing at a stabilized flow rate of 900 STB/day and wellbore pressure of 1000 psi. The current average reservoir pressure in 2000 psi. Generate the IPR data of this horizontal well by using Cheng's method.

REFERENCES

1. Beggs, D., "Gas Production Operations," *OGCI,* Tulsa, Oklahoma, 1984.

2. Beggs, D., "Production Optimization Using NODAL Analysis," *OGCI,* Tulsa, Oklahoma, 1991

3. Bendakhlia, H., and Aziz, K., "IPR for Solution-Gas Drive Horizontal Wells," SPE Paper 19823, presented at the 64th Annual Meeting in San Antonio, Texas, October 8–11, 1989.

4. Borisov, Ju. P., "Oil Production Using Horizontal and Multiple Deviation Wells," Nedra, Moscow, 1964. Translated by J. Strauss. S. D. Joshi (ed.). Bartlesville, OK: Phillips Petroleum Co., the R & D Library Translation, 1984.

5. Cheng, A. M., "IPR For Solution Gas-Drive Horizontal Wells," SPE Paper 20720, presented at the 65th Annual SPE meeting held in New Orleans, September 23–26, 1990.

6. Fetkovich, M. J., "The Isochronal Testing of Oil Wells," SPE Paper 4529, presented at the SPE 48th Annual Meeting, Las Vegas, Sept. 30–Oct. 3, 1973.

7. Giger, F. M., Reiss, L. H., and Jourdan, A. P., "The Reservoir Engineering Aspect of Horizontal Drilling," SPE Paper 13024, presented at the SPE 59th Annual Technical Conference and Exhibition, Houston, Texas, Sept. 16–19, 1984.

8. Golan, M., and Whitson, C. H., *Well Performance,* 2nd ed. Englewood Cliffs, NJ: Prentice-Hall, 1986.

9. Joshi, S., *Horizontal Well Technology.* Tulsa, OK: PennWell Publishing Company, 1991.

10. Klins, M., and Clark, L., "An Improved Method to Predict Future IPR Curves," *SPE Reservoir Engineering,* November 1993, pp. 243–248.

11. Muskat, M., and Evinger, H. H., "Calculations of Theoretical Productivity Factor," *Trans. AIME,* 1942, pp. 126–139, 146.

12. Renard, G. I., and Dupuy, J. M., "Influence of Formation Damage on the Flow Efficiency of Horizontal Wells," SPE Paper 19414, presented at the Formation Damage Control Symposium, Lafayette, Louisiana, Feb. 22–23, 1990.

13. Standing, M. B., "Inflow Performance Relationships for Damaged Wells Producing by Solution-Gas Drive," *JPT,* (Nov. 1970, pp. 1399–1400.

14. Vogel, J. V., "Inflow Performance Relationships for Solution-Gas Drive Wells," *JPT,* Jan. 1968, pp. 86–92; *Trans. AIME,* p. 243.

15. Wiggins, M. L., "Generalized Inflow Performance Relationships for Three-Phase Flow," SPE Paper 25458, presented at the SPE Production Operations Symposium, Oklahoma City, March 21–23, 1993.

GAS WELL PERFORMANCE

Determination of the flow capacity of a gas well requires a relationship between the inflow gas rate and the sand-face pressure or flowing bottom-hole pressure. This inflow performance relationship may be established by the proper solution of Darcy's equation. Solution of Darcy's Law depends on the conditions of the flow existing in the reservoir or the flow regime.

When a gas well is first produced after being shut-in for a period of time, the gas flow in the reservoir follows an unsteady-state behavior until the pressure drops at the drainage boundary of the well. Then the flow behavior passes through a short transition period, after which it attains a steady-state or semisteady (pseudosteady)-state condition. The objective of this chapter is to describe the empirical as well as analytical expressions that can be used to establish the inflow performance relationships under the pseudosteady-state flow condition.

VERTICAL GAS WELL PERFORMANCE

The exact solution to the differential form of Darcy's equation for compressible fluids under the pseudosteady-state flow condition was given previously by Equation 6-150 as:

$$Q_g = \frac{kh\left[\overline{\psi}_r - \psi_{wf}\right]}{1422\,T\left[\ln\left(\dfrac{r_e}{r_w}\right) - 0.75 + s\right]} \tag{8-1}$$

where Q_g = gas flow rate, Mscf/day
  k = permeability, md
  $\overline{\Psi}_r$ = average reservoir real gas pseudo-pressure, psi²/cp
  T = temperature, °R
  s = skin factor
  h = thickness
  r_e = drainage radius
  r_w = wellbore radius

The productivity index J for a gas well can be written analogous to that for oil wells as:

$$J = \frac{Q_g}{\overline{\Psi}_r - \Psi_{wf}} = \frac{kh}{1422\,T\left[\ln\left(\dfrac{r_e}{r_w}\right) - 0.75 + s\right]} \tag{8-2}$$

or

$$Q_g = J(\overline{\Psi}_r - \Psi_{wf}) \tag{8-3}$$

with the absolute open flow potential (AOF), i.e., maximum gas flow rate $(Q_g)_{max}$, as calculated by:

$$(Q_g)_{max} = J\,\overline{\Psi}_r \tag{8-4}$$

where J = productivity index, Mscf/day/psi²/cp
  $(Q_g)_{max}$ = AOF

Equation 8-3 can be expressed in a linear relationship as:

$$\Psi_{wf} = \overline{\Psi}_r - \left(\frac{1}{J}\right)Q_g \tag{8-5}$$

Equation 8-5 indicates that a plot of Ψ_{wf} vs. Q_g would produce a straight line with a slope of (1/J) and intercept of $\overline{\Psi}_r$, as shown in Figure 8-1. If two different stabilized flow rates are available, the line can be extrapolated and the slope is determined to estimate AOF, J, and $\overline{\Psi}_r$.

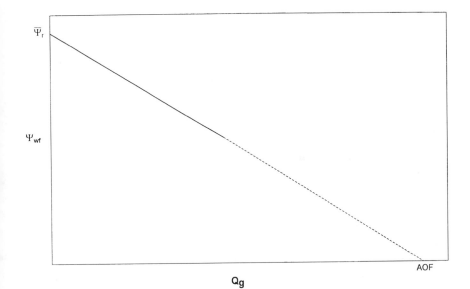

Figure 8-1. Steady-state gas well flow.

Equation 8-1 can be alternatively written in the following integral form:

$$Q_g = \frac{kh}{1422\ T\left[\ln\left(\dfrac{r_e}{r_w}\right) - 0.75 + s\right]} \int_{p_{wf}}^{\bar{p}_r}\left(\frac{2p}{\mu_g\ z}\right)dp \qquad (8\text{-}6)$$

Note that $(p/\mu_g\ z)$ is directly proportional to $(1/\mu_g\ B_g)$ where B_g is the gas formation volume factor and defined as:

$$B_g = 0.00504\ \frac{zT}{p} \qquad (8\text{-}7)$$

where B_g = gas formation volume factor, bbl/scf
$\quad\quad$ z = gas compressibility factor
$\quad\quad$ T = temperature, °R

Equation 8-6 can then be written in terms of B_g as:

$$Q_g = \left[\frac{7.08(10^{-6})kh}{\ln\left(\frac{r_e}{r_w}\right) - 0.75 + s} \right] \int_{p_{wf}}^{\bar{p}_r} \left(\frac{1}{\mu_g B_g} \right) dp \qquad (8\text{-}8)$$

where Q_g = gas flow rate, Mscf/day
μ_g = gas viscosity, cp
k = permeability, md

Figure 8-2 shows a typical plot of the gas pressure functions $(2p/\mu_g z)$ and $(1/\mu_g B_g)$ versus pressure. The integral in Equations 8-6 and 8-8 represents the area under the curve between $\bar{p}_r$ and p_{wf}.

As illustrated in Figure 8-2, the pressure function exhibits the following three distinct pressure application regions:

Region I. High-Pressure Region

When both p_{wf} and $\bar{p}_r$ are higher than 3000 psi, the pressure functions $(2p/\mu_g z)$ and $(1/\mu_g B_g)$ are nearly constants. This observation suggests that the pressure term $(1/\mu_g B_g)$ in Equation 8-8 can be treated as a constant and removed outside the integral, to give the following approximation to Equation 8-6:

$$Q_g = \frac{7.08(10^{-6})kh(\bar{p}_r - p_{wf})}{(\mu_g B_g)_{avg}\left[\ln\left(\frac{r_e}{r_w}\right) - 0.75 + s\right]} \qquad (8\text{-}9)$$

where Q_g = gas flow rate , Mscf/day
B_g = gas formation volume factor, bbl/scf
k = permeability, md

The gas viscosity μ_g and formation volume factor B_g should be evaluated at the average pressure p_{avg} as given by:

$$p_{avg} = \frac{\bar{p}_r + p_{wf}}{2} \qquad (8\text{-}10)$$

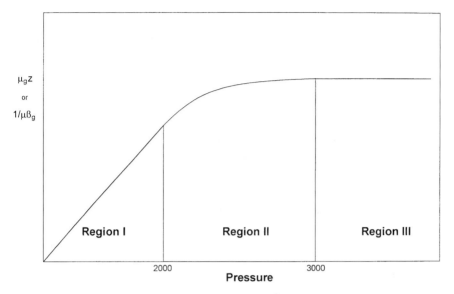

Figure 8-2. Gas PVT data.

The method of determining the gas flow rate by using Equation 8-9 commonly called the **pressure-approximation method.**

It should be pointed out the concept of the productivity index J cannot be introduced into Equation 8-9 since Equation 8-9 is only valid for applications when both p_{wf} and $\bar{p}_r$ are above 3000 psi.

Region II. Intermediate-Pressure Region

Between 2000 and 3000 psi, the pressure function shows distinct curvature. When the bottom-hole flowing pressure and average reservoir pressure are both between 2000 and 3000 psi, the pseudopressure gas pressure approach (i.e., Equation 8-1) should be used to calculate the gas flow rate.

Region III. Low-Pressure Region

At low pressures, usually less than 2000 psi, the pressure functions $(2p/\mu_g z)$ and $(1/\mu_g B_g)$ exhibit a linear relationship with pressure. Golan and Whitson (1986) indicated that the product $(\mu_g z)$ is essentially constant when evaluating any pressure below 2000 psi. Implementing this observation in Equation 8-6 and integrating gives:

$$Q_g = \frac{kh(\bar{p}_r^2 - p_{wf}^2)}{1422\,T\,(\mu_g\,z)_{avg}\left[\ln\left(\dfrac{r_e}{r_w}\right)-0.75+s\right]} \tag{8-11}$$

where Q_g = gas flow rate, Mscf/day
        k = permeability, md
        T = temperature, °R
        z = gas compressibility factor
        μ_g = gas viscosity, cp

It is recommended that the z-factor and gas viscosity be evaluated at the average pressure p_{avg} as defined by:

$$p_{avg} = \sqrt{\frac{\bar{p}_r^2 + p_{wf}^2}{2}}$$

The method of calculating the gas flow rate by Equation 8-11 is called the **pressure-squared approximation method.**

If both $\bar{p}_r$ and p_{wf} are lower than 2000 psi, Equation 8-11 can be expressed in terms of the productivity index J as:

$$Q_g = J\,(\bar{p}_r^2 - p_{wf}^2) \tag{8-12}$$

with

$$(Q_g)_{max} = AOF = J\,\bar{p}_r^2 \tag{8-13}$$

where

$$J = \frac{kh}{1422\,T\,(\mu_g\,z)_{avg}\left[\ln\left(\dfrac{r_e}{r_w}\right)-0.75+s\right]} \tag{8-14}$$

Example 8-1

The PVT properties of a gas sample taken from a dry gas reservoir are given in the following table:

p, psi	μ_g, cp	Z	ψ, psi²/cp	B_g, bbl/scf
0	0.01270	1.000	0	—
400	0.01286	0.937	13.2×10^6	0.007080
1200	0.01530	0.832	113.1×10^6	0.00210
1600	0.01680	0.794	198.0×10^6	0.00150
2000	0.01840	0.770	304.0×10^6	0.00116
3200	0.02340	0.797	678.0×10^6	0.00075
3600	0.02500	0.827	816.0×10^6	0.000695
4000	0.02660	0.860	950.0×10^6	0.000650

The reservoir is producing under the pseudosteady-state condition. The following additional data is available:

$k = 65$ md $h = 15'$ $T = 600°R$
$r_e = 1000'$ $r_w = 0.25'$ $s = 0.4$

Calculate the gas flow rate under the following conditions:

a. $\bar{p}_r = 4000$ psi, $p_{wf} = 3200$ psi
b. $\bar{p}_r = 2000$ psi, $p_{wf} = 1200$ psi

Use the appropriate approximation methods and compare results with the exact solution.

Solution

a. Calculate Q_g at $\bar{p}_r = 4000$ and $p_{wf} = 3200$ psi:

Step 1. Select the approximation method. Because $\bar{p}_r$ and p_{wf} are both > 3000, the pressure-approximation method is used, i.e., Equation 8-9.

Step 2. Calculate average pressure and determine the corresponding gas properties.

$$\bar{p} = \frac{4000 + 3200}{2} = 3600 \text{ psi}$$

$\mu_g = 0.025$ $B_g = 0.000695$

Step 3. Calculate the gas flow rate by applying Equation 8-9.

$$Q_g = \frac{7.08\,(10^{-6})\,(65)\,(15)\,(4000 - 3200)}{(0.025)\,(0.000695)\left[\ln\left(\dfrac{1000}{0.25}\right) - 0.75 - 0.4\right]}$$

$$= 44,490\,\text{Mscf/day}$$

Step 4. Recalculate Q_g by using the pseudopressure equation, i.e., Equation 8-1.

$$Q_g = \frac{(65)\,(15)\,(950.0 - 678.0)\,(65)\,(15)10^6}{(1422)\,(600)\left[\ln\left(\dfrac{1000}{0.25}\right) - 0.75 - 0.4\right]} = 43,509\,\text{Mscf/day}$$

b. Calculate Q_g at $\bar{p}_r = 2000$ and $p_{wf} = 1058$:

Step 1. Select the appropriate approximation method. Because $\bar{p}_r$ and p_{wf} " 2000, use the pressure-squared approximation.

Step 2. Calculate average pressure and the corresponding μ_g and z.

$$\bar{p} = \sqrt{\frac{2000^2 + 1200^2}{2}} = 1649\,\text{psi}$$

$$\mu_g = 0.017 \qquad z = 0.791$$

Step 3. Calculate Q_g by using the pressure-squared equation, i.e., Equation 8-11.

$$Q_g = \frac{(65)\,(15)\,(2000^2 - 1200^2)}{1422\,(600)\,(0.017)\,(0.791)\left[\ln\left(\dfrac{1000}{0.25}\right) - 0.75 - 0.4\right]}$$

$$= 30,453\,\text{Mscf/day}$$

Step 4. Compare Q_g with the exact value from Equation 8-1:

$$Q_g = \frac{(65)\,(15)\,(304.0 - 113.1)\,10^6}{(1422)\,(600)\left[\ln\left(\dfrac{1000}{0.25}\right) - 0.75 - 0.4\right]}$$

$$= 30,536 \text{ Mscf/day}$$

All of the mathematical formulations presented thus far in this chapter are based on the assumption that laminar (viscous) flow conditions are observed during the gas flow. During radial flow, the flow velocity increases as the wellbore is approached. This increase of the gas velocity might cause the development of a turbulent flow around the wellbore. If turbulent flow does exist, it causes an additional pressure drop similar to that caused by the mechanical skin effect.

As presented in Chapter 6 by Equations 6-164 through 6-166, the semisteady-state flow equation for compressible fluids can be modified to account for the additional pressure drop due the turbulent flow by including the rate-dependent skin factor DQ_g. The resulting pseudosteady-state equations are given in the following three forms:

First Form : Pressure-Squared Approximation Form

$$Q_g = \frac{kh\left(\bar{p}_r^2 - p_{wf}^2\right)}{1422\,T\,(\mu_g z)_{avg}\left[\ln\left(\dfrac{r_e}{r_w}\right) - 0.75 + s + DQ_g\right]} \tag{8-15}$$

where D is the **inertial** or **turbulent flow** factor and is given by Equation 6-160 as:

$$D = \frac{FKh}{1422\,T} \tag{8-16}$$

where the non-Darcy flow coefficient F is defined by Equation 6-156 as:

$$F = 3.161\,(10^{-12})\left[\frac{\beta T \gamma_g}{\mu_g\, h^2\, r_w}\right] \tag{8-17}$$

where F = non-Darcy flow coefficient
           k = permeability, md
           T = temperature, °R
           γ_g = gas gravity
           r_w = wellbore radius, ft
           h = thickness, ft
           β = turbulence parameter as given by Equation 6-157 as
           $\beta = 1.88 \, (10^{-10}) \, k^{-1.47} \, \phi^{-0.53}$

Second Form: Pressure-Approximation Form

$$Q_g = \frac{7.08 \, (10^{-6}) \, kh \, (\overline{p}_r - p_{wf})}{(\mu_g \beta_g)_{avg} \, T \left[\ln \left(\dfrac{r_e}{r_w} \right) - 0.75 + s + DQ_g \right]} \tag{8-18}$$

Third Form: Real Gas Potential (Pseudopressure) Form

$$Q_g = \frac{kh \, (\overline{\psi}_r - \psi_{wf})}{1422 \, T \left[\ln \left(\dfrac{r_e}{r_w} \right) - 0.75 + s + DQ_g \right]} \tag{8-19}$$

Equations 8-15, 8-18, and 8-19 are essentially quadratic relationships in Q_g and, thus, they do not represent explicit expressions for calculating the gas flow rate. There are two separate empirical treatments that can be used to represent the turbulent flow problem in gas wells. Both treatments, with varying degrees of approximation, are directly derived and formulated from the three forms of the pseudosteady-state equations, i.e., Equations 8-15 through 8-17. These two treatments are called:

• Simplified treatment approach
• Laminar-inertial-turbulent (LIT) treatment

The above two empirical treatments of the gas flow equation are presented on the following pages.

The Simplified Treatment Approach

Based on the analysis for flow data obtained from a large member of gas wells, Rawlins and Schellhardt (1936) postulated that the relationship between the gas flow rate and pressure can be expressed as:

$$Q_g = C(\overline{p}_r^2 - p_{wf}^2)^n \tag{8-20}$$

where Q_g = gas flow rate, Mscf/day
$\overline{p}_r$ = average reservoir pressure, psi
n = exponent
C = performance coefficient, Mscf/day/psi^2

The exponent n is intended to account for the additional pressure drop caused by the high-velocity gas flow, i.e., turbulence. Depending on the flowing conditions, the exponent n may vary from 1.0 for completely laminar flow to 0.5 for fully turbulent flow. The performance coefficient C in Equation 8-20 is included to account for:

• Reservoir rock properties
• Fluid properties
• Reservoir flow geometry

Equation 8-20 is commonly called the **deliverability** or **back-pressure equation.** If the coefficients of the equation (i.e., n and C) can be determined, the gas flow rate Q_g at any bottom-hole flow pressure p_{wf} can be calculated and the IPR curve constructed.

Taking the logarithm of both sides of Equation 8-20 gives:

$$\log(Q_g) = \log(C) + n \log(\overline{p}_r^2 - p_{wf}^2) \tag{8-21}$$

Equation 8-22 suggests that a plot of Q_g versus $(\overline{p}_r^2 - p_{wf}^2)$ on log-log scales should yield a straight line having a slope of n. In the natural gas industry the plot is traditionally reversed by plotting $(\overline{p}_r^2 - p_{wf}^2)$ versus Q_g on the logarithmic scales to produce a straight line with a slope of (1/n). This plot as shown schematically in Figure 8-3 is commonly referred to as the *deliverability graph* or the *back-pressure plot*.

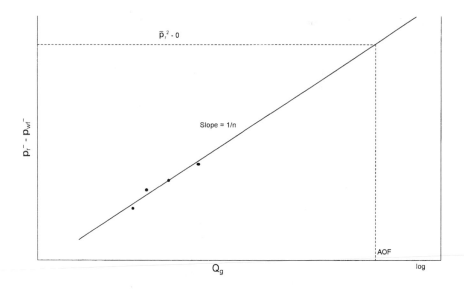

Figure 8-3. Well deliverability graph.

The deliverability exponent n can be determined from any two points on the *straight line,* i.e., $(Q_{g1}, \Delta p_1^2)$ and $(Q_{g2}, \Delta p_2^2)$, according to the flowing expression:

$$n = \frac{\log(Q_{g1}) - \log(Q_{g2})}{\log(\Delta p_1^2) - \log(\Delta p_2^2)} \tag{8-22}$$

Given n, any point on the straight line can be used to compute the performance coefficient C from:

$$C = \frac{Q_g}{(\overline{p}_r^2 - p_{wf}^2)^n} \tag{8-23}$$

The coefficients of the back-pressure equation or any of the other empirical equations are traditionally determined from analyzing gas well testing data. Deliverability testing has been used for more than sixty years by the petroleum industry to characterize and determine the flow potential of gas wells. There are essentially three types of deliverability tests and these are:

- Conventional deliverability (back-pressure) test
- Isochronal test
- Modified isochronal test

These tests basically consist of flowing wells at multiple rates and measuring the bottom-hole flowing pressure as a function of time. When the recorded data is properly analyzed, it is possible to determine the flow potential and establish the inflow performance relationships of the gas well. The deliverability test is discussed later in this chapter for the purpose of introducing basic techniques used in analyzing the test data.

The Laminar-Inertial-Turbulent (LIT) Approach

The three forms of the semisteady-state equation as presented by Equations 8-15, 8-18, and 8-19 can be rearranged in quadratic forms for the purpose of separating the *laminar* and *inertial-turbulent* terms composing these equations as follows:

a. Pressure-Squared Quadratic Form

Equation 8-15 can be written in a more simplified form as:

$$\bar{p}_r^2 - p_{wf}^2 = a Q_g + b Q_g^2 \tag{8-24}$$

with

$$a = \left(\frac{1422 \, T \, \mu_g \, z}{kh}\right)\left[\ln\left(\frac{r_e}{r_w}\right) - 0.75 + s\right] \tag{8-25}$$

$$b = \left(\frac{1422 \, T \, \mu_g \, z}{kh}\right) D \tag{8-26}$$

where a = laminar flow coefficient
  b = inertial-turbulent flow coefficient
  Q_g = gas flow rate, Mscf/day
  z = gas deviation factor
  k = permeability, md
  μ_g = gas viscosity, cp

The term (a Q_g) in Equation 8-26 represents the pressure-squared drop due to laminar flow while the term (b Q_g^2) accounts for the pressure-squared drop due to inertial-turbulent flow effects.

Equation 8-24 can be linearized by dividing both sides of the equation by Q_g to yield:

$$\frac{\overline{p}_r^2 - p_{wf}^2}{Q_g} = a + b\,Q_g \qquad (8\text{-}27)$$

The coefficients a and b can be determined by plotting $\left(\dfrac{\overline{p}_r^2 - p_{wf}^2}{Q_g} \right)$ versus Q_g on a Cartesian scale and should yield a straight line with a slope of b and intercept of a. As presented later in this chapter, data from deliverability tests can be used to construct the linear relationship as shown schematically in Figure 8-4.

Given the values of a and b, the quadratic flow equation, i.e., Equation 8-24, can be solved for Q_g at any p_{wf} from:

$$Q_g = \frac{-a + \sqrt{a^2 + 4b\left(\overline{p}_r^2 - p_{wf}^2\right)}}{2b} \qquad (8\text{-}28)$$

Furthermore, by assuming various values of p_{wf} and calculating the corresponding Q_g from Equation 8-28, the current IPR of the gas well at the current reservoir pressure $\overline{p}_r$ can be generated.

It should be pointed out the following assumptions were made in developing Equation 8-24:

• Single phase flow in the reservoir
• Homogeneous and isotropic reservoir system
• Permeability is independent of pressure
• The product of the gas viscosity and compressibility factor, i.e., (μ_g z) is constant.

This method is recommended for applications at pressures below 2000 psi.

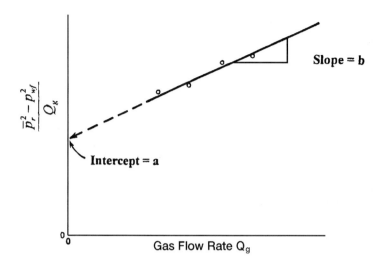

Figure 8-4. Graph of the pressure-squared data.

b. Pressure-Quadratic Form

The pressure-approximation equation, i.e., Equation 8-18, can be rearranged and expressed in the following quadratic form.

$$\bar{p}_r - p_{wf} = a_1 Q_g + b_1 Q_g^2 \tag{8-29}$$

where

$$a_1 = \frac{141.2 \,(10^{-3})\,(\mu_g B_g)}{kh} \left[\ln\left(\frac{r_e}{r_w}\right) - 0.75 + s \right] \tag{8-30}$$

$$b_1 = \left[\frac{141.2 \,(10^{-3})\,(\mu_g B_g)}{kh} \right] D \tag{8-31}$$

The term $(a_1 Q_g)$ represents the pressure drop due to laminar flow, while the term $(b_1 Q_g^2)$ accounts for the additional pressure drop due to the turbulent flow condition. In a linear form, Equation 8-17 can be expressed as:

$$\frac{\bar{p}_r - p_{wf}}{Q_g} = a_1 + b_1 \, Q_g \qquad (8\text{-}32)$$

The laminar flow coefficient a_1 and inertial-turbulent flow coefficient b_1 can be determined from the linear plot of the above equation as shown in Figure 8-5.

Having determined the coefficient a_1 and b_1, the gas flow rate can be determined at any pressure from:

$$Q_g = \frac{-a_1 + \sqrt{a_1^2 + 4b_1 \left(\bar{p}_r - p_{wf}\right)}}{2b_1} \qquad (8\text{-}33)$$

The application of Equation 8-29 is also restricted by the assumptions listed for the pressure-squared approach. However, the pressure method is applicable at pressures higher than 3000 psi.

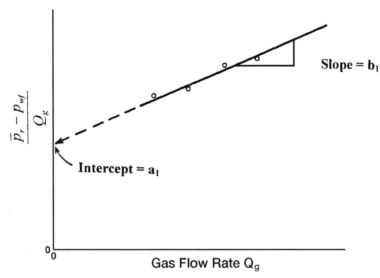

Figure 8-5. Graph of the pressure-method data.

c. Pseudopressure Quadratic Approach

Equation 8-19 can be written as:

$$\overline{\psi}_r - \psi_{wf} = a_2 Q_g + b_2 Q_g^2 \qquad (8\text{-}34)$$

where

$$a_2 = \left(\frac{1422}{kh}\right)\left[\ln\left(\frac{r_e}{r_w}\right) - 0.75 + s\right] \qquad (8\text{-}35)$$

$$b_2 = \left(\frac{1422}{kh}\right)D \qquad (8\text{-}36)$$

The term $(a_2 Q_g)$ in Equation 8-34 represents the pseudopressure drop due to laminar flow while the term $(b_2 Q_g^2)$ accounts for the pseudopressure drop due to inertial-turbulent flow effects.

Equation 8-34 can be linearized by dividing both sides of the equation by Q_g to yield:

$$\frac{\overline{\psi}_r - \psi_{wf}}{Q_g} \, a_2 + b_2 Q_g \qquad (8\text{-}37)$$

The above expression suggests that a plot of $\left(\dfrac{\overline{\psi}_r - \psi_{wf}}{Q_g}\right)$ versus Q_g on a Cartesian scale should yield a straight line with a slope of b_2 and intercept of a_2 as shown in Figure 8-6.

Given the values of a_2 and b_2, the gas flow rate at any p_{wf} is calculated from:

$$Q_g = \frac{-a_2 + \sqrt{a_2^2 + 4b_2\,(\overline{\psi}_r - \psi_{wf})}}{2b_2} \qquad (8\text{-}38)$$

It should be pointed out that the pseudopressure approach is more rigorous than either the pressure-squared or pressure-approximation method and is applicable to all ranges of pressure.

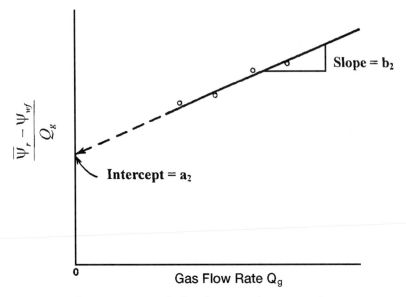

Figure 8-6. Graph of real gas pseudo-pressure data.

In the next section, the back-pressure test is introduced. The material, however, is intended only to be an introduction. There are several excellent books by the following authors that address transient flow and well testing in great detail:

- Earlougher (1977)
- Matthews and Russell (1967)
- Lee (1982)
- Canadian Energy Resources Conservation Board (1975).

The Back-Pressure Test

Rawlins and Schellhardt (1936) proposed a method for testing gas wells by gauging the ability of the well to flow against various back pressures. This type of flow test is commonly referred to as the *conventional deliverability test*. The required procedure for conducting this back-pressure test consists of the following steps:

Step 1. Shut in the gas well sufficiently long for the formation pressure to equalize at the volumetric average pressure $\bar{p}_r$.

Step 2. Place the well on production at a constant flow rate Q_{g1} for a sufficient time to allow the bottom-hole flowing pressure to stabilize at p_{wf1}, i.e., to reach the pseudosteady state.

Step 3. Repeat Step 2 for several rates and the stabilized bottom-hole flow pressure is recorded at each corresponding flow rate. If three or four rates are used, the test may be referred to as a three-point or four-point flow test.

The rate and pressure history of a typical four-point test is shown in Figure 8-7. The figure illustrates a normal sequence of rate changes where the rate is increased during the test. Tests may be also run, however, using a reverse sequence. Experience indicates that a normal rate sequence gives better data in most wells.

The most important factor to be considered in performing the conventional deliverability test is the length of the flow periods. It is required that each rate be maintained sufficiently long for the well to stabilize, i.e., to reach the pseudosteady state. The stabilization time for a well in the center of a circular or square drainage area may be estimated from:

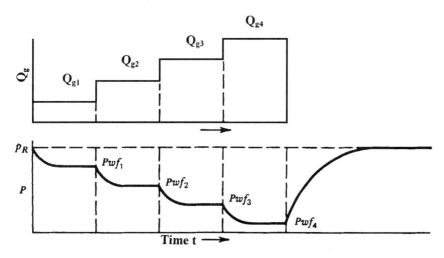

Figure 8-7. Conventional back-pressure test.

$$t_s = \frac{1200 \, \phi \, S_g \, \mu_g \, r_e^2}{k \, \overline{p}_r} \qquad (8\text{-}39)$$

where t_s = stabilization time, hr
ϕ = porosity, fraction
μ_g = gas viscosity, cp
S_g = gas saturation, fraction
k = gas effective permeability, md
$\overline{p}_r$ = average reservoir pressure, psia
r_e = drainage radius, ft

The application of the back-pressure test data to determine the coefficients of any of the empirical flow equations is illustrated in the following example.

Example 8-2

A gas well was tested using a three-point conventional deliverability test. Data recorded during the test are given below:

p_{wf}, psia	ψ_{wf}, psi^2/cp	Q_g, Mscf/day
$\overline{p}_r = 1952$	316×10^6	0
1700	245×10^6	2624.6
1500	191×10^6	4154.7
1300	141×10^6	5425.1

Figure 8-8 shows the gas pseudopressure ψ as a function of pressure. Generate the current IPR by using the following methods.

a. Simplified back-pressure equation
b. Laminar-inertial-turbulent (LIT) methods:

   i. Pressure-squared approach, Equation 8-29
   ii. Pressure-approach, Equation 8-33
   iii. Pseudopressure approach, Equation 8-26

c. Compare results of the calculation.

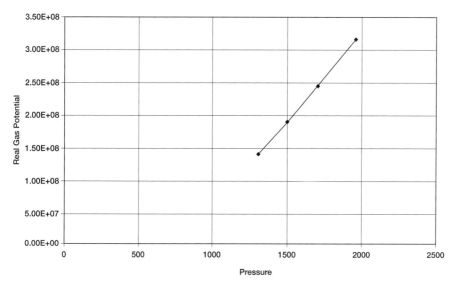

Figure 8-8. Real gas potential vs. pressure.

Solution

a. Back-Pressure Equation:

Step 1. Prepare the following table:

p_{wf}	p^2_{wf}, psi² × 10³	$(\bar{p}^2_r - p^2_{wf})$, psi² × 10³	Q_g, Mscf/day
$\bar{p}_r = 1952$	3810	0	0
1700	2890	920	2624.6
1500	2250	1560	4154.7
1300	1690	2120	5425.1

Step 2. Plot $(\bar{p}^2_r - p^2_{wf})$ versus Q_g on a log-log scale as shown in Figure 8-9. Draw the best straight line through the points.

Step 3. Using any two points on the straight line, calculate the exponent n from Equation 8-22, as

$$n = \frac{\log(4000) - \log(1800)}{\log(1500) - \log(600)} = 0.87$$

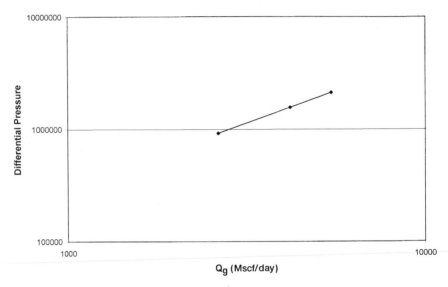

Figure 8-9. Back-pressure curve.

Step 4. Determine the performance coefficient C from Equation 8-23 by using the coordinate of any point on the straight line, or:

$$C = \frac{1800}{(600,000)^{0.87}} = 0.0169 \text{ Mscf/psi}^2$$

Step 5. The back-pressure equation is then expressed as:

$$Q_g = 0.0169 \, (3,810,000 - p_{wf}^2)^{0.87}$$

Step 6. Generate the IPR data by assuming various values of p_{wf} and calculate the corresponding Q_g.

p_{wf}	Q_g, Mscf/day
1952	0
1800	1720
1600	3406
1000	6891
500	8465
0	8980 = AOF = $(Q_g)_{max}$

b. **LIT Method**

i. *Pressure-squared method*

Step 1. Construct the following table:

p_{wf}	$(\bar{p}_r^2 - p_{wf}^2)$, psi$^2 \times 10^3$	Q_g, Mscf/day	$(\bar{p}_r^2 - p_{wf}^2)/Q_g$
$\bar{p}_r = 1952$	0	0	—
1700	920	2624.6	351
1500	1560	4154.7	375
1300	2120	5425.1	391

Step 2. Plot $(\bar{p}_r^2 - p_{wf}^2)/Q_g$ versus Q_g on a Cartesian scale and draw the best straight line as shown in Figure 8-10.

Step 3. Determine the intercept and the slope of the straight line to give:

intercept a = 318
slope b = 0.01333

Step 4. The quadratic form of the pressure-squared approach can be expressed as:

$$(3,810,000 - p_{wf}^2) = 318\ Q_g + 0.01333\ Q_g^2$$

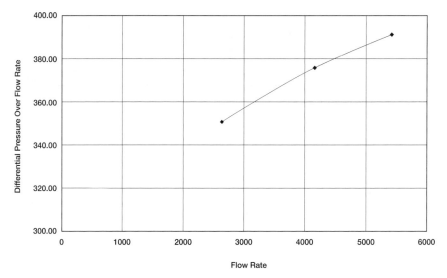

Figure 8-10. Pressure-squared method.

Step 5. Construct the IPR data by assuming various values of p_{wf} and solving for Q_g by using Equation 8-28.

p_{wf}	$(\bar{p}_r^2 - p_{wf}^2)$, psi$^2 \times 10^3$	Q_g, Mscf/day
$\bar{p}_r = 1952$	0	0
1800	570	1675
1600	1250	3436
1000	2810	6862
500	3560	8304
0	3810	$8763 = AOF = (Q_g)_{max}$

ii. *Pressure-approximation method*

Step 1. Construct the following table:

P_{wf}	$(\bar{p}_r - p_{wf})$	Q_g, Mscf/day	$(\bar{p}_r - p_{wf})/Q_g$
$\bar{p}_r = 1952$	0	0	—
1700	252	262.6	0.090
1500	452	4154.7	0.109
1300	652	5425.1	0.120

Step 2. Plot $(\bar{p}_r - p_{wf})/Q_g$ versus Q_g on a Cartesian scale as shown in Figure 8-11.

Draw the best straight line and determine the intercept and slope as:

intercept $a_1 = 0.06$
slope $\quad b_1 = 1.111 \times 10^{-5}$

Step 3. The quadratic form of the pressure-approximation method is then given by:

$$(1952 - p_{wf}) = 0.06\, Q_g + 1.111\,(10^{-5})\, Q_g^2$$

Step 4. Generate the IPR data by applying Equation 8-33:

p_{wf}	$(\bar{p}_r - p_{wf})$	Q_g, Mscf/day
1952	0	0
1800	152	1879
1600	352	3543
1000	952	6942
500	1452	9046
0	1952	10827

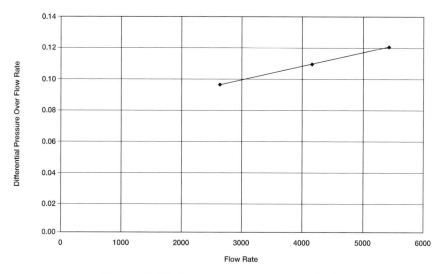

Figure 8-11. Pressure-approximation method.

iii. *Pseudopressure approach*

Step 1. Construct the following table:

p_{wf}	ψ, psi²/cp	$(\bar{\psi}_r - \psi_{wf})$	Q_g, Mscf/day	$(\bar{\psi}_r - \psi_{wf})/Q_g$
$\bar{p}_r = 1952$	316×10^6	0	0	—
1700	245×10^6	71×10^6	262.6	27.05×10^3
1500	191×10^6	125×10^6	4154.7	30.09×10^3
1300	141×10^6	175×10^6	5425.1	32.26×10^3

Step 2. Plot $(\bar{\psi}_r - \psi_{wf})/Q_g$ on a Cartesian scale as shown in Figure 8-12 and determine the intercept a_2 and slope b_2, or:

$$a_2 = 22.28 \times 10^3$$
$$b_2 = 1.727$$

Step 3. The quadratic form of the gas pseudopressure method is given by:

$$(316 \times 10^6 - \psi_{wf}) = 22.28 \times 10^3 \, Q_g + 1.727 \, Q_g^2$$

Step 4. Generate the IPR data by assuming various values of p_{wf}, i.e., ψ_{wf}, and calculate the corresponding Q_g from Equation 8-38.

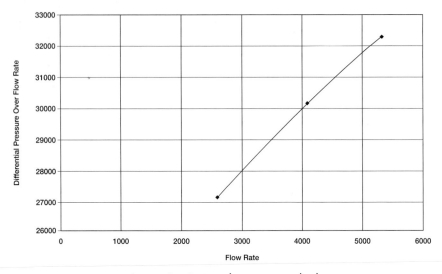

Figure 8-12. Pseudopressure method.

p_{wf}	ψ	$\bar{\psi}_r - \psi_{wf}$	Q_g, Mscf/day
1952	316×10^6	0	0
1800	270×10^6	46×10^6	1794
1600	215×10^6	101×10^6	3503
1000	100×10^6	216×10^6	6331
500	40×10^6	276×10^6	7574
0	0	316×10^6	8342 = AOF $(Q_g)_{max}$

c. Compare the gas flow rates as calculated by the four different methods. Results of the IPR calculation are documented below:

	Gas Flow Rate, Mscf/day			
Pressure	Back-pressure	p^2-Approach	p-Approach	ψ-Approach
19520	0	0	0	0
1800	1720	1675	1879	1811
1600	3406	3436	3543	3554
1000	6891	6862	6942	6460
500	8465	8304	9046	7742
0	8980	8763	10827	8536
	6.0%	5.4%	11%	—

Since the pseudo-pressure analysis is considered more accurate and rigorous than the other three methods, the accuracy of each of the methods in predicting the IPR data is compared with that of the ψ-approach. Figure 8-13 compares graphically the performance of each method with that of ψ-approach. Results indicate that the pressure-squared equation generated the IPR data with an absolute average error of 5.4% as compared with 6% and 11% for the back-pressure equation and the pressure-approximation method, respectively.

It should be noted that the pressure-approximation method is limited to applications for pressures greater than 3000 psi.

Future Inflow Performance Relationships

Once a well has been tested and the appropriate deliverability or inflow performance equation established, it is essential to predict the IPR data as a function of average reservoir pressure. The gas viscosity μ_g and gas compressibility z-factor are considered the parameters that are subject to the greatest change as reservoir pressure $\bar{p}_r$ changes.

Assume that the current average reservoir pressure is $\bar{p}_r$, with gas viscosity of μ_g and a compressibility factor of z_1. At a selected future aver-

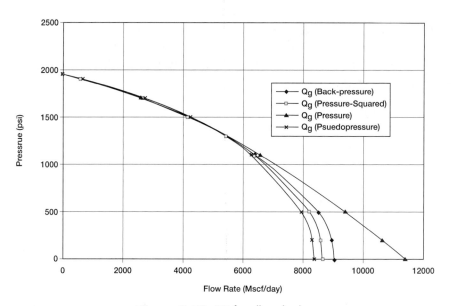

Figure 8-13. IPR for all methods.

age reservoir pressure $\bar{p}_{r2}$, μ_{g2} and z_2 represent the corresponding gas properties. To approximate the effect of reservoir pressure changes, i.e. from $\bar{p}_{r1}$ to $\bar{p}_{r2}$, on the coefficients of the deliverability equation, the following methodology is recommended:

Back-Pressure Equation

The performance coefficient C is considered a pressure-dependent parameter and adjusted with each change of the reservoir pressure according to the following expression:

$$C_2 = C_1 \left[\frac{\mu_{g1}\, z_1}{\mu_{g2}\, z_2} \right] \qquad (8\text{-}40)$$

The value of n is considered essentially constant.

LIT Methods

The laminar flow coefficient a and the inertial-turbulent flow coefficient b of any of the previous LIT methods, i.e., Equations 8-24, 8-29, and 8-34, are modified according to the following simple relationships:

• Pressure-Squared Method

The coefficients a and b of pressure-squared are modified to account for the change of the reservoir pressure from $\bar{p}_{r1}$ to $\bar{p}_{r2}$ by adjusting the coefficients as follows:

$$a_2 = a_1 \left[\frac{\mu_{g2}\, z_2}{\mu_{g1}\, z_1} \right] \qquad (8\text{-}41)$$

$$b_2 = b_1 \left[\frac{\mu_{g2}\, z_2}{\mu_{g1}\, z_1} \right] \qquad (8\text{-}42)$$

where the subscripts 1 and 2 represent conditions at reservoir pressure $\bar{p}_{r1}$ to $\bar{p}_{r2}$, respectively.

• **Pressure-Approximation Method**

$$a_2 = a_1 \left[\frac{\mu_{g2} \, \beta_{g2}}{\mu_{g1} \, \beta_{g1}} \right] \tag{8-43}$$

$$b_2 = b_1 \left[\frac{\mu_{g2} \, \beta_{g2}}{\mu_{g1} \, \beta_{g1}} \right] \tag{8-44}$$

where B_g is the gas formation volume factor

• **Pseudopressure Approach**
    The coefficients a and b of the pseudo-pressure approach are essentially independent of the reservoir pressure and they can be treated as constants.

Example 8-3

In addition to the data given in Example 8-2, the following information is available:

• $(\mu_g \, z) = 0.01206$ at 1952 psi
• $(\mu_g \, z) = 0.01180$ at 1700 psi

Using the following methods:

a. Back-pressure method
b. Pressure-squared method
c. Pseudo-pressure method

Generate the IPR data for the well when the reservoir pressure drops from 1952 to 1700 psi.

Solution

Step 1. Adjust the coefficients a and b of each equation. For the:
    • **Back-pressure equation:**
        Using Equation 8-40, adjust C:

$$C = 0.0169 \left(\frac{0.01206}{0.01180} \right) = 0.01727$$

$$Q_g = 0.01727 \, (1700^2 - p_{wf}^2)0.87$$

• **Pressure-squared method:**
Adjust a and b by applying Equations 8-41 and 8-42

$$a = 318 \left(\frac{0.01180}{0.01206} \right) = 311.14$$

$$b = 0.01333 \left(\frac{0.01180}{0.01206} \right) = 0.01304$$

$$(1700^2 - p_{wf}^2) = 311.14 \, Q_g + 0.01304 \, Q_g^2$$

• **Pseudopressure method:**
No adjustments are needed.

$$(245 \times 10^6) - \psi_{wf} = (22.28 \times 10^3) \, Q_g + 1.727 \, Q_g^2$$

Step 2. Generate the IPR data:

	Gas Flow rate Q_g, Mscf/day		
p_{wf}	Back-Pressure	p^2-Method	ψ-Method
$\bar{p}_r = 1700$	0	0	0
1600	1092	1017	1229
1000	4987	5019	4755
500	6669	6638	6211
0	7216	7147	7095

Figure 8-14 compares graphically the IPR data as predicted by the above three methods.

HORIZONTAL GAS WELL PERFORMANCE

Many low permeability gas reservoirs are historically considered to be noncommercial due to low production rates. Most vertical wells drilled in tight gas reservoirs are stimulated using hydraulic fracturing and/or

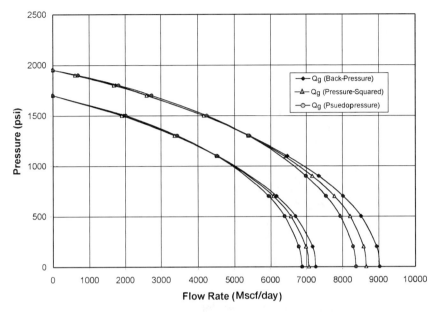

Figure 8-14. IPR comparison.

acidizing treatments to attain economical flow rates. In addition, to deplete a tight gas reservoir, vertical wells must be drilled at close spacing to efficiently drain the reservoir. This would require a large number of vertical wells. In such reservoirs, horizontal wells provide an attractive alternative to effectively deplete tight gas reservoirs and attain high flow rates. Joshi (1991) points out those horizontal wells are applicable in both low-permeability reservoirs as well as in high-permeability reservoirs.

An excellent reference textbook by Sada Joshi (1991) gives a comprehensive treatment of horizontal wells performance in oil and gas reservoirs.

In calculating the gas flow rate from a horizontal well, Joshi introduced the concept of the **effective wellbore radius** r'_w into the gas flow equation. The effective wellbore radius is given by:

$$r'_w = \frac{r_{eh}\,(L/2)}{a[1+\sqrt{1-(L/2a)^2}\,[h/(2r_w)]^{h/L}}$$

(8-45)

with

$$a = \left(\frac{L}{2}\right) \left[0.5 + \sqrt{0.25 + (2r_{eh}/L)^4}\right]^{0.5} \tag{8-46}$$

and

$$r_{eh} = \sqrt{\frac{43,560\ A}{\pi}} \tag{8-47}$$

where L = length of the horizontal well, ft
      h = thickness, ft
      r_w = wellbore radius, ft
      r_{eh} = horizontal well drainage radius, ft
      a = half the major axis of drainage ellipse, ft
      A = drainage area, acres

Methods of calculating the horizontal well drainage area A are presented in Chapter 7 by Equations 7-45 and 7-46.

For a pseudosteady-state flow, Joshi expressed Darcy's equation of a laminar flow in the following two familiar forms:

Pressure-Squared Form

$$Q_g = \frac{kh\ (\bar{p}_r^2 - p_{wf}^2)}{1422\ T\,(\mu_g\, z)_{avg}\left[\ln(r_{eh}/r_w') - 0.75 + s\right]} \tag{8-48}$$

where Q_g = gas flow rate, Mscf/day
      s = skin factor
      k = permeability, md
      T = temperature, °R

Pseudo-Pressure Form

$$Q_g = \frac{kh\,(\bar{\psi}_r - \psi_{wf})}{1422\ T\left[\ln\left(\dfrac{r_{eh}}{r_w}\right) - 0.75 + s\right]}$$

Example 8-4

A 2,000-foot-long horizontal gas well is draining an area of approximately 120 acres. The following data are available:

$\bar{p}_r = 2000$ psi $\quad\quad \bar{\psi}_r = 340 \times 10^6$ psi²/cp

$p_{wf} = 1200$ psi $\quad \psi_{wf} = 128 \times 10^6$ psi²/cp

$(\mu_g z)_{avg} = 0.011826 \quad\quad r_w = 0.3$ ft $\quad\quad s = 0.5$
$\quad\quad\quad\quad h = 20$ ft $\quad\quad\quad\quad\quad T = 180°F \quad\quad k = 1.5$ md

Assuming a pseudosteady-state flow, calculate the gas flow rate by using the pressure-squared and pseudopressure methods.

Solution

Step 1. Calculate the drainage radius of the horizontal well:

$$r_{eh} = \sqrt{\frac{(43,560)(120)}{\pi}} = 1290 \text{ ft}$$

Step 2. Calculate half the major axis of drainage ellipse by using Equation 8-46:

$$a = \left[\frac{2000}{2}\right]\left[0.5 + \sqrt{0.25 + \left[\frac{(2)(1290)}{2000}\right]^4}\right]^{0.5} = 1495.8$$

Step 3. Calculate the effective wellbore radius r'_w from Equation 8-45:

$$(h/2r_w)^{h/L} = \left[\frac{20}{(2)(0.3)}\right]^{20/2000} = 1.0357$$

$$1 + \sqrt{1 - \left(\frac{L}{2a}\right)^2} = 1 + \sqrt{1 - \left(\frac{2000}{2(1495.8)}\right)^2} = 1.7437$$

Applying Equation 8-45, gives:

$$r'_w = \frac{1290(200/2)}{1495.8\,(1.7437)\,(1.0357)} = 477.54\,\text{ft}$$

Step 4. Calculate the flow rate by using the pressure-squared approximation and ψ-approach.

- **Pressure-squared**

$$Q_g = \frac{(1.5)\,(20)\,(2000^2 - 1200^2)}{(1422)\,(640)\,(0.011826)\left[\ln\left(\dfrac{1290}{477.54}\right) - 0.75 + 0.5\right]}$$

$$= 9,594\,\text{Mscf/day}$$

- **ψ-Method**

$$Q_g = \frac{(1.5)\,(20)\,(340 - 128)\,(10^6)}{(1422)\,(640)\left[\ln\left(\dfrac{1290}{477.54}\right) - 0.75 + 0.5\right]}$$

$$= 9396\,\text{Mscf/day}$$

For turbulent flow, Darcy's equation must be modified to account for the additional pressure caused by the non-Darcy flow by including the rate-dependent skin factor DQ_g. In practice, the back-pressure equation and the LIT approach are used to calculate the flow rate and construct the IPR curve for the horizontal well. Multirate tests, i.e., deliverability tests, must be performed on the horizontal well to determine the coefficients of the selected flow equation.

PROBLEMS

1. A gas well is producing under a constant bottom-hole flowing pressure of 1000 psi. The specific gravity of the produced gas is 0.65, given:

$p_i = 1500$ psi	$r_w = 0.33$ ft	$r_e = 1000$ ft	$k = 20$ md
$h = 20$ ft	$T = 140°F$	$s = 0.40$	

Calculate the gas flow rate by using:

a. Real gas pseudopressure approach
b. Pressure-squared approximation

2. The following data[1] were obtained from a back-pressure test on a gas well.

Q_g, Mscf/day	p_{wf}, psi
0	481
4928	456
6479	444
8062	430
9640	415

a. Calculate values of C and n
b. Determine AOF
c. Generate the IPR curves at reservoir pressures of 481 and 300 psi.

3. The following back-pressure test data are available:

Q_g, Mscf/day	p_{wf}, psi
0	5240
1000	4500
1350	4191
2000	3530
2500	2821

Given:

gas gravity $= 0.78$
porosity $= 12\%$
$s_{wi} = 15\%$
$T = 281°F$

a. Generate the current IPR curve by using:

i. Simplified back-pressure equation
ii. Laminar-inertial-turbulent (LIT) methods:

• Pressure-squared approach

---

[1]Chi Ikoku, *Natural Gas Reservoir Engineering,* John Wiley and Sons, 1984.

• Pressure-approximation approach
• Pseudopressure approach

b. Repeat part a for a future reservoir pressure of 4000 psi.

4. A 3,000-foot horizontal gas well is draining an area of approximately 180 acres, given:

$p_i = 2500$ psi $p_{wf} = 1500$ psi $k = 25$ md
$T = 120°F$ $r_w = 0.25$ $h = 20$ ft
$\upsilon_g = 0.65$

Calculate the gas flow rate.

REFERENCES

1. Earlougher, Robert C., Jr., *Advances in Well Test Analysis.* Mongraph Vol. 5, Society of Petroleum Engineers of AIME. Dallas, TX: Millet the Printer, 1977.

2. ERCB. *Theory and Practice of the Testing of Gas Wells,* 3rd ed. Calgary: Energy Resources Conservation Board, 1975.

3. Fetkovich, M. J., "Multipoint Testing of Gas Wells," SPE Mid-continent Section Continuing Education Course of Well Test Analysis, March 17, 1975.

4. Golan, M., and Whitson, C., *Well Performance.* International Human Resources Development Corporation, 1986.

5. Joshi, S., *Horizontal Well Technology.* Tulsa, OK: PennWell Publishing Company, 1991.

6. Lee, J., *Well Testing.* Dallas: Society of Petroleum Engineers of AIME, 1982.

7. Matthews, C., and Russell, D., "Pressure Buildup and Flow Tests in Wells." Dallas: SPE Monograph Series, 1967.

8. Rawlins, E. L., and Schellhardt, M. A., "Back-Pressure Data on Natural Gas Wells and Their Application to Production Practices." U.S. Bureau of Mines Monograph 7, 1936.

GAS AND WATER CONING

Coning is a term used to describe the mechanism underlying the upward movement of water and/or the down movement of gas into the perforations of a producing well. Coning can seriously impact the well productivity and influence the degree of depletion and the overall recovery efficiency of the oil reservoirs. The specific problems of water and gas coning are listed below.

• Costly added water and gas handling
• Gas production from the original or secondary gas cap reduces pressure without obtaining the displacement effects associated with gas drive
• Reduced efficiency of the depletion mechanism
• The water is often corrosive and its disposal costly
• The afflicted well may be abandoned early
• Loss of the total field overall recovery

Delaying the encroachment and production of gas and water are essentially the controlling factors in maximizing the field's ultimate oil recovery. Since coning can have an important influence on operations, recovery, and economics, it is the objective of this chapter to provide the theoretical analysis of coning and outline many of the practical solutions for calculating water and gas coning behavior.

CONING

Coning is primarily the result of movement of reservoir fluids in the direction of least resistance, balanced by a tendency of the fluids to maintain gravity equilibrium. The analysis may be made with respect to either gas or water. Let the original condition of reservoir fluids exist as shown schematically in Figure 9-1, water underlying oil and gas overlying oil. For the purposes of discussion, assume that a well is partially penetrating the formation (as shown in Figure 9-1) so that the production interval is halfway between the fluid contacts.

Production from the well would create pressure gradients that tend to lower the gas-oil contact and elevate the water-oil contact in the immediate vicinity of the well. Counterbalancing these flow gradients is the tendency of the gas to remain above the oil zone because of its lower density and of the water to remain below the oil zone because of its higher density. These counterbalancing forces tend to deform the gas-oil and water-oil contacts into a bell shape as shown schematically in Figure 9-2.

There are essentially three forces that may affect fluid flow distributions around the well bores. These are:

• Capillary forces
• Gravity forces
• Viscous forces

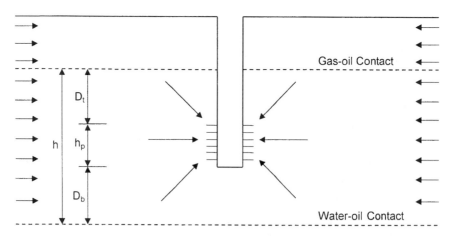

Figure 9-1. Original reservoir static condition.

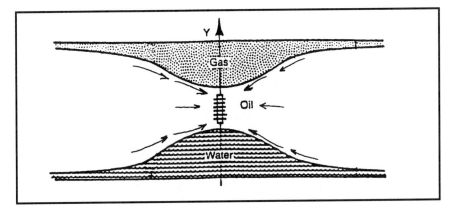

Figure 9-2. Gas and water coning.

Capillary forces usually have negligible effect on coning and will be neglected. Gravity forces are directed in the vertical direction and arise from fluid density differences. The term *viscous forces* refers to the pressure gradients associated fluid flow through the reservoir as described by Darcy's Law. Therefore, at any given time, there is a balance between gravitational and viscous forces at points on and away from the well completion interval. When the dynamic (viscous) forces at the wellbore exceed gravitational forces, a "cone" will ultimately break into the well.

We can expand on the above basic visualization of coning by introducing the concepts of:

• Stable cone
• Unstable cone
• Critical production rate

If a well is produced at a constant rate and the pressure gradients in the drainage system have become constant, a steady-state condition is reached. If at this condition the dynamic (viscous) forces at the well are less than the gravity forces, then the water or gas cone that has formed will not extend to the well. Moreover, the cone will neither advance nor recede, thus establishing what is known as a *stable cone*. Conversely, if the pressure in the system is an unsteady-state condition, then an *unstable cone* will continue to advance until steady-state conditions prevail.

If the flowing pressure drop at the well is sufficient to overcome the gravity forces, the unstable cone will grow and ultimately break into the well. It is important to note that in a realistic sense, stable system cones may only be "pseudo-stable" because the drainage system and pressure distributions generally change. For example, with reservoir depletion, the water-oil contact may advance toward the completion interval, thereby increasing chances for coning. As another example, reduced productivity due to well damage requires a corresponding increase in the flowing pressure drop to maintain a given production rate. This increase in pressure drop may force an otherwise stable cone into a well.

The *critical production rate* is the rate above which the flowing pressure gradient at the well causes water (or gas) to cone into the well. It is, therefore, the maximum rate of oil production without concurrent production of the displacing phase by coning. At the critical rate, the built-up cone is stable but is at a position of incipient breakthrough.

Defining the conditions for achieving the maximum water-free and/or gas-free oil production rate is a difficult problem to solve. Engineers are frequently faced with the following specific problems:

1. Predicting the maximum flow rate that can be assigned to a completed well without the simultaneous production of water and/or free-gas.
2. Defining the optimum length and position of the interval to be perforated in a well in order to obtain the maximum water and gas-free production rate.

Calhoun (1960) pointed out that the rate at which the fluids can come to an equilibrium level in the rock may be so slow, due to the low permeability or to capillary properties, that the gradient toward the wellbore overcomes it. Under these circumstances, the water is lifted into the wellbore and the gas flows downward, creating a cone as illustrated in Figure 9-2. Not only is the direction of gradients reversed with gas and oil cones, but the rapidity with which the two levels will balance will differ. Also, the rapidity with which any fluid will move is inversely proportional to its viscosity, and, therefore, the gas has a greater tendency to cone than does water. For this reason, the amount of coning will depend upon the viscosity of the oil compared to that of water.

It is evident that the degree or rapidity of coning will depend upon the rate at which fluid is withdrawn from the well and upon the permeability in the vertical direction k_v compared to that in the horizontal direction k_h. It will also depend upon the distance from the wellbore withdrawal point to the gas-oil or oil-water discontinuity.

The elimination of coning could be aided by shallower penetration of wells where there is a water zone or by the development of better horizontal permeability. Although the vertical permeability could not be lessened, the ratio of horizontal to vertical flow can be increased by such techniques as acidizing or pressure parting the formation. The application of such techniques needs to be controlled so that the effect occurs above the water zone or below the gas zone, whichever is the desirable case. This permits a more uniform rise of a water table.

Once either gas coning or water coning has occurred, it is possible to shut in the well and permit the contacts to restabilize. Unless conditions for rapid attainment of gravity equilibrium are present, restabilization will not be extremely satisfactory. Fortunately, bottom water is found often where favorable conditions for gravity separation do exist. Gas coning is more difficult to avoid because gas saturation, once formed, is difficult to eliminate.

There are essentially three categories of correlation that are used to solve the coning problem. These categories are:

• Critical rate calculations
• Breakthrough time predictions
• Well performance calculations after breakthrough

The above categories of calculations are applicable in evaluating the coning problem in vertical and horizontal wells.

CONING IN VERTICAL WELLS

Vertical Well Critical Rate Correlations

Critical rate Q_{oc} is defined as the maximum allowable oil flow rate that can be imposed on the well to avoid a cone breakthrough. The critical rate would correspond to the development of a stable cone to an elevation just below the bottom of the perforated interval in an oil-water system or to an elevation just above the top of the perforated interval in a gas-oil system. There are several empirical correlations that are commonly used to predict the oil critical rate, including the correlations of:

• Meyer-Garder
• Chierici-Ciucci
• Hoyland-Papatzacos-Skjaeveland

- Chaney et al.
- Chaperson
- Schols

The practical applications of these correlations in predicting the critical oil flow rate are presented over the following pages.

The Meyer-Garder Correlation

Meyer and Garder (1954) suggest that coning development is a result of the radial flow of the oil and associated pressure sink around the well-bore. In their derivations, Meyer and Garder assume a homogeneous system with a uniform permeability throughout the reservoir, i.e., $k_h = k_v$. It should be pointed out that the ratio k_h/k_v is the most critical term in evaluating and solving the coning problem. They developed three separate correlations for determining the critical oil flow rate:

- Gas coning
- Water coning
- Combined gas and water coning

Gas coning

Consider the schematic illustration of the gas-coning problem shown in Figure 9-3.

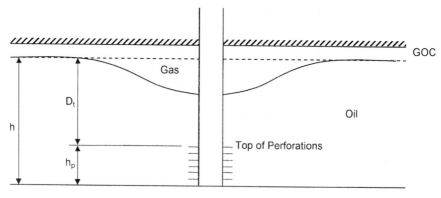

Figure 9-3. Gas coning.

Meyer and Garder correlated the critical oil rate required to achieve a stable gas cone with the following well penetration and fluid parameters:

• Difference in the oil and gas density
• Depth D_t from the original gas-oil contact to the top of the perforations
• The oil column thickness h

The well perforated interval h_p, in a gas-oil system, is essentially defined as

$$h_p = h - D_t$$

Meyer and Garder propose the following expression for determining the oil critical flow rate in a gas-oil system:

$$Q_{oc} = 0.246 \times 10^{-4} \left[\frac{\rho_o - \rho_g}{\ln (r_e/r_w)} \right] \left(\frac{k_o}{\mu_o B_o} \right) \left[h^2 - (h - D_t)^2 \right] \qquad (9\text{-}1)$$

where Q_{oc} = critical oil rate, STB/day
ρ_g, ρ_o = density of gas and oil, respectively, lb/ft^3
k_o = effective oil permeability, md
r_e, r_w = drainage and wellbore radius, respectively, ft
h = oil column thickness, ft
D_t = distance from the gas-oil contact to the *top* of the perforations, ft

Water coning

Meyer and Garder propose a similar expression for determining the critical oil rate in the water coning system shown schematically in Figure 9-4. The proposed relationship has the following form:

$$Q_{oc} = 0.246 \times 10^{-4} \left[\frac{\rho_w - \rho_o}{\ln (r_e/r_w)} \right] \left(\frac{k_o}{\mu_o B_o} \right) (h^2 - h_p^2) \qquad (9\text{-}2)$$

where ρ_w = water density, lb/ft^3
h_p = perforated interval, ft

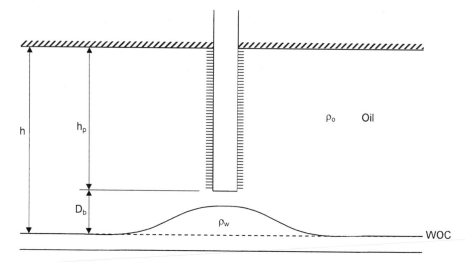

Figure 9-4. Water coning.

Simultaneous gas and water coning

If the effective oil-pay thickness h is comprised between a gas cap and a water zone (Figure 9-5), the completion interval h_p must be such as to permit maximum oil-production rate without having gas and water simultaneously produced by coning, gas breaking through at the top of the interval and water at the bottom.

This case is of particular interest in the production from a thin column underlaid by bottom water and overlaid by gas.

For this combined gas and water coning, Pirson (1977) combined Equations 9-1 and 9-2 to produce the following simplified expression for determining the maximum oil-flow rate without gas and water coning:

$$Q_{oc} = 0.246 \times 10^{-4} \left[\frac{k_o}{\mu_o \, B_o} \right] \frac{h^2 - h_p^2}{\ln \, (r_e / r_w)}$$

$$\times \left[(\rho_w - \rho_o) \left(\frac{\rho_o - \rho_g}{\rho_w - \rho_g} \right)^2 + (\rho_o - \rho_g) \left(1 - \frac{\rho_o - \rho_g}{\rho_w - \rho_g} \right)^2 \right] \qquad (9 \text{-} 3)$$

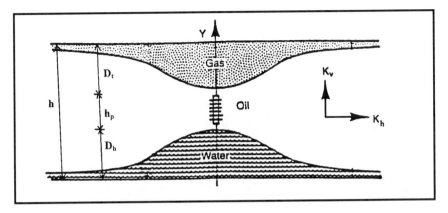

Figure 9-5. The development of gas and water coning.

Example 9-1

A vertical well is drilled in an oil reservoir overlaid by a gas cap. The related well and reservoir data are given below:

horizontal and vertical permeability, i.e., k_h, k_v = 110 md
oil relative permeability, k_{ro} = 0.85
oil density, ρ_o = 47.5 lb/ft^3
gas density, ρ_g = 5.1 lb/ft^3
oil viscosity, μ_o = 0.73 cp
oil formation volume factor, B_o = 1.1 bbl/STB
oil column thickness, h   = 40 ft
perforated interval, h_p = 15 ft
depth from GOC to top of perforations, D_t = 25 ft
wellbore radius, r_w = 0.25 ft
drainage radius, r_e = 660 ft

Using the Meyer and Garder relationships, calculate the critical oil flow rate.

Solution

The critical oil flow rate for this gas-coning problem can be determined by applying Equation 9-1. The following two steps summarize Meyer-Garder methodology:

Step 1. Calculate effective oil permeability k_o

$$k_o = k_{ro}\, k = (0.85)\,(110) = 93.5 \text{ md}$$

Step 2. Solve for Q_{oc} by applying Equation 9-1

$$Q_{oc} = 0.246 \times 10^{-4}\,\frac{47.5 - 5.1}{\ln\,(660/0.25)}\,\frac{93.5}{(0.73)(1.1)}\,[40^2 - (40 - 25)^2]$$
$$= 21.20 \text{ STB/day}$$

Example 9-2

Resolve Example 9-1 assuming that the oil zone is underlaid by bottom water. The water density is given as 63.76 lb/ft^3. The well completion interval is 15 feet as measured from the top of the formation (no gas cap) to the bottom of the perforations.

Solution

The critical oil flow rate for this water-coning problem can be estimated by applying Equation 9-2. The equation is designed to determine the critical rate at which the water cone "touches" the bottom of the well to give

$$Q_{oc} = 0.246 \times 10^{-4}\left[\frac{(63.76 - 47.5)}{\ln\,(660/0.25)}\right]\left(\frac{93.5}{(0.73)(1.1)}\right)[40^2 - 15^2]$$

$$Q_{oc} = 8.13 \text{ STB/day}$$

The above two examples signify the effect of the fluid density differences on critical oil flow rate.

Example 9-3

A vertical well is drilled in an oil reservoir that is overlaid by a gas cap and underlaid by bottom water. Figure 9-6 shows an illustration of the simultaneous gas and water coning.

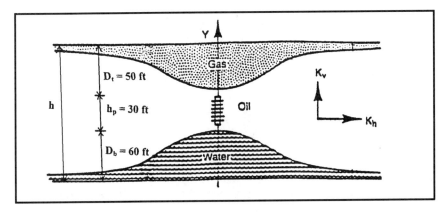

Figure 9-6. Gas and water coning problem (Example 9-3).

The following data are available:

oil density	$\rho_o = 47.5 \text{ lb/ft}^3$
water density	$\rho_w = 63.76 \text{ lb/ft}^3$
gas density	$\rho_g = 5.1 \text{ lb/ft}^3$
oil viscosity	$\mu_o = 0.73 \text{ cp}$
oil FVF	$B_o = 1.1 \text{ bbl/STB}$
oil column thickness	$h = 65 \text{ ft}$
depth from GOC to top of perforations	$D_t = 25 \text{ ft}$
well perforated interval	$h_p = 15 \text{ ft}$
wellbore radius	$r_w = 0.25 \text{ ft}$
drainage radius	$r_e = 660 \text{ ft}$
oil effective permeability	$k_o = 93.5 \text{ md}$
horizontal and vertical permeability, i.e., k_h, $k_v = 110 \text{ md}$
oil relative permeability	$k_{ro} = 0.85$

Calculate the maximum permissible oil rate that can be imposed to avoid cones breakthrough, i.e., water and gas coning.

Solution

Apply Equation 9-3 to solve for the simultaneous gas- and water-coning problem, to give:

$$Q_{oc} = 0.246 \times 10^{-4} \frac{93.5}{(0.73)(1.1)} \left[\frac{65^2 - 15^2}{\text{Ln}(660/0.25)} \right]$$

$$\times \left[(63.76 - 47.5) \left(\frac{47.5 - 5.1}{63.76 - 5.1} \right)^2 \right.$$

$$\left. + (47.5 - 5.1) \left(1 - \frac{47.5 - 5.1}{63.76 - 5.1} \right)^2 \right] = 17.1 \text{ STB/day}$$

Pirson (1977) derives a relationship for determining the optimum placement of the desired h_p feet of perforation in an oil zone with a gas cap above and a water zone below. Pirson proposes that the optimum distance D_t from the GOC to the top of the perforations can determined from the following expression:

$$D_t = (h - h_p) \left[1 - \frac{\rho_o - \rho_g}{\rho_w - \rho_g} \right] \tag{9-4}$$

where the distance D_t is expressed in feet.

Example 9-4

Using the data given in Example 9-3, calculate the optimum distance for the placement of the 15-foot perforations.

Solution

Applying Equation 9-4 gives

$$D_t = (65 - 15) \left[1 - \frac{47.5 - 5.1}{63.76 - 5.1} \right] = 13.9 \text{ ft}$$

Slider (1976) presented an excellent overview of the coning problem and the above-proposed predictive expressions. Slider points out that Equations 9-1 through 9-4 are not based on realistic assumptions. One of the biggest difficulties is in the assumption that the permeability is the same in all directions. As noted, this assumption is seldom realistic. Since sedimentary formations were initially laid down in thin, horizontal sheets, it is natural for the formation permeability to vary from one sheet to another vertically.

Therefore, there is generally quite a difference between the permeability measured in a vertical direction and the permeability measured in a horizontal direction. Furthermore, the permeability in the horizontal direction is normally considerably greater than the permeability in the vertical direction. This also seems logical when we recognize that very thin, even microscopic sheets of impermeable material, such as shale, may have been periodically deposited. These permeability barriers have a great effect on the vertical flow and have very little effect on the horizontal flow, which would be parallel to the plane of the sheets.

The Chierici-Ciucci Approach

Chierici and Ciucci (1964) used a potentiometric model to predict the coning behavior in vertical oil wells. The results of their work are presented in dimensionless graphs that take into account the vertical and horizontal permeability. The diagrams can be used for solving the following two types of problems:

a. Given the reservoir and fluid properties, as well as the position of and length of the perforated interval, determine the maximum oil production rate without water and/or gas coning.
b. Given the reservoir and fluids characteristics only, determine the optimum position of the perforated interval.

The authors introduced four dimensionless parameters that can be determined from a graphical correlation to determine the critical flow rates. The proposed four dimensionless parameters are shown in Figure 9-7 and defined as follows:

Effective dimensionless radius r_{De}:

The first dimensionless parameter that the authors used to correlate results of potentiometric model is called the *effective dimensionless radius* and is defined by:

$$r_{De} = \frac{r_e}{h} \sqrt{\frac{k_h}{k_v}} \qquad (9\text{-}5)$$

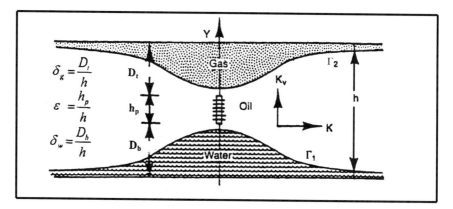

Figure 9-7. Water and gas coning in a homogeneous formation. (*After Chierici, Ciucci, and Pizzi, courtesy JPT, August 1964.*)

Meyer and Garder stated that the proposed graphical correlation is valid in the following range of r_{De} values:

$$5 \text{ "} r_{De} \text{ "} 80$$

where   h = oil column thickness, ft

       r_e = drainage radius, ft

       k_v, k_h = vertical and horizontal permeability, respectively

Dimensionless perforated length ε:

The second dimensionless parameter that the authors used in developing their correlation is termed the *dimensionless perforated length* and is defined by:

$$\varepsilon = h_p / h \tag{9-6}$$

The authors pointed out that the proposed graphical correlation is valid when the value of the dimensionless perforated length is in the following range:

$$0 \text{ "} \varepsilon \text{ "} 0.75$$

Dimensionless gas cone ratio δ_g:

The authors introduced the *dimensionless gas cone ratio* as defined by the following relationship:

$$\delta_g = D_t/h \qquad (9\text{-}7)$$

with

$$0.070 '' \ \delta_g \ '' \ 0.9$$

where D_t is the distance from the original GOC to the top of perforations, ft.

Dimensionless water cone ratio δ_w:

The last dimensionless parameter that Chierici et al. proposed in developing their correlation is called the *dimensionless water-cone ratio* and is defined by:

$$\delta_w = D_b/h \qquad (9\text{-}8)$$

with

$$0.07 '' \ \delta_w \ '' \ 0.9$$

where D_b = distance from the original WOC to the bottom of the perforations, ft

Chierici and coauthors proposed that the oil-water and gas-oil contacts are stable only if the oil production rate of the well is not higher than the following rates:

$$Q_{ow} = 0.492 \times 10^{-4} \ \frac{h^2 \ (\rho_w - \rho_o)}{B_o \ \mu_o} \ (k_{ro} \ k_h) \ \Psi_w (r_{De}, \ \varepsilon, \ \delta_w) \qquad (9\text{-}9)$$

$$Q_{og} = 0.492 \times 10^{-4} \ \frac{h^2 \ (\rho_o - \rho_g)}{B_o \ \mu_o} \ (k_{ro} \ k_h) \ \Psi_g (r_{De}, \ \varepsilon, \ \delta_g) \qquad (9\text{-}10)$$

where Q_{ow} = critical oil flow rate in oil-water system, STB/day
$\qquad Q_{og}$ = critical oil flow rate in gas-oil system, STB/day
ρ_o, ρ_w, ρ_g = densities in lb/ft^3
$\qquad \psi_w$ = water dimensionless function
$\qquad \psi_g$ = gas dimensionless function
$\qquad k_h$ = horizontal permeability, md

The authors provided a set of working graphs for determining the dimensionless function ψ from the calculated dimensionless parameters r_{De}, ε, and δ. These graphs are shown in Figures 9-8 through 9-14. *This set of curves should be only applied to homogeneous formations.*

It should be noted that if a gas cap and an aquifer are present together, the following conditions must be satisfied in order to avoid water and free-gas production.

$$Q_o \,''\, Q_{ow}$$

and

$$Q_o \,''\, Q_{og}$$

(*text continued on page 588*)

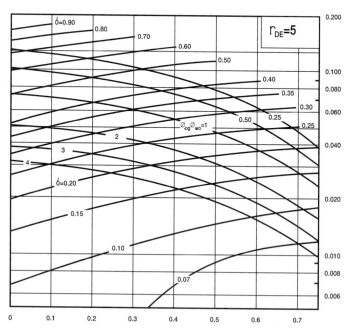

Figure 9-8. Dimensionless functions for $r_{De} = 5$. (*After Chierici, Ciucci, and Pizzi, courtesy JPT, August 1964.*)

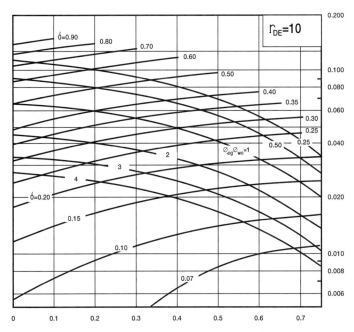

Figure 9-9. Dimensionless functions for $r_{De} = 10$. (*After Chierici, Ciucci, and Pizzi, courtesy JPT, August 1964.*)

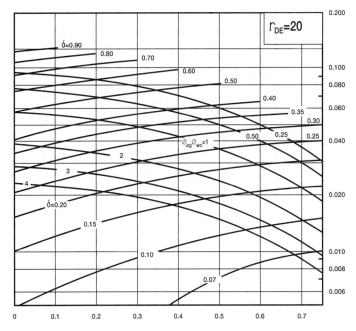

Figure 9-10. Dimensionless functions for $r_{De} = 20$. (*After Chierici, Ciucci, and Pizzi, courtesy JPT, August 1964.*)

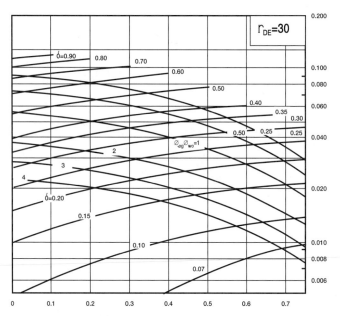

Figure 9-11. Dimensionless functions for $r_{De} = 30$. (*After Chierici, Ciucci, and Pizzi, courtesy JPT, August 1964.*)

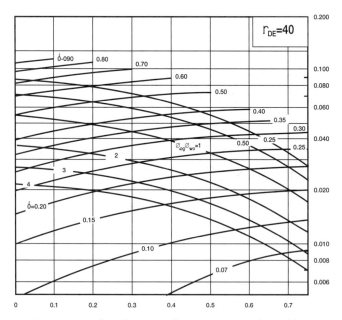

Figure 9-12. Dimensionless functions for $r_{De} = 40$. (*After Chierici, Ciucci, and Pizzi, courtesy JPT, August 1964.*)

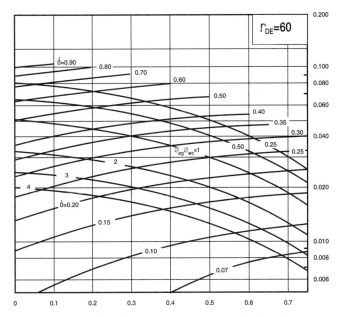

Figure 9-13. Dimensionless functions for $r_{De} = 60$. (*After Chierici, Ciucci, and Pizzi, courtesy JPT, August 1964.*)

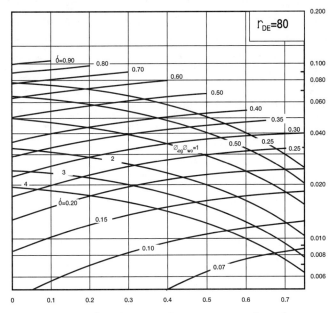

Figure 9-14. Dimensionless functions for $r_{De} = 80$. (*After Chierici, Ciucci, and Pizzi, courtesy JPT, August 1964.*)

(*text continued from page 584*)

Example 9-5

A vertical well is drilled on a regular 40-acre spacing in an oil reservoir that is overlaid by a gas cap and underlaid by an aquifer. The following data are available:

Oil pay thickness	$h = 140$ ft
Distance from the GOC to the top of perforations	$D_t = 50$ ft
Length of the perforated interval	$h_p = 30$ ft
Horizontal permeability	$k_h = 300$ md
Relative oil permeability	$k_{ro} = 1.00$
Vertical permeability	$k_v = 90$ md
Oil density	$\rho_o = 46.24$ lb/ft^3
Water density	$\rho_w = 68.14$ lb/ft^3
Gas density	$\rho_g = 6.12$ lb/ft^3
Oil FVF	$B_o = 1.25$ bbl/STB
Oil viscosity	$\mu_o = 1.11$ cp

A schematic representation of the given data is shown in Figure 9-15. Calculate the maximum allowable oil-flow rate without water and free-gas production.

Solution

Step 1. Calculate the drainage radius r_e:

$$\pi r_e^2 = (40)(43{,}560)$$

$$r_e = 745 \text{ ft}$$

Step 2. Compute the distance from the WOC to the bottom of the perforations D_b:

$$D_b = h - D_t - h_p$$

$$D_b = 140 - 50 - 30 = 60 \text{ ft}$$

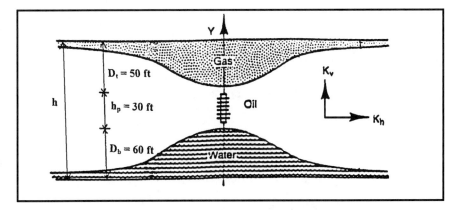

Figure 9-15. Gas and water coning problem (Example 9-5).

Step 3. Find the dimensionless radius r_{De} from Equation 9-5:

$$r_{De} = \frac{745}{140} \sqrt{\frac{300}{90}} = 9.72$$

Step 4. Calculate the dimensionless perforated length ε by applying Equation 9-6:

$$\varepsilon = \frac{30}{140} = 0.214$$

Step 5. Calculate the gas cone ratio δ_g from Equation 9-7:

$$\delta_g = \frac{50}{140} = 0.357$$

Step 6. Determine the water cone ratio δ_w by applying Equation 9-8:

$$\delta_w = \frac{60}{140} = 0.429$$

Step 7. Calculate the oil-gas and water-oil density differences:

$$\Delta\rho_{ow} = \rho_w - \rho_o = 68.14 - 46.24 = 21.90 \text{ lb/ft}^3$$

$$\Delta\rho_{og} = \rho_o - \rho_g = 46.24 - 6.12 = 40.12 \text{ lb/ft}^3$$

Step 8. Find the density differences ratio:

$$\Delta\rho_{og}/\Delta\rho_{ow} = \frac{40.12}{21.90} = 1.83$$

Step 9. From Figure 9-10, which corresponds to $r_{De} = 10$; approximate the dimensionless functions ψ_g and ψ_w:

for $\varepsilon = 0.214$ and $\delta_g = 0.357$ to give $\psi_g = 0.051$

and

for $\varepsilon = 0.214$ and $\delta_w = 0.429$ to give $\psi_w = 0.065$

Step 10. Estimate the oil critical rate by applying Equations 9-9 and 9-10:

$$Q_{ow} = 0.492 \times 10^{-4} \frac{140^2 \ (21.90)}{(1.25)\ (1.11)} [(1)\ (300)]\ 0.065 = 297 \text{ STB/day}$$

$$Q_{og} = 0.492 \times 10^{-4} \frac{140^2 \ (40.12)}{(1.25)\ (1.11)} [(1)\ (300)]\ 0.051 = 426 \text{ STB/day}$$

These calculations show that the water coning is the limiting condition for the oil-flow rate. The maximum oil rate without water or free-gas production is, therefore, 297 STB/day.

Chierici and Ciucci (1964) proposed a methodology for determining the optimum completion interval in coning problems. The method is basically based on the "trial and error" approach.

For a given dimensionless radius r_{De} and knowing GOC, WOC, and fluids density, the specific steps of the proposed methodology are summarized below:

Step 1. Assume the length of the perforated interval h_p.

Step 2. Calculate the dimensionless perforated length $\varepsilon = h_p/h$.

Step 3. Select the appropriate family of curves that corresponds to r_{De}, interpolate if necessary, and enter the working charts with ε on the x-axis and move vertically to the calculated ratio $\Delta\rho_{og}/\Delta\rho_{ow}$.

Estimate the corresponding δ and ψ. Designate these two dimensionless parameters as the optimum gas cone ratio $\delta_{g,opt}$ and optimum dimensionless function ψ_{opt}.

Step 4. Calculate the distance from GOC to the top of the perforation,

$$D_t = (h)\,(\delta_{g,opt})$$

Step 5. Calculate the distance from the WOC to the bottom of the perforation, h_w

$$D_b = h - D_t - h_p$$

Step 6. Using the optimum dimensionless function ψ_{opt} in Equation 9-9; calculate the maximum allowable oil-flow rate Q_{ow}.

Step 7. Repeat Steps 1 through 6.

Step 8. The calculated values of Q_{ow} at different assumed perforated intervals should be compared with those obtained from flow-rate equations, e.g., Darcy's equation, using the maximum drawdown pressure.

Example 9-6

Example 9-5 indicates that a vertical well is drilled in an oil reservoir that is overlaid by a gas cap and underlaid by an aquifer. Assuming that the pay thickness h is 200 feet and the rock and fluid properties are identical to those given in Example 9-5, calculate length and position of the perforated interval.

Solution

Step 1. Using the available data, calculate

$$r_{De} = \frac{745}{200}\sqrt{\frac{300}{90}} = 6.8$$

and

$$\Delta\rho_{og}/\Delta\rho_{wo} = 40.12\,/\,21.90 = 1.83$$

Step 2. Assume the length of the perforated interval is 40 feet; therefore,

$$h_p = 40'$$

$$\varepsilon = 40/200 = 0.2$$

Step 3. To obtain the values of ψ_{opt} and $\delta_{g,opt}$ for $r_{De} = 6.8$, interpolate between Figures 9-8 and 9-9 to give

$$\psi_{opt} = 0.043$$

$$\delta_{g,opt} = 0.317$$

Step 4. Calculate the distance from GOC to the top of the perforations.

$$D_t = (200)\,(0.317) = 63 \text{ ft}$$

Step 5. Determine the distance from the WOC to the bottom of the perforations.

$$D_b = 200 - 63 - 40 = 97 \text{ ft}$$

Step 6. Calculate the optimum oil-flow rate.

$$(Q_o)_{opt} = 0.492 \times 10^{-4} \,\frac{200^2\,(40.12)\,(300)\,(0.043)}{(1.25)\,(1.11)}$$
$$= 740 \text{ STB/day}$$

Step 7. Repeat Steps 2 through 6 with the results of the calculation as shown below. The oil-flow rates as calculated from appropriate flow equations are also included.

h_P	20	40	60	80	100
ε	0.1	0.2	0.3	0.4	0.5
ψ_{opt}	0.0455	0.0430	0.0388	0.0368	0.0300
$\delta_{g,opt}$	0.358	0.317	0.271	0.230	0.190
D_t	72	63	54	46	38
D_b	108	97	86	74	62
$(Q_o)_{opt}$	786	740	669	600	516
Expected Q_o	525	890	1320	1540	1850

The maximum oil production rate that can be obtained from this well without coning breakthrough is 740 STB/day. This indicates that the optimum distance from the GOC to the top of the perforations is 63 ft and the optimum distance from the WOC to the bottom of the perforations is 97 ft. The total length of the perforated interval is $200 - 63 - 97 = 40$ ft.

The Hoyland-Papatzacos-Skjaeveland Methods

Hoyland, Papatzacos, and Skjaeveland (1989) presented two methods for predicting critical oil rate for bottom water coning in anisotropic, homogeneous formations with the well completed from the top of the formation. The first method is an analytical solution, and the second is a numerical solution to the coning problem. A brief description of the methods and their applications are presented below.

The Analytical Solution Method

The authors presented an analytical solution that is based on the Muskat-Wyckoff (1953) theory. In a steady-state flow condition, the solution takes a simple form when it is combined with the method of images to give the boundary conditions as shown in Figure 9-16.

To predict the critical rate, the authors superimpose the same criteria as those of Muskat and Wyckoff on the single-phase solution and, therefore, neglect the influence of cone shape on the potential distribution. Hoyland and his coworkers presented their analytical solution in the following form:

$$Q_{oc} = 0.246 \times 10^{-4} \left[\frac{h^2 \, (\rho_w - \rho_o) \, k_h}{\mu_o \, B_o} \right] q_{CD} \tag{9-11}$$

where Q_{oc} = critical oil rate, STB/day
h = total thickness of the oil zone, ft
ρ_w, ρ_o = water and oil density, lb/ft^3
k_h = horizontal permeability, md
q_{CD} = dimensionless critical flow rate

The authors correlated the dimensionless critical rate q_{CD} with the dimensionless radius r_D and the fractional well penetration ratio h_P/h as shown in Figure 9-17.

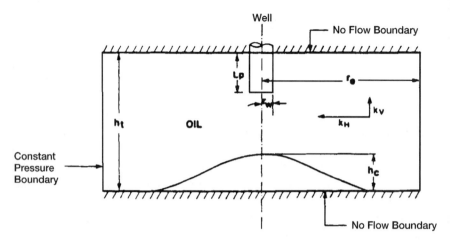

Figure 9-16. Illustration of the boundary condition for analytical solution. (*After Hoyland, A. et al., courtesy* SPE Reservoir Engineering, *November 1989.*)

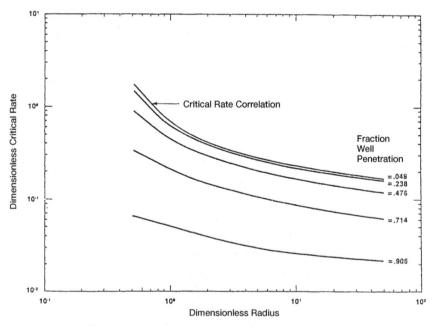

Figure 9-17. Critical rate correlation. (*After Hoyland, A. et al., courtesy* SPE Reservoir Engineering, *November 1989.*)

$$r_D = \frac{r_e}{h} \sqrt{\frac{k_v}{k_h}} \tag{9-12}$$

where r_e = drainage radius, ft
k_v = vertical permeability, md
k_h = horizontal permeability, md

The Numerical Solution Method

Based on a large number of simulation runs with more than 50 critical rate values, the authors used a regression analysis routine to develop the following relationships:

• For isotropic reservoirs with $k_h = k_v$, the following expression is proposed:

$$Q_{oc} = 0.924 \times 10^{-4} \frac{k_o (\rho_w - \rho_o)}{\mu_o B_o} \left[1 - \left(\frac{h_p}{h} \right)^2 \right]^{1.325}$$

$$\times h^{2.238} \left[\ln (r_e) \right]^{-1.99} \tag{9-13}$$

• For anisotropic reservoirs, the authors correlated the dimensionless critical rate with the dimensionless radius r_D and five different fractional well penetrations. The correlation is presented in a graphical form as shown in Figure 9-18.

The authors illustrated their methodology through the following example.

Example 9-7

Given the following data, determine the oil critical rate:

Density differences (water/oil), lbm/ft³ = 17.4
Oil FVF, RB/STB   = 1.376
Oil viscosity, cp   = 0.8257
Horizontal permeability, md   = 1,000
Vertical permeability, md   = 640
Total oil thickness, ft   = 200
Perforated thickness, ft   = 50
External radius, ft   = 500

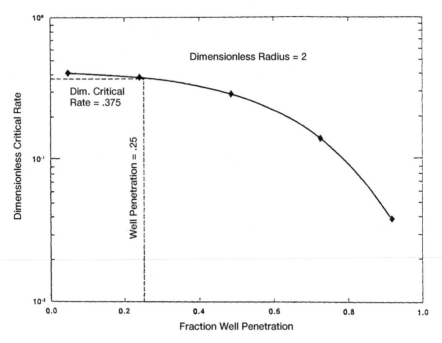

Figure 9-18. Critical rate calculation for Example 9-7. (*After Hoyland, A. et al., courtesy* SPE Reservoir Engineering, *November 1989.*)

Solution

Step 1. Calculate the dimensionless radius r_D by applying Equation 9-12.

$$r_D = \frac{r_e}{h}(k_v/k_h)^{0.5} = \frac{500}{200}(40/1000)^{0.5} = 2$$

Step 2. Determine dimensionless critical rate for several fractional well penetrations from Figure 9-17 for a dimensionless radius of 2.

Step 3. Plot dimensionless critical rate as a function of well penetration. The plot is shown in section A of Figure 9-17.

Step 4. Calculate fractional well penetration, $h_p/h = 50/200 = 0.25$.

Step 5. Interpolate in the plot in section A of Figure 9-17 to find dimensionless critical rate q_{Dc} equal to 0.375.

Step 6. Use Equation 9-11 and find the critical rate.

$$Q_{oc} = 0.246 \times 10^{-4} \left[\frac{200^2 (17.4)}{(1.376)(0.8257)} \right] (1000)(0.375)$$
$$= 5,651 \, STB/day$$

Critical Rate Curves by Chaney et al.

Chaney et al. (1956) developed a set of working curves for determining oil critical flow rate. The authors proposed a set of working graphs that were generated by using a potentiometric analyzer study and applying the water coning mathematical theory as developed by Muskat-Wyckoff (1935).

The graphs, as shown in Figures 9-19 through 9-23, were generated using the following fluid and sand characteristics:

Drainage radius $r_e = 1000$ ft
Wellbore radius $r_w = 3''$
Oil column thickness $h = 12.5, 25, 50, 75$ and 100 ft
Permeability $k = 1000$ md
Oil viscosity $\mu_o = 1$ cp
$\rho_o - \rho_w$ $= 18.72$ lb/ft^3
$\rho_o - \rho_g$ $= 37.44$ lb/ft^3

The graphs are designed to determine the critical flow rate in oil-water, gas-oil, and gas-water systems with fluid and rock properties as listed above. The hypothetical rates as determined from the Chaney et al. curves (designated as Q_{curve}), are corrected to account for the actual reservoir rock and fluid properties by applying the following expressions:

In oil-water systems

$$Q_{oc} = 0.5288 \times 10^{-4} \left[\frac{k_o (\rho_w - \rho_o)}{\mu_o B_o} \right] Q_{curve} \qquad (9-14)$$

(*text continued on page 603*)

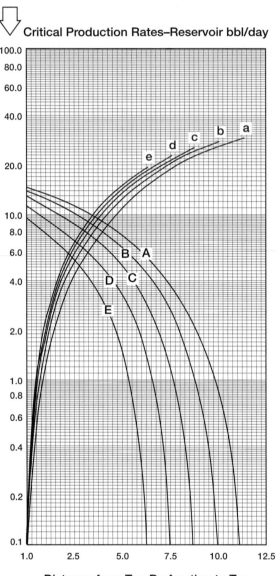

Critical Production Rates–Reservoir bbl/day

Distance from Top Perforation to Top
of Sand or Gas-Oil Contact–In Feet

Critical-production-rate curves for sand thickness of 12.5 ft., well radius of 3 in., and drainage radius of 1,000 ft. Water coning curves: A, 1.25 ft. perforated interval; B, 2.5 ft.; C, 3.75 ft.; D, 5.00 ft.; and E, 6.25 ft. Gas coning curves: a, 1.25 ft. perforated interval; b, 2.5 ft.; c, 3.75 ft.; d, 5.00 ft., and e, 6.25 ft.

Figure 9-19. Critical production rate curves. *(After Chaney et al., courtesy OGJ, May 1956.)*

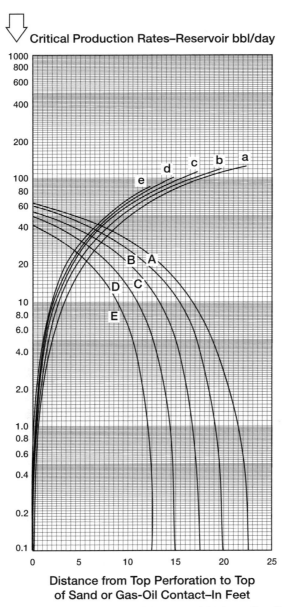

Critical Production Rates–Reservoir bbl/day

Distance from Top Perforation to Top
of Sand or Gas-Oil Contact–In Feet

Critical-production-rate curves for sand thickness of 25 ft., well radius of 3 in., and drainage radius of 1,000 ft. Water coning curves: A, 2.5 ft. perforated interval; B, 5 ft.; C, 7.5 ft.; D, 10 ft.; and E, 12.5 ft. Gas coning curves: a, 2.5 ft. perforated interval; b, 5 ft.; c, 7.5 ft.; d, 10 ft., and e, 12.5 ft.

Figure 9-20. Critical production rate curves. (*After Chaney et al., courtesy OGJ, May 1956.*)

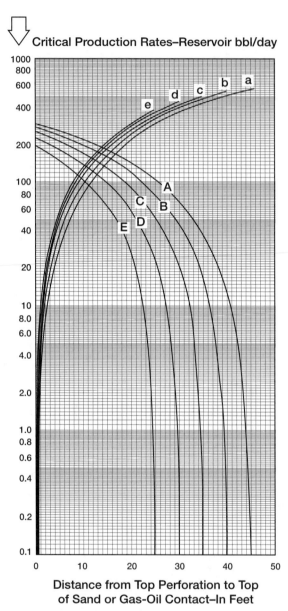

Critical Production Rates–Reservoir bbl/day

Distance from Top Perforation to Top
of Sand or Gas-Oil Contact–In Feet

Critical-production-rate curves for sand thickness of 50 ft., well radius of 3 in., and drainage radius of 1,000 ft. Water coning curves: A, 5 ft. perforated interval; B, 10 ft.; C, 15 ft.; D, 20 ft.; and E, 25 ft. Gas coning curves: a, 5 ft. perforated interval; b, 10 ft.; c, 15 ft.; d, 20 ft., and e, 25 ft.

Figure 9-21. Critical production rate curves. (*After Chaney et al., courtesy OGJ, May 1956.*)

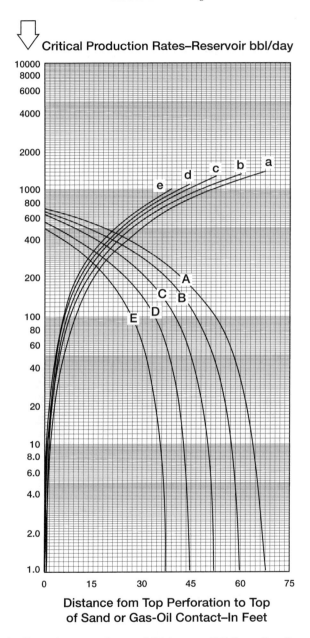

Critical Production Rates–Reservoir bbl/day

Distance fom Top Perforation to Top
of Sand or Gas-Oil Contact–In Feet

Critical-production-rate curves for sand thickness of 75 ft., well radius of 3 in., and drainage radius of 1,000 ft. Water coning curves: A, 7.5 ft. perforated interval; B, 15 ft.; C, 22.5 ft.; D, 30 ft.; and E, 37.5 ft. Gas coning curves: a, 7.5 ft. perforated interval; b, 15 ft.; c, 22.5 ft.; d, 30 ft., and e, 37.5 ft.

Figure 9-22. Critical production rate curves. (*After Chaney et al., courtesy OGJ, May 1956.*)

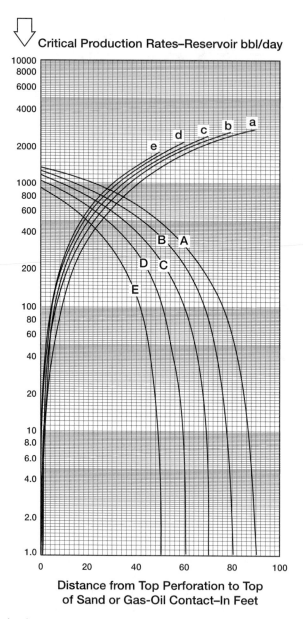

Critical-production-rate curves for sand thickness of 100 ft., well radius of 3 in., and drainage radius of 1,000 ft. Water coning curves: A, 10 ft. perforated interval; B, 20 ft.; C, 30 ft.; D, 40 ft.; and E, 50 ft. Gas coning curves: a, 10 ft. perforated interval; b, 20 ft.; c, 30 ft.; d, 40 ft., and e, 50 ft.

Figure 9-23. Critical production rate curves. (*After Chaney et al., courtesy OGJ, May 1956.*)

(*text continued from page 597*)

where ρ_o = oil density, lb/ft^3
ρ_w = water density, lb/ft^3
Q_{oc} = critical oil flow rate, STB/day
k_o = effective oil permeability, md

In gas-water systems

$$Q_{gc} = 0.5288 \times 10^{-4} \left[\frac{k_g \ (\rho_w - \rho_g)}{\mu_g \ B_g} \right] Q_{curve} \qquad (9\text{-}15)$$

where ρ_g = gas density, lb/ft^3
ρ_w = water density, lb/ft^3
Q_{gc} = critical gas flow rate, Mscf/day
B_g = gas FVF, bbl/Mscf
k_g = effective gas permeability, md

In gas-oil systems

$$Q_{oc} = 0.2676 \times 10^{-4} \left[\frac{k_o \ (\rho_o - \rho_g)}{\mu_o \ B_o} \right] Q_{curve} \qquad (9\text{-}16)$$

Example 9-8

In an oil-water system, the following fluid and sand data are available:

h = 50'	h_p = 15'
ρ_o = 47.5 lb/ft^3	ρ_w = 63.76 lb/ft^3
μ_o = 0.73 cp	B_o = 1.1 bbl/STB
r_w = 3"	r_e = 1000'
k_o = 93.5 md	

Calculate the oil critical rate.

Solution

Step 1. Distance from the top of the perforations to top of the sand = 0'

Step 2. Using Figure 9-20, for h = 50, enter the graph with 0′ and move vertically to curve C to give:

$$Q_{curve} = 270 \text{ bbl/day}$$

Step 3. Calculate critical oil rate from Equation 9-14.

$$Q_{oc} = 0.5288 \times 10^{-4} \left[\frac{93.5 \ (63.76 - 47.5)}{(1.1) \ (0.73)} \right] 270 = 27 \text{ STB/day}$$

The above method can be used through the trial-and-error procedure to optimize the location of the perforated interval in two-cone systems. It should be pointed out that Chaney's method was developed for a homogeneous, isotropic reservoir with $k_v = k_h$.

Chaperson's Method

Chaperson (1986) proposed a simple relationship to estimate the critical rate of a vertical well in an anisotropic formation ($k_v \neq k_h$). The relationship accounts for the distance between the production well and boundary. The proposed correlation has the following form:

$$Q_{oc} = 0.0783 \times 10^{-4} \frac{k_h \ (h - h_p)^2}{\mu_o \ B_o} [\Delta \rho] \ q_c^* \qquad (9\text{-}17)$$

where Q_{oc} = critical oil rate, STB/day
  k_h = horizontal permeability, md
  $\Delta \rho = \rho_w - \rho_o$, density difference, lb/ft^3
  h = oil column thickness, ft
  h_p = perforated interval, ft

Joshi (1991) correlated the coefficient q_c^* with the parameter α'' as

$$q_c^* = 0.7311 + (1.943/\alpha'') \qquad (9\text{-}18)$$

$$\alpha'' = (r_e/h) \sqrt{k_v / k_h} \qquad (9\text{-}19)$$

Example 9-9

The following data are available on an oil-water system:

$h = 50'$	$r_e = 1000'$	$\mu_o = 0.73$ cp
$B_o = 1.1$ bbl/STB	$\rho_w = 63.76$ lb/ft^3	$k_h = 100$ md
$\rho_o = 47.5$ lb/ft^3	$k_v = 10$ md	$h_P = 15'$

Calculate the critical rate.

Solution

Step 1. Calculate α'' from Equation 9-19.

$$\alpha'' = (1000/50) \sqrt{10/100} = 6.324$$

Step 2. Solve for q_c^* by applying Equation 9-18.

$$q_c^* = 0.7311 + (1.943/6.324) = 1.0383$$

Step 3. Solve for the critical oil rate Q_{oc} by using Equation 9-17.

$$Q_{oc} = 0.0783 \times 10^{-4} \frac{(100)\ (50-15)^2}{(0.73)\ (1.1)} [63.76 - 47.5]\ (1.0383)$$

$$Q_{oc} = 20.16 \text{ STB/day}$$

Schols' Method

Schols (1972) developed an empirical equation based on results obtained from numerical simulator and laboratory experiments. His critical rate equation has the following form:

$$Q_{oc} = 0.0783 \times 10^{-4} \left[\frac{(\rho_w - \rho_o)\ k_o\ (h^2 - h_p^2)}{u_o\ B_o} \right]$$

$$\times \left[0.432 + \frac{3.142}{\ln (r_e/r_w)} \right] (h/r_e)^{0.14} \qquad (9\text{-}20)$$

where k_o = effective oil permeability, md
  r_w = wellbore radius, ft
  h_p = perforated interval, ft
  ρ = density, lb/ft³

Schols' equation is only valid for isotropic formation, i.e., $k_h = k_v$.

Example 9-10

In an oil-water system, the following fluid and rock data are available:

$h = 50'$ $h_p = 15'$ $\rho_o = 47.5$ lb/ft³ $\rho_w = 63.76$ lb/ft³
$\mu_o = 0.73$ cp $B_o = 1.1$ bbl/STB $r_e = 1000'$ $r_w = 0.25'$
$k_o = k = 93.5$ md

Calculate the critical oil flow rate.

Solution

Applying Equation 9-20, gives

$$Q_{oc} = 0.0783 \times 10^{-4} \left[\frac{(63.76 - 47.5)\,(93.5)\,(50^2 - 15^2)}{(0.73)\,(1.1)} \right]$$

$$\times \left[0.432 + \frac{3.142}{\ln(1000/.25)} \right] (50/1000)^{0.14}$$

$$Q_{oc} = 18 \text{ STB/day}$$

BREAKTHROUGH TIME IN VERTICAL WELLS

Critical flow rate calculations frequently show low rates that, for economic reasons, cannot be imposed on production wells. Therefore, if a well produces above its critical rate, the cone will break through after a given time period. This time is called *time to breakthrough* t_{BT}. Two of the most widely used correlations are documented below.

The Sobocinski-Cornelius Method

Sobocinski and Cornelius (1965) developed a correlation for predicting water breakthrough time based on laboratory data and modeling results. The authors correlated the breakthrough time with two dimensionless parameters, the dimensionless cone height and the dimensionless breakthrough time. Those two dimensionless parameters are defined by the following expressions:

Dimensionless cone height Z

$$Z = 0.492 \times 10^{-4} \frac{(\rho_w - \rho_o)\,k_h\,h\,(h - h_p)}{\mu_o\,B_o\,Q_o} \tag{9-21}$$

where ρ = density, lb/ft^3
$\quad\quad k_h$ = horizontal permeability, md
$\quad\quad Q_o$ = oil production rate, STB/day
$\quad\quad h_p$ = perforated interval, ft
$\quad\quad h$ = oil column thickness, ft

Dimensionless breakthrough time $(t_D)_{BT}$

$$(t_D)_{BT} = \frac{4Z + 1.75\,Z^2 - 0.75\,Z^3}{7 - 2\,Z} \tag{9-22}$$

The authors proposed the following expression for predicting time to breakthrough from the calculated value of the dimensionless breakthrough time $(t_D)_{BT}$:

$$t_{BT} = \frac{20{,}325\,\mu_o\,h\,\phi\,(t_D)_{BT}}{(\rho_w - \rho_o)\,k_v\,(1 + M^\alpha)} \tag{9-23}$$

where t_{BT} = time to breakthrough, days
$\quad\quad \phi$ = porosity, fraction
$\quad\quad k_v$ = vertical permeability, md
$\quad\quad M$ = water-oil mobility and is defined by:

$$M = \left[\frac{(k_{rw})_{sor}}{(k_{ro})_{swc}}\right]\left(\frac{\mu_o}{\mu_w}\right) \tag{9-24}$$

with $\quad (k_{ro})_{swc}$ = oil relative permeability at connate water saturation
$\quad\quad\quad (k_{rw})_{sor}$ = water relative permeability at residual oil saturation
$\quad\quad\quad\quad \alpha = 0.5$ for M ″ 1
$\quad\quad\quad\quad \alpha = 0.6$ for $1 < M$ ″ 10

Joshi (1991) observed by examining Equation 9-22 that if $Z = 3.5$ or greater, there will be no water breakthrough. This observation can be imposed on Equation 9-21 with Z = 3.5 to give an expression for calculating the critical oil flow rate, or

$$Q_{oc} = 0.141 \times 10^{-4} \frac{(\rho_w - \rho_o) \, k_h \, h \, (h - h_p)}{\mu_o \, B_o} \qquad (9\text{-}25)$$

Example 9-11

Calculate the water breakthrough using the Sobocinski-Cornelius method for a vertical well producing at 250 STB/day. The following reservoir data are available:

$Q_o = 250$ STB/day $h = 50$ ft $hp = 15$ ft $\rho_w = 63.76$ lb/ft^3
$\rho_o = 47.5$ lb/ft^3 $\mu_o = 0.73$ cp $B_o = 1.1$ bbl/STB $k_v = 9$ md
$k_h = 93$ md $\phi = 13\%$ $M = 3$

Solution

Step 1. Solve for the dimensionless cone height Z from Equation 9-21 to give

$$Z = 0.492 \times 10^{-4} \left[\frac{(63.76 - 47.5) \, (93) \, (50) \, (50 - 15)}{(0.73) \, (1.1) \, (250)} \right] = 0.6486$$

Step 2. Calculate the dimensionless breakthrough time by using Equation 9-22.

$$(t_D)_{BT} = \frac{(4) \, (0.64866) + 1.75 \, (0.6486)^2 - 0.75 \, (0.6486)^3}{7 - 2 \, (0.6486)}$$

$$= 0.5481$$

Step 3. Estimate time to breakthrough from Equation 9-23.

$$t_{BT} = \frac{20,325 \, (0.73) \, (0.13) \, (50) \, (0.5481)}{(63.76 - 47.5) \, (9) \, (1 + 3^{.6})} = 123 \text{ days}$$

Example 9-12

Using the data given in Example 9-11, approximate the critical oil flow rate by using Equation 9-25.

Solution

$$Q_{oc} = 0.141 \times 10^{-4} \frac{(63.76 - 47.5)\,(93)\,(50)\,(50 - 15)}{(0.73)\,(1.1)}$$

$$= 46.3 \text{ STB/day}$$

The Bournazel-Jeanson Method

Based on experimental data, Bournazel and Jeanson (1971) developed a methodology that uses the same dimensionless groups proposed in the Sobocinski-Cornelius method. The procedure of calculating the time to breakthrough is given below.

Step 1. Calculate the dimensionless core height Z from Equation 9-21.

Step 2. Calculate the dimensionless breakthrough time by applying the following expression:

$$(t_D)_{BT} = \frac{Z}{3 - 0.7\,Z} \qquad (9\text{-}26)$$

Step 3. Solve for the time to breakthrough t_{BT} by substituting the above-calculated dimensionless breakthrough time into Equation 9-23, i.e.,

$$t_{BT} = \frac{20{,}325\,\mu_o\,h\,\phi\,(t_D)_{BT}}{(\rho_w - \rho_o)\,k_v\,(1 + M^\alpha)}$$

As pointed out by Joshi (1991), Equation 9-26 indicates that no breakthrough occurs if $Z \geq 4.286$. Imposing this value on Equation 9-21 gives a relationship for determining Q_{oc}.

$$Q_{oc} = 0.1148 \times 10^{-4} \frac{(\rho_w - \rho_o)\,k_h\,h\,(h - h_p)}{\mu_o\,B_o} \qquad (9\text{-}27)$$

Example 9-13

Resolve Example 9-11 by using the Bournazel-Jeanson method.

Step 1. Solve for the dimensionless cone height Z = 0.6486

Step 2. Calculate the dimensionless breakthrough time from Equation 9-26.

$$(t_D)_{BT} = \frac{0.6486}{3 - 0.7\,(.6486)} = 0.2548$$

Step 3. Calculate the time to breakthrough by apply Equation 9-23 to give

$$t_{BT} = \frac{20,325\,(0.73)\,(0.13)\,(50)\,(0.2548)}{(63.76 - 47.5)\,(9)\,(1 + 3^{.6})} = 57.2 \text{ days}$$

Step 4. From Equation 9-27, the critical oil rate is

$$Q_{oc} = 37.8 \text{ STB/day}$$

AFTER BREAKTHROUGH PERFORMANCE

Once the water breakthrough occurs, it is important to predict the performance of water production as a function of time. Normally, using numerical radial models solves such a problem. Currently, no simple analytical solution exists to predict the performance of the vertical well after breakthrough. Kuo and Desbrisay (1983) applied the material balance equation to predict the rise in the oil-water contact in a homogeneous reservoir and correlated their numerical results in terms of the following dimensionless parameters:

- Dimensionless water cut $(f_w)_D$.
- Dimensionless breakthrough time t_{DBT}
- Dimensionless limiting water cut $(WC)_{limit}$

The specific steps of the proposed procedure are given below.

Step 1. Calculate the time to breakthrough t_{BT} by using the Sobocinski-Cornelius method or the Bournazel-Jeanson correlation.

Step 2. Assume any time t after breakthrough.

Step 3. Calculate the dimensionless breakthrough time ratio t_{DBT} from:

$$t_{DBT} = t/t_{BT} \tag{9-28}$$

Step 4. Compute the dimensionless limiting water cut from:

$$(WC)_{limit} = \frac{M}{M + (h/h_w)} \tag{9-29}$$

With the parameters in Equation 9-29 as defined below:

$$M = \left[\frac{(k_{rw})_{sor}}{(k_{ro})_{swc}}\right] \frac{\mu_o}{\mu_w} \tag{9-30}$$

$$h = H_o (1 - R) \tag{9-31}$$

$$h_w = H_w + H_o R \tag{9-32}$$

$$R = (N_p / N) \left[\frac{1 - S_{wc}}{1 - S_{or} - S_{wc}}\right] \tag{9-33}$$

where $(WC)_{limit}$ = current limiting value for water cut
$\qquad M$ = mobility ratio
$\qquad (k_{rw})_{sor}$ = relative permeability for the water and residual oil saturation (S_{or})
$\qquad (k_{ro})_{swc}$ = relative permeability for the oil at the connate water saturation (S_{wc})
$\qquad \mu_o, \mu_w$ = oil and water viscosities, cp
$\qquad H_o$ = initial oil zone thickness, ft
$\qquad H_w$ = initial water zone thickness, ft
$\qquad h$ = current oil zone thickness, ft
$\qquad h_w$ = current water zone thickness, ft
$\qquad N_p$ = cumulative oil production, STB
$\qquad N$ = initial oil in place, STB

Step 5. Calculate the dimensionless water cut $(f_w)_D$ based upon the dimensionless breakthrough time ratio as given by the following relationships:

$$(f_w)_D = 0 \quad \text{for } t_{DBT} < 0.5 \tag{9-34}$$

$$(f_w)_D = 0.29 + 0.94 \log(t_{DBT}) \quad \text{for } 0.5 \text{ '' } t_{DBT} \text{ '' } 5.7 \qquad (9\text{-}35)$$

$$(f_w)_D = 1.0 \quad \text{for } t_{DBT} > 5.7 \qquad (9\text{-}36)$$

Step 6. Calculate the actual water cut f_w from the expression:

$$f_w = (f_w)_D \, (WC)_{limit} \qquad (9\text{-}37)$$

Step 7. Calculate water and oil flow rate by using the following expressions:

$$Q_w = (f_w) \, Q_T \qquad (9\text{-}38)$$

$$Q_o = Q_T - Q_w \qquad (9\text{-}39)$$

where Q_w, Q_o, Q_T are the water, oil, and total flow rates, respectively.

It should be pointed out that as oil is recovered, the oil-water contact will rise and the limiting value for water cut will change. It also should be noted the limiting water cut value $(WC)_{limit}$ lags behind one time step when calculating future water cut.

Example 9-14

The rock, fluid, and the related reservoir properties of a bottom-water drive reservoir are given below:

well spacing = 80 acres
initial oil column thickness = 80 ft

$h_p = 20'$	$\rho_o = 47 \text{ lb/ft}^3$	$\rho_w = 63 \text{ lb/ft}^3$	$r_e = 1053'$
$r_w = 0.25'$	$M = 3.1$	$\phi = 14\%$	$S_{or} = 0.35$
$S_{wc} = 0.25$	$B_o = 1.2 \text{ bbl/STB}$	$\mu_o = 1.6 \text{ cp}$	$\mu_w = 0.82 \text{ cp}$
$k_h = 60 \text{ md}$	$k_v = 6 \text{ md}$		

Calculate the water cut behavior of a vertical well in the reservoir assuming a total production rate of 500, 1000, and 1500 STB/day.

Solution

Step 1. Calculate the dimensionless cone height Z by using Equation 9-21.

$$Z = 0.492 \times 10^{-4} \frac{(63-47)\,(60)\,(80)\,(80-20)}{(1.6)\,(1.2)\,Q_o}$$

Q_o	500	1000	1500
Z	0.2362	0.1181	0.0787

Step 2. Calculate the dimensionless breakthrough time by applying Equation 9-26.

Q	500	1000	1500
Z	0.2362	0.1181	0.0787
$(t_D)_{BT}$	0.08333	0.04048	0.02672

Step 3. Calculate the time to breakthrough from Equation 9-23.

$$t_{BT} = \left[\frac{(20,325)\,(1.6)\,(0.14)\,(80)}{(63-47)\,(6)\,(1+3.1)^{.6}} \right] (t_D)_{BT}$$

$$t_{BT} = 1276.76\,(t_D)_{BT}$$

Q	500	1000	1500
$(t_D)_{BT}$	0.08333	0.04048	0.02672
t_{BT}	106.40	51.58	34.11

Step 4. Calculate initial oil in place N.

$$N = 7758\,A\phi\,h\,(1 - S_{wi})/B_o$$

$$N = 7758\,(80)\,(0.14)\,(80)\,(1 - 0.25)\,/\,1.2 = 4,344,480 \text{ STB}$$

Step 5. Calculate the parameter R by applying Equation 9-33.

$$R = [N_p/(4,344,480)] \frac{1-0.25}{1-0.35-0.25} = 4.3158 \times 10^{-7} N_p$$

Step 6. Calculate the limiting water cut at breakthrough.

Q_o	500	1000	1500
t_{BT}	106.4	51.58	34.11
N_p	53,200	51,580	51,165
R	0.02296	0.022261	0.022082
h	78.16	78.22	78.23
h_w	21.84	21.78	21.77
$(WC)_{limit}$	0.464	0.463	0.463

Step 7. The water cut calculations after an assumed elapsed time of 120 days at a fixed total flow rate of 500 STB/days are given below:

• From Equation 9-28, calculate t_{DBT}

$$t_{DBT} = 120/106.4 = 1.1278$$

• Apply Equation 9-36 to find $(f_w)_D$:

$$(f_w)_D = 0.29 + 0.96 \log (1.1278) = 0.3391$$

• Solve for the present water cut from Equation 9-37:

$$f_w = (0.3391)(0.464) = 0.1573$$

Step 8. Calculate water and oil flow rate:

$$Q_w = (0.1573)(500) = 78.65 \text{ STB/day}$$

$$Q_o = 500 - 78.65 = 421.35 \text{ STB/day}$$

Step 9. Calculate cumulative oil produced from breakthrough to 120 days:

$$\Delta N_p = \left[\frac{500 + 421.35}{2} \right] (120 - 106.4) = 6,265.18 \text{ STB}$$

Step 10. Calculate cumulative oil produced after 120 days:

$$N_p = 53,200 + 6,265.18 = 59,465.18 \text{ STB}$$

Step 11. Find the recovery factory (RF):

$$RF = 59,465.18/4,344,480 = 0.0137$$

Step 12. Assume an elapsed time of 135 days, repeat the above steps at the same total rate of 500 STB/day:

- $R = 4.3158 \times 10^{-7} (59,465.18) = 0.020715$
- $h_w = 21.66$
- $h = 78.34$
- $(W_c)_{limit} = 0.4615$
- $(f_w)_D = 0.29 + 0.94 \log (135/106.4) = 0.3872$
- $f_w = (0.3872) (0.4615) = 0.1787$
- $Q_w = (500) (0.1787) = 89.34$ STB/day
  $Q_o = 500 - 89.34 = 410.66$ STB/day

$$\Delta N_p = \left[\frac{410.66 + 421.34}{2} \right] (135 - 120) = 6,240.0 \text{ STB}$$

- $N_p = 59,465.18 + 6,240.0 = 65,705.22$
- $RF = 0.0151$

Tables 9-1 through 9-3 summarize the calculations for water cut versus time for total flow rates of 500, 100, and 1500 STB/day, respectively.

CONING IN HORIZONTAL WELLS

The applications of horizontal well technology in developing hydrocarbon reservoirs have been widely used in recent years. One of the main objectives of using this technology is to improve hydrocarbon recovery from water and/or gas-cap drive reservoirs. The advantages of using a horizontal well over a conventional vertical well are their larger capacity to produce oil at the same drawdown and a longer breakthrough time at a given production rate.

Many correlations to predict coning behavior in horizontal wells are available in the literature. Joshi (1991) provides a detailed treatment of the coning problem in horizontal wells. As in vertical wells, the coning problem in horizontal wells involves the following calculations:

- Determination of the critical flow rate
- Breakthrough time predictions
- Well performance calculations after breakthrough

Table 9-1
Results of Example 9-14
Total Production Rate is 500 STB/day

Time days	Oil Rate STB/day	Water Rate STB/day	Water Cut Fraction	Cum. Oil MSTB	Oil Rec. %
120.	406.5	93.5	0.187	59.999	1.38
135.	379.9	120.1	0.240	65.897	1.52
150.	355.8	144.2	0.288	71.415	1.64
165.	333.8	166.2	0.332	76.587	1.76
180.	313.5	186.5	0.373	81.442	1.87
195.	294.5	205.5	0.411	86.002	1.98
210.	276.9	223.1	0.446	90.287	2.08
765.	239.3	260.7	0.521	177.329	4.08
1020.	226.4	273.6	0.547	236.676	5.45
1035.	225.6	274.4	0.549	240.066	5.53
1575.	202.4	297.6	0.595	355.425	8.18
2145.	182.4	317.6	0.635	464.927	10.70
2415.	174.2	325.8	0.652	513.062	11.81
2430.	173.8	326.2	0.652	515.672	11.87
2445.	173.4	326.6	0.653	518.276	11.93
3300.	151.5	348.5	0.697	656.768	15.12
3615.	144.6	355.4	0.711	703.397	16.19
3630.	144.3	355.7	0.711	705.564	16.24
3645.	144.0	356.0	0.712	707.727	16.29

Horizontal Well Critical Rate Correlations

The following four correlations for estimating critical flow rate in horizontal wells are discussed:

• Chaperson's Method
• Efros' Method
• Karcher's Method
• Joshi's Method

Chaperson's Method

Chaperson (1986) provides a simple and practical estimate or the critical rate under steady-state or pseudosteady-state flowing conditions for an isotropic formation. The author proposes the following two relationships for predicting water and gas coning:

Table 9-2
Results of Example 9-14
Total Production Rate is 1000 STB/day

Time days	Oil Rate STB/day	Water Rate STB/day	Water Cut Fraction	Cum. Oil MSTB	Oil Rec. %
80.	674.7	325.3	0.325	64.14	1.48
95.	594.2	405.8	0.406	73.66	1.70
110.	524.8	475.2	0.475	82.05	1.89
125.	463.2	536.8	0.537	89.46	2.06
140.	407.8	592.2	0.592	95.99	2.21
335.	494.8	505.2	0.505	145.68	3.35
905.	390.8	609.2	0.609	395.80	9.11
1115.	362.4	637.6	0.638	474.81	10.93
1130.	360.5	639.5	0.639	480.23	11.05
1475.	321.8	678.2	0.678	597.70	13.76
1835.	288.8	711.2	0.711	707.42	16.28
1850.	287.5	712.5	0.712	711.74	16.38
1865.	286.3	713.7	0.714	716.04	16.48
1880.	285.1	714.9	0.715	720.33	16.58
1895.	283.8	716.2	0.716	724.59	16.68
2615.	234.4	765.6	0.766	910.20	20.95
2630.	233.5	766.5	0.766	913.71	21.03
3065.	210.4	789.6	0.790	1010.11	23.25
3080.	209.6	790.4	0.790	1013.26	23.32
3620.	185.8	814.2	0.814	1119.83	25.78
3635.	185.2	814.8	0.815	1122.61	25.84
3650.	184.6	815.4	0.815	1125.39	25.90

Water coning

$$Q_{oc} = 0.0783 - 10^{-4} \left(\frac{L q_c^*}{y_e} \right) (\rho_w - \rho_o) \frac{k_h \left[h - (h - D_b) \right]^2}{\mu_o B_o} \qquad (9\text{-}40)$$

Gas coning

$$Q_{oc} = 0.0783 - 10^{-4} \left(\frac{L q_c^*}{y_e} \right) (\rho_o - \rho_g) \frac{k_h \left[h - (h - D_t) \right]^2}{\mu_o B_o} \qquad (9\text{-}41)$$

The above two equations are applicable under the following constraint:

$$1 \leq \alpha'' < 70 \quad \text{and} \quad 2y_e < 4L \qquad (9\text{-}42)$$

Table 9-3
Results of Example 9-14
Total Production Rate is 1500 STB/day

Time days	Oil Rate STB/day	Water Rate STB/day	Water Cut Fraction	Cum. Oil MSTB	Oil Rec. %
80.	742.1	757.9	0.505	67.98	1.56
260.	734.6	765.4	0.510	160.34	3.69
275.	727.1	772.9	0.515	171.31	3.94
290.	719.6	780.4	0.520	182.16	4.19
770.	541.8	958.2	0.639	481.00	11.07
785.	537.6	962.4	0.642	489.10	11.26
800.	533.5	966.5	0.644	497.13	11.44
1295.	423.5	1076.5	0.718	732.08	16.85
1310.	420.8	1079.2	0.719	738.42	17.00
1325.	418.1	1081.9	0.721	744.71	17.14
2060.	315.7	1184.3	0.790	1011.42	23.28
2075.	314.0	1186.0	0.791	1016.14	23.39
2090.	312.4	1187.6	0.792	1020.84	23.50
2105.	310.8	1189.2	0.793	1025.51	23.60
2120.	309.2	1190.8	0.794	1030.16	23.71
2135.	307.6	1192.4	0.795	1034.79	23.82
2705.	255.5	1244.5	0.830	1194.54	27.50
3545.	200.1	1299.9	0.867	1384.48	31.87
3650.	194.4	1305.6	0.870	1405.19	32.34

where

$$\alpha'' = \left(\frac{y_e}{h}\right)\sqrt{\frac{k_v}{k_h}} \qquad (9\text{-}43)$$

D_b = distance between the WOC and the horizontal well
D_t = distance between the GOC and the horizontal well
Q_{oc} = critical oil rate, STB/day
ρ = density, lb/ft^3
k_h = horizontal permeability, md
h = oil column thickness, ft
y_e = half distance between two lines of horizontal wells
     (half drainage length perpendicular to the horizontal well)
L = length of the horizontal well
q_c^* = dimensionless function

Joshi (1991) correlated the dimensionless function F with the parameter α'':

$$q_c^* = 3.9624955 + 0.0616438\alpha'' - 0.000504(\alpha'')^2 \qquad (9\text{-}44)$$

Example 9-15

A 1,640-ft.-long horizontal well is drilled in the top elevation of the pay zone in a water-drive reservoir. The following data are available:

$h = 50$ ft	$k_h = 60$ md	$k_v = 15$ md	$B_o = 1.1$ bbL/STB
$\mu_o = 0.73$ cp	$r_w = 0.3$ ft	$D_b = 50$ ft	$\rho_o = 47.5$ lb/ft^3
$\rho_w = 63.76$ lb/ft^3	$y_e = 1320$ ft		

Using the Chaperson method, calculate:

a. The oil critical flow rate for the horizontal well.
b. Repeat the calculation assuming a vertical well with $h_p = 15'$ and $r_e = 1489$ ft.

Solution

Critical rate for a horizontal well:

Step 1. Solve for α'' by applying Equation 9-43.

$$\alpha'' = \frac{1320}{50}\sqrt{\frac{15}{60}} = 13.20$$

Step 2. Solve for the dimensionless function q_c^* by applying Equation 9-44.

$$q_c^* = 4.6821$$

Step 3. Calculate the critical rate from Equation 9-41.

$$Q_{oc} = 0.0783 \times 10^{-4} \left(\frac{1640 \times 4.6821}{1320}\right)(63.76 - 47.5)\left(\frac{60 \times 50^2}{.73 \times 1.1}\right)$$

$$= 138.4 \text{ STB/day}$$

Critical rate for a vertical well:

Step 1. Solve for α'' by using Equation 9-19.

$$\alpha'' = 14.89$$

Step 2. Solve for q_c^* by applying Equation 9-18.

$$q_c^* = 0.8616$$

Step 3. Calculate the critical rate for the vertical well from Equation 9-17.

$$Q_{oc} = 0.0783 \times 10^{-4} \, \frac{60 \times (50-15)^2}{(0.73) \, (1.1)} \, (63.76 - 47.5) \, (0.8616)$$
$$= 10 \text{ STB/day}$$

The ratio of the two critical oil rates is

$$\text{Rate ratio} = \frac{138.4}{10} \cong 14$$

This rate ratio clearly shows the *critical rate improvement* in the case of the horizontal well over that of the vertical well.

Efros' Method

Efros (1963) proposed a critical flow rate correlation that is based on the assumption that the critical rate is nearly independent of drainage radius. The correlation does not account for the effect of the vertical permeability. Efros developed the following two relationships that are designed to calculate the critical rate in oil-water and gas-oil systems:

Water coning

$$Q_{oc} = 0.0783 \times 10^{-4} \, \frac{k_h \, (\rho_w - \rho_o) \, [h - (h - D_b)]^2 L}{\mu_o B_o \left[y_e + \sqrt{y_e^2 + (h^2/3)} \right]} \qquad (9\text{-}45)$$

Gas coning

$$Q_{oc} = 0.0783 \times 10^{-4} \frac{k_h \, (\rho_o - \rho_g) \, [h - (h - D_t)]^2 \, L}{\mu_o B_o \left[y_e + \sqrt{y_e^2 + (h^2 / 3)} \right]} \qquad (9\text{-}46)$$

where L = length of the horizontal well, ft
$\quad\;\; y_e$ = half distance between two lines of horizontal wells
$\quad\;\; \rho$ = density, lb/ft^3
$\quad\;\; h$ = net pay thickness
$\quad\;\; k$ = permeability, md

Example 9-16

Using the horizontal well data given in Example 9-15, solve for the horizontal well critical flow rate by using Efros' correlation.

Solution

Step 1. Calculate the critical oil flow rate by applying Equation 9-45 to give

$$Q_{oc} = 0.0783 \times 10^{-4} \frac{60 \, (63.76 - 47.5) \, 50^2 \, (1640)}{(1.1) \, (0.73) \left[1320 + \sqrt{1320^2 + \dfrac{50^2}{3}} \right]}$$

$$\cong 15 \; STB/day$$

Karcher's Method

Karcher (1986) proposed a correlation that produces a critical oil flow rate value similar to that of Efros' equation. Again, the correlation does not account for the vertical permeability.

Water coning

$$Q_{oc} = 0.0783 \times 10^{-4} \frac{k_h (\rho_w - \rho_o) \, (h - B)^2 \, L}{\mu_o \, B_o \, (2 \, y_e)}$$

$$\times \left[1 - \left(\frac{h - B}{y_e} \right)^2 (1 / 24) \right] \qquad (9\text{-}47)$$

where $B = h - D_b$
D_b = distance between WOC and horizontal well, ft

Gas coning

$$Q_{oc} = 0.0783 \times 10^{-4} \frac{k_h (\rho_o - \rho_g) (h - T)^2 \, L}{\mu_o \, \beta_o \, (2 \, y_e)}$$

$$\times \left[1 - \left(\frac{h - T}{y_e} \right)^2 (1/24) \right] \tag{9-48}$$

where $T = h - D_t$
D_t = distance between GOC and horizontal well, ft

Example 9-17

Resolve example by using Karcher's method.

Solution

$$Q_{oc} = 0.0783 \times 10^{-4} \frac{60 \, (63.76 - 47.5) \, (50)^2 \, 1640}{(1.1) \, (0.73) \, (2 \times 1320)}$$

$$\times \left[1 - \left(\frac{50}{1320} \right)^2 (1/24) \right] \cong 15 \text{ STB/day}$$

Joshi's Method

Joshi (1988) suggests the following relationships for determining the critical oil flow rate in horizontal wells by defining the following parameters:

• Horizontal well drainage radius r_{eh}

$$r_{eh} = \sqrt{\frac{43,560 \, A}{\pi}}$$

where A is the horizontal well drainage area in acres.

- Half the major axis of drainage ellipse a

$$a = (L/2) \left[0.5 + \sqrt{0.25 + (2r_{eh}/L)^4} \right]^{0.5} \tag{9-49}$$

- Effective wellbore radius r_w'

$$r_w' = \frac{r_{eh} \left[\dfrac{L}{2a} \right]}{\left[1 + \sqrt{1 - [L/(2a)]^2} \right] [h/(2r_w)]^{h/L}} \tag{9-50}$$

For oil-water systems:

$$Q_{oc} = 0.0246 \times 10^{-3} \frac{(\rho_w - \rho_o) \, k_h \, [h^2 - (h - D_b)^2]}{\mu_o \, B_o \, \ln{(r_{eh}/r_w')}} \tag{9-51}$$

For oil-gas systems:

$$Q_{oc} = 0.0246 \times 10^{-3} \frac{(\rho_o - \rho_g) \, k_h \, [h^2 - (h - D_t)^2]}{\mu_o \, B_o \, \ln{(r_{eh}/r_w')}} \tag{9-52}$$

where ρ = density, lb/ft^3
  k_h = horizontal density, md
  D_b = distance between the horizontal well and the WOC, ft
  D_t = distance between the horizontal well and GOC, ft
  r_w = wellbore radius, ft

Example 9-18

Resolve Example 9-17 by applying Joshi's approach.

Solution

Step 1. Solve for a by applying Equation 9-49

$$a = (1640/2) \left[0.5 \sqrt{0.25 + (2 \times 1489/1640)^4} \right]^{0.5} = 1,606 \text{ ft}$$

Step 2. Calculate r'_w from Equation 9-50.

$$r'_w = \frac{1489 \left[\dfrac{1640}{2(1606)} \right]}{\left[1 + \sqrt{1 - [1640/(2 \times 1606)]^2} \right] [50/(2 \times 0.3)]^{50/1640}} = 357 \text{ ft}$$

Step 3. Estimate the critical flow rate from Equation 9-51.

$$Q_{oc} = 0.0246 \times 10^{-3} \frac{(63.76 - 47.5)\,(60)\,[50^2 - (50 - 50)^2]}{(0.73)\,(1.1)\,\ln\,(1489/357)}$$
$$= 52 \text{ STB/day}$$

HORIZONTAL WELL BREAKTHROUGH TIME

Several authors have proposed mathematical expressions for determining the time to breakthrough in horizontal wells. The following two methodologies are presented in the following sections:

• The Ozkan-Raghavan Method
• Papatzacos' Method

The Ozkan-Raghavan Method

Ozkan and Raghavan (1988) proposed a theoretical correlation for calculating time to breakthrough in a bottom-water-drive reservoir. The authors introduced the following dimensionless parameters:

$$L_D = \text{dimensionless well length} = [L/(2h)] \sqrt{k_v/k_h} \qquad (9\text{-}53)$$

$$z_{WD} = \text{dimensionless vertical distance} = D_b/h \qquad (9\text{-}54)$$

where   L = well length, ft
D_b = distance between WOC and horizontal well
H = formation thickness, ft
k_v = vertical permeability, md
k_h = horizontal permeability, md

Ozkan and Raghavan expressed the water breakthrough time by the following equation:

$$t_{BT} = \left[\frac{f_d \, h^3 \, E_s}{5.615 \, Q_o \, B_o} \right] (k_h / k_v) \qquad (9-55)$$

with the parameter f_d as defined by:

$$f_d = \phi \, (1 - S_{wc} - S_{or}) \qquad (9-56)$$

where t_{BT} = time to breakthrough, days
$\quad k_v$ = vertical permeability, md
$\quad k_h$ = horizontal permeability, md
$\quad \phi$ = porosity, fraction
$\quad S_{wc}$ = connate water saturation, fraction
$\quad S_{or}$ = residual oil saturation, fraction
$\quad Q_o$ = oil flow rate, STB/day
$\quad E_s$ = sweep efficiency, dimensionless

Ozkan and Raghavan graphically correlated the sweep efficiency with the dimensionless well length L_D and dimensionless vertical distance Z_{WD} as shown in Figure 9-24.

Example 9-19

A 1,640-foot-long horizontal well is drilled in a bottom-water-drive reservoir. The following data are available:

$$
\begin{array}{llll}
h = 50 \text{ ft} & k_h = 60 \text{ md} & k_v = 15 \text{ md} & B_o = 1.1 \text{ bbl/STB} \\
\mu_o = 0.73 \text{ cp} & r_w = 0.3 \text{ ft} & \rho_o = 47.5 \text{ lb/ft}^3 & \rho_w = 63.76 \text{ lb/ft}^3 \\
Z_{WD} = 1 & \phi = 15\% & S_{wc} = 0.25 & S_{or} = 0.3
\end{array}
$$

The well is producing at 1000 STB/day. Calculate time to breakthrough.

Solution

Step 1. Solve for L_D by using Equation 9-53.

$$L_D = \left[\frac{1640}{2(50)} \right] \sqrt{15/60} = 8.2$$

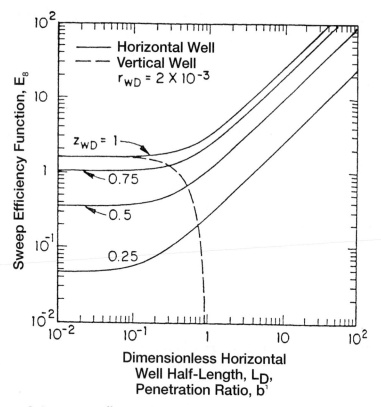

Figure 9-24. Sweep efficiency for horizontal and vertical wells. (*After Ozkan, E., and Raghavan, R., courtesy SPE, 1988.*)

Step 2. Calculate the parameter f_d from Equation 9-56.

$$f_d = 0.15 \, (1 - 0.25 - 0.30) = 0.0675$$

Step 3. Estimate the sweep efficiency E_s from Figure 9-24.

$$E_s \cong 21$$

Step 4. Solve for time to breakthrough by applying Equation 9-55.

$$t_{BT} = \left[\frac{0.0675 \, (50)^3 \, 21}{5.615 \times 1000 \times 1.1} \right] (60 / 15) = 114.7 \text{ days}$$

Papatzacos' Method

Papatzacos et al. (1989) proposed a methodology that is based on semianalytical solutions for time development of a gas or water cone and simultaneous gas and water cones in an anisotropic, infinite reservoir with a horizontal well placed in the oil column.

Water coning

Step 1. Calculate the dimensionless rate q_D from the following expression:

$$q_D = 20,333.66 \, \mu_o \, B_o \, Q_o / \left[L \, h \, (\rho_w - \rho_o) \, \sqrt{k_V \, k_h} \right] \qquad (9\text{-}57)$$

where ρ = density, lb/ft^3
k_v = vertical permeability, md
k_h = horizontal permeability, md
h = oil zone thickness, ft
L = length of horizontal well

Step 2. Solve for the dimensionless breakthrough time t_{DBT} by applying the following relationship:

$$t_{DBT} = 1 - (3q_D - 1) \ln \left[\frac{3q_D}{3q_D - 1} \right] \qquad (9\text{-}58)$$

Step 3. Estimate the time to the water breakthrough t_{BT} by using the water and oil densities in the following expression:

$$t_{BT} = \frac{22,758.528 \, h \, \phi \, \mu_o \, t_{DBT}}{k_v (\rho_w - \rho_o)} \qquad (9\text{-}59)$$

where t_{BT} = time to water breakthrough as expressed in days
ρ_o = oil density, lb/ft^3
ρ_w = water density, lb/ft^3

Gas coning

Step 1. Calculate the dimensionless flow rate q_D.

$$q_D = 20,333.66 \, \mu_o \, B_o \, Q_o / \left[L \, h \, (\rho_o - \rho_g) \, \sqrt{k_V k_h} \right] \qquad (9\text{-}60)$$

Step 2. Solve for t_{DBT} by applying Equation 9-58.

Step 3. Estimate the time to the gas breakthrough t_{BT} by using the gas and oil densities in the following expression:

$$t_{BT} = \frac{22,758.528\,h\,\phi\,\mu_o\,t_{DBT}}{k_v\,(\rho_o - \rho_g)} \tag{9-61}$$

where t_{BT} = time to gas breakthrough as expressed in days
ρ_o = oil density, lb/ft³
ρ_g = gas density, lb/ft³

Water and gas coning

For the two-cone case, the authors developed two graphical correlations for determining the time to breakthrough and optimum placement of the horizontal well. The proposed method is summarized below:

Step 1. Calculate the gas coning dimensionless flow rate by applying Equation 9-60.

Step 2. Calculate the density difference ratio.

$$\psi = \frac{\rho_w - \rho_o}{\rho_o - \rho_g} \tag{9-62}$$

Step 3. Solve for the dimensionless breakthrough time by using Figure 9-25 or applying the following polynomial:

$$\ln(t_{DBT}) = c_0 + c_1 U + c_2 U^2 + c_3 U^3 \tag{9-63}$$

where $U = \ln(q_D)$

The coefficients $c_0 - c_3$ are tabulated in Table 9-4.

Step 4. Solve for the time to breakthrough by applying the gas-coning Equation 9-61.

Step 5. Solve for the optimum placement of the horizontal above the WOC by applying the following expression:

$$D_b^{opt} = h\,\beta_{opt} \tag{9-64}$$

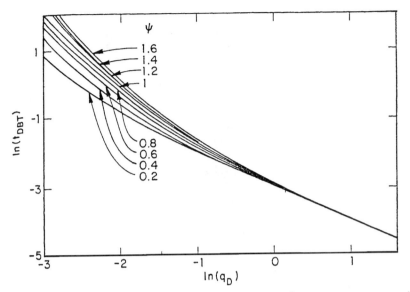

Figure 9-25. Dimensionless time for two-cone case. (*After Paptzacos, P. et. al., courtesy SPE, 1989.*)

where D_b^{opt} = optimum distance above the WOC, ft

h = oil thickness, ft

β_{opt} = optimum fractional well placement

The fractional well placement β_{opt} is determined from Figure 9-26 or the following relationship:

$$\beta_{opt} = c_0 + c_1 U + c_2 U^2 + c_3 U^3 \tag{9-65}$$

The coefficients of the above polynomial are given in Table 9-5.

Example 9-20

Resolve Example 9-18 by using Papatzacos' method.

Solution

Step 1. Solve for the dimensionless flow rate by using Equation 9-57.

$$q_D = \frac{20,333.66 \times 0.73 \times 1.1 \times 1000}{\left[1640 \times 50 \,(63.76 - 47.5)\, \sqrt{60 \times 15}\right]} = 0.408$$

Table 9-4
Coefficients for Breakthrough Time, t_{DBT} (Equation 4-64)
(*After Papatzacos, P. et al., SPE Paper 19822, 1989*)

ψ	C_0	C_1	C_2	C_3
0.2	−2.9494	−0.94654	−0.0028369	−0.029879
0.4	−2.9473	−0.93007	0.016244	−0.049687
0.6	−2.9484	−0.9805	0.050875	−0.046258
0.8	−2.9447	−1.0332	0.075238	−0.038897
1.0	−2.9351	−1.0678	0.088277	−0.034931
1.2	−2.9218	−1.0718	0.091371	−0.040743
1.4	−2.9162	−1.0716	0.093986	−0.042933
1.6	−2.9017	−1.0731	0.094943	−0.048212
1.8	−2.8917	−1.0856	0.096654	−0.046621
2.0	−2.8826	−1.1103	0.10094	−0.040963

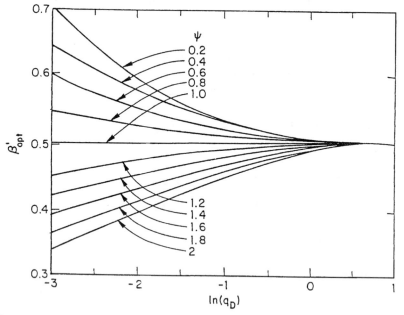

Figure 9-26. Optimum well placement for two-cone case. (*After Paptzacos, P. et. al., courtesy SPE, 1989.*)

Table 9-5
Coefficients for Optimum Well Placement (Equation 4-66)
(After Papatzacos, P. et al., SPE Paper 19822, 1989)

ψ	C_0	C_1	C_2	C_3
0.2	0.507	−0.0126	0.01055	−0.002483
0.4	0.504	−0.0159	0.01015	−0.000096
0.6	0.503	−0.0095	0.00624	−0.000424
0.8	0.502	−0.0048	0.00292	−0.000148
1.0	0.500	−0.0001	0.00004	0.000009
1.2	0.497	0.0042	−0.00260	0.000384
1.4	0.495	0.0116	−0.00557	−0.000405
1.6	0.493	0.0178	−0.00811	−0.000921
1.8	0.490	0.0231	−0.01020	−0.001242
2.0	0.488	0.0277	−0.01189	−0.001467

Step 2. Calculate the dimensionless breakthrough time from Equation 9-58.

$$t_{DBT} = 1 - (3 \times 0.408 - 1) \ln \left[\frac{3 \times 0.408}{3 \times 0.408 - 1} \right] = 0.6191$$

Step 3. Estimate the time to breakthrough from Equation 9-59

$$t_{BT} = \frac{22,758.528 \times 50 \times .15 \times 0.73 \times 0.6191}{[15 \, (63.74 - 47.5)]} = 316 \text{ days}$$

Example 9-21

A 1,640-foot-long horizontal well is drilled in an oil reservoir with developing gas and water cones. The following data are available:

$$h = 50 \text{ ft} \quad k_h = 60 \text{ md} \quad k_v = 15 \text{ md} \quad B_o = 1.1 \text{ bbl/STB}$$
$$\mu_o = 0.73 \text{ cp} \quad r_w = 0.3 \text{ ft} \quad \rho_o = 47.5 \text{ lb/ft}^3 \quad \rho_w = 63.76 \text{ lb/ft}^3$$
$$\rho_g = 9.1 \text{ lb/ft}^3 \quad \phi = 15\% \quad Q_o = 1000 \text{ STB/day}$$

Determine the optimum well placement and calculate the corresponding breakthrough time.

Solution

Step 1. Calculate the dimensionless flow rate from Equation 9-60.

$$q_D = \frac{20{,}333.66 \times 0.73 \times 1.1 \times 1000}{\left[1640 \times 50 \, (47.5 - 9.1) \, \sqrt{60 \times 15}\right]} = 0.1728$$

Step 2. Calculate the density difference ratio from Equation 9-62.

$$\psi = \frac{63.76 - 47.5}{47.5 - 9.1} = 0.4234$$

Step 3. Read the fraction well placement β_{opt} from Figure 9-26 by using the calculated values of ψ and q_D to give:

$$B_{opt} \cong 0.565$$

Step 4. Calculate the optimum well placement above the WOC from Equation 9-64.

$$D_b^{opt} = (0.565)\,(50) = 28.25 \text{ ft}$$

Step 5. From Figure 9-25, for $q_D = 0.1728$ and $\psi = 0.4234$, find the dimensionless breakthrough time t_{DBT}:

$$Ln\,(t_{DBT}) = -.8 \quad \text{(from Figure 9-25)}$$
$$t_{DBT} = 0.449$$

Step 6. Estimate the time to breakthrough by applying Equation 9-61.

$$t_{BT} = 22{,}758.528 \times 50 \times 0.15 \times 0.73 \times 0.449 / [15\,(47.3 - 9.1)]$$
$$= 97.71 \text{ days}$$

PROBLEMS

1. In an oil-water system, the following fluid and rock data are available:

$h = 60'$ $h_p = 25'$ $\rho_o = 47.5 \text{ lb/ft}^3$ $\rho_w = 63.76 \text{ lb/ft}^3$

$\mu_o = 0.85 \text{ cp}$ $B_o = 1.2 \text{ bbl/STB}$ $r_e = 660$ $r_w = 0.25'$

$k_o = k = 90.0 \text{ md}$

Calculate the critical oil flow rate, by using the following methods:

- Meyer-Garder
- Chierici-Ciucci
- Hoyland-Papatzacos-Skjaeveland
- Chaney
- Chaperson
- Schols

2. Given:

$Q_o = 400$ STB/day $h = 60$ ft $h_p = 25$ ft $\rho_w = 63.76$ lb/ft^3
$\rho_o = 47.5$ lb/ft^3 $\mu_o = 0.85$ cp $B_o = 1.2$ bbl/STB
$k_v = 9$ md $k_h = 90$ md $\phi = 15\%$ $M = 3.5$

Calculate the water breakthrough time by using the:

a. Sobocinski-Cornelius method
b. Bournazel-Jeanson correlation

3. The rock, fluid, and the related reservoir properties of a bottom-water drive reservoir are given below:

well spacing = 80 acres
initial oil column thickness = 100 ft

$h_p = 40'$ $\rho_o = 48$ lb/ft^3 $\rho_w = 63$ lb/ft^3 $r_e = 660'$
$r_w = 0.25'$ $M = 3.0$ $\phi = 14\%$ $S_{or} = 0.25$
$S_{wc} = 0.25$ $B_o = 1.2$ bbl/STB $\mu_o = 2.6$ cp $\mu_w = 1.00$ cp
$k_h = 80$ md $k_v = 16$ md

Calculate the water-cut behavior of a vertical well in the reservoir assuming a total production rate of 500, 1000, and 1500 STB/day.

4. A 2,000-ft-long horizontal well is drilled in the top elevation of the pay zone in a water-drive reservoir. The following data are available:

$h = 50$ ft $k_h = 80$ md $k_v = 25$ md $B_o = 1.2$ bbl/STB
$\mu_o = 2.70$ cp $r_w = 0.3$ ft $D_b = 50$ ft $\rho_o = 48.5$ lb/ft^3
$\rho_w = 62.50$ lb/ft^3 $y_e = 1320$ ft

Calculate the critical flow rate by using:

a. Chaperson's method
b. Efros' correlation
c. Karcher's equation
d. Joshi's method

5. A 2,000-foot-long horizontal well is producing at 1500 STB/day. The following data are available:

$h = 60$ ft $\quad$ $k_h = 80$ md $\quad$ $k_v = 15$ md $\quad$ $B_o = 1.2$ bbl/STB
$\mu_o = 2.70$ cp $\quad$ $r_w = 0.3$ ft $\quad$ $\rho_o = 47.5$ lb/ft^3 $\quad$ $\rho_w = 63.76$ lb/ft^3
$z_{wD} = 1$ $\quad$ $\phi = 15\%$ $\quad$ $S_{wc} = 0.25$ $\quad$ $S_{or} = 0.25$

Calculate the time to breakthrough by using the:

a. Ozkan-Raghavan method
b. Papatzacos' method

REFERENCES

1. Bournazel, C., and Jeanson, B., "Fast Water Coning Evaluation," SPE Paper 3628 presented at the SPE 46th Annual Fall Meeting, New Orleans, Oct. 3–6, 1971.

2. Calhoun, John, *Fundamentals of Reservoir Engineering.* Norman, OK: The University of Oklahoma Press, 1960.

3. Chaney, P. E. et al., "How to Perforate Your Well to Prevent Water and Gas Coning," *OGJ,* May 1956, p. 108.

4. Chaperson, I., "Theoretical Study of Coning Toward Horizontal and Vertical Wells in Anisotrophic Formations: Subcritical and Critical Rates," SPE Paper 15377, SPE 61st Annual Fall Meeting, New Orleans, LA, Oct. 5–8, 1986.

5. Chierici, G. L., Ciucci, G. M., and Pizzi, G., "A Systematic Study of Gas and Water Coning by Potentiometric Models," *JPT,* Aug. 1964, p. 923.

6. Efros, D. A., "Study of Multiphase Flows in Porous Media" (in Russian), *Gastoptexizdat,* Leningrad, 1963.

7. Hoyland, L. A., Papatzacos, P., and Skjaeveland, S. M., "Critical Rate for Water Coning: Correlation and Analytical Solution," *SPERE,* Nov. 1989, p. 495.

8. Joshi, S. D., "Augmentation of Well Productivity Using Slant and Horizontal Wells," *J. of Petroleum Technology,* June 1988, pp. 729–739.

9. Joshi, S., *Horizontal Well Technology.* Tulsa, OK: Pennwell Publishing Company, 1991.

10. Karcher, B., Giger, F., and Combe, J., "Some Practical Formulas to Predict Horizontal Well Behavior," SPE Paper 15430, presented at the SPE 61st Annual Conference, New Orleans, Oct. 5–8, 1986.

11. Kuo, C. T., and Desbrisay, C. L., "A Simplified Method for Water Coning Predictions," SPE Paper 12067, presented at the Annual SPE Technical Conference, San Francisco, Oct. 5–8, 1983.

12. Meyer, H. I., and Garder, A. O., "Mechanics of Two Immiscible Fluids in Porous Media," *J. Applied Phys.,* Nov. 1954, No. 11, p. 25.

13. Muskat, M., and Wyckoff, R. D., "An Approximate Theory of Water Coning in Oil Production," *Trans. AIME,* 1953, pp. 114, 144.

14. Ozkan, E., and Raghavan, R., "Performance of Horizontal Wells Subject to Bottom Water Drive," SPE Paper 18545, presented at the SPE Eastern Regional Meeting, Charleston, West Virginia, Nov. 2–4, 1988.

15. Papatzacos, P., Herring, T. U., Martinsen, R., and Skjaeveland, S. M., "Cone Breakthrough Time for Horizontal Wells," SPE Paper 19822, presented at the 64th SPE Annual Conference and Exhibition, San Antonio, TX, Oct. 8–11, 1989.

16. Pirson, S. J., *Oil Reservoir Engineering.* Huntington, NY: Robert E. Krieger Publishing Company, 1977.

17. Schols, R. S., "An Empirical Formula for the Critical Oil Production Rate," *Erdoel Erdgas,* A., January 1972, Vol. 88, No. 1, pp. 6–11.

18. Slider, H. C., *Practical Petroleum Reservoir Engineering Methods.* Tulsa, OK: Petroleum Publishing Company, 1976.

19. Sobocinski, D. P., and Cornelius, A. J., "A Correlation for Predicting Water Coning Time," *JPT,* May 1965, pp. 594–600.

WATER INFLUX

Nearly all hydrocarbon reservoirs are surrounded by water-bearing rocks called *aquifers*. These aquifers may be substantially larger than the oil or gas reservoirs they adjoin as to appear infinite in size, or they may be so small in size as to be negligible in their effect on reservoir performance.

As reservoir fluids are produced and reservoir pressure declines, a pressure differential develops from the surrounding aquifer into the reservoir. Following the basic law of fluid flow in porous media, the aquifer reacts by encroaching across the original hydrocarbon-water contact. In some cases, water encroachment occurs due to hydrodynamic conditions and recharge of the formation by surface waters at an outcrop.

In many cases, the pore volume of the aquifer is not significantly larger than the pore volume of the reservoir itself. Thus, the expansion of the water in the aquifer is negligible relative to the overall energy system, and the reservoir behaves volumetrically. In this case, the effects of water influx can be ignored. In other cases, the aquifer permeability may be sufficiently low such that a very large pressure differential is required before an appreciable amount of water can encroach into the reservoir. In this instance, the effects of water influx can be ignored as well.

This chapter focuses on those those reservoir-aquifer systems in which the size of the aquifer is large enough and the permeability of the rock is high enough that water influx occurs as the reservoir is depleted. This chapter also provides various water influx calculation models and a detailed description of the computational steps involved in applying these models.

CLASSIFICATION OF AQUIFERS

Many gas and oil reservoirs produced by a mechanism termed *water drive*. Often this is called *natural water drive* to distinguish it from *artificial water drive* that involves the injection of water into the formation. Hydrocarbon production from the reservoir and the subsequent pressure drop prompt a response from the aquifer to offset the pressure decline. This response comes in a form of *water influx,* commonly called *water encroachment,* which is attributed to:

• Expansion of the water in the aquifer
• Compressibility of the aquifer rock
• Artesian flow where the water-bearing formation outcrop is located structurally higher than the pay zone

Reservoir-aquifer systems are commonly classified on the basis of:

• Degree of pressure maintenance   • Flow regimes
• Outer boundary conditions   • Flow geometries

Degree of Pressure Maintenance

Based on the degree of the reservoir pressure maintenance provided by the aquifer, the natural water drive is often qualitatively described as:

• Active water drive
• Partial water drive
• Limited water drive

The term *active* water drive refers to the water encroachment mechanism in which the rate of water influx equals the reservoir *total* production rate. Active water-drive reservoirs are typically characterized by a gradual and slow reservoir pressure decline. If, during any long period, the production rate and reservoir pressure remain reasonably constant, the reservoir voidage rate must be equal to the water influx rate.

$$\begin{bmatrix} \text{water influx} \\ \text{rate} \end{bmatrix} = \begin{bmatrix} \text{oil flow} \\ \text{rate} \end{bmatrix} + \begin{bmatrix} \text{free gas} \\ \text{flow rate} \end{bmatrix} + \begin{bmatrix} \text{water production} \\ \text{rate} \end{bmatrix}$$

or

$$e_w = Q_o \, B_o + Q_g \, B_g + Q_w \, B_w \qquad (10\text{-}1)$$

where e_w = water influx rate, bbl/day
  Q_o = oil flow rate, STB/day
  B_o = oil formation volume factor, bbl/STB
  Q_g = **free** gas flow rate, scf/day
  B_g = gas formation volume factor, bbl/scf
  Q_w = water flow rate, STB/day
  B_w = water formation volume factor, bbl/STB

Equation 10-1 can be equivalently expressed in terms of cumulative production by introducing the following derivative terms:

$$e_w = \frac{dW_e}{dt} = B_o \frac{dN_p}{dt} + (GOR - R_s) \frac{dN_p}{dt} B_g + \frac{dW_p}{dt} B_w \qquad (10\text{-}2)$$

where W_e = cumulative water influx, bbl
  t = time, days
  N_p = cumulative oil production, STB
  GOR = current gas-oil ratio, scf/STB
  R_s = current gas solubility, scf/STB
  B_g = gas formation volume factor, bbl/scf
  W_p = cumulative water production, STB
  dN_p/dt = daily oil flow rate Q_o, STB/day
  dW_p/dt = daily water flow rate Q_w, STB/day
  dW_e/dt = daily water influx rate e_w, bbl/day
  $(GOR - R_s)dN_p/dt$ = daily free gas flow rate, scf/day

Example 10-1

Calculate the water influx rate e_w in a reservoir whose pressure is stabilized at 3000 psi.

Given: initial reservoir pressure = 3500 psi
  dN_p/dt = 32,000 STB/day
  B_o = 1.4 bbl/STB
  GOR = 900 scf/STB
  R_s = 700 scf/STB
  B_g = 0.00082 bbl/scf
  dW_p/dt = 0
  B_w = 1.0 bbl/STB

Solution

Applying Equation 10-1 or 10-2 gives:

e_w = (1.4) (32,000) + (900 − 700) (32,000) (0.00082) + 0
   = 50,048 bbl/day

Outer Boundary Conditions

The aquifer can be classified as infinite or finite (bounded). Geologically all formations are finite, but may act as infinite if the changes in the pressure at the oil-water contact are not "felt" at the aquifer boundary. Some aquifers outcrop and are infinite acting because of surface replenishment. In general, the outer boundary governs the behavior of the aquifer and, therefore:

a. Infinite system indicates that the effect of the pressure changes at the oil/aquifer boundary can never be felt at the outer boundary. This boundary is for all intents and purposes at a constant pressure equal to initial reservoir pressure.
b. Finite system indicates that the aquifer outer limit is affected by the influx into the oil zone and that the pressure at this outer limit changes with time.

Flow Regimes

There are basically three flow regimes that influence the rate of water influx into the reservoir. As previously described in Chapter 6, those flow regimes are:

a. Steady-state
b. Semisteady (pseudosteady)-state
c. Unsteady-state

Flow Geometries

Reservoir-aquifer systems can be classified on the basis of flow geometry as:

a. Edge-water drive
b. Bottom-water drive
c. Linear-water drive

In edge-water drive, as shown in Figure 10-1, water moves into the flanks of the reservoir as a result of hydrocarbon production and pressure drop at the reservoir-aquifer boundary. The flow is essentially radial with negligible flow in the vertical direction.

Bottom-water drive occurs in reservoirs with large areal extent and gentle dip where the reservoir-water contact completely underlies the reservoir. The flow is essentially radial and, in contrast to the edge-water drive, the bottom-water drive has significant vertical flow.

In linear-water drive, the influx is from one flank of the reservoir. The flow is strictly linear with a constant cross-sectional area.

RECOGNITION OF NATURAL WATER INFLUX

Normally very little information is obtained during the exploration-development period of a reservoir concerning the presence or characteristics of an aquifer that could provide a source of water influx during the depletion period. Natural water drive may be assumed by analogy with nearby producing reservoirs, but early reservoir performance trends can provide clues. A comparatively low, and decreasing, rate of reservoir pressure decline with increasing cumulative withdrawals is indicative of fluid influx.

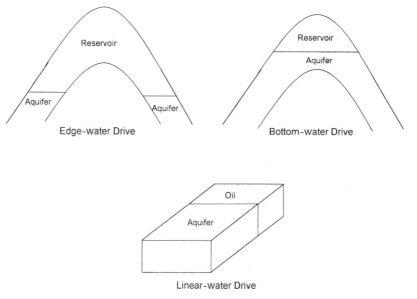

Figure 10-1. Flow geometries.

Successive calculations of barrels withdrawn per psi change in reservoir pressure can supplement performance graphs. If the reservoir limits have not been delineated by developed dry holes, however, the influx could be from an undeveloped area of the reservoir not accounted for in averaging reservoir pressure. If the reservoir pressure is below the oil saturation pressure, a low rate of increase in produced gas-oil ratio is also indicative of fluid influx.

Early water production from edge wells is indicative of water encroachment. Such observations must be tempered by the possibility that the early water production is due to formation fractures; thin, high permeability streaks; or to coning in connection with a limited aquifer. The water production may be due to casing leaks.

Calculation of increasing original oil-in-place from successive reservoir pressure surveys by using the material balance assuming no water influx is also indicative of fluid influx.

WATER INFLUX MODELS

It should be appreciated that in reservoir engineering there are more uncertainties attached to this subject than to any other. This is simply because one seldom drills wells into an aquifer to gain the necessary information about the porosity, permeability, thickness and fluid properties. Instead, these properties frequently have to be inferred from what has been observed in the reservoir. Even more uncertain, however, is the geometry and areal continuity of the aquifer itself.

Several models have been developed for estimating water influx that are based on assumptions that describe the characteristics of the aquifer. Due to the inherent uncertainties in the aquifer characteristics, all of the proposed models require historical reservoir performance data to evaluate constants representing aquifer property parameters since these are rarely known from exploration-development drilling with sufficient accuracy for direct application. The material balance equation can be used to determine historical water influx provided original oil-in-place is known from pore volume estimates. This permits evaluation of the constants in the influx equations so that future water influx rate can be forecasted.

The mathematical water influx models that are commonly used in the petroleum industry include:

• Pot aquifer
• Schilthuis' steady-state

- Hurst's modified steady-state
- The Van Everdingen-Hurst unsteady-state
  - Edge-water drive
  - Bottom-water drive
- The Carter-Tracy unsteady-state
- Fetkovich's method
  - Radial aquifer
  - Linear aquifer

The following sections describe these models and their practical applications in water influx calculations.

The Pot Aquifer Model

The simplest model that can be used to estimate the water influx into a gas or oil reservoir is based on the basic definition of compressibility. A drop in the reservoir pressure, due to the production of fluids, causes the aquifer water to expand and flow into the reservoir. The compressibility is defined mathematically as:

$$e_w = \frac{dW_e}{dt} = B_o \frac{dN_p}{dt} + (GOR - R_s)\frac{dN_p}{dt} B_g + \frac{dW_p}{dt} B_w \qquad (10\text{-}2)$$

or

$$\Delta V = c\,V\,\Delta p$$

Applying the above basic compressibility definition to the aquifer gives:

Water influx = (aquifer compressibility) (initial volume of water) (pressure drop)

or

$$W_e = (c_w + c_f)\,W_i\,(p_i - p) \qquad (10\text{-}3)$$

where W_e = cumulative water influx, bbl
c_w = aquifer water compressibility, psi^{-1}
c_f = aquifer rock compressibility, psi^{-1}
W_i = initial volume of water in the aquifer, bbl

p_i = initial reservoir pressure, psi

p = current reservoir pressure (pressure at oil-water contact), psi

Calculating the initial volume of water in the aquifer requires the knowledge of aquifer dimension and properties. These, however, are seldom measured since wells are not deliberately drilled into the aquifer to obtain such information. For instance, if the aquifer shape is radial, then:

$$W_i = \left[\frac{\pi \left(r_a^2 - r_e^2 \right) h \phi}{5.615} \right] \qquad (10\text{-}4)$$

where r_a = radius of the aquifer, ft

r_e = radius of the reservoir, ft

h = thickness of the aquifer, ft

ϕ = porosity of the aquifer

Equation 10-3 suggests that water is encroaching in a radial form from all directions. Quite often, water does not encroach on all sides of the reservoir, or the reservoir is not circular in nature.

To account for these cases, a modification to Equation 10-2 must be made in order to properly describe the flow mechanism. One of the simplest modifications is to include the fractional encroachment angle f in the equation, as illustrated in Figure 10-2, to give:

$$W_e = (c_w + c_f) \, W_i \, f \, (p_i - p) \qquad (10\text{-}5)$$

where the fractional encroachment angle f is defined by:

$$f = \frac{(\text{encoachment angle})^{\circ}}{360^{\circ}} = \frac{\theta}{360^{\circ}} \qquad (10\text{-}6)$$

The above model is only applicable to a small aquifer, i.e., pot aquifer, whose dimensions are of the same order of magnitude as the reservoir itself. Dake (1978) points out that because the aquifer is considered relatively small, a pressure drop in the reservoir is instantaneously transmitted throughout the entire reservoir-aquifer system. Dake suggests that for large aquifers, a mathematical model is required which includes time dependence to account for the fact that it takes a finite time for the aquifer to respond to a pressure change in the reservoir.

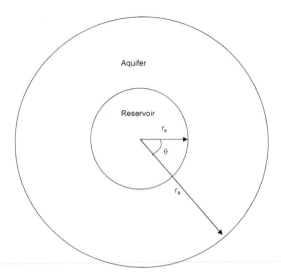

Figure 10-2. Radial aquifer geometries.

Example 10-2

Calculate the cumulative water influx that results from a pressure drop of 200 psi at the oil-water contact with an encroachment angle of 80°. The reservoir-aquifer system is characterized by the following properties:

	Reservoir	Aquifer
radius, ft	2600	10,000
porosity	0.18	0.12
c_f, psi^{-1}	4×10^{-6}	3×10^{-6}
c_w, psi^{-1}	5×10^{-6}	4×10^{-6}
h, ft	20	25

Solution

Step 1. Calculate the initial volume of water in the aquifer from Equation 10-4.

$$W_i = \left(\frac{\pi (10,000^2 - 2600^2)(25)(0.12)}{5.615} \right) = 156.5 \, \text{MMbbl}$$

Step 2. Determine the cumulative water influx by applying Equation 10-5.

$$W_e = (4+3)\,10^{-6}\ (156.5 \times 10^6)\left(\frac{80}{360}\right)(200) = 48,689\ \text{bbl}$$

Schilthuis' Steady-State Model

Schilthuis (1936) proposed that for an aquifer that is flowing under the steady-state flow regime, the flow behavior could be described by Darcy's equation. The rate of water influx e_w can then be determined by applying Darcy's equation:

$$\frac{dW_e}{dt} = e_w = \left[\frac{0.00708\,kh}{\mu_w\,\ln\left(\dfrac{r_a}{r_e}\right)}\right](p_i - p) \tag{10-7}$$

The above relationship can be more conveniently expressed as:

$$\frac{dW_e}{dt} = e_w = C\,(p_i - p) \tag{10-8}$$

where e_w = rate of water influx, bbl/day
$\quad\quad\ k$ = permeability of the aquifer, md
$\quad\quad\ h$ = thickness of the aquifer, ft
$\quad\quad\ r_a$ = radius of the aquifer, ft
$\quad\quad\ r_e$ = radius of the reservoir
$\quad\quad\ t$ = time, days

The parameter C is called the water influx constant and is expressed in bbl/day/psi. This water influx constant C may be calculated from the reservoir historical production data over a number of selected time intervals, provided that the rate of water influx e_w has been determined independently from a different expression. For instance, the parameter C may be estimated by combining Equation 10-1 with 10-8. Although the influx constant can only be obtained in this manner when the reservoir pressure stabilizes, once it has been found, it may be applied to both stabilized and changing reservoir pressures.

Example 10-3

The data given in Example 10-1 is used in this example:

p_i = 3500 psi   p = 3000 psi Q_o = 32,000 STB/day
B_o = 1.4 bbl/STB   GOR = 900 scf/STB R_s = 700 scf/STB
B_g = 0.00082 bbl/scf Q_w = 0 B_w = 1.0 bbl/STB

Calculate Schilthuis' water influx constant.

Solution

Step 1. Solve for the rate of water influx e_w by using Equation 10-1.

$$e_w = (1.4)\,(32{,}000) + (900 - 700)\,(32{,}000)\,(0.00082) + 0$$
$$= 50{,}048 \text{ bbl/day}$$

Step 2. Solve for the water influx constant from Equation 10-8.

$$C = \frac{50{,}048}{(3500 - 3000)} = 100 \ \text{bbl/day/psi}$$

If the steady-state approximation adequately describes the aquifer flow regime, the calculated water influx constant C values will be constant over the historical period.

Note that the pressure drops contributing to influx are the cumulative pressure drops from the initial pressure.

In terms of the cumulative water influx W_e, Equation 10-8 is integrated to give the common Schilthuis expression for water influx as:

$$\int_0^{W_e} dW_e = \int_0^t C\,(p_i - p)\,dt$$

or

$$W_e = C \int_0^t (p_i - p)\,dt \tag{10-9}$$

where W_e = cumulative water influx, bbl
$\quad\quad$ C = water influx constant, bbl/day/psi
$\quad\quad$ t = time, days
$\quad\quad$ p_i = initial reservoir pressure, psi
$\quad\quad$ p = pressure at the oil-water contact at time t, psi

When the pressure drop $(p_i - p)$ is plotted versus the time t, as shown in Figure 10-3, the area under the curve represents the integral $\int_0^t (p_i - p) = dt$.

This area at time t can be determined numerically by using the trapezoidal rule (or any other numerical integration method), as:

$$\int_0^t (p_i - p) = dt = area_I + area_{II} + area_{III} + etc. = \left(\frac{p_i - p_1}{2}\right)(t_1 - 0)$$

$$+ \frac{(p_i - p_1) + (p_i - p_2)}{2}(t_2 - t_1) + \frac{(p_i - p_2) + (p_i - p_3)}{2}(t_3 - t_2)$$

$$+ etc.$$

Equation 10-9 can then be written as:

$$W_e = C \sum_0^t (\Delta p)\, \Delta t \tag{10-10}$$

Example 10-4

The pressure history of a water-drive oil reservoir is given below:

t, days	p, psi
0	3500 (p_i)
100	3450
200	3410
300	3380
400	3340

The aquifer is under a steady-state flowing condition with an estimated water influx constant of 130 bbl/day/psi. Calculate the cumulative water influx after 100, 200, 300, and 400 days using the steady-state model.

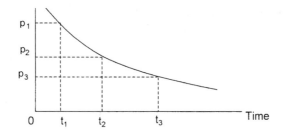

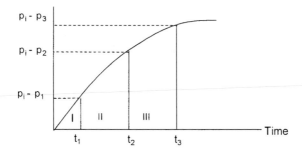

Figure 10-3. Calculating the area under the curve.

Solution

Step 1. Calculate the total pressure drop at each time t.

t, days	p	$p_i - p$
0	3500	0
100	3450	50
200	3410	90
300	3380	120
400	3340	160

Step 2. Calculate the cumulative water influx after 100 days:

$$W_e = 130\left(\frac{50}{2}\right)(100 - 0) = 325,000 \text{ bbl}$$

Step 3. Determine W_e after 200 days.

$$W_e = 130 \left[\left(\frac{50}{2} \right)(100 - 0) + \left(\frac{50+90}{2} \right)(200 - 100) \right] = 1,235,000 \, \text{bbl}$$

Step 4. W_e after 300 days.

$$W_e = 130 \left[\left(\frac{50}{2} \right)(100) + \left(\frac{50+90}{2} \right)(200 - 100) \right.$$

$$\left. + \left(\frac{120+90}{2} \right)(300 - 200) \right] = 2,600,000 \, \text{bbl}$$

Step 5. Calculate W_e after 400 days.

$$W_e = 130 \left[2500 + 7000 + 10,500 + \left(\frac{160+120}{2} \right)(400 - 300) \right]$$

$$= 4,420,000 \, \text{bbl}$$

Hurst's Modified Steady-State Model

One of the problems associated with the Schilthuis' steady-state model is that as the water is drained from the aquifer, the aquifer drainage radius r_a will increase as the time increases. Hurst (1943) proposed that the "apparent" aquifer radius r_a would increase with time and, therefore the dimensionless radius r_a/r_e may be replaced with a time dependent function, as:

$$r_a/r_e = at \qquad (10\text{-}11)$$

Substituting Equation 10-11 into Equation 10-7 gives:

$$e_w = \frac{dW_e}{dt} = \frac{0.00708 \, kh \, (p_i - p)}{\mu_w \ln (at)} \qquad (10\text{-}12)$$

The Hurst modified steady-state equation can be written in a more simplified form as:

$$e_w = \frac{dW_e}{dt} = \frac{C(p_i - p)}{\ln(at)} \qquad (10\text{-}13)$$

and in terms of the cumulative water influx

$$W_e = C \int_0^t \left[\frac{p_i - p}{\ln(at)} \right] dt \qquad (10\text{-}14)$$

or

$$W_e = C \sum_0^t \left[\frac{\Delta p}{\ln(at)} \right] \Delta t \qquad (10\text{-}15)$$

The Hurst modified steady-state equation contains two unknown constants, i.e., a and C, that must be determined from the reservoir-aquifer pressure and water influx historical data. The procedure of determining the constants a and C is based on expressing Equation 10-13 as a linear relationship.

$$\left(\frac{p_i - p}{e_w} \right) = \frac{1}{C} \ln(at)$$

or

$$\frac{p_i - p}{e_w} = \left(\frac{1}{C} \right) \ln(a) + \left(\frac{1}{C} \right) \ln(t) \qquad (10\text{-}16)$$

Equation 10-16 indicates that a plot of $(p_i - p)/e_w$ versus $\ln(t)$ will be a straight line with a slope of $1/C$ and intercept of $(1/C)\ln(a)$, as shown schematically in Figure 10-4.

Example 10-5

The following data, as presented by Craft and Hawkins (1959), documents the reservoir pressure as a function of time for a water-drive reservoir. Using the reservoir historical data, Craft and Hawkins calculated the water influx by applying the material balance equation (see Chapter 11). The rate of water influx was also calculated numerically at each time period.

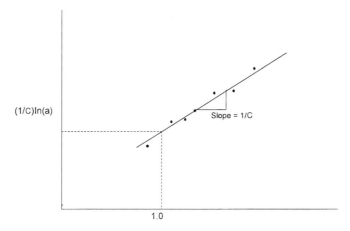

Figure 10-4. Graphical determination of C and a.

Time days	Pressure psi	W_e M bbl	e_w bbl/day	$p_i - p$ psi
0	3793	0	0	0
182.5	3774	24.8	389	19
365.0	3709	172.0	1279	84
547.5	3643	480.0	2158	150
730.0	3547	978.0	3187	246
912.5	3485	1616.0	3844	308
1095.0	3416	2388.0	4458	377

Assuming that the boundary pressure would drop to 3379 psi after 1186.25 days of production, calculate cumulative water influx at that time.

Solution

Step 1. Construct the following table:

t, days	ln(t)	$p_i - p$	e_w, bbl/day	$(p_i - p)/e_w$
0	—	0	0	—
182.5	5.207	19	389	0.049
365.0	5.900	84	1279	0.066
547.5	6.305	150	2158	0.070
730.0	6.593	246	31.87	0.077
912.5	6.816	308	3844	0.081
1095.0	6.999	377	4458	0.085

Step 2. Plot the term $(p_i - p)/e_w$ versus $\ln(t)$ and draw the best straight line through the points as shown in Figure 10-5, and determine the slope of the line to give:

slope $= 0.020$

Step 3. Determine the coefficient C of the Hurst equation from the slope to give:

$C = 1/0.02 = 50$

Step 4. Using any point on the straight line, solve for the parameter a by applying Equation 10-13 to give:

$a = 0.064$

Step 5. The Hurst equation is represented by:

$$W_e = 50 \int_o^t \left[\frac{p_i - p}{\ln(0.064\, t)} \right] dt$$

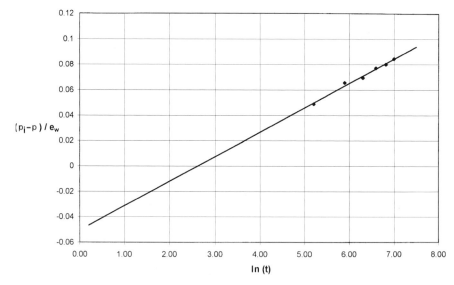

Figure 10-5. Determination of C and n for Example 10-5.

Step 6. Calculate the cumulative water influx after 1186.25 days from:

$$W_e = 2388 \times 10^3 + \int_{1095}^{1186.25} 50 \left[\frac{p_i - p}{\ln(0.064\,t)} \right] dt$$

$$W_e = 2388 \times 10^3 + 50 \left[\frac{\frac{3793 - 3379}{\ln(0.064 \times 1186.25)} + \frac{3793 - 3416}{\ln(0.064 \times 1095)}}{2} \right]$$

$$\times (1186.25 - 1095)$$

$$W_e = 2388 \times 10^3 + 420.508 \times 10^3 = 2809 \text{ Mbbl}$$

The Van Everdingen-Hurst Unsteady-State Model

The mathematical formulations that describe the flow of crude oil system into a wellbore are identical in form to those equations that describe the flow of water from an aquifer into a cylindrical reservoir, as shown schematically in Figure 10-6.

When an oil well is brought on production at a constant flow rate after a shut-in period, the pressure behavior is essentially controlled by the

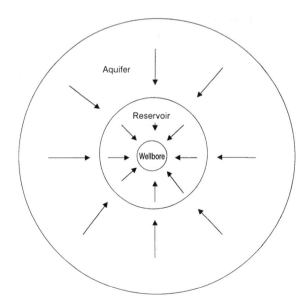

Figure 10-6. Water influx into a cylindrical reservoir.

transient (unsteady-state) flowing condition. This flowing condition is defined as the time period during which the boundary has no effect on the pressure behavior.

The dimensionless form of the diffusivity equation, as presented in Chapter 6 by Equation 6-90, is basically the general mathematical equation that is designed to model the transient flow behavior in reservoirs or aquifers. In a dimensionless form, the diffusivity equation takes the form:

$$\frac{\partial^2 P_D}{\partial r_D^2} + \frac{1}{r_D}\frac{\partial P_D}{\partial r_D} = \frac{\partial P_D}{\partial t_D}$$

Van Everdingen and Hurst (1949) proposed solutions to the dimensionless diffusivity equation for the following two reservoir-aquifer boundary conditions:

• Constant terminal rate
• Constant terminal pressure

For the constant-terminal-rate boundary condition, the rate of water influx is assumed constant for a given period; and the pressure drop at the reservoir-aquifer boundary is calculated.

For the constant-terminal-pressure boundary condition, a boundary pressure drop is assumed constant over some finite time period, and the water influx rate is determined.

In the description of water influx from an aquifer into a reservoir, there is greater interest in calculating the influx rate rather than the pressure. This leads to the determination of the water influx as a function of a given pressure drop at the inner boundary of the reservoir-aquifer system.

Van Everdingen and Hurst solved the diffusivity equation for the aquifer-reservoir system by applying the Laplace transformation to the equation. The authors' solution can be used to determine the water influx in the following systems:

• Edge-water-drive system (radial system)
• Bottom-water-drive system
• Linear-water-drive system

Edge-Water Drive

Figure 10-7 shows an idealized radial flow system that represents an edge-water-drive reservoir. The inner boundary is defined as the interface

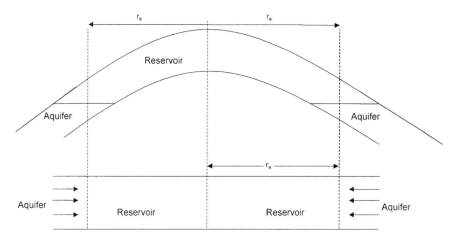

Figure 10-7. Idealized radial flow model.

between the reservoir and the aquifer. The flow across this inner boundary is considered horizontal and encroachment occurs across a cylindrical plane encircling the reservoir. With the interface as the inner boundary, it is possible to impose a constant terminal pressure at the inner boundary and determine the rate of water influx across the interface.

Van Everdingen and Hurst proposed a solution to the dimensionless diffusivity equation that utilizes the constant terminal pressure condition in addition to the following initial and outer boundary conditions:

Initial conditions:

$p = p_i$ for all values of radius r

Outer boundary conditions

• For an infinite aquifer

$p = p_i$ at $r = \infty$

• For a bounded aquifer

$\dfrac{\partial p}{\partial r} = 0$ at $r = r_a$

Van Everdingen and Hurst assumed that the aquifer is characterized by:

• Uniform thickness
• Constant permeability

- Uniform porosity
- Constant rock compressibility
- Constant water compressibility

The authors expressed their mathematical relationship for calculating the water influx in a form of a dimensionless parameter that is called *dimensionless water influx* W_{eD}. They also expressed the dimensionless water influx as a function of the *dimensionless time* t_D and *dimensionless radius* r_D, thus they made the solution to the diffusivity equation generalized and applicable to any aquifer where the flow of water into the reservoir is essentially radial.

The solutions were derived for cases of bounded aquifers and aquifers of infinite extent. The authors presented their solution in tabulated and graphical forms as reproduced here in Figures 10-8 through 10-11 and Tables 10-1 and 10-2.

The two dimensionless parameters t_D and r_D are given by:

$$t_D = 6.328 \times 10^{-3} \frac{kt}{\phi \mu_w c_t r_e^2} \qquad (10\text{-}17)$$

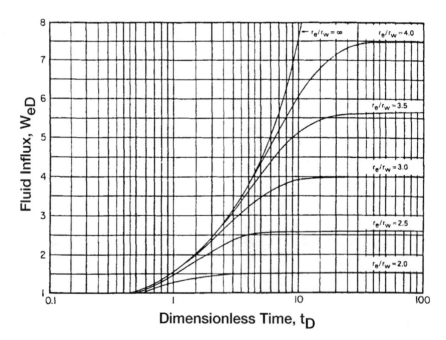

Figure 10-8. Dimensionless water influx W_{eD} for several values of r_e/r_R, i.e. r_a/r_e. (*Van Everdingen and Hurst W_{eD}. Permission to publish by the SPE.*)

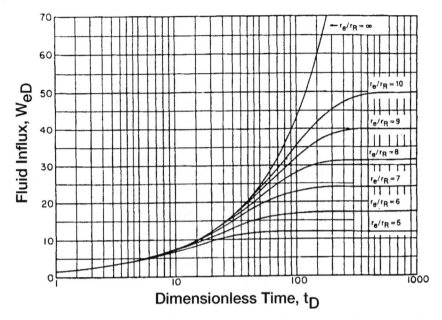

Figure 10-9. Dimensionless water influx W_{eD} for several values of r_e/r_R, i.e. r_a/r_e. (*Van Everdingen and Hurst W_{eD}. Permission to publish by the SPE.*)

$$r_D = \frac{r_a}{r_e} \qquad (10\text{-}18)$$

$$c_t = c_w + c_f \qquad (10\text{-}19)$$

where t = time, days
k = permeability of the aquifer, md
ϕ = porosity of the aquifer
μ_w = viscosity of water in the aquifer, cp
r_a = radius of the aquifer, ft
r_e = radius of the reservoir, ft
c_w = compressibility of the water, psi^{-1}
c_f = compressibility of the aquifer formation, psi^{-1}
c_t = total compressibility coefficient, psi^{-1}

The water influx is then given by:

$$W_e = B \, \Delta p \, W_{eD} \qquad (10\text{-}20)$$

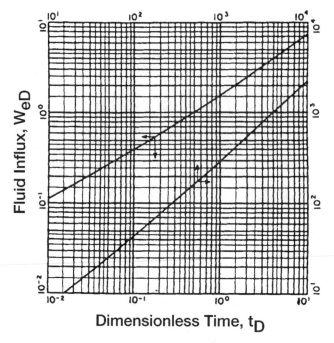

Dimensionless Time, t_D

Figure 10-10. Dimensionless water influx W_{eD} for infinite aquifer. (*Van Everdingen and Hurst W_{eD}. Permission to publish by the SPE.*)

with

$$B = 1.119 \phi c_t r_e^2 h \qquad (10\text{-}21)$$

where W_e = cumulative water influx, bbl
$\quad B$ = water influx constant, bbl/psi
$\quad \Delta p$ = pressure drop at the boundary, psi
$\quad W_{eD}$ = dimensionless water influx

Equation 10-21 assumes that the water is encroaching in a radial form. Quite often, water dies not encroach on all sides of the reservoir, or the reservoir is not circular in nature. In these cases, some modifications must be made in Equation 10-21 to properly describe the flow mechanism. One of the simplest modifications is to introduce the encroachment angle to the water influx constant B as:

$$f = \frac{\theta}{360} \qquad (10\text{-}22)$$

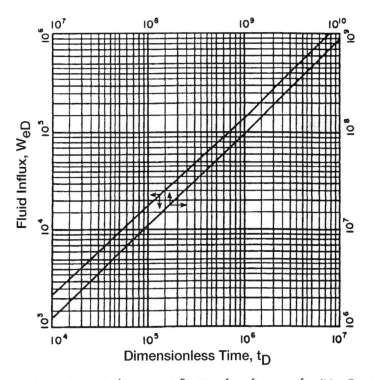

Figure 10-11. Dimensionless water influx W_{eD} for infinite aquifer. (*Van Everdingen and Hurst W_{eD}. Permission to publish by the SPE.*)

$$B = 1.119 \, \phi \, c_t \, r_e^2 \, h \, f \tag{10-23}$$

θ is the angle subtended by the reservoir circumference, i.e., for a full circle $\theta = 360°$ and for semicircle reservoir against a fault $\theta = 180°$, as shown in Figure 10-12.

Example 10-6[1]

Calculate water influx at the end of 1, 2, and 5 years into a circular reservoir with an aquifer of infinite extent. The initial and current reservoir pressures are 2500 and 2490 psi, respectively. The reservoir-aquifer system has the following properties.

(*text continued on page 668*)

---

[1]Data of this example was reported by Cole, F., *Reservoir Engineering Manual,* Gulf Publishing Company, 1969.

Table 10-1
Dimensionless Water Influx W_{eD} for Infinite Aquifer
(Van Everdingen and Hurst W_{eD}. Permission to publish by the SPE.)

Dimensionless time t_D	Fluid influx W_{eD}	Dimensionless time t_D	Fluid influx W_{eD}	Dimensionless time t_D	Fluid influx W_{eD}	Dimensionless time t_D	Fluid influx W_{eD}	Dimensionless time t_D	Fluid influx W_{eD}	Dimensionless time t_D	Fluid influx W_{eD}
0.00	0.000	79	35.697	455	150.249	1190	340.843	3250	816.090	35.000	6780.247
0.01	0.112	80	36.058	460	151.640	1200	343.308	3300	827.088	40.000	7650.096
0.05	0.278	81	36.418	465	153.029	1210	345.770	3350	838.067	50.000	9363.099
0.10	0.404	82	36.777	470	154.416	1220	348.230	3400	849.028	60.000	11,047.299
0.15	0.520	83	37.136	475	155.801	1225	349.460	3450	859.974	70.000	12,708.358
0.20	0.606	84	37.494	480	157.184	1230	350.688	3500	870.903	75.000	13,531.457
0.25	0.689	85	37.851	485	158.565	1240	353.144	3550	881.816	80.000	14,350.121
0.30	0.758	86	38.207	490	159.945	1250	355.597	3600	892.712	90.000	15,975.389
0.40	0.898	87	38.563	495	161.322	1260	358.048	3650	903.594	100.000	17,586.284
0.50	1.020	88	38.919	500	162.698	1270	360.496	3700	914.459	125.000	21,560.732
0.60	1.140	89	39.272	510	165.444	1275	361.720	3750	925.309	$1.5(10)^5$	$2.538(10)^4$
0.70	1.251	90	39.626	520	168.183	1280	362.942	3800	936.144	2.0″	3.308″
0.80	1.359	91	39.979	525	169.549	1290	365.386	3850	946.966	2.5″	4.066″
0.90	1.469	92	40.331	530	170.914	1300	367.828	3900	957.773	3.0″	4.817″
1	1.569	93	40.684	540	173.639	1310	370.267	3950	968.566	4.0″	6.267″
2	2.447	94	41.034	550	176.357	1320	372.704	4000	979.344	5.0″	7.699″
3	3.202	95	41.385	560	179.069	1325	373.922	4050	990.108	6.0″	9.113″
4	3.893	96	41.735	570	181.774	1330	375.139	4100	1000.858	7.0″	$1.051(10)^5$
5	4.539	97	42.084	575	183.124	1340	377.572	4150	1011.595	8.0″	1.189″
6	5.153	98	42.433	580	184.473	1350	380.003	4200	1022.318	9.0″	1.326″
7	5.743	99	42.781	590	187.166	1360	382.432	4250	1033.028	$1.0(10)^6$	1.462″
8	6.314	100	43.129	600	189.852	1370	384.859	4300	1043.724	1.5″	2.126″
9	6.869	105	44.858	610	192.533	1375	386.070	4350	1054.409	2.0″	2.781″

10	7.411	110	46.574	620	195.208	1380	387.283	4400	1065.082	2.5"	3.427"
11	7.940	115	48.277	625	196.544	1390	389.705	4450	1075.743	3.0"	4.064"
12	8.457	120	49.968	630	197.878	1400	392.125	4500	1086.390	4.0"	5.313"
13	8.964	125	51.648	640	200.542	1410	394.543	4550	1097.024	5.0"	6.544"
14	9.461	130	53.317	650	203.201	1420	396.959	4600	1107.646	6.0"	7.761"
15	9.949	135	54.976	660	205.854	1425	398.167	4650	1118.257	7.0"	8.965"
16	10.434	140	56.625	670	208.502	1430	399.373	4700	1128.854	8.0"	$1.016(10)^6$
17	10.913	145	58.265	675	209.825	1440	401.786	4750	1139.439	9.0"	1.134"
18	11.386	150	59.895	680	211.145	1450	404.197	4800	1150.012	$1.0(10)^7$	1.252"
19	11.855	155	61.517	690	213.784	1460	406.606	4850	1160.574	1.5"	1.828"
20	12.319	160	63.131	700	216.417	1470	409.013	4900	1171.125	2.0"	2.398"
21	12.778	165	64.737	710	219.046	1475	410.214	4950	1181.666	2.5"	2.961"
22	13.233	170	66.336	720	221.670	1480	411.418	5000	1192.198	3.0"	3.517"
23	13.684	175	67.928	725	222.980	1490	413.820	5100	1213.222	4.0"	4.610"
24	14.131	180	69.512	730	224.289	1500	416.220	5200	1234.203	5.0"	5.689"
25	14.573	185	71.090	740	226.904	1525	422.214	5300	1255.141	6.0"	6.758"
26	15.013	190	72.661	750	229.514	1550	428.196	5400	1276.037	7.0"	7.816"
27	15.450	195	74.226	760	232.120	1575	434.168	5500	1296.893	8.0"	8.866"
28	15.883	200	75.785	770	234.721	1600	440.128	5600	1317.709	9.0"	9.911"
29	16.313	205	77.338	775	236.020	1625	446.077	5700	1338.486	$1.0(10)^8$	$1.095(10)^7$
30	16.742	210	78.886	780	237.318	1650	452.016	5800	1359.225	1.5"	1.604"
31	17.167	215	80.428	790	239.912	1675	457.945	5900	1379.927	2.0"	2.108"
32	17.590	220	81.965	800	242.501	1700	463.863	6000	1400.593	2.5"	2.607"
33	18.011	225	83.497	810	245.086	1725	469.771	6100	1421.224	3.0"	3.100"
34	18.429	230	85.023	820	247.668	1750	475.669	6200	1441.820	4.0"	4.071"
35	18.845	235	86.545	825	248.957	1775	481.558	6300	1462.383	5.0"	5.032"
36	19.259	240	88.062	830	250.245	1800	487.437	6400	1482.912	6.0"	5.984"
37	19.671	245	89.575	840	252.819	1825	493.307	6500	1503.408	7.0"	6.928"
38	20.080	250	91.084	850	255.388	1850	499.167	6600	1523.872	8.0"	7.865"
39	20.488	255	92.589	860	257.953	1875	505.019	6700	1544.305	9.0"	8.797"
40	20.894	260	94.090	870	260.515	1900	510.861	6800	1564.706	$1.0(10)^9$	9.725"
41	21.298	265	95.588	875	261.795	1925	516.695	6900	1585.077	1.5"	$1.429(10)^8$

(table continued on next page)

Table 10-1 (continued)

Dimensionless time t_D	Fluid influx W_{eD}	Dimensionless time t_D	Fluid influx W_{eD}	Dimensionless time t_D	Fluid influx W_{eD}	Dimensionless time t_D	Fluid influx W_{eD}	Dimensionless time t_D	Fluid influx W_{eD}	Dimensionless time t_D	Fluid influx W_{eD}
42	21.701	270	97.081	880	263.073	1950	522.520	7000	1605.418	2.0"	1.880"
43	22.101	275	98.571	890	265.629	1975	528.337	7100	1625.729	2.5"	2.328"
44	22.500	280	100.057	900	268.181	2000	534.145	7200	1646.011	3.0"	2.771"
45	22.897	285	101.540	910	270.729	2025	539.945	7300	1666.265	4.0"	3.645"
46	23.291	290	103.019	920	273.274	2050	545.737	7400	1686.490	5.0"	4.510"
47	23.684	295	104.495	925	274.545	2075	551.522	7500	1706.688	6.0"	5.368"
48	24.076	300	105.968	930	275.815	2100	557.299	7600	1726.859	7.0"	6.220"
49	24.466	305	107.437	940	278.353	2125	563.068	7700	1747.002	8.0"	7.066"
50	24.855	310	108.904	950	280.888	2150	568.830	7800	1767.120	9.0"	7.909"
51	25.244	315	110.367	960	283.420	2175	574.585	7900	1787.212	$1.0(10)^{10}$	8.747"
52	25.633	320	111.827	970	285.948	2200	580.332	8000	1807.278	1.5"	$1.288"(10)^{9}$
53	26.020	325	113.284	975	287.211	2225	586.072	8100	1827.319	2.0"	1.697"
54	26.406	330	114.738	980	288.473	2250	591.806	8200	1847.336	2.5"	2.103"
55	26.791	335	116.189	990	290.995	2275	597.532	8300	1867.329	3.0"	2.505"
56	27.174	340	117.638	1000	293.514	2300	603.252	8400	1887.298	4.0"	3.299"
57	27.555	345	119.083	1010	296.030	2325	608.965	8500	1907.243	5.0"	4.087"
58	27.935	350	120.526	1020	298.543	2350	614.672	8600	1927.166	6.0"	4.868"

59	28.314	355	121.966	1025	299.799	2375	620.372	8700	1947.065	7.0″	5.643″
60	28.691	360	123.403	1030	301.053	2400	626.066	8800	1966.942	8.0″	6.414″
61	29.068	365	124.838	1040	303.560	2425	631.755	8900	1986.796	9.0″	7.183″
62	29.443	370	126.720	1050	306.065	2450	637.437	9000	2006.628	$1.0(10)^{11}$	7.948″
63	29.818	375	127.699	1060	308.567	2475	643.113	9100	2026.438	1.5″	$1.17(10)^{10}$
64	30.192	380	129.126	1070	311.066	2500	648.781	9200	2046.227	2.0″	1.55″
65	30.565	385	130.550	1075	312.314	2550	660.093	9300	2065.996	2.5″	1.92″
66	30.937	390	131.972	1080	313.562	2600	671.379	9400	2085.744	3.0″	2.29″
67	31.308	395	133.391	1090	316.055	2650	682.640	9500	2105.473	4.0″	3.02″
68	31.679	400	134.808	1100	318.545	2700	693.877	9600	2125.184	5.0″	3.75″
69	32.048	405	136.223	1110	321.032	2750	705.090	9700	2144.878	6.0″	4.47″
70	32.417	410	137.635	1120	323.517	2800	716.280	9800	2164.555	7.0″	5.19″
71	32.785	415	139.045	1125	324.760	2850	727.449	9900	2184.216	8.0″	5.89″
72	33.151	420	140.453	1130	326.000	2900	738.598	10,000	2203.861	9.0″	6.58″
73	33.517	425	141.859	1140	328.480	2950	749.725	12,500	2688.967	$1.0(10)^{12}$	7.28″
74	33.883	430	143.262	1150	330.958	3000	760.833	15,000	3164.780	1.5″	$1.08(10)^{11}$
75	34.247	435	144.664	1160	333.433	3050	771.922	17,500	3633.368	2.0″	1.42″
76	34.611	440	146.064	1170	335.906	3100	782.992	20,000	4095.800		
77	34.974	445	147.461	1175	337.142	3150	794.042	25,000	5005.726		
78	35.336	450	148.856	1180	338.376	3200	805.075	30,000	5899.508		

Table 10-2
Dimensionless Water Influx W_{eD} for Several Values of r_e/r_R, i.e. r_a/r_e
(Van Everdingen and Hurst W_{eD}. Permission to publish by the SPE.)

$r_e/r_R = 1.5$ Dimensionless time t_D	Fluid influx W_{eD}	$r_e/r_R = 2.0$ Dimensionless time t_D	Fluid fux W_{eD}	$r_e/r_R = 2.5$ Dimensionless time t_D	Fluid influx W_{eD}	$r_e/r_R = 3.0$ Dimensionless time t_D	Fluid influx W_{eD}	$r_e/r_R = 3.5$ Dimensionless time t_D	Fluid influx W_{eD}	$r_e/r_R = 4.0$ Dimensionless time t_D	Fluid flux W_{eD}	$r_e/r_R = 4.5$ Dimensionless time t_D	Fluid influx W_{eD}
$5.0(10)^{-2}$	0.276	$5.0(10)^{-2}$	0.278	$1.0(10)^{-1}$	0.408	$3.0(10)^{-1}$	0.755	1.00	1.571	2.00	2.442	2.5	2.835
6.0"	0.304	7.5"	0.345	1.5"	0.509	4.0"	0.895	1.20	1.761	2.20	2.598	3.0	3.196
7.0"	0.330	$1.0(10)^{-1}$	0.404	2.0"	0.599	5.0"	1.023	1.40	1.940	2.40	2.748	3.5	3.537
8.0"	0.354	1.25"	0.458	2.5"	0.681	6.0"	1.143	1.60	2.111	2.60	2.893	4.0	3.859
9.0"	0.375	1.50"	0.507	3.0"	0.758	7.0"	1.256	1.80	2.273	2.80	3.034	4.5	4.165
$1.0(10)^{-1}$	0.395	1.75"	0.553	3.5"	0.829	8.0"	1.363	2.00	2.427	3.00	3.170	5.0	4.454
1.1"	0.414	2.00"	0.597	4.0"	0.897	9.0"	1.465	2.20	2.574	3.25	3.334	5.5	4.727
1.2"	0.431	2.25"	0.638	4.5"	0.962	1.00	1.563	2.40	2.715	3.50	3.493	6.0	4.986
1.3"	0.446	2.50"	0.678	5.0"	1.024	1.25	1.791	2.60	2.849	3.75	3.645	6.5	5.231
1.4"	0.461	2.75"	0.715	5.5"	1.083	1.50	1.997	2.80	2.976	4.00	3.792	7.0	5.464
1.5"	0.474	3.00"	0.751	6.0"	1.140	1.75	2.184	3.00	3.098	4.25	3.932	7.5	5.684
1.6"	0.486	3.25"	0.785	6.5"	1.195	2.00	2.353	3.25	3.242	4.50	4.068	8.0	5.892
1.7"	0.497	3.50"	0.817	7.0"	1.248	2.25	2.507	3.50	3.379	4.75	4.198	8.5	6.089
1.8"	0.507	3.75"	0.848	7.5"	1.299	2.50	2.646	3.75	3.507	5.00	4.323	9.0	6.276
1.9"	0.517	4.00"	0.877	8.0"	1.348	2.75	2.772	4.00	3.628	5.50	4.560	9.5	6.453
2.0"	0.525	4.25"	0.905	8.5"	1.395	3.00	2.886	4.25	3.742	6.00	4.779	10	6.621
2.1"	0.533	4.50"	0.932	9.0"	1.440	3.25	2.990	4.50	3.850	6.50	4.982	11	6.930
2.2"	0.541	4.75"	0.958	9.5"	1.484	3.50	3.084	4.75	3.951	7.00	5.169	12	7.208

2.3″	0.548	5.00″	0.993	1.0	1.526	3.75	3.170	5.00	4.047	7.50	5.343	13	7.457
2.4″	0.554	5.50″	1.028	1.1	1.605	4.00	3.247	5.50	4.222	8.00	5.504	14	7.680
2.5″	0.559	6.00″	1.070	1.2	1.679	4.25	3.317	6.00	4.378	8.50	5.653	15	7.880
2.6″	0.565	6.50″	1.108	1.3	1.747	4.50	3.381	6.50	4.516	9.00	5.790	16	8.060
2.8″	0.574	7.00″	1.143	1.4	1.811	4.75	3.439	7.00	4.639	9.50	5.917	18	8.365
3.0″	0.582	7.50″	1.174	1.5	1.870	5.00	3.491	7.50	4.749	10	6.035	20	8.611
3.2″	0.588	8.00″	1.203	1.6	1.924	5.50	3.581	8.00	4.846	11	6.246	22	8.809
3.4″	0.594	9.00″	1.253	1.7	1.975	6.00	3.656	8.50	4.932	12	6.425	24	8.968
3.6″	0.599	1.00″	1.295	1.8	2.022	6.50	3.717	9.00	5.009	13	6.580	26	9.097
3.8″	0.603	1.1	1.330	2.0	2.106	7.00	3.767	9.50	5.078	14	6.712	28	9.200
4.0″	0.606	1.2	1.358	2.2	2.178	7.50	3.809	10.00	5.138	15	6.825	30	9.283
4.5″	0.613	1.3	1.382	2.4	2.241	8.00	3.843	11	5.241	16	6.922	34	9.404
5.0″	0.617	1.4	1.402	2.6	2.294	9.00	3.894	12	5.321	17	7.004	38	9.481
6.0″	0.621	1.6	1.432	2.8	2.340	10.00	3.928	13	5.385	18	7.076	42	9.532
7.0″	0.623	1.7	1.444	3.0	2.380	11.00	3.951	14	5.435	20	7.189	46	9.565
8.0″	0.624	1.8	1.453	3.4	2.444	12.00	3.967	15	5.476	22	7.272	50	9.586
		2.0	1.468	3.8	2.491	14.00	3.985	16	5.506	24	7.332	60	9.612
		2.5	1.487	4.2	2.525	16.00	3.993	17	5.531	26	7.377	70	9.621
		3.0	1.495	4.6	2.551	18.00	3.997	18	5.551	30	7.434	80	9.623
		4.0	1.499	5.0	2.570	20.00	3.999	20	5.579	34	7.464	90	9.624
		5.0	1.500	6.0	2.599	22.00	3.999	25	5.611	38	7.481	100	9.625
				7.0	2.613	24.00	4.000	30	5.621	42	7.490		
				8.0	2.619			35	5.624	46	7.494		
				9.0	2.622			40	5.625	50	7.499		
				10.0	2.624								

Table 10-2 (continued)

$r_e/r_R = 5.0$		$r_e/r_R = 6.0$		$r_e/r_R = 7.0$		$r_e/r_R = 8.0$		$r_e/r_R = 9.0$		$r_e/r_R = 10.0$	
Dimensionless time t_D	Fluid influx W_{eD}	Dimensionless time t_D	Fluid influx W_{eD}	Dimensionless time t_D	Fluid influx W_{eD}	Dimensionless time t_D	Fluid influx W_{eD}	Dimensionless time t_D	Fluid influx W_{eD}	Dimensionless time t_D	Fluid influx W_{eD}
3.0	3.195	6.0	5.148	9.00	6.861	9	6.861	10	7.417	15	9.965
3.5	3.542	6.5	5.440	9.50	7.127	10	7.398	15	9.945	20	12.32
4.0	3.875	7.0	5.724	10	7.389	11	7.920	20	12.26	22	13.22
4.5	4.193	7.5	6.002	11	7.902	12	8.431	22	13.13	24	14.95
5.0	4.499	8.0	6.273	12	8.397	13	8.930	24	13.98	26	14.95
5.5	4.792	8.5	6.537	13	8.876	14	9.418	26	14.79	28	15.78
6.0	5.074	9.0	6.795	14	9.341	15	9.895	28	15.59	30	16.59
6.5	5.345	9.5	7.047	15	9.791	16	10.361	30	16.35	32	17.38
7.0	5.605	10.0	7.293	16	10.23	17	10.82	32	17.10	34	18.16
7.5	5.854	10.5	7.533	17	10.65	18	11.26	34	17.82	36	18.91
8.0	6.094	11	7.767	18	11.06	19	11.70	36	18.52	38	19.65
8.5	6.325	12	8.220	19	11.46	20	12.13	38	19.19	40	20.37
9.0	6.547	13	8.651	20	11.85	22	12.95	40	19.85	42	21.07
9.5	6.760	14	9.063	22	12.58	24	13.74	42	20.48	44	21.76
10	6.965	15	9.456	24	13.27	26	14.50	44	21.09	46	22.42
11	7.350	16	9.829	26	13.92	28	15.23	46	21.69	48	23.07
12	7.706	17	10.19	28	14.53	30	15.92	48	22.26	50	23.71

t_D	W_{eD}	t_D	W_{eD}	t_D	W_{eD}	t_D	W_{eD}	t_D	W_{eD}	t_D	W_{eD}
13	8.035	18	10.53	30	15.11	34	17.22	50	22.82	52	24.33
14	8.339	19	10.85	35	16.39	38	18.41	52	23.36	54	24.94
15	8.620	20	11.16	40	17.49	40	18.97	54	23.89	56	25.53
16	8.879	22	11.74	45	18.43	45	20.26	56	24.39	58	26.11
18	9.338	24	12.26	50	19.24	50	21.42	58	24.88	60	26.67
20	9.731	25	12.50	60	20.51	55	22.46	60	25.36	65	28.02
22	10.07	31	13.74	70	21.45	60	23.40	65	26.48	70	29.29
24	10.35	35	14.40	80	22.13	70	24.98	70	27.52	75	30.49
26	10.59	39	14.93	90	22.63	80	26.26	75	28.48	80	31.61
28	10.80	51	16.05	100	23.00	90	27.28	80	29.36	85	32.67
30	10.98	60	16.56	120	23.47	100	28.11	85	30.18	90	33.66
34	11.26	70	16.91	140	23.71	120	29.31	90	30.93	95	34.60
38	11.46	80	17.14	160	23.85	140	30.08	95	31.63	100	35.48
42	11.61	90	17.27	180	23.92	160	30.58	100	32.27	120	38.51
46	11.71	100	17.36	200	23.96	180	30.91	120	34.39	140	40.89
50	11.79	110	17.41	500	24.00	200	31.12	140	35.92	160	42.75
60	11.91	120	17.45			240	31.34	160	37.04	180	44.21
70	11.96	130	17.46			280	31.43	180	37.85	200	45.36
80	11.98	140	17.48			320	31.47	200	38.44	240	46.95
90	11.99	150	17.49			360	31.49	240	39.17	280	47.94
100	12.00	160	17.49			400	31.50	280	39.56	320	48.54
120	12.00	180	17.50			500	31.50	320	39.77	360	48.91
		200	17.50					360	39.88	400	49.14
		220	17.50					400	39.94	440	49.28
								440	39.97	480	49.36
								480	39.98		

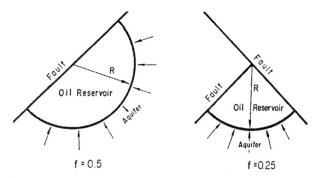

Figure 10-12. Gas cap drive reservoir. *(After Cole, F., Reservoir Engineering Manual, Gulf Publishing Company, 1969.)*

(text continued from page 659)

	Reservoir	Aquifer
radius, ft	2000	∞
h, ft	20	25
k, md	50	100
ϕ, %	15	20
μ_w, cp	0.5	0.8
c_w, psi^{-1}	1×10^{-6}	0.7×10^{-6}
c_f, psi^{-1}	2×10^{-6}	0.3×10^{-6}

Solution

Step 1. Calculate the total compressibility coefficient c_t.

$$c_t = 0.7\,(10^{-6}) + 0.3\,(10^{-3}) = 1 \times 10^{-6}\ \text{psi}^{-1}$$

Step 2. Determine the water influx constant from Equation 10-23.

$$B = 1.119\,(0.2)\,(\,1 \times 10^{-6})\,(2000)^2\,(25)\,(360/360) = 22.4$$

Step 3. Calculate the corresponding dimensionless time after 1, 2, and 5 years.

$$t_D = 6.328 \times 10^{-3}\,\frac{100\,t}{(0.8)\,(0.2)\,(1 \times 10^{-6})\,(2000)^2}$$

$$t_D = 0.9888t$$

t, days	$t_D = 0.9888\ t$
365	361
730	722
1825	1805

Step 4. Using Table 10-1, determine the dimensionless water influx W_{eD}.

t, days	t_D	W_{eD}
365	361	123.5
730	722	221.8
1825	1805	484.6

Step 5. Calculate the cumulative water influx by applying Equation 10-20.

t, days	W_{eD}	$W_e = (20.4)\ (2500 - 2490)\ W_{eD}$
365	123.5	25,200 bbl
730	221.8	45,200 bbl
1825	484.6	98,800 bbl

Example 10-6 shows that, for a given pressure drop, doubling the time interval will not double the water influx. This example also illustrates how to calculate water influx as a result of a single pressure drop. As there will usually be many of these pressure drops occurring throughout the prediction period, it is necessary to analyze the procedure to be used where these multiple pressure drops are present.

Consider Figure 10-13, which illustrates the decline in the boundary pressure as a function of time for a radial reservoir-aquifer system. If the boundary pressure in the reservoir shown in Figure 10-13 is suddenly reduced at time t, from p_i to p_1, a pressure drop of $(p_i - p_1)$ will be imposed across the aquifer. Water will continue to expand and the new reduced pressure will continue to move outward into the aquifer. Given a sufficient length of time the pressure at the outer edge of the aquifer will finally be reduced to p_1.

If some time after the boundary pressure has been reduced to p_1, a second pressure p_2 is suddenly imposed at the boundary, and a new pressure wave will begin moving outward into the aquifer. This new pressure wave will also cause water expansion and therefore encroachment into the reservoir. This new pressure drop, however, will not be $p_i - p_2$, but will be $p_1 - p_2$. This second pressure wave will be moving behind the

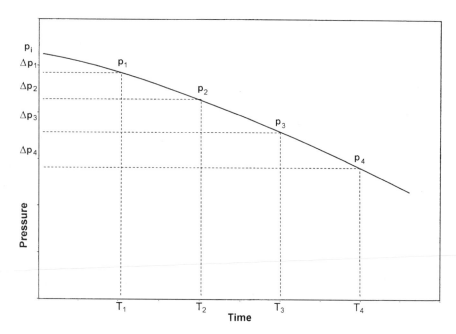

Figure 10-13. Boundary pressure versus time.

first pressure wave. Just ahead of the second pressure wave will be the pressure at the end of the first pressure drop, p_1.

Since these pressure waves are assumed to occur at different times, they are entirely independent of each other. Thus, water expansion will continue to take place as a result of the first pressure drop, even though additional water influx is also taking place as a result of one or more later pressure drops. This is essentially an application of the **principle of superposition.** In order to determine the total water influx into a reservoir at any given time, it is necessary to determine the water influx as a result of each successive pressure drop that has been imposed on the reservoir and aquifer.

In calculating cumulative water influx into a reservoir at successive intervals, it is necessary to calculate the total water influx from the beginning. This is required because of the different times during which the various pressure drops have been effective.

The van Everdingen-Hurst computational steps for determining the water influx are summarized below in conjunction with Figure 10-14:

Step 1. Assume that the boundary pressure has declined from its initial value of p_i to p_1 after t_1 days. To determine the cumulative water

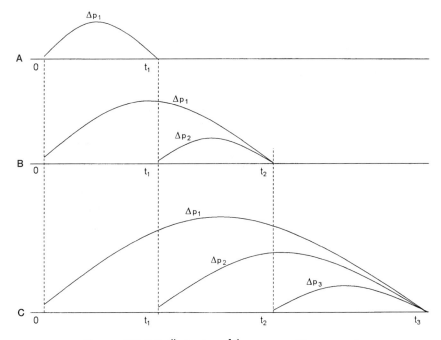

Figure 10-14. Illustration of the superposition concept.

influx in response to this first pressure drop, $\Delta p_1 = p_i - p_1$ can be simply calculated from Equation 10-20, or:

$$W_e = B \, \Delta p_1 \, (W_{eD})_{t_1}$$

Where W_e is the cumulative water influx due to the first pressure drop Δp_1. The dimensionless water influx $(W_{eD})_{t_1}$ is evaluated by calculating the dimensionless time at t_1 days. This simple calculation step is shown in section A of Figure 10-14.

Step 2. Let the boundary pressure decline again to p_2 after t_2 days with a pressure drop of $\Delta p_2 = p_1 - p_2$. The cumulative (total) water influx after t_2 days will result from the first pressure drop Δp_1 and the second pressure drop Δp_2, or:

$$W_e = \text{water influx due to } \Delta p_1 + \text{water influx due to } \Delta p_2$$

$$W_e = (W_e)_{\Delta p_1} + (W_e)_{\Delta p_2}$$

where

$$(W_e)_{\Delta p_1} = B \, \Delta p_1 \, (W_{eD})_{t_2}$$

$$(W_e)_{\Delta p_2} = B \, \Delta p_2 \, (W_{eD})_{t_2 - t_1}$$

The above relationships indicate that the effect of the first pressure drop Δp_1 will continue for the entire time t_2, while the effect of the second pressure drop will continue only for $(t_2 - t_1)$ days as shown in section B of Figure 10-14.

Step 3. A third pressure drop of $\Delta p_3 = p_2 - p_3$ would cause an additional water influx as illustrated in section C of Figure 10-14. The cumulative (total) cumulative water influx can then be calculated from:

$$W_e = (W_e)_{\Delta p_1} + (W_e)_{\Delta p_2} + (W_e)_{\Delta p_3}$$

where

$$(W_e)_{\Delta p_1} = B \, \Delta p_1 \, (W_{eD})_{t_3}$$

$$(W_e)_{\Delta p_2} = B \, \Delta p_2 \, (W_{eD})_{t_3 - t_1}$$

$$(W_e)_{\Delta p_3} = B \, \Delta p_3 \, (W_{eD})_{t_3 - t_2}$$

The van Everdingen-Hurst water influx relationship can then be expressed in a more generalized form as:

$$W_e = B \, \Sigma \, \Delta p \, W_{eD} \tag{10-24}$$

The authors also suggested that instead of using the entire pressure drop for the first period, a better approximation is to consider that one-half of the pressure drop, $\frac{1}{2} (p_i - p_1)$, is effective during the entire first period. For the second period, the effective pressure drop then is one-half of the pressure drop during the first period, $\frac{1}{2} (p_i - p_2)$, which simplifies to:

$$\frac{1}{2} (p_i - p_1) + \frac{1}{2} (p_1 - p_2) = \frac{1}{2} (p_i - p_2)$$

Similarly, the effective pressure drop for use in the calculations for the third period would be one-half of the pressure drop during the second period, $\frac{1}{2} (p_1 - p_2)$, plus one-half of the pressure drop during the third period, $\frac{1}{2} (p_2 - p_3)$, which simplifies to $\frac{1}{2} (p_1 - p_3)$. The time intervals must all be equal in order to preserve the accuracy of these modifications.

Example 10-7

Using the data given in Example 10-6, calculate the cumulative water influx at the end of 6, 12, 18, and 24 months. The predicted boundary pressure at the end of each specified time period is given below:

Time, months	Boundary pressure, psi
0	2500
6	2490
12	2472
18	2444
24	2408

Solution

Water influx at the end of 6 months

Step 1. Determine water influx constant B:

$$B = 22.4 \text{ bbl/psi}$$

Step 2. Calculate the dimensionless time t_D at 182.5 days.

$$t_D = 0.9888t$$
$$= 0.9888 \,(182.5) = 180.5$$

Step 3. Calculate the first pressure drop Δp_1. This pressure is taken as ½ of the actual pressure drop, or:

$$\Delta p_1 = \frac{p_i - p_1}{2}$$

$$\Delta p_1 = \frac{2500 - 2490}{2} = 5\,\text{psi}$$

Step 4. Determine the dimension water influx W_{eD} from Table 10-1 at $t_D = 180.5$ to give:

$$W_{eD} = 69.46$$

Step 5. Calculate the cumulative water influx at the end of 182.5 days due to the first pressure drop of 5 psi by using the van Everdingen-Hurst equation, or:

$$W_e = (20.4) \, (5) \, (69.46) = 7080 \text{ bbl}$$

Cumulative water influx after 12 months

Step 1. After an additional six months, the pressure has declined from 2490 psi to 2472 psi. This second pressure Δp_2 is taken as one-half the *actual* pressure drop during the first period, plus one-half the *actual* pressure drop during the second period, or:

$$\Delta p_2 = \frac{p_i - p_2}{2}$$

$$= \frac{2500 - 2472}{2} = 14 \, \text{psi}$$

Step 2. The cumulative (total) water influx at the end of 12 months would result from the first pressure drop Δp_1 and the second pressure drop Δp_2.

   The first pressure drop Δp_1 has been effective for one year, but the second pressure drop, Δp_2, has been effective only 6 months, as shown in Figure 10-15.

   Separate calculations must be made for the two pressure drops because of this time difference and the results added in order to determine the total water influx, i.e.:

$$W_e = (W_e)_{\Delta p_1} + (W_e)_{\Delta p_2}$$

Figure 10-15. Duration of the pressure drop in Example 10-7.

Step 3. Calculate the dimensionless time at 365 days as:

$t_D = 0.9888t$
$= 0.9888\,(365) = 361$

Step 4. Determine the dimensionless water influx at $t_D = 361$ from Table 10-1 to give:

$W_{eD} = 123.5$

Step 5. Calculate the water influx due to the first and second pressure drop, i.e., $(W_e)_{\Delta p_1}$ and $(W_e)_{\Delta p_2}$, or:

$(W_e)_{\Delta p_1} = (20.4)(5)(123.5) = 12{,}597$ bbl

$(W_e)_{\Delta p_2} = (20.4)(14)(69.46) = 19{,}838$

Step 6. Calculate total (cumulative) water influx after one year.

$W_e = 12{,}597 + 19{,}938 = 32{,}435$ bbl

Water influx after 18 months

Step 1. Calculate the third pressure drop Δp_3 which is taken as ½ of the *actual* pressure drop during the second period plus ½ of the *actual* pressure drop during the third period, or:

$$\Delta p_3 = \frac{p_1 - p_3}{2}$$

$$\Delta p_3 = \frac{2490 - 2444}{2} = 23\,\text{psi}$$

Step 2. Calculate the dimensionless time after 6 months.

$t_D = 0.9888\,t$
$= 0.9888\,(547.5) = 541.5$

Step 3. Determine the dimensionless water influx at:

$t_D = 541.5$ from Table 10-1

$W_{eD} = 173.7$

Step 4. The first pressure drop will have been effective the entire 18 months, the second pressure drop will have been effective for 12 months, and the last pressure drop will have been effective only 6 months, as shown in Figure 10-16. Therefore, the cumulative water influx is calculated below:

Time, days	t_D	Δp	W_{eD}	$B\Delta p\,W_{eD}$
547.5	541.5	5	173.7	17,714
365	361	14	123.5	35,272
182.5	180.5	23	69.40	32,291

$$W_e = 85,277 \text{ bbl}$$

Water influx after two years

The first pressure drop has now been effective for the entire two years, the second pressure drop has been effective for 18 months, the third pressure drop has been effective for 12 months, and the fourth pressure drop has been effective only 6 months. Summary of the calculations is given below:

Time, days	t_D	Δp	W_{eD}	$B\Delta p\,W_{eD}$
730	722	5	221.8	22,624
547.5	541.5	14	173.7	49,609
365	631	23	123.5	57,946
182.5	180.5	32	69.40	45,343

$$W_e = 175,522 \text{ bbl}$$

Edwardson and coworkers (1962) developed three sets of simple polynomial expressions for calculating the dimensionless water influx W_{eD} for

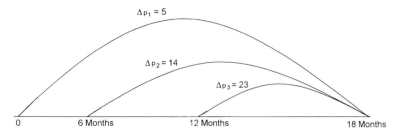

Figure 10-16. Pressure drop data for Example 10-7.

infinite-acting aquifers. The proposed three expressions essentially approximate the W_{eD} values in three different dimensionless time regions.

• For $t_D < 0.01$

$$W_{eD} = 2\left(\frac{t_D}{\pi}\right)^{0.5} \qquad (10\text{-}25)$$

• For $0.01 < t_D < 200$

$$W_{eD} = \frac{1.2838\sqrt{t_D} + 1.19328\,t_D + 0.269872\,(t_D)^{3/2} + 0.00855294\,(t_D)^2}{1 + 0.616599\sqrt{t_D} + 0.0413008\,t_D}$$

$$(10\text{-}26)$$

• For $t_D > 200$

$$W_{eD} = \frac{-4.29881 + 2.02566\,t_D}{\ln(t_D)} \qquad (10\text{-}27)$$

Bottom-Water Drive

The van Everdingen-Hurst solution to the *radial* diffusivity equation is considered the most rigorous aquifer influx model to date. The proposed solution technique, however, is not adequate to describe the vertical water encroachment in bottom-water-drive system. Coats (1962) presented a mathematical model that takes into account the vertical flow effects from bottom-water aquifers. He correctly noted that in many cases reservoirs are situated on top of an aquifer with a continuous horizontal interface between the reservoir fluid and the aquifer water and with a significant aquifer thickness. He stated that in such situations significant bottom-water drive would occur. Coats modified the diffusivity equation to account for the vertical flow by including an additional term in the equation, to give:

$$\frac{\partial^2 p}{\partial r^2} + \frac{1}{r}\frac{\partial p}{\partial r} + F_k \frac{\partial^2 p}{\partial z^2} = \frac{\mu \phi c}{k}\frac{\partial p}{\partial t} \qquad (10\text{-}28)$$

where F_k is the ratio of vertical to horizontal permeability, or:

$$F_k = k_v/k_h \qquad (10\text{-}29)$$

where k_v = vertical permeability

k_h = horizontal permeability

Allard and Chen (1988) pointed out that there are an infinite number of solutions to Equation 10-28, representing all possible reservoir-aquifer configurations. They suggested that it is possible to derive a general solution that is applicable to a variety of systems by the solution to Equation 10-28 in terms of the dimensionless time t_D, dimensionless radius r_D, and a newly introduced dimensionless variable z_D.

$$z_D = \frac{h}{r_e \sqrt{F_k}} \tag{10-30}$$

where z_D = dimensionless vertical distance

h = aquifer thickness, ft

Allen and Chen used a numerical model to solve Equation 10-28. The authors developed a solution to the bottom-water influx that is comparable in form with that of van Everdingen and Hurst.

$$W_e = B \, \Delta p \, W_{eD} \tag{10-31}$$

They defined the water influx constant B identical to that of Equation 10-21, or

$$B = 1.119 \, \phi \, c_t \, r_e^2 \, h \tag{10-32}$$

Notice that the water influx constant B does not include the encroachment angle θ.

The actual values of W_{eD} are different from those of the van Everdingen-Hurst model because W_{eD} for the bottom-water drive is also a function of the vertical permeability. Allard and Chen tabulated the values of W_{eD} as a function of r_D, t_D, and z_D. These values are presented in Tables 10-3 through 10-7.

The solution procedure of a bottom-water influx problem is identical to the edge-water influx problem outlined in Example 10-7. Allard and Chen illustrated results of their method in the following example.

Example 10-8

An infinite-acting bottom-water aquifer is characterized by the following properties:

$r_a = \infty$ $k_h = 50$ md $F_k = 0.04$
$\phi = 0.1$ $\mu_w = 0.395$ cp $c_t = 8 \times 10^{-6}$ psi^{-1}
$h = 200'$ $r_e = 2000'$ $\theta = 360°$

The boundary pressure history is given below:

Time, days	p, psi
0	3000
30	2956
60	2917
90	2877
120	2844
150	2811
180	2791
210	2773
240	2755

Calculate the cumulative water influx as a function of time by using the bottom-water-drive solution and compare with the edge-water-drive approach.

Solution

Step 1. For an infinite-acting aquifer:

$$r_D = \infty$$

Step 2. Calculate z_D from Equation 10-30.

$$z_D = \frac{200}{2000\sqrt{0.04}} = 0.5$$

Step 3. Calculate the water influx constant B.

$$B = 1.119\,(0.1)\,(200)\,(8 \times 10^{-6})\,(2000)^2 = 716 \text{ bbl/psi}$$

(text continued on page 702)

Table 10-3
Dimensionless Water Influx, W_{eD}, for Infinite Aquifer
(Permission to publish by the SPE)

t_D				z_D'			
	0.05	0.1	0.3	0.5	0.7	0.9	1.0
0.1	0.700	0.677	0.508	0.349	0.251	0.195	0.176
0.2	0.793	0.786	0.696	0.547	0.416	0.328	0.295
0.3	0.936	0.926	0.834	0.692	0.548	0.440	0.396
0.4	1.051	1.041	0.952	0.812	0.662	0.540	0.486
0.5	1.158	1.155	1.059	0.918	0.764	0.631	0.569
0.6	1.270	1.268	1.167	1.021	0.862	0.721	0.651
0.7	1.384	1.380	1.270	1.116	0.953	0.806	0.729
0.8	1.503	1.499	1.373	1.205	1.039	0.886	0.803
0.9	1.621	1.612	1.477	1.286	1.117	0.959	0.872
1	1.743	1.726	1.581	1.347	1.181	1.020	0.932
2	2.402	2.393	2.288	2.034	1.827	1.622	1.509
3	3.031	3.018	2.895	2.650	2.408	2.164	2.026
4	3.629	3.615	3.477	3.223	2.949	2.669	2.510
5	4.217	4.201	4.048	3.766	3.462	3.150	2.971
6	4.784	4.766	4.601	4.288	3.956	3.614	3.416
7	5.323	5.303	5.128	4.792	4.434	4.063	3.847
8	5.829	5.808	5.625	5.283	4.900	4.501	4.268
9	6.306	6.283	6.094	5.762	5.355	4.929	4.680
10	6.837	6.816	6.583	6.214	5.792	5.344	5.080
11	7.263	7.242	7.040	6.664	6.217	5.745	5.468
12	7.742	7.718	7.495	7.104	6.638	6.143	5.852
13	8.196	8.172	7.943	7.539	7.052	6.536	6.231
14	8.648	8.623	8.385	7.967	7.461	6.923	6.604
15	9.094	9.068	8.821	8.389	7.864	7.305	6.973
16	9.534	9.507	9.253	8.806	8.262	7.682	7.338

17	7.699	8.056	8.656	9.218	9.679	9.942	9.969
18	8.057	8.426	9.046	9.626	10.100	10.371	10.399
19	8.411	8.793	9.432	10.029	10.516	10.794	10.823
20	8.763	9.156	9.815	10.430	10.929	11.211	11.241
21	9.111	9.516	10.194	10.826	11.339	11.633	11.664
22	9.457	9.874	10.571	11.219	11.744	12.045	12.075
23	9.801	10.229	10.944	11.609	12.147	12.454	12.486
24	10.142	10.581	11.315	11.996	12.546	12.861	12.893
25	10.481	10.931	11.683	12.380	12.942	13.264	13.297
26	10.817	11.279	12.048	12.761	13.336	13.665	13.698
27	11.152	11.625	12.411	13.140	13.726	14.062	14.097
28	11.485	11.968	12.772	13.517	14.115	14.458	14.493
29	11.816	12.310	13.131	13.891	14.501	14.850	14.886
30	12.145	12.650	13.488	14.263	14.884	15.241	15.277
31	12.473	12.990	13.843	14.634	15.266	15.628	15.666
32	12.799	13.324	14.196	15.002	15.645	16.015	16.053
33	13.123	13.659	14.548	15.368	16.023	16.398	16.437
34	13.446	13.992	14.897	15.732	16.398	16.780	16.819
35	13.767	14.324	15.245	16.095	16.772	17.160	17.200
36	14.088	14.654	15.592	16.456	17.143	17.538	17.579
37	14.406	14.983	15.937	16.815	17.513	17.915	17.956
38	14.724	15.311	16.280	17.173	17.882	18.289	18.331
39	15.040	15.637	16.622	17.529	18.249	18.662	18.704
40	15.356	15.963	16.964	17.886	18.620	19.045	19.088
41	15.671	16.288	17.305	18.240	18.982	19.407	19.450
42	15.985	16.611	17.644	18.592	19.344	19.777	19.821
43	16.297	16.933	17.981	18.943	19.706	20.144	20.188
44	16.608	17.253	18.317	19.293	20.065	20.510	20.555
45	16.918	17.573	18.651	19.641	20.424	20.874	20.920
46	17.227	17.891	18.985	19.988	20.781	21.237	21.283
47	17.535	18.208	19.317	20.333	21.137	21.598	21.645
48	17.841	18.524	19.648	20.678	21.491	21.958	22.006
49	18.147	18.840	19.978	21.021	21.844	22.317	22.365
50	18.452	19.154	20.307	21.363	22.196	22.674	22.722
51	18.757	19.467	20.635	21.704	22.547	23.032	23.081

(table continued on next page)

Table 10-3 *(continued)*

t_D	0.05	0.1	0.3	0.5	0.7	0.9	1.0
				z_D			
52	23.436	23.387	22.897	22.044	20.962	19.779	19.060
53	23.791	23.741	23.245	22.383	21.288	20.091	19.362
54	24.145	24.094	23.593	22.721	21.613	20.401	19.664
55	24.498	24.446	23.939	23.058	21.937	20.711	19.965
56	24.849	24.797	24.285	23.393	22.260	21.020	20.265
57	25.200	25.147	24.629	23.728	22.583	21.328	20.564
58	25.549	25.496	24.973	24.062	22.904	21.636	20.862
59	25.898	25.844	25.315	24.395	23.225	21.942	21.160
60	26.246	26.191	25.657	24.728	23.545	22.248	21.457
61	26.592	26.537	25.998	25.059	23.864	22.553	21.754
62	26.938	26.883	26.337	25.390	24.182	22.857	22.049
63	27.283	27.227	26.676	25.719	24.499	23.161	22.344
64	27.627	27.570	27.015	26.048	24.816	23.464	22.639
65	27.970	27.913	27.352	26.376	25.132	23.766	22.932
66	28.312	28.255	27.688	26.704	25.447	24.068	23.225
67	28.653	28.596	28.024	27.030	25.762	24.369	23.518
68	28.994	28.936	28.359	27.356	26.075	24.669	23.810
69	29.334	29.275	28.693	27.681	26.389	24.969	24.101
70	29.673	29.614	29.026	28.006	26.701	25.268	24.391
71	30.011	29.951	29.359	28.329	27.013	25.566	24.681
72	30.349	30.288	29.691	28.652	27.324	25.864	24.971
73	30.686	30.625	30.022	28.974	27.634	26.161	25.260
74	31.022	30.960	30.353	29.296	27.944	26.458	25.548
75	31.357	31.295	30.682	29.617	28.254	26.754	25.836
76	31.692	31.629	31.012	29.937	28.562	27.049	26.124
77	32.026	31.963	31.340	30.257	28.870	27.344	26.410
78	32.359	32.296	31.668	30.576	29.178	27.639	26.697
79	32.692	32.628	31.995	30.895	29.485	27.933	26.983
80	33.024	32.959	32.322	31.212	29.791	28.226	27.268
81	33.355	33.290	32.647	31.530	30.097	28.519	27.553

82	27.837	28.812	30.402	31.846	32.973	33.621	33.686
83	28.121	29.104	30.707	32.163	33.297	33.950	34.016
84	28.404	29.395	31.011	32.478	33.622	34.279	34.345
85	28.687	29.686	31.315	32.793	33.945	34.608	34.674
86	28.970	29.976	31.618	33.107	34.268	34.935	35.003
87	29.252	30.266	31.921	33.421	34.590	35.263	35.330
88	29.534	30.556	32.223	33.735	34.912	35.589	35.657
89	29.815	30.845	32.525	34.048	35.233	35.915	35.984
90	30.096	31.134	32.826	34.360	35.554	36.241	36.310
91	30.376	31.422	33.127	34.672	35.874	36.566	36.636
92	30.656	31.710	33.427	34.983	36.194	36.890	36.960
93	30.935	31.997	33.727	35.294	36.513	37.214	37.285
94	31.215	32.284	34.026	35.604	36.832	37.538	37.609
95	31.493	32.570	34.325	35.914	37.150	37.861	37.932
96	31.772	32.857	34.623	36.223	37.467	38.183	38.255
97	32.050	33.142	34.921	36.532	37.785	38.505	38.577
98	32.327	33.427	35.219	36.841	38.101	38.826	38.899
99	32.605	33.712	35.516	37.149	38.417	39.147	39.220
100	32.881	33.997	35.813	37.456	38.733	39.467	39.541
105	34.260	35.414	37.290	38.987	40.305	41.062	41.138
110	35.630	36.821	38.758	40.508	41.865	42.645	42.724
115	36.993	38.221	40.216	42.018	43.415	44.218	44.299
120	38.347	39.612	41.666	43.520	44.956	45.781	45.864
125	39.694	40.995	43.107	45.012	46.487	47.334	47.420
130	41.035	42.372	44.541	46.497	48.009	48.879	48.966
135	42.368	43.741	45.967	47.973	49.523	50.414	50.504
140	43.696	45.104	47.386	49.441	51.029	51.942	52.033
145	45.017	46.460	48.798	50.903	52.528	53.462	53.555
150	46.333	47.810	50.204	52.357	54.019	54.974	55.070
155	47.643	49.155	51.603	53.805	55.503	56.479	56.577
160	48.947	50.494	52.996	55.246	56.981	57.977	58.077
165	50.247	51.827	54.384	56.681	58.452	59.469	59.570
170	51.542	53.156	55.766	58.110	59.916	60.954	61.058
175	52.832	54.479	57.143	59.534	61.375	62.433	62.539
180	54.118	55.798	58.514	60.952	62.829	63.906	64.014

(table continued on next page)

Table 10-3 (continued)

z'_D

t_D	0.05	0.1	0.3	0.5	0.7	0.9	1.0
185	65.484	65.374	64.276	62.365	59.881	57.112	55.399
190	66.948	66.836	65.718	63.773	61.243	58.422	56.676
195	68.406	68.293	67.156	65.175	62.600	59.727	57.949
200	69.860	69.744	68.588	66.573	63.952	61.028	59.217
205	71.309	71.191	70.015	67.967	65.301	62.326	60.482
210	72.752	72.633	71.437	69.355	66.645	63.619	61.744
215	74.191	74.070	72.855	70.740	67.985	64.908	63.001
220	75.626	75.503	74.269	72.120	69.321	66.194	64.255
225	77.056	76.931	75.678	73.496	70.653	67.476	65.506
230	78.482	78.355	77.083	74.868	71.981	68.755	66.753
235	79.903	79.774	78.484	76.236	73.306	70.030	67.997
240	81.321	81.190	79.881	77.601	74.627	71.302	69.238
245	82.734	82.602	81.275	78.962	75.945	72.570	70.476
250	84.144	84.010	82.664	80.319	77.259	73.736	71.711
255	85.550	85.414	84.050	81.672	78.570	75.098	72.943
260	86.952	86.814	85.432	83.023	79.878	76.358	74.172
265	88.351	88.211	86.811	84.369	81.182	77.614	75.398
270	89.746	89.604	88.186	85.713	82.484	78.868	76.621
275	91.138	90.994	89.558	87.053	83.782	80.119	77.842
280	92.526	92.381	90.926	88.391	85.078	81.367	79.060
285	93.911	93.764	92.292	89.725	86.371	82.612	80.276
290	95.293	95.144	93.654	91.056	87.660	83.855	81.489
295	96.672	96.521	95.014	92.385	88.948	85.095	82.700
300	98.048	97.895	96.370	93.710	90.232	86.333	83.908
305	99.420	99.266	97.724	95.033	91.514	87.568	85.114

310	100.79	100.64	99.07	96.35	92.79	88.80	86.32
315	102.16	102.00	100.42	97.67	94.07	90.03	87.52
320	103.52	103.36	101.77	98.99	95.34	91.26	88.72
325	104.88	104.72	103.11	100.30	96.62	92.49	89.92
330	106.24	106.08	104.45	101.61	97.89	93.71	91.11
335	107.60	107.43	105.79	102.91	99.15	94.93	92.30
340	108.95	108.79	107.12	104.22	100.42	96.15	93.49
345	110.30	110.13	108.45	105.52	101.68	97.37	94.68
350	111.65	111.48	109.78	106.82	102.94	98.58	95.87
355	113.00	112.82	111.11	108.12	104.20	99.80	97.06
360	114.34	114.17	112.43	109.41	105.45	101.01	98.24
365	115.68	115.51	113.76	110.71	106.71	102.22	99.42
370	117.02	116.84	115.08	112.00	107.96	103.42	100.60
375	118.36	118.18	116.40	113.29	109.21	104.63	101.78
380	119.69	119.51	117.71	114.57	110.46	105.83	102.95
385	121.02	120.84	119.02	115.86	111.70	107.04	104.13
390	122.35	122.17	120.34	117.14	112.95	108.24	105.30
395	123.68	123.49	121.65	118.42	114.19	109.43	106.47
400	125.00	124.82	122.94	119.70	115.43	110.63	107.64
405	126.33	126.14	124.26	120.97	116.67	111.82	108.80
410	127.65	127.46	125.56	122.25	117.90	113.02	109.97
415	128.97	128.78	126.86	123.52	119.14	114.21	111.13
420	130.28	130.09	128.16	124.79	120.37	115.40	112.30
425	131.60	131.40	129.46	126.06	121.60	116.59	113.46
430	132.91	132.72	130.75	127.33	122.83	117.77	114.62
435	134.22	134.03	132.05	128.59	124.06	118.96	115.77
440	135.53	135.33	133.34	129.86	125.29	120.14	116.93
445	136.84	136.64	134.63	131.12	126.51	121.32	118.08
450	138.15	137.94	135.92	132.38	127.73	122.50	119.24
455	139.45	139.25	137.20	133.64	128.96	123.68	120.39
460	140.75	140.55	138.49	134.90	130.18	124.86	121.54
465	142.05	141.85	139.77	136.15	131.39	126.04	122.69
470	143.35	143.14	141.05	137.40	132.61	127.21	123.84
475	144.65	144.44	142.33	138.66	133.82	128.38	124.98
480	145.94	145.73	143.61	139.91	135.04	129.55	126.13

(table continued on next page)

Table 10-3 (continued)

t_D	0.05	0.1	0.3	0.5	0.7	0.9	1.0
				z'_D			
485	147.24	147.02	144.89	141.15	136.25	130.72	127.27
490	148.53	148.31	146.16	142.40	137.46	131.89	128.41
495	149.82	149.60	147.43	143.65	138.67	133.06	129.56
500	151.11	150.89	148.71	144.89	139.88	134.23	130.70
510	153.68	153.46	151.24	147.38	142.29	136.56	132.97
520	156.25	156.02	153.78	149.85	144.70	138.88	135.24
530	158.81	158.58	156.30	152.33	147.10	141.20	137.51
540	161.36	161.13	158.82	154.79	149.49	143.51	139.77
550	163.91	163.68	161.34	157.25	151.88	145.82	142.03
560	166.45	166.22	163.85	159.71	154.27	148.12	144.28
570	168.99	168.75	166.35	162.16	156.65	150.42	146.53
580	171.52	171.28	168.85	164.61	159.02	152.72	148.77
590	174.05	173.80	171.34	167.05	161.39	155.01	151.01
600	176.57	176.32	173.83	169.48	163.76	157.29	153.25
610	179.09	178.83	176.32	171.92	166.12	159.58	155.48
620	181.60	181.34	178.80	174.34	168.48	161.85	157.71
630	184.10	183.85	181.27	176.76	170.83	164.13	159.93
640	186.60	186.35	183.74	179.18	173.18	166.40	162.15
650	189.10	188.84	186.20	181.60	175.52	168.66	164.37
660	191.59	191.33	188.66	184.00	177.86	170.92	166.58
670	194.08	193.81	191.12	186.41	180.20	173.18	168.79
680	196.57	196.29	193.57	188.81	182.53	175.44	170.99
690	199.04	198.77	196.02	191.21	184.86	177.69	173.20
700	201.52	201.24	198.46	193.60	187.19	179.94	175.39
710	203.99	203.71	200.90	195.99	189.51	182.18	177.59

720	179.78	184.42	191.83	198.37	203.34	206.17	206.46
730	181.97	186.66	194.14	200.75	205.77	208.63	208.92
740	184.15	188.89	196.45	203.13	208.19	211.09	211.38
750	186.34	191.12	198.76	205.50	210.62	213.54	213.83
760	188.52	193.35	201.06	207.87	213.04	215.99	216.28
770	190.69	195.57	203.36	210.24	215.45	218.43	218.73
780	192.87	197.80	205.66	212.60	217.86	220.87	221.17
790	195.04	200.01	207.95	214.96	220.27	223.31	223.61
800	197.20	202.23	210.24	217.32	222.68	225.74	226.05
810	199.37	204.44	212.53	219.67	225.08	228.17	228.48
820	201.53	206.65	214.81	222.02	227.48	230.60	230.91
830	203.69	208.86	217.09	224.36	229.87	233.02	233.33
840	205.85	211.06	219.37	226.71	232.26	235.44	235.76
850	208.00	213.26	221.64	229.05	234.65	237.86	238.18
860	210.15	215.46	223.92	231.38	237.04	240.27	240.59
870	212.30	217.65	226.19	233.72	239.42	242.68	243.00
880	214.44	219.85	228.45	236.05	241.80	245.08	245.41
890	216.59	222.04	230.72	238.37	244.17	247.49	247.82
900	218.73	224.22	232.98	240.70	246.55	249.89	250.22
910	220.87	226.41	235.23	243.02	248.92	252.28	252.62
920	223.00	228.59	237.49	245.34	251.28	254.68	255.01
930	225.14	230.77	239.74	247.66	253.65	257.07	257.41
940	227.27	232.95	241.99	249.97	256.01	259.46	259.80
950	229.39	235.12	244.24	252.28	258.36	261.84	262.19
960	231.52	237.29	246.48	254.59	260.72	264.22	264.57

(table continued on next page)

Table 10-3 (continued)

t_D	z'_D						
	0.05	0.1	0.3	0.5	0.7	0.9	1.0
970	266.95	266.60	263.07	256.89	248.72	239.46	233.65
980	269.33	268.98	265.42	259.19	250.96	241.63	235.77
990	271.71	271.35	267.77	261.49	253.20	243.80	237.89
1,000	274.08	273.72	270.11	263.79	255.44	245.96	240.00
1,010	276.35	275.99	272.35	265.99	257.58	248.04	242.04
1,020	278.72	278.35	274.69	268.29	259.81	250.19	244.15
1,030	281.08	280.72	277.03	270.57	262.04	252.35	246.26
1,040	283.44	283.08	279.36	272.86	264.26	254.50	248.37
1,050	285.81	285.43	281.69	275.15	266.49	256.66	250.48
1,060	288.16	287.79	284.02	277.43	268.71	258.81	252.58
1,070	290.52	290.14	286.35	279.71	270.92	260.95	254.69
1,080	292.87	292.49	288.67	281.99	273.14	263.10	256.79
1,090	295.22	294.84	290.99	284.26	275.35	265.24	258.89
1,100	297.57	297.18	293.31	286.54	277.57	267.38	260.98
1,110	299.91	299.53	295.63	288.81	279.78	269.52	263.08
1,120	302.26	301.87	297.94	291.07	281.98	271.66	265.17
1,130	304.60	304.20	300.25	293.34	284.19	273.80	267.26
1,140	306.93	306.54	302.56	295.61	286.39	275.93	269.35
1,150	309.27	308.87	304.87	297.87	288.59	278.06	271.44
1,160	311.60	311.20	307.18	300.13	290.79	280.19	273.52
1,170	313.94	313.53	309.48	302.38	292.99	282.32	275.61
1,180	316.26	315.86	311.78	304.64	295.19	284.44	277.69
1,190	318.59	318.18	314.08	306.89	297.38	286.57	279.77
1,200	320.92	320.51	316.38	309.15	299.57	288.69	281.85
1,210	323.24	322.83	318.67	311.39	301.76	290.81	283.92
1,220	325.56	325.14	320.96	313.64	303.95	292.93	286.00
1,230	327.88	327.46	323.25	315.89	306.13	295.05	288.07
1,240	330.19	329.77	325.54	318.13	308.32	297.16	290.14
1,250	332.51	332.08	327.83	320.37	310.50	299.27	292.21
1,260	334.82	334.39	330.11	322.61	312.68	301.38	294.28

1,270	337.13	336.70	332.39	324.85	314.85	303.49	296.35
1,280	339.44	339.01	334.67	327.08	317.03	305.60	298.41
1,290	341.74	341.31	336.95	329.32	319.21	307.71	300.47
1,300	344.05	343.61	339.23	331.55	321.38	309.81	302.54
1,310	346.35	345.91	341.50	333.78	323.55	311.92	304.60
1,320	348.65	348.21	343.77	336.01	325.72	314.02	306.65
1,330	350.95	350.50	346.04	338.23	327.89	316.12	308.71
1,340	353.24	352.80	348.31	340.46	330.05	318.22	310.77
1,350	355.54	355.09	350.58	342.68	332.21	320.31	312.82
1,360	357.83	357.38	352.84	344.90	334.38	322.41	314.87
1,370	360.12	359.67	355.11	347.12	336.54	324.50	316.92
1,380	362.41	361.95	357.37	349.34	338.70	326.59	318.97
1,390	364.69	364.24	359.63	351.56	340.85	328.68	321.02
1,400	366.98	366.52	361.88	353.77	343.01	330.77	323.06
1,410	369.26	368.80	364.14	355.98	345.16	332.86	325.11
1,420	371.54	371.08	366.40	358.19	347.32	334.94	327.15
1,430	373.82	373.35	368.65	360.40	349.47	337.03	329.19
1,440	376.10	375.63	370.90	362.61	351.62	339.11	331.23
1,450	378.38	377.90	373.15	364.81	353.76	341.19	333.27
1,460	380.65	380.17	375.39	367.02	355.91	343.27	335.31
1,470	382.92	382.44	377.64	369.22	358.06	345.35	337.35
1,480	385.19	384.71	379.88	371.42	360.20	347.43	339.38
1,490	387.46	386.98	382.13	373.62	362.34	349.50	341.42
1,500	389.73	389.25	384.37	375.82	364.48	351.58	343.45
1,525	395.39	394.90	389.96	381.31	369.82	356.76	348.52
1,550	401.04	400.55	395.55	386.78	375.16	361.93	353.59
1,575	406.68	406.18	401.12	392.25	380.49	367.09	358.65
1,600	412.32	411.81	406.69	397.71	385.80	372.24	363.70
1,625	417.94	417.42	412.24	403.16	391.11	377.39	368.74
1,650	423.55	423.03	417.79	408.60	396.41	382.53	373.77
1,675	429.15	428.63	423.33	414.04	401.70	387.66	378.80
1,700	434.75	434.22	428.85	419.46	406.99	392.78	383.82
1,725	440.33	439.79	434.37	424.87	412.26	397.89	388.83
1,750	445.91	445.37	439.89	430.28	417.53	403.00	393.84
1,775	451.48	450.93	445.39	435.68	422.79	408.10	398.84

(table continued on next page)

Table 10-3 (continued)

z'_D

t_D	0.05	0.1	0.3	0.5	0.7	0.9	1.0
1,800	457.04	456.48	450.88	441.07	428.04	413.20	403.83
1,825	462.59	462.03	456.37	446.46	433.29	418.28	408.82
1,850	468.13	467.56	461.85	451.83	438.53	423.36	413.80
1,875	473.67	473.09	467.32	457.20	443.76	428.43	418.77
1,900	479.19	478.61	472.78	462.56	448.98	433.50	423.73
1,925	484.71	484.13	478.24	467.92	454.20	438.56	428.69
1,950	490.22	489.63	483.69	473.26	459.41	443.61	433.64
1,975	495.73	495.13	489.13	478.60	464.61	448.66	438.59
2,000	501.22	500.62	494.56	483.93	469.81	453.70	443.53
2,025	506.71	506.11	499.99	489.26	475.00	458.73	448.47
2,050	512.20	511.58	505.41	494.58	480.18	463.76	453.40
2,075	517.67	517.05	510.82	499.89	485.36	468.78	458.32
2,100	523.14	522.52	516.22	505.19	490.53	473.80	463.24
2,125	528.60	527.97	521.62	510.49	495.69	478.81	468.15
2,150	534.05	533.42	527.02	515.78	500.85	483.81	473.06
2,175	539.50	538.86	532.40	521.07	506.01	488.81	477.96
2,200	544.94	544.30	537.78	526.35	511.15	493.81	482.85
2,225	550.38	549.73	543.15	531.62	516.29	498.79	487.74
2,250	555.81	555.15	548.52	536.89	521.43	503.78	492.63
2,275	561.23	560.56	553.88	542.15	526.56	508.75	497.51
2,300	566.64	565.97	559.23	547.41	531.68	513.72	502.38
2,325	572.05	571.38	564.58	552.66	536.80	518.69	507.25
2,350	577.46	576.78	569.92	557.90	541.91	523.65	512.12
2,375	582.85	582.17	575.26	563.14	547.02	528.61	516.98
2,400	588.24	587.55	580.59	568.37	552.12	533.56	521.83
2,425	593.63	592.93	585.91	573.60	557.22	538.50	526.68
2,450	599.01	598.31	591.23	578.82	562.31	543.45	531.53
2,475	604.38	603.68	596.55	584.04	567.39	548.38	536.37
2,500	609.75	609.04	601.85	589.25	572.47	553.31	541.20
2,550	620.47	619.75	612.45	599.65	582.62	563.16	550.86

2,600	560.50	572.99	592.75	610.04	623.03	630.43	631.17
2,650	570.13	582.80	602.86	620.40	633.59	641.10	641.84
2,700	579.73	592.60	612.95	630.75	644.12	651.74	652.50
2,750	589.32	602.37	623.02	641.07	654.64	662.37	663.13
2,800	598.90	612.13	633.07	651.38	665.14	672.97	673.75
2,850	608.45	621.88	643.11	661.67	675.61	683.56	684.34
2,900	617.99	631.60	653.12	671.94	686.07	694.12	694.92
2,950	627.52	641.32	663.13	682.19	696.51	704.67	705.48
3,000	637.03	651.01	673.11	692.43	706.94	715.20	716.02
3,050	646.53	660.69	683.08	702.65	717.34	725.71	726.54
3,100	656.01	670.36	693.03	712.85	727.73	736.20	737.04
3,150	665.48	680.01	702.97	723.04	738.10	746.68	747.53
3,200	674.93	689.64	712.89	733.21	748.45	757.14	758.00
3,250	684.37	699.27	722.80	743.36	758.79	767.58	768.45
3,300	693.80	708.87	732.69	753.50	769.11	778.01	778.89
3,350	703.21	718.47	742.57	763.62	779.42	788.42	789.31
3,400	712.62	728.05	752.43	773.73	789.71	798.81	799.71
3,450	722.00	737.62	762.28	783.82	799.99	809.19	810.10
3,500	731.38	747.17	772.12	793.90	810.25	819.55	820.48
3,550	740.74	756.72	781.94	803.97	820.49	829.90	830.83
3,600	750.09	766.24	791.75	814.02	830.73	840.24	841.18
3,650	759.43	775.76	801.55	824.06	840.94	850.56	851.51
3,700	768.76	785.27	811.33	834.08	851.15	860.86	861.83
3,750	778.08	794.76	821.10	844.09	861.34	871.15	872.13
3,800	787.38	804.24	830.86	854.09	871.51	881.43	882.41
3,850	796.68	813.71	840.61	864.08	881.68	891.70	892.69
3,900	805.96	823.17	850.34	874.05	891.83	901.95	902.95
3,950	815.23	832.62	860.06	884.01	901.96	912.19	913.20
4,000	824.49	842.06	869.77	893.96	912.09	922.41	923.43
4,050	833.74	851.48	879.47	903.89	922.20	932.62	933.65
4,100	842.99	860.90	889.16	913.82	932.30	942.82	943.86
4,150	852.22	870.30	898.84	923.73	942.39	953.01	954.06
4,200	861.44	879.69	908.50	933.63	952.47	963.19	964.25
4,250	870.65	889.08	918.16	943.52	962.53	973.35	974.42
4,300	879.85	898.45	927.80	953.40	972.58	983.50	984.58

(table continued on next page)

Table 10-3 (continued)

t_D	0.05	0.1	0.3	z'_D 0.5	0.7	0.9	1.0
4,350	994.73	993.64	982.62	963.27	937.43	907.81	889.04
4,400	1,004.9	1,003.8	992.7	973.1	947.1	917.2	898.2
4,450	1,015.0	1,013.9	1,002.7	983.0	956.7	926.5	907.4
4,500	1,025.1	1,024.0	1,012.7	992.8	966.3	935.9	916.6
4,550	1,035.2	1,034.1	1,022.7	1,002.6	975.9	945.2	925.7
4,600	1,045.3	1,044.2	1,032.7	1,012.4	985.5	954.5	934.9
4,650	1,055.4	1,054.2	1,042.6	1,022.2	995.0	963.8	944.0
4,700	1,065.5	1,064.3	1,052.6	1,032.0	1,004.6	973.1	953.1
4,750	1,075.5	1,074.4	1,062.6	1,041.8	1,014.1	982.4	962.2
4,800	1,085.6	1,084.4	1,072.5	1,051.6	1,023.7	991.7	971.4
4,850	1,095.6	1,094.4	1,082.4	1,061.4	1,033.2	1,000.9	980.5
4,900	1,105.6	1,104.5	1,092.4	1,071.1	1,042.8	1,010.2	989.5
4,950	1,115.7	1,114.5	1,102.3	1,080.9	1,052.3	1,019.4	998.6
5,000	1,125.7	1,124.5	1,112.2	1,090.6	1,061.8	1,028.7	1,007.7
5,100	1,145.7	1,144.4	1,132.0	1,110.0	1,080.8	1,047.2	1,025.8
5,200	1,165.6	1,164.4	1,151.7	1,129.4	1,099.7	1,065.6	1,043.9
5,300	1,185.5	1,184.3	1,171.4	1,148.8	1,118.6	1,084.0	1,062.0
5,400	1,205.4	1,204.1	1,191.1	1,168.2	1,137.5	1,102.4	1,080.0
5,500	1,225.3	1,224.0	1,210.7	1,187.5	1,156.4	1,120.7	1,098.0
5,600	1,245.1	1,243.7	1,230.3	1,206.7	1,175.2	1,139.0	1,116.0
5,700	1,264.9	1,263.5	1,249.9	1,226.0	1,194.0	1,157.3	1,134.0
5,800	1,284.6	1,283.2	1,269.4	1,245.2	1,212.8	1,175.5	1,151.9
5,900	1,304.3	1,302.9	1,288.9	1,264.4	1,231.5	1,193.8	1,169.8
6,000	1,324.0	1,322.6	1,308.4	1,283.5	1,250.2	1,211.9	1,187.7
6,100	1,343.6	1,342.2	1,327.9	1,302.6	1,268.9	1,230.1	1,205.5
6,200	1,363.2	1,361.8	1,347.3	1,321.7	1,287.5	1,248.3	1,223.3
6,300	1,382.8	1,381.4	1,366.7	1,340.8	1,306.2	1,266.4	1,241.1
6,400	1,402.4	1,400.9	1,386.0	1,359.8	1,324.7	1,284.5	1,258.9
6,500	1,421.9	1,420.4	1,405.3	1,378.8	1,343.3	1,302.5	1,276.6
6,600	1,441.4	1,439.9	1,424.6	1,397.8	1,361.9	1,320.6	1,294.3

6,700	1,460.9	1,459.4	1,443.9	1,416.7	1,380.4	1,338.6	1,312.0
6,800	1,480.3	1,478.8	1,463.1	1,435.6	1,398.9	1,356.6	1,329.7
6,900	1,499.7	1,498.2	1,482.4	1,454.5	1,417.3	1,374.5	1,347.4
7,000	1,519.1	1,517.5	1,501.5	1,473.4	1,435.8	1,392.5	1,365.0
7,100	1,538.5	1,536.9	1,520.7	1,492.3	1,454.2	1,410.4	1,382.6
7,200	1,557.8	1,556.2	1,539.8	1,511.1	1,472.6	1,428.3	1,400.2
7,300	1,577.1	1,575.5	1,559.0	1,529.9	1,491.0	1,446.2	1,417.8
7,400	1,596.4	1,594.8	1,578.1	1,548.6	1,509.3	1,464.1	1,435.3
7,500	1,615.7	1,614.0	1,597.1	1,567.4	1,527.6	1,481.9	1,452.8
7,600	1,634.9	1,633.2	1,616.2	1,586.1	1,545.9	1,499.7	1,470.3
7,700	1,654.1	1,652.4	1,635.2	1,604.8	1,564.2	1,517.5	1,487.8
7,800	1,673.3	1,671.6	1,654.2	1,623.5	1,582.5	1,535.3	1,505.3
7,900	1,692.5	1,690.7	1,673.1	1,642.2	1,600.7	1,553.0	1,522.7
8,000	1,711.6	1,709.9	1,692.1	1,660.8	1,619.0	1,570.8	1,540.1
8,100	1,730.8	1,729.0	1,711.0	1,679.4	1,637.2	1,588.5	1,557.6
8,200	1,749.9	1,748.1	1,729.9	1,698.0	1,655.3	1,606.2	1,574.9
8,300	1,768.9	1,767.1	1,748.8	1,716.6	1,673.5	1,623.9	1,592.3
8,400	1,788.0	1,786.2	1,767.7	1,735.2	1,691.6	1,641.5	1,609.7
8,500	1,807.0	1,805.2	1,786.5	1,753.7	1,709.8	1,659.2	1,627.0
8,600	1,826.0	1,824.2	1,805.4	1,772.2	1,727.9	1,676.8	1,644.3
8,700	1,845.0	1,843.2	1,824.2	1,790.7	1,746.0	1,694.4	1,661.6
8,800	1,864.0	1,862.1	1,842.9	1,809.2	1,764.0	1,712.0	1,678.9
8,900	1,883.0	1,881.1	1,861.7	1,827.7	1,782.1	1,729.6	1,696.2
9,000	1,901.9	1,900.0	1,880.5	1,846.1	1,800.1	1,747.1	1,713.4
9,100	1,920.8	1,918.9	1,899.2	1,864.5	1,818.1	1,764.7	1,730.7
9,200	1,939.7	1,937.4	1,917.9	1,882.9	1,836.1	1,782.2	1,747.9
9,300	1,958.6	1,956.6	1,936.6	1,901.3	1,854.1	1,799.7	1,765.1
9,400	1,977.4	1,975.4	1,955.2	1,919.7	1,872.0	1,817.2	1,782.3
9,500	1,996.3	1,994.3	1,973.9	1,938.0	1,890.0	1,834.7	1,799.4
9,600	2,015.1	2,013.1	1,992.5	1,956.4	1,907.9	1,852.1	1,816.6
9,700	2,033.9	2,031.9	2,011.1	1,974.7	1,925.8	1,869.6	1,833.7
9,800	2,052.7	2,050.6	2,029.7	1,993.0	1,943.7	1,887.0	1,850.9
9,900	2,071.5	2,069.4	2,048.3	2,011.3	1,961.6	1,904.4	1,868.0

(table continued on next page)

Table 10-3 (continued)

t_D	$z'_D = 0.05$	0.1	0.3	0.5	0.7	0.9	1.0
1.00×10^4	2.090×10^3	2.088×10^3	2.067×10^3	2.029×10^3	1.979×10^3	1.922×10^3	1.885×10^3
1.25×10^4	2.553×10^3	2.551×10^3	2.526×10^3	2.481×10^3	2.421×10^3	2.352×10^3	2.308×10^3
1.50×10^4	3.009×10^3	3.006×10^3	2.977×10^3	2.925×10^3	2.855×10^3	2.775×10^3	2.724×10^3
1.75×10^4	3.457×10^3	3.454×10^3	3.421×10^3	3.362×10^3	3.284×10^3	3.193×10^3	3.135×10^3
2.00×10^4	3.900×10^3	3.897×10^3	3.860×10^3	3.794×10^3	3.707×10^3	3.605×10^3	3.541×10^3
2.50×10^4	4.773×10^3	4.768×10^3	4.724×10^3	4.646×10^3	4.541×10^3	4.419×10^3	4.341×10^3
3.00×10^4	5.630×10^3	5.625×10^3	5.574×10^3	5.483×10^3	5.361×10^3	5.219×10^3	5.129×10^3
3.50×10^4	6.476×10^3	6.470×10^3	6.412×10^3	6.309×10^3	6.170×10^3	6.009×10^3	5.906×10^3
4.00×10^4	7.312×10^3	7.305×10^3	7.240×10^3	7.125×10^3	6.970×10^3	6.790×10^3	6.675×10^3
4.50×10^4	8.139×10^3	8.132×10^3	8.060×10^3	7.933×10^3	7.762×10^3	7.564×10^3	7.437×10^3
5.00×10^4	8.959×10^3	8.951×10^3	8.872×10^3	8.734×10^3	8.548×10^3	8.331×10^3	8.193×10^3
6.00×10^4	1.057×10^4	1.057×10^4	1.047×10^4	1.031×10^4	1.010×10^4	9.846×10^3	9.684×10^3
7.00×10^4	1.217×10^4	1.217×10^4	1.206×10^4	1.188×10^4	1.163×10^4	1.134×10^4	1.116×10^4
8.00×10^4	1.375×10^4	1.375×10^4	1.363×10^4	1.342×10^4	1.315×10^4	1.283×10^4	1.262×10^4
9.00×10^4	1.532×10^4	1.531×10^4	1.518×10^4	1.496×10^4	1.465×10^4	1.430×10^4	1.407×10^4
1.00×10^5	1.687×10^4	1.686×10^4	1.672×10^4	1.647×10^4	1.614×10^4	1.576×10^4	1.551×10^4
1.25×10^5	2.071×10^4	2.069×10^4	2.052×10^4	2.023×10^4	1.982×10^4	1.936×10^4	1.906×10^4
1.50×10^5	2.448×10^4	2.446×10^4	2.427×10^4	2.392×10^4	2.345×10^4	2.291×10^4	2.256×10^4
2.00×10^5	3.190×10^4	3.188×10^4	3.163×10^4	3.119×10^4	3.059×10^4	2.989×10^4	2.945×10^4
2.50×10^5	3.918×10^4	3.916×10^4	3.885×10^4	3.832×10^4	3.760×10^4	3.676×10^4	3.622×10^4
3.00×10^5	4.636×10^4	4.633×10^4	4.598×10^4	4.536×10^4	4.452×10^4	4.353×10^4	4.290×10^4
4.00×10^5	6.048×10^4	6.044×10^4	5.999×10^4	5.920×10^4	5.812×10^4	5.687×10^4	5.606×10^4
5.00×10^5	7.436×10^4	7.431×10^4	7.376×10^4	7.280×10^4	7.150×10^4	6.998×10^4	6.900×10^4
6.00×10^5	8.805×10^4	8.798×10^4	8.735×10^4	8.623×10^4	8.471×10^4	8.293×10^4	8.178×10^4
7.00×10^5	1.016×10^5	1.015×10^5	1.008×10^5	9.951×10^4	9.777×10^4	9.573×10^4	9.442×10^4
8.00×10^5	1.150×10^5	1.149×10^5	1.141×10^5	1.127×10^5	1.107×10^6	1.084×10^5	1.070×10^5
9.00×10^5	1.283×10^5	1.282×10^5	1.273×10^5	1.257×10^5	1.235×10^5	1.210×10^5	1.194×10^5

1.00 × 10⁶	1.415 × 10⁶	1.412 × 10⁵	1.404 × 10⁵	1.387 × 10⁵	1.363 × 10⁵	1.335 × 10⁵	1.317 × 10⁵
1.50 × 10⁶	2.059 × 10⁵	2.060 × 10⁵	2.041 × 10⁵	2.016 × 10⁵	1.982 × 10⁵	1.943 × 10⁵	1.918 × 10⁵
2.00 × 10⁶	2.695 × 10⁵	2.695 × 10⁵	2.676 × 10⁵	2.644 × 10⁵	2.601 × 10⁵	2.551 × 10⁵	2.518 × 10⁵
2.50 × 10⁶	3.320 × 10⁵	3.319 × 10⁵	3.296 × 10⁵	3.254 × 10⁵	3.202 × 10⁵	3.141 × 10⁵	3.101 × 10⁵
3.00 × 10⁶	3.937 × 10⁵	3.936 × 10⁵	3.909 × 10⁵	3.864 × 10⁵	3.803 × 10⁵	3.731 × 10⁵	3.684 × 10⁵
4.00 × 10⁶	5.154 × 10⁵	5.152 × 10⁵	5.118 × 10⁵	5.060 × 10⁵	4.981 × 10⁵	4.888 × 10⁵	4.828 × 10⁵
5.00 × 10⁶	6.352 × 10⁵	6.349 × 10⁵	6.308 × 10⁵	6.238 × 10⁵	6.142 × 10⁵	6.029 × 10⁵	5.956 × 10⁵
6.00 × 10⁶	7.536 × 10⁵	7.533 × 10⁵	7.485 × 10⁵	7.402 × 10⁵	7.290 × 10⁵	7.157 × 10⁵	7.072 × 10⁵
7.00 × 10⁶	8.709 × 10⁵	8.705 × 10⁵	8.650 × 10⁵	8.556 × 10⁵	8.427 × 10⁵	8.275 × 10⁵	8.177 × 10⁵
8.00 × 10⁶	9.972 × 10⁵	9.867 × 10⁵	9.806 × 10⁵	9.699 × 10⁵	9.555 × 10⁵	9.384 × 10⁵	9.273 × 10⁵
9.00 × 10⁶	1.103 × 10⁶	1.102 × 10⁶	1.095 × 10⁶	1.084 × 10⁶	1.067 × 10⁶	1.049 × 10⁶	1.036 × 10⁶
1.00 × 10⁷	1.217 × 10⁶	1.217 × 10⁶	1.209 × 10⁶	1.196 × 10⁶	1.179 × 10⁶	1.158 × 10⁶	1.144 × 10⁶
1.50 × 10⁷	1.782 × 10⁶	1.781 × 10⁶	1.771 × 10⁶	1.752 × 10⁶	1.727 × 10⁶	1.697 × 10⁶	1.678 × 10⁶
2.00 × 10⁷	2.337 × 10⁶	2.336 × 10⁶	2.322 × 10⁶	2.298 × 10⁶	2.266 × 10⁶	2.227 × 10⁶	2.202 × 10⁶
2.50 × 10⁷	2.884 × 10⁶	2.882 × 10⁶	2.866 × 10⁶	2.837 × 10⁶	2.797 × 10⁶	2.750 × 10⁶	2.720 × 10⁶
3.00 × 10⁷	3.425 × 10⁶	3.423 × 10⁶	3.404 × 10⁶	3.369 × 10⁶	3.323 × 10⁶	3.268 × 10⁶	3.232 × 10⁶
4.00 × 10⁷	4.493 × 10⁶	4.491 × 10⁶	4.466 × 10⁶	4.422 × 10⁶	4.361 × 10⁶	4.290 × 10⁶	4.224 × 10⁶
5.00 × 10⁷	5.547 × 10⁶	5.544 × 10⁶	5.514 × 10⁶	5.460 × 10⁶	5.386 × 10⁶	5.299 × 10⁶	5.243 × 10⁶
6.00 × 10⁷	6.590 × 10⁶	6.587 × 10⁶	6.551 × 10⁶	6.488 × 10⁶	6.401 × 10⁶	6.299 × 10⁶	6.232 × 10⁶
7.00 × 10⁷	7.624 × 10⁶	7.620 × 10⁶	7.579 × 10⁶	7.507 × 10⁶	7.407 × 10⁶	7.290 × 10⁶	7.213 × 10⁶
8.00 × 10⁷	8.651 × 10⁶	8.647 × 10⁶	8.600 × 10⁶	8.519 × 10⁶	8.407 × 10⁶	8.274 × 10⁶	8.188 × 10⁶
9.00 × 10⁷	9.671 × 10⁶	9.666 × 10⁶	9.615 × 10⁶	9.524 × 10⁶	9.400 × 10⁶	9.252 × 10⁶	9.156 × 10⁶
1.00 × 10⁸	1.069 × 10⁷	1.067 × 10⁷	1.062 × 10⁷	1.052 × 10⁷	1.039 × 10⁷	1.023 × 10⁷	1.012 × 10⁷
1.50 × 10⁸	1.567 × 10⁷	1.567 × 10⁷	1.555 × 10⁷	1.541 × 10⁷	1.522 × 10⁷	1.499 × 10⁷	1.483 × 10⁷
2.00 × 10⁸	2.059 × 10⁷	2.059 × 10⁷	2.048 × 10⁷	2.029 × 10⁷	2.004 × 10⁷	1.974 × 10⁷	1.954 × 10⁷
2.50 × 10⁸	2.546 × 10⁷	2.545 × 10⁷	2.531 × 10⁷	2.507 × 10⁷	2.476 × 10⁷	2.439 × 10⁷	2.415 × 10⁷
3.00 × 10⁸	3.027 × 10⁷	3.026 × 10⁷	3.010 × 10⁷	2.984 × 10⁷	2.947 × 10⁷	2.904 × 10⁷	2.875 × 10⁷
4.00 × 10⁸	3.979 × 10⁷	3.978 × 10⁷	3.958 × 10⁷	3.923 × 10⁷	3.875 × 10⁷	3.819 × 10⁷	3.782 × 10⁷
5.00 × 10⁸	4.920 × 10⁷	4.918 × 10⁷	4.894 × 10⁷	4.851 × 10⁷	4.793 × 10⁷	4.724 × 10⁷	4.679 × 10⁷
6.00 × 10⁸	5.852 × 10⁷	5.850 × 10⁷	5.821 × 10⁷	5.771 × 10⁷	5.702 × 10⁷	5.621 × 10⁷	5.568 × 10⁷

(table continued on next page)

Table 10-3 (continued)

| | | | | z_D' | | | |
t_D	0.05	0.1	0.3	0.5	0.7	0.9	1.0
7.00×10^8	6.777×10^7	6.774×10^7	6.741×10^7	6.684×10^7	6.605×10^7	6.511×10^7	6.450×10^7
8.00×10^8	7.700×10^7	7.693×10^7	7.655×10^7	7.590×10^7	7.501×10^7	7.396×10^7	7.327×10^7
9.00×10^8	8.609×10^7	8.606×10^7	8.564×10^7	8.492×10^7	8.393×10^7	8.275×10^7	8.199×10^7
1.00×10^9	9.518×10^7	9.515×10^7	9.469×10^7	9.390×10^7	9.281×10^7	9.151×10^7	9.066×10^7
1.50×10^9	1.401×10^8	1.400×10^8	1.394×10^8	1.382×10^8	1.367×10^8	1.348×10^8	1.336×10^8
2.00×10^9	1.843×10^8	1.843×10^8	1.834×10^8	1.819×10^8	1.799×10^8	1.774×10^8	1.758×10^8
2.50×10^9	2.281×10^8	2.280×10^8	2.269×10^8	2.251×10^8	2.226×10^8	2.196×10^8	2.177×10^8
3.00×10^9	2.714×10^8	2.713×10^8	2.701×10^8	2.680×10^8	2.650×10^8	2.615×10^8	2.592×10^8
4.00×10^9	3.573×10^8	3.572×10^8	3.556×10^8	3.528×10^8	3.489×10^8	3.443×10^8	3.413×10^8
5.00×10^9	4.422×10^8	4.421×10^8	4.401×10^8	4.367×10^8	4.320×10^8	4.263×10^8	4.227×10^8
6.00×10^9	5.265×10^8	5.262×10^8	5.240×10^8	5.199×10^8	5.143×10^8	5.077×10^8	5.033×10^8
7.00×10^9	6.101×10^8	6.098×10^8	6.072×10^8	6.025×10^8	5.961×10^8	5.885×10^8	5.835×10^8
8.00×10^9	6.932×10^8	6.930×10^8	6.900×10^8	6.847×10^8	6.775×10^8	6.688×10^8	6.632×10^8
9.00×10^9	7.760×10^8	7.756×10^8	7.723×10^8	7.664×10^8	7.584×10^8	7.487×10^8	7.424×10^8
1.00×10^{10}	8.583×10^8	8.574×10^8	8.543×10^8	8.478×10^8	8.389×10^8	8.283×10^8	8.214×10^8
1.50×10^{10}	1.263×10^9	1.264×10^9	1.257×10^9	1.247×10^9	1.235×10^9	1.219×10^9	1.209×10^9
2.00×10^{10}	1.666×10^9	1.666×10^9	1.659×10^9	1.646×10^9	1.630×10^9	1.610×10^9	1.596×10^9
2.50×10^{10}	2.065×10^9	2.063×10^9	2.055×10^9	2.038×10^9	2.018×10^9	1.993×10^9	1.977×10^9
3.00×10^{10}	2.458×10^9	2.458×10^9	2.447×10^9	2.430×10^9	2.405×10^9	2.376×10^9	2.357×10^9
4.00×10^{10}	3.240×10^9	3.239×10^9	3.226×10^9	3.203×10^9	3.171×10^9	3.133×10^9	3.108×10^9

5.00×10^{10}	4.014×10^{9}	4.013×10^{9}	3.997×10^{9}	3.968×10^{9}	3.929×10^{9}	3.883×10^{9}	3.852×10^{9}
6.00×10^{10}	4.782×10^{9}	4.781×10^{9}	4.762×10^{9}	4.728×10^{9}	4.682×10^{9}	4.627×10^{9}	4.591×10^{9}
7.00×10^{10}	5.546×10^{9}	5.544×10^{9}	5.522×10^{9}	5.483×10^{9}	5.430×10^{9}	5.366×10^{9}	5.325×10^{9}
8.00×10^{10}	6.305×10^{9}	6.303×10^{9}	6.278×10^{9}	6.234×10^{9}	6.174×10^{9}	6.102×10^{9}	6.055×10^{9}
9.00×10^{10}	7.060×10^{9}	7.058×10^{9}	7.030×10^{9}	6.982×10^{9}	6.914×10^{9}	6.834×10^{9}	6.782×10^{9}
1.00×10^{11}	7.813×10^{9}	7.810×10^{9}	7.780×10^{9}	7.726×10^{9}	7.652×10^{9}	7.564×10^{9}	7.506×10^{9}
1.50×10^{11}	1.154×10^{10}	1.153×10^{10}	1.149×10^{10}	1.141×10^{10}	1.130×10^{10}	1.118×10^{10}	1.109×10^{10}
2.00×10^{11}	1.522×10^{10}	1.521×10^{10}	1.515×10^{10}	1.505×10^{10}	1.491×10^{10}	1.474×10^{10}	1.463×10^{10}
2.50×10^{11}	1.886×10^{10}	1.885×10^{10}	1.878×10^{10}	1.866×10^{10}	1.849×10^{10}	1.828×10^{10}	1.814×10^{10}
3.00×10^{11}	2.248×10^{10}	2.247×10^{10}	2.239×10^{10}	2.224×10^{10}	2.204×10^{10}	2.179×10^{10}	2.163×10^{10}
4.00×10^{11}	2.965×10^{10}	2.964×10^{10}	2.953×10^{10}	2.934×10^{10}	2.907×10^{10}	2.876×10^{10}	2.855×10^{10}
5.00×10^{11}	3.677×10^{10}	3.675×10^{10}	3.662×10^{10}	3.638×10^{10}	3.605×10^{10}	3.566×10^{10}	3.540×10^{10}
6.00×10^{11}	4.383×10^{10}	4.381×10^{10}	4.365×10^{10}	4.337×10^{10}	4.298×10^{10}	4.252×10^{10}	4.221×10^{10}
7.00×10^{11}	5.085×10^{10}	5.082×10^{10}	5.064×10^{10}	5.032×10^{10}	4.987×10^{10}	4.933×10^{10}	4.898×10^{10}
8.00×10^{11}	5.783×10^{10}	5.781×10^{10}	5.760×10^{10}	5.723×10^{10}	5.673×10^{10}	5.612×10^{10}	5.572×10^{10}
9.00×10^{11}	6.478×10^{10}	6.476×10^{10}	6.453×10^{10}	6.412×10^{10}	6.355×10^{10}	6.288×10^{10}	6.243×10^{10}
1.00×10^{12}	7.171×10^{10}	7.168×10^{10}	7.143×10^{10}	7.098×10^{10}	7.035×10^{10}	6.961×10^{10}	6.912×10^{10}
1.50×10^{12}	1.060×10^{11}	1.060×10^{11}	1.056×10^{11}	1.050×10^{11}	1.041×10^{11}	1.030×10^{11}	1.022×10^{11}
2.00×10^{12}	1.400×10^{11}	1.399×10^{11}	1.394×10^{11}	1.386×10^{11}	1.374×10^{11}	1.359×10^{11}	1.350×10^{11}

Table 10-4
Dimensionless Water Influx, W_{eD}, for $r'_D = 4$
(*Permission to publish by the SPE*)

t_D	z'_D						
	0.05	0.1	0.3	0.5	0.7	0.9	1.0
2	2.398	2.389	2.284	2.031	1.824	1.620	1.507
3	3.006	2.993	2.874	2.629	2.390	2.149	2.012
4	3.552	3.528	3.404	3.158	2.893	2.620	2.466
5	4.053	4.017	3.893	3.627	3.341	3.045	2.876
6	4.490	4.452	4.332	4.047	3.744	3.430	3.249
7	4.867	4.829	4.715	4.420	4.107	3.778	3.587
8	5.191	5.157	5.043	4.757	4.437	4.096	3.898
9	5.464	5.434	5.322	5.060	4.735	4.385	4.184
10	5.767	5.739	5.598	5.319	5.000	4.647	4.443
11	5.964	5.935	5.829	5.561	5.240	4.884	4.681
12	6.188	6.158	6.044	5.780	5.463	5.107	4.903
13	6.380	6.350	6.240	5.983	5.670	5.316	5.113
14	6.559	6.529	6.421	6.171	5.863	5.511	5.309
15	6.725	6.694	6.589	6.345	6.044	5.695	5.495
16	6.876	6.844	6.743	6.506	6.213	5.867	5.671
17	7.014	6.983	6.885	6.656	6.371	6.030	5.838
18	7.140	7.113	7.019	6.792	6.523	6.187	5.999
19	7.261	7.240	7.140	6.913	6.663	6.334	6.153
20	7.376	7.344	7.261	7.028	6.785	6.479	6.302
22	7.518	7.507	7.451	7.227	6.982	6.691	6.524
24	7.618	7.607	7.518	7.361	7.149	6.870	6.714
26	7.697	7.685	7.607	7.473	7.283	7.026	6.881
28	7.752	7.752	7.674	7.563	7.395	7.160	7.026
30	7.808	7.797	7.741	7.641	7.484	7.283	7.160
34	7.864	7.864	7.819	7.741	7.618	7.451	7.350
38	7.909	7.909	7.875	7.808	7.719	7.585	7.496
42	7.931	7.931	7.909	7.864	7.797	7.685	7.618
46	7.942	7.942	7.920	7.898	7.842	7.752	7.697
50	7.954	7.954	7.942	7.920	7.875	7.808	7.764
60	7.968	7.968	7.965	7.954	7.931	7.898	7.864
70	7.976	7.976	7.976	7.968	7.965	7.942	7.920
80	7.982	7.982	7.987	7.976	7.976	7.965	7.954
90	7.987	7.987	7.987	7.984	7.983	7.976	7.965
100	7.987	7.987	7.987	7.987	7.987	7.983	7.976
120	7.987	7.987	7.987	7.987	7.987	7.987	7.987

Table 10-5
Dimensionless Water Influx, W_{eD}, for $r'_D = 6$
(*Permission to publish by the SPE*)

				z'_D			
t_D	0.05	0.1	0.3	0.5	0.7	0.9	1.0
6	4.780	4.762	4.597	4.285	3.953	3.611	3.414
7	5.309	5.289	5.114	4.779	4.422	4.053	3.837
8	5.799	5.778	5.595	5.256	4.875	4.478	4.247
9	6.252	6.229	6.041	5.712	5.310	4.888	4.642
10	6.750	6.729	6.498	6.135	5.719	5.278	5.019
11	7.137	7.116	6.916	6.548	6.110	5.648	5.378
12	7.569	7.545	7.325	6.945	6.491	6.009	5.728
13	7.967	7.916	7.719	7.329	6.858	6.359	6.067
14	8.357	8.334	8.099	7.699	7.214	6.697	6.395
15	8.734	8.709	8.467	8.057	7.557	7.024	6.713
16	9.093	9.067	8.819	8.398	7.884	7.336	7.017
17	9.442	9.416	9.160	8.730	8.204	7.641	7.315
18	9.775	9.749	9.485	9.047	8.510	7.934	7.601
19	10.09	10.06	9.794	9.443	8.802	8.214	7.874
20	10.40	10.37	10.10	9.646	9.087	8.487	8.142
22	10.99	10.96	10.67	10.21	9.631	9.009	8.653
24	11.53	11.50	11.20	10.73	10.13	9.493	9.130
26	12.06	12.03	11.72	11.23	10.62	9.964	9.594
28	12.52	12.49	12.17	11.68	11.06	10.39	10.01
30	12.95	12.92	12.59	12.09	11.46	10.78	10.40
35	13.96	13.93	13.57	13.06	12.41	11.70	11.32
40	14.69	14.66	14.33	13.84	13.23	12.53	12.15
45	15.27	15.24	14.94	14.48	13.90	13.23	12.87
50	15.74	15.71	15.44	15.01	14.47	13.84	13.49
60	16.40	16.38	16.15	15.81	15.34	14.78	14.47
70	16.87	16.85	16.67	16.38	15.99	15.50	15.24
80	17.20	17.18	17.04	16.80	16.48	16.06	15.83
90	17.43	17.42	17.30	17.10	16.85	16.50	16.29
100	17.58	17.58	17.49	17.34	17.12	16.83	16.66
110	17.71	17.69	17.63	17.50	17.34	17.09	16.93
120	17.78	17.78	17.73	17.63	17.49	17.29	17.17
130	17.84	17.84	17.79	17.73	17.62	17.45	17.34
140	17.88	17.88	17.85	17.79	17.71	17.57	17.48
150	17.92	17.91	17.88	17.84	17.77	17.66	17.58
175	17.95	17.95	17.94	17.92	17.87	17.81	17.76
200	17.97	17.97	17.96	17.95	17.93	17.88	17.86
225	17.97	17.97	17.97	17.96	17.95	17.93	17.91
250	17.98	17.98	17.98	17.97	17.96	17.95	17.95
300	17.98	17.98	17.98	17.98	17.98	17.97	17.97
350	17.98	17.98	17.98	17.98	17.98	17.98	17.98
400	17.98	17.98	17.98	17.98	17.98	17.98	17.98
450	17.98	17.98	17.98	17.98	17.98	17.98	17.98
500	17.98	17.98	17.98	17.98	17.98	17.98	17.98

Table 10-6
Dimensionless Water Influx, W_{eD}, for $r'_D = 8$
(*Permission to publish by the SPE*)

t_D	z'_D 0.05	0.1	0.3	0.5	0.7	0.9	1.0
9	6.301	6.278	6.088	5.756	5.350	4.924	4.675
10	6.828	6.807	6.574	6.205	5.783	5.336	5.072
11	7.250	7.229	7.026	6.650	6.204	5.732	5.456
12	7.725	7.700	7.477	7.086	6.621	6.126	5.836
13	8.173	8.149	7.919	7.515	7.029	6.514	6.210
14	8.619	8.594	8.355	7.937	7.432	6.895	6.578
15	9.058	9.032	8.783	8.351	7.828	7.270	6.940
16	9.485	9.458	9.202	8.755	8.213	7.634	7.293
17	9.907	9.879	9.613	9.153	8.594	7.997	7.642
18	10.32	10.29	10.01	9.537	8.961	8.343	7.979
19	10.72	10.69	10.41	9.920	9.328	8.691	8.315
20	11.12	11.08	10.80	10.30	9.687	9.031	8.645
22	11.89	11.86	11.55	11.02	10.38	9.686	9.280
24	12.63	12.60	12.27	11.72	11.05	10.32	9.896
26	13.36	13.32	12.97	12.40	11.70	10.94	10.49
28	14.06	14.02	13.65	13.06	12.33	11.53	11.07
30	14.73	14.69	14.30	13.68	12.93	12.10	11.62
34	16.01	15.97	15.54	14.88	14.07	13.18	12.67
38	17.21	17.17	16.70	15.99	15.13	14.18	13.65
40	17.80	17.75	17.26	16.52	15.64	14.66	14.12
45	19.15	19.10	18.56	17.76	16.83	15.77	15.21
50	20.42	20.36	19.76	18.91	17.93	16.80	16.24
55	21.46	21.39	20.80	19.96	18.97	17.83	17.24
60	22.40	22.34	21.75	20.91	19.93	18.78	18.19
70	23.97	23.92	23.36	22.55	21.58	20.44	19.86
80	25.29	25.23	24.71	23.94	23.01	21.91	21.32
90	26.39	26.33	25.85	25.12	24.24	23.18	22.61
100	27.30	27.25	26.81	26.13	25.29	24.29	23.74
120	28.61	28.57	28.19	27.63	26.90	26.01	25.51
140	29.55	29.51	29.21	28.74	28.12	27.33	26.90
160	30.23	30.21	29.96	29.57	29.04	28.37	27.99
180	30.73	30.71	30.51	30.18	29.75	29.18	28.84
200	31.07	31.04	30.90	30.63	30.26	29.79	29.51
240	31.50	31.49	31.39	31.22	30.98	30.65	30.45
280	31.72	31.71	31.66	31.56	31.39	31.17	31.03
320	31.85	31.84	31.80	31.74	31.64	31.49	31.39
360	31.90	31.90	31.88	31.85	31.78	31.68	31.61
400	31.94	31.94	31.93	31.90	31.86	31.79	31.75
450	31.96	31.96	31.95	31.94	31.91	31.88	31.85
500	31.97	31.97	31.96	31.96	31.95	31.93	31.90
550	31.97	31.97	31.97	31.96	31.96	31.95	31.94
600	31.97	31.97	31.97	31.97	31.97	31.96	31.95
700	31.97	31.97	31.97	31.97	31.97	31.97	31.97
800	31.97	31.97	31.97	31.97	31.97	31.97	31.97

Table 10-7
Dimensionless Water Influx W_{eD} for $r_D = 10$
(Permission to publish by the SPE)

t_D				z'_D			
	0.05	0.1	0.3	0.5	0.7	0.9	1.0
22	12.07	12.04	11.74	11.21	10.56	9.865	9.449
24	12.86	12.83	12.52	11.97	11.29	10.55	10.12
26	13.65	13.62	13.29	12.72	12.01	11.24	10.78
28	14.42	14.39	14.04	13.44	12.70	11.90	11.42
30	15.17	15.13	14.77	14.15	13.38	12.55	12.05
32	15.91	15.87	15.49	14.85	14.05	13.18	12.67
34	16.63	16.59	16.20	15.54	14.71	13.81	13.28
36	17.33	17.29	16.89	16.21	15.35	14.42	13.87
38	18.03	17.99	17.57	16.86	15.98	15.02	14.45
40	18.72	18.68	18.24	17.51	16.60	15.61	15.02
42	19.38	19.33	18.89	18.14	17.21	16.19	15.58
44	20.03	19.99	19.53	18.76	17.80	16.75	16.14
46	20.67	20.62	20.15	19.36	18.38	17.30	16.67
48	21.30	21.25	20.76	19.95	18.95	17.84	17.20
50	21.92	21.87	21.36	20.53	19.51	18.38	17.72
52	22.52	22.47	21.95	21.10	20.05	18.89	18.22
54	23.11	23.06	22.53	21.66	20.59	19.40	18.72
56	23.70	23.64	23.09	22.20	21.11	19.89	19.21
58	24.26	24.21	23.65	22.74	21.63	20.39	19.68
60	24.82	24.77	24.19	23.26	22.13	20.87	20.15
65	26.18	26.12	25.50	24.53	23.34	22.02	21.28
70	27.47	27.41	26.75	25.73	24.50	23.12	22.36
75	28.71	28.55	27.94	26.88	25.60	24.17	23.39
80	29.89	29.82	29.08	27.97	26.65	25.16	24.36
85	31.02	30.95	30.17	29.01	27.65	26.10	25.31
90	32.10	32.03	31.20	30.00	28.60	27.03	26.25
95	33.04	32.96	32.14	30.95	29.54	27.93	27.10
100	33.94	33.85	33.03	31.85	30.44	28.82	27.98
110	35.55	35.46	34.65	33.49	32.08	30.47	29.62
120	36.97	36.90	36.11	34.98	33.58	31.98	31.14
130	38.28	38.19	37.44	36.33	34.96	33.38	32.55
140	39.44	39.37	38.64	37.56	36.23	34.67	33.85
150	40.49	40.42	39.71	38.67	37.38	35.86	35.04
170	42.21	42.15	41.51	40.54	39.33	37.89	37.11
190	43.62	43.55	42.98	42.10	40.97	39.62	38.90
210	44.77	44.72	44.19	43.40	42.36	41.11	40.42
230	45.71	45.67	45.20	44.48	43.54	42.38	41.74
250	46.48	46.44	46.01	45.38	44.53	43.47	42.87
270	47.11	47.06	46.70	46.13	45.36	44.40	43.84
290	47.61	47.58	47.25	46.75	46.07	45.19	44.68

(*table continued on next page*)

Table 10-7 *(continued)*

t_D	0.05	0.1	0.3	0.5	0.7	0.9	1.0
				z_D'			
310	48.03	48.00	47.72	47.26	46.66	45.87	45.41
330	48.38	48.35	48.10	47.71	47.16	46.45	46.03
350	48.66	48.64	48.42	48.08	47.59	46.95	46.57
400	49.15	49.14	48.99	48.74	48.38	47.89	47.60
450	49.46	49.45	49.35	49.17	48.91	48.55	48.31
500	49.65	49.64	49.58	49.45	49.26	48.98	48.82
600	49.84	49.84	49.81	49.74	49.65	49.50	49.41
700	49.91	49.91	49.90	49.87	49.82	49.74	49.69
800	49.94	49.94	49.93	49.92	49.90	49.85	49.83
900	49.96	49.96	49.94	49.94	49.93	49.91	49.90
1,000	49.96	49.96	49.96	49.96	49.94	49.93	49.93
1,200	49.96	49.96	49.96	49.96	49.96	49.96	49.96

(text continued from page 679)

Step 4. Calculate the dimensionless time t_D.

$$t_D = 6.328 \times 10^{-3} \left[\frac{50}{(0.1)(0.395)(8 \times 10^{-6})(2000)^2} \right] t$$

$$t_D = 0.2503 \, t$$

Step 5. Calculate the water influx.

t Days	t_D	Δp psi	Bottom-Water Model W_{eD}	W_e, Mbbl	Edge-Water Model W_{eD}	W_e, Mbbl
0	0	0	—	—	—	—
30	7.5	22	5.038	79	6.029	95
60	15.0	41.5	8.389	282	9.949	336
90	22.5	39.5	11.414	572	13.459	678
120	30.0	36.5	14.994	933	16.472	1,103
150	37.5	33.0	16.994	1,353	19.876	1,594
180	45.0	26.5	19.641	1,810	22.897	2,126
210	52.5	19.0	22.214	2,284	25.827	2,676
240	60.0	18.0	24.728	2,782	28.691	3,250

The Carter-Tracy Water Influx Model

Van Everdingen-Hurst methodology provides the exact solution to the radial diffusivity equation and therefore is considered the correct technique for calculating water influx. However, because superposition of solutions is required, their method involves tedious calculations. To reduce the complexity of water influx calculations, Carter and Tracy (1960) proposed a calculation technique that does not require superposition and allows direct calculation of water influx.

The primary difference between the Carter-Tracy technique and the van Everdingen-Hurst technique is that the Carter-Tracy technique assumes constant water influx rates over each finite time interval. Using the Carter-Tracy technique, the cumulative water influx at any time, t_n, can be calculated directly from the previous value obtained at t_{n-1}, or:

$$(W_e)_n = (W_e)_{n-1} + [(t_D)_n - (t_D)_{n-1}]$$

$$\left[\frac{B \, \Delta p_n - (W_e)_{n-1} (p'_D)_n}{(p_D)_n - (t_D)_{n-1} (p'_D)_n} \right] \tag{10-33}$$

where   B = the van Everdingen-Hurst water influx constant as defined
            by Equation 10-23
      t_D = the dimensionless time as defined by Equation 10-17
      n = refers to the *current* time step
    n − 1 = refers to the *previous* time step
      Δp_n = total pressure drop, $p_i - p_n$, psi
      p_D = dimensionless pressure
      p'_D = dimensionless pressure derivative

Values of the dimensionless pressure p_D as a function of t_D and r_D are tabulated in Chapter 6, Table 6-2. In addition to the curve-fit equations given in Chapter 6 (Equations 6-91 through 6-96), Edwardson and coauthors (1962) developed the following approximation of p_D for an infinite-acting aquifer.

$$p_D = \frac{370.529\sqrt{t_D} + 137.582\,t_D + 5.69549\,(t_D)^{1.5}}{328.834 + 265.488\sqrt{t_D} + 45.2157\,t_D + (t_D)^{1.5}} \tag{10-34}$$

The dimensionless pressure derivative can then be approximated by

$$p'_D = \frac{E}{F}$$

(10 - 35)

where $E = 716.441 + 46.7984 \, (t_D)^{0.5} + 270.038 \, t_D + 71.0098 \, (t_D)^{1.5}$
$F = 1296.86 \, (t_D)^{0.5} + 1204.73 \, t_D + 618.618 \, (t_d)^{1.5}$
$\quad + 538.072 \, (t_D)^2 + 142.41 \, (t_D)^{2.5}$

The following approximation could also be used between $t_D > 100$:

$p_D = 0.5 \, [\text{Ln} \, (t_D) + 0.80907]$

with the derivative as given by:

$p'_D = 1/(2t_D)$

It should be noted that the Carter-Tracy method is not an exact solution to the diffusivity equation and should be considered an approximation.

Example 10-9

Rework Example 10-7 by using the Carter-Tracy method.

Solution

Example 10-7 shows the following preliminary results:

- Water influx constant $B = 20.4$ bbl/psi
- $t_D = 0.9888 \, t$

Step 1. For each time step n, calculate the *total* pressure drop $\Delta p_n = p_i - p_n$ and the corresponding t_D

N	t, days	p_n	Δp_n	t_D
0	0	2500	0	0
1	182.5	2490	10	180.5
2	365.0	2472	28	361.0
3	547.5	2444	56	541.5
4	730.0	2408	92	722.0

Step 2. Since values of t_D are greater than 100, use Equation 6-92 to calculate p_D and its derivative p'_D, i.e.,

$p_D = 0.5 \, [\text{Ln} \, (t_D) + 0.80907]$

$p'_D = 1/(2 \, t_D)$

N	t	t_D	p_D	p_D'
0	0	0	—	—
1	182.5	180.5	3.002	2.770×10^{-3}
2	365	361.0	3.349	1.385×10^{-3}
3	547.5	541.5	3.552	0.923×10^{-3}
4	730.0	722.0	3.696	0.693×10^{-3}

Step 3. Calculate cumulative water influx by applying Equation 10-33.

- **W_e after 182.5 days:**

$$W_e = 0 + [180.5 - 0] \left[\frac{(20.4)\,(10) - (0)\,(2.77 \times 10^{-3})}{3.002 - (0)\,(2.77 \times 10^{-3})} \right]$$

$$W_e = 12{,}266 \text{ bbl}$$

- **W_e after 365 days:**

$$W_e = 12{,}266 + [361 - 180.5]$$

$$\left[\frac{(20.4)\,(28) - (12{,}266)\,(1.385 \times 10^{-3})}{3.349 - (180.5)\,(1.385 \times 10^{-3})} \right]$$

$$= 42{,}546 \text{ bbl}$$

- **W_e after 547.5 days:**

$$W_e = 42{,}546 + [541.5 - 361]$$

$$\left[\frac{(20.4)\,(56) - (42{,}546)\,(0.923 \times 10^{-3})}{3.552 - (361)\,(0.923 \times 10^{-3})} \right]$$

$$W_e = 104{,}406$$

- **W_e after 720 days:**

$$W_e = 104{,}406 + [722 - 541.5]$$

$$\left[\frac{(20.4)\,(92) - (104{,}406)\,(0.693 \times 10^{-3})}{3.696 - (541.5)\,(0.693 \times 10^{-3})} \right]$$

$$W_e = 202{,}477 \text{ bbl}$$

The following table compares results of the Carter-Tracy water influx calculations with those of the van Everdingen-Hurst method.

Time, month	Carter-Tracy W_e, bbl	Van Everdingen-Hurst W_e, bbl
0	0	0
6	12,266	7,080
12	42,546	32,435
18	104,400	85,277
24	202,477	175,522

The above comparison indicates that the Carter-Tracy method considerably overestimates the water influx. This is due, however, to the fact that a large time-step of 6 months was used in the Carter-Tracy method to determine the water influx. Accuracy of the Carter-Tracy method can be increased substantially by restricting the time step used in performing the water influx calculations to less than 30 days, i.e. $\Delta t = 30$ days. Recalculating the water influx on monthly basis produces an excellent match with the van Everdingen-Hurst method as shown below.

time months	time days	p psi	Δp psi	t_D	p_D	p_D'	Carter-Tracy W_e bbl	van Everdingen-Hurst W_e bbl
0	0	2500.0	0.00	0	0.00	0	0.0	0
1	30	2498.9	1.06	30.0892	2.11	0.01661	308.8	
2	61	2497.7	2.31	60.1784	2.45	0.00831	918.3	
3	91	2496.2	3.81	90.2676	2.66	0.00554	1860.3	
4	122	2494.4	5.56	120.357	2.80	0.00415	3171.7	
5	152	2492.4	7.55	150.446	2.91	0.00332	4891.2	
6	183	2490.2	9.79	180.535	3.00	0.00277	7057.3	7088.9
7	213	2487.7	12.27	210.624	3.08	0.00237	9709.0	
8	243	2485.0	15.00	240.713	3.15	0.00208	12884.7	
9	274	2482.0	17.98	270.802	3.21	0.00185	16622.8	
10	304	2478.8	21.20	300.891	3.26	0.00166	20961.5	
11	335	2475.3	24.67	330.981	3.31	0.00151	25938.5	
12	365	2471.6	28.38	361.070	3.35	0.00139	31591.5	32438.0
13	396	2467.7	32.34	391.159	3.39	0.00128	37957.8	
14	426	2463.5	36.55	421.248	3.43	0.00119	45074.5	
15	456	2459.0	41.00	451.337	3.46	0.00111	52978.6	
16	487	2454.3	45.70	481.426	3.49	0.00104	61706.7	
17	517	2449.4	50.64	511.516	3.52	0.00098	71295.3	
18	547	2444.3	55.74	541.071	3.55	0.00092	81578.8	85552.0
19	578	2438.8	61.16	571.130	3.58	0.00088	92968.2	
20	608	2433.2	66.84	601.190	3.60	0.00083	105323.	
21	638	2427.2	72.75	631.249	3.63	0.00079	118681.	
22	669	2421.1	78.92	661.309	3.65	0.00076	133076.	
23	699	2414.7	85.32	691.369	3.67	0.00072	148544.	
24	730	2408.0	91.98	721.428	3.70	0.00069	165119.	175414.0

Fetkovich's Method

Fetkovich (1971) developed a method of describing the approximate water influx behavior of a finite aquifer for radial and linear geometries. In many cases, the results of this model closely match those determined using the van Everdingen-Hurst approach. The Fetkovich theory is much simpler, and, like the Carter-Tracy technique, this method does not require the use of superposition. Hence, the application is much easier, and this method is also often utilized in numerical simulation models.

Fetkovich's model is based on the premise that the productivity index concept will adequately describe water influx from a finite aquifer into a hydrocarbon reservoir. That is, the water influx rate is directly proportional to the pressure drop between the average aquifer pressure and the pressure at the reservoir/aquifer boundary. The method neglects the effects of any transient period. Thus, in cases where pressures are changing rapidly at the aquifer/reservoir interface, predicted results may differ somewhat from the more rigorous van Everdingen-Hurst or Carter-Tracy approaches. In many cases, however, pressure changes at the waterfront are gradual and this method offers an excellent approximation to the two methods discussed above.

This approach begins with two simple equations. The first is the productivity index (PI) equation for the aquifer, which is analogous to the PI equation used to describe an oil or gas well:

$$e_w = \frac{dW_e}{dt} = J(\overline{p}_a - p_r) \qquad (10\text{-}36)$$

where e_w = water influx rate from aquifer, bbl/day
$\quad\quad J$ = productivity index for the aquifer, bbl/day/psi
$\quad\quad \overline{p}_a$ = average aquifer pressure, psi
$\quad\quad p_r$ = inner aquifer boundary pressure, psi

The second equation is an aquifer material balance equation for a constant compressibility, which states that the amount of pressure depletion in the aquifer is directly proportional to the amount of water influx from the aquifer, or:

$$W_e = c_t \, W_i \, (p_i - \overline{p}_a) f \qquad (10\text{-}37)$$

where W_i = initial volume of water in the aquifer, bbl
$\quad\quad c_t$ = total aquifer compressibility, $c_w + c_f$, psi^{-1}
$\quad\quad p_i$ = initial pressure of the aquifer, psi
$\quad\quad f = \theta/360$

Equation 10-37 suggests that the maximum possible water influx occurs if $p_a = 0$, or:

$$W_{ei} = c_t \, W_i \, p_i \, f \tag{10-38}$$

Combining Equation 10-38 with 10-37 gives:

$$\bar{p}_a = p_i \left(1 - \frac{W_e}{c_t \, W_i \, p_i} \right) = p_i \left(1 - \frac{W_e}{W_{ei}} \right) \tag{10-39}$$

Equation 10-37 provides a simple expression to determine the average aquifer pressure $\bar{p}_a$ after removing W_e bbl of water from the aquifer to the reservoir, i.e., cumulative water influx.

Differentiating Equation 10-39 with respect to time gives:

$$\frac{dW_e}{dt} = -\frac{W_{ei}}{p_i} \frac{d\bar{p}_a}{dt} \tag{10-40}$$

Fetkovich combined Equation 10-40 with 10-36 and integrated to give the following form:

$$W_e = \frac{W_{ei}}{p_i} (p_i - p_r) \exp\left(\frac{-J p_i \, t}{W_{ei}} \right) \tag{10-41}$$

where W_e = cumulative water influx, bbl

p_r = reservoir pressure, i.e. pressure at the oil- or gas-water contact

t = time, days

Equation 10-41 has no practical applications since it was derived for a constant inner boundary pressure. To use this solution in the case in which the boundary pressure is varying continuously as a function of time, the superposition technique must be applied. Rather than using superposition, Fetkovich suggested that, if the reservoir-aquifer boundary pressure history is divided into a finite number of time intervals, the incremental water influx during the nth interval is:

$$(\Delta W_e)_n = \frac{W_{ei}}{p_i} \left[(\bar{p}_a)_{n-1} - (\bar{p}_r)_n \right] \left[1 - \exp\left(-\frac{J p_i \, \Delta t_n}{W_{ei}} \right) \right] \tag{10-42}$$

where $(\bar{p}_a)_{n-1}$ is the average aquifer pressure at the end of the previous time step. This average pressure is calculated from Equation 10-39 as:

$$(\bar{p}_a)_{n-1} = p_i \left(1 - \frac{(W_e)_{n-1}}{W_{ei}} \right) \qquad (10\text{-}43)$$

The average reservoir boundary pressure $(\bar{p}_r)_n$ is estimated from:

$$(\bar{p}_r)_n = \frac{(P_r)_n + (p_r)_{n-1}}{2} \qquad (10\text{-}44)$$

The productivity index J used in the calculation is a function of the geometry of the aquifer. Fetkovich calculated the productivity index from Darcy's equation for bounded aquifers. Lee and Wattenbarger (1996) pointed out that Fetkovich's method can be extended to infinite-acting aquifers by requiring that the ratio of water influx rate to pressure drop to be approximately constant throughout the productive life of the reservoir. The productivity index J of the aquifer is given by the following expressions.

Type of Outer Aquifer Boundary	J for Radial flow, bbl/day/psi	J for Linear Flow, bbl/day/psi	Equation #
Finite, no flow	$J = \dfrac{0.00708\,kh\,f}{\mu_w\,[\ln\,r_D - 0.75]}$	$J = \dfrac{0.003381\,kwh}{\mu_w\,L}$	(10-45)
Finite, constant pressure	$J = \dfrac{0.00708\,kh\,f}{\mu_w\,[\ln\,(r_D)]}$	$J = \dfrac{0.001127\,k\,wh}{\mu_w\,L}$	(10-46)
Infinite	$J = \dfrac{0.00708\,kh\,f}{\mu_w\,\ln\,(a/r_e)}$ $a = \sqrt{0.0142\,kt\,/\,(f\mu_w\,c_t)}$	$J = \dfrac{0.001\,k\,wh}{\mu_w\,\sqrt{0.0633\,kt/(f\,\mu_w c_t)}}$	(10-47)

where w = width of the linear aquifer
  L = length of the linear aquifer
  r_D = dimensionless radius, r_a/r_e
  k = permeability of the aquifer, md
  t = time, days
  θ = encroachment angle
  h = thickness of the aquifer
  $f = \theta/360$

The following steps describe the methodology of using the Fetkovich's model in predicting the cumulative water influx.

Step 1. Calculate initial volume of water in the aquifer from:

$$W_i = \frac{\pi}{5.615}(r_a^2 - r_e^2)h\phi$$

Step 2. Calculate the maximum possible water influx W_{ei} by applying Equation 10-38, or:

$$W_{ei} = c_t\, W_i\, p_i\, f$$

Step 3. Calculate the productivity index J based on the boundary conditions and aquifer geometry.

Step 4. Calculate the incremental water influx $(\Delta W_e)_n$ from the aquifer during the nth time interval by using Equation 10-42. For example, during the first time interval Δt_1:

$$(\Delta W_e)_1 = \frac{W_{ei}}{p_i}\left[p_i - (\overline{p}_r)_1\right]\left[1 - \exp\left(\frac{-Jp_i\,\Delta t_1}{W_{ei}}\right)\right]$$

with

$$(\overline{p}_r)_1 = \frac{p_i + (p_r)_1}{2}$$

For the second time interval Δt_2

$$(\Delta W_e)_2 = \frac{W_{ei}}{p_i}\left[(\overline{p}_a)_1 - (\overline{p}_r)_2\right]\left[1 - \exp\left(\frac{-Jp_i\,\Delta t_2}{W_{ei}}\right)\right]$$

where $(\overline{p}_a)_1$ is the average aquifer pressure at the end of the first period and removing $(\Delta W_e)_1$ barrels of water from the aquifer to the reservoir. From Equation 10-43:

$$(\overline{p}_a)_1 = p_i\left(1 - \frac{(\Delta W_e)_1}{W_{ei}}\right)$$

Step 5. Calculate the cumulative (total) water influx at the end of any time period from:

$$W_e = \sum_{t=1}^{n} (\Delta W_e)_i$$

Example 10-10[2]

Using Fetkovich's method, calculate the water influx as a function of time for the following reservoir-aquifer and boundary pressure data:

$p_i = 2740$ psi $h = 100'$ $c_t = 7 \times 10^{-6}$ psi
$\mu_w = 0.55$ cp $k = 200$ md $\theta = 140°$
reservoir area $= 40{,}363$ acres   aquifer area $= 1{,}000{,}000$ acres.

Time, days	p_r, psi
0	2740
365	2500
730	2290
1095	2109
1460	1949

Figure 10-17 shows the wedge reservoir-aquifer system with an encroachment angle of 140°.

Solution

Step 1. Calculate the reservoir radius r_e:

$$r_e = \left(\frac{140}{360}\right)\sqrt{\frac{(2374)(43,560)}{\pi}} = 9200\,\text{ft}$$

Step 2. Calculate the equivalent aquifer radius r_a:

$$r_a = \left(\frac{140}{360}\right)\sqrt{\frac{(1,000,000)(43,560)}{\pi}} = 46,000\,\text{ft}$$

---

[2]Data of this example is given by L. P. Dake, *Fundamentals of Reservoir Engineering,* Elsevier Publishing Company, 1978.

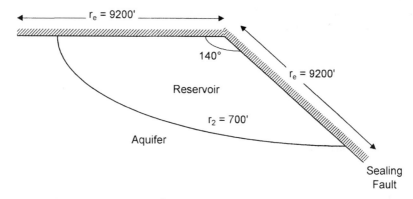

Figure 10-17. Aquifer-reservoir geometry for Example 10-10.

Step 3. Calculate the dimensionless radius r_D.

$$r_D = r_a/r_e$$
$$= 46{,}000/9{,}200 = 5$$

Step 4. Calculate initial water in place W_i.

$$W_i = \pi(r_a^2 - r_e^2)h\phi/5.615$$
$$= \frac{\pi(46{,}000^2 - 9{,}200^2)(100)\,(0.25)}{5.615} = 28.41\,\text{MMM bbl}$$

Step 5. Calculate W_{ei} from Equation 10-38.

$$W_{ei} = c_t\,W_i\,p_i\,f$$

$$W_{ei} = 7 \times 10^{-6}\,(28.41 \times 10^9)\,(2740)\left(\frac{140}{360}\right) = 211.9\,\text{MM bbl}$$

Step 6. Calculate the productivity index J of the radial aquifer from Equation 10-45.

$$J = \frac{0.00708(200)\,(100)\left(\dfrac{140}{360}\right)}{0.55\,\text{Ln}(5)} = 116.5\,\text{bbl/day/psi}$$

Therefore, $Jp_i/W_{ei} = (116.5 \times 2740)/(211.9 \times 10^6) = 1.506 \times 10^{-3}$

Since the time step Δt is fixed at 365 days, then

$$1 - e^{-Jp_i \Delta t / W_{ei}} = 1 - e^{-1.506 \times 10^{-3} \times 365} = 0.4229$$

Equation 10-42 can be reduced to:

$$(\Delta W_e)_n = \frac{211.9 \times 10^6}{2740} [(\overline{p}_a)_{n-1} - (\overline{p}_r)_n]^{(0.4229)}$$

$$(\Delta W_e)_n = 32705 [(\overline{p}_a)_{n-1} - (\overline{p}_r)_n]$$

Step 7. Calculate cumulative water influx as shown in the following table:

n	t days	p_r	$(\overline{p}_r)_n$	$(\overline{p}_a)_{n-1}$	$(\overline{p}_a)_{n-1} - (\overline{p}_r)_n$	$(\Delta W_e)_n$ MM bbl	(W_e) MM bbl
0	0	2740	2740	2740	0	0	0
1	365	2500	2620	2740	120	3.925	3.925
2	730	2290	2395	2689	294	9.615	13.540
3	1095	2109	2199	2565	366	11.970	25.510
4	1460	1949	2029	2409	381	12.461	37.971

PROBLEMS

1. Calculate the cumulative water influx that results from a pressure drop of 200 psi at the oil-water contact with an encroachment angle of 50°. The reservoir-aquifer system is characterized by the following properties:

	Reservoir	Aquifer
radius, ft	6000	20,000
porosity	0.18	0.15
c_f, psi^{-1}	4×10^{-6}	3×10^{-6}
c_w, psi^{-1}	5×10^{-6}	4×10^{-6}
h, ft	25	20

2. An active water drive oil reservoir is producing under the steady-state flowing conditions. The following data is available:

$p_i = 4000$ psi $Q_w = 0$ $R_s = 500$ scf/STB
$Q_o = 40,000$ STB/day $p = 3000$ psi $T = 140°F$
GOR $= 700$ scf/STB $B_o = 1.3$ bbl/STB $B_w = 1.0$ bbl/STB
$z = 0.82$

Calculate Schilthuis' water influx constant.

3. The pressure history of a water-drive oil reservoir is given below:

t, days	p, psi
0	4000
120	3950
220	3910
320	3880
420	3840

The aquifer is under a steady-state flowing condition with an estimated water influx constant of 80 bbl/day/psi. Using the steady-state model, calculate and plot the cumulative water influx as a function of time.

4. A water-drive reservoir has the following boundary-pressure history:

Time, months	Boundary, pressure, psi
0	2610
6	2600
12	2580
18	2552
24	2515

The aquifer-reservoir system is characterized by the following data:

	Reservoir	Aquifer
radius, ft	2000	∞
h, ft	25	30
k, md	60	80
ϕ, %	17	18
μ_w, cp	0.55	0.85
c_w, psi^{-1}	0.7×10^{-6}	0.8×10^{-6}
c_f, psi^{-1}	0.2×10^{-6}	0.3×10^{-6}

If the encroachment angle is 360°, calculate the water influx as a function of time by using:

a. The van Everdingen-Hurst Method
b. The Carter-Tracy Method

5. The following table summarizes the original data available on the West Texas water-drive reservoir:

	Oil Zone	Aquifer
Geometry	Circle	Semi-circle
Area, acres	640	Infinite
Initial reservoir pressure, psia	4000	4000
Initial oil saturation	0.80	0
Porosity, %	22	—
B_{oi}, bbl/STB	1.36	—
B_{wi}, bbl/STB	1.00	1.05
c_o, psi^{-1}	6×10^{-6}	—
c_w, psi^{-1}	3×10^{-6}	7×10^{-6}

The aquifer geological data estimates the water influx constant at 551 bbl/psi. After 1120 days of production, the reservoir average pressure has dropped to 3800 psi and the field has produced 860,000 STB of oil. The field condition after 1120 days of production is given below:

$p = 3800$ psi
$N_p = 860,000$ STB
$B_o = 1.34$ bbl/STB
$B_w = 1.05$ bbl/STB
$W_e = 991,000$ bbl
$t_D = 32.99$ (dimensionless time after 1120 days)
$W_p = 0$ bbl

It is expected that the average reservoir pressure will drop to 3400 psi after 1,520 days (i.e., from the start of production). Calculate the cumulative water influx after 1,520 days.

6. A wedge reservoir-aquifer system with an encroachment angle of 60° has the following boundary pressure history:

Time, days	Boundary pressure, psi
0	2850
365	2610
730	2400
1095	2220
1460	2060

Given:

$$h = 120' \qquad c_f = 5 \times 10^{-6} \text{ psi}^{-1} \qquad c_w = 4 \times 10^{-6} \text{ psi}^{-1}$$
$$\mu_w = 0.7 \text{ cp} \qquad k = 60 \text{ md} \qquad \phi = 12\%$$

reservoir area = 40,000 acres   aquifer area = 980,000 acres   T = 140°F

Calculate the cumulative influx as a function of time by using Fetkovich's Method.

REFERENCES

1. Allard, D. R., and Chen, S. M., "Calculation of Water Influx for Bottom Water Drive Reservoirs," *SPE Reservoir Engineering,* May 1988, pp. 369–379.

2. Carter, R. D., and Tracy, G. W., "An Improved Method for Calculations Water Influx," *Trans. AIME,* 1960.

3. Chatas, A., "A Practical Treatment of Nonsteady-State Flow Problems in Reservoir Systems," *Petroleum Engineering,* May 1953, 25, No. 5, B-42; No. 6, June, p. B-38; No. 8, August, p. B-44.

4. Coats, K., "A Mathematical Model for Water Movement about Bottom-Water-Drive Reservoirs," *SPE Jour.,* March 1962, pp. 44–52; *Trans. AIME,* p. 225.

5. Craft, B., and Hawkins, M., *Applied Reservoir Engineering.* Prentice Hall, 1959.

6. Craft, B., Hawkins, M., and Terry, R., *Applied Petroleum Reservoir Engineering,* 2nd ed. Prentice Hall, 1991.

7. Dake, L. P., *Fundamentals of Reservoir Engineering.* Amsterdam: Elsevier, 1978.

8. Dake, L., *The Practice of Reservoir Engineering.* Amsterdam: Elsevier, 1994.

9. Edwardson, M. et al., "Calculation of Formation Temperature Disturbances Caused by Mud Circulation," *JPT,* April 1962, pp. 416–425; *Trans. AIME,* p. 225.

10. Fetkovich, M. J., "A Simplified Approach to Water Influx Calculations-Finite Aquifer Systems," *JPT,* July 1971, pp. 814–828.

11. Hurst, W., "Water Influx into a Reservoir and its Application to the Equation of Volumetric Balance," *Trans. AIME,* Vol. 151, pp. 57, 1643.

12. Lee, J., and Wattenbarger, R., *Gas Reservoir Engineering.* SPE Textbook Series, Vol. 5, SPE, Dallas, TX, 1996.

13. Schilthuis, R., "Active Oil and Reservoir Energy," *Trans. AIME,* 1936, pp. 37, 118.

14. Van Everdingen, A., and Hurst, W., "The Application of the Laplace Transformation to Flow Problems in Reservoirs," *Trans. AIME,* 1949, pp. 186, 305.

OIL RECOVERY MECHANISMS AND THE MATERIAL BALANCE EQUATION

Each reservoir is composed of a unique combination of geometric form, geological rock properties, fluid characteristics, and primary drive mechanism. Although no two reservoirs are identical in all aspects, they can be grouped according to the primary recovery mechanism by which they produce. It has been observed that each drive mechanism has certain typical performance characteristics in terms of:

• Ultimate recovery factor
• Pressure decline rate
• Gas-oil ratio
• Water production

The recovery of oil by any of the natural drive mechanisms is called **primary recovery.** The term refers to the production of hydrocarbons from a reservoir without the use of any process (such as fluid injection) to supplement the natural energy of the reservoir.

The two main objectives of this chapter are to:

1. Introduce and give a detailed discussion of the various primary recovery mechanisms and their effects on the overall performance of oil reservoirs.

2. Provide the basic principles of the material balance equation and other governing relationships that can be used to predict the volumetric performance of oil reservoirs.

PRIMARY RECOVERY MECHANISMS

For a proper understanding of reservoir behavior and predicting future performance, it is necessary to have knowledge of the driving mechanisms that control the behavior of fluids within reservoirs. The overall performance of oil reservoirs is largely determined by the nature of the energy, i.e., driving mechanism, available for moving the oil to the wellbore. There are basically six driving mechanisms that provide the natural energy necessary for oil recovery:

• Rock and liquid expansion drive
• Depletion drive
• Gas cap drive
• Water drive
• Gravity drainage drive
• Combination drive

These driving mechanisms are discussed as follows.

Rock and Liquid Expansion

When an oil reservoir initially exists at a pressure higher than its bubble-point pressure, the reservoir is called an **undersaturated oil reservoir.** At pressures above the bubble-point pressure, crude oil, connate water, and rock are the only materials present. As the reservoir pressure declines, the rock and fluids expand due to their individual compressibilities. The reservoir rock compressibility is the result of two factors:

• Expansion of the individual rock grains
• Formation compaction

Both of the above two factors are the results of a decrease of fluid pressure within the pore spaces, and both tend to reduce the pore volume through the reduction of the porosity.

As the expansion of the fluids and reduction in the pore volume occur with decreasing reservoir pressure, the crude oil and water will be forced out of the pore space to the wellbore. Because liquids and rocks are only slightly compressible, the reservoir will experience a rapid pressure decline. The oil reservoir under this driving mechanism is characterized by a constant gas-oil ratio that is equal to the gas solubility at the bubble point pressure.

This driving mechanism is considered the least efficient driving force and usually results in the recovery of only a small percentage of the total oil in place.

The Depletion Drive Mechanism

This driving form may also be referred to by the following various terms:

• Solution gas drive
• Dissolved gas drive
• Internal gas drive

In this type of reservoir, the principal source of energy is a result of gas liberation from the crude oil and the subsequent expansion of the solution gas as the reservoir pressure is reduced. As pressure falls below the bubble-point pressure, gas bubbles are liberated within the microscopic pore spaces. These bubbles expand and force the crude oil out of the pore space as shown conceptually in Figure 11-1.

Cole (1969) suggests that a depletion-drive reservoir can be identified by the following characteristics:

• **Reservoir pressure:** The reservoir pressure declines rapidly and continuously. This reservoir pressure behavior is attributed to the fact that no extraneous fluids or gas caps are available to provide a replacement of the gas and oil withdrawals.
• **Water production:** The absence of a water drive means there will be little or no water production with the oil during the entire producing life of the reservoir.
• **Gas-oil ratio:** A depletion-drive reservoir is characterized by a rapidly increasing gas-oil ratio from all wells, regardless of their structural position. After the reservoir pressure has been reduced below the bubble-point pressure, gas evolves from solution throughout the reservoir. Once the gas saturation exceeds the critical gas saturation, free gas begins to flow toward the wellbore and gas-oil ratio increases. The gas

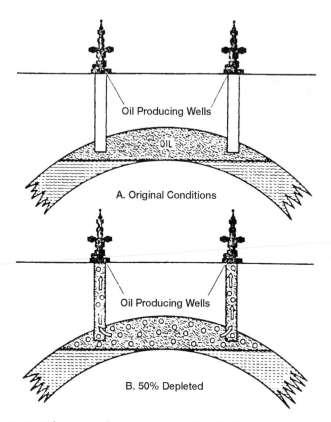

Oil Producing Wells

OIL

A. Original Conditions

Oil Producing Wells

B. 50% Depleted

Figure 11-1. Solution gas drive reservoir. (*After Clark, N. J.,* Elements of Petroleum Reservoirs, *SPE, 1969.*)

will also begin a vertical movement due to the gravitational forces, which may result in the formation of a secondary gas cap. Vertical permeability is an important factor in the formation of a secondary gas cap.

- **Ultimate Oil Recovery:** Oil production by depletion drive is usually the least efficient recovery method. This is a direct result of the formation of gas saturation throughout the reservoir. Ultimate oil recovery from depletion-drive reservoirs may vary from less than 5% to about 30%. The low recovery from this type of reservoirs suggests that large quantities of oil remain in the reservoir and, therefore, depletion-drive reservoirs are considered the best candidates for secondary recovery applications.

The above characteristic trends occurring during the production life of depletion-drive reservoirs are shown in Figure 11-2 and summarized below:

Characteristics	Trend
Reservoir pressure	Declines rapidly and continuously
Gas-oil ratio	Increases to maximum and then declines
Water production	None
Well behavior	Requires pumping at early stage
Oil recovery	5 to 30%

Gas Cap Drive

Gas-cap-drive reservoirs can be identified by the presence of a gas cap with little or no water drive as shown in Figure 11-3.

Due to the ability of the gas cap to expand, these reservoirs are characterized by a slow decline in the reservoir pressure. The natural energy available to produce the crude oil comes from the following two sources:

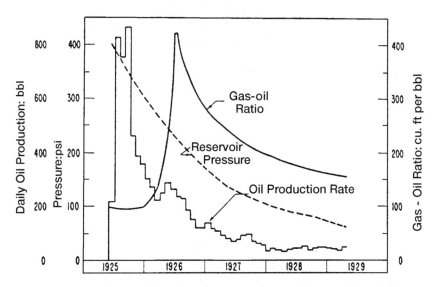

Figure 11-2. Production data of a solution-gas-drive reservoir. (*After Clark, N. J.*, Elements of Petroleum Reservoirs, *SPE, 1969.*)

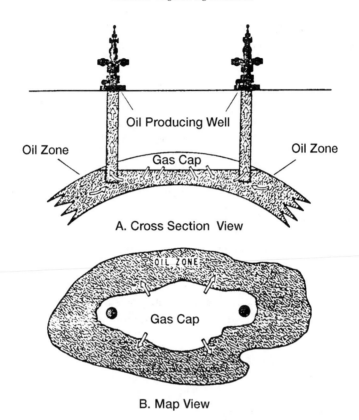

Figure 11-3. Gas-cap-drive reservoir. (*After Clark, N. J.*, Elements of Petroleum Reservoirs, *SPE, 1969.*)

• Expansion of the gas-cap gas
• Expansion of the solution gas as it is liberated

Cole (1969) and Clark (1969) presented a comprehensive review of the characteristic trends associated with gas-cap-drive reservoirs. These characteristic trends are summarized below:

• **Reservoir pressure:** The reservoir pressure falls slowly and continuously. Pressure tends to be maintained at a higher level than in a depletion drive reservoir. The degree of pressure maintenance depends upon the volume of gas in the gas cap compared to the oil volume.
• **Water production:** Absent or negligible water production.

- **Gas-oil ratio:** The gas-oil ratio rises continuously in up-structure wells. As the expanding gas cap reaches the producing intervals of upstructure wells, the gas-oil ratio from the affected wells will increase to high values.
- **Ultimate oil recovery:** Oil recovery by gas-cap expansion is actually a frontal drive displacing mechanism that, therefore, yields a considerably larger recovery efficiency than that of depletion-drive reservoirs. This larger recovery efficiency is also attributed to the fact that no gas saturation is being formed throughout the reservoir at the same time. Figure 11-4 shows the relative positions of the gas-oil contact at different times in the producing life of the reservoir. The expected oil recovery ranges from 20% to 40%.

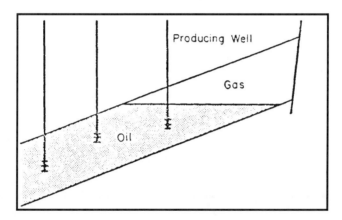

A. Initial fluid distribution.

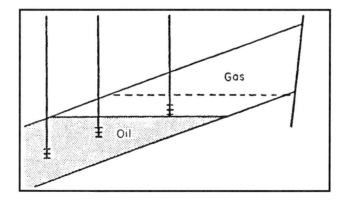

Figure 11-4. Gas cap drive reservoir. (*After Cole, F.,* Reservoir Engineering Manual, *Gulf Publishing Company, 1969.*)

• **Well behavior:** Because of effects of gas-cap expansion on maintaining reservoir pressure and the effect of decreased liquid column weight as it is produced out the well, gas-cap-drive reservoirs tend to flow longer than depletion-drive reservoirs.

The ultimate oil recovery from a gas-cap-drive reservoir will vary depending largely on the following six important parameters:

Size of the Original Gas Cap

As shown graphically in Figure 11-5, the ultimate oil recovery increases with increasing the size of the gas cap.

Vertical Permeability

Good vertical permeability will permit the oil to move downward with less bypassing of gas.

Oil Viscosity

As the oil viscosity increases, the amount of gas bypassing will also increase, which leads to a lower oil recovery.

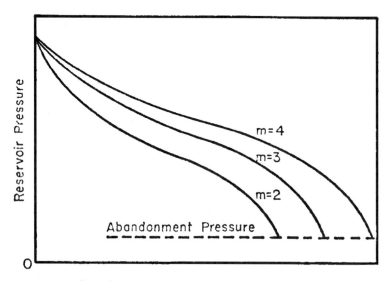

Figure 11-5. Effect of gas cap size on ultimate oil recovery. (*After Cole, F.,* Reservoir Engineering Manual, *Gulf Publishing Company, 1969.*)

Degree of Conservation of the Gas

In order to conserve gas, and thereby increase ultimate oil recovery, it is necessary to shut in the wells that produce excessive gas.

Oil Production Rate

As the reservoir pressure declines with production, solution gas evolves from the crude oil and the gas saturation increases continuously. If the gas saturation exceeds the critical gas saturation, the evolved gas begins to flow in the oil zone. As a result of creating a mobile gas phase in the oil zone, the following two events will occur:

• The effective permeability to oil will be decreased as a result of the increased gas saturation.
• The effective permeability to gas will be increased, thereby increasing the flow of gas.

The formation of the free gas saturation in the oil zone cannot be prevented without resorting to pressure maintenance operations. Therefore, in order to achieve maximum benefit from a gas-cap drive-producing mechanism, gas saturation in the oil zone must be kept to an absolute minimum. This can be accomplished by taking advantage of gravitational segregation of the fluids. In fact, an efficiently operated gas-cap-drive reservoir must also have an efficient gravity segregation drive. As the gas saturation is formed in the oil zone it must be allowed to migrate upstructure to the gas cap. Thus, a gas-cap-drive reservoir is in reality a combination-driving reservoir, although it is not usually considered as such.

Lower producing rates will permit the maximum amount of free gas in the oil zone to migrate to the gas cap. Therefore, gas-cap-drive reservoirs are rate sensitive, as lower producing rates will usually result in increased recovery.

Dip Angle

The size of the gas cap, a measure of reservoir energy available to produce the oil, will in large part determine the recovery percent to be expected. Such recovery normally will be 20 to 40 percent of the original oil in place; if some other features are present to assist, however, such as a steep angle of dip that allows good oil drainage to the bottom of the structure, considerably higher recoveries (up to 60 percent or greater) may be obtained. Conversely, extremely thin oil columns (where early break-through of the advancing gas cap occurs in producing wells) may limit oil recovery to lower figures regardless of the size of the gas cap. Figure 11-6 a typical production and pressure data for a gas-cap-drive reservoir.

The Water-Drive Mechanism

Many reservoirs are bounded on a portion or all of their peripheries by water bearing rocks called aquifers. The aquifers may be so large compared to the reservoir they adjoin as to appear infinite for all practical purposes, and they may range down to those so small as to be negligible in their effects on the reservoir performance.

The aquifer itself may be entirely bounded by impermeable rock so that the reservoir and aquifer together form a closed (volumetric) unit. On the other hand, the reservoir may be outcropped at one or more places where it may be replenished by surface water as shown schematically in Figure 11-7.

It is common to speak of edge water or bottom water in discussing water influx into a reservoir. Bottom water occurs directly beneath the oil and edge water occurs off the flanks of the structure at the edge of the oil

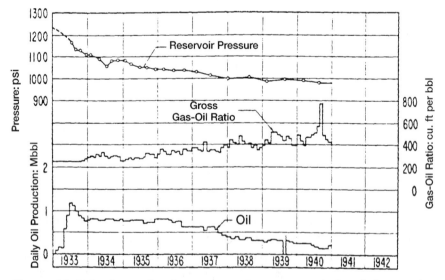

Figure 11-6. Production data for a gas-cap-drive reservoir. (*After Clark, N. J. Elements of Petroleum Reservoirs, SPE, 1969. Courtesy of API.*)

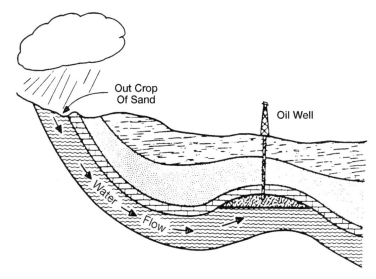

Figure 11-7. Reservoir having artesian water drive. (*After Clark, N. J.,* Elements of Petroleum Reservoirs, *SPE, 1969.*)

as illustrated in Figure 11-8. Regardless of the source of water, the water drive is the result of water moving into the pore spaces originally occupied by oil, replacing the oil and displacing it to the producing wells.

Cole (1969) presented the following discussion on the characteristics that can be used for identification of the water-driving mechanism:

Reservoir Pressure

The reservoir pressure decline is usually very gradual. Figure 11-9 shows the pressure-production history of a typical water-drive reservoir.

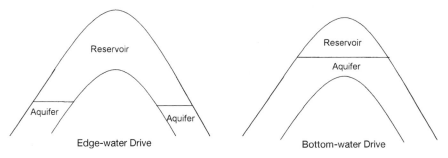

Figure 11-8. Aquifer geometries.

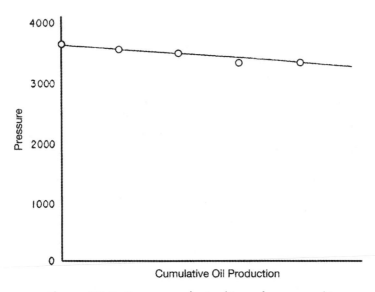

Figure 11-9. Pressure-production history for a water-drive reservoir.

It is not uncommon for many thousands of barrels of oil to be produced for each pound per square inch drop in reservoir pressure. The reason for the small decline in reservoir pressure is that oil and gas withdrawals from the reservoir are replaced almost volume for volume by water encroaching into the oil zone.

Several large oil reservoirs in the Gulf Coast areas of the U.S. have such active water drives that the reservoir pressure has declined only about one psi per million barrels of oil produced. Although pressure history is normally plotted versus cumulative oil production, it should be understood that total reservoir fluid withdrawals are the really important criteria in the maintenance of reservoir pressure. In a water-drive reservoir, only a certain number of barrels of water can move into the reservoir as a result of a unit pressure drop within the reservoir.

Since the principal income production is from oil, if the withdrawals of water and gas can be minimized, then the withdrawal of oil from the reservoir can be maximized with minimum pressure decline. Therefore, it is extremely important to reduce water and gas production to an absolute minimum. This can usually be accomplished by shutting in wells producing large quantities of these fluids and, where possible, transferring their allowables to other wells producing with lower water-oil or gas-oil ratios.

Water Production

Early excess water production occurs in structurally low wells. This is characteristic of a water-drive reservoir, and, provided the water is encroaching in a uniform manner, nothing can or should be done to restrict this encroachment, as the water will probably provide the most efficient displacing mechanism possible.

If the reservoir has one or more lenses of very high permeability, then the water may be moving through this more permeable zone. In this case, it may be economically feasible to perform remedial operations to shut off this permeable zone producing water. It should be realized that in most cases the oil that is being recovered from a structurally low well will be recovered from wells located higher on the structure and any expenses involved in remedial work to reduce the water-oil ratio of structurally low wells may be needless expenditures.

Gas-Oil Ratio

There is normally little change in the producing gas-oil ratio during the life of the reservoir. This is especially true if the reservoir does not have an initial free gas cap. Pressure will be maintained as a result of water encroachment and therefore there will be relatively little gas released from this solution.

Ultimate Oil Recovery

Ultimate recovery from water-drive reservoirs is usually much larger than recovery under any other producing mechanism. Recovery is dependent upon the efficiency of the flushing action of the water as it displaces the oil. In general, as the reservoir heterogeneity increases, the recovery will decrease, due to the uneven advance of the displacing water.

The rate of water advance is normally faster in the zones of high permeability. This results in earlier high water-oil ratios and consequent earlier economic limits. Where the reservoir is more or less homogeneous, the advancing waterfront will be more uniform, and when the economic limit, due primarily to high water-oil ratio, has been reached, a greater portion of the reservoir will have been contacted by the advancing water.

Ultimate oil recovery is also affected by the degree of activity of the water drive. In a very active water drive where the degree of pressure maintenance is good, the role of solution gas in the recovery process is reduced to almost zero, with maximum advantage being taken of the water as a displacing force. This should result in maximum oil recovery from the reservoir. The ultimate oil recovery normally ranges from 35% to 75% of the original oil in place.

The characteristic trends of a water drive reservoir is shown graphically in Figure 11-10 and is summarized below:

Characteristics	Trends
Reservoir pressure	Remains high
Surface gas-oil ratio	Remains low
Water production	Starts early and increases to appreciable amounts
Well behavior	Flow until water production gets excessive
Expected oil recovery	35 to 75 percent

The Gravity-Drainage-Drive Mechanism

The mechanism of gravity drainage occurs in petroleum reservoirs as a result of differences in densities of the reservoir fluids. The effects of gravitational forces can be simply illustrated by placing a quantity of crude oil and a quantity of water in a jar and agitating the contents. After agitation, the jar is placed at rest, and the more denser fluid (normally

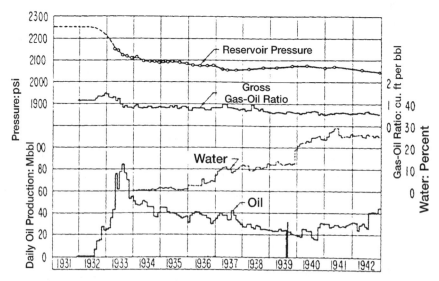

Figure 11-10. Production data for a water-drive reservoir. (*After Clark, N. J.,* Elements of Petroleum Reservoirs, *SPE, 1969. Courtesy of API.*)

water) will settle to the bottom of the jar, while the less dense fluid (normally oil) will rest on top of the denser fluid. The fluids have separated as a result of the gravitational forces acting on them.

The fluids in petroleum reservoirs have all been subjected to the forces of gravity, as evidenced by the relative positions of the fluids, i.e., gas on top, oil underlying the gas, and water underlying oil. The relative positions of the reservoir fluids are shown in Figure 11-11. Due to the long periods of time involved in the petroleum accumulation-and-migration process, it is generally assumed that the reservoir fluids are in equilibrium. If the reservoir fluids are in equilibrium, then the gas-oil and oil-water contacts should be essentially horizontal. Although it is difficult to determine precisely the reservoir fluid contacts, best available data indicate that, in most reservoirs, the fluid contacts actually are essentially horizontal.

Gravity segregation of fluids is probably present to some degree in all petroleum reservoirs, but it may contribute substantially to oil production in some reservoirs.

Cole (1969) stated that reservoir operating largely under a gravity drainage producing mechanism are characterized by:

Reservoir Pressure

Variable rates of pressure decline, depending principally upon the amount of gas conservation. Strictly speaking, where the gas is conserved

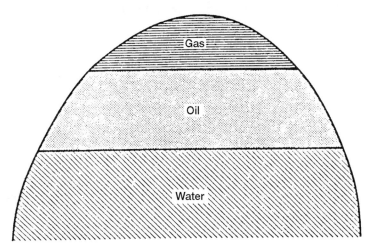

Figure 11-11. Initial fluids distribution in an oil reservoir.

and reservoir pressure is maintained, the reservoir would be operating under combined gas-cap drive and gravity-drainage mechanisms. Therefore, for the reservoir to be operating solely as a result of gravity drainage, the reservoir would show a rapid pressure decline. This would require the upstructure migration of the evolved gas where it later would be produced from structurally high wells, resulting in rapid loss of pressure.

Gas-Oil Ratio

Low gas-oil ratio from structurally low wells. This is caused by migration of the evolved gas upstructure due to gravitational segregation of the fluids. On the other hand, the structurally high wells will experience an increasing gas-oil ratio as a result of the upstructure migration of the gas released from the crude oil.

Secondary Gas Cap

Formation of a secondary gas cap in reservoirs that initially were undersaturated. Obviously the gravity-drainage mechanism does not become operative until reservoir pressure has declined below the saturation pressure, since above the saturation pressure there will be no free gas in the reservoir.

Water Production

Little or no water production. Water production is indicative of a water drive.

Ultimate Oil Recovery

Ultimate recovery from gravity-drainage reservoirs will vary widely, due primarily to the extent of depletion by gravity drainage alone. Where gravity drainage is good, or where producing rates are restricted to take maximum advantage of the gravitational forces, recovery will be high. There are reported cases where recovery from gravity-drainage reservoirs has exceeded 80% of the initial oil in place. In other reservoirs where depletion drive also plays an important role in the oil recovery process, the ultimate recovery will be less.

In operating a gravity-drainage reservoir, it is essential that the oil saturation in the vicinity of the wellbore must be maintained as high as possible. There are two basic reasons for this requirement:

• A high oil saturation means a higher oil flow rate
• A high oil saturation means a lower gas flow rate

If the evolved gas migrates upstructure instead of toward the wellbore, then a high oil saturation in the vicinity of the wellbore can be maintained.

In order to take maximum advantage of the gravity-drainage-producing mechanism, wells should be located as structurally low as possible. This will result in maximum conservation of the reservoir gas. A typical gravity-drainage reservoir is shown in Figure 11-12.

Factors that affect ultimate recovery from gravity-drainage reservoirs are:

• Permeability in the direction of dip
• Dip of the reservoir
• Reservoir producing rates
• Oil viscosity
• Relative permeability characteristics

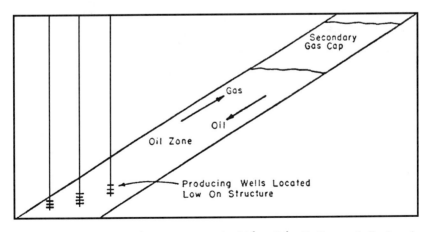

Figure 11-12. Gravity-drainage reservoir. (*After Cole, F.,* Reservoir Engineering Manual, *Gulf Publishing Company, 1969.*)

Cole (1969) presented the following complete treatment of the above listed factors.

Permeability in the Direction of Dip

Good permeability in the direction of migration of the oil is a prerequisite for efficient gravity drainage. For example, a reservoir with little structural relief that also contained many more or less continuous shale "breaks" could probably not be operated under gravity drainage because the oil could not flow to the base of the structure.

Dip of the Reservoir

In most reservoirs, the permeability in the direction of dip is considerably larger than the permeability transverse to the direction of dip. Therefore, as the dip of the reservoir increases, the oil and gas can flow along the direction of dip (which is also the direction of greatest permeability) and still achieve their desired structural position.

Reservoir-Producing Rates

Since the gravity-drainage rate is limited, the reservoir-producing rates should be limited to the gravity-drainage rate, and then maximum recovery will result. If the reservoir-producing rate exceeds the gravity-drainage rate, the depletion-drive-producing mechanism will become more significant with a consequent reduction in ultimate oil recovery.

Oil Viscosity

Oil viscosity is important because the gravity-drainage rate is dependent upon the viscosity of the oil. In the fluid flow equations, the flow rate increases as the viscosity decreases. Therefore, the gravity-drainage rate will increase as the reservoir oil viscosity decreases.

Relative Permeability Characteristics

For an efficient gravity-drive mechanism to be operative, the gas must flow upstructure while the oil flows downstructure. Although this situation involves counterflow of the oil and gas, both fluids are flowing and, therefore, relative permeability characteristics of the formation are very important.

The Combination-Drive Mechanism

The driving mechanism most commonly encountered is one in which both water and free gas are available in some degree to displace the oil toward the producing wells. The most common type of drive encountered, therefore, is a combination-drive mechanism as illustrated in Figure 11-13.

Two combinations of driving forces can be present in combination-drive reservoirs. These are (1) depletion drive and a weak water drive and; (2) depletion drive with a small gas cap and a weak water drive. Then, of course, gravity segregation can play an important role in any of the aforementioned drives.

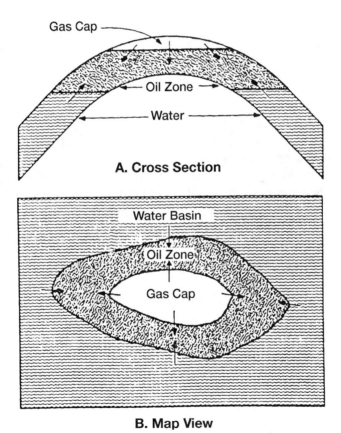

Figure 11-13. Combination-drive reservoir. (*After Clark, N. J.,* Elements of Petroleum Reservoirs, *SPE, 1969.*)

Combination-drive reservoirs can be recognized by the occurrence of a combination of some of the following factors:

a. Relatively rapid pressure decline. Water encroachment and/or external gas-cap expansion are insufficient to maintain reservoir pressures.

b. Water encroaching slowly into the lower part of the reservoir. Structurally low producing wells will exhibit slowly increasing water producing rates.

c. If a small gas cap is present the structurally high wells will exhibit continually increasing gas-oil ratios, provided the gas cap is expanding. It is possible that the gas cap will shrink due to production of excess free gas, in which case the structurally high wells will exhibit a **decreasing** gas-oil ratio. This condition should be avoided whenever possible, as large volumes of oil can be lost as a result of a shrinking gas cap.

d. A substantial percentage of the total oil recovery may be due to the depletion-drive mechanism. The gas-oil ratio of structurally low wells will also continue to increase due to evolution of solution gas throughout the reservoir, as pressure is reduced.

e. Ultimate recovery from combination-drive reservoirs is usually greater than recovery from depletion-drive reservoirs but less than recovery from water-drive or gas-cap-drive reservoirs. Actual recovery will depend upon the degree to which it is possible to reduce the magnitude of recovery by depletion drive. In most combination-drive reservoirs, it will be economically feasible to institute some type of pressure maintenance operation, either gas injection, water injection, or both gas and water injection, depending upon the availability of the fluids.

THE MATERIAL BALANCE EQUATION

The material balance equation (MBE) has long been recognized as one of the basic tools of reservoir engineers for interpreting and predicting reservoir performance. The MBE, when properly applied, can be used to:

- Estimate initial hydrocarbon volumes in place
- Predict future reservoir performance
- Predict ultimate hydrocarbon recovery under various types of primary driving mechanisms

The equation is structured to simply keep inventory of all materials entering, leaving and accumulating in the reservoir. The concept the material balance equation was presented by Schilthuis in 1941. In its simplest form, the equation can be written on volumetric basis as:

Initial volume = volume remaining + volume removed

Since oil, gas, and water are present in petroleum reservoirs, the material balance equation can be expressed for the total fluids or for any one of the fluids present.

Before deriving the material balance, it is convenient to denote certain terms by symbols for brevity. The symbols used conform where possible to the standard nomenclature adopted by the Society of Petroleum Engineers.

p_i	Initial reservoir pressure, psi
p	Volumetric average reservoir pressure
Δp	Change in reservoir pressure = $p_i - p$, psi
p_b	Bubble point pressure, psi
N	Initial (original) oil in place, STB
N_p	Cumulative oil produced, STB
G_p	Cumulative gas produced, scf
W_p	Cumulative water produced, bbl
R_p	Cumulative gas-oil ratio, scf/STB
GOR	Instantaneous gas-oil ratio, scf/STB
R_{si}	Initial gas solubility, scf/STB
R_s	Gas solubility, scf/STB
B_{oi}	Initial oil formation volume factor, bbl/STB
B_o	Oil formation volume factor, bbl/STB
B_{gi}	Initial gas formation volume factor, bbl/scf
B_g	Gas formation volume factor, bbl/scf
W_{inj}	Cumulative water injected, STB
G_{inj}	Cumulative gas injected, scf
W_e	Cumulative water influx, bbl
m	Ratio of initial gas-cap-gas reservoir volume to initial reservoir oil volume, bbl/bbl
G	Initial gas-cap gas, scf
P.V	Pore volume, bbl
c_w	Water compressibility, psi^{-1}
c_f	Formation (rock) compressibility, psi^{-1}

Several of the material balance calculations require the total pore volume (P.V) as expressed in terms of the initial oil volume N and the volume of the gas cap. The expression for the total pore volume can be derived by conveniently introducing the parameter m into the relationship as follows:

Defining the ratio m as:

$$m = \frac{\text{Initial volume of gas cap}}{\text{Volume of oil initially in place}} = \frac{G B_{gi}}{N B_{oi}}$$

Solving for the volume of the gas cap gives:

Initial volume of the gas cap $= G B_{gi} = m N B_{oi}$

The total volume of the hydrocarbon system is then given by:

Initial oil volume + initial gas cap volume $= (P.V) (1 - S_{wi})$

$$N B_{oi} + m N B_{oi} = (P.V) (1 - S_{wi})$$

or

$$P.V = \frac{N B_{oi} (1 + m)}{1 - S_{wi}} \tag{11-1}$$

where S_{wi} = initial water saturation
$\quad N$ = initial oil in place, STB
$\quad P.V$ = total pore volume, bbl
$\quad m$ = ratio of initial gas-cap-gas reservoir volume to initial reservoir oil volume, bbl/bbl

Treating the reservoir pore as an idealized container as illustrated in Figure 11-14, volumetric balance expressions can be derived to account for all volumetric changes which occurs during the natural productive life of the reservoir.

The MBE can be written in a generalized form as follows:

Pore volume occupied by the oil initially in place at p_i

+

Pore volume occupied by the gas in the gas cap at p_i

=

Pore volume occupied by the remaining oil at p

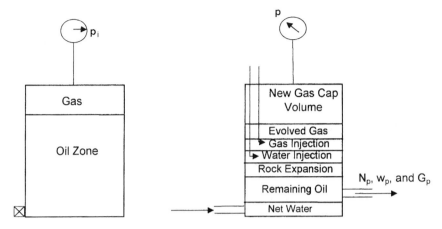

Figure 11-14. Tank-model concept.

$+$

Pore volume occupied by the gas in the gas cap at p

$+$

Pore volume occupied by the evolved solution gas at p

$+$

Pore volume occupied by the net water influx at p

$+$

Change in pore volume due to connate water expansion and pore volume reduction due to rock expansion

$+$

Pore volume occupied by the injected gas at p

$+$

Pore volume occupied by the injected water at p $\hspace{2cm}$ (11-2)

The above nine terms composing the MBE can be separately determined from the hydrocarbon PVT and rock properties, as follows:

Pore Volume Occupied by the Oil Initially in Place

Volume occupied by initial oil in place = $N\,B_{oi}$ $\hspace{2cm}$ (11-3)

where N = oil initially in place, STB
B_{oi} = oil formation volume factor at initial reservoir pressure p_i, bbl/STB

Pore Volume Occupied by the Gas in the Gas Cap

$$\text{Volume of gas cap} = m \, N \, B_{oi} \tag{11-4}$$

where m is a dimensionless parameter and defined as the ratio of gas-cap volume to the oil zone volume.

Pore Volume Occupied by the Remaining Oil

$$\text{Volume of the remaining oil} = (N - N_p) \, B_o \tag{11-5}$$

where N_p = cumulative oil production, STB

B_o = oil formation volume factor at reservoir pressure p, bbl/STB

Pore Volume Occupied by the Gas Cap at Reservoir Pressure p

As the reservoir pressure drops to a new level p, the gas in the gas cap expands and occupies a larger volume. Assuming no gas is produced from the gas cap during the pressure decline, the new volume of the gas cap can be determined as:

$$\text{Volume of the gas cap at p} = \left[\frac{m \, N \, B_{oi}}{B_{gi}} \right] B_g \tag{11-6}$$

where B_{gi} = gas formation volume factor at initial reservoir pressure, bbl/scf

B_g = current gas formation volume factor, bbl/scf

Pore Volume Occupied by the Evolved Solution Gas

This volumetric term can be determined by applying the following material balance on the solution gas:

$$\begin{bmatrix} \text{volume of the evolved} \\ \text{solution gas} \end{bmatrix} = \begin{bmatrix} \text{volume of gas initially} \\ \text{in solution} \end{bmatrix}$$

$$- \begin{bmatrix} \text{volume of gas} \\ \text{produced} \end{bmatrix}$$

$$- \begin{bmatrix} \text{volume of gas} \\ \text{remaining in solution} \end{bmatrix}$$

or

$$\begin{bmatrix} \text{volume of the evolved} \\ \text{solution gas} \end{bmatrix} = [N R_{si} - N_p R_p - (N - N_p)R_s]B_g \quad (11\text{-}7)$$

where N_p = cumulative oil produced, STB
R_p = net cumulative produced gas-oil ratio, scf/STB
R_s = current gas solubility factor, scf/STB
B_g = current gas formation volume factor, bbl/scf
R_{si} = gas solubility at initial reservoir pressure, scf/STB

Pore Volume Occupied by the Net Water Influx

net water influx $= W_e - W_p B_w$ $\qquad\qquad\qquad\qquad$ (11-8)

where W_e = cumulative water influx, bbl
W_p = cumulative water produced, STB
B_w = water formation volume factor, bbl/STB

Change in Pore Volume Due to Initial Water and Rock Expansion

The component describing the reduction in the hydrocarbon pore volume due to the expansion of initial (connate) water and the reservoir rock cannot be neglected for an undersaturated oil reservoir. The water compressibility c_w and rock compressibility c_f are generally of the same order of magnitude as the compressibility of the oil. The effect of these two components, however, can be generally neglected for gas-cap-drive reservoir or when the reservoir pressure drops below the bubble-point pressure.

The compressibility coefficient c which describes the changes in the volume (expansion) of the fluid or material with changing pressure is given by:

$$c = \frac{-1}{V}\frac{\partial V}{\partial p}$$

or

$$\Delta V = V c \Delta p$$

where ΔV represents the net changes or expansion of the material as a result of changes in the pressure. Therefore, the reduction in the pore volume due to the expansion of the connate water in the oil zone and the gas cap is given by:

Connate water expansion = [(pore volume) S_{wi}] c_w Δp

Substituting for the pore volume (P.V) with Equation 11-1 gives:

$$\text{Expansion of connate water} = \frac{N\,B_{oi}\,(1+m)}{1-S_{wi}}\,S_{wi}\,c_w\,\Delta p \qquad (11\text{-}9)$$

where Δp = change in reservoir pressure, $(p_i - p)$
c_w = water compressibility coefficient, psi^{-1}
m = ratio of the volume of the gas-cap gas to the reservoir oil volume, bbl/bbl

Similarly, the reduction in the pore volume due to the expansion of the reservoir rock is given by:

$$\text{Change in pore volume} = \frac{N\,B_{oi}\,(1+m)}{1-S_{wi}}\,c_f\,\Delta p \qquad (11\text{-}10)$$

Combining the expansions of the connate water and formation as represented by Equations 11-9 and 11-10 gives:

Total changes in the pore volume

$$= N\,B_{oi}\,(1+m)\left(\frac{S_{wi}\,c_w + c_f}{1-S_{wi}}\right)\,\Delta p \qquad (11\text{-}11)$$

Pore Volume Occupied by the Injection Gas and Water

Assuming that G_{inj} volumes of gas and W_{inj} volumes of water have been injected for pressure maintenance, the total pore volume occupied by the two injected fluids is given by:

$$\text{Total volume} = G_{inj}\,B_{ginj} + W_{inj}\,B_w \qquad (11\text{-}12)$$

where G_{inj} = cumulative gas injected, scf
B_{ginj} = injected gas formation volume factor, bbl/scf
W_{inj} = cumulative water injected, STB
B_w = water formation volume factor, bbl/STB

Combining Equations 11-3 through 11-12 with Equation 11-2 and rearranging gives:

$$N = \frac{N_p B_o + (G_p - N_p R_s) B_g - (W_e - W_p B_w) - G_{inj} B_{ginj} - W_{inj} B_w}{(B_o - B_{oi}) + (R_{si} - R_s) B_g + mB_{oi} \left[\dfrac{B_g}{B_{gi}} - 1\right] + B_{oi}(1+m)\left[\dfrac{S_{wi}c_w + c_f}{1 - S_{wi}}\right]\Delta p} \quad (11\text{-}13)$$

where N = initial oil in place, STB
G_p = cumulative gas produced, scf
N_p = cumulative oil produced, STB
R_{si} = gas solubility at initial pressure, scf/STB
m = ratio of gas-cap gas volume to oil volume, bbl/bbl
B_{gi} = gas formation volume factor at p_i, bbl/scf
B_{ginj} = gas formation volume factor of the injected gas, bbl/scf

The cumulative gas produced G_p can be expressed in terms of the cumulative gas-oil ratio R_p and cumulative oil produced N_p by:

$$G_p = R_p N_p \quad (11\text{-}14)$$

Combining Equation 11-14 with Equation 11-13 gives:

$$N = \frac{N_p [B_o + (R_p - R_s) B_g] - (W_e - W_p B_w) - G_{inj} B_{ginj} - W_{inj} B_{wi}}{(B_o - B_{oi}) + (R_{si} - R_s) B_g + m B_{oi} \left[\dfrac{B_g}{B_{gi}} - 1\right] + B_{oi}(1+m)\left[\dfrac{S_{wi}c_w + c_f}{1 - S_{wi}}\right]\Delta p} \quad (11\text{-}15)$$

The above relationship is referred to as the **material balance equation** (MBE). A more convenient form of the MBE can be determined by introducing the concept of the total (two-phase) formation volume factor B_t into the equation. This oil PVT property is defined as:

$$B_t = B_o + (R_{si} - R_s) B_g \quad (11\text{-}16)$$

Introducing B_t into Equation 11-15 and assuming, for sake of simplicity, no water or gas injection gives:

$$N = \frac{N_p\left[B_t + (R_p - R_{si})B_g\right] - (W_e - W_pB_w)}{(B_t - B_{ti}) + mB_{ti}\left[\dfrac{B_g}{B_{gi}} - 1\right] + B_{ti}(1+m)\left[\dfrac{S_{wi}c_w + c_f}{1 - S_{wi}}\right]\Delta p} \qquad (11\text{-}17)$$

where S_{wi} = initial water saturation
R_p = cumulative produced gas-oil ratio, scf/STB
Δp = change in the volumetric average reservoir pressure, psi

In a combination drive reservoir where all the driving mechanisms are simultaneously present, it is of practical interest to determine the relative magnitude of each of the driving mechanisms and its contribution to the production.

Rearranging Equation 11-17 gives:

$$\frac{N(B_t - B_{ti})}{A} + \frac{NmB_{ti}(B_g - B_{gi})/B_{gi}}{A} + \frac{W_e - W_pB_w}{A}$$

$$+ \frac{NB_{oi}(1+m)\left[\dfrac{c_wS_{wi} + c_f}{1 - S_{wi}}\right](p_i - p)}{A} = 1 \qquad (11\text{-}18)$$

with the parameter A as defined by:

$$A = N_p\left[B_t + (R_p - R_{si})B_g\right] \qquad (11\text{-}19)$$

Equation 11-18 can be abbreviated and expressed as:

$$DDI + SDI + WDI + EDI = 1.0 \qquad (11\text{-}20)$$

where DDI = depletion-drive index
SDI = segregation (gas-cap)-drive index
WDI = water-drive index
EDI = expansion (rock and liquid)-depletion index

The four terms of the left-hand side of Equation 11-20 represent the major primary driving mechanisms by which oil may be recovered from oil reservoirs. As presented earlier in this chapter, these driving forces are:

a. Depletion Drive. Depletion drive is the oil recovery mechanism wherein the production of the oil from its reservoir rock is achieved by the expansion of the original oil volume with all its original dissolved gas. This driving mechanism is represented mathematically by the first term of Equation 11-18 or:

$$DDI = N \, (B_t - B_{ti})/A \qquad (11\text{-}21)$$

where DDI is termed the **depletion-drive index.**

b. Segregation Drive. Segregation drive (gas-cap drive) is the mechanism wherein the displacement of oil from the formation is accomplished by the expansion of the original free gas cap. This driving force is described by the second term of Equation 11-18, or:

$$SDI = [N \, m \, B_{ti} \, (B_g - B_{gi})/B_{gi}]/A \qquad (11\text{-}22)$$

where SDI is termed the **segregation-drive index.**

c. Water Drive. Water drive is the mechanism wherein the displacement of the oil is accomplished by the net encroachment of water into the oil zone. This mechanism is represented by the third term of Equation 11-18 or:

$$WDI = (W_e - W_p \, B_w)/A \qquad (11\text{-}23)$$

where WDI is termed the **water-drive index.**

d. Expansion Drive. For undersaturated oil reservoirs with no water influx, the principle source of energy is a result of the rock and fluid expansion. Where all the other three driving mechanisms are contributing to the production of oil and gas from the reservoir, the contribution of the rock and fluid expansion to the oil recovery is too small and essentially negligible and can be ignored.

Cole (1969) pointed out that since the sum of the driving indexes is equal to one, it follows that if the magnitude of one of the index terms is reduced, then one or both of the remaining terms must be correspondingly increased. An effective water drive will usually result in maximum recovery from the reservoir. Therefore, if possible, the reservoir should be operated to yield a maximum water-drive index and minimum values for the depletion-drive index and the gas-cap-drive index. Maximum advantage should be taken of the most efficient drive available, and where the water drive is too weak to provide an effective displacing force, it may be possible to utilize the displacing energy of the gas cap. In any event, the depletion-drive index should be maintained as low as possible at all times, as this is normally the most inefficient driving force available.

Equation 11-20 can be solved at any time to determine the magnitude of the various driving indexes. The forces displacing the oil and gas from the reservoir are subject to change from time to time and for this reason Equation 11-20 should be solved periodically to determine whether there has been any change in the driving indexes. Changes in fluid withdrawal rates are primarily responsible for changes in the driving indexes. For example, reducing the oil-producing rate could result in an increased water-drive index and a correspondingly reduced depletion-drive index in a reservoir containing a weak water drive. Also, by shutting in wells producing large quantities of water, the water-drive index could be increased, as the net water influx (gross water influx minus water production) is the important factor.

When the reservoir has a very weak water drive but a fairly large gas cap, the most efficient reservoir producing mechanism may be the gas cap, in which case a large gas-cap-drive index is desirable. Theoretically, recovery by gas-cap drive is independent of producing rate, as the gas is readily expansible. Low vertical permeability could limit the rate of expansion of the gas cap, in which case the gas-cap-drive index would be rate sensitive. Also, gas coning into producing wells will reduce the effectiveness of the gas-cap expansion due to the production of free gas. Gas coning is usually a rate sensitive phenomenon, the higher the producing rates, the greater the amount of coning.

An important factor in determining the effectiveness of a gas-cap drive is the degree of conservation of the gas-cap gas. As a practical matter, it will often be impossible, because of royalty owners or lease agreements, to completely eliminate gas-cap gas production. Where free gas is being produced, the gas-cap-drive index can often be markedly increased by shutting in high gas-oil ratio wells and, if possible, transferring their allowables to other low gas-oil ratio wells.

Figure 11-15 shows a set of plots that represents various driving indexes for a combination-drive reservoir. At point A, some of the structurally low wells are reworked to reduce water production. This resulted in an effective increase in the water-drive index. At point B, workover operations are complete; water-, gas-, and oil-producing rates are relatively stable; and the driving indexes show no change. At point C, some of the wells which have been producing relatively large, but constant, volumes of water are shut in, which results in an increase in the water-drive index. At the same time, some of the upstructure, high gas-oil ratio wells have been shut in and their allowables transferred to wells lower on the structure producing with normal gas-oil ratios. At point D, gas is being returned to the reservoir, and the gas-cap-drive index is exhibiting a decided increase.

The water-drive index is relatively constant, although it is decreasing somewhat, and the depletion-drive index is showing a marked decline. This is indicative of a more efficient reservoir operation, and, if the depletion-drive index can be reduced to zero, relatively good recovery can be expected from the reservoir. Of course, to achieve a zero-depletion-drive index would require the complete maintenance of reservoir pressure, which is often difficult to accomplish. It can be noted from Figure 11-15 that the sum of the various indexes of drive is always equal to one.

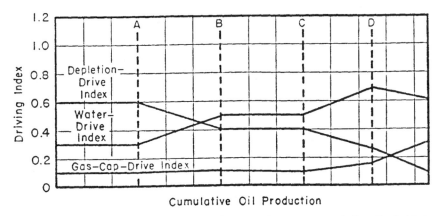

Figure 11-15. Driving indexes in a combination-drive reservoir. (*After Clark, N. J.,* Elements of Petroleum Reservoirs, *SPE, 1969.*)

Example 11-1

A combination-drive reservoir contains 10 MMSTB of oil initially in place. The ratio of the original gas-cap volume to the original oil volume, i.e., m, is estimated as 0.25. The initial reservoir pressure is 3000 psia at 150°F. The reservoir produced 1 MMSTB of oil, 1100 MMscf of 0.8 specific gravity gas, and 50,000 STB of water by the time the reservoir pressure dropped to 2800 psi. The following PVT is available:

	3000 psi	2800 psi
B_o, bbl/STB	1.58	1.48
R_s, scf/STB	1040	850
B_g, bbl/scf	0.00080	0.00092
B_t, bbl/STB	1.58	1.655
B_w, bbl/STB	1.000	1.000

The following data is also available:

$$S_{wi} = 0.20 \qquad c_w = 1.5 \times 10^{-6} \text{ psi}^{-1} \qquad c_f = 1 \times 10^{-6} \text{ psi}^{-1}$$

Calculate:

a. Cumulative water influx
b. Net water influx
c. Primary driving indexes at 2800 psi

Solution

Because the reservoir contains a gas cap, the rock and fluid expansion can be neglected, i.e., set c_f and $c_w = 0$. For illustration purposes, however, the rock and fluid expansion term will be included in the calculations.

Part A. Cumulative water influx

Step 1. Calculate cumulative gas-oil ratio R_p:

$$R_p = \frac{1100 \times 10^6}{1 \times 10^6} = 1100 \text{ scf/STB}$$

Step 2. Arrange Equation 11-17 to solve for W_e:

$$W_e = N_p [B_t + (R_p - R_{si})B_g] - N\left[(B_g - B_{ti}) + mB_{ti}\left(\frac{B_g}{B_{gi}} - 1\right)\right.$$

$$\left. + B_{ti}(1+m)\left(\frac{S_{wi}c_w + c_f}{1 - S_{wi}}\right)\Delta p\right] + W_p B_{wp}$$

$$W_e = 10^6 [1.655 + (1100 - 1040)\,0.00092]$$

$$- 10^7 \left[(1.655 - 1.58) + 0.25\,(1.58)\left(\frac{0.00092}{0.00080} - 1\right)\right.$$

$$\left. + 1.58\,(1 + 0.25)\left(\frac{0.2\,(1.5 \times 10^{-6})}{1 - 0.2}\right)(3000 - 2800)\right] + 50,000$$

$$= 411,281\,\text{bbl}$$

Neglecting the rock and fluid expansion term, the cumulative water influx is 417,700 bbl.

Part B. Net water influx

Net water influx $= W_e - W_p B_w = 411,281 - 50,000 = 361,281\,\text{bbl}$

Part C. Primary recovery indexes

Step 1. Calculate the parameter A by using Equation 11-19:

$$A = 10^6 [1.655 + (1100 - 1040)\,0.00092] = 1,710,000$$

Step 2. Calculate DDI, SDI, and WDI by applying Equations 11-21 through 11-23, respectively:

$$DDI = \frac{10 \times 10^6\,(1.655 - 1.58)}{1,710,000} = 0.4385$$

$$SDI = \frac{10 \times 10^6\,(0.25)\,(1.58)\,(0.00092 - 0.0008)/0.0008}{1,710,000} = 0.3465$$

$$WDI = \frac{411,281 - 50,000}{1,710,000} = 0.2112$$

$$EDI = 1 - 0.4385 - 0.3465 - 0.2112 = 0.0038$$

where EDI is termed the **expansion-drive index.**

These calculations show that the 43.85% of the recovery was obtained by depletion drive, 34.65% by gas-cap drive, 21.12% by water drive, and only 0.38% by connate water and rock expansion. The results suggest that the expansion-drive index (EDI) term can be neglected in the presence of a gas cap or when the reservoir pressure drops below the bubble-point pressure. In high pore volume compressibility reservoirs, such as chalks and unconsolidated sands, however, the energy contribution of the rock and water expansion cannot be ignored even at high gas saturations.

Example 11-2

The Big Butte field is a combination-drive reservoir. The current reservoir pressure is estimated at 2500 psi. The reservoir production data and PVT information are given below:

	Initial reservoir condition	Current reservoir condition
p, psi	3000	2500
B_o, bbl/STB	1.35	1.33
R_s, scf/STB	600	500
N_p, MMSTB	0	5
G_p, MMMscf		5.5
B_w, bbl/STB	1.00	1.00
W_e, MMbbl	0	3
W_p, MMbbl	0	0.2
B_g, bbl/scf	0.0011	0.0015
c_f, c_w	0	0

The following additional information is available:

Volume of bulk oil zone = 100,000 ac-ft
Volume of bulk gas zone = 20,000 ac-ft

Calculate the initial oil in place.

Solution

Step 1. Assuming the same porosity and connate water for the oil and gas zones, calculate m:

$$m = \frac{20,000}{100,000} = 0.2$$

Step 2. Calculate the cumulative gas-oil ratio R_p:

$$R_p = \frac{5.5 \times 10^9}{5 \times 10^6} = 1100 \text{ scf/STB}$$

Step 3. Solve for the initial oil-in-place by applying Equation 11-15:

$$N = \frac{5 \times 10^6 [1.33 + (1100 - 500)0.0015] - (3 \times 10^6 - 0.2 \times 10^6)}{(1.35 - 1.33) + (600 - 500)0.0015 + (0.2)(1.35)\left[\dfrac{0.0015}{0.0011} - 1\right]}$$

$$= 31.14 \text{ MMSTB}$$

Basic Assumptions in the MBE

The material balance equation calculation is based on changes in reservoir conditions over discrete periods of time during the production history. The calculation is most vulnerable to many of its underlying assumptions early in the depletion sequence when fluid movements are limited and pressure changes are small. Uneven depletion and partial reservoir development compound the accuracy problem.

The basic assumptions in the material balance equation (MBE) are as follows:

• **Constant temperature.** Pressure-volume changes in the reservoir are assumed to occur without any temperature changes. If any temperature changes occur, they are usually sufficiently small to be ignored without significant error.
• **Pressure equilibrium.** All parts of the reservoir have the same pressure, and fluid properties are therefore constant throughout. Minor variations in the vicinity of the well bores may usually be ignored. Substantial pressure variation across the reservoir may cause excessive calculation error.

It is assumed that the PVT samples or data sets represent the actual fluid compositions and that reliable and representative laboratory procedures have been used. Notably, the vast majority of material balances assume that differential depletion data represent reservoir flow and that separator flash data may be used to correct for the wellbore transition to surface conditions. Such "black oil" PVT treatments relate volume changes to temperature and pressure only. They lose validity in cases of volatile oil or gas condensate reservoirs where compositions are also important. Special laboratory procedures may be used to improve PVT data for volatile fluid situations.

Constant reservoir volume. Reservoir volume is assumed to be constant except for those conditions of rock and water expansion or water influx that are specifically considered in the equation. The formation is considered to be sufficiently competent that no significant volume change will occur through movement or reworking of the formation due to overburden pressure as the internal reservoir pressure is reduced. The constant volume assumption is also related to an area of interest to which the equation is applied. If the focus is on some part of a reservoir system, except for specific exterior flow terms it is assumed that the particular portion is encased in no-flow boundaries.

Reliable production data. All production data should be recorded with respect to the same time period. If possible, gas-cap- and solution-gas production records should be maintained separately.

Gas and oil gravity measurements should be recorded in conjunction with the fluid volume data. Some reservoirs require a more detailed analysis and that the material balance be solved for volumetric segments. The produced fluid gravities will aid in the selection of the volumetric segments and also in the averaging of fluid properties. There are essentially three types of production data that must be recorded in order to use the MBE in performing reliable reservoir calculations. These are:

- **Oil-production data,** even for non-interest properties, which can usually be obtained from various sources and is usually fairly reliable.
- **Gas-production data,** which are becoming more available and reliable as the market value of this commodity increases; unfortunately, these data will often be more questionable where gas is flared.
- **The water-production term,** which need represent only the net withdrawals of water; therefore, where subsurface disposal of produced brine is to the same source formation, most of the error due to poor data will be eliminated.

The MBE as an Equation of a Straight Line

An insight into the general MBE, i.e., Equation 11-15, may be gained by considering the physical significance of the following groups of terms of which it is comprised:

- $N_p [B_o + (R_p - R_s) B_g]$ Represents the reservoir volume of cumulative oil and gas produced.
- $[W_e - W_p B_w]$ Refers to the net water influx that is retained in the reservoir.
- $[G_{inj} B_{ginj} + W_{inj} B_w]$ This pressure maintenance term represents cumulative fluid injection in the reservoir.
- $[m B_{oi} (B_g/B_{gi} - 1)]$ Represents the net expansion of the gas cap that occurs with the production of N_p stock-tank barrels of oil (as expressed in bbl/STB of original oil in place).

There are essentially three unknowns in Equation 11-15:

a. The original oil in place N
b. The cumulative water influx W_e
c. The original size of the gas cap as compared to the oil zone size m

In developing a methodology for determining the above three unknowns, Havlena and Odeh (1963) expressed Equation 11-15 in the following form:

$$N_p [B_o + (R_p - R_s)B_g] + W_p B_w = N[(B_o - B_{oi}) + (R_{si} - R_s)B_g]$$

$$+ m N B_{oi} \left(\frac{B_g}{B_{gi}} - 1 \right) + N (1+m) B_{oi} \left[\frac{c_w S_{wi} + c_f}{1 + S_{wi}} \right] \Delta p$$

$$+ W_e + W_{inj} B_w + G_{inj} B_{ginj} \qquad (11\text{-}24)$$

Havlena and Odeh further expressed Equation 11-24 in a more condensed form as:

$$F = N [E_o + m E_g + E_{f,w}] + (W_e + W_{inj} B_w + G_{inj} B_{ginj})$$

Assuming, for the purpose of simplicity, that no pressure maintenance by gas or water injection is being considered, the above relationship can be further simplified and written as:

$$F = N [E_o + m E_g + E_{f,w}] + W_e \qquad (11\text{-}25)$$

In which the terms F, E_o, E_g, and $E_{f,w}$ are defined by the following relationships:

• F represents the underground withdrawal and given by:

$$F = N_p [B_o + (R_p - R_s) B_g] + W_p B_w \tag{11-26}$$

In terms of the two-phase formation volume factor B_t, the underground withdrawal F can be written as:

$$F = N_p [B_t + (R_p - R_{si}) B_g] + W_p B_w \tag{11-27}$$

• E_o describes the expansion of oil and its originally dissolved gas and is expressed in terms of the oil formation volume factor as:

$$E_o = (B_o - B_{oi}) + (R_{si} - R_s) B_g \tag{11-28}$$

Or equivalently, in terms of B_t:

$$E_o = B_t - B_{ti} \tag{11-29}$$

• E_g is the term describing the expansion of the gas-cap gas and is defined by the following expression:

$$E_g = B_{oi} [(B_g/B_{gi}) - 1] \tag{11-30}$$

In terms of the two-phase formation volume factor B_t, essentially $B_{ti} = B_{oi}$ or:

$$E_g = B_{ti} [(B_g/B_{gi}) - 1]$$

• $E_{f,w}$ represents the expansion of the initial water and the reduction in the pore volume and is given by:

$$E_{f,w} = (1 + m) B_{oi} \left[\frac{c_w S_{wi} + c_f}{1 - S_{wi}} \right] \Delta p \tag{11-31}$$

Havlena and Odeh examined several cases of varying reservoir types with Equation 11-25 and pointed out that the relationship can be rearranged into the form of a straight line. For example, in the case of a reservoir which has no initial gas cap (i.e., m = 0) or water influx (i.e., $W_e = 0$), and negligible formation and water compressibilities (i.e., c_f and $c_w = 0$); Equation 11-25 reduces to:

$$F = N E_o$$

The above expression suggests that a plot of the parameter F as a function of the oil expansion parameter E_o would yield a straight line with a slope N and intercept equal to zero.

The Straight-Line Solution Method to the MBE

The straight-line solution method requires the plotting of a variable group versus another variable group, with the variable group selection depending on the mechanism of production under which the reservoir is producing. The most important aspect of this method of solution is that it attaches significance the sequence of the plotted points, the direction in which they plot, and to the shape of the resulting plot.

The significance of the straight-line approach is that the sequence of plotting is important and if the plotted data deviates from this straight line there is some reason for it. This significant observation will provide the engineer with valuable information that can be used in determining the following unknowns:

- Initial oil in place   N
- Size of the gas cap   m
- Water influx W_e
- Driving mechanism

The remainder of this chapter is devoted to illustrations of the use of the straight-line solution method in determining N, m, and W_e for different reservoir mechanisms.

Case 1. Volumetric Undersaturated-Oil Reservoirs

Assuming no water or gas injection, the linear form of the MBE as expressed by Equation 11-25 can be written as:

$$F = N [E_o + m E_g + E_{f,w}] + W_e \qquad (11\text{-}32)$$

Several terms in the above relationship may disappear when imposing the conditions associated with the assumed reservoir driving mechanism. For a volumetric and undersaturated reservoir, the conditions associated with driving mechanism are:

- $W_e = 0$, since the reservoir is volumetric
- $m = 0$, since the reservoir is undersaturated
- $R_s = R_{si} = R_p$, since all produced gas is dissolved in the oil

Applying the above conditions on Equation 11-32 gives:

$$F = N (E_o + E_{f,w}) \tag{11-33}$$

or

$$N = \frac{F}{E_o + E_{f,w}} \tag{11-34}$$

where N = initial oil in place, STB
$$F = N_p B_o + W_p B_w \tag{11-35}$$
$$E_o = B_o - B_{oi} \tag{11-36}$$

$$E_{f,w} = B_{oi} \left[\frac{c_w S_w + c_f}{1 - S_{wi}} \right] \Delta p \tag{11-37}$$

$\Delta p = p_i - \bar{p}_r$
p_i = initial reservoir pressure
$\bar{p}_r$ = volumetric average reservoir pressure

When a new field is discovered, one of the first tasks of the reservoir engineer is to determine if the reservoir can be classified as a volumetric reservoir, i.e., $W_e = 0$. The classical approach of addressing this problem is to assemble all the necessary data (i.e., production, pressure, and PVT) that are required to evaluate the right-hand side of Equation 11-36. The term $F/(E_o + E_{f,w})$ for each pressure and time observation is plotted versus cumulative production N_p or time, as shown in Figure 11-16. Dake (1994) suggests that such a plot can assume two various shapes, which are:

• All the calculated points of $F/(E_o + E_{f,w})$ lie on a horizontal straight line (see Line A in Figure 11-16). Line A in the plot implies that the reservoir can be classified as a volumetric reservoir. This defines a purely depletion-drive reservoir whose energy derives solely from the expansion of the rock, connate water, and the oil. Furthermore, the ordinate value of the plateau determines the initial oil in place N.
• Alternately, the calculated values of the term $F/(E_o + E_{f,w})$ rise, as illustrated by the curves B and C, indicating that the reservoir has been energized by water influx, abnormal pore compaction, or a combination of

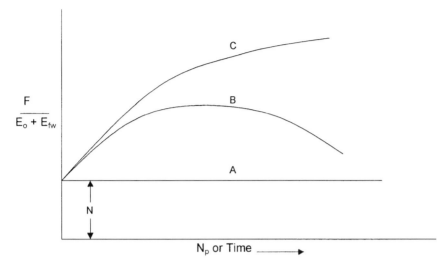

Figure 11-16. Classification of the reservoir.

these two. Curve C in Figure 11-16 might be for a strong water-drive field in which the aquifer is displacing an infinite acting behavior, whereas B represents an aquifer whose outer boundary has been felt and the aquifer is depleting in unison with the reservoir itself. The downward trend in points on curve B as time progresses denotes the diminishing degree of energizing by the aquifer. Dake (1994) points out that in water-drive reservoirs, the shape of the curve, i.e., $F/(E_o + E_{f,w})$ vs. time, is highly rate dependent. For instance, if the reservoir is producing at higher rate than the water-influx rate, the calculated values of $F/(E_o + E_{f,w})$ will dip downward revealing a lack of energizing by the aquifer, whereas, if the rate is decreased, the reverse happens and the points are elevated.

Similarly, Equation 11-33 could be used to verify the characteristic of the reservoir-driving mechanism and to determine the initial oil in place. A plot of the underground withdrawal F versus the expansion term $(E_o + E_{f,w})$ should result in a straight line going through the origin with N being the slope. It should be noted that the origin is a "must" point; thus, one has a fixed point to guide the straight-line plot (as shown in Figure 11-17).

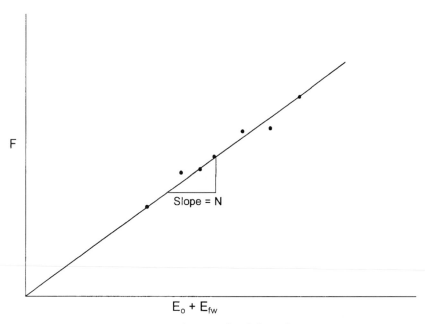

Figure 11-17. Underground withdrawal vs. $E_o + E_{fw}$.

This interpretation technique is useful in that, if the linear relationship is expected for the reservoir and yet the actual plot turns out to be non-linear, then this deviation can itself be diagnostic in determining the actual drive mechanisms in the reservoir.

A linear plot of the underground withdrawal F versus $(E_o + E_{f,w})$ indicates that the field is producing under volumetric performance, i.e., no water influx, and strictly by pressure depletion and fluid expansion. On the other hand, a nonlinear plot indicates that the reservoir should be characterized as a water-drive reservoir.

Example 11-3

The Virginia Hills Beaverhill Lake field is a volumetric undersaturated reservoir. Volumetric calculations indicate the reservoir contains 270.6

MMSTB of oil initially in place. The initial reservoir pressure is 3685 psi. The following additional data is available:

$$S_{wi} = 24\% \qquad c_w = 3.62 \times 10^{-6} \text{ psi}^{-1} \qquad c_f = 4.95 \times 10^{-6} \text{ psi}^{-1}$$
$$B_w = 1.0 \text{ bbl/STB} \qquad p_b = 1500 \text{ psi}$$

The field production and PVT data are summarized below:

Volumetric Average Pressure	No. of producing wells	B_o bbl/STB	N_p MSTB	W_p MSTB
3685	1	1.3102	0	0
3680	2	1.3104	20.481	0
3676	2	1.3104	34.750	0
3667	3	1.3105	78.557	0
3664	4	1.3105	101.846	0
3640	19	1.3109	215.681	0
3605	25	1.3116	364.613	0
3567	36	1.3122	542.985	0.159
3515	48	1.3128	841.591	0.805
3448	59	1.3130	1273.530	2.579
3360	59	1.3150	1691.887	5.008
3275	61	1.3160	2127.077	6.500
3188	61	1.3170	2575.330	8.000

Calculate the initial oil in place by using the MBE and compare with the volumetric estimate of N.

Solution

Step 1. Calculate the initial water and rock expansion term $E_{f,w}$ from Equation 11-37:

$$E_{f,w} = 1.3102 \left[\frac{3.62 \times 10^{-6} (0.24) + 4.95 \times 10^{-6}}{1 - 0.24} \right] \Delta p$$

$$E_{f,w} = 10.0 \times 10^{-6} (3685 - \bar{p}_r)$$

Step 2. Construct the following table:

$\bar{p}_r$, psi	F, Mbbl Equation 10-35	E_o, bbl/STB Equation 10-36	Δp	$E_{f, w}$	$E_o + E_{f, w}$
3685	—	—	0	0	—
3680	26.84	0.0002	5	50×10^{-6}	0.00025
3676	45.54	0.0002	9	90×10^{-6}	0.00029
3667	102.95	0.0003	18	180×10^{-6}	0.00048
3664	133.47	0.0003	21	210×10^{-6}	0.00051
3640	282.74	0.0007	45	450×10^{-6}	0.00115
3605	478.23	0.0014	80	800×10^{-6}	0.00220
3567	712.66	0.0020	118	1180×10^{-6}	0.00318
3515	1,105.65	0.0026	170	1700×10^{-6}	0.00430
3448	1,674.72	0.0028	237	2370×10^{-6}	0.00517
3360	2,229.84	0.0048	325	3250×10^{-6}	0.00805
3275	2,805.73	0.0058	410	4100×10^{-6}	0.00990
3188	3,399.71	0.0068	497	4970×10^{-6}	0.01170

Step 3. Plot the underground withdrawal term F against the expansion term $(E_o + E_{f,w})$ on a Cartesian scale, as shown in Figure 11-18).

Step 4. Draw the best straight line through the points and determine the slope of the line and the volume of the active initial oil in place as:

$$N = 257 \text{ MMSTB}$$

It should be noted that the value of the initial oil in place as determined from the MBE is referred to as the effective or active initial oil in place. This value is usually smaller than that of the volumetric estimate due to oil being trapped in undrained fault compartments or low-permeability regions of the reservoir.

Case 2. Volumetric Saturated-Oil Reservoirs

An oil reservoir that originally exists at its bubble-point pressure is referred to as a saturated oil reservoir. The main driving mechanism in this type of reservoir results from the liberation and expansion of the solution gas as the pressure drops below the bubble-point pressure. The only unknown in a volumetric saturated-oil reservoir is the initial oil in place N. Assuming that the water and rock expansion term $E_{f,w}$ is negligi-

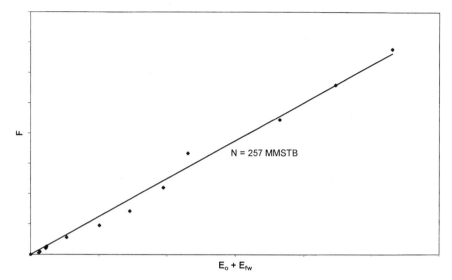

Figure 11-18. F vs. $(E_o + E_{fw})$ for Example 11-3.

ble in comparison with the expansion of solution gas, Equation 11-32 can be simplified as:

$$F = N E_o \qquad (11\text{-}38)$$

where the underground withdrawal F and the oil expansion E_o were defined previously by Equations 11-26 and 11-28 or Equations 11-27 and 11-29 to give:

$$F = N_p [B_t + (R_p - R_{si}) B_g] + W_p B_w$$

$$E_o = B_t - B_{ti}$$

Equation 11-38 indicates that a plot of the underground withdrawal F, evaluated by using the actual reservoir production data, as a function of the fluid expansion term E_o, should result in a straight line going through the origin with a slope of N.

The above interpretation technique is useful in that, if a simple linear relationship such as Equation 11-38 is expected for a reservoir and yet the actual plot turns out to be nonlinear, then this deviation can itself be diagnostic in determining the actual drive mechanisms in the reservoir. For instance, Equation 11-38 may turn out to be nonlinear because there is an unsuspected water influx into the reservoir helping to maintain the pressure.

Case 3. Gas-Cap-Drive Reservoirs

For a reservoir in which the expansion of the gas-cap gas is the predominant driving mechanism and assuming that the natural water influx is negligible ($W_e = 0$), the effect of water and pore compressibilities can be considered negligible. Under these conditions, the Havlena-Odeh material balance can be expressed as:

$$F = N [E_o + m E_g] \qquad (11\text{-}39)$$

where E_g is defined by Equation 11-30 as:

$$E_g = B_{oi} [(B_g/B_{gi}) - 1]$$

The way in which Equation 11-39 can be used depends on the number of unknowns in the equation. There are three possible unknowns in Equation 11-39:

- N is unknown, m is known
- m is unknown, N is known
- N and m are unknown

The practical use of Equation 11-39 in determining the three possible unknowns is presented below:

a. Unknown N, known m:

Equation 11-39 indicates that a plot of F versus ($E_o + m E_g$) on a Cartesian scale would produce a straight line through the origin with a slope of N, as shown in Figure 11-19. In making the plot, the underground withdrawal F can be calculated at various times as a function of the production terms N_p and R_p.
Conclusion: N = Slope

b. Unknown m, known N:

Equation 11-39 can be rearranged as an equation of straight line, to give:

$$\left(\frac{F}{N} - E_o\right) = m E_g \qquad (11\text{-}40)$$

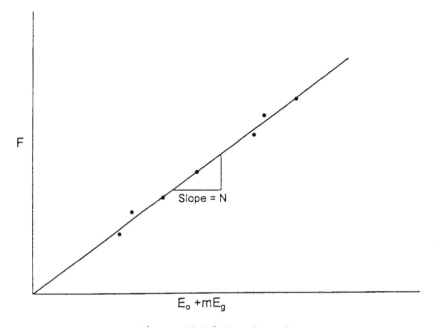

Figure 11-19. F vs. $E_o + mE_g$.

The above relationship shows that a plot of the term $(F/N - E_o)$ versus E_g would produce a straight line with a slope of m. One advantage of this particular arrangement is that the straight line must pass through the origin which, therefore, acts as a control point. Figure 11-20 shows an illustration of such a plot.

Conclusion: m = Slope

c. N and m are Unknown

If there is uncertainty in both the values of N and m, Equation 11-39 can be re-expressed as:

$$\frac{F}{E_o} = N + mN\left(\frac{E_g}{E_o}\right) \qquad (11\text{-}41)$$

A plot of F/E_o versus E_g/E_o should then be linear with intercept N and slope mN. This plot is illustrated in Figure 11-21.

Conclusions: N = intercept
                   mN = slope
                   m = slope/intercept

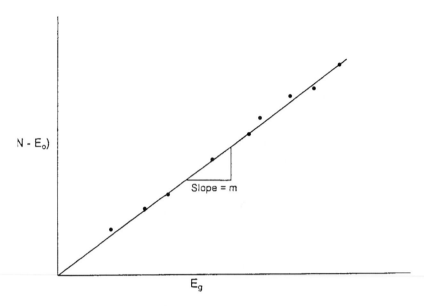

Figure 11-20. $(F/N - E_o)$ vs. E_g.

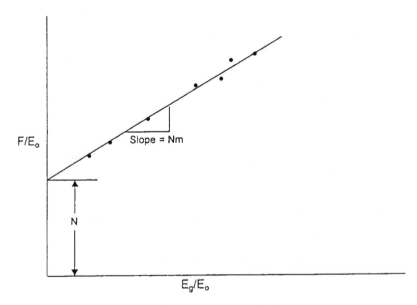

Figure 11-21. F/E_o vs. E_g/E_o.

Example 11-4[1]

The production history and the PVT data of a gas-cap-drive reservoir are given below:

Date	$\bar{p}$ psi	N_p MSTB	G_p Mscf	B_t bbl/STB	B_g bbl/scf
5/1/89	4415	—	—	1.6291	0.00077
1/1/91	3875	492.5	751.3	1.6839	0.00079
1/1/92	3315	1015.7	2409.6	1.7835	0.00087
1/1/93	2845	1322.5	3901.6	1.9110	0.00099

The initial gas solubility R_{si} is 975 scf/STB. Estimate the initial oil and gas in place.

Solution

Step 1. Calculate the cumulative produced gas-oil ratio R_p

$\bar{p}$	G_p Mscf	N_p MSTB	$R_p = G_p/N_p$ scf/STB
4415	—	—	—
3875	751.3	492.5	1525
3315	2409.6	1015.7	2372
2845	3901.6	1322.5	2950

Step 2. Calculate F, E_o, and E_g:

p	F	E_o	E_g
3875	2.04×10^6	0.0548	0.0529
3315	8.77×10^6	0.1540	0.2220
2845	17.05×10^6	0.2820	0.4720

Step 3. Calculate F/E_o and E_g/E_o

p	F/E_o	E_g/E_o
3875	3.72×10^7	0.96
3315	5.69×10^7	1.44
2845	6.00×10^7	1.67

---

[1]After Economides, M., and Hill, D., *Petroleum Production Systems,* Prentice Hall, 1993.

Step 4. Plot (F/E_o) versus (E_g/E_o) as shown in Figure 11-22 to give:

- Intercept = N = 9 MMSTB
- Slope = N m = 3.1 × 107

Step 5. Calculate m:

$$m = 3.1 \times 10^7/(9 \times 10^6) = 3.44$$

Step 6. Calculate initial gas in place G:

$$m = \frac{G B_{gi}}{N B_{oi}}$$

$$G = \frac{(3.44)(9 \times 10^6)(1.6291)}{0.00077} = 66\,\text{MMMscf}$$

Case 4. Water-Drive Reservoirs

In a water-drive reservoir, identifying the type of the aquifer and characterizing its properties are perhaps the most challenging tasks involved in conducting a reservoir engineering study. Yet, without an accurate

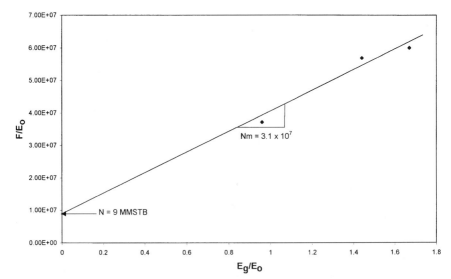

Figure 11-22. Calculation of m and N for Example 11-4.

description of the aquifer, future reservoir performance and management cannot be properly evaluated.

The full MBE can be expressed again as:

$$F = N \, (E_o + m \, E_g + E_{f,w}) + W_e$$

Dake (1978) points out that the term $E_{f,w}$ can frequently be neglected in water-drive reservoirs. This is not only for the usual reason that the water and pore compressibilities are small, but also because a water influx helps to maintain the reservoir pressure and, therefore, the Δp appearing in the $E_{f,w}$ term is reduced, or

$$F = N \, (E_o + m \, E_g) + W_e \qquad (11\text{-}42)$$

If, in addition, the reservoir has initial gas cap, then Equation 11-42 can be further reduced to:

$$F = N \, E_o + W_e \qquad (11\text{-}43)$$

Dake (1978) points out that in attempting to use the above two equations to match the production and pressure history of a reservoir, the greatest uncertainty is always the determination of the water influx W_e. In fact, in order to calculate the influx the engineer is confronted with what is inherently the greatest uncertainty in the whole subject of reservoir engineering. The reason is that the calculation of W_e requires a mathematical model which itself relies on the knowledge of aquifer properties. These, however, are seldom measured since wells are not deliberately drilled into the aquifer to obtain such information.

For a water-drive reservoir with no gas cap, Equation 11-43 can be rearranged and expressed as:

$$\frac{F}{E_o} = N + \frac{W_e}{E_o} \qquad (11\text{-}44)$$

Several water influx models have been described in Chapter 10, including the:

• Pot-aquifer model
• Schilthuis steady-state method
• Van Everdingen-Hurst model

The use of these models in connection with Equation 11-44 to simultaneously determine N and W_e is described below.

The Pot-Aquifer Model in the MBE

Assume that the water influx could be properly described using the simple pot aquifer model given by Equation 10-5 as:

$$W_e = (c_w + c_f) \, W_i \, f \, (p_i - p) \tag{11-45}$$

$$f = \frac{(\text{encroachment angle})^\circ}{360^\circ} = \frac{\theta}{360^\circ}$$

$$W_i = \left[\frac{\pi (r_a^2 - r_e^2) h \phi}{5.615} \right]$$

where r_a = radius of the aquifer, ft
r_e = radius of the reservoir, ft
h = thickness of the aquifer, ft
ϕ = porosity of the aquifer
θ = encroachment angle
c_w = aquifer water compressibility, psi^{-1}
c_f = aquifer rock compressibility, psi^{-1}
W_i = initial volume of water in the aquifer, bbl

Since the aquifer properties c_w, c_f, h, r_a, and θ are seldom available, it is convenient to combine these properties and treated as one unknown K. Equation 11-45 can be rewritten as:

$$W_e = K \, \Delta p \tag{11-46}$$

Combining Equation 11-46 with Equation 11-44 gives:

$$\frac{F}{E_o} = N + K \left(\frac{\Delta p}{E_o} \right) \tag{11-47}$$

Equation 11-47 indicates that a plot of the term (F/E_o) as a function of ($\Delta p/E_o$) would yield a straight line with an intercept of N and slope of K, as illustrated in Figure 11-23.

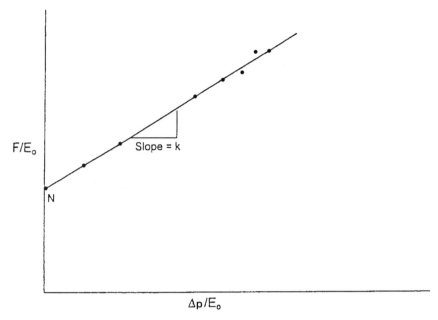

Figure 11-23. F/E_o vs. $\Delta p/E_o$.

The Steady-State Model in the MBE

The steady-state aquifer model as proposed by Schilthuis (1936) is given by:

$$W_e = C \int_0^t (p_i - p)\, dt \tag{11-48}$$

where W_e = cumulative water influx, bbl
$\quad$ C = water influx constant, bbl/day/psi
$\quad$ t = time, days
$\quad$ p_i = initial reservoir pressure, psi
$\quad$ p = pressure at the oil-water contact at time t, psi

Combining Equation 11-48 with Equation 11-44 gives:

$$\frac{F}{E_o} = N + C \left(\frac{\displaystyle\int_0^t (p_i - p)\,dt}{E_o} \right) \qquad (11\text{-}49)$$

Plotting (F/E_o) versus $\displaystyle\int_0^t (p_i - p)\,dt/E_o$ results in a straight line with an intercept that represents the initial oil in place N and a slope that describes the water influx C as shown in Figure 11-24.

The Unsteady-State Model in the MBE

The van Everdingen-Hurst unsteady-state model is given by:

$$W_e = B \, \Sigma \, \Delta p \, W_{eD} \qquad (11\text{-}50)$$

with

$$B = 1.119 \, \phi \, c_t \, r_e^2 \, h \, f$$

Van Everdingen and Hurst presented the dimensionless water influx W_{eD} as a function of the dimensionless time t_D and dimensionless radius r_D that are given by:

$$t_D = 6.328 \times 10^{-3} \, \frac{kt}{\phi \mu_w c_t r_e^2}$$

$$r_D = \frac{r_a}{r_e}$$

$$c_t = c_w + c_f$$

where t = time, days
k = permeability of the aquifer, md
ϕ = porosity of the aquifer
μ_w = viscosity of water in the aquifer, cp

r_a = radius of the aquifer, ft
r_e = radius of the reservoir, ft
c_w = compressibility of the water, psi^{-1}

Combining Equation 11-50 with Equation 11-44 gives:

$$\frac{F}{E_o} = N + B\left(\frac{\sum \Delta p\, W_{eD}}{E_o}\right) \qquad (11\text{-}51)$$

The proper methodology of solving the above linear relationship is summarized in the following steps.

Step 1. From the field past production and pressure history, calculate the underground withdrawal F and oil expansion E_o.

Step 2. Assume an aquifer configuration, i.e., linear or radial.

Step 3. Assume the aquifer radius r_a and calculate the dimensionless radius r_D.

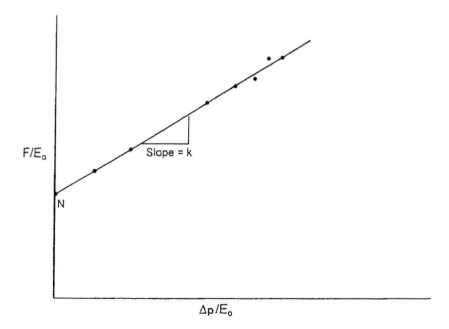

Figure 11-24. Graphical determination of N and c.

Step 4. Plot (F/E_o) versus $(\Sigma \ \Delta p \ W_{eD})/E_o$ on a Cartesian scale. If the assumed aquifer parameters are correct, the plot will be a straight line with N being the intercept and the water influx constant B being the slope. It should be noted that four other different plots might result. These are:

- Complete random scatter of the individual points, which indicates that the calculation and/or the basic data are in error.
- A systematically upward curved line, which suggests that the assumed aquifer radius (or dimensionless radius) is too small.
- A systematically downward curved line, indicating that the selected aquifer radius (or dimensionless radius) is too large.
- An s-shaped curve indicates that a better fit could be obtained if a linear water influx is assumed.

Figure 11-25 shows a schematic illustration of Havlena-Odeh (1963) methodology in determining the aquifer fitting parameters.

Example 11-5

The material balance parameters, the underground withdrawal F, and oil expansion E_o of a saturated-oil reservoir (i.e., m = o) are given below:

p	F	E_o
3500	—	—
3488	2.04×10^6	0.0548
3162	8.77×10^6	0.1540
2782	17.05×10^6	0.2820

Assuming that the rock and water compressibilities are negligible, calculate the initial oil in place.

Solution

Step 1. The most important step in applying the MBE is to verify that no water influx exists. Assuming that the reservoir is volumetric, calculate the initial oil in place N by using every individual production data point in Equation 11-38, or:

$$N = F/E_o$$

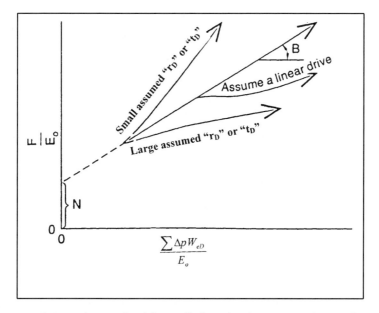

Figure 11-25. Havlena and Odeh straight-line plot. (*Source: Havlena and Odeh, 1963.*)

F	E_o	$N = F/E_o$
2.04×10^6	0.0548	37 MMSTB
8.77×10^6	0.1540	57 MMSTB
17.05×10^6	0.2820	60 MMSTB

Step 2. The above calculations show the calculated values of the initial oil in place are increasing (as shown graphically in Figure 11-26), which indicates a water encroachment, i.e., water-drive reservoir.

Step 3. For simplicity, select the pot-aquifer model to represent the water encroachment calculations in the MBE as given by Equation 11-47, or:

$$\frac{F}{E_o} = N + K\left(\frac{\Delta p}{E_o}\right)$$

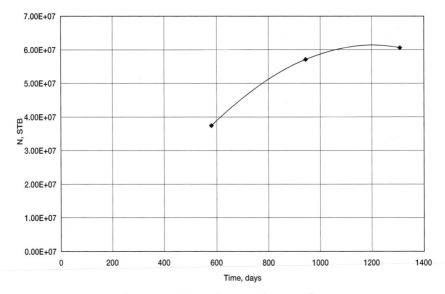

Figure 11-26. Indication of water influx.

Step 4. Calculate the terms (F/E_o) and $(\Delta p/E_o)$ of Equation 11-47.

p	Δp	F	E_o	F/E_o	$\Delta p/E_o$
3500	0	—	—	—	—
3488	12	2.04×10^6	0.0548	37.23×10^6	219.0
3162	338	8.77×10^6	0.1540	56.95×10^6	2194.8
2782	718	17.05×10^6	0.2820	60.46×10^6	2546

Step 5. Plot (F/E_o) versus $(\Delta p/E_o)$, as shown in Figure 11-27, and determine the intercept and the slope.

Intercept = N = 35 MMSTB

Slope = K = 9983

Tracy's Form of the Material Balance Equation

Neglecting the formation and water compressibilities, the general material balance equation as expressed by Equation 11-13 can be reduced to the following:

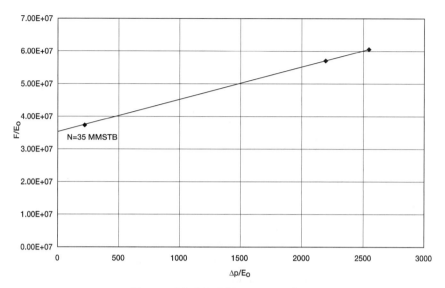

Figure 11-27. F/E_o versus $\Delta p/E_o$.

$$N = \frac{N_p B_o + (G_p - N_p R_s) B_g - (W_e - W_p B_w)}{(B_o - B_{oi}) + (R_{si} - R_s) B_g + m B_{oi} \left[\dfrac{B_g}{B_{gi}} - 1 \right]} \qquad (11\text{-}52)$$

Tracy (1955) suggested that the above relationship can be rearranged into a more usable form as:

$$N = N_p \Phi_o + G_p \Phi_g + (W_p B_w - W_e) \Phi_w \qquad (11\text{-}53)$$

where Φ_o, Φ_g, and Φ_w are considered PVT related properties that are functions of pressure and defined by:

$$\Phi_o = \frac{B_o - R_s B_g}{\text{Den}} \qquad (11\text{-}54)$$

$$\Phi_g = \frac{B_g}{\text{Den}} \qquad (11\text{-}55)$$

$$\phi_w = \frac{1}{Den} \tag{11-56}$$

with

$$Den = (B_o - B_{oi}) + (R_{si} - R_s)B_g + mB_{oi}\left[\frac{B_g}{B_{gi}} - 1\right] \tag{11-57}$$

where Φ_o = oil PVT function

$\quad\quad\quad \Phi_g$ = gas PVT function

$\quad\quad\quad \Phi_w$ = water PVT function

Figure 11-28 gives a graphical presentation of the behavior of Tracy's PVT functions with changing pressure.

Notice that Φ_o is negative at low pressures and all Φ functions are approaching infinity at bubble-point pressure. Tracy's form is valid only for initial pressures equal to bubble-point pressure and cannot be used at pressures above bubble point. Furthermore, the shape of the Φ function curves illustrate that small errors in pressure and/or production can cause large errors in calculated oil in place at pressures near the bubble point.

Steffensen (1992), however, pointed out the Tracy's equation uses the oil formation volume factor at the bubble-point pressure B_{ob} for the initial B_{oi} which causes all the PVT functions to become infinity at the bubble-point pressure. Steffensen suggested that Tracy's equation could be extended for applications above the bubble-point pressure, i.e., for under-saturated-oil reservoirs, by simply using the value of B_o at the initial reservoir pressure. He concluded that Tracy's methodology could predict reservoir performance for the entire pressure range from any initial pressure down to abandonment.

The following example is given by Tracy (1955) to illustrate his proposed approach.

Example 11-6

The production history of a saturated-oil reservoir is as follows:

Pressure, psia	Cumulative Oil, MSTB	Cumulative Gas, MMscf
1690	0	0
1600	398	38.6
1500	1570	155.8
1100	4470	803

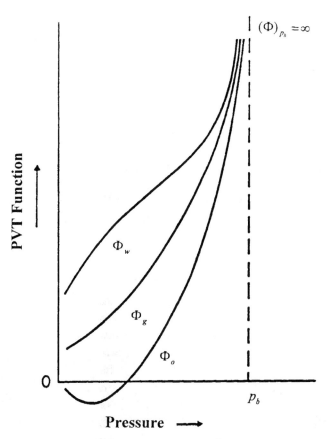

Figure 11-28. Tracy's PVT functions.

The calculated values of the PVT functions are given below:

Pressure, psia	Φ_o	Φ_g
1600	36.60	0.4000
1500	14.30	0.1790
1100	2.10	0.0508

Calculate the oil in place N.

Solution

The calculations can be conveniently performed in following table:

p, psia	N_p, MSTB	G_p, MMscf	$(N_p \, \Phi_o)$	$(G_p \, \Phi_g)$	N, STB
1600	398	38.6	14.52×10^6	15.42×10^6	29.74×10^6
1500	155.8	155.8	22.45×10^6	27.85×10^6	50.30×10^6
1100	803.0	803.0	9.39×10^6	40.79×10^6	50.18×10^6

The above results show that the original oil in place in this reservoir is approximately 50 MMSTB of oil. The calculation at 1600 psia is a good example of the sensitivity of such a calculation near the bubble-point pressure. Since the last two values of the original oil in place agree so well, the first calculation is probably wrong.

PROBLEMS

1. Given the following data on an oil reservoir:

	Oil	Aquifer
Geometry	circle	semi-circle
Encroachment angle	—	180°
Radius, ft	4000	80,000
Flow regime	semisteady-state	unsteady-state
Porosity	—	0.20
Thickness, ft	—	30
Permeability, md	200	50
Viscosity, cp	1.2	0.36
Original pressure	3800	3800
Current pressure	3600	—
Original volume factor	1.300	1.04
Current volume factor	1.303	1.04
Bubble-point pressure	3000	—

The field has been on production for 1120 days and has produced 800,000 STB of oil and 60,000 STB of water. Water and formation compressibilities are estimated to 3×10^{-6} and 3.5×10^{-6} psi^{-1}, respectively. Calculate the original oil in place.

2. The following rock- and fluid-properties data are available on the Nameless Fields:

Reservoir area = 1000 acres   porosity = 10%   thickness = 20′

T = 140°F s_{wi} = 20%

p_i = 4000 psi p_b = 4000 psi

The gas compressibility factor and relative permeability ratio are given by the following expressions:

$$z = 0.8 - 0.00002\,(p - 4000)$$

$$\frac{k_{rg}}{k_{ro}} = 0.00127\,e^{17.269\,Sg}$$

The production history of the field is given below:

	4000 psi	3500 psi	3000 psi
μ_o, cp	1.3	1.25	1.2
μ_g, cp	—	0.0125	0.0120
B_o, bbl/STB	1.4	1.35	1.30
R_s, scf/STB	—	—	450
GOR, scf/STB	600	—	1573

Subsurface information indicates that there is no aquifer and has been no water production.

Calculate:

a. Remaining oil in place at 3000 psi

b. Cumulative gas produced at 3000 psi

3. The following PVT and production history data are available on an oil reservoir in West Texas:

Original oil in place   = 10 MMSTB

Initial water saturation   = 22%

Initial reservoir pressure = 2496 psia

Bubble-point pressure   = 2496 psi

Pressure psi	B_o bbl/STB	R_s scf/STB	B_g bbl/scf	μ_o cp	μ_g cp	GOR scf/STB
2496	1.325	650	0.000796	0.906	0.016	650
1498	1.250	486	0.001335	1.373	0.015	1360
1302	1.233	450	0.001616	1.437	0.014	2080

The cumulative gas-oil ratio at 1302 psi is recorded at 953 scf/STB. Calculate:

a. Oil saturation at 1302 psia
b. Volume of the free gas in the reservoir at 1302 psia
c. Relative permeability ratio (k_g/k_o) at 1302 psia

4. The Nameless Field is an undersaturated-oil reservoir. The crude oil system and rock type indicates that the reservoir is highly compressible. The available reservoir and production data are given below:

$S_{wi} = 0.25$ $\qquad$ $\phi = 20\%$ $\qquad$ Area = 1,000 acres
$h = 70'$ $\qquad$ $T = 150°F$

Bubble-point pressure = 3500 psia

	Original condition	Current conditions
Pressure, psi	5000	4500
B_o, bbl/STB	1.905	1.920
R_s, scf/STB	700	700
N_P, MSTB	0	610.9

Calculate the cumulative oil production at 3900 psi. The PVT data show that the oil formation volume factor is equal to 1.938 bbl/STB at 3900 psia.

5. The following data[2] is available on a gas-cap-drive reservoir:

Pressure (psi)	N_p (MMSTB)	R_p (scf/STB)	B_o (RB/STB)	R_s (scf/STB)	B_g (RB/scf)
3,330			1.2511	510	0.00087
3,150	3.295	1,050	1.2353	477	0.00092
3,000	5.903	1,060	1.2222	450	0.00096
2,850	8.852	1,160	1.2122	425	0.00101
2,700	11.503	1,235	1.2022	401	0.00107
2,550	14.513	1,265	1.1922	375	0.00113
2,400	17.730	1,300	1.1822	352	0.00120

Calculate the initial oil and free gas volumes.

6. The Wildcat Reservoir was discovered in 1980. This reservoir had an initial reservoir pressure of 3,000 psia, and laboratory data indicated a

---

[2]Dake, L. P., *Fundamentals of Reservoir Engineering*, Elsevier Publishing Co., Amsterdam, 1978.

bubble-point pressure of 2,500 psi. The following additional data are available:

Area   = 700 acres
Thickness   = 35 ft
Porosity   = 20%
Temperature   = 150°F
API gravity   = 50°
Specific gravity of gas = 0.72
Initial water saturation = 25%

Average isothermal oil compressibility above the bubble point = 18 × 10^{-6} psi^{-1}

Calculate the volume of oil initially in place at 3,000 psi as expressed in STB.

REFERENCES

1. Clark, N., *Elements of Petroleum Reservoirs.* SPE, Dallas, TX 1969.

2. Cole, F., *Reservoir Engineering Manual.* Gulf Publishing Co., Houston, TX 1969.

3. Craft, B. C., and Hawkins, M. (Revised by Terry, R. E.), *Applied Petroleum Reservoir Engineering,* 2nd ed. Englewood Cliffs, NJ: Prentice Hall, 1991.

4. Dake, L. P., *Fundamentals of Reservoir Engineering.* Amsterdam: Elsevier. 1978.

5. Dake, L., *The Practice of Reservoir Engineering,* Amsterdam: Elsevier. 1994.

6. Economides, M., and Hill, D., *Petroleum Production System.* Prentice Hall, 1993.

7. Havlena, D., and Odeh, A. S., "The Material Balance as an Equation of a Straight Line," *JPT,* August 1963, pp. 896–900.

8. Havlena, D., and Odeh, A. S., "The Material Balance as an Equation of a Straight Line, Part II—Field Cases," *JPT,* July 1964, pp. 815–822.

9. Schilthuis, R., "Active Oil and Reservoir Energy," *Trans. AIME,* 1936, Vol. 118, p. 33.

10. Steffensen, R., "Solution-Gas-Drive Reservoirs," *Petroleum Engineering Handbook,* Chapter 37. Dallas: SPE, 1992.

11. Tracy, G., "Simplified Form of the MBE," *Trans. AIME,* 1955, Vol. 204, pp. 243–246.

12. Van Everdingen, A., and Hurst, W., "The Application of the Laplace Transformation to Flow Problems in Reservoirs," *Trans. AIME,* 1949, p. 186.

PREDICTING OIL RESERVOIR PERFORMANCE

Most reservoir engineering calculations involve the use of the material balance equation. Some of the most useful applications of the MBE require the concurrent use of fluid flow equations, e.g., Darcy's equation. Combining the two concepts would enable the engineer to predict the reservoir future production performance as a function of time. Without the fluid flow concepts, the MBE simply provides performance as a function of the average reservoir pressure. Prediction of the reservoir future performance is ordinarily performed in the following two phases:

Phase 1. Predicting cumulative hydrocarbon production as a function of declining reservoir pressure. This stage is accomplished without regard to:

• Actual number of wells
• Location of wells
• Production rate of individual wells
• Time required to deplete the reservoir

Phase 2. The second stage of prediction is the time-production phase. In these calculations, the reservoir performance data, as calculated from Phase One, are correlated with time. It is necessary in this phase to account for the number of wells and the productivity of each well.

PHASE 1. RESERVOIR PERFORMANCE PREDICTION METHODS

The material balance equation in its various mathematical forms as presented in Chapter 11 is designed to provide with estimates of the initial oil in place N, size of the gas cap m, and water influx W_e. To use the MBE to predict the reservoir future performance, it requires two additional relations:

• Equation of producing (instantaneous) gas-oil ratio
• Equation for relating saturations to cumulative oil production

These auxiliary mathematical expressions are presented as follows:

Instantaneous Gas-Oil Ratio

The produced gas-oil ratio (GOR) at any particular time is the ratio of the standard cubic feet of *total* gas being produced at any time to the stock-tank barrels of oil being produced at that same instant. Hence, the name *instantaneous gas-oil ratio*. Equation 6-54 in Chapter 6 describes the GOR mathematically by the following expression:

$$GOR = R_s + \left(\frac{k_{rg}}{k_{ro}}\right)\left(\frac{\mu_o B_o}{\mu_g B_g}\right) \qquad (12\text{-}1)$$

where GOR = instantaneous gas-oil ratio, scf/STB
  R_s = gas solubility, scf/STB
  k_{rg} = relative permeability to gas
  k_{ro} = relative permeability to oil
  B_o = oil formation volume factor, bbl/STB
  B_g = gas formation volume factor, bbl/scf
  μ_o = oil viscosity, cp
  μ_g = gas viscosity, cp

The instantaneous GOR equation is of fundamental importance in reservoir analysis. The importance of Equation 12-1 can appropriately be discussed in conjunction with Figures 12-1 and 12-2.

These illustrations show the history of the gas-oil ratio of a hypothetical depletion-drive reservoir that is typically characterized by the following points:

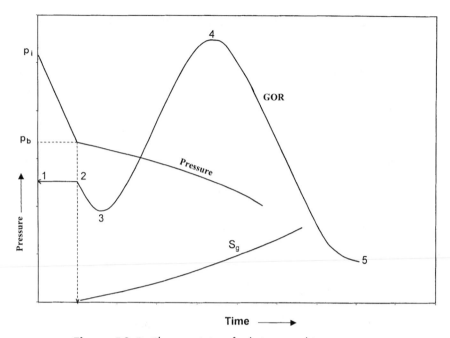

Figure 12-1. Characteristics of solution-gas-drive reservoirs.

Point 1. When the reservoir pressure p is above the bubble-point pressure p_b, there is no free gas in the formation, i.e., $k_{rg} = 0$, and therefore:

$$GOR = R_{si} = R_{sb} \qquad (12\text{-}2)$$

The gas-oil ratio remains constant at R_{si} until the pressure reaches the bubble-point pressure at Point 2.

Point 2. As the reservoir pressure declines below p_b, the gas begins to evolve from solution and its saturation increases. This free gas, however, cannot flow until the gas saturation S_g reaches the critical gas saturation S_{gc} at Point 3. From Point 2 to Point 3, the instantaneous GOR is described by a decreasing gas solubility as:

$$GOR = R_s \qquad (12\text{-}3)$$

Point 3. At Point 3, the free gas begins to flow with the oil and the values of GOR are progressively increasing with the declining reservoir pressure to Point 4. During this pressure decline period, the GOR is described by Equation 12-1, or:

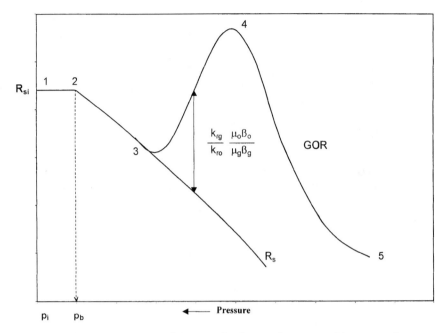

Figure 12-2. History of GOR and R_s for a solution-gas-drive reservoir.

$$GOR = R_s + \left(\frac{k_{rg}}{k_{ro}}\right)\left(\frac{\mu_o\,B_o}{\mu_g\,B_g}\right)$$

Point 4. At Point 4, the maximum GOR is reached due to the fact that the supply of gas has reached a maximum and marks the beginning of the *blow-down* period to Point 5.

Point 5. This point indicates that all the producible free gas has been produced and the GOR is essentially equal to the gas solubility and continues to Point 6.

There are three types of gas-oil ratios, all expressed in scf/STB, which must be clearly distinguished from each other. These are:

• Instantaneous GOR (defined by Equation 12-1)
• Solution GOR
• Cumulative GOR

The solution gas-oil ratio is a PVT property of the crude oil system. It is commonly referred to as *gas solubility* and denoted by R_s. It measures the tendency of the gas to dissolve in or evolve from the oil with changing pressures. It should be pointed out that as long as the evolved gas remains immobile, i.e., gas saturation S_g is less than the critical gas saturation, the instantaneous GOR is equal to the gas solubility, i.e.:

$$GOR = R_s$$

The cumulative gas-oil ratio R_p, as defined previously in the material balance equation, should be clearly distinguished from the producing (instantaneous) gas-oil ratio (GOR). The cumulative gas-oil ratio is defined as:

$$R_p = \frac{\text{cumulative (TOTAL) gas produced}}{\text{cumulative oil produced}}$$

or

$$R_p = \frac{G_p}{N_p} \tag{12-4}$$

where R_p = cumulative gas-oil ratio, scf/STB
$\quad\quad\quad G_p$ = cumulative gas produced, scf
$\quad\quad\quad N_p$ = cumulative oil produced, STB

The cumulative gas produced G_p is related to the instantaneous GOR and cumulative oil production by the expression:

$$G_p = \int_0^{N_p} (GOR) \, dN_p \tag{12-5}$$

Equation 12-5 simply indicates that the cumulative gas production at any time is essentially the area under the curve of the GOR versus N_p relationship, as shown in Figure 12-3.

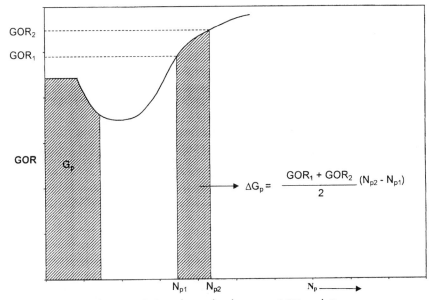

Figure 12-3. Relationship between GOR and G_p.

The incremental cumulative gas produced ΔG_p between N_{p1}, and N_{p2} is then given by:

$$\Delta G_p = \int_{N_{p1}}^{N_{p2}} (GOR)\, dN_p \qquad (12\text{-}6)$$

The above integral can be approximated by using the trapezoidal rule, to give:

$$\Delta G_p = \left[\frac{(GOR)_1 + (GOR)_2}{2} \right] (N_{p2} - N_{p1})$$

or

$$\Delta G_p = (GOR)_{avg}\, \Delta N_p$$

Equation 12-5 can then be approximated as:

$$G_p = \sum_o (GOR)_{avg} \Delta N_p \qquad (12\text{-}7)$$

Example 12-1

The following production data are available on a depletion-drive reservoir:

p psi		GOR scf/STB	N$_p$ MMSTB
2925	(p$_i$)	1340	0
2600		1340	1.380
2400		1340	2.260
2100	(p$_i$)	1340	3.445
1800		1936	7.240
1500		3584	12.029
1200		6230	15.321

Calculate cumulative gas produced G$_p$ and cumulative gas-oil ratio at each pressure.

Solution

Step 1. Construct the following table:

p psi	GOR scf/STB	(GOR)$_{avg}$ scf/STB	N$_p$ MMSTB	ΔN$_p$ MMSTB	ΔG$_p$ MMscf	G$_p$ MMscf	R$_p$ scf/STB
2925	1340	1340	0	0	0	0	—
2600	1340	1340	1.380	1.380	1849	1849	1340
2400	1340	1340	2.260	0.880	1179	3028	1340
2100	1340	1340	3.445	1.185	1588	4616	1340
1800	1936	1638	7.240	3.795	6216	10,832	1496
1500	3584	2760	12.029	4.789	13,618	24,450	2033
1200	6230	4907	15.321	3.292	16,154	40,604	2650

The Reservoir Saturation Equations

The saturation of a fluid (gas, oil, or water) in the reservoir is defined as the volume of the fluid divided by the pore volume, or:

$$S_o = \frac{\text{oil volume}}{\text{pore volume}} \tag{12-8}$$

$$S_w = \frac{\text{water volume}}{\text{pore volume}} \tag{12-9}$$

$$S_g = \frac{\text{gas volume}}{\text{pore volume}} \tag{12-10}$$

$$S_o + S_w + S_g + 1.0 \tag{12-11}$$

Consider a volumetric oil reservoir with no gas cap that contains N stock-tank barrels of oil at the initial reservoir pressure p_i. Assuming no water influx gives:

$$S_{oi} = 1 - S_{wi}$$

where the subscript i indicates initial reservoir condition. From the definition of oil saturation:

$$1 - S_{wi} = \frac{N\,B_{oi}}{\text{pore volume}}$$

or

$$\text{pore volume} = \frac{N\,B_{oi}}{1 - S_{wi}} \tag{12-12}$$

If the reservoir has produced N_p stock-tank barrels of oil, the remaining oil volume is given by:

$$\text{remaining oil volume} = (N - N_p)\,B_o \tag{12-13}$$

Substituting Equations 12-13 and 12-12 into Equation 12-8 gives:

$$S_o = \frac{(N - N_p) B_o}{\left(\dfrac{N B_{oi}}{1 - S_{wi}}\right)} \tag{12-14}$$

or

$$S_o = (1 - S_{wi})\left(1 - \frac{N_p}{N}\right)\frac{B_o}{B_{oi}} \tag{12-15}$$

$$S_g = 1 - S_o - S_{wi} \tag{12-16}$$

Example 12-2

A volumetric solution-gas-drive reservoir has an initial water saturation of 20%. The initial oil formation volume factor is reported at 1.5 bbl/STB. When 10% of the initial oil was produced, the value of B_o decreased to 1.38. Calculate the oil saturation and gas saturation.

Solution

From Equation 12-5

$$S_o = (1 - 0.2)(1 - 0.1)\left(\frac{1.38}{1.50}\right) = 0.662$$

$$S_g = 1 - 0.662 - 0.20 = 0.138$$

It should be pointed out that the values of the relative permeability ratio k_{rg}/k_{ro} as a function of oil saturation can be generated by using the actual field production as expressed in terms of N_p, GOR, and PVT data. The proposed methodology involves the following steps:

Step 1. Given the actual field cumulative oil production N_p and the PVT data as a function of pressure, calculate the oil and gas saturations from Equations 12-15 and 12-16, i.e.:

$$S_o = (1 - S_{wi})\left(1 - \frac{N_p}{N}\right)\frac{B_o}{B_{oi}}$$

$$S_g = 1 - S_o - S_{wi}$$

Step 2. Using the actual field instantaneous GORs, solve Equation 12-1 for the relative permeability ratio as:

$$\frac{k_{rg}}{k_{ro}} = (GOR - R_s)\left(\frac{\mu_g B_g}{\mu_o B_o}\right) \tag{12-17}$$

Step 3. Plot (k_{rg}/k_{ro}) versus S_o on a semilog paper.

Equation 12-15 suggests that all the remaining oil saturation be distributed uniformly throughout the reservoir. If water influx, gas-cap expansion, or gas-cap shrinking has occurred, the oil saturation equation, i.e., Equation 12-15, must be adjusted to account for oil trapped in the invaded regions.

Oil saturation adjustment for water influx

The proposed oil saturation adjustment methodology is illustrated in Figure 12-4 and described by the following steps:

Step 1. Calculate the pore volume in the water-invaded region, as:

$$W_e - W_p B_w = (P.V)_{water} (1 - S_{wi} - S_{orw})$$

Solving for the pore volume of water-invaded zone $(P.V)_{water}$ gives:

$$(P.V)_{water} = \frac{W_e - W_p B_w}{1 - S_{wi} - S_{orw}} \tag{12-18}$$

where $(P.V)_{water}$ = pore volume in water-invaded zone, bbl
S_{orw} = residual oil saturated in the imbibition water-oil system.

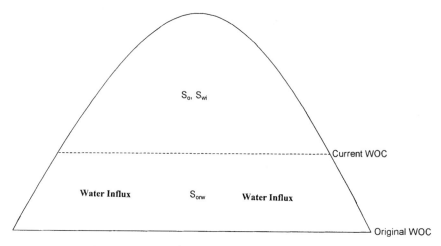

Figure 12-4. Oil saturation adjusted for water influx.

Step 2. Calculate oil volume in the water-invaded zone, or:

$$\text{volume of oil} = (P.V)\text{water } S_{orw} \qquad (12\text{-}19)$$

Step 3. Adjust Equation 12-14 to account for the trapped oil by using Equations 12-18 and 12-19:

$$S_o = \frac{(N - N_p)B_o - \left[\dfrac{W_e - W_p\, B_w}{1 - S_{wi} - S_{orw}}\right] S_{orw}}{\left(\dfrac{N\,B_{oi}}{1 - S_{wi}}\right) - \left[\dfrac{W_e - W_p\, B_w}{1 - S_{wi} - S_{orw}}\right]} \qquad (12\text{-}20)$$

Oil saturation adjustment for gas-cap expansion

The oil saturation adjustment procedure is illustrated in Figure 12-5 and summarized below:

Step 1. Assuming no gas is produced from the gas cap, calculate the net expansion of the gas cap, from:

$$\text{Expansion of the gas cap} = m\,N\,B_{oi}\left(\frac{B_g}{B_{gi}} - 1\right) \qquad (12\text{-}21)$$

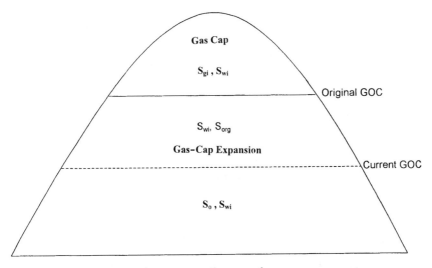

Figure 12-5. Oil saturation adjustment for gas-cap expansion.

Step 2. Calculate the pore volume of the gas-invaded zone, $(P.V)_{gas}$, by solving the following simple material balance:

$$m\,N\,B_{oi}\left(\frac{B_g}{B_{gi}}-1\right)=(P.V)_{gas}\,(1-S_{wi}-S_{org})$$

or

$$(P.V)_{gas}=\frac{m\,N\,B_{oi}\left(\dfrac{B_g}{B_{gi}}-1\right)}{1-S_{wi}-S_{org}} \qquad (12\text{-}22)$$

where $(P.V)_{gas}$ = pore volume of the gas-invaded zone
S_{org} = residual oil saturation in gas-oil system

Step 3. Calculate the volume of oil in the gas-invaded zone.

$$\text{oil volume} = (P.V)_{gas}\,S_{org} \qquad (12\text{-}23)$$

Step 4. Adjust Equation 12-14 to account for the trapped oil in the gas expansion zone by using Equations 12-22 and 12-23, to give:

$$S_o = \frac{(N - N_p)B_o - \left[\dfrac{mNB_{oi}\left(\dfrac{B_g}{B_{gi}} - 1\right)}{1 - S_{wi} - S_{org}}\right]S_{org}}{\left(\dfrac{NB_{oi}}{1 - S_{wi}}\right) - \left[\dfrac{mNB_{oi}}{1 - S_{wi} - S_{org}}\right]\left(\dfrac{B_g}{B_{gi}} - 1\right)} \qquad (12\text{-}24)$$

Oil saturation adjustment for combination drive

For a combination-drive reservoir, i.e., water influx and gas cap, the oil-saturation equation as given by Equation 12-14 can be adjusted to account for both driving mechanisms, as:

$$S_o = \frac{(N - N_p)B_o - \left[\dfrac{mNB_{oi}\left(\dfrac{B_g}{B_{gi}} - 1\right)S_{org}}{1 - S_{wi} - S_{org}} + \dfrac{(W_e - W_p)S_{orw}}{1 - S_{wi} - S_{orw}}\right]}{\dfrac{NB_{oi}}{1 - S_{wi}} - \left[\dfrac{mNB_{oi}\left(\dfrac{B_g}{B_{gi}} - 1\right)}{1 - S_{wi} - S_{org}} + \dfrac{W_e - W_p\,B_w}{1 - S_{wi} - S_{orw}}\right]} \qquad (12\text{-}25)$$

Oil saturation adjustment for shrinking gas cap

Cole (1969) points out that the control of the gas cap size is very often a reliable guide to the efficiency of reservoir operations. A shrinking gas cap will cause the loss of substantial amount of oil, which might otherwise be recovered. Normally, there is little or no oil saturation in the gas cap, and if the oil migrates into the original gas zone there will necessarily be some residual oil saturation remaining in this portion of the gas cap at abandonment. The magnitude of this loss may be quite large, depending upon the:

• Area of the gas-oil contact
• Rate of gas-cap shrinkage
• Relative permeability characteristics
• Vertical permeability.

A shrinking gas cap can be controlled by either shutting in wells which are producing large quantities of gas-cap gas or by returning some of the produced gas back the gas cap portion of the reservoir. In many cases, the shrinkage cannot be completely eliminated by shutting in wells, as there is a practical limit to the number of wells that can be shut in. The amount of oil lost by the shrinking gas cap can be very well the engineer's most important economic justification for the installation of gas return facilities.

The difference between the original volume of the gas cap and the volume occupied by the gas cap at any subsequent time is a measure of the volume of oil that has migrated into the gas cap. If the size of the original gas cap is $m \, N \, B_{oi}$, then the expansion of the original free gas resulting from reducing the pressure from p_i to p is:

$$\text{Expansion of the original gas cap} = m \, N \, B_{oi} \, [(B_g/B_{gi}) - 1]$$

where $m \, N \, B_{oi}$ = original gas-cap volume, bbl
       B_g = gas FVF, bbl/scf

If the gas cap is shrinking, then the volume of the produced gas must be larger than the gas-cap expansion. All of the oil that moves into the gas cap will not be lost, as this oil will also be subject to the various driving mechanisms. Assuming no original oil saturation in the gas zone, the oil that will be lost is essentially the residual oil saturation remaining at abandonment. If the cumulative gas production from the gas cap is G_{pc} scf, the volume of the gas-cap shrinkage as expressed in barrels is equal to:

$$\text{Gas-cap shrinkage} = G_{pc} \, B_g - m \, N \, B_{oi} \, [(B_g/B_{gi}) - 1]$$

From the volumetric equation:

$$G_{pc} B_g - m N B_{oi} [(B_g/B_{gi}) - 1] = 7758 A h \phi (1 - S_{wi} - S_{gr})$$

where A = average cross-sectional area of the gas-oil contact, acres
   h = average change in depth of the gas-oil contact, feet
   S_{gr} = residual gas saturation in the shrinking zone

The volume of oil lost as a result of oil migration to the gas cap can also be calculated from the volumetric equation as follows:

$$\text{Oil lost} = 7758 A h \phi S_{org}/B_{oa}$$

where S_{org} = residual oil saturation in the gas-cap shrinking zone
   B_{oa} = oil FVF at abandonment

Combining the above relationships and eliminating the term $7758 A h \phi$, give the following expression for estimating the volume of oil in barrels lost in the gas cap:

$$\text{Oil lost} = \frac{\left[G_{pc} B_g - m N B_{oi} \left(\dfrac{B_g}{B_{gi}} - 1 \right) \right] S_{org}}{(1 - S_{wi} - S_{gr}) B_{oa}}$$

where G_{pc} = cumulative gas production for the gas cap, scf
   B_g = gas FVF, bbl/scf

All the methodologies that have been developed to predict the future reservoir performance are essentially based on employing and combining the above relationships that include the:

• MBE
• Saturation equations
• Instantaneous GOR
• Equation relating the cumulative gas-oil ratio to the instantaneous GOR

Using the above information, it is possible to predict the field primary recovery performance with declining reservoir pressure. There are three methodologies that are widely used in the petroleum industry to perform a reservoir study, these are:

• Tracy's method
• Muskat's method
• Tarner's method

These methods yield essentially the same results when small intervals of pressure or time are used. The methods can be used to predict the performance of a reservoir under any driving mechanism, including:

• Solution-gas drive
• Gas-cap drive
• Water drive
• Combination drive

The practical use of all the techniques is illustrated in predicting the primary recovery performance of a volumetric solution-gas-drive reservoir. Using the appropriate saturation equation, e.g., Equation 12-20 for a water-drive reservoir, any of the available reservoir prediction techniques could be applied to other reservoirs operating under different driving mechanisms.

The following two cases of the solution-gas-drive reservoir are considered:

• Undersaturated-oil reservoirs
• Saturated-oil reservoirs

Undersaturated-Oil Reservoirs

When the reservoir pressure is above the bubble-point pressure of the crude oil system, the reservoir is considered an undersaturated. The general material balance is expressed in Chapter 11 by Equation 11-15.

$$N = \frac{N_p [B_o + (R_p - R_s) B_g] - (W_e - W_p B_w) - G_{inj} B_{ginj} - W_{inj} B_{wi}}{(B_o - B_{oi}) + (R_{si} - R_s) B_g + m B_{oi} \left[\dfrac{B_g}{B_{gi}} - 1 \right] + B_{oi} (1 + m) \left[\dfrac{S_{wi} c_w + c_f}{1 - S_{wi}} \right] \Delta p}$$

For a volumetric undersaturated reservoir with no fluid injection, the following conditions are observed:

$$m = 0$$
$$W_e = 0$$
$$R_s = R_{si} = R_p$$

Imposing the above conditions on the MBE reduces the equation to the following simplified form:

$$N = \frac{N_p \, B_o}{(B_o - B_{oi}) + B_{oi} \left[\dfrac{S_{wi} \, c_w + c_f}{1 - S_{wi}} \right] \Delta p} \tag{12-26}$$

with

$$\Delta p = p_i - p$$

where p_i = initial reservoir pressure
       p = current reservoir pressure

Hawkins (1955) introduced the oil compressibility c_o into the MBE to further simplify the equation. The oil compressed is defined in Chapter 2 by:

$$c_o = \frac{1}{B_{oi}} \frac{B_o - B_{oi}}{\Delta p}$$

rearranging, gives:

$$B_o - B_{oi} = c_o \, B_{oi} \, \Delta p$$

Combining the above expression with Equation 12-26 gives:

$$N = \frac{N_p \, B_o}{c_o \, B_{oi} \, \Delta p + B_{oi} \left[\dfrac{S_{wi} \, c_w + c_f}{1 - S_{wi}} \right] \Delta p} \tag{12-27}$$

The denominator of the above equation can be written as:

$$B_{oi} \left[c_o + \frac{S_{wi} c_w}{1 - S_{wi}} + \frac{c_f}{1 - S_{wi}} \right] \Delta p \qquad (12\text{-}28)$$

Since there are only two fluids in the reservoir, i.e., oil and water, then:

$$S_{oi} + S_{wi} = 1$$

Equation 12-28 can then be expressed as:

$$B_{oi} \left[\frac{S_{oi} c_o + S_{wi} c_w + c_f}{1 - S_{wi}} \right] \Delta p$$

The term between the two brackets is called the effective compressibility and defined by Hawkins (1955) as:

$$c_e = \frac{S_{oi} c_o + S_{wi} c_w + c_f}{1 - S_{wi}} \qquad (12\text{-}29)$$

Combining Equations 12-27, 12-28, and 12-29, the MBE above the bubble-point pressure becomes:

$$N = \frac{N_p B_o}{B_{oi} c_e \Delta p} = \frac{N_p B_o}{B_{oi} c_e \left(P_i - P \right)} \qquad (12\text{-}30)$$

Equation 12-30 can be expressed as an equation of a straight line by:

$$P = P_i - \left[\frac{1}{N B_{oi} c_e} \right] N_p B_o \qquad (12\text{-}31)$$

Figure 12-6 indicates that the reservoir pressure will decrease linearly with cumulative reservoir voidage $N_p B_o$.

Rearranging Equation 12-31 and solving for the cumulative oil production N_p gives:

$$N_p = N c_e \left(\frac{B_o}{B_{oi}} \right) \Delta p \qquad (12\text{-}32)$$

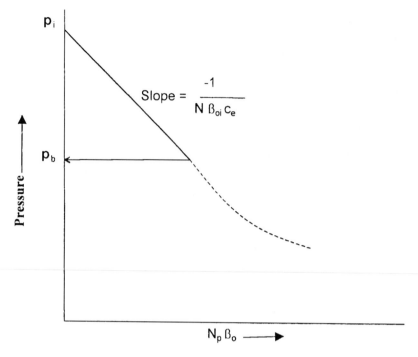

Figure 12-6. Pressure voidage relationship.

The calculation of future reservoir production, therefore, does not require a trial-and-error procedure, but can be obtained directly from the above expression.

Example 12-3

The following data are available on a volumetric undersaturated-oil reservoir:

p_i = 4000 psi p_b = 3000 psi   N = 85 MMSTB
c_f = 5×10^{-6} psi^{-1} c_o = 15×10^{-6} psi^{-1} c_w = 3×10^{-6} psi^{-1}
S_{wi} = 30% B_{oi} = 1.40 bbl/STB

Estimate cumulative oil production when the reservoir pressure drops to 3500 psi. The oil formation volume factor at 3500 psi is 1.414 bbl/STB.

Solution

Step 1. Determine the effective compressibility from Equation 12-29.

$$c_e = \frac{(0.7)(15 \times 10^{-6}) + (0.3)(3 \times 10^{-6}) + 5 \times 10^{-6}}{1 - 0.3}$$

$$= 23.43 \times 10^{-6} \, psi^{-1}$$

Step 2. Estimate N_p from Equation 12-32.

$$N_p = (85 \times 10^{-6})(23.43 \times 10^{-6})\left(\frac{1.411}{1.400}\right)(4000 - 3500)$$

$$= 985.18 \, MSTB$$

Saturated-Oil Reservoirs

If the reservoir originally exists at its bubble-point pressure, the reservoir is referred to as a saturated-oil reservoir. This is considered as the second type of the solution-gas-drive-reservoir. As the reservoir pressure declines below the bubble-point, the gas begins to evolve from solution. The general MBE may be simplified by assuming that the expansion of the gas is much greater than the expansion of rock and initial water and, therefore, can be neglected. For a volumetric and saturated-oil reservoir with no fluid injection, the MBE can be expressed by:

$$N = \frac{N_p B_o + (G_p - N_p R_s) B_g}{(B_o - B_{oi}) + (R_{si} - R_s) B_g} \tag{12-33}$$

The above material balance equation contains two unknowns, which are:

• Cumulative oil production N_p
• Cumulative gas production G_p

The following reservoir and PVT data must be available in order to predict the primary recovery performance of a depletion-drive reservoir in terms of N_p and G_p:

a. Initial oil-in-place N
Generally the volumetric estimate of in-place oil is used in calculating the performance. Where there is sufficient solution-gas-drive history,

however, this estimate may be checked by calculating a material-balance estimate.

b. Hydrocarbon PVT data

Since differential gas liberation is assumed to best represent the conditions in the reservoir, differential laboratory PVT data should be used in reservoir material balance. The flash PVT data are then used to convert from reservoir conditions to stock-tank conditions.

If laboratory data are not available, reasonable estimates may sometimes be obtained from published correlations. If differential data are not available, the flash data may be used instead; however, this may result in large errors for high-solubility crude oils.

c. Initial fluid saturations

Initial fluid saturations obtained from a laboratory analysis of core data are preferred; however, if these are not available, estimates in some cases may be obtained from a well-log analysis or may be obtained from other reservoirs in the same or similar formations.

d. Relative permeability data

Generally, laboratory-determined k_g/k_o and k_{ro} data are averaged to obtain a single representative set for the reservoir. If laboratory data are not available, estimates in some cases may be obtained from other reservoirs in the same or similar formations.

Where there is sufficient solution-gas-drive history for the reservoir, calculate (k_{rg}/k_{ro}) values versus saturation from Equations 12-15 and 12-17, i.e.:

$$S_o = (1 - S_{wi}) (1 - N_p/N) (B_o/B_{oi})$$

$$k_{rg}/k_{ro} = (GOR - R_s) (\mu_g B_g/\mu_o B_o)$$

The above results should be compared with the averaged laboratory relative permeability data. This may indicate a needed adjustment in the early data and possibly an adjustment in the overall data.

All the techniques that are used to predict the future performance of a reservoir are based on combining the appropriate MBE with the instantaneous GOR using the proper saturation equation. The calculations are repeated at a series of assumed reservoir pressure drops. These calculations are usually based on one stock-tank barrel of oil in place at the bubble-point pressure, i.e., N = 1. This avoids carrying large numbers in the calculation procedure and permits calculations to be made on the basis of the fractional recovery of initial oil in place.

There are several widely used techniques that were specifically developed to predict the performance of solution-gas-drive reservoirs, including:

• Tracy's method
• Muskat's method
• Tarner's method

These methodologies are presented below.

Tracy's Method

Tracy (1955) suggests that the general material balance equation can be rearranged and expressed in terms of three functions of PVT variables. Tracy's arrangement is given in Chapter 11 by Equation 11-53 and is repeated here for convenience:

$$N = N_p \, \Phi_o + G_p \, \Phi_g + (W_p \, B_w - W_e) \, \Phi_w \qquad (12\text{-}34)$$

where Φ_o, Φ_g, and Φ_w are considered PVT-related properties that are functions of pressure and defined by:

$$\Phi_o = \frac{B_o - R_s B_g}{\text{Den}}$$

$$\Phi_g = \frac{B_g}{\text{Den}}$$

$$\Phi_w = \frac{1}{\text{Den}}$$

with

$$\text{Den} = \left(B_o - B_{oi}\right) + \left(R_{si} - R_s\right)B_g + m B_{oi}\left[\frac{B_g}{B_{gi}} - 1\right] \qquad (12\text{-}35)$$

For a solution-gas-drive reservoir, Equations 12-34 and 12-35 are reduced to the following expressions, respectively:

$$N = N_p \, \Phi_o + G_p \, \Phi_g \qquad (12\text{-}36)$$

and

$$Den = (B_o - B_{oi}) + (R_{si} - R_s) \, B_g \qquad (12\text{-}37)$$

Tracy's calculations are performed in series of pressure drops that proceed from known reservoir condition at the previous reservoir pressure p* to the new assumed lower pressure p. The calculated results at the new reservoir pressure become "known" at the next assumed lower pressure.

In progressing from the conditions at any pressure p* to the lower reservoir pressure p, consider that the incremental oil and gas production are ΔN_p and ΔG_p, or:

$$N_p = N_p^* + \Delta N_p \qquad (12\text{-}38)$$

$$G_p = G_p^* + \Delta G_p \qquad (12\text{-}39)$$

where N_p^*, G_p^* = "known" cumulative oil and gas production at previous pressure level p*

N_p, G_p = "unknown" cumulative oil and gas at new pressure level p

Replacing N_p and G_p in Equation 12-36 with those of Equations 12-38 and 12-39 gives:

$$N = (N_p^* + \Delta N_p) \, \Phi_o + (G_p^* + \Delta G_p) \, \Phi_g \qquad (12\text{-}40)$$

Define the average instantaneous GOR between the two pressure p* and p by:

$$(GOR)_{avg} = \frac{GOR * + GOR}{2} \qquad (12\text{-}41)$$

The incremental cumulative gas production ΔG_p can be approximated by Equation 12-7 as:

$$\Delta G_p = (GOR)_{avg} \, \Delta N_p \qquad (12\text{-}42)$$

Replacing ΔG_p in Equation 12-40 with that of 12-41 gives:

$$N = [N_p^* + \Delta N_p] \, \Phi_o + [G_p^* + \Delta N_p \, (GOR)_{avg}] \, \Phi_g \qquad (12\text{-}43)$$

If Equation 12-43 is expressed for $N = 1$, the cumulative oil production N_p and cumulative gas production G_p become fractions of initial oil in place. Rearranging Equation 12-43 gives:

$$\Delta N_p = \frac{1 - (N_p^* \, \Phi_o + G_p^* \, \Phi_g)}{\Phi_o + (GOR)_{avg} \, \Phi_g} \qquad (12\text{-}44)$$

Equation 12-44 shows that there are essentially two unknowns, the incremental cumulative oil production ΔN_P and the average gas oil ratio $(GOR)_{avg}$.

Tracy suggested the following alternative technique for solving Equation 12-44.

Step 1. Select an average reservoir pressure p.

Step 2. Calculate the values of the PVT functions Φ_o and Φ_g.

Step 3. Estimate the GOR at p.

Step 4. Calculate the average instantaneous GOR $(GOR)_{avg} = (GOR^* + GOR)/2$.

Step 5. Calculate the incremental cumulative oil production ΔN_p from Equation 12-44 as:

$$\Delta N_P = \frac{1 - (N_p^* \, \Phi_o + G_p^* \, \Phi_g)}{\Phi_o + (GOR)_{avg} \, \Phi_g}$$

Step 6. Calculate cumulative oil production N_p:

$$N_p = N_p^* + \Delta N_p$$

Step 7. Calculate the oil and gas saturations at selected average reservoir pressure by using Equations 12-15 and 12-16, as:

$$S_o = (1 - S_{wi})\,(1 - N_p)\,(B_o/B_{oi})$$

$$S_g = 1 - S_o - S_{wi}$$

Step 8. Obtain relative permeability ratio k_{rg}/k_{ro} at S_g.

Step 9. Calculate the instantaneous GOR from Equation 12-1.

$$GOR = R_s + (k_{rg}/k_{ro})\,(\mu_o\,B_o/\mu_g\,B_g)$$

Step 10. Compare the estimated GOR in Step 3 with the calculated GOR in Step 9. If the values are within acceptable tolerance, proceed to next step. If not within the tolerance, set the estimated GOR equal to the calculated GOR and repeat the calculations from Step 3.

Step 11. Calculate the cumulative gas production.

$$G_p = G_p^* + \Delta N_p\,(GOR)_{avg}$$

Step 12. Since results of the calculations are based on 1 STB of oil initially in place, a final check on the accuracy of the prediction should be made on the MBE, or:

$$N_p\,\Phi_o + G_p\,\Phi_g = 1 \pm \text{tolerance}$$

Step 13. Repeat from Step 1.

As the calculation progresses, a plot of GOR versus pressure can be maintained and extrapolated as an aid in estimating GOR at each new pressure.

Example 12-4[1]

The following PVT data characterize a solution-gas-drive reservoir.

---

[1]The example data and solution are given by Economides, M., Hill, A., and Economides, C., *Petroleum Production System,* Prentice Hall Petroleum Engineering series, 1994.

The relative permeability data are shown in Figure 12-7.

p psi	B_o bbl/STB	B_g bbl/scf	R_s scf/STB
4350	1.43	6.9×10^{-4}	840
4150	1.420	7.1×10^{-4}	820
3950	1.395	7.4×10^{-4}	770
3750	1.380	7.8×10^{-4}	730
3550	1.360	8.1×10^{-4}	680
3350	1.345	8.5×10^{-4}	640

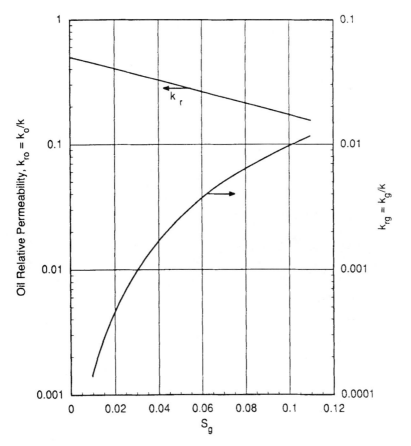

Figure 12-7. Relative permeability data for Example 12-3. (*After Economides, M., et al.,* Petroleum Production Systems, *Prentice Hall Petroleum Engineers Series, 1994.*)

The following additional data are available:

$$p_i = p_b = 4350 \text{ psi} \quad S_{wi} = 30\% \quad N = 15 \text{ MMSTB}$$

Predict the cumulative oil and gas production to 3350 psi.

Solution

A sample of the Tracy's calculation procedure is performed at 4150 psi.

Step 1. Calculate Tracy's PVT functions at 4150

   • Calculate the term Den from Equation 12-37

$$\text{Den} = (B_o - B_{oi}) + (R_{si} - R_s) B_g$$

$$\text{Den} = (1.42 - 1.43) + (840 - 820) (7.1 \times 10^{-4}) = 0.0042$$

   • Calculate Φ_o and Φ_g

$$\Phi_o = (B_o - R_s B_g)/\text{Den}$$

$$\Phi_o = [1.42 - (820) (7.1 \times 10^{-4})]/0.0042 = 199$$

$$\Phi_g = B_g/\text{Den}$$

$$= 7.1 \times 10^{-4}/0.0042 = 0.17$$

   Similarly, these PVT variables are calculated for all other pressures to give:

p	Φ_o	Φ_g
4350	—	—
4150	199	0.17
3950	49	0.044
3750	22.6	0.022
3550	13.6	0.014
3350	9.42	0.010

Step 2. Assume a value for the GOR at 4150 psi as 850 scf/STB.

Step 3. Calculate the average GOR.

$$(GOR)_{avg} = \frac{840 + 850}{2} = 845 \, scf \, / \, STB$$

Step 4. Calculate the incremental cumulative oil production ΔN_p.

$$\Delta N_p = \frac{1 - 0}{199 + (845)(0.17)} = 0.00292 \, STB$$

Step 5. Calculate the cumulative oil production N_p.

$$N_p = N_p^* + \Delta N_p$$

$$N_p = 0 + 0.00292 = 0.00292$$

Step 6. Calculate oil and gas saturations.

$$S_o = (1 - N_p) \, (B_o/B_{oi}) \, (1 - S_{wi})$$

$$S_o = (1 - 0.00292) \, (1.42/1.43) \, (1 - 0.3) = 0.693$$

$$S_g = 1 - S_{wi} - S_o$$

$$S_g = 1 - 0.3 - 0.693 = 0.007$$

Step 7. Determine the relative permeability ratio k_{rg}/k_{ro} from Figure 12-7, to give:

$$k_{rg}/k_{ro} = 8 \times 10^{-5}$$

Step 8. Using $\mu_o = 1.7$ cp and $\mu_g = 0.023$ cp, calculate the instantaneous GOR.

$$GOR = 820 + (1.7 \times 10^{-4}) \, \frac{(1.7) \, (1.42)}{(0.023)(7.1 \times 10^{-4})} = 845 \, scf \, / STB$$

which agrees with the assumed value.

Step 9. Calculate cumulative gas production.

$$G_p = 0 + (0.00292)(850) = 2.48$$

Complete results of the method are shown below:

$\bar{p}$	ΔN_p	N_p	$(GOR)_{avg}$	ΔG_p	G_p scf/STB	$N_p = 15 \times 10^6$ N STB	$G_p = 15 \times 10^6$ N scf
4350	—	—	—	—	—	—	—
4150	0.00292	0.00292	845	2.48	2.48	0.0438×10^6	37.2×10^6
3950	0.00841	0.0110	880	7.23	9.71	0.165×10^6	145.65×10^6
3750	0.0120	0.0230	1000	12	21.71	0.180×10^6	325.65×10^6
3550	0.0126	0.0356	1280	16.1	37.81	0.534×10^6	567.15×10^6
3350	0.011	0.0460	1650	18.2	56.01	0.699×10^6	840×10^6

Muskat's Method

Muskat (1945) expressed the material balance equation for a depletion-drive reservoir in the following differential form:

$$\frac{dS_o}{dp} = \frac{\dfrac{S_o B_g}{B_o}\dfrac{dR_s}{dp} + \dfrac{S_o}{B_o}\dfrac{k_{rg}}{k_{ro}}\dfrac{\mu_o}{\mu_g}\dfrac{dB_o}{dp} + (1 - S_o - S_{wc})B_g\dfrac{d(1/B_g)}{dp}}{1 + \dfrac{\mu_o}{\mu_g}\dfrac{k_{rg}}{k_{ro}}} \qquad (12\text{-}45)$$

with

$$\Delta S_o = S_o^* - S_o$$

$$\Delta p = p^* - p$$

where S_o^*, p^* = oil saturation and average reservoir pressure at the
beginning of the pressure step

S_o, p = oil saturation and average reservoir pressure at the end
of the time step

R_s = gas solubility, scf/STB

B_g = gas formation volume factor, bbl/scf

Craft, Hawkins, and Terry (1991) suggested the calculations can be greatly facilitated by computing and preparing in advance in graphical form the following pressure dependent groups:

$$X(p) = \frac{B_g}{B_o} \frac{dR_s}{dp} \qquad (12\text{-}46)$$

$$Y(p) = \frac{1}{B_o} \frac{\mu_o}{\mu_g} \frac{dB_o}{dp} \qquad (12\text{-}47)$$

$$Z(p) = B_g \frac{d(1/B_g)}{dp} \qquad (12\text{-}48)$$

Introducing the above pressure dependent terms into Equation 12-45, gives:

$$\left(\frac{\Delta S_o}{\Delta p}\right) = \frac{S_o X(p) + S_o \dfrac{k_{rg}}{k_{ro}} Y(p) + (1 - S_o - S_{wc}) Z(p)}{1 + \dfrac{\mu_o}{\mu_g} \dfrac{k_{rg}}{k_{ro}}} \qquad (12\text{-}49)$$

Craft, Hawkins, and Terry (1991) proposed the following procedure for solving Muskat's equation for a given pressure drop Δp, i.e., $(p^* - p)$:

Step 1. Prepare a plot of k_{rg}/k_{ro} versus gas saturation.

Step 2. Plot R_s, B_o and $(1/B_g)$ versus pressure and determine the slope of each plot at selected pressures, i.e., dB_o/dp, dR_s/dp, and $d(1/B_g)/dp$.

Step 3. Calculate the pressure dependent terms $X(p)$, $Y(p)$, and $Z(p)$ that correspond to the selected pressures in Step 2.

Step 4. Plot the pressure dependent terms as a function of pressure, as illustrated in Figure 12-8.

Step 5. Graphically determine the values of $X(p)$, $Y(p)$, and $Z(p)$ that correspond to the pressure p.

Step 6. Solve Equation 12-49 for $(\Delta S_o/\Delta p)$ by using the oil saturation S_o^* at the beginning of the pressure drop interval p^*.

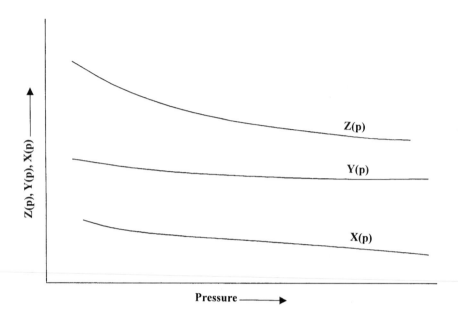

Figure 12-8. Pressure dependant terms vs. p.

Step 7. Determine the oil saturation S_o at the average reservoir pressure p, from:

$$S_o = S_o^* - (p^* - p)\left(\frac{\Delta S_o}{\Delta p}\right)$$
(12-50)

Step 8. Using the S_o from Step 7 and the pressure p, recalculate $(\Delta S_o/\Delta p)$ from Equation 12-49.

Step 9. Calculate the average value for $(\Delta S_o/\Delta p)$ from the two values obtained in Steps 6 and 8, or:

$$\left(\frac{\Delta S_o}{\Delta p}\right)_{avg} = \frac{1}{2}\left[\left(\frac{\Delta S_o}{\Delta p}\right)_{step\,6} + \left(\frac{\Delta S_o}{\Delta p}\right)_{step\,8}\right]$$

Step 10. Using $(\Delta S_o/\Delta p)_{avg}$, solve for the oil saturation S_o from:

$$S_o = S_o^* - (p^* - p)\left(\frac{\Delta S_o}{\Delta p}\right)_{avg} \qquad (12\text{-}51)$$

This value of S_o becomes S_o^* for the next pressure drop interval.

Step 11. Calculate gas saturation S_g by:

$$S_g = 1 - S_{wi} - S_o$$

Step 12. Using the saturation equation, i.e., Equation 12-15, solve for the cumulative oil production.

$$N_P = N\left[1 - \left(\frac{B_{oi}}{B_o}\right)\left(\frac{S_o}{1 - S_{wi}}\right)\right] \qquad (12\text{-}52)$$

Step 13. Calculate the cumulative gas production by using Equations 12-40 and 12-41.

Step 14. Repeat Steps 5 through 13 for all pressure drops of interest.

Example 12-5[2]

A volumetric depletion-drive reservoir exists at its bubble-point pressure of 2500 psi. Detailed fluid property data are listed by Craft and his coauthors and given here at only two pressures.

Fluid property	p_b = 2500 psi	p = 2300 psi
B_o, bbl/STB	1.498	1.463
R_s, scf/STB	721	669
B_g, bbl/scf	0.001048	0.001155
μ_o, cp	0.488	0.539
μ_g, cp	0.0170	0.0166
X (p)	0.00018	0.00021
Y (p)	0.00328	0.00380
Z (p)	0.00045	0.00050

[2]Craft, B. C., Hawkins, M., Terry, R., *Applied Petroleum Reservoir Engineering*, 2nd ed. Prentice Hall, 1991.

The following additional information is available:

$N = 56$ MMSTB $S_{wi} = 20\%$ $S_{oi} = 80\%$

S_g	k_{rg}/k_{ro}
0.10	0.010
0.20	0.065
0.30	0.200
0.50	2.000
0.55	3.000
0.57	5.000

Calculate the cumulative oil production for a pressure drop of 200 psi, i.e., at 2300 psi.

Solution

Step 1. Using the oil saturation at the beginning of the pressure interval, i.e., $S_o^* = 0.8$, calculate (k_{rg}/k_{ro}) to give:

$k_{rg}/k_{ro} = 0.0$ (No free gas initially in place.)

Step 2. Evaluate $(\Delta S_o/\Delta p)$ by applying Equation 12-49.

$$\left(\frac{\Delta S_o}{\Delta p} \right) = \frac{(0.8)(0.00018) + 0 + 0}{1 + 0} = 0.000146$$

Step 3. Estimate the oil saturation at $p = 2300$ psi from Equation 12-51.

$$S_o = 0.8 - 200(0.000146) = 0.7709$$

Step 4. Recalculate $(\Delta S_o/\Delta p)$ by using $S_o = 0.7709$ and the pressure dependent terms at 2300 psi.

$$\left(\frac{\Delta S_o}{\Delta p} \right) = \frac{0.7709(0.00021) + 0.7709(0.00001)0.0038 + (1 - 0.2 - 0.7709)0.0005}{1 + \left(\dfrac{0.539}{0.0166} \right)(0.00001)}$$

$$\left(\frac{\Delta S_o}{\Delta p} \right) = 0.000173$$

Step 5. Calculate the average $(\Delta S_o/\Delta p)$.

$$\left(\frac{\Delta S_o}{\Delta p}\right)_{avg} = \frac{0.000146 + 0.000173}{2} = 0.000159$$

Step 6. Calculate $S_o = 0.8 - (2500 - 2300)(0.000159) = 0.7682$.

Step 7. Calculate gas saturation.

$$S_g = 1 - 0.2 - 0.7682 + 0.0318$$

Step 8. Calculate cumulative oil production at 2300 psi by using Equation 12-52.

$$N_p = 56 \times 10^6 \left[1 - \left(\frac{1.498}{1.463}\right)\left(\frac{0.7682}{1 - 0.2}\right)\right] = 939,500\,\text{STB}$$

Step 9. Calculate k_{rg}/k_{ro} at 2300 psi, to give $k_{rg}/k_{ro} = 0.00001$.

Step 10. Calculate the instantaneous GOR at 2300 psi.

$$GOR = 669 + 0.00001\frac{(0.539)(1.463)}{(0.0166)(0.001155)} = 669\,\text{scf/STB}$$

Step 11. Calculate cumulative gas production.

$$G_p = \left(\frac{669 + 669}{2}\right)939,500 = 629\,\text{MMscf}$$

It should be stressed that this method is based on the assumption of uniform oil saturation in the whole reservoir and that the solution will therefore break down when there is appreciable gas segregation in the formation. It is therefore applicable only when permeabilities are relatively low.

Tarner's Method

Tarner (1944) suggests an iterative technique for predicting cumulative oil production N_p and cumulative gas production G_p as a function of reservoir pressure. The method is based on solving the material-balance equation and the instantaneous gas-oil ratio equation simultaneously for a given reservoir pressure drop from p_1 to p_2. It is accordingly assumed that the cumulative oil and gas production has increased from N_{p1} and G_{p1} to N_{p2} and G_{p2}. To simplify the description of the proposed iterative procedure, the stepwise calculation is illustrated for a volumetric saturated-oil reservoir. It should be pointed out that Tarner's method could be extended to predict the volumetric behavior of reservoirs under different driving mechanisms.

Step 1. Select a future reservoir pressure p_2 below the initial (current) reservoir pressure p_1 and obtain the necessary PVT data. Assume that the cumulative oil production has increased from N_{p1} to N_{p2}. It should be pointed out that N_{p1} and G_{p1} are set equal to zero at the initial reservoir pressure, i.e., bubble-point pressure.

Step 2. Estimate or guess the cumulative oil production N_{p2} at p_2.

Step 3. Calculate the cumulative gas production G_{p2} by rearranging the MBE, i.e., Equation 12-33, to give:

$$G_{p2} = N\left[(R_{si} - R_s) - \frac{B_{oi} - B_o}{B_g}\right] - N_{p2}\left[\frac{B_o}{B_g} - R_s\right] \qquad (12\text{-}53)$$

Equivalently, the above relationship can be expressed in terms of the two-phase (total) formation volume factor B_t as:

$$G_{p2} = \frac{N(B_t - B_{ti}) - N_{p2}(B_t - R_{si}B_g)}{B_g} \qquad (12\text{-}54)$$

where B_g = gas formation volume factor at p_2, bbl/scf
B_o = oil formation volume factor at p_2, bbl/STB
B_t = two-phase formation volume factor at p_2, bbl/STB
N = initial oil in place, STB

Step 4. Calculate the oil and gas saturations at the assumed cumulative oil production N_{p2} and the selected reservoir pressure p_2 by applying Equations 12-15 and 12-16 respectively, or:

$$S_o = (1 - S_{wi}) \left[1 - \frac{N_{p2}}{N} \right] \left(\frac{B_o}{B_{oi}} \right)$$

$$S_g = 1 - S_o - S_{wi}$$

where B_o = initial oil formation volume factor at p_i, bbl/STB
B_o = oil formation volume factor at p_2, bbl/STB
S_g = gas saturation at p_2
B_o = oil saturation at p_2

Step 5. Using the available relative permeability data, determine the relative permeability ratio k_{rg}/k_{ro} that corresponds to the gas saturation at p_2 and compute the instantaneous $(GOR)_2$ at p_2 from Equation 12-1, as:

$$(GOR)_2 = R_s + \left(\frac{k_{rg}}{k_{ro}} \right) \left(\frac{\mu_o B_o}{\mu_g B_g} \right) \qquad (12\text{-}55)$$

It should be noted that all the PVT data in the expression must be evaluated at the assumed reservoir pressure p_2.

Step 6. Calculate again the cumulative gas production G_{p2} at p_2 by applying Equation 12-7, or:

$$G_{p2} = (G_{p1}) + \left[\frac{(GOR)_1 + (GOR)_2}{2} \right] [N_{p2} - N_{p1}] \qquad (12\text{-}56)$$

in which $(GOR)_1$ represents the instantaneous GOR at p_1. If p_1 represents the initial reservoir pressure, then set $(GOR)_1 = R_{si}$.

Step 7. The total gas produced G_{p2} during the first prediction period as calculated by the material balance equation is compared to the total gas produced as calculated by the GOR equation. These two equations provide with two independent methods required for determining the total gas produced. Therefore, if the cumulative gas production G_{p2} as calculated from Step 3 agrees with the value of Step 6, the assumed value of N_{p2} is correct and a new pressure may be selected and Steps 1 through 6 are repeated. Otherwise, assume another value of N_{p2} and repeat Steps 2 through 6.

Step 8. In order to simplify this iterative process, three values of N_p can be assumed, which yield three different solutions of cumulative gas production for each of the equations (i.e., MBE and GOR equation).When the computed values of G_{p2} are plotted versus the assumed values of N_{p2}, the resulting two curves (one representing results of Step 3 and the one representing Step 5) will intersect. This intersection indicates the cumulative oil and gas production that will satisfy both equations.

It should be pointed out that it may be more convenient to assume values of N_P as a fraction of the initial oil in place N. For instance, N_p could be assumed as 0.01 N, rather than as 10,000 STB. In this method, a true value of N is not required. Results of the calculations would be, therefore, in terms of STB of oil produced per STB of oil initially in place and scf of gas produced per STB of oil initially in place.

To illustrate the application of Tarner's method, Cole (1969) presented the following example:

Example 12-6

A saturated-oil reservoir has a bubble-point pressure of 2100 psi at 175°F. The initial reservoir pressure is 2925 psi. The following data summarizes the rock and fluid properties of the field:

Original oil in place = 10 MMSTB
Connate-water saturation = 15%
Porosity = 12 %
$c_w = 3.6 \times 10^{-6} \text{ psi}^{-1}$
$c_f = 4.9 \times 10^{-6} \text{ psi}^{-1}$

Basic PVT Data

p, psi	B_o, bbl/STB	B_t, bbl/STB	R_s, scf/STB	B_g, bbl/scf	μ_o/μ_g
2925	1.429	1.429	1340	—	—
2100	1.480	1.480	1340	0.001283	34.1
1800	1.468	1.559	1280	0.001518	38.3
1500	1.440	1.792	1150	0.001853	42.4

Relative Permeability Ratio

S_o, %	k_{rg}/k_{ro}
81	0.018
76	0.063
60	0.85
50	3.35
40	10.2

Predict cumulative oil and gas production at 2100, 1800, and 1500 psi.

Solution

The required calculations will be performed under the following two different driving mechanisms:

- During the reservoir pressure declines from the initial reservoir pressure of 2925 to the bubble-point pressure of 2100 psi, the reservoir is considered undersaturated and, therefore, the MBE can be used directly to cumulative production without restoring to the iterative technique.
- For reservoir pressures below the bubble-point pressure, the reservoir is treated as a saturated-oil reservoir and Tarner's method may be applied.

Phase 1: Oil recovery prediction above the bubble-point pressure

Step 1. Arrange the MBE (Equation 11-32) and solve for the cumulative oil as:

$$N_p = \frac{N[E_o + E_{f,w}]}{B_o} \qquad (12\text{-}57)$$

where

$$E_{f,w} = B_{oi}\left[\frac{c_w S_w + c_f}{1 - S_{wi}}\right](p_i - p)$$

$$E_o = B_o - B_{oi}$$

Step 2. Calculate the two expansion factors E_o and $E_{f,w}$ for the pressure declines from 2925 to 2100 psi:

$$E_o = 1.480 - 1.429 = 0.051$$

$$E_{f,w} = 1.429\left[\frac{(3.6\times10^{-6})(0.15)+(4.9\times10^{-6})}{1-0.15}\right] = 9.1456\times10^{-6}$$

Step 3. Calculate cumulative oil and gas production when the reservoir pressure declines from 2925 to 2100 psi by applying Equation 12-57, to give:

$$N_p = \frac{10\times10^6[0.051+9.1456\times10^{-6}]}{1.48} = 344,656 \text{ STB}$$

At or above the bubble-point pressure, the producing gas-oil ratio is equal to the gas solubility at the bubble point and, therefore, the cumulative gas production is given by:

$$G_p = N_p R_{si}$$

$$G_p = (344,656)(1340) = 462 \text{ MMscf}$$

Step 4. Determine remaining oil in place at 2100 psi.

Remaining oil in place $= 10,000,000 - 344,656 = 9,655,344$ STB

This remaining oil in place is considered as the initial oil in place during the reservoir performance below the saturation pressure, i.e.:

$N = 9,655,344$ STB
$N_p = 0.0$ STB
$G_p = 0.0$ scf
$R_{si} = 1340$ scf/STB
$B_{oi} = 1.489$ bbl/STB
$B_{ti} = 1.489$ bbl/STB
$B_{gi} = 0.001283$ bbl/scf

Phase 2: Oil recovery prediction above the bubble-point pressure

First prediction period at 1800 psi:

1. Assume $N_p = 0.01 \, N$ and apply Equation 12-54 to solve for G_p.

$$G_p = \frac{N(1.559 - 1.480) - (0.01N)(1.559 - 1340 \times 0.001518)}{0.001518} = 55.17 \, N$$

2. Calculate the oil saturation, to give:

$$S_o = (1 - S_{wi})\left(1 - \frac{N_p}{N}\right)\frac{B_o}{B_{oi}} = (1 - 0.15)\left(1 - \frac{0.01N}{N}\right)\frac{1.468}{1.480} = 0.835$$

3. Determine the relative permeability ratio k_{rg}/k_{ro} from the available data to give:

$$k_{rg}/k_{ro} = 0.0100$$

4. Calculate the instantaneous GOR at 1800 psi by appying Equation 12-55 to give:

$$GOR = 1280 + 0.0100\,(38.3)\left(\frac{1.468}{0.001518}\right) = 1650 \, scf \, / \, STB$$

5. Solve again for the cumulative gas production by using the average GOR and applying Equation 12-56 to yield:

$$G_p = 0 + \frac{1340 + 1650}{2}(0.01N - 0) = 14.95N$$

6. Since the cumulative gas production as calculated by the two independent methods (Step 1 and Step 5) do not agree, the calculations must be repeated by assuming a different value for N_p and plotting results of the calculation. The final results as summarized below show the cumulative gas and oil production as the pressure declines from the bubble-point pressure. It should be pointed out that the cumulative production above the bubble-point pressure must be included when reporting the *total* cumulative oil and gas production.

Pressure	N_p	Actual N_p, STB	G_p	Actual G_p, MMscf
1800	0.0393 N	379,455	64.34 N	621.225
1500	0.0889 N	858,360	136.6 N	1318.92

PHASE 2. RELATING RESERVOIR PERFORMANCE TO TIME

All reservoir performance techniques show the relationship of cumulative oil production and the instantaneous GOR as a function of average reservoir pressure. These techniques, however, do not relate the cumulative oil production N_p and cumulative gas production G_p with time. Figure 12-9 shows a schematic illustration of the predicted cumulative oil production with reservoir pressure.

The time required for production can be calculated by applying the concept of the inflow performance relation (IPR) in conjunction with the

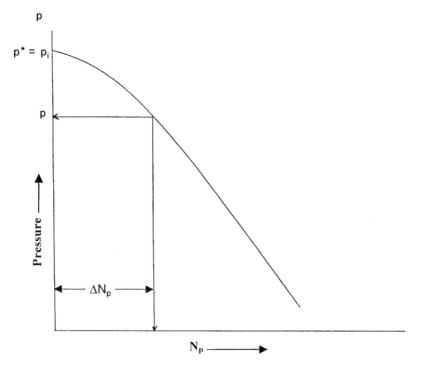

Figure 12-9. Cumulative production as a function of average reservoir pressure.

MBE predictions. Vogel (1969) expressed the well's inflow performance relationship by Equation 7-9, or:

$$Q_o = (Q_o)_{max}\left[1 - 0.2\left(\frac{p_{wf}}{\bar{p}_r}\right) - 0.8\left(\frac{p_{wf}}{\bar{p}_r}\right)^2\right]$$

The following methodology can be employed to correlate the predicted cumulative field production with time t.

Step 1. Plot the predicted cumulative oil production N_p as a function of average reservoir pressure p as shown in Figure 12-9.

Step 2. Construct the IPR curve for each well in field at the initial average reservoir pressure p*. Calculate the oil flow rate for the entire field by taking the summation of the flow rates. Plot the flow rates as shown schematically in Figure 12-10 for two hypothetical wells and establish the IPR for the field.

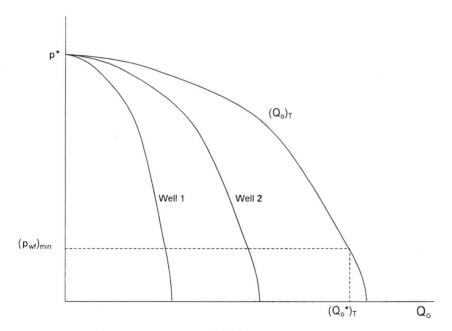

Figure 12-10. Overall field IPR at current pressure.

Step 3. Using the minimum bottom-hole flowing pressure $(p_{wf})_{min}$, determine the total field flow rate $(Q_o)_T^*$.

Step 4. Select a future average reservoir pressure p and determine the future IPR for each well in field. Construct the field IPR curve as shown in Figure 12-11.

Step 5. Using the minimum p_{wf}, determine the field total oil flow rate $(Q_o)_T$.

Step 6. Calculate the average field production rate $(\overline{Q}_o)_T$.

$$(\overline{Q}_o)_T = \frac{(Q_o)_T + (Q_o)_T^*}{2}$$

Step 7. Calculate the time Δt required for the incremental oil production ΔN_p during the first pressure drop interval, i.e., from p* to p, by:

$$\Delta t = \frac{\Delta N_p}{(\overline{Q}_o)_T}$$

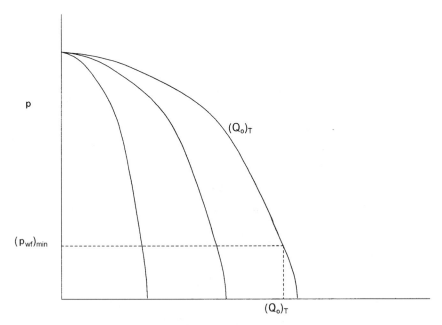

Figure 12-11. Future field IPR.

Step 8. Repeat the above steps and calculate the total time t to reach an average reservoir pressure p by:

$$t = \Sigma \, \Delta t$$

PROBLEMS

1. Determine the fractional oil recovery, during depletion down to bubble-point pressure, for the reservoir whose PVT parameters are listed in Table 3-7 and for which:

$$c_w = 3.5 \times 10^{-6} \text{ psi}^{-1} \qquad c_f = 3.5 \times 10^{-6} \text{ psi}^{-1} \qquad S_{wc} = 0.20$$

2. The Big Butte field is a depletion drive reservoir that contains 25 MMSTB of oil initially in place. Tables 3-4 through 3-7 show the experimental PVT data of the crude oil system. The initial reservoir pressure is recorded as 1936 psi at 247°F. The relative permeability ratio k_{rg}/k_{ro} is given by:

$$k_{rg}/k_{ro} = 0.007 \, e^{11.513 \, Sg}$$

Given

$$S_{or} = 35\% \qquad S_{gc} = 3\% \qquad S_{wi} = 25\%$$

Using a pressure drop increment of 200 psi, predict the reservoir future performance in terms of:

- Cumulative oil production N_p
- Cumulative gas production G_p
- Oil saturation S_o
- Gas saturation S_g
- Instantaneous GOR
- Cumulative producing gas-oil ratio R_p

Plot results of the calculations to an abandonment pressure of 500 psi. Use the following three methods:

1. Tracy's method
2. Muskat's method
3. Tarner's method

REFERENCES

1. Cole, F., *Reservoir Engineering Manual*. Houston: Gulf Publishing Company, 1969.

2. Craft, B. C., and Hawkins, M. (revised by Terry, R. E.), *Applied Petroleum Reservoir Engineering,* 2nd ed. Englewood Cliffs, NJ: Prentice Hall, 1991.

3. Dake, L. P., *Fundamentals of Reservoir Engineering*. Amsterdam, Elsevier, 1978.

4. Economides, M., Hill, A., and Economides, C., *Petroleum Production Systems*. Englewood Cliffs, NJ: Prentice Hall, 1994.

5. Havlena, D., and Odeh, A. S., "The Material Balance as an Equation of a Straight Line," *JPT,* August 1963, pp. 896–900.

6. Havlena, D., and Odeh, A. S., "The Material Balance as an Equation of a Straight Line. Part II—Field Cases," *JPT,* July 1964, pp. 815–822.

7. Hawkins, M., "Material Balances in Expansion Type Reservoirs Above Bubble-Point," SPE Transactions Reprint Series No. 3, 1955, pp. 36–40.

8. Muskat, M., "The Production Histories of Oil Producing Gas-Drive Reservoirs," *Journal of Applied Physics,* 1945, Vol. 16, p. 167.

9. Tarner, J., "How Different Size Gas Caps and Pressure Maintenance Programs Affect Amount of Recoverable Oil," *Oil Weekly,* June 12, 1944, Vol. 144.

10. Tracy, G., Simplified Form of the MBE," *Trans. AIME,* 1955, Vol. 204, pp. 243–246.

11. Vogel, J. V., "Inflow Performance Performance Relationships for Three-Phase Flow," SPE Paper 25458 presented at the SPE Production Operations Symposium, Oklahoma City, March 21–23, 1993.

GAS RESERVOIRS

Reservoirs containing only free gas are termed gas reservoirs. Such a reservoir contains a mixture of hydrocarbons, which exists wholly in the gaseous state. The mixture may be a *dry, wet,* or *condensate* gas, depending on the composition of the gas, along with the pressure and temperature at which the accumulation exists.

Gas reservoirs may have water influx from a contiguous water-bearing portion of the formation or may be volumetric (i.e., have no water influx).

Most gas engineering calculations involve the use of gas formation volume factor B_g and gas expansion factor E_g. Both factors are defined in Chapter 2 by Equations 2-52 through 2-56. Those equations are summarized below for convenience:

• Gas formation volume factor B_g is defined is defined as the actual volume occupied by n moles of gas at a specified pressure and temperature, divided by the volume occupied by the same amount of gas at standard conditions. Applying the real gas equation-of-state to both conditions gives:

$$B_g = \frac{p_{sc}}{T_{sc}} \frac{zT}{p} = 0.02827 \frac{zT}{p} \tag{13-1}$$

• The gas expansion factor is simply the reciprocal of B_g, or:

$$E_g = \frac{T_{sc}}{p_{sc}} \frac{p}{zT} = 35.37 \frac{p}{zT} \tag{13-2}$$

where B_g = gas formation volume factor, ft³/scf

E_g = gas expansion factor, scf/ft³

This chapter presents two approaches for estimating initial gas in place G, gas reserves, and the gas recovery for volumetric and water-drive mechanisms:

• Volumetric method
• Material balance approach

THE VOLUMETRIC METHOD

Data used to estimate the gas-bearing reservoir PV include, but are not limited to, well logs, core analyses, bottom-hole pressure (BHP) and fluid sample information, along with well tests. This data typically is used to develop various subsurface maps. Of these maps, structural and stratigraphic cross-sectional maps help to establish the reservoir's areal extent and to identify reservoir discontinuities, such as pinch-outs, faults, or gas-water contacts. Subsurface contour maps, usually drawn relative to a known or marker formation, are constructed with lines connecting points of equal elevation and therefore portray the geologic structure. Subsurface isopachous maps are constructed with lines of equal net gas-bearing formation thickness. With these maps, the reservoir PV can then be estimated by planimetering the areas between the isopachous lines and using an approximate volume calculation technique, such as the pyramidal or trapezoidal method.

The volumetric equation is useful in reserve work for estimating gas in place at any stage of depletion. During the development period before reservoir limits have been accurately defined, it is convenient to calculate gas in place per acre-foot of bulk reservoir rock. Multiplication of this unit figure by the best available estimate of bulk reservoir volume then gives gas in place for the lease, tract, or reservoir under consideration. Later in the life of the reservoir, when the reservoir volume is defined and performance data are available, volumetric calculations provide valuable checks on gas in place estimates obtained from material balance methods.

The equation for calculating gas in place is:

$$G = \frac{43,560 \, Ah\phi(1-S_{wi})}{B_{gi}} \qquad (13-3)$$

where   G = gas in place, scf
  A = area of reservoir, acres
  h = average reservoir thickness, ft
  ϕ = porosity
  S_{wi} = water saturation, and
  B_{gi} = gas formation volume factor, ft³/scf

This equation can be applied at both initial and abandonment conditions in order to calculate the recoverable gas.

Gas produced = Initial gas − Remaining gas

or

$$G_p = 43,560 \, Ah\phi \, (1 - S_{wi}) \left(\frac{1}{B_{gi}} - \frac{1}{B_{ga}} \right) \qquad (13\text{-}4)$$

where B_{ga} is evaluated at abandonment pressure. Application of the volumetric method assumes that the pore volume occupied by gas is constant. If water influx is occurring, A, h, and S_w will change.

Example 13-1

A gas reservoir has the following characteristics:

A = 3000 acres   h = 30 ft ϕ = 0.15 S_{wi} = 20%
T = 150°F p_i = 2600 psi

p	z
2600	0.82
1000	0.88
400	0.92

Calculate cumulative gas production and recovery factor at 1000 and 400 psi.

Solution

Step 1. Calculate the reservoir pore volume P.V

$$P.V = 43,560 \, Ah\phi$$

$$P.V = 43,560 \, (3000) \, (30) \, (0.15) = 588.06 \, MMft^3$$

Step 2. Calculate B_g at every given pressure by using Equation 13-1.

p	z	B_g, ft^3/scf
2600	0.82	0.0054
1000	0.88	0.0152
400	0.92	0.0397

Step 3. Calculate initial gas in place at 2600 psi

$$G = 588.06 \, (10^6) \, (1 - 0.2)/0.0054 = 87.12 \, MMMscf$$

Step 4. Since the reservoir is assumed volumetric, calculate the remaining gas at 1000 and 400 psi.

• Remaining gas at 1000 psi

$$G_{1000 \, psi} = 588.06(10^6) \, (1 - 0.2)/0.0152 = 30.95 \, MMMscf$$

• Remaining gas at 400 psi

$$G_{400 \, psi} = 588.06(10^6) \, (1 - 0.2)/0.0397 = 11.95 \, MMMscf$$

Step 5. Calculate cumulative gas production G_p and the recovery factor RF at 1000 and 400 psi.

• At 1000 psi:

$$G_p = (87.12 - 30.95) \times 10^9 = 56.17 \, MMM \, scf$$

$$RF = \frac{56.17 \times 10^9}{87.12 \times 10^9} = 64.5\%$$

• At 400 psi:

$$G_p = (87.12 - 11.95) \times 10^9 = 75.17 \text{ MMMscf}$$

$$RF = \frac{75.17 \times 10^9}{87.12 \times 10^9} = 86.3\%$$

The recovery factors for volumetric gas reservoirs will range from 80 to 90%. If a strong water drive is present, trapping of residual gas at higher pressures can reduce the recovery factor substantially, to the range of 50 to 80%.

THE MATERIAL BALANCE METHOD

If enough production-pressure history is available for a gas reservoir, the initial gas in place G, the initial reservoir pressure p_i, and the gas reserves can be calculated without knowing A, h, ϕ, or S_w. This is accomplished by forming a mass or mole balance on the gas as:

$$n_p = n_i - n_f \tag{13-5}$$

where n_p = moles of gas produced
n_i = moles of gas initially in the reservoir
n_f = moles of gas remaining in the reservoir

Representing the gas reservoir by an idealized gas container, as shown schematically in Figure 13-1, the gas moles in Equation 13-5 can be replaced by their equivalents using the real gas law to give:

$$\frac{p_{sc}G_p}{RT_{sc}} = \frac{p_i V}{z_i RT} - \frac{p[V - (W_e - W_p)]}{zRT} \tag{13-6}$$

where p_i = initial reservoir pressure
G_p = cumulative gas production, scf
p = current reservoir pressure
V = original gas volume, ft^3
z_i = gas deviation factor at p_i
z = gas deviation factor at p
T = temperature, °R
W_e = cumulative water influx, ft^3
W_p = cumulative water production, ft^3

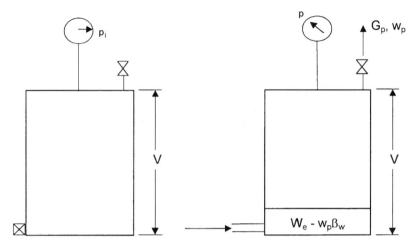

Figure 13-1. Idealized water-drive gas reservoir.

Equation 13-6 is essentially the general material balance equation (MBE). Equation 13-6 can be expressed in numerous forms depending on the type of the application and the driving mechanism. In general, dry gas reservoirs can be classified into two categories:

• Volumetric gas reservoirs
• Water-drive gas reservoirs

The remainder of this chapter is intended to provide the basic background in natural gas engineering. There are several excellent textbooks that comprehensively address this subject, including the following:

• Ikoku, C., *Natural Gas Reservoir Engineering,* 1984
• Lee, J. and Wattenbarger, R., *Gas Reservoir Engineering,* SPE, 1996

Volumetric Gas Reservoirs

For a volumetric reservoir and assuming no water production, Equation 13-6 is reduced to:

$$\frac{p_{sc}G_p}{T_{sc}} = \left(\frac{p_i}{z_i T}\right)V - \left(\frac{p}{zT}\right)V \tag{13-7}$$

Equation 13-7 is commonly expressed in the following two forms:

Form 1. In terms of p/z

Rearranging Equation 13-7 and solving for p/z gives:

$$\frac{p}{z} = \frac{p_i}{z_i} - \left(\frac{p_{sc}T}{T_{sc}V}\right)G_p \tag{13-8}$$

Equation 13-8 is an equation of a straight line when (p/z) is plotted versus the cumulative gas production G_p, as shown in Figure 13-2. This straight-line relationship is perhaps one of the most widely used relationships in gas-reserve determination.

The straight-line relationship provides the engineer with the reservoir characteristics:

• Slope of the straight line is equal to:

$$\text{slope} = \frac{p_{sc}T}{T_{sc}V} \tag{13-9}$$

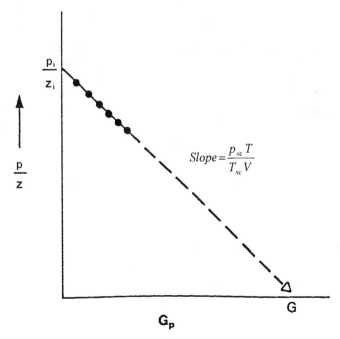

Figure 13-2. Gas material balance equation.

The original gas volume V can be calculated from the slope and used to determine the areal extend of the reservoir from:

$$V = 43,560 \, Ah \, \phi \, (1 - S_{wi}) \tag{13-10}$$

where A is the reservoir area in acres.

- Intercept at $G_p = 0$ gives p_i/z_i
- Intercept at $p/z = 0$ gives the gas initially in place G in scf
- Cumulative gas production or gas recovery at any pressure

Example 13-2[1]

A volumetric gas reservoir has the following production history.

Time, t years	Reservoir pressure, p psia	z	Cumulative production, G_p MMMscf
0.0	1798	0.869	0.00
0.5	1680	0.870	0.96
1.0	1540	0.880	2.12
1.5	1428	0.890	3.21
2.0	1335	0.900	3.92

The following data is also available:

$$\phi = 13\%$$
$$S_{wi} = 0.52$$
$$A = 1060 \text{ acres}$$
$$h = 54 \text{ ft.}$$
$$T = 164°F$$

Calculate the gas initially in place volumetrically and from the MBE.

Solution

Step 1. Calculate B_{gi} from Equation 13-1

$$B_{gi} = 0.02827 \frac{(0.869) \, (164 + 460)}{1798} = 0.00853 \text{ ft}^3/\text{scf}$$

---

[1]After Ikoku, C., *Natural Gas Reservoir Engineering,* John Wiley & Sons, 1984.

Step 2. Calculate the gas initially in place volumetrically by applying Equation 13-3.

G = 43,560 (1060) (54) (0.13) (1 − 0.52)/0.00853 = 18.2 MMMscf

Step 3. Plot p/z versus G_p as shown in Figure 13-3 and determine G.

G = 14.2 MMMscf

This checks the volumetric calculations.

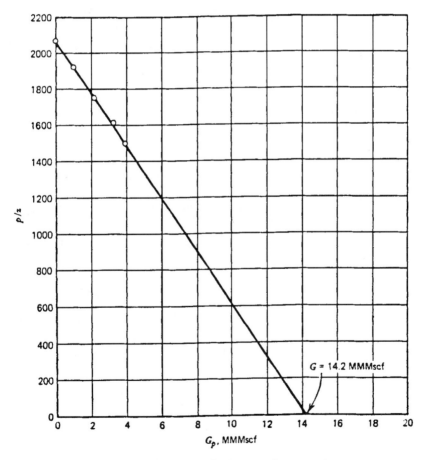

Figure 13-3. Relationship of p/z vs. G_p for Example 13-2.

The initial reservoir gas volume V can be expressed in terms of the volume of gas at standard conditions by:

$$V = B_g \, G = \left(\frac{p_{sc}}{T_{sc}} \frac{z_i \, T}{p_i} \right) G$$

Combining the above relationship with that of Equation 13-8 gives:

$$\frac{p}{z} = \frac{p_i}{z_i} - \left[\left(\frac{p_i}{z_i} \right) \frac{1}{G} \right] G_p \qquad (13\text{-}11)$$

Again, Equation 13-11 shows that for a volumetric reservoir, the relationship between (p/z) and G_p is essentially linear. This popular equation indicates that by extrapolation of the straight line to abscissa, i.e., at $p/z = 0$, will give the value of the gas initially in place as $G = G_p$.

The graphical representation of Equation 13-11 can be used to detect the presence of water influx, as shown graphically in Figure 13-4. When the plot of (p/z) versus G_p deviates from the linear relationship, it indicates the presence of water encroachment.

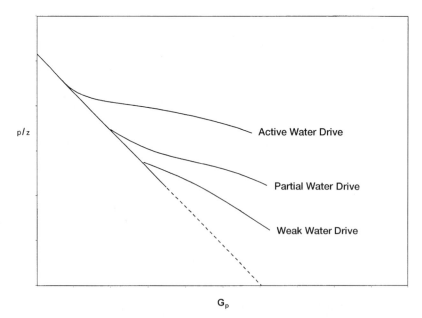

Figure 13-4. Effect of water drive on p/z vs. G_p relationship.

Many other graphical methods have been proposed for solving the gas MBE that are useful in detecting the presence of water influx. One such graphical technique is called the **energy plot,** which is based on arranging Equation 13-11 and taking the logarithm of both sides to give:

$$\log\left[1 - \frac{z_i\, p}{p_i\, z}\right] = \log G_p - \log G \qquad (13\text{-}12)$$

Figure 13-5 shows a schematic illustration of the plot.

From Equation 13-12, it is obvious that a plot of $[1 - (z_i\, p)/(p_i\, z)]$ versus G_p on log-log coordinates will yield a straight line with a slope of one (45° angle). An extrapolation to one on the vertical axis (p = 0) yields a value for initial gas in place, G. The graphs obtained from this type of analysis have been referred to as *energy plots.* They have been found to be useful in detecting water influx early in the life of a reservoir. If W_e is not zero, the slope of the plot will be less than one, and will also decrease with time, since W_e increases with time. An increasing slope can only occur as a result of either gas leaking from the reservoir or bad data,

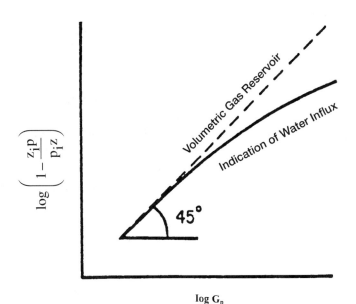

Figure 13-5. An energy plot.

since the increasing slope would imply that the gas-occupied pore volume was increasing with time.

Form 2. In terms of B_g

From the definition of the gas formation volume factor, it can be expressed as:

$$B_{gi} = \frac{V}{G}$$

Combining the above expression with Equation 13-1 gives:

$$\frac{p_{sc}}{T_{sc}} \frac{z_i\,T}{p_i} = \frac{V}{G} \tag{13-13}$$

where V = volume of gas originally in place, ft^3
G = volume of gas originally in place, scf
p_i = original reservoir pressure
z_i = gas compressibility factor at p_i

Equation 13-13 can be combined with Equation 13-7, to give:

$$G = \frac{G_p\,B_g}{B_g - B_{gi}} \tag{13-14}$$

Equation 13-14 suggests that to calculate the initial gas volume, the only information required is production data, pressure data, gas specific gravity for obtaining z-factors, and reservoir temperature. Early in the producing life of a reservoir, however, the denominator of the right-hand side of the material balance equation is very small, while the numerator is relatively large. A small change in the denominator will result in a large discrepancy in the calculated value of initial gas in place. Therefore, the material balance equation should not be relied on early in the producing life of the reservoir.

Material balances on volumetric gas reservoirs are simple. Initial gas in place may be computed from Equation 13-14 by substituting cumulative gas produced and appropriate gas formation volume factors at corresponding reservoir pressures during the history period. If successive calculations at various times during the history give consistent values for initial gas in place, the reservoir is operating under volumetric control and computed G is reliable, as shown in Figure 13-6. Once G has been determined and the absence of water influx established in this fashion, the same equation can be used to make future predictions of cumulative gas production function of reservoir pressure.

Ikoku (1984) points out that successive application of Equation 13-14 will normally result in increasing values of the gas initially in place G with time if water influx is occurring. If there is gas leakage to another zone due to bad cement jobs or casing leaks, however, the computed value of G may decrease with time.

Example 13-3

After producing 360 MMscf of gas from a volumetric gas reservoir, the pressure has declined from 3200 psi to 3000 psi, given:

$B_{gi} = 0.005278$ ft^3/scf
$B_g = 0.005390$ ft^3/scf

a. Calculate the gas initially in place.

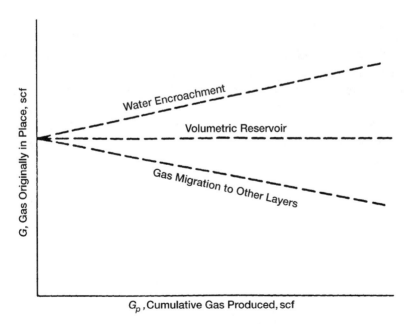

Figure 13-6. Graphical determination of the gas initially in place G.

b. Recalculate the gas initially in place assuming that the pressure measurements were incorrect and the true average pressure is 2900 psi. The gas formation volume factor at this pressure is 0.00558 ft³/scf.

Solution

a. Using Equation 13-14, calculate G.

$$G = \frac{360 \times 10^6\,(0.00539)}{0.00539 - 0.005278} = 17.325\,\text{MMMscf}$$

b. Recalculate G by using the correct value of B_g.

$$G = \frac{360 \times 10^6\,(0.00668)}{0.00558 - 0.005278} = 6.652\,\text{MMMscf}$$

Thus, an error of 100 psia, which is only 3.5% of the total reservoir pressure, resulted in an increase in calculated gas in place of approximately 160%, a 2½-fold increase. Note that a similar error in reservoir pressure later in the producing life of the reservoir will not result in an error as large as that calculated early in the producing life of the reservoir.

Water-Drive Gas Reservoirs

If the gas reservoir has a water drive, then there will be two unknowns in the material balance equation, even though production data, pressure, temperature, and gas gravity are known. These two unknowns are initial gas in place and cumulative water influx. In order to use the material balance equation to calculate initial gas in place, some independent method of estimating W_e, the cumulative water influx, must be developed as discussed in Chapter 11.

Equation 13-14 can be modified to include the cumulative water influx and water production to give:

$$G = \frac{G_p B_g - (W_e - W_p B_w)}{B_g - B_{gi}} \tag{13-15}$$

The above equation can be arranged and expressed as:

$$G + \frac{W_e}{B_g - B_{gi}} = \frac{G_p B_g + W_p B_w}{B_g - B_{gi}} \tag{13-16}$$

Equation 13-16 reveals that for a volumetric reservoir, i.e., $W_e = 0$, the right-hand side of the equation will be constant regardless of the amount of gas G_p which has been produced. For a water-drive reservoir, the values of the right-hand side of Equation 13-16 will continue to increase because of the $W_e/(B_g - B_{gi})$ term. A plot of several of these values at successive time intervals is illustrated in Figure 13-7. Extrapolation of the line formed by these points back to the point where $G_p = 0$ shows the true value of G, because when $G_p = 0$, then $W_e/(B_g - B_{gi})$ is also zero.

This graphical technique can be used to estimate the value of We, because at any time the difference between the horizontal line (i.e., true value of G) and the sloping line $[G + (W_e)/(B_g - B_{gi})$ will give the value of $W_e/(B_g - B_{gi})$.

Because gas often is bypassed and trapped by the encroaching water, recovery factors for gas reservoirs with water drive can be significantly lower than for volumetric reservoirs produced by simple gas expansion. In addition, the presence of reservoir heterogeneities, such as low-permeability stringers or layering, may reduce gas recovery further. As noted previously, ultimate recoveries of 80% to 90% are common in volumetric gas reservoirs, while typical recovery factors in water-drive gas reservoirs can range from 50% to 70%.

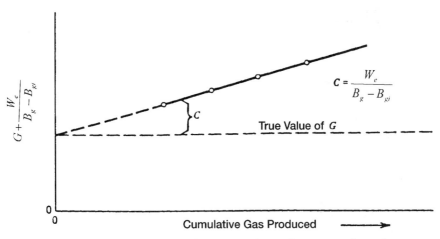

Figure 13-7. Effect of water influx on calculating the gas initially in place.

MATERIAL BALANCE EQUATION AS A STRAIGHT LINE

Havlena and Odeh (1963) expressed the material balance in terms of gas production, fluid expansion, and water influx as:

$$
\begin{array}{ccccccc}
\text{Underground} & = & \text{Gas} & + & \text{Water expansion/} & + & \text{Water} \\
\text{withdrawal} & & \text{expansion} & & \text{pore compaction} & & \text{influx}
\end{array}
$$

or

$$
G_p\,B_g + W_p\,B_w = G\,(B_g - B_{gi}) + G\,B_{gi}\,\frac{(c_w\,S_{wi} + c_f)}{1 - S_{wi}}\,\Delta p
$$
$$
+ W_e\,B_w \tag{13-17}
$$

Using the nomenclature of Havlena and Odeh, as described in Chapter 11, gives:

$$
F = G\,(E_g + E_{f,w}) + W_e\,B_w \tag{13-18}
$$

with the terms F, E_g, and $E_{f,w}$ as defined by:

• Underground fluid withdrawal F:

$$
F = G_p\,B_g + W_p\,B_w \tag{13-19}
$$

• Gas expansion E_g:

$$
E_g = B_g - B_{gi} \tag{13-20}
$$

• Water and rock expansion $E_{f,w}$:

$$
E_{f,w} = B_{gi}\,\frac{(c_w\,S_{wi} + c_f)}{1 - S_{wi}} \tag{13-21}
$$

Assuming that the rock and water expansion term $E_{f,w}$ is negligible in comparison with the gas expansion E_g, Equation 13-18 is reduced to:

$$
F = G\,E_g + W_e\,B_w \tag{13-22}
$$

Finally, dividing both sides of the equation by E_g gives:

$$\frac{F}{E_g} = G + \frac{W_e\, B_w}{E_g} \qquad (13\text{-}23)$$

Using the production, pressure and PVT data, the left-hand side of this expression should be plotted as a function of the cumulative gas production, G_p. This is simply for display purposes to inspect its variation during depletion. Plotting F/E_g versus production time or pressure decline, Δp, can be equally illustrative.

Dake (1994) presented an excellent discussion of the strengths and weaknesses of the MBE as a straight line. He points out that the plot will have one of the three shapes depicted in Figure 13-8. If the reservoir is of the volumetric depletion type, $W_e = 0$, then the values of F/E_g evaluated,

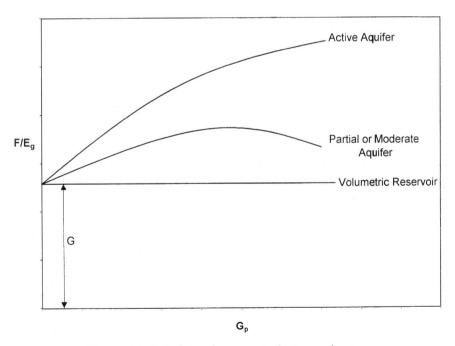

Figure 13-8. Defining the reservoir-driving mechanism.

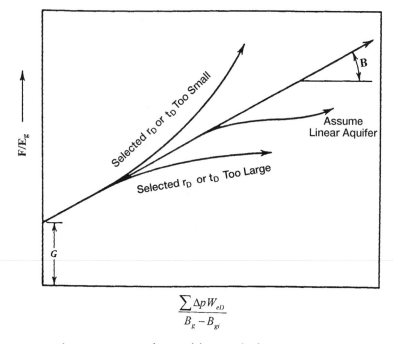

$$\frac{\sum \Delta p \, W_{eD}}{B_g - B_{gi}}$$

Figure 13-9. Havlena-Odeh MBE plot for a gas reservoir.

say, at six monthly intervals, should plot as a straight line parallel to the abscissa—whose ordinate value is the GIIP.

Alternatively, if the reservoir is affected by natural water influx then the plot of F/E_g will usually produce a concave downward shaped arc whose exact form is dependent upon the aquifer size and strength and the gas off-take rate. Backward extrapolation of the F/E_g trend to the ordinate should nevertheless provide an estimate of the GIIP ($W_e \sim 0$); however, the plot can be highly nonlinear in this region yielding a rather uncertain result. The main advantage in the F/E_g versus G_p plot is that it is much more sensitive than other methods in establishing whether the reservoir is being influenced by natural water influx or not.

The graphical presentation of Equation 13-23 is illustrated by Figure 13-9. A graph of F/E_g vs. $\Sigma \Delta p \, W_{eD}/E_g$ yields a straight line, provided the unsteady-state influx summation, $\Sigma \Delta p \, W_{eD}$, is accurately assumed. The resulting straight line intersects the y-axis at the initial gas in place G and has a slope equal to the water influx constant B.

Nonlinear plots will result if the aquifer is improperly characterized. A systematic upward or downward curvature suggests that the summation term is too small or too large, respectively, while an S-shaped curve indicates that a linear (instead of a radial) aquifer should be assumed. The points should plot sequentially from left to right. A reversal of this plotting sequence indicates that an unaccounted aquifer boundary has been reached and that a smaller aquifer should be assumed in computing the water influx term.

A linear infinite system rather than a radial system might better represent some reservoirs, such as reservoirs formed as fault blocks in salt domes. The van Everdingen-Hurst dimensionless water influx W_{eD} is replaced by the square root of time as:

$$W_e = C \sum \Delta p_n \sqrt{t - t_n} \tag{13-24}$$

where C = water influx constant ft^3/psi
 t = time (any convenient units, i.e., days, year)

The water influx constant C must be determined by using the past production and pressure of the field in conjunction with Havlena-Odeh methodology. For the linear system, the underground withdrawal F is plotted versus $[\Sigma \Delta p_n \; Zt - t_n/(B - B_{gi})]$ on a Cartesian coordinate graph. The plot should result in a straight line with G being the intercept and the water influx constant C being the slope of the straight line.

To illustrate the use of the linear aquifer model in the gas MBE as expressed as an equation of straight line, i.e., Equation 13-23, Havlena and Odeh proposed the following problem.

Example 13-4

The volumetric estimate of the gas initially in place for a dry-gas reservoir ranges from 1.3 to 1.65×10^{12} scf. Production, pressures and pertinent gas expansion term, i.e., $E_g = B_g - B_{gi}$, are presented in Table 13-1. Calculate the original gas in place G.

Solution

Step 1. Assume volumetric gas reservoir.

Step 2. Plot (p/z) versus G_p or $G_p B_g/(B_g - B_{gi})$ versus G_p.

Step 3. A plot of $G_p B_g/(B_g - B_{gi})$ vs. $G_p B_g$ showed an upward curvature, as shown in Figure 13-10, indicating water influx.

Table 13-1
Havlena-Odeh Dry-Gas Reservoir Data for Example 13-4

Time (months)	Average Reservoir Pressure (psi)	$E_g =$ $(B_g - B_{gi}) \times 10^{-6}$ (ft^3/scf)	$F =$ $(G_p B_g) \times 10^6$ (ft^3)	$\dfrac{\Sigma \Delta p_n \overline{zt - t_n}}{B_g - B_{gi}}$ (10^6)	$\dfrac{F/E_g =}{G_p B_g}$ $\dfrac{}{B_g - B_{gi}}$ (10^{12})
0	2,883	0.0	—	—	—
2	2,881	4.0	5.5340	0.3536	1.3835
4	2,874	18.0	24.5967	0.4647	1.3665
6	2,866	34.0	51.1776	0.6487	1.5052
8	2,857	52.0	76.9246	0.7860	1.4793
10	2,849	68.0	103.3184	0.9306	1.5194
12	2,841	85.0	131.5371	1.0358	1.5475
14	2,826	116.5	180.0178	1.0315	1.5452
16	2,808	154.5	240.7764	1.0594	1.5584
18	2,794	185.5	291.3014	1.1485	1.5703
20	2,782	212.0	336.6281	1.2426	1.5879
22	2,767	246.0	392.8592	1.2905	1.5970
24	2,755	273.5	441.3134	1.3702	1.6136
26	2,741	305.5	497.2907	1.4219	1.6278
28	2,726	340.0	556.1110	1.4672	1.6356
30	2,712	373.5	613.6513	1.5174	1.6430
32	2,699	405.0	672.5969	1.5714	1.6607
34	2,688	432.5	723.0868	1.6332	1.6719
36	2,667	455.5	771.4902	1.7016	1.6937

Step 4. Assuming a linear water influx, plot $G_p B_g/(B_g - B_{gi})$ versus $\left[\Sigma \Delta p_n \sqrt{t - t_n}\right]/(B_g - B_{gi})$ as shown in Figure 13-11.

Step 5. As evident from Figure 13-11, the necessary straight-line relationship is regarded as satisfactory evidence of the presence of linear aquifer.

Step 6. From Figure 13-11, determine the original gas in place G and the linear water influx constant C as:

$$G = 1.325 \times 10^{12} \text{ scf}$$

$$C = 212.7 \times 10^3 \text{ ft}^3/\text{psi}$$

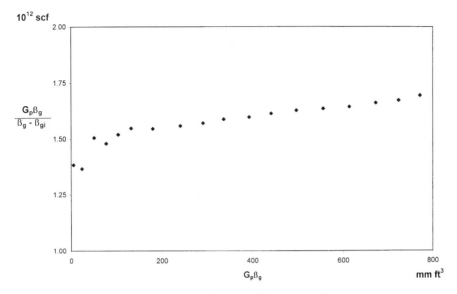

Figure 13-10. Indication of the water influx.

ABNORMALLY PRESSURED GAS RESERVOIRS

Hammerlindl (1971) pointed out that in abnormally high-pressure volumetric gas reservoirs, two distinct slopes are evident when the plot of p/z versus G_p is used to predict reserves because of the formation and fluid compressibility effects as shown in Figure 13-12. The final slope of the p/z plot is steeper than the initial slope; consequently, reserve estimates based on the early life portion of the curve are erroneously high. The initial slope is due to gas expansion and significant pressure maintenance brought about by formation compaction, crystal expansion, and water expansion. At approximately normal pressure gradient, the formation compaction is essentially complete and the reservoir assumes the characteristics of a normal gas expansion reservoir. This accounts for the second slope. Most early decisions are made based on the early life extrapolation of the p/z plot; therefore, the effects of hydrocarbon pore volume change on reserve estimates, productivity, and abandonment pressure must be understood.

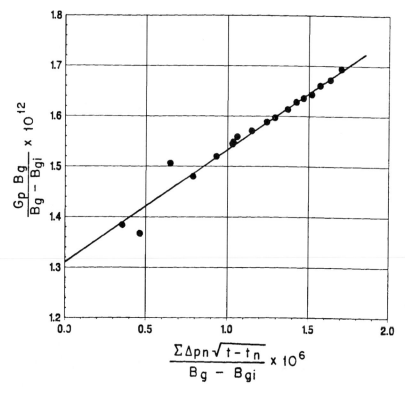

Figure 13-11. Havlena-Odeh MBE plot for Example 13-4.

All gas reservoir performance is related to effective compressibility, not gas compressibility. When the pressure is abnormal and high, effective compressibility may equal two or more times that of gas compressibility. If effective compressibility is equal to twice the gas compressibility, then the first cubic foot of gas produced is due to 50% gas expansion and 50% formation compressibility and water expansion. As the pressure is lowered in the reservoir, the contribution due to gas expansion becomes greater because gas compressibility is approaching effective compressibility. Using formation compressibility, gas production, and shut-in bottom-hole pressures, two methods are presented for correcting the reserve estimates from the early life data (assuming no water influx).

Roach (1981) proposed a graphical technique for analyzing abnormally pressured gas reservoirs. The MBE as expressed by Equation 13-17 may be written in the following form for a volumetric gas reservoir:

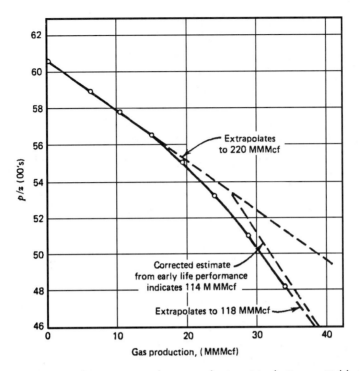

Figure 13-12. P/z versus cumulative production. North Ossum Field, Lafayette Parish, Louisiana NS2B Reservoir. (*After Hammerlindl.*)

$$(p/z)c_t = (p_i/z_i) - \left[1 - \frac{G_p}{G}\right] \tag{13-25}$$

where

$$c_t = 1 - \frac{(c_f + c_w S_{wi})(p_i - p)}{1 - S_{wi}} \tag{13-26}$$

Defining the rock expansion term E_R as:

$$E_R = \frac{c_f + c_w S_{wi}}{1 - S_{wi}} \tag{13-27}$$

Equation 13-26 can be expressed as:

$$c_t = 1 - E_R (p_i - p) \tag{13-28}$$

Equation 13-25 indicates that plotting $(p/z)c_t$ versus cumulative gas production on Cartesian coordinates results in a straight line with an x-intercept at the original gas in place and a y-intercept at the original p/z. Since c_t is unknown and must be found by choosing the compressibility values resulting in the best straight-line fit, this method is a trial-and-error procedure.

Roach used the data published by Duggan (1972) for the Mobil-David Anderson gas field to illustrate the application of Equations 13-25 and 13-28 to determine graphically the gas initially in place. Duggan reported that the reservoir had an initial pressure of 9507 psig at 11,300 ft. Volumetric estimates of original gas in place indicated that the reservoir contains 69.5 MMMscf. The historical p/z versus G_p plot produced an initial gas in place of 87 MMMscf, as shown in Figure 13-13.

Using the trial-and-error approach, Roach showed that a value of the rock expansion term E_R of 18.5×10^{-6} would result in a straight line with a gas initially in place of 75 MMMscf, as shown in Figure 13-13.

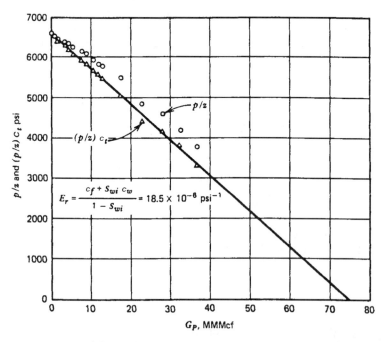

Figure 13-13. Mobil-David Anderson "L" p/z versus cumulative production. (*After Roach.*)

To avoid the trial-and-error procedure, Roach proposed that Equations 13-25 and 13-28 can be combined and expressed in a linear form by:

$$\alpha = \left(\frac{1}{G}\right)\beta - E_R \qquad (13\text{-}29)$$

with

$$\alpha = \frac{[\,(p_i/z_i)/(p/z)\,] - 1}{(p_i - p)} \qquad (13\text{-}30)$$

$$\beta = \frac{(p_i/z_i)\,(p/z)}{(p_i - p)} \qquad (13\text{-}31)$$

where   G = initial gas in place, scf
$\qquad$ E_R = rock expansion term, psi^{-1}
$\qquad$ S_{wi} = initial water saturation

Roach (1981) shows that a plot of α versus β will yield a straight line with slope 1/G and y-intercept = $-E_R$. To illustrate his proposed methodology, he applied Equation 13-29 to the Mobil-David gas field as shown in Figure 13-14. The slope of the straight line gives G = 75.2 MMMscf and the intercept gives $E_R = 18.5 \times 10^{-6}$.

Begland and Whitehead (1989) proposed a method to predict the percent recovery of volumetric, high-pressured gas reservoirs from the initial pressure to the abandonment pressure with only initial reservoir data. The proposed technique allows the pore volume and water compressibilities to be pressure-dependent. The authors derived the following form of the MBE for a volumetric gas reservoir:

$$r = \frac{G_p}{G} = \frac{B_g - B_{gi}}{B_g} + \frac{\dfrac{B_{gi}\,S_{wi}}{1 - S_{wi}}\left[\dfrac{B_{tw}}{B_{twi}} - 1 + \dfrac{c_f\,(p_i - p)}{S_{wi}}\right]}{B_g} \qquad (13\text{-}32)$$

where   r = recovery factor
$\qquad$ B_g = gas formation volume factor, bbl/scf
$\qquad$ c_f = formation compressibility, psi^{-1}
$\qquad$ B_{tw} = two-phase water formation volume factor, bbl/STB
$\qquad$ B_{twi} = initial two-phase water formation volume factor, bbl/STB

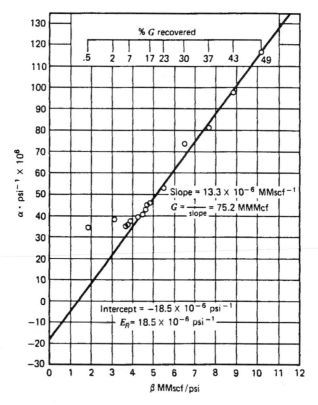

Figure 13-14. Mobil-David Anderson "L" gas material-balance. (*After Roach.*)

The water two-phase FVF is determined from:

$$B_{tw} = B_w + B_g (R_{swi} - R_{sw}) \tag{13-33}$$

where R_{sw} = gas solubility in the water phase, scf/STB
B_w = water FVF, bbl/STB

The following three assumptions are inherent in Equation 13-32:

- A volumetric, single-phase gas reservoir
- No water production
- The formation compressibility c_f remains constant over the pressure drop $(p_i - p)$.

The authors point out that the changes in water compressibility c_w are implicit in the change of B_{tw} with pressure as determined by Equation 13-33.

Begland and Whitehead suggest that because c_f is pressure dependent, Equation 13-32 is not correct as reservoir pressure declines from the initial pressure to some value several hundred psi lower. The pressure dependence of c_f can be accounted for in Equation 13-32 is solved in an incremental manner.

Effect of Gas Production Rate on Ultimate Recovery

Volumetric gas reservoirs are essentially depleted by expansion and, therefore, the ultimate gas recovery is independent of the field production rate. The gas saturation in this type of reservoir is never reduced; only the number of pounds of gas occupying the pore spaces is reduced. Therefore, it is important to reduce the abandonment pressure to the lowest possible level. In closed-gas reservoirs, it is not uncommon to recover as much as 90 percent of the initial gas in place.

Cole (1969) points out that for water-drive gas reservoirs, recovery may be rate dependent. There are two possible influences which producing rate may have on ultimate recovery. First, in an active water-drive reservoir, the abandonment pressure may be quite high, sometimes only a few psi below initial pressure. In such a case, the number of pounds of gas remaining in the pore spaces at abandonment will be relatively great.

The encroaching water, however, reduces the initial gas saturation. Therefore, the high abandonment pressure is somewhat offset by the reduction in initial gas saturation. If the reservoir can be produced at a rate greater than the rate of water influx rate, without water coning, then a high producing rate could result in maximum recovery by taking advantage of a combination of reduced abandonment pressure and reduction in initial gas saturation. Second, the water coning problems may be very severe in gas reservoirs, in which case it will be necessary to restrict withdrawal rates to reduce the magnitude of this problem.

Cole suggests that the recovery from water-drive gas reservoirs is substantially less than recovery from closed-gas reservoirs. As a rule of thumb, recovery from a water-drive reservoir will be approximately 50 to 80 percent of the initial gas in place. The structural location of producing wells and the degree of water coning are important considerations in determining ultimate recovery.

A set of circumstances could exist—such as the location of wells very high on the structure with very little coning tendencies—where water-drive recovery would be greater than depletion-drive recovery. Abandonment pressure is a major factor in determining recovery efficiency, and permeability is usually the most important factor in determining the magnitude of the abandonment pressure. Reservoirs with low permeability will have higher abandonment pressures than reservoirs with high permeability. A certain minimum flow rate must be sustained, and a higher permeability will permit this minimum flow rate at a lower pressure.

PROBLEMS

1. The following information is available on a volumetric gas reservoir:

Initial reservoir temperature, T_i = 155°F
Initial reservoir pressure, p_i = 3500 psia
Specific gravity of gas, γ_g = 0.65 (air = 1)
Thickness of reservoir, h   = 20 ft
Porosity of the reservoir, ϕ = 10%
Initial water saturation, S_{wi} = 25%

After producing 300 MMscf, the reservoir pressure declined to 2500 psia. Estimate the areal extent of this reservoir.

2. The following pressures and cumulative production data[2] are available for a natural gas reservoir:

Reservoir pressure, psia	Gas deviation factor, z	Cumulative production, MMMscf
2080	0.759	0
1885	0.767	6.873
1620	0.787	14.002
1205	0.828	23.687
888	0.866	31.009
645	0.900	36.207

a. Estimate the initial gas in place.
b. Estimate the recoverable reserves at an abandonment pressure of 500 psia. Assume z_a = 1.00.

---

[2]Ikoku, C., *Natural Gas Reservoir Engineering,* John Wiley and Sons, 1984.

c. What is the recovery factor at the abandonment pressure of 500 psia?

3. A gas field with an active water drive showed a pressure decline from 3000 to 2000 psia over a 10-month period. From the following production data, match the past history and calculate the original hydrocarbon gas in the reservoir. Assume $z = 0.8$ in the range of reservoir pressures and $T = 140°F$.

Data				
t, months	0	2.5	5.0	7.5	10.0
p, psia	3000	2750	2500	2250	2000
G_p, MMscf	0	97.6	218.9	355.4	500.0

4. A volumetric gas reservoir produced 600 MMscf of 0.62 specific gravity gas when the reservoir pressure declined from 3600 to 2600 psi. The reservoir temperature is reported at 140°F. Calculate:

a. Gas initially in place
b. Remaining reserves to an abandonment pressure of 500 psi
c. Ultimate gas recovery at abandonment

5. The following information on a water-drive gas reservoir is given:

Bulk volume = 100,000 acre-ft
Gas Gravity = 0.6
Porosity = 15%
$S_{wi} = 25\%$
$T = 140°F$
$p_i = 3500$ psi

Reservoir pressure has declined to 3000 psi while producing 30 MMMscf of gas and no water production. Calculate cumulative water influx.

6. The pertinent data for the Mobil-David field is given below.

$G = 70$ MMMscf $p_i = 9507$ psi $\phi = 24\%$ $S_{wi} = 35\%$
$c_w = 401 \times 10^{-6}$ psi^{-1} $c_f = 3.4 \times 10^{-6}$ psi^{-1} $\gamma_g = 0.94$ $T = 266°F$

For this volumetric abnormally-pressured reservoir, calculate and plot cumulative gas production as a function of pressure.

7. The Big Butte field is a volumetric dry-gas reservoir with a recorded initial pressure of 3,500 psi at 140°F. The specific gravity of the pro-

duced gas is measured at 0.65. The following reservoir data are available from logs and core analysis:

Reservoir area   = 1500 acres
Thickness   = 25 ft
Porosity   = 15%
Initial water saturation = 20%

Calculate:

a. Initial gas in place as expressed in scf
b. Gas viscosity at 3,500 psi and 140°F

REFERENCES

1. Begland, T., and Whitehead, W., "Depletion Performance of Volumetric High-Pressured Gas Reservoirs," *SPE Reservoir Engineering,* August 1989, pp. 279–282.

2. Cole, F. W., *Reservoir Engineering Manual.* Houston: Gulf Publishing Co., 1969.

3. Dake, L., *The Practice of Reservoir Engineering.* Amsterdam: Elsevier Publishing Company, 1994.

4. Duggan, J. O., "The Anderson 'L'—An Abnormally Pressured Gas Reservoir in South Texas," *Journal of Petroleum Technology,* February 1972, Vol. 24, No. 2, pp. 132–138.

5. Hammerlindl, D. J., "Predicing Gas Reserves in Abnormally Pressure Reservoirs." SPE Paper 3479 presented at the 46th Annual Fall Meeting of SPE, New Orleans, October 1971.

6. Havlena, D., and Odeh, A. S., "The Material Balance as an Equation of a Straight Line," *Trans. AIME,* Part 1: 228 I-896 (1963); Part 2: 231 I-815 (1964).

7. Ikoku, C., *Natural Gas Reservoir Engineering.* John Wiley & Sons, Inc., 1984.

8. Roach, R. H., "Analyzing Geopressured Reservoirs—A Material Balance Technique," SPE Paper 9968, Society of Petroleum Engineers of AIME, Dallas, December 1981.

9. Van Everdingen, A. F., and Hurst, W., "Application of Laplace Transform to Flow Problems in Reservoirs," *Trans. AIME,* 1949, Vol. 186, pp. 305–324B.

PRINCIPLES OF WATERFLOODING

The terms primary oil recovery, secondary oil recovery, and tertiary (enhanced) oil recovery are traditionally used to describe hydrocarbons recovered according to the method of production or the time at which they are obtained.

Primary oil recovery describes the production of hydrocarbons under the natural driving mechanisms present in the reservoir without supplementary help from injected fluids such as gas or water. In most cases, the natural driving mechanism is a relatively inefficient process and results in a low overall oil recovery. The lack of sufficient natural drive in most reservoirs has led to the practice of supplementing the natural reservoir energy by introducing some form of artificial drive, the most basic method being the injection of gas or water.

Secondary oil recovery refers to the additional recovery that results from the conventional methods of water injection and immiscible gas injection. Usually, the selected secondary recovery process follows the primary recovery but it can also be conducted concurrently with the primary recovery. Waterflooding is perhaps the most common method of secondary recovery. However, before undertaking a secondary recovery project, it should be clearly proven that the natural recovery processes are insufficient; otherwise there is a risk that the substantial capital investment required for a secondary recovery project may be wasted.

Tertiary (enhanced) oil recovery is that additional recovery over and above what could be recovered by primary and secondary recovery methods. Various methods of enhanced oil recovery (EOR) are essentially designed to recover oil, commonly described as residual oil, left in the reservoir after both primary and secondary recovery methods have been exploited to their respective economic limits. Figure 14-1 illustrates the concept of the three oil recovery categories.

FACTORS TO CONSIDER IN WATERFLOODING

Thomas, Mahoney, and Winter (1989) pointed out that in determining the suitability of a candidate reservoir for waterflooding, the following reservoir characteristics must be considered:

• Reservoir geometry
• Fluid properties
• Reservoir depth
• Lithology and rock properties
• Fluid saturations

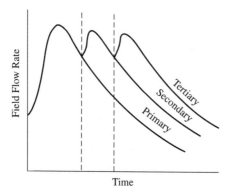

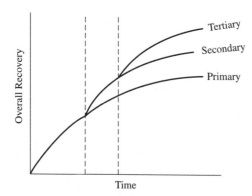

Figure 14-1. Oil recovery categories.

• Reservoir uniformity and pay continuity
• Primary reservoir driving mechanisms

Each of these topics is discussed in detail in the following subsections.

Reservoir Geometry

The areal geometry of the reservoir will influence the location of wells and, if offshore, will influence the location and number of platforms required. The reservoir's geometry will essentially dictate the methods by which a reservoir can be produced through water-injection practices.

An analysis of reservoir geometry and past reservoir performance is often important when defining the presence and strength of a natural water drive and, thus, when defining the need to supplement the natural injection. If a water-drive reservoir is classified as an active water drive, injection may be unnecessary.

Fluid Properties

The physical properties of the reservoir fluids have pronounced effects on the suitability of a given reservoir for further development by water-flooding. The viscosity of the crude oil is considered the most important fluid property that affects the degree of success of a waterflooding project. The oil viscosity has the important effect of determining the mobility ratio that, in turn, controls the sweep efficiency.

Reservoir Depth

Reservoir depth has an important influence on both the technical and economic aspects of a secondary or tertiary recovery project. Maximum injection pressure will increase with depth. The costs of lifting oil from very deep wells will limit the maximum economic water–oil ratios that can be tolerated, thereby reducing the ultimate recovery factor and increasing the total project operating costs. On the other hand, a shallow reservoir imposes a restraint on the injection pressure that can be used, because this must be less than fracture pressure. In waterflood operations, there is a critical pressure (approximately 1 psi/ft of depth) that, if exceeded, permits the injecting water to expand openings along fractures or to create fractures. This results in the channeling of the injected water or the bypassing of large portions of the reservoir matrix. Consequently, an operational pressure gradient of 0.75 psi/ft of depth normally is allowed to provide a sufficient margin of safety to prevent pressure parting.

Lithology and Rock Properties

Thomas et al. (1989) pointed out that lithology has a profound influence on the efficiency of water injection in a particular reservoir. Reservoir lithology and rock properties that affect flood ability and success are:

- Porosity
- Permeability
- Clay content
- Net thickness

In some complex reservoir systems, only a small portion of the total porosity, such as fracture porosity, will have sufficient permeability to be effective in water-injection operations. In these cases, a water-injection program will have only a minor impact on the matrix porosity, which might be crystalline, granular, or vugular in nature.

Although evidence suggests that the clay minerals present in some sands may clog the pores by swelling and deflocculating when water-flooding is used, no exact data are available as to the extent to which this may occur.

Tight (low-permeability) reservoirs or reservoirs with thin net thickness possess water-injection problems in terms of the desired water-injection rate or pressure. Note that the water-injection rate and pressure are roughly related by the following expression:

$$p_{inj} \propto \frac{i_w}{hk}$$

where p_{inj} = water-injection pressure
   i_w = water-injection rate
   h = net thickness
   k = absolute permeability

The above relationship suggests that to deliver a desired daily injection rate of i_w in a tight or thin reservoir, the required injection pressure might exceed the formation fracture pressure.

Fluid Saturations

In determining the suitability of a reservoir for waterflooding, a high oil saturation that provides a sufficient supply of recoverable oil is the primary criterion for successful flooding operations. Note that higher oil saturation at the beginning of flood operations increases the oil mobility that, in turn, gives higher recovery efficiency.

Reservoir Uniformity and Pay Continuity

Substantial reservoir uniformity is one of the major physical criterions for successful waterflooding. For example, if the formation contains a stratum of limited thickness with a very high permeability (i.e., **thief zone**), rapid channeling and bypassing will develop. Unless this zone can be located and shut off, the producing water–oil ratios will soon become too high for the flooding operation to be considered profitable.

The lower depletion pressure that may exist in the highly permeable zones will also aggravate the water-channeling tendency due to the high-permeability variations. Moreover, these thief zones will contain less residual oil than the other layers, and their flooding will lead to relatively lower oil recoveries than other layers.

Areal continuity of the pay zone is also a prerequisite for a successful waterflooding project. Isolated lenses may be effectively depleted by a single well completion, but a flood mechanism requires that both the injector and producer be present in the lens. Breaks in pay continuity and reservoir anisotropy caused by depositional conditions, fractures, or faulting need to be identified and described before determining the proper well spanning and the suitable flood pattern orientation.

Primary Reservoir Driving Mechanisms

As described in Chapter 11, six driving mechanisms basically provide the natural energy necessary for oil recovery:

• Rock and liquid expansion
• Solution gas drive
• Gas cap drive
• Water drive
• Gravity drainage drive
• Combination drive

The recovery of oil by any of the above driving mechanisms is called *primary recovery*. The term refers to the production of hydrocarbons from a reservoir without the use of any process (such as water injection) to supplement the natural energy of the reservoir. The primary drive mechanism and anticipated ultimate oil recovery should be considered when reviewing possible waterflood prospects. The approximate oil recovery range is tabulated below for various driving mechanisms. Note that these calculations are approximate and, therefore, oil recovery may fall outside these ranges.

Driving Mechanism	Oil Recovery Range, %
Rock and liquid expansion	3–7
Solution gap	5–30
Gas cap	20–40
Water drive	35–75
Gravity drainage	<80
Combination drive	30–60

Water-drive reservoirs that are classified as strong water-drive reservoirs are not usually considered to be good candidates for waterflooding because of the natural ongoing water influx. However, in some instances a natural water drive could be supplemented by water injection in order to:

• Support a higher withdrawal rate
• Better distribute the water volume to different areas of the field to achieve more uniform areal coverage
• Better balance voidage and influx volumes.

Gas-cap reservoirs are not normally good waterflood prospects because the primary mechanism may be quite efficient without water injection. In these cases, gas injection may be considered in order to help maintain pressure. Smaller gas-cap drives may be considered as waterflood prospects, but the existence of the gas cap will require greater care to prevent migration of displaced oil into the gas cap. This migration would result in a loss of recoverable oil due to the establishment of residual oil saturation in pore volume, which previously had none. If a gas cap is repressured with water, a substantial volume may be required for this purpose, thereby lengthening the project life and requiring a higher volume of water. However, the presence of a gas cap does not always mean that an effective gas-cap drive is functioning. If the vertical communication between the gas cap and the oil zone is considered poor due to low vertical permeability, a waterflood may be appropriate in this case. Analysis of past performance, together with reservoir geology studies, can provide insight as to the degree of effective communication. Natural permeability barriers can often restrict the migration of fluids to the gas cap. It may also be possible to use selective plugging of input wells to restrict the loss of injection fluid to the gas cap.

Solution gas-drive mechanisms generally are considered the best candidates for waterfloods. Because the primary recovery will usually be low, the potential exists for substantial additional recovery by water injection. In effect, we hope to create an artificial water-drive mechanism. The typical range of water-drive recovery is approximately double that of solution gas drive. As a general guideline, waterfloods in solution gas-drive reservoirs frequently will recover an additional amount of oil equal to primary recovery.

Volumetric undersaturated oil reservoirs producing above the bubble-point pressure must depend on rock and liquid expansion as the main driving mechanism. In most cases, this mechanism will not recover more than about 5% of the original oil in place. These reservoirs will offer an opportunity for greatly increasing recoverable reserves if other conditions are favorable.

OPTIMUM TIME TO WATERFLOOD

The most common procedure for determining the optimum time to start waterflooding is to calculate:

- Anticipated oil recovery
- Fluid production rates
- Monetary investment
- Availability and quality of the water supply
- Costs of water treatment and pumping equipment
- Costs of maintenance and operation of the water installation facilities
- Costs of drilling new injection wells or converting existing production wells into injectors

These calculations should be performed for several assumed times and the net income for each case determined. The scenario that maximizes the profit and perhaps meets the operator's desirable goal is selected.

Cole (1969) lists the following factors as being important when determining the reservoir pressure (or time) to initiate a secondary recovery project:

- **Reservoir oil viscosity.** Water injection should be initiated when the reservoir pressure reaches its bubble-point pressure since the oil viscosity reaches its minimum value at this pressure. The mobility of the oil will increase with decreasing oil viscosity, which in turns improves the sweeping efficiency.
- **Free gas saturation.** (1) In **water injection projects**. It is desirable to have initial gas saturation, possibly as much as 10%. This will occur at a pressure that is below the bubble point pressure. (2) In **gas injection projects.** Zero gas saturation in the oil zone is desired. This occurs while reservoir pressure is at or above bubble-point pressure.
- **Cost of injection equipment.** This is related to reservoir pressure, and at higher pressures, the cost of injection equipment increases. Therefore, a low reservoir pressure at initiation of injection is desirable.
- **Productivity of producing wells.** A high reservoir pressure is desirable to increase the productivity of producing wells, which prolongs the flowing period of the wells, decreases lifting costs, and may shorten the overall life of the project.
- **Effect of delaying investment on the time value of money.** A delayed investment in injection facilities is desirable from this standpoint.
- **Overall life of the reservoir.** Because operating expenses are an important part of total costs, the fluid injection process should be started as early as possible.

Some of these six factors act in opposition to others. Thus the actual pressure at which a fluid injection project should be initiated will require optimization of the various factors in order to develop the most favorable overall economics.

The principal requirement for a successful fluid injection project is that sufficient oil must remain in the reservoir after primary operations have ceased to render economic the secondary recovery operations. This high residual oil saturation after primary recovery is essential not only because there must be a sufficient volume of oil left in the reservoir, but also because of relative permeability considerations. A high oil relative permeability, i.e., high oil saturation, means more oil recovery with less production of the displacing fluid. On the other hand, low oil saturation means a low oil relative permeability with more production of the displacing fluid at a given time.

EFFECT OF TRAPPED GAS ON WATERFLOOD RECOVERY

Numerous experimental and field studies have been conducted to study the effect of the presence of initial gas saturation on waterflood recovery. Early research indicated that the waterflooding of a linear system results in the formation of an oil bank, or zone of increased oil saturation, ahead of the injection water. The moving oil bank will displace a portion of the free water ahead of it, trapping the rest as a residual gas. An illustration of the water saturation profile is shown schematically in Figure 14-2. Several authors have shown through experiments that oil recovery by water is improved as a result of the establishment of **trapped gas saturation,** S_{gt}, in the reservoir.

The theory of this phenomenon of improving overall oil recovery when initial gas exists at the start of the flood is not well established; however, Cole (1969) proposed the following two different theories that perhaps provide insight to this phenomenon.

First Theory

Cole (1969) postulates that since the interfacial tension of a gas–oil system is less than the interfacial tension of a gas–water system, in a three-phase system containing gas, water, and oil, the reservoir fluids will tend to arrange themselves in a minimum energy relationship. In this

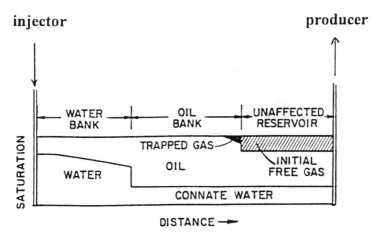

Figure 14-2. Water saturation profile during a waterflood.

case, this would dictate that the gas molecules enclose themselves in an oil "blanket." This increases the effective size of any oil globules, which have enclosed some gas. When the oil is displaced by water, the oil globules are reduced to some size dictated by the flow mechanics. If a gas bubble existed on the inside of the oil globule, the amount of residual oil left in the reservoir would be reduced by the size of the gas bubble within the oil globule. As illustrated in Figure 14-3, the external diameters of the residual oil globules are the same in both views. However, in view b, the center of the residual oil globule is not oil, but gas. Therefore, in view b, the actual residual oil saturation is reduced by the size of the gas bubble within the oil globule.

Second Theory

Cole (1969) points out that reports on other laboratory experiments have noted the increased recovery obtained by flooding cores with air after waterflooding. These cores were classified as water-wet at the time the laboratory experiments were conducted. On the basis of these experiments, it was postulated that the residual oil saturation was located in the larger pore spaces, since the water would be preferentially pulled into the smaller pore spaces by capillary action in the water-wet sandstone. At a later time, when air was flooded through the core, it moved preferentially

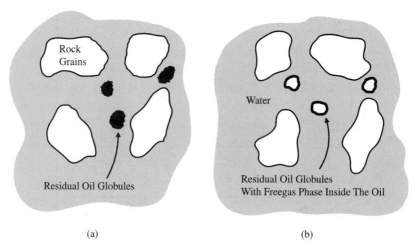

(a)   (b)

Figure 14-3. Effect of free gas saturation on S_{or} (first theory). *(After Cole, F., 1969.)*

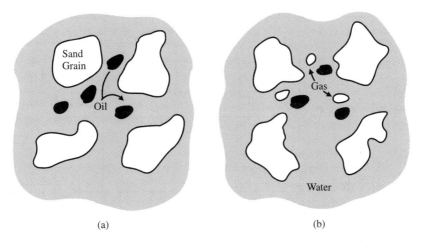

(a)   (b)

Figure 14-4. Effect of free gas saturation on S_{or} (second theory). *(After Cole, F., 1969.)*

through the larger pore spaces since it was nonwetting. However, in passing through these large pore spaces, the air displaced some of the residual oil left by water displacement.

This latter theory is more nearly compatible with fluid flow observations, because the gas saturation does not have to exist inside the oil phase. If this theory were correct, the increased recovery due to the presence of free gas saturation could be explained quite simply for water-wet porous media. As the gas saturation formed, it displaced oil from the larger pore spaces, because it is more nonwetting to the reservoir rock than the oil. Then, as water displaced the oil from the reservoir rock, the amount of residual oil left in the larger pore spaces would be reduced because of occupancy of a portion of this space by gas. This phenomenon is illustrated in Figure 14-4. In view a, there is no free gas saturation and the residual oil occupies the larger pore spaces. In view b, free gas saturation is present and this free gas now occupies a portion of the space originally occupied by the oil. The combined residual saturations of oil and gas in view b are approximately equal to the residual oil saturation of view a.

Craig (1971) presented two graphical correlations that are designed to account for the reduction in the residual oil saturation due to the presence of the trapped gas. The first graphical correlation, shown in Figure 14-5, correlates the trapped gas saturation (S_{gt}) as a function of the initial gas saturation (S_{gi}). The second correlation as presented in Figure 14-6

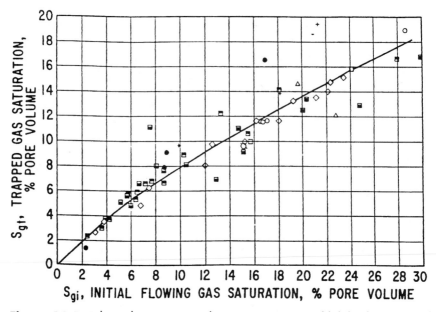

Figure 14-5. Relation between S_{gi} and S_{gt}. *(Permission to publish by the Society of Petroleum Engineers.)*

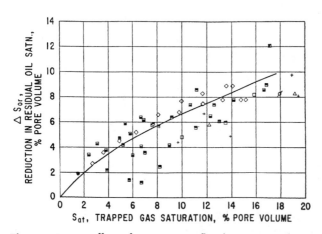

Figure 14-6. Effect of S_{gt} on waterflood recovery. *(Permission to publish by the Society of Petroleum Engineers.)*

illustrates the effect of the trapped gas saturation on the reduction in residual oil saturation (ΔS_{or}) for preferentially water-wet rock. The two graphic correlations can be expressed mathematically by the following two expressions:

$$S_{gt} = a_1 + a_2 S_{gi} + a_3 S_{gi}^2 + a_4 S_{gi}^3 + \frac{a_5}{S_{gi}} \qquad (14\text{-}1)$$

and

$$\Delta S_{or} = a_1 + a_2 S_{gt} + a_3 S_{gt}^2 + a_5 S_{gt}^3 + \frac{a_5}{S_{gt}} \qquad (14\text{-}2)$$

where S_{gi} = initial gas saturation
   S_{gt} = trapped gas saturation
   ΔS_{or} = reduction in residual oil saturation

Values of coefficients a_1 through a_5 for both expressions are tabulated below:

Coefficients	Equation 14-1	Equation 14-2
a_1	0.030517211	0.026936065
a_2	0.4764700	0.41062853
a_3	0.69469046	0.29560322
a_4	−1.8994762	−1.4478797
a_5	−4.1603083 × 10^{-4}	−3.0564771 × 10^{-4}

Example 14-1

An oil reservoir is being considered for further development by initiating a waterflooding project. The oil–water relative permeability data indicate that the residual oil saturation is 35%. It is projected that the initial gas saturation at the start of the flood is approximately 10%. Calculate the anticipated reduction in residual oil, ΔS_{or}, due to the presence of the initial gas at the start of the flood.

Solution

Step 1. From Figure 14-5 or Equation 14-1, determine the trapped gas saturation, to give:

$$S_{gt} = 8\%$$

Step 2. Estimate the reduction in the residual oil saturation from Figure 14-6 or Equation 14-2, to give:

$$\Delta S_{or} = 5.7\%$$

Therefore, new residual oil saturation is:

$$S_{or} = 33\%$$

Khelil (1983) suggests that waterflood recovery can possibly be improved if a so-called "optimum gas saturation" is present at the start of the flood. This optimum gas saturation is given by:

$$(S_g)_{opt} = \frac{0.001867\,k^{0.634}\,B_o^{0.902}}{\left(\frac{S_o}{\mu_o}\right)^{0.352}\left(\frac{S_{wi}}{\mu_w}\right)^{0.166}\phi^{1.152}} \tag{14-3}$$

where $(S_g)_{opt}$ = optimum gas saturation, fraction
S_o , S_{wi} = oil and initial water saturations, fraction
μ_o , μ_w = oil and water viscosities, cp
k = absolute permeability, md
B_o = oil formation volume factor, bbl/STB
ϕ = porosity, fraction

The above correlation is not explicit and must be used in conjunction with the *material balance equation* (MBE). The proposed methodology of determining $(S_g)_{opt}$ is based on calculating the gas saturation as a function of reservoir pressure (or time) by using both the MBE and Equation 14-3. When the gas saturation as calculated by the two equations is identical, this gas saturation is identified as $(S_g)_{opt}$.

Example 14-2

An absolute permeability of 33 md, porosity of 25%, and an initial water saturation of 30% characterize a saturated oil reservoir that exists at its bubble-point pressure of 1925 psi. The water viscosity is treated as a constant with a value of 0.6 cp. Results of the material balance calculations are given below:

Pressure, psi	Bo, bbl/STB	μ_o, cp	S_o	$S_g = 1 - S_o - S_{wi}$
1925	1.333	0.600	0.700	0.000
1760	1.287	0.625	0.628	0.072
1540	1.250	0.650	0.568	0.132
1342	1.221	0.700	0.527	0.173

Using the above data, calculate the optimum gas saturation.

Solution

Pressure, psi	Bo, bbl/STB	μ_o, cp	MBE, S_o	S_g	Equation 14-3 $(S_g)_{opt}$
1925	1.333	0.600	0.700	0.000	—
1760	1.287	0.625	0.628	0.072	0.119
1540	1.250	0.650	0.568	0.132	0.122
1342	1.221	0.700	0.527	0.173	

The calculated value of $(S_g)_{opt}$ at 1540 psi agrees with the value of S_g as calculated from the MBE. Thus, to obtain the proposed additional recovery benefit, the primary depletion should be terminated at a pressure of 1540 psi and water injection initiated.

The injection into a solution gas-drive reservoir usually occurs at injection rates that cause repressurization of the reservoir. If pressure is high enough, the trapped gas will dissolve in the oil with no effect on subsequent residual oil saturations. It is of interest to estimate what pressure increases would be required in order to dissolve the trapped gas in the oil system. The pressure is essentially defined as the "new" bubble-point pressure (P_b^{new}). As the pressure increases to the new bubble-point pressure, the trapped gas will dissolve in the oil phase with a subsequent increase in the gas solubility from R_s to R_s^{new}. As illustrated in Figure 14-7, the new gas solubility can be estimated as the sum of the volumes of the dissolved gas and the trapped gas in the reservoir divided by the volume of stock-tank oil in the reservoir, or:

$$R_S^{new} = \frac{\left[\dfrac{(S_o)(\text{Pore volume})}{B_o}\right]R_S + \left[\dfrac{(S_{gt})(\text{Pore volume})}{B_g}\right]}{\dfrac{(S_o)(\text{Pore volume})}{B_o}}$$

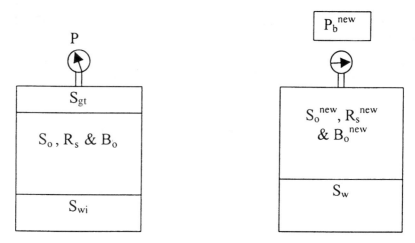

Figure 14-7. Variable bubble-point pressures.

Simplifying gives:

$$R_S^{new} = R_S + \left(\frac{S_{gt}}{S_o}\right)\left(\frac{B_o}{B_g}\right)$$ (14-4)

where R_s^{new} = gas solubility at the "new" bubble-point pressure, scf/STB

  R_s = gas solubility at current pressure p, scf/STB
  B_g = gas formation volume factor, bbl/scf
  B_o = oil formation volume factor, bbl/STB
  S_{gt} = trapped gas saturation

The pressure that corresponds to the new gas solubility (R_s^{new}) on the R_s vs. p relationship is then identified as the pressure at which the trapped gas will completely dissolve in the oil phase.

Example 14-3

The Big Butte Field is a solution gas-drive reservoir that is under consideration for a waterflood project. The volumetric calculations of the field indicate that the areal extent of the field is 1612.6 acres. The field is characterized by the following properties:

Thickness h	= 25 ft
Porosity ϕ	= 15%
Initial water saturation S_{wi}	= 20%
Initial pressure p_i	= 2377 psi

Results from the MBE in terms of cumulative oil production N_p as a function of reservoir pressure p are given below:

Pressure, psi	N_p, MMSTB
2377	0
2250	1.10
1950	1.76
1650	2.64
1350	3.3

The PVT properties of the crude oil system are tabulated below:

Pressure, psi	B_o, bbl/STB	R_s, scf/STB	B_g, bbl/scf
2377	1.706	921	—
2250	1.678	872	0.00139
1950	1.555	761	0.00162
1650	1.501	657	0.00194
1350	1.448	561	0.00240
1050	1.395	467	0.00314
750	1.336	375	0.00448
450	1.279	274	0.00754

Assume that the waterflood will commence when the reservoir pressure declines to 1650 psi; find the pressure that is required to dissolve the trapped gas.

Solution

Step 1. Calculate initial oil in place N:

$$N = 7758 \, A \, h \, \phi \, (1 - S_{wi})/B_{oi}$$

$$N = 7758 \, (1612.6) \, (25) \, (0.15) \, (1 - 0.2)/1.706 = 22 \, \text{MMSTB}$$

Step 2. Calculate remaining oil saturation by applying Equation 12-5 at 1650 psi:

$$S_o = (1 - S_{wi}) \left(1 - \frac{N_p}{N} \right) \left(\frac{B_o}{B_{oi}} \right)$$

$$S_o = (1 - 0.2) \left(1 - \frac{2.64}{22} \right) \left(\frac{1.501}{1.706} \right) = 0.619$$

Step 3. Calculate gas saturation at 1650 psi:

$$S_g = 1 - S_o - S_{wi}$$

$$S_g = 1 - 0.619 - 0.2 = 0.181$$

Step 4. Calculate the trapped gas saturation from Figure 14-5 or Equation 14-1, to give:

$$S_{gt} = 12.6\%$$

Step 5. Calculate the gas solubility when all the trapped gas is dissolved in the oil by applying Equation 14-4:

$$R_S^{new} = 657 + \left(\frac{0.126}{0.619} \right) \left(\frac{1.501}{0.00194} \right) = 814 \, \text{scf/STB}$$

Step 6. Enter the tabulated PVT data with the new gas solubility of 814 scf/STB and find the corresponding pressure of approximately 2140 psi. This pressure is identified as the pressure that is required to dissolve the trapped gas.

SELECTION OF FLOODING PATTERNS

One of the first steps in designing a waterflooding project is flood pattern selection. The objective is to select the proper pattern that will provide the injection fluid with the maximum possible contact with the crude oil system. This selection can be achieved by (1) converting existing production wells into injectors or (2) drilling infill injection wells. When making the selection, the following factors must be considered:

• Reservoir heterogeneity and directional permeability
• Direction of formation fractures
• Availability of the injection fluid (gas or water)
• Desired and anticipated flood life
• Maximum oil recovery
• Well spacing, productivity, and injectivity

In general, the selection of a suitable flooding pattern for the reservoir depends on the number and location of existing wells. In some cases, producing wells can be converted to injection wells while in other cases it may be necessary or desirable to drill new injection wells. Essentially four types of well arrangements are used in fluid injection projects:

• Irregular injection patterns
• Peripheral injection patterns
• Regular injection patterns
• Crestal and basal injection patterns

Irregular Injection Patterns

Willhite (1986) points out that surface or subsurface topology and/or the use of slant-hole drilling techniques may result in production or injection wells that are not uniformly located. In these situations, the region affected by the injection well could be different for every injection well. Some small reservoirs are developed for primary production with a limited number of wells and when the economics are marginal, perhaps only few production wells are converted into injectors in a nonuniform pattern. Faulting and localized variations in porosity or permeability may also lead to irregular patterns.

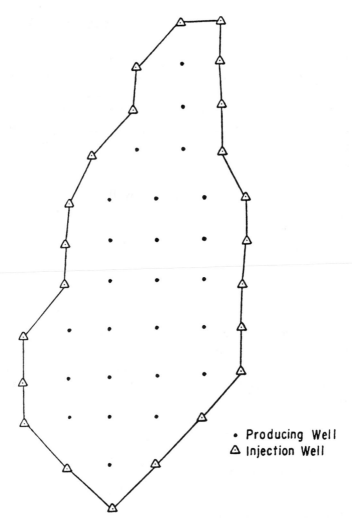

Figure 14-8. Typical peripheral waterflood. *(After Cole, F., 1969.)*

Peripheral Injection Patterns

In peripheral flooding, the injection wells are located at the external boundary of the reservoir and the oil is displaced toward the interior of the reservoir, as shown in Figure 14-8. Craig (1971), in an excellent review of the peripheral flood, points out the following main characteristics of the flood:

- The peripheral flood generally yields a maximum oil recovery with a minimum of produced water.
- The production of significant quantities of water can be delayed until only the last row of producers remains.
- Because of the unusually small number of injectors compared with the number of producers, it takes a long time for the injected water to fill up the reservoir gas space. The result is a delay in the field response to the flood.
- For a successful peripheral flood, the formation permeability must be large enough to permit the movement of the injected water at the desired rate over the distance of several well spacings from injection wells to the last line of producers.
- To keep injection wells as close as possible to the waterflood front without bypassing any movable oil, watered-out producers may be

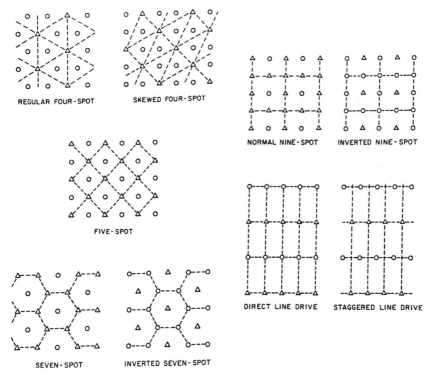

Figure 14-9. Flood patterns. *(Permission to publish by the Society of Petroleum Engineers.)*

converted into injectors. However, moving the location of injection wells frequently requires laying longer surface water lines and adding costs.

- Results from peripheral flooding are more difficult to predict. The displacing fluid tends to displace the oil bank past the inside producers, which are thus difficult to produce.
- Injection rates are generally a problem because the injection wells continue to push the water greater distances.

Regular Injection Patterns

Due to the fact that oil leases are divided into square miles and quarter square miles, fields are developed in a very regular pattern. A wide variety of injection-production well arrangements have been used in injection projects. The most common patterns, as shown in Figure 14-9, are the following:

- **Direct line drive**. The lines of injection and production are directly opposed to each other. The pattern is characterized by two parameters: a = distance between wells of the same type, and d = distance between lines of injectors and producers.
- **Staggered line drive**. The wells are in lines as in the direct line, but the injectors and producers are no longer directly opposed but laterally displaced by a distance of a/2.
- **Five spot**. This is a special case of the staggered line drive in which the distance between all like wells is constant, i.e., a = 2d. Any four injection wells thus form a square with a production well at the center.
- **Seven spot**. The injection wells are located at the corner of a hexagon with a production well at its center.
- **Nine spot**. This pattern is similar to that of the five spot but with an extra injection well drilled at the middle of each side of the square. The pattern essentially contains eight injectors surrounding one producer.

The patterns termed **inverted** have only one injection well per pattern. This is the difference between **normal** and **inverted** well arrangements. Note that the four-spot and inverted seven-spot patterns are identical.

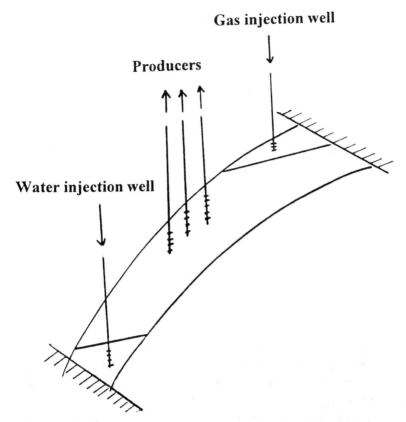

Figure 14-10. Well arrangements for dipping reservoirs.

Crestal and Basal Injection Patterns

In crestal injection, as the name implies, the injection is through wells located at the top of the structure. Gas injection projects typically use a crestal injection pattern. In basal injection, the fluid is injected at the bottom of the structure. Many water-injection projects use basal injection patterns with additional benefits being gained from gravity segregation. A schematic illustration of the two patterns is shown in Figure 14-10.

OVERALL RECOVERY EFFICIENCY

The overall recovery factor (efficiency) RF of any secondary or tertiary oil recovery method is the product of a combination of three individual efficiency factors as given by the following generalized expression:

$$RF = E_D E_A E_V \qquad (14\text{-}5)$$

In terms of cumulative oil production, Equation 14-5 can be written as:

$$N_P = N_S E_D E_A E_V \qquad (14\text{-}6)$$

where   RF = overall recovery factor
N_S = initial oil in place at the start of the flood, STB
N_P = cumulative oil produced, STB
E_D = displacement efficiency
E_A = areal sweep efficiency
E_V = vertical sweep efficiency

The displacement efficiency E_D is the fraction of movable oil that has been displaced from the swept zone at any given time or pore volume injected. Because an immiscible gas injection or waterflood will always leave behind some residual oil, E_D will always be less than 1.0.

The areal sweep efficiency E_A is the fractional area of the pattern that is swept by the displacing fluid. The major factors determining areal sweep are:

• Fluid mobilities
• Pattern type
• Areal heterogeneity
• Total volume of fluid injected

The vertical sweep efficiency E_V is the fraction of the vertical section of the pay zone that is contacted by injected fluids. The vertical sweep efficiency is primarily a function of:

• Vertical heterogeneity
• Degree of gravity segregation
• Fluid mobilities
• Total volume injection

Note that the product of $E_A E_V$ is called the **volumetric sweep efficiency** and represents the overall fraction of the flood pattern that is contacted by the injected fluid.

All three efficiency factors (i.e., E_D, E_A, and E_V) are variables that increase during the flood and reach maximum values at the economic limit of the injection project. Each of the three efficiency factors is discussed individually and methods of estimating these efficiencies are presented.

I. DISPLACEMENT EFFICIENCY

As defined previously, displacement efficiency is the fraction of movable oil that has been recovered from the swept zone at any given time. Mathematically, the displacement efficiency is expressed as:

$$E_D = \frac{\text{Volume of oil at start of flood } - \text{ Remaining oil volume}}{\text{Volume of oil at start of flood}}$$

$$E_D = \frac{(\text{Pore volume})\left(\frac{S_{oi}}{B_{oi}}\right) - (\text{Pore volume})\left(\frac{\overline{S}_o}{B_o}\right)}{(\text{Pore volume})\left(\frac{S_{oi}}{B_{oi}}\right)}$$

or

$$E_D = \frac{\dfrac{S_{oi}}{B_{oi}} - \dfrac{\overline{S}_o}{B_o}}{\dfrac{S_{oi}}{B_{oi}}} \qquad (14\text{-}7)$$

where S_{oi} = initial oil saturation at start of flood
B_{oi} = oil FVF at start of flood, bbl/STB
$\overline{S}_o$ = average oil saturation in the flood pattern at a particular point during the flood

Assuming a constant oil formation volume factor during the flood life, Equation 14-7 is reduced to:

$$E_D = \frac{S_{oi} - \overline{S}_o}{S_{oi}} \qquad (14\text{-}8)$$

where the initial oil saturation S_{oi} is given by:

$$S_{oi} = 1 - S_{wi} - S_{gi}$$

However, in the swept area, the gas saturation is considered zero, thus:

$$\overline{S}_o = 1 - \overline{S}_w$$

The displacement efficiency E_D can be expressed more conveniently in terms of water saturation by substituting the above relationships into Equation 14-8, to give:

$$E_D = \frac{\overline{S}_w - S_{wi} - S_{gi}}{1 - S_{wi} - S_{gi}} \qquad (14\text{-}9)$$

where $\overline{S}_w$ = average water saturation in the swept area
S_{gi} = initial gas saturation at the start of the flood
S_{wi} = initial water saturation at the start of the flood

If no initial gas is present at the start of the flood, Equation 14-9 is reduced to:

$$E_D = \frac{\overline{S}_w - S_{wi}}{1 - S_{wi}} \qquad (14\text{-}10)$$

The displacement efficiency E_D will continually increase at different stages of the flood, i.e., with increasing $\overline{S}_w$. Equation 14-8 or 14-10 suggests that E_D reaches its maximum when the average oil saturation in the area of the flood pattern is reduced to the residual oil saturation S_{or} or, equivalently, when $\overline{S}_w = 1 - S_{or}$.

Example 14-4

A saturated oil reservoir is under consideration to be waterflooded immediately after drilling and completion. Core analysis tests indicate that the initial and residual oil saturations are 70 and 35%, respectively.

Calculate the displacement efficiency when the oil saturation is reduced to 65, 60, 55, 50, and 35%. Assume that B_o will remain constant throughout the project life.

Solution

Step 1. Calculate initial water saturation:

$$S_{wi} = 1 - 0.7 = 0.3$$

Step 2. Calculate E_D from Equation 14-10:

$$E_D = \frac{\overline{S}_w - S_{wi}}{1 - S_{wi}}$$

$\overline{S}_o$	$\overline{S}_w = 1 - \overline{S}_o$	$E_D = \dfrac{\overline{S}_w - S_{wi}}{1 - S_{wi}}$
0.65	0.35	0.071
0.60	0.40	0.142
0.55	0.45	0.214
0.50	0.50	0.286
$S_{or} = 0.35$	0.65	0.500 (maximum)

Example 14-4 shows that E_D will continually increase with increasing water saturation in the reservoir. The problem, of course, lies with developing an approach for determining the increase in the average water saturation in the swept area as a function of cumulative water injected (or injection time). Buckley and Leverett (1942) developed a well-established theory, called the *frontal displacement theory,* which provides the basis for establishing such a relationship. This classic theory consists of two equations:

- Fractional flow equation
- Frontal advance equation

The frontal displacement theory and its main two components are discussed next.

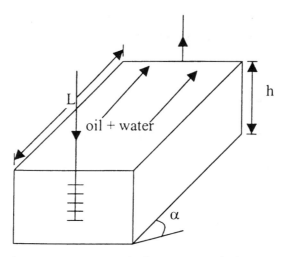

Figure 14-11. Linear displacement in a tilted system.

Fractional Flow Equation

The development of the fractional flow equation is attributed to Leverett (1941). For two immiscible fluids, oil and water, the fractional flow of water, f_w (or any immiscible displacing fluid), is defined as the water flow rate divided by the total flow rate, or:

$$f_w = \frac{q_w}{q_t} = \frac{q_w}{q_w + q_o} \qquad (14\text{-}11)$$

where f_w = fraction of water in the flowing stream, i.e., water cut, bbl/bbl

           q_t = total flow rate, bbl/day

           q_w = water flow rate, bbl/day

           q_o = oil flow rate, bbl/day

Consider the steady-state flow of two immiscible fluids (oil and water) through a tilted-linear porous media as shown in Figure 14-11. Assuming a homogeneous system, Darcy's equation can be applied for each of the fluids:

$$q_o = \frac{-k_o A}{\mu_o} \left[\frac{\partial P_o}{\partial x} + g\rho_o \sin(\alpha) \right] \qquad (14\text{-}12)$$

$$q_w = \frac{-k_w A}{\mu_w} \left[\frac{\partial P_w}{\partial x} + g\rho_w \sin(\alpha) \right] \qquad (14\text{-}13)$$

where subscripts o, w = oil and water
k_o, k_w = effective permeability
μ_o, μ_w = viscosity
p_o, p_w = pressure
ρ_o, ρ_w = density
A = cross-sectional area
x = distance
α = dip angle
$\sin(\alpha)$ = **positive** for updip flow and
negative for downdip flow

Rearranging Equations 14-12 and 14-13 gives:

$$\frac{q_o \mu_o}{A k_o} = -\frac{\partial p_o}{\partial x} - g\rho_o \sin(\alpha)$$

$$\frac{q_w \mu_w}{A k_w} = -\frac{\partial p_w}{\partial x} - g\rho_w \sin(\alpha)$$

Subtracting the above two equations yields:

$$\frac{q_w \mu_w}{A k_w} - \frac{q_o \mu_{ow}}{A k_o} = \left(\frac{\partial p_o}{\partial x} - \frac{\partial p_w}{\partial x} \right) - g(\rho_w - \rho_o)\sin\alpha \qquad (14\text{-}14)$$

From the definition of the capillary pressure p_c:

$$P_c = p_o - p_w$$

Differentiating the above expression with respect to the distance x gives:

$$\frac{\partial p_c}{\partial x} = \frac{\partial p_o}{\partial x} - \frac{\partial p_w}{\partial x} \qquad (14\text{-}15)$$

Combining Equation 14-15 with 14-16 gives:

$$\frac{q_w\mu_w}{Ak_w} - \frac{q_o\mu_o}{Ak_o} = \frac{\partial p_c}{\partial x} - g\Delta\rho\,\sin(\alpha) \tag{14-16}$$

where $\Delta\rho = \rho_w - \rho_o$. From the water cut equation, i.e., Equation 14-11:

$$q_w = f_w q_t \text{ and } q_o = (1 - f_w)q_t \tag{14-17}$$

Replacing q_o and q_w in Equation 14-16 with those of Equation 14-17 gives:

$$f_w = \frac{1 + \left(\dfrac{k_o A}{\mu_o q_t}\right)\left[\dfrac{\partial p_c}{\partial x} - g\Delta\rho\,\sin(\alpha)\right]}{1 + \dfrac{k_o}{k_w}\dfrac{\mu_w}{\mu_o}}$$

In field units, the above equation can be expressed as:

$$f_w = \frac{1 + \left(\dfrac{0.001127 k_o A}{\mu_o q_t}\right)\left[\dfrac{\partial p_c}{\partial x} - 0.433\Delta\rho\,sin(\alpha)\right]}{1 + \dfrac{k_o}{k_w}\dfrac{\mu_w}{\mu_o}} \tag{14-18}$$

where f_w = fraction of water (water cut), bbl/bbl
           k_o = effective permeability of oil, md
           k_w = effective permeability of water, md
           $\Delta\rho$ = water–oil density differences, g/cm^3
           k_w = effective permeability of water, md
           q_t = total flow rate, bbl/day
           μ_o = oil viscosity, cp
           μ_w = water viscosity, cp
           A = cross-sectional area, ft^2

Noting that the relative permeability ratios $k_{ro}/k_{rw} = k_o/k_w$ and, for two-phase flow, the total flow rate q_t are essentially equal to the water injection rate, i.e., $i_w = q_t$, Equation 14-18 can be expressed more conveniently in terms of k_{ro}/k_{rw} and i_w as:

$$f_w = \frac{1 + \left(\dfrac{0.001127(kk_{ro})A}{\mu_o i_w}\right)\left[\dfrac{\partial p_c}{\partial x} - 0.433\Delta\rho\sin(\alpha)\right]}{1 + \dfrac{k_{ro}}{k_{rw}}\dfrac{\mu_w}{\mu_o}} \qquad (14\text{-}19)$$

where i_w = water injection rate, bbl/day
        f_w = water cut, bbl/bbl
        k_{ro} = relative permeability to oil
        k_{rw} = relative permeability to water
        k = absolute permeability, md

The fractional flow equation as expressed by the above relationship suggests that for a given rock–fluid system, all the terms in the equation are defined by the characteristics of the reservoir, except:

- water injection rate, i_w
- water viscosity, μ_w
- direction of the flow, i.e., updip or downdip injection

Equation 14-19 can be expressed in a more generalized form to describe the fractional flow of any displacement fluid as:

$$f_D = \frac{1 + \left(\dfrac{0.001127(kk_{rD})A}{\mu_o i_D}\right)\left[\dfrac{\partial p_c}{\partial x} - 0.433\Delta\rho\sin(\alpha)\right]}{1 + \dfrac{k_{ro}}{k_{rD}}\dfrac{\mu_D}{\mu_o}} \qquad (14\text{-}20)$$

where the subscript D refers to the displacement fluid and $\Delta\rho$ is defined as:

$$\Delta\rho = \rho_D - \rho_o$$

For example, when the displacing fluid is immiscible gas, then:

$$f_g = \frac{1 - \left(\dfrac{0.001127(kk_{rg})A}{\mu_o i_g}\right)\left[\dfrac{\partial p_c}{\partial x} - 0.433(\rho_g - \rho_o)\sin(\alpha)\right]}{1 + \dfrac{k_{ro}}{k_{rg}}\dfrac{\mu_g}{\mu_o}} \qquad (14\text{-}21)$$

The effect of capillary pressure is usually neglected because the capillary pressure gradient is generally small and, thus, Equations 14-19 and 14-21 are reduced to:

$$f_w = \frac{1 - \left(\dfrac{0.001127(kk_{ro})A}{\mu_o i_w}\right)\left[0.433(\rho_w - \rho_o)\sin(\alpha)\right]}{1 + \dfrac{k_{ro}}{k_{rw}}\dfrac{\mu_w}{\mu_o}} \qquad (14\text{-}22)$$

and

$$f_g = \frac{1 - \left(\dfrac{0.001127(kk_{ro})A}{\mu_o i_g}\right)\left[0.433(\rho_g - \rho_o)\sin(\alpha)\right]}{1 + \dfrac{k_{ro}}{k_{rg}}\dfrac{\mu_g}{\mu_o}}$$

where i_g = gas injection rate, bbl/day
           μ_g = gas viscosity, cp
           ρ_g = gas density, g/cm^3

From the definition of water cut, i.e., $f_w = q_w/(q_w + q_o)$, we can see that the limits of the water cut are 0 and 100%. At the irreducible (connate) water saturation, the water flow rate q_w is zero and, therefore, the water cut is 0%. At the residual oil saturation point, S_{or}, the oil flow rate is zero and the water cut reaches its upper limit of 100%. The shape of the water cut versus water saturation curve is characteristically S-shaped, as shown in Figure 14-12. The limits of the f_w curve (0 and 1) are defined by the end points of the relative permeability curves.

The implications of the above discussion are also applied to defining the relationship that exists between fg and gas saturation, as shown in Figure 14-12.

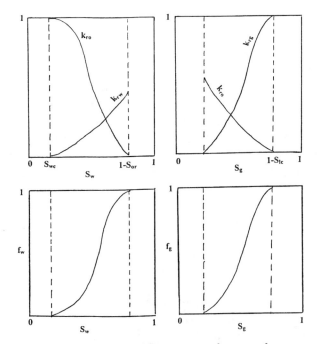

Figure 14-12. Fractional flow curves a function of saturations.

Note that, in general, any influences that cause the fractional flow curve to shift upward (i.e., increase in f_w or f_g) will result in a *less efficient displacement* process. It is essential, therefore, to determine the effect of various component parts of the fractional flow equation on the displacement efficiency. Note that for any two immiscible fluids, e.g., water and oil, the fraction of the oil (oil cut) f_o flowing at any point in the reservoir is given by:

$$f_o + f_w = 1 \quad \text{or} \quad f_o = 1 - f_w$$

The above expression indicates that during the displacement of oil by waterflood, an **increase in f_w** at any point in the reservoir will cause a proportional **decrease in f_o and oil mobility.** Therefore, the objective is to select the proper injection scheme that could possibly reduce the water fractional flow. This can be achieved by investigating the effect of the injected water viscosity, formation dip angle, and water-injection rate on the water cut. The overall effect of these parameters on the water fractional flow curve are discussed next.

Effect of Water and Oil Viscosities

Figure 14-13 shows the general effect of oil viscosity on the fractional flow curve for both water-wet and oil-wet rock systems. This illustration reveals that regardless of the system wettability, a higher oil viscosity results in an upward shift (an increase) in the fractional flow curve. The apparent effect of the water viscosity on the water fractional flow is clearly indicated by examining Equation 14-22. Higher injected water viscosities will result in an increase in the value of the denominator of Equation 14-22 with an overall reduction in f_w (i.e., a downward shift).

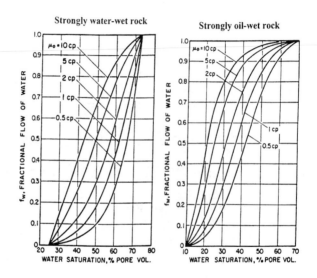

Figure 14-13. Effect of oil viscosity on f_w.

Effect of Dip Angle and Injection Rate

To study the effect of the formation dip angle α and the injection rate on the displacement efficiency, consider the water fractional flow equation as represented by Equation 14-22. Assuming a constant injection rate and realizing that $(\rho_w - \rho_o)$ is always positive and in order to isolate the effect of the dip angle and injection rate on f_w, Equation 14-22 is expressed in the following simplified form:

$$f_w = \frac{1 - \left[X \dfrac{\sin(\alpha)}{i_w} \right]}{1 + Y} \qquad (14\text{-}23)$$

where the variables X and Y are a collection of different terms that are all considered positives and given by:

$$X = \frac{(0.001127)(0.433)(k\,k_{ro})A(\rho_w - \rho_o)}{\mu_o}$$

$$Y = \frac{k_{ro}}{k_{rw}}\frac{\mu_w}{\mu_o}$$

- Updip flow, i.e., $\sin(\alpha)$ is positive. Figure 14-14 shows that when the water displaces oil *updip* (i.e., injection well is located downdip), a more efficient performance is obtained. This improvement is due to the fact that the term $X \sin(\alpha)/i_w$ will always remain positive, which leads to a decrease (downward shift) in the f_w curve. Equation 14-23 also reveals that a *lower water-injection rate* i_w is desirable since the nominator $1 - [X \sin(\alpha)/i_w]$ of Equation 14-23 will decrease with a lower injection rate i_w, resulting in an overall downward shift in the f_w curve.
- Downdip flow, i.e., $\sin(\alpha)$ is negative. When the oil is displaced *downdip* (i.e., injection well is located updip), the term $X \sin(\alpha)/i_w$ will always remain negative and, therefore, the numerator of Equation 14-23 will be $1+[X \sin(\alpha)/i_w]$, i.e.:

$$f_w = \frac{1 + \left[X \dfrac{\sin(\alpha)}{i_w} \right]}{1 + Y}$$

which causes an increase (upward shift) in the f_w curve. It is beneficial, therefore, when injection wells are located at the top of the structure to inject the water at a higher injection rate to improve the displacement efficiency.

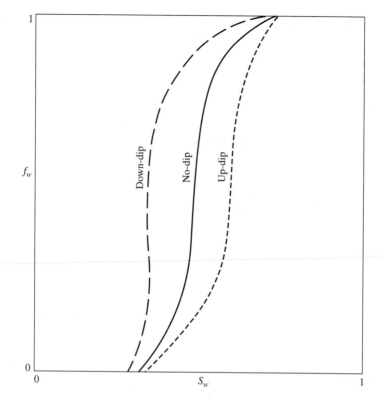

Figure 14-14. Effect of dip angle on f_w.

It is interesting to reexamine Equation 14-23 when displacing the oil *downdip*. Combining the product $X \sin(\alpha)$ as C, Equation 14-23 can be written:

$$f_w = \frac{1 + \left(\dfrac{C}{i_w} \right)}{1 + Y}$$

The above expression shows that the possibility exists that the water cut f_w could reach a value greater than unity ($f_w > 1$) if:

$$\frac{C}{i_w} > Y$$

This could only occur when displacing the oil downdip at a low water-injection rate i_w. The resulting effect of this possibility is called a

counterflow, where the oil phase is moving in a direction opposite to that of the water (i.e., oil is moving upward and the water downward). When the water injection wells are located at the top of a tilted formation, the injection rate must be high to avoid oil migration to the top of the formation.

Note that for a horizontal reservoir, i.e., $\sin(\alpha) = 0$, the injection rate has no effect on the fractional flow curve. When the dip angle α is zero, Equation 14-22 is reduced to the following simplified form:

$$f_w = \frac{1}{1 + \left(\dfrac{k_{ro}}{k_{rw}} \dfrac{\mu_w}{\mu_o} \right)} \qquad (14\text{-}24)$$

In waterflooding calculations, the reservoir water cut f_w and the water–oil ratio WOR are both traditionally expressed in two different units: bb/bbl and STB/STB. The interrelationships that exist between these two parameters are conveniently presented below, where

$$
\begin{aligned}
Q_o &= \text{oil flow rate, STB/day} \\
q_o &= \text{oil flow rate, bbl/day} \\
Q_w &= \text{water flow rate, STB/day} \\
q_w &= \text{water flow rate, bbl/day} \\
WOR_s &= \text{surface water–oil ratio, STB/STB} \\
WOR_r &= \text{reservoir water–oil ratio, bbl/bbl} \\
f_{ws} &= \text{surface water cut, STB/STB} \\
f_w &= \text{reservoir water cut, bbl/bbl}
\end{aligned}
$$

i) Reservoir f_w – Reservoir WOR_r Relationship

$$f_w = \frac{q_w}{q_w + q_o} = \frac{\left(\dfrac{q_w}{q_o} \right)}{\left(\dfrac{q_w}{q_o} \right) + 1}$$

Substituting for WOR gives:

$$f_w = \frac{WOR_r}{WOR_r + 1} \qquad \left(14\text{-}25\right)$$

Solving for WOR_r gives:

$$WOR_r = \frac{1}{\dfrac{1}{f_w} - 1} = \frac{f_w}{1 - f_w} \qquad (14\text{-}26)$$

ii) Reservoir f_w – Surface WOR_s Relationship

By definition:

$$f_w = \frac{q_w}{q_w + q_o} = \frac{Q_w B_w}{Q_w B_w + Q_o B_o} = \frac{\left(\dfrac{Q_w}{Q_o}\right) B_w}{\left(\dfrac{Q_w}{Q_o}\right) B_w + B_o}$$

Introducing the surface WOR_s into the above expression gives:

$$f_w = \frac{B_w WOR_s}{B_w WOR_s + B_o} \qquad (14\text{-}27)$$

Solving for WOR_s yields:

$$WOR_s = \frac{B_o}{B_w\left(\dfrac{1}{f_w} - 1\right)} = \frac{B_o f_w}{B_w(1 - f_w)} \qquad (14\text{-}28)$$

iii) Reservoir WOR_r – Surface WOR_s Relationship

From the definition of WOR:

$$WOR_r = \frac{q_w}{q_o} = \frac{Q_w B_w}{Q_o B_o} = \frac{\left(\dfrac{Q_w}{Q_o}\right) B_w}{B_o}$$

Introducing the surface WOR_s into the above expression gives:

$$WOR_r = (WOR)_s \left(\frac{B_w}{B_o}\right) \qquad (14\text{-}29)$$

or

$$WOR_s = (WOR)_r \left(\frac{B_o}{B_w} \right)$$

iv) Surface f_{ws} – Surface WOR_s Relationship

$$f_{ws} = \frac{Q_w}{Q_w + Q_o} = \frac{\left(\dfrac{Q_w}{Q_o} \right)}{\left(\dfrac{Q_w}{Q_o} \right) + 1}$$

or

$$f_{ws} = \frac{WOR_s}{WOR_s + 1} \qquad\qquad (14\text{-}30)$$

v) Surface f_{ws} – Reservoir f_w Relationship

$$f_{ws} = \frac{B_o}{B_w \left(\dfrac{1}{f_w} - 1 \right) + B_o} \qquad\qquad (14\text{-}31)$$

Example 14-5

Use the relative permeability as shown in Figure 14-15 to plot the fractional flow curve for a linear reservoir system with the following properties:

Dip angle	$= 0$	Absolute permeability	$= 50$ md
B_o	$= 1.20$ bbl/STB	B_w	$= 1.05$ bbl/STB
ρ_o	$= 45$ lb/ft^3	ρ_w	$= 64.0$ lb/ft^3
μ_w	$= 0.5$ cp	Cross-sectional area A	$= 25,000$ ft^2

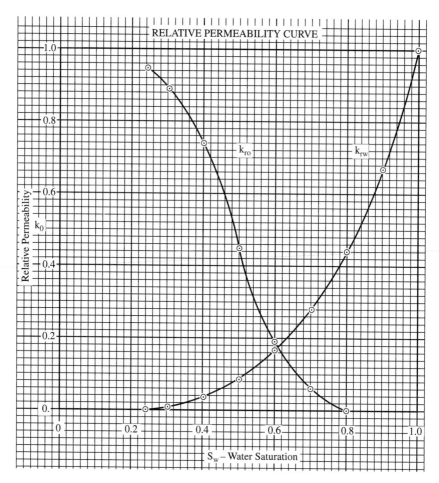

Figure 14-15. Relative permeability curves for Example 14-5.

Perform the calculations for the following values of oil viscosity: $\mu_0 = 0.5$, 1.0, 5, and 10 cp.

Solution

For a horizontal system, Equation 14-24 can be used to calculate f_w as a function of saturation.

S_w	k_{ro}	k_{rw}	k_{ro}/k_{rw}	$f_w = \dfrac{1}{1 + \dfrac{k_{ro}}{k_{rw}}\dfrac{\mu_w}{\mu_o}}$			
				$\mu_o=0.5$	$\mu_o=1.0$	$\mu_o=5$	$\mu_o=10$
0.24	0.95	0.00	00	0	0	0	0
0.30	0.89	0.01	89.0	0.011	0.022	0.101	0.183
0.40	0.74	0.04	18.5	0.051	0.098	0.351	0.519
0.50	0.45	0.09	5.0	0.17	0.286	0.667	0.800
0.60	0.19	0.17	1.12	0.47	0.641	0.899	0.947
0.65	0.12	0.28	0.43	0.70	0.823	0.459	0.979
0.70	0.06	0.22	0.27	0.79	0.881	0.974	0.987
0.75	0.03	0.36	0.08	0.93	0.962	0.992	0.996
0.78	0.00	0.41	0	1.00	1.000	1.000	1.000

Results of the above example are documented graphically in Figure 14-16, which shows the apparent effect of oil viscosity on the fractional flow curve.

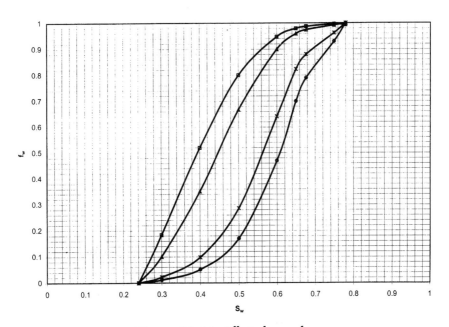

Figure 14-16. Effect of μ_o on f_w.

Example 14-6

The linear system in Example 14-5 is under consideration for a water-flooding project with a water injection rate of 1000 bbl/day. The oil viscosity is considered constant at 1.0 cp. Calculate the fractional flow curve for the reservoir dip angles of 10, 20, and 30°, assuming (a) updip displacement and (b) downdip displacement.

Solution

Step 1. Calculate the density difference $(\rho_w - \rho_o)$ in g/cm^3:

$$(\rho_w - \rho_o) = (64 - 45) / 62.4 = 0.304 \text{ g/cm}^3$$

Step 2. Simplify Equation 14-22 by using the given fixed data:

$$f_w = \frac{1 - \dfrac{0.001127(50\,k_{ro})(25,000)}{(1)(1000)}\left[0.433(0.304)\sin(\alpha)\right]}{1 + \left(\dfrac{0.5}{1}\right)\left(\dfrac{k_{ro}}{k_{rw}}\right)}$$

$$f_w = \frac{1 - 0.185\,k_{ro}\left[\sin(\alpha)\right]}{1 + 0.5\left(\dfrac{k_{ro}}{k_{rw}}\right)}$$

For updip displacement, **$\sin(\alpha)$ is positive,** therefore:

$$f_w = \frac{1 - 0.185\,k_{ro}\sin(\alpha)}{1 + 0.5\left(\dfrac{k_{ro}}{k_{rw}}\right)}$$

For downdip displacement, **$\sin(\alpha)$ is negative,** therefore:

$$f_w = \frac{1 + 0.185\,k_{ro}\sin(\alpha)}{1 + 0.5\left(\dfrac{k_{ro}}{k_{rw}}\right)}$$

Step 3. Perform the fractional flow calculations in the following tabulated form:

S_w	k_{ro}	k_{ro}/k_{rw}	f_w, Updip Displacement			f_w, Downdip Displacement		
			10°	20°	30°	10°	20°	30°
0.24	0.95	00	0	0	0	0	0	0
0.30	0.89	89	0.021	0.021	0.020	0.023	0.023	0.024
0.40	0.74	18.5	0.095	0.093	0.091	0.100	0.102	0.104
0.50	0.45	5.0	0.282	0.278	0.274	0.290	0.294	0.298
0.60	0.19	1.12	0.637	0.633	0.630	0.645	0.649	0.652
0.65	0.12	0.43	0.820	0.817	0.814	0.826	0.830	0.832
0.70	0.06	0.27	0.879	0.878	0.876	0.883	0.884	0.886
0.75	0.03	0.08	0.961	0.960	0.959	0.962	0.963	0.964
0.78	0.00	0	1.000	1.000	1.000	1.000	1.000	1.000

The fractional flow equation, as discussed in the previous section, is used to determine the water cut f_w at any point in the reservoir, assuming that the water saturation at the point is known. The question, however, is how to determine the water saturation at this particular point. The answer is to use the **frontal advance equation.** The frontal advance equation is designed to determine the water saturation profile in the reservoir at any give time during water injection.

Frontal Advance Equation

Buckley and Leverett (1942) presented what is recognized as the basic equation for describing two-phase, immiscible displacement in a linear system. The equation is derived based on developing a material balance for the displacing fluid as it flows through any given element in the porous media:

Volume entering the element − Volume leaving the element
= change in fluid volume

Consider a differential element of porous media, as shown in Figure 14-17, having a differential length dx, an area A, and a porosity ϕ. During a differential time period dt, the total volume of water entering the element is given by:

Volume of water entering the element $= q_t \, f_w \, d_t$

The volume of water leaving the element has a differentially smaller water cut $(f_w - df_w)$ and is given by:

Volume of water leaving the element $= q_t \, (f_w - df_w) \, dt$

Subtracting the above two expressions gives the accumulation of the water volume within the element in terms of the differential changes of the saturation df_w:

$$q_t\, f_w\, d_t - q_t\, (f_w - df_w)\, dt = A\phi\, (dx)\, (dS_w)/5.615$$

Simplifying:

$$q_t\, df_w\, dt = A\, \phi\, (dx)\, (dS_w)/5.615$$

Separating the variables gives:

$$\left(\frac{dx}{dt}\right)_{Sw} = (\upsilon)_{Sw} = \left(\frac{5.615 q_t}{\phi A}\right)\left(\frac{df_w}{dS_w}\right)_{Sw} \tag{14-32}$$

where $(\upsilon)_{Sw}$ = velocity of any specified value of S_w, ft/day
    A = cross-sectional area, ft^2
    q_t = total flow rate (oil + water), bbl/day
    $(df_w/dS_w)_{Sw}$ = slope of the f_w vs. S_w curve at S_w

The above relationship suggests that the velocity of any specific water saturation S_w is directly proportional to the value of the slope of the f_w vs. S_w curve, evaluated at S_w. Note that for two-phase flow, the total flow rate q_t is essentially equal to the injection rate i_w, or:

$$\left(\frac{dx}{dt}\right)_{Sw} = (\upsilon)_{Sw} = \left(\frac{5.615\ i_w}{\phi A}\right)\left(\frac{df_w}{dS_w}\right)_{Sw} \tag{14-33}$$

where i_w = water injection rate, bbl/day.

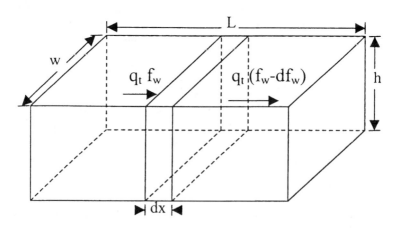

Figure 14-17. Water flow through a linear differential element.

To calculate the total distance any specified water saturation will travel during a total time t, Equation 14-33 must be integrated:

$$\int_0^X dx = \left(\frac{5.615 i_w}{\phi A} \right) \left(\frac{df_w}{dt} \right)_{Sw} \int_0^t dt$$

or

$$\left(x \right)_{Sw} = \left(\frac{5.615 i_w t}{\phi A} \right) \left(\frac{df_w}{dt} \right)_{Sw} \tag{14-34}$$

Equation 14-34 can also be expressed in terms of total volume of water injected by recognizing that under a constant water-injection rate, the cumulative water injected is given by:

$$W_{inj} = t\, i_w$$

or

$$\left(x \right)_{sw} = \frac{5.615 W_{inj}}{\phi A} \left(\frac{df_w}{dt} \right)_{Sw} \tag{14-35}$$

where　　i_w = water injection rate, bbl/day
　　　　W_{inj} = cumulative water injected, bbl
　　　　t = time, day
　　　　$(x)_{Sw}$ = distance from the injection for any given saturation S_w, ft

Equation 14-35 also suggests that the position of any value of water saturation S_w at given cumulative water injected W_{inj} is proportional to the slope (df_w/dS_w) for this particular S_w. At any given time t, the water saturation profile can be plotted by simply determining the slope of the f_w curve at each selected saturation and calculating the position of S_w from Equation 14-35.

Figure 14-18 shows the typical S shape of the f_w curve and its derivative curve. However, a mathematical difficulty arises when using the derivative curve to construct the water saturation profile at any given time. Suppose we want to calculate the positions of two different saturations (shown in Figure 14-18 as saturations A and B) after W_{inj} barrels of water have been injected in the reservoir. Applying Equation 14-35 gives:

$$\left(x\right)_A = \frac{5.615W_{inj}}{\phi A}\left(\frac{df_w}{dS_w}\right)_A$$

$$\left(x\right)_B = \frac{5.615W_{inj}}{\phi A}\left(\frac{df_w}{dS_w}\right)_B$$

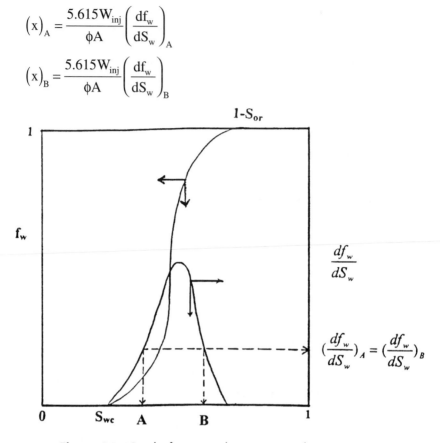

Figure 14-18. The f_w curve with its saturation derivative curve.

Figure 14-18 indicates that both derivatives are identical, i.e., $(df_w/dS_w)_A = (df_w/dS_w)_B$, which implies that multiple water saturations can coexist at the same position—but this is physically impossible. Buckley and Leverett (1942) recognized the physical impossibility of such a condition. They pointed out that this apparent problem is due to the neglect of the capillary pressure gradient term in the fractional flow equation. This capillary term is given by:

$$\text{Capillary term} = \left(\frac{0.001127k_oA}{\mu_o i_w}\right)\left(\frac{dP_c}{dx}\right)$$

Including the above capillary term when constructing the fractional flow curve would produce a graphical relationship that is characterized by the following two segments of lines, as shown in Figure 14-19:

- A straight line segment with a constant slope of $(df_w/dS_w)_{Swf}$ from S_{wc} to S_{wf}
- A concaving curve with decreasing slopes from S_{wf} to $(1 - S_{or})$

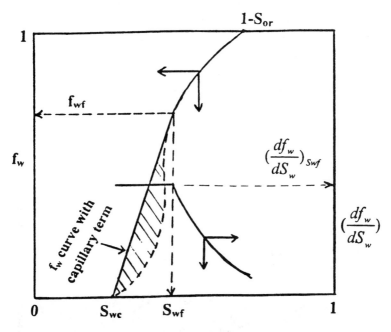

Figure 14-19. Effect of the capillary term on the f_w curve.

Terwilliger et al. (1951) found that at the lower range of water saturations between S_{wc} and S_{wf}, all saturations move at the same velocity as a function of time and distance. Notice that all saturations in that range have the same value for the slope and, therefore, the same velocity as given by Equation 14-33:

$$(\upsilon)_{Sw<Swf} = \left(\frac{5.615i_w}{\phi A}\right)\left(\frac{df_w}{dS_w}\right)_{Swf}$$

We can also conclude that all saturations in this particular range will travel the same distance x at any particular time, as given by Equation 14-34 or 14-35:

$$(x)_{Sw<Swf} = \left(\frac{5.615i_w t}{\phi A}\right)\left(\frac{df_w}{dS_w}\right)_{Swf}$$

The result is that the water saturation profile will maintain a constant shape over the range of saturations between S_{wc} and S_{wf} with time. Terwilliger and his coauthors termed the reservoir-flooded zone with this range of saturations the **stabilized zone.** They define the stabilized zone as that particular saturation interval (i.e., S_{wc} to S_{wf}) where all points of saturation travel at the same velocity. Figure 14-20 illustrates the concept of the stabilized zone. The authors also identified another saturation zone between S_{wf} and $(1 - S_{or})$, where the velocity of any water saturation is variable. They termed this zone the **nonstabilized zone.**

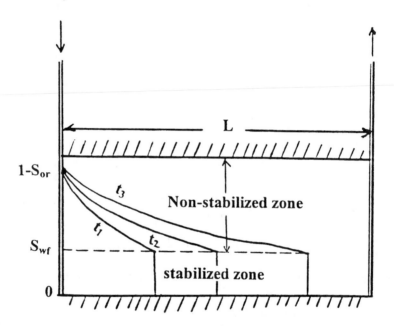

Figure 14-20. Water saturation profile as a function of distance and time.

Experimental core flood data show that the actual water saturation profile during water flooding is similar to that of Figure 14-20. There is a distinct front, or **shock front,** at which the water saturation abruptly increases from S_{wc} to S_{wf}. Behind the flood front there is a gradual increase in saturations from S_{wf} up to the maximum value of $1 - S_{or}$. Therefore, the saturation S_{wf} is called the water saturation at the front or, alternatively, the water saturation of the stabilized zone.

Welge (1952) showed that by drawing a straight line from S_{wc} (or from S_{wi} if it is different from S_{wc}) *tangent* to the fractional flow curve, the saturation value at the tangent point is equivalent to that at the front S_{wf}.

The coordinate of the point of tangency represents also the value of the water cut at the leading edge of the water front f_{wf}.

From the above discussion, the water saturation profile at any given time t_1 can be easily developed as follows:

Step 1. Ignoring the capillary pressure term, construct the fractional flow curve, i.e., f_w vs. S_w.

Step 2. Draw a straight line tangent from S_{wi} to the curve.

Step 3. Identify the point of tangency and read off the values of S_{wf} and f_{wf}.

Step 4. Calculate graphically the slope of the tangent as $(df_w/dS_w)_{Swf}$.

Step 5. Calculate the distance of the leading edge of the water front from the injection well by using Equation 14-34, or:

$$\left(x\right)_{Swf} = \left(\frac{5.615 i_w t_1}{\phi A}\right)\left(\frac{df_w}{dS_w}\right)_{Swf}$$

Step 6. Select several values for water saturation S_w *greater than* S_{wf} and determine $(df_w/dS_w)_{Sw}$ by graphically drawing a tangent to the f_w curve at each selected water saturation (as shown in Figure 14-21).

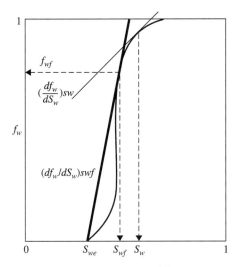

Figure 14-21. Fractional flow curve.

Step 7. Calculate the distance from the injection well to each selected saturation by applying Equation 14-36, or:

$$(x)_{Sw} = \left(\frac{5.615 i_w t_1}{\phi A} \right) \left(\frac{df_w}{dS_w} \right)_{S_w}$$

Step 8. Establish the water saturation profile after t_1 days by plotting results obtained in step 7.

Step 9. Select a new time t_2 and repeat steps 5 through 7 to generate a family of water saturation profiles as shown schematically in Figure 14-20.

Some erratic values of $(df_w/dS_w)_{S_w}$ might result when determining the slope graphically at different saturations. A better way is to determine the derivative mathematically by recognizing that the relative permeability ratio (k_{ro}/k_{rw}) can be expressed by Equation 5-29 of Chapter 5 as:

$$\frac{k_{ro}}{k_{rw}} = a \, e^{bS_w} \tag{14-36}$$

Notice that the slope b in the above expression has a negative value. The above expression can be substituted into Equation 14-26 to give:

$$f_w = \frac{1}{1 + \left(\dfrac{\mu_w}{\mu_o} \right) a \, e^{bS_w}} \tag{14-37}$$

The derivative of $(df_w/dS_w)_{S_w}$ may be obtained mathematically by differentiating the above equation with respect to S_w to give:

$$\left(\frac{df_w}{dS_w} \right)_{Sw} = \frac{-\left(\dfrac{\mu_w}{\mu_o} \right) a \, b \, e^{bS_w}}{\left[1 + \left(\dfrac{\mu_w}{\mu_o} \right) a \, e^{bS_w} \right]^2} \tag{14-38}$$

The data in the following example, as given by Craft and Hawkins (1959), are used to illustrate one of the practical applications of the frontal displacement theory.

Example 14-7

The following data are available for a linear-reservoir system:

S_w	0.25	0.30	0.35	0.40	0.45	0.50	0.55	0.60	0.65	0.70	0.75
k_{ro}/k_{rw}	30.23	17.00	9.56	5.38	3.02	1.70	0.96	0.54	0.30	0.17	0.10

Oil formation volume factor B_o	= 1.25 bbl/STB
Water formation volume factor B_w	= 1.02 bbl/STB
Formation thickness h	= 20 ft
Cross-sectional area A	= 26,400 ft
Porosity ϕ	= 25%
Injection rate i_w	= 900 bbl/day
Distance between producer and injector L	= 600 ft
Oil viscosity μ_o	= 2.0 cp
Water viscosity μ_w	= 1.0 cp
Dip angle α	= 0°
Connate water saturation S_{wc}	= 20%
Initial water saturation S_{wi}	= 20%
Residual water saturation S_{or}	= 20%

Calculate and plot the water saturation profile after 60, 120, and 240 days.

Solution

Step 1. Plot the relative permeability ratio k_{ro}/k_{rw} vs. water saturation on a semi-log paper and determine the coefficients a and b of Equation 14-36, as shown in Figure 14-22, to give:

$a = 537.59$ and $b = -11.51$

Therefore,

$$\frac{k_{ro}}{k_{rw}} = 537.59e^{-11.51S_w}$$

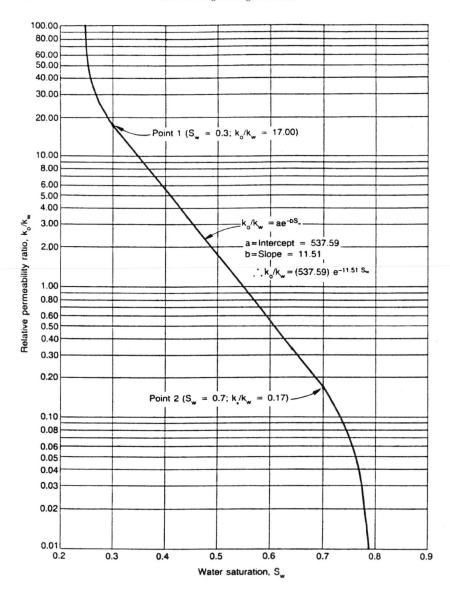

Figure 14-22. Relative permeability ratio.

Step 2. Assume several values of water saturation and calculate the fractional flow curve at its derivatives by applying Equations 14-37 and 14-38:

S_w	k_{ro}/k_{rw}	f_w, Equation 14-37	(df_w/dS_w), Equation 14-38
0.25	30.23	0.062	0.670
0.30	17.00	0.105	1.084
0.35	9.56	0.173	1.647
0.40	5.38	0.271	2.275
0.45	3.02	0.398	2.759
0.50	1.70	0.541	2.859
0.55	0.96	0.677	2.519
0.60	0.54	0.788	1.922
0.65	0.30	0.869	1.313
0.70	0.17	0.922	0.831
0.75	0.10	0.956	0.501

Step 3. Plot f_w and (df_w/S_w) vs. S_w on a Cartesian scale as shown in Figure 14-23. Draw a straight line from S_{wc} and tangent to the f_w curve. Determine the coordinates of point of tangency and the slope of the tangent $(df_w/dS_w)_{Swf}$, to give:

$$\left(S_{wf}, f_{wf}\right) = (0.596, 0.48) \quad \text{and} \quad \left(\frac{df_w}{dS_w}\right)_{S_{wf}} = 1.973$$

This means that the leading edge of the waterfront (stabilized zone) has a constant saturation of 0.596 and water cut of 78%.

Step 4. When constructing the water saturation profile, it should be noted that **no water saturation with a value less than S_{wf}, i.e., 59.6%, exists behind the leading edge of the water bank.** Assume water saturation values in the range of S_{wf} to $(1 - S_{or})$, i.e., 59.6 to 75%, and calculate the water saturation profile as a function of time by using Equation 14-36:

$$\left(x\right)_{S_w} = \left(\frac{5.615 i_w t}{\phi A}\right)\left(\frac{df_w}{dS_w}\right)_{S_w}$$

$$\left(x\right)_{S_w} = \left(\frac{(5.615)(900)t}{(0.25)(26,400)}\right)\left(\frac{df_w}{dS_w}\right)_{S_w}$$

$$\left(x\right)_{S_w} = (0.77t)\left(\frac{df_w}{dS_w}\right)_{S_w}$$

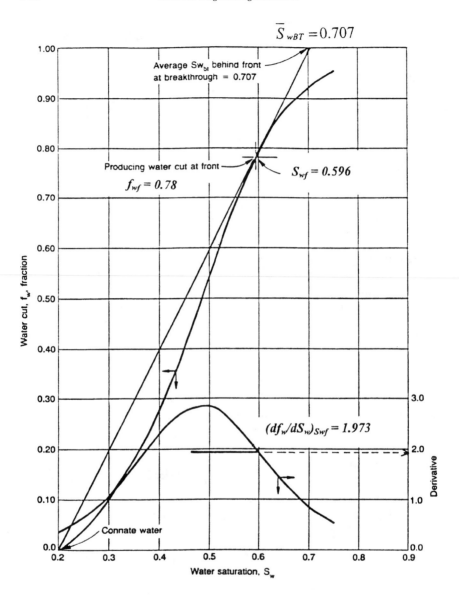

Figure 14-23. Water cut curve and its derivative.

Assumed S_w	(df_w/dS_w)	$x = 0.77t(df/dS_w)$ $t = 60$ days	$x = 0.77t(df/dS_w)$ $t = 120$ days	$x = 0.77t(df/dS_w)$ $t = 240$days
0.596	1.973	91	182	365
0.60	1.922	88	177	353
0.65	1.313	60	121	241
0.70	0.831	38	76	153
0.75	0.501	23	46	92

Step 5. Plot the water saturation profile as a function of distance and time, as shown in Figure 14-24.

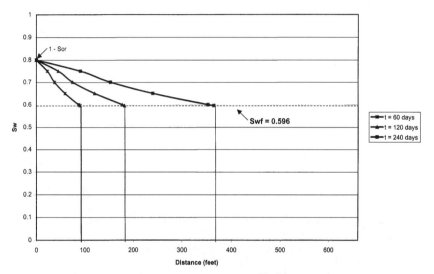

Figure 14-24. Water saturation profile for Example 14-7.

The above example shows that after 240 days of water injection, the leading edge of the water front has moved 365 feet from the injection well (235 feet from the producer). The water front (leading edge) will eventually reach the production well and water breakthrough occurs.

The example also indicates that at water breakthrough, the leading edge of the water front would have traveled exactly the entire distance between the two wells, i.e., 600 feet. Therefore, to determine the time to breakthrough, t_{BT}, simply set $(x)S_{wf}$ equal to the distance between the injector and producer L in Equation 14-34 and solve for the time:

$$L = \left(\frac{5.615 i_w t_{BT}}{\phi A} \right) \left(\frac{df_w}{dS_w} \right)_{S_{wf}}$$

Note that the pore volume (PV) is given by:

$$(PV) = \frac{\phi A L}{5.615}$$

Combining the above two expressions and solving for the time to breakthrough t_{BT} gives:

$$t_{BT} = \left[\frac{(PV)}{i_w} \right] \frac{1}{\left(\frac{df_w}{dS_w} \right)_{S_{wf}}} \tag{14-39}$$

where t_{BT} = time to breakthrough, day
            PV = total flood pattern pore volume, bbl
            L = distance between the injector and producer, ft

Assuming a constant water-injection rate, the cumulative water injected at breakthrough is calculated from Equation 14-39 as:

$$W_{iBT} = i_w t_{BT} = \frac{(PV)}{\left(\frac{df_w}{dS_w} \right)_{S_{wf}}} \tag{14-40}$$

where W_{iBT} = cumulative water injected at breakthrough, bbl

$$\left(\frac{\phi A L}{5.615} \right) = \text{total flood pattern pore volume, bbl}$$

It is convenient to express the cumulative water injected in terms of pore volumes injected, i.e., dividing W_{inj} by the reservoir total pore volume. Conventionally, Q_i refers to the total pore volumes of water injected. From Equation 14-40, Q_i at breakthrough is:

$$Q_{iBT} = \frac{W_{iBT}}{(PV)} = \frac{1}{\left(\frac{df_w}{dS_w} \right)_{S_{wf}}} \tag{14-41}$$

where Q_{iBT} = cumulative pore volumes of water injected at breakthrough
           PV = total flood pattern pore volume, bbl

Example 14-8

Using the data given in Example 14-7, calculate:

- Time to breakthrough
- Cumulative water injected at breakthrough
- Total pore volumes of water injected at breakthrough

Solution

Step 1. Calculate the reservoir pore volume:

$$(PV) = \frac{(0.25)(26,400)(660)}{5.615} = 775,779 \text{ bbl}$$

Step 2. Calculate the time to breakthrough from Equation 14-39:

$$t_{BT} = \frac{(PV)}{i_w} \frac{1}{\left(\dfrac{df_w}{dS_w} \right)_{S_{wf}}}$$

$$t_{BT} = \left(\frac{775,779}{900} \right) \left(\frac{1}{1.973} \right) = 436.88 \text{ days}$$

Step 3. Determine cumulative water injected at breakthrough:

$$W_{iBT} = i_w t_{BT}$$
$$W_{iBT} = (900)(436.88) = 393,198 \text{ bbl}$$

Step 4. Calculate total pore volumes of water injected at breakthrough:

$$Q_{iBT} = \frac{1}{\left(\dfrac{df_w}{dS_w} \right)_{S_{wf}}}$$

$$Q_{iBT} = \frac{1}{1.973} = 0.507 \text{ pore volumes}$$

A further discussion of Equation 14-40 is needed to better understand the significance of the Buckley and Leverett (1942) frontal advance theory. Equation 14-40, which represents cumulative water injected at breakthrough, is given by:

$$W_{iBT} = (PV)\dfrac{1}{\left(\dfrac{df_w}{dS_w}\right)_{S_{wf}}} = (PV)Q_{iBT}$$

If the tangent to the fractional flow curve is extrapolated to $f_w = 1$ with a corresponding water saturation of S_w^* (as shown in Figure 14-25), then the slope of the tangent can be calculated numerically as:

$$\left(\dfrac{df_w}{dS_w}\right)_{S_{wf}} = \dfrac{1-0}{S_w^* - S_{wi}}$$

Combining the above two expressions gives:

$$W_{iBT} = (PV)(S_w^* - S_{wi}) = (PV)Q_{iBT}$$

The above equation suggests that the water saturation value denoted as S_w^* must be the average water saturation at breakthrough, or:

$$W_{iBT}(PV)(\overline{S}_{wBT} - S_{wi}) = (PV)Q_{iBT} \qquad (14\text{-}42)$$

where $\overline{S}_{wBT}$ = average water saturation in the reservoir at breakthrough

PV = flood pattern pore volume, bbl

W_{iBT} = cumulative water injected at breakthrough, bbl

S_{wi} = initial water saturation

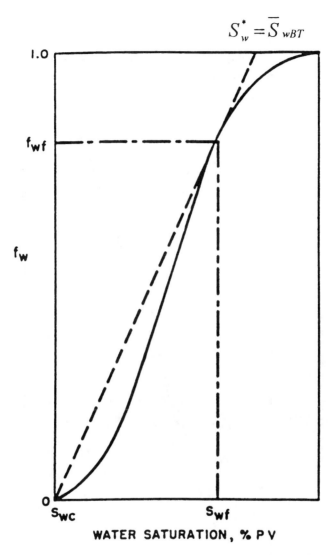

$$S_{w}^{*} = \overline{S}_{wBT}$$

Figure 14-25. Average water saturation at breakthrough.

Two important points **must be** considered when determining $\overline{S}_{wBT}$:

1. When drawing the tangent, the line must be originated from the initial water saturation S_{wi} if it is different from the connate water saturation S_{wc}, as shown in Figure 14-26.

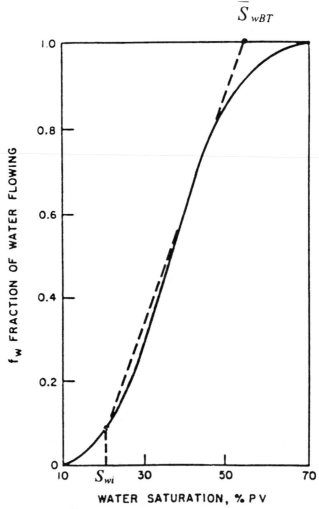

Figure 14-26. Tangent from S_{wi}.

2. When considering the areal sweep efficiency E_A and vertical sweep efficiency E_V, Equation 14-42 should be expressed as:

$$W_{iBT} = (PV)(\overline{S}_{wBT} - S_{wi})E_{ABT}E_{VBT} \qquad (14\text{-}43)$$

or equivalently as:

$$W_{iBT} = (PV)Q_{iBT}E_{ABT}E_{VBT} \qquad (14\text{-}44)$$

where E_{ABT} and E_{VBT} are the areal and vertical sweep efficiencies at breakthrough (as discussed later in the chapter). Note that the average water saturation in the swept area would remain constant with a value of $\overline{S}_{wBT}$ until breakthrough occurs, as illustrated in Figure 14-27. At the time of breakthrough, the flood front saturation S_{wf} reaches the producing well and the water cut increases suddenly from zero to f_{wf}. At breakthrough, S_{wf} and f_{wf} are designated $\overline{S}_{wBT}$ and f_{wBT}.

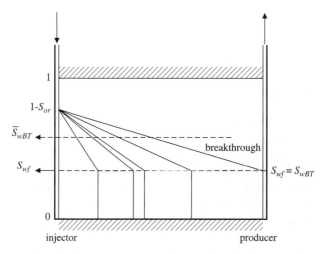

Figure 14-27. Average water saturation before breakthrough.

After breakthrough, the water saturation and the water cut at the producing well gradually increase with continuous injection of water, as shown in Figure 14-28. Traditionally, the produced well is designated as well 2 and, therefore, the water saturation and water cut at the producing well are denoted as S_{w2} and f_{w2}, respectively.

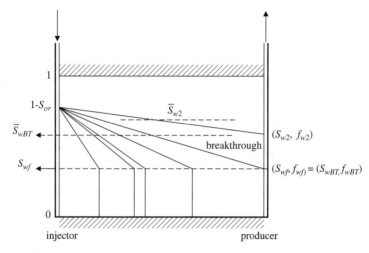

Figure 14-28. Average water saturation after breakthrough.

Welge (1952) illustrated that when the water saturation at the producing well reaches any assumed value S_{w2} after breakthrough, the fractional flow curve can be used to determine:

- Producing water cut f_{w2}
- Average water saturation in the reservoir $\overline{S}_{w2}$
- Cumulative water injected in pore volumes, i.e., Q_i

As shown in Figure 14-29, the author pointed out that drawing a tangent to the fractional flow curve at **any** assumed value of S_{w2} greater than S_{wf} has the following properties:

1. The value of the fractional flow at the point of tangency corresponds to the well producing water cut f_{w2}, as expressed in bbl/bbl.
2. The saturation at which the tangent intersects $f_w = 1$ is the average water saturation $\overline{S}_{w2}$ in the swept area. Mathematically, the average water saturation is determined from:

$$\overline{S}_{w2} = S_{w2} + \frac{1 - f_{w2}}{\left(\dfrac{df_w}{dS_w} \right)_{S_{w2}}} \tag{14-45}$$

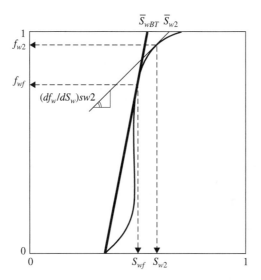

Figure 14-29. Determination of average water saturation after breakthrough.

3. The reciprocal of the slope of the tangent is defined as the cumulative pore volumes of water injected Q_i at the time when the water saturation reaches S_{w2} at the producing well, or:

$$Q_i = \frac{1}{\left(\dfrac{df_w}{dS_w}\right)_{S_{w2}}} \tag{14-46}$$

4. The cumulative water injected when the water saturation at the producing well reaches S_{w2} is given by:

$$W_{inj} = (PV)Q_i E_A E_V \tag{14-47}$$

or equivalently as:

$$W_{inj} = (PV)(\overline{S}_{w2} - S_{wi})E_A E_V \tag{14-48}$$

where: W_{inj} = cumulative water injected, bbl
   (PV) = pattern pore volume, bbl
   E_A = areal sweep efficiency
   E_V = vertical sweep efficiency

5. For a constant injection rate i_w, the total time t to inject W_{inj} barrels of water is given by:

$$t = \frac{W_{inj}}{i_w} \qquad (14\text{-}49)$$

Example 14-9

Using the data given in Example 14-7 for the linear reservoir system, calculate the following when the water saturation at the producing well reaches 0.70 (i.e., $S_{w2} = 0.7$):

a. reservoir water cut in bbl/bbl
b. surface water cut in STB/STB
c. reservoir water–oil ratio in bbl/bbl
d. surface water–oil ratio in STB/STB
e. average water saturation in the swept area
f. pore volumes of water injected
g. cumulative water injected in bbl

Assume that the areal and vertical sweep efficiency are 100%, i.e., $E_A = 1.0$ and $E_V = 1.0$.

Solution

a. Results of Example 14-7 indicate that at a water saturation value of 70%, the corresponding water cut f_w is 0.922, therefore:

$f_{w2} = 0.922$ bbl/bbl

b. Calculate the surface water cut by applying Equation 14-31:

$$f_{ws} = \frac{B_o}{B_w \left(\dfrac{1}{f_w} - 1 \right) + B_o}$$

$$f_{ws} = \frac{1.25}{1.02 \left(\dfrac{1}{0.922} - 1 \right) + 1.25} = 0.935 \ \text{STB/STB}$$

c. Determine the producing water–oil ratio by using Equation 14-26:

$$WOR_r = \frac{f_w}{1-f_w}$$

$$WOR_r = \frac{0.922}{1-0.922} = 11.82 \ \text{bbl/bbl}$$

d. Apply Equation 14-29 to determine the surface water–oil ratio:

$$WOR_S = \frac{B_o WOR_r}{B_w}$$

$$WOR_S = \frac{(1.25)(11.82)}{1.02} = 14.49 \ \text{STB/STB}$$

e. Draw a tangent to the fractional flow curve at the coordinate of the point $(S_w, f_w) = (0.7, 0.922)$ and extrapolate to $f_w = 1.0$ to give a corresponding $\bar{S}_{w2} = 0.794$. Equivalently, the average water saturation can be calculated by determining the slope of the tangent and applying Equation 14-45, to give:

$$\left(\frac{df_w}{dS_w} \right)_{S_{w2}} = 0.831$$

$$\bar{S}_{W2} = 0.70 + \frac{1-0.922}{0.831} = 0.794$$

f. From Equation 14-46, the cumulative pore volume of water injected is the reciprocal of the slope of the tangent line:

$Q_i = 1/0.831 = 1.203$

g. Calculate cumulative water injected by applying Equation 14-47:

$$W_{inj} = (PV)(Q_i)E_A E_V$$
$$W_{inj} = (775,779)(1.203)(1)(1) = 933,262 \ \text{bbl}$$

Oil Recovery Calculations

The main objective of performing oil recovery calculations is to generate a set of performance curves under a specific water-injection scenario.

A set of performance curves is defined as the graphical presentation of the time-related oil recovery calculations in terms of:

- Oil production rate, Q_o
- Water production rate, Q_w
- Surface water–oil ratio, WOR_s
- Cumulative oil production, N_p
- Recovery factor, RF
- Cumulative water production, W_p
- Cumulative water injected, W_{inj}
- Water-injection pressure, p_{inj} (discussed later in the chapter)
- Water-injection rate, i_w (discussed later in the chapter)

In general, oil recovery calculations are divided into two parts: (1) before breakthrough calculations and (2) after breakthrough calculations. Regardless of the stage of the waterflood, i.e., before or after breakthrough, the cumulative oil production is given previously by Equation 14-6 as:

$$N_p = N_S E_D E_A E_V$$

where N_p = cumulative oil production, STB
             N_S = initial oil in place at start of the flood, STB
             E_D = displacement efficiency
             E_A = areal sweep efficiency
             E_V = vertical sweep efficiency

As defined by Equation 14-10 when $S_{gi} = 0$, the displacement efficiency is given by:

$$E_D = \frac{\overline{S}_w - S_{wi}}{1 - S_{wi}}$$

At breakthrough, the E_D can be calculated by determining the average water saturation at breakthrough:

$$E_{DBT} = \frac{\overline{S}_{wBT} - S_{wi}}{1 - S_{wi}} \qquad (14\text{-}50)$$

where E_{DBT} = displacement efficiency at breakthrough
$\overline{S}_{wBT}$ = average water saturation at breakthrough

The cumulative oil production at breakthrough is then given by:

$$\left(N_p\right)_{BT} = N_S E_{DBT} E_{ABT} E_{VBT} \qquad (14\text{-}51)$$

where $(N_p)_{BT}$ = cumulative oil production at breakthrough, STB
E_{ABT}, E_{VBT} = areal and vertical sweep efficiencies at breakthrough

Assuming E_A and E_V are 100%, Equation 14-51 is reduced to:

$$\left(N_p\right)_{BT} = N_S E_{DBT} \qquad (14\text{-}52)$$

Before breakthrough occurs, the oil recovery calculations are simple when assuming that **no free gas exists at the start of the flood, i.e., S_{gi} = 0.** The cumulative oil production is simply equal to the volume of water injected with no water production during this phase ($W_p = 0$ and $Q_w = 0$).

Oil recovery calculations after breakthrough are based on determining E_D at various assumed values of water saturations at the producing well. The specific steps of performing complete oil recovery calculations are composed of three stages:

1. Data preparation
2. Recovery performance to breakthrough
3. Recovery performance after breakthrough

Stage 1: Data Preparation

Step 1. Express the relative permeability data as relative permeability ratio k_{ro}/k_{rw} and plot their values versus their corresponding water saturations on a semi-log scale.

Step 2. Assuming that the resulting plot of relative permeability ratio, k_{ro}/k_{rw} vs. S_w, forms a straight-line relationship, determine values of the coefficients a and b of the straight line (see Example 14-7).

Express the straight-line relationship in the form given by Equation 14-36, or:

$$\frac{k_{ro}}{k_{rw}} = ae^{bS_w}$$

Step 3. Calculate and plot the fractional flow curve f_w, allowing for gravity effects if necessary, but neglecting the capillary pressure gradient.

Step 4. Select several values of water saturations between S_{wf} and $(1 - S_{or})$ and determine the slope (df_w/dS_w) at each saturation. The numerical calculation of each slope as expressed by Equation 14-38 provides consistent values as a function of saturation, or:

$$\left(\frac{df_w}{dS_w}\right) = \frac{-\left(\frac{\mu_w}{\mu_o}\right)abe^{bS_w}}{\left[1+\left(\frac{\mu_w}{\mu_o}\right)ae^{bS_w}\right]^2}$$

Step 5. Prepare a plot of the calculated values of the slope (df_w/dS_w) versus S_w on a Cartesian scale and draw a smooth curve through the points.

Stage 2: Recovery Performance to Breakthrough ($S_{gi} = 0$, E_A, $E_V = 100\%$)

Step 1. Draw a tangent to the fractional flow curve as originated from S_{wi} and determine:

- Point of tangency with the coordinate (S_{wf}, f_{wf})
- Average water saturation at breakthrough $\overline{S}_{wBT}$ by extending the tangent line to $f_w = 1.0$
- Slope of the tangent line $\left(\dfrac{df_w}{dS_w}\right)_{S_{wf}}$

Step 2. Calculate pore volumes of water injected at breakthrough by using Equation 14-41:

$$Q_{iBT} = \frac{1}{\left(\dfrac{df_w}{dS_w}\right)_{S_{wf}}} = \left(\overline{S}_{wBT} - S_{wi}\right)$$

Step 3. Assuming E_A and E_V are 100%, calculate cumulative water injected at breakthrough by applying Equation 14-42:

$$W_{iBT} = (PV)(\overline{S}_{wBT} - S_{wi})$$

or equivalently:

$$W_{iBT} = (PV)Q_{iBT}$$

Step 4. Calculate the displacement efficiency at breakthrough by applying Equation 14-50:

$$E_{DBT} = \frac{\overline{S}_{wBT} - S_{wi}}{1 - S_{wi}}$$

Step 5. Calculate cumulative oil production at breakthrough from Equation 14-52:

$$\left(N_p\right)_{BT} = N_S E_{DBT}$$

Step 6. Assuming a constant water-injection rate, calculate time to breakthrough from Equation 14-40:

$$t_{BT} = \frac{W_{iBT}}{i_w}$$

Step 7. Select several values of injection time less than the breakthrough time, i.e., $t < t_{BT}$, and set:

$$
\begin{aligned}
W_{inj} &= i_w t \\
Q_o &= i_w/B_o \\
WOR &= 0 \\
W_p &= 0 \\
N_p &= \frac{i_w t}{B_o} = \frac{W_{inj}}{B_o}
\end{aligned}
$$

Step 8. Calculate the surface water–oil ratio WOR_s exactly at breakthrough by using Equation 14-28:

$$WOR_s = \frac{B_o}{B_w\left(\dfrac{1}{f_{wBT}} - 1\right)}$$

where f_{wBT} is the water cut at breakthrough (notice that $f_{wBT} = f_{wf}$).

Note that WOR$_s$ as calculated from the above expression is only correct when **both** the areal sweep efficiency E$_A$ and vertical sweep efficiency E$_V$ are 100%.

Stage 3: Recovery Performance After Breakthrough (S_{gi} = 0, E_A, E_V = 100%)

The recommended methodology of calculating recovery performance after breakthrough is based on selecting several values of water saturations around the producing well, i.e., S$_{w2}$, and determining the corresponding average reservoir water saturation $\overline{S}_{w2}$ for each S$_{w2}$. The specific steps that are involved are summarized below:

Step 1. Select six to eight different values of S$_{w2}$ (i.e., S$_w$ at the producing well) between S$_{wBT}$ and (1 − S$_{or}$) and determine (df$_w$/dS$_w$) values corresponding to these S$_{w2}$ points.

Step 2. For each selected value of S$_{w2}$, calculate the corresponding reservoir water cut and average water saturation from Equations 14-37 and 14-45:

$$f_{w2} = \frac{1}{1 + \left(\dfrac{\mu_w}{\mu_o}\right)ae^{bS_{w2}}}$$

$$\overline{S}_{w2} = S_{w2} + \frac{1 - f_{w2}}{\left(\dfrac{df_w}{dS_w}\right)_{S_{w2}}}$$

Step 3. Calculate the displacement efficiency E$_D$ for each selected value of S$_{w2}$:

$$E_D = \frac{\overline{S}_{w2} - S_{wi}}{1 - S_{wi}}$$

Step 4. Calculate cumulative oil production N$_p$ for each selected value of S$_{w2}$ from Equation 14-6, or:

$$N_P = N_s\, E_D\, E_A\, E_V$$

Assuming E_A and E_V are equal to 100%, then:

$$N_P = N_s\, E_D$$

Step 5. Determine pore volumes of water injected, Q_i, for each selected value of S_{w2} from Equation 14-46:

$$Q_i = \dfrac{1}{\left(\dfrac{df_w}{dS_w}\right)_{S_{w2}}}$$

Step 6. Calculate cumulative water injected for each selected value of S_{w2} by applying Equation 14-47 or 14-48:

$$W_{inj} = (PV)Q_i \quad \text{or} \quad W_{inj} = (PV)(\overline{S}_{w2} - S_{wi})$$

Notice that E_A and E_V are set equal to 100%

Step 7. Assuming a constant water-injection rate i_w, calculate the time t to inject W_{inj} barrels of water by applying Equation 14-49:

$$t = \dfrac{W_{inj}}{i_w}$$

Step 8. Calculate cumulative water production W_P at any time t from the material balance equation, which states that the cumulative water injected at any time will displace an equivalent volume of oil and water, or:

$$W_{inj} = N_p B_o + W_p B_w$$

Solving for W_p gives:

$$W_p \dfrac{W_{inj} - N_p B_o}{B_w} \tag{14-53}$$

or equivalently in a more generalized form:

$$W_p = \dfrac{W_{inj} - (\overline{S}_{w2} - S_{wi})(PV)E_A E_V}{B_w} \tag{14-54}$$

We should emphasize that all of the above derivations are based on the assumption that **no free gas exists from the start of the flood till abandonment.**

Step 9. Calculate the surface water–oil ratio WOR_s that corresponds to each value of f_{w2} (as determined in step 2) from Equation 14-28:

$$WOR_s = \frac{B_o}{B_w \left(\dfrac{1}{f_{w2}} - 1 \right)}$$

Step 10. Calculate the oil and water flow rates from the following derived relationships:

$$i_w = Q_o B_o + Q_w B_w$$

Introducing the surface water–oil ratio into the above expression gives:

$$i_w = Q_o B_o + Q_o WOR_s B_w$$

Solving for Q_o gives:

$$Q_o = \frac{i_w}{B_o + B_w WOR_s} \tag{14-55}$$

and

$$Q_w = Q_o WOR_s \tag{14-56}$$

where Q_o = oil flow rate, STB/day
        Q_w = water flow rate, STB/day
        i_w = water injection rate, bbl/day

Step 11. The preceding calculations as described in steps 1 through 10 can be organized in the following tabulated form:

S_{w2}	F_{w2}	(df_w/dS_w)	$\bar{S}_{w2}$	E_D	N_p	Q_i	W_{inj}	t	W_p	WOR$_s$	Q_o	Q_w
S_{wBT}	f_{wBT}	•	S_{wBT}	E_{DBT}	N_{PBT}	Q_{iBT}	W_{iBT}	t_{BT}	0	•	•	•
•	•	•	•	•	•	•	•	•	•	•	•	•
•	•	•	•	•	•	•	•	•	•	•	•	•
•	•	•	•	•	•	•	•	•	•	•	•	•
$(1-S_{or})$	1.0	•	•	•	•	•	•	•	•	100%	0	•

Step 12. Express the results in a graphical form.

Example 14-10

The data of Example 14-7 are reproduced here for convenience:

S_w	0.25	0.30	0.35	0.40	0.45	0.50	0.55	0.60	0.65	0.70	0.75
k_{ro}/k_{ro}	30.23	17.00	9.56	5.38	3.02	1.70	0.96	0.54	0.30	0.17	0.10
f_w	0.062	0.105	0.173	0.271	0.398	0.541	0.677	0.788	0.869	0.922	0.956
df_w/dS_w	0.670	10.84	1.647	2.275	2.759	2.859	2.519	1.922	1.313	0.831	0.501

$$\mu_o = 2.0 \text{ cp} \qquad \mu_w = 1.0 \text{ cp}$$
$$B_o = 1.25 \text{ bbl/STB} \qquad B_w = 1.02 \text{ bbl/STB}$$
$$\phi = 25\% \qquad h = 20 \text{ ft}$$
$$S_{wi} = 20\% \qquad S_{or} = 20\%$$
$$i_w = 900 \text{ bbl/day} \qquad (PV) = 775{,}779 \text{ bbl}$$
$$N_s = 496{,}449 \text{ STB} \qquad E_A = 100\%$$
$$E_V = 100\%$$

Predict the waterflood performance to abandonment at a WOR$_s$ of 45 STB/STB.

Solution

Step 1. Plot f_w vs. S_w as shown in Figure 14-30 and construct the tangent to the curve. Extrapolate the tangent to $f_w=1.0$ and determine:

$$S_{wf} = S_{wBT} = 0.596$$
$$f_{wf} = f_{wBT} = 0.780$$
$$(df_w/dS_w)_{swf} = 1.973$$
$$Q_{iBT} = 1/1.973 = 0.507$$
$$\bar{S}_{wBT} = 0.707$$

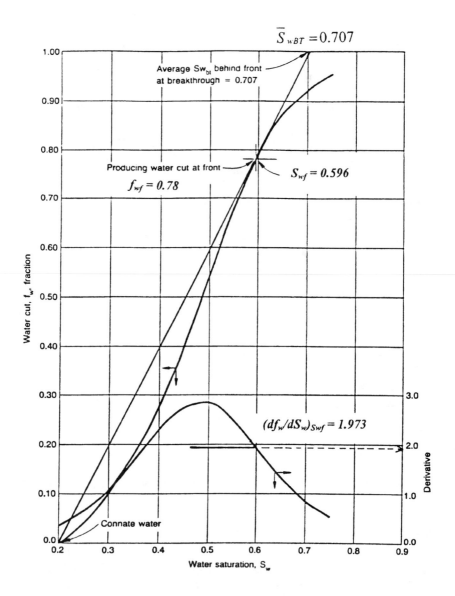

Figure 14-30. Fractional flow curve for Example 14-10.

Step 2. Calculate E_{DBT} by using Equation 14-50:

$$E_{DBT} = \frac{0.707 - 0.20}{1 - 0.20} = 0.634$$

Step 3. Calculate $(N_p)_{BT}$ by applying Equation 14-52:

$$\left(N_p\right)_{BT} = 496,499 \ (0.634) = 314,780 \ \text{STB}$$

Step 4. Calculate cumulative water injected at breakthrough from Equation 14-42:

$$W_{iBT} = 775,779(0.507) = 393,198 \ \text{bbl}$$

Step 5. Calculate the time to breakthrough:

$$t_{BT} = \frac{393,198}{900} = 436.88 \ \text{days}$$

Step 6. Calculate WOR_s exactly at breakthrough by applying Equation 14-28:

$$WOR_s = \frac{1.25}{1.02\left(\dfrac{1}{0.78} - 1\right)} = 4.34 \ \text{STB/STB}$$

Step 7. Describe the recovery performance to breakthrough in the following tabulated form:

t, days	$W_{inj} = 900 \ t$	$N_p = \dfrac{W_{inj}}{B_o}$	$Q_o = \dfrac{i_w}{B_o}$	WOR_s	$Q_w = Q_o \ WOR_s$	W_p
0	0	0	0	0	0	0
100.0	90,000	72,000	720	0	0	0
200.0	180,000	144,000	720	0	0	0
300.0	270,000	216,000	720	0	0	0
400.0	360,000	288,000	720	0	0	0
436.88	393,198	314,780	720	4.34	3125	0

Step 8. Following the computational procedure as outlined for recovery performance after breakthrough, construct the following table:

S_{w2}	f_{w2}	df_w/dS_w	Q_i	$\bar{S}_{w2}$	E_D	N_p	W_{inj}	t, days	W_p	WOR_s	Q_o	Q_w
0.598	0.784	1.948	0.513	0.709	0.636	315,773	397,975	442	82,202	4.45	155	690
0.600	0.788	1.922	0.520	0.710	0.638	316,766	403,405	448	86,639	4.56	153	698
0.700	0.922	0.831	1.203	0.794	0.743	368,899	933,262	1,037	564,363	14.49	56	814
0.800	0.974	0.293	3.407	0.889	0.861	427,486	2,643,079	2,937	2,215,593	45.91	19	859

Step 9. Express graphically results of the calculations as a set of performance curves, as shown in Figure 14-31.

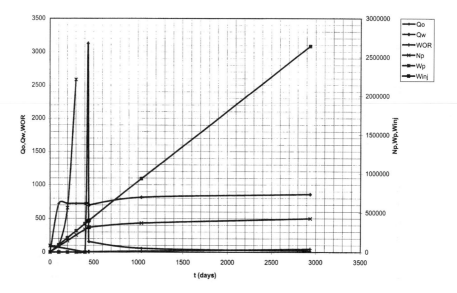

Figure 14-31. Performance curves of Example 14-10.

II. AREAL SWEEP EFFICIENCY

The areal sweep efficiency E_A is defined as the fraction of the total
flood pattern that is contacted by the displacing fluid. It increases steadily
with injection from zero at the start of the flood until breakthrough
occurs, after which E_A continues to increase at a slower rate.

The areal sweep efficiency depends basically on the following three
main factors:

1. Mobility ratio M
2. Flood pattern
3. Cumulative water injected W_{inj}

Mobility Ratio

In general, the mobility of any fluid λ is defined as the ratio of the effective permeability of the fluid to the fluid viscosity, i.e.:

$$\lambda_o = \frac{k_o}{\mu_o} = \frac{k \, k_{ro}}{\mu_o} \qquad (14\text{-}57)$$

$$\lambda_w = \frac{k_w}{\mu_w} = \frac{k \, k_{rw}}{\mu_w} \qquad (14\text{-}58)$$

$$\lambda_g = \frac{k_g}{\mu_g} = \frac{k \, k_{rg}}{\mu_g} \qquad (14\text{-}59)$$

where $\lambda_o, \lambda_w, \lambda_g$ = mobility of oil, water, and gas, respectively
$\quad\quad k_o, k_w, k_g$ = effective permeability to oil, water, and gas, respectively
$\quad\quad k_{ro}, k_{rw}$ = relative permeability to oil, water, and gas, respectively
$\quad\quad k$ = absolute permeability

The fluid mobility as defined mathematically by the above three relationships indicates that λ is a strong function of the fluid saturation. The mobility ratio M is defined as the mobility of the *displacing fluid* to the mobility of the *displaced fluid*, or:

$$M = \frac{\lambda_{\text{displacing}}}{\lambda_{\text{displaced}}}$$

For waterflooding then:

$$M = \frac{\lambda_w}{\lambda_0}$$

Substituting for λ:

$$M = \frac{k \, k_{rw}}{\mu_w} \frac{\mu_o}{k \, k_{ro}}$$

Simplifying gives:

$$M = \frac{k_{rw}}{k_{ro}} \frac{\mu_o}{\mu_w} \qquad (14\text{-}60)$$

Muskat (1946) points out that in calculating M by applying Equation 14-60, the following concepts must be employed in determining k_{ro} and k_{rw}:

- **Relative permeability of oil k_{ro}.** Because the displaced oil is moving ahead of the water front in the noninvaded portion of the pattern, as shown schematically in Figure 14-32, k_{ro} must be evaluated at the initial water saturation S_{wi}.

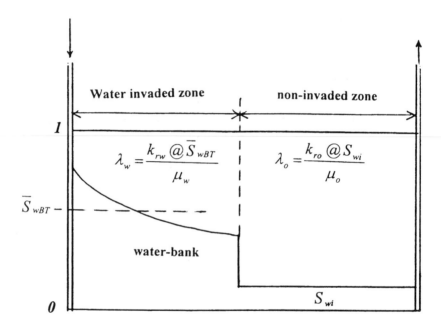

Figure 14-32. Oil and water mobilities to breakthrough.

- **Relative permeability of water k_{rw}.** The displacing water will form a water bank that is characterized by an average water saturation of $\overline{S}_{wBT}$ in the swept area. This average saturation will remain constant until breakthrough, after which the average water saturation will continue to increase (as denoted by $\overline{S}_{w2}$). The mobility ratio, therefore, can be expressed more explicitly under two different stages of the flood:

  From the start to breakthrough:

$$M = \frac{k_{rw} @ \overline{S}_{wBT}}{k_{ro} @ S_{wi}} \frac{\mu_0}{\mu_w} \tag{14-61}$$

  where $k_{rw} @ \overline{S}_{wBT}$ = relative permeability of water at $\overline{S}_{wBT}$

$$k_{ro} \,@\, S_{wi} = \text{relative permeability of oil at } S_{wi}$$

The above relationship indicates that the mobility ratio will remain constant from the start of the flood until breakthrough occurs.

After breakthrough:

$$M = \frac{k_{rw} \,@\, \overline{S}_{w2}}{k_{ro} \,@\, S_{wi}} \frac{\mu_0}{\mu_w} \qquad (14\text{-}62)$$

Equation 14-62 indicates that the mobility of the water krw/μw will increase after breakthrough due to the continuous increase in the average water saturation $\overline{S}_{w2}$. This will result in a proportional increase in the mobility ratio M after breakthrough, as shown in Figure 14-33.

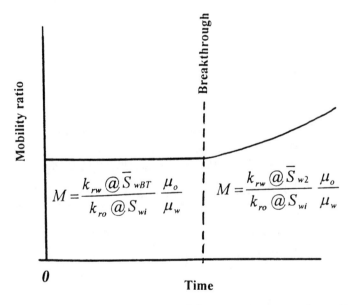

Figure 14-33. Mobility ratio versus time relationship.

In general, if no further designation is applied, the term **mobility ratio** refers to the mobility ratio before breakthrough.

Flood Patterns

In designing a waterflood project, it is common practice to locate injection and producing wells in a regular geometric pattern so that a symmetrical and interconnected network is formed. As shown previously in Figure 14-9, regular flood patterns include these:

- Direct line drive
- Staggered line drive
- Five spot
- Seven spot
- Nine spot

By far the most used pattern is the five spot and, therefore, most of the discussion in the remainder of the chapter will focus on this pattern.

Craig et al. (1955) performed experimental studies on the influence of fluid mobilities on the areal sweep efficiency resulting from water or gas injection. Craig and his co-investigators used horizontal laboratory models representing a quadrant of five spot patterns. Areal sweep efficiencies were determined from x-ray shadowgraphs taken during various stages of the displacement as illustrated in Figure 14-34. Two mobility ratios, 1.43 and 0.4, were used in the study.

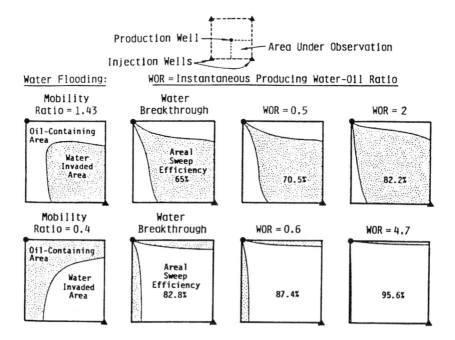

Figure 14-34. X-ray shadowgraphs of flood progress. *(Permission to publish by the Society of Petroleum Engineers.)*

Figure 14-34 shows that at the start of the flood, the water front takes on a cylindrical form around the injection point (well). As a result of the continuous injection, pressure distribution and corresponding streamlines are developed between the injection and production wells. However, various streamlines have different lengths with the shortest streamline being the direct line between the injector and producer. The pressure gradient along this line is the highest that causes the injection fluid to flow faster along the shortest streamline than the other lines. The water front gradually begins to deform from the cylindrical form and cusp into the production well as water breakthrough occurs. The effect of the mobility ratio on the areal sweep efficiency is apparent by examining Figure 14-34. This figure shows that at breakthrough, only 65% of the flood pattern area has been contacted (swept) by the injection fluid with a mobility ratio of 1.43 and 82.8% when the mobility ratio is 0.4. This contacted fraction when water breakthrough occurs is defined as the **areal sweep efficiency at breakthrough**, as denoted by E_{ABT}. In general, **lower mobility ratios would increase the areal sweep efficiency and higher mobility ratios would decrease the E_A**. Figure 14-34 also shows that with continued injection after breakthrough, the areal sweep efficiency continues to increase until it eventually reaches 100%.

Cumulative Water Injected

Continued injection after breakthrough can result in substantial increases in recovery, especially in the case of an adverse mobility ratio. The work of Craig et al. (1955) has shown that significant quantities of oil may be swept by water after breakthrough. It should be pointed out that the higher the mobility ratio, the more important is the "after-breakthrough" production.

Areal Sweep Prediction Methods

Methods of predicting the areal sweep efficiency are essentially divided into the following three phases of the flood:

- Before breakthrough
- At breakthrough
- After breakthrough

Phase 1: Areal Sweep Efficiency Before Breakthrough

The areal sweep efficiency before breakthrough is simply proportional to the volume of water injected and is given by:

$$E_A = \frac{W_{inj}}{(PV)(\overline{S}_{wBT} - S_{wi})}$$

(14-63)

where W_{inj} = cumulative water injected, bbl
(PV) = flood pattern pore volume, bbl

Phase 2: Areal Sweep Efficiency at Breakthrough

Craig (1955) proposed a graphical relationship that correlates the areal sweep efficiency at breakthrough E_{ABT} with the mobility ratio for the five spot pattern. The correlation, as shown in Figure 14-35, closely simulates flooding operations and is probably the most representative of actual waterfloods. The graphical illustration of areal sweep efficiency as a strong function of mobility ratio shows that a change in the mobility ratio from 0.15 to 10.0 would change the breakthrough areal sweep efficiency from 100 to 50%. Willhite (1986) presented the following mathematical correlation, which closely approximates the graphical relationship presented in Figure 14-35:

$$E_{ABT} = 0.54602036 + \frac{0.03170817}{M} + \frac{0.30222997}{e^M} - 0.00509693\,M$$

(14-64)

where E_{ABT} = areal sweep efficiency at breakthrough
M = mobility ratio

Phase 3: Areal Sweep Efficiency After Breakthrough

In the same way that displacement efficiency E_D increases after breakthrough, the areal sweep efficiency also increases due to the gradual increase in the total swept area with continuous injection. Dyes et al. (1954) correlated the increase in the areal sweep efficiency after breakthrough with the ratio of water volume injected at any time after breakthrough, W_{inj}, to water volume injected at breakthrough, W_{iBT}, as given by:

$$E_A = E_{ABT} + 0.633 \log\left(\frac{W_{inj}}{W_{iBT}}\right)$$

(14-65)

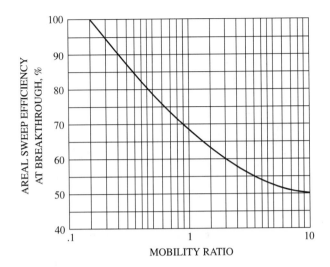

Figure 14-35. Areal sweep efficiency at breakthrough. *(Permission to publish by the Society of Petroleum Engineers.)*

or

$$E_A = E_{ABT} + 0.2749 \ln\left(\frac{W_{inj}}{W_{iBT}}\right) \tag{14-66}$$

where $\quad E_A$ = areal sweep efficiency after breakthrough
$\quad\quad\quad W_{inj}$ = cumulative water injected
$\quad\quad\quad W_{iBT}$ = Cumulative water injected at breakthrough

The authors also presented a graphical relationship that relates the areal sweep efficiency with the reservoir water cut f_w and the reciprocal of mobility ratio $1/M$ as shown in Figure 14-36. Fassihi (1986) used a nonlinear regression model to reproduce the data of Figure 14-36 by using the following expression:

$$E_A = \frac{1}{1+A} \tag{14-67}$$

with

$$A = \left[a_1 \ln(M + a_2) + a_3\right]f_w + a_4 \ln(M + a_5) + a_6$$

The coefficient of Equation 14-67 for patterns such as the five spot, staggered line drive, and direct line drive are given below:

Coefficients in Areal Sweep Efficiency Correlations

Coefficient	Five Spot	Direct Line	Staggered Line
a_1	−0.2062	−0.3014	−0.2077
a_2	−0.0712	−0.1568	−0.1059
a_3	−0.511	−0.9402	−0.3526
a_4	0.3048	0.3714	0.2608
a_5	0.123	−0.0865	0.2444
a_6	0.4394	0.8805	0.3158

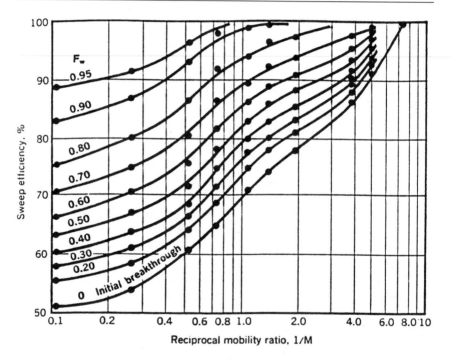

Figure 14-36. Areal sweep efficiency as a function of 1/M and f_w. *(Permission to publish by the Society of Petroleum Engineers.)*

Craig (1971) proposed that for a given value of E_{ABT} for a five-spot flood pattern, the ratio Q_i/Q_{iBT} that corresponds to W_{inj}/W_{iBT} could be determined mathematically by evaluating the following expression:

$$\frac{Q_i}{Q_{iBT}} = 1 + E_{ABT} \int\limits_{1}^{x} \left(\frac{1}{E_A}\right) dx$$

with

$$x = \frac{W_{inj}}{W_{iBT}}$$

where Q_i = total pore volumes of water injected any time after breakthrough = $1/(df_w/dS_w)_{Sw2}$
Q_{iBT} = total pore volumes of water injected at water breakthrough = $1/(df_w/dS_w)_{Swf}$

Craig tabulated the values of Q_i/Q_{iBT} as a function of W_{inj}/W_{iBT} and E_{ABT}. The author listed the values for a wide range of W_{inj}/W_{iBT} with E_{ABT} ranging from 50 to 90% as shown in Table 14-1. The value of Q_i/Q_{iBT} is read from the table for any particular value of E_{ABT} and the value of W_{inj}/W_{iBT} using interpolation if necessary. For example, if E_{ABT} = 70% and W_{inj}/W_{iBT} = 2.00, the value of the ratio Q_i/Q_{iBT} is read from Table 14-1 as 1.872, i.e., Q_i/Q_{iBT} = 1.872.

Table 14-1.
Q_i/Q_{iBT} values for various values of E_{ABT}.
(Permission to publish by the Society of Petroleum Engineers)

W_i/W_{iBT}					E_{ABT} percent					
	50.	51.	52.	53.	54.	55.	56.	57.	58.	59.
1.0	1.000	1.000	1.000	1.000	1.000	1.000	1.000	1.000	1.000	1.000
1.2	1.190	1.191	1.191	1.191	1.191	1.191	1.191	1.191	1.192	1.192
1.4	1.365	1.366	1.366	1.367	1.368	1.368	1.369	1.369	1.370	1.370
1.6	1.529	1.530	1.531	1.532	1.533	1.535	1.536	1.536	1.537	1.538
1.8	1.684	1.686	1.688	1.689	1.691	1.693	1.694	1.696	1.697	1.699
2.0	1.832	1.834	1.837	1.839	1.842	1.844	1.846	1.849	1.851	1.853
2.2	1.974	1.977	1.981	1.984	1.987	1.990	1.993	1.996	1.999	2.001
2.4	2.111	2.115	2.119	2.124	2.127	2.131	2.135	2.139	2.142	2.146
2.6	2.244	2.249	2.254	2.259	2.264	2.268	2.273	2.277	2.282	2.286
2.8	2.373	2.379	2.385	2.391	2.397	2.402	2.407	2.413	2.418	2.422

Table 14-1
Continued

E_{ABT} percent

3.0	2.500	2.507	2.513	2.520	2.526	2.533	2.539	2.545	2.551	2.556
3.2	2.623	2.631	2.639	2.646	2.653	2.660	2.667	2.674	2.681	2.687
3.4	2.744	2.752	2.761	2.770	2.778	2.786	2.793	2.801	2.808	2.816
3.6	2.862	2.872	2.881	2.891	2.900	2.909	2.917	2.926	2.934	2.942
3.8	2.978	2.989	3.000	3.010	3.020	3.030	3.039	3.048	3.057	3.066
4.0	3.093	3.105	3.116	3.127	3.138	3.149	3.159	3.169	3.179	3.189
4.2	3.205	3.218	3.231	3.243	3.254	3.266	3.277	3.288	3.299	3.309
4.4	3.316	3.330	3.343	3.357	3.369	3.382	3.394	3.406	3.417	3.428
4.6	3.426	3.441	3.455	3.469	3.483	3.496	3.509	3.521	3.534	3.546
4.8	3.534	3.550	3.565	3.580	3.594	3.609	3.622	3.636	3.649	
5.0	3.641	3.657	3.674	3.689	3.705	3.720	3.735			
5.2	3.746	3.764	3.781	3.798	3.814	3.830				
5.4	3.851	3.869	3.887	3.905	3.922					
5.6	3.954	3.973	3.993	4.011						
5.8	4.056	4.077	4.097							
6.0	4.157	4.179								
6.2	4.257									

Values of W_i/W_{iBT} at which $E_A = 100$ percent

6.164	5.944	5.732	5.527	5.330	5.139	4.956	4.779	4.608	4.443

E_{ABT} percent

W_i/W_{iBT}	60.	61.	62.	63.	64.	65.	66.	67.	68.	69.
1.0	1.000	1.000	1.000	1.000	1.000	1.000	1.000	1.000	1.000	1.000
1.2	1.192	1.192	1.192	1.192	1.192	1.192	1.193	1.193	1.193	1.193
1.4	1.371	1.371	1.371	1.372	1.372	1.373	1.373	1.373	1.374	1.374
1.6	1.539	1.540	1.541	1.542	1.543	1.543	1.544	1.545	1.546	1.546
1.8	1.700	1.702	1.703	1.704	1.706	1.707	1.708	1.709	1.710	1.711
2.0	1.855	1.857	1.859	1.861	1.862	1.864	1.866	1.868	1.869	1.871
2.2	2.004	2.007	2.009	2.012	2.014	2.016	2.019	2.021	2.023	2.025
2.4	2.149	2.152	2.155	2.158	2.161	2.164	2.167	2.170	2.173	2.175
2.6	2.290	2.294	2.298	2.301	2.305	2.308	2.312	2.315	2.319	2.322
2.8	2.427	2.432	2.436	2.441	2.445	2.449	2.453	2.457	2.461	2.465
3.0	2.562	2.567	2.572	2.577	2.582	2.587	2.592	2.597	2.601	2.606
3.2	2.693	2.700	2.705	2.711	2.717	2.723	2.728	2.733	2.738	2.744
3.4	2.823	2.830	2.836	2.843	2.849	2.855	2.862	2.867	2.873	
3.6	2.950	2.957	2.965	2.972	2.979	2.986	2.993			
3.8	3.075	3.083	3.091	3.099	3.107					
4.0	3.198	3.207	3.216	3.225						
4.2	3.319	3.329								
4.4	3.439									

Values of W_i/W_{iBT} at which E_A = 100 percent

| 4.235 | 4.132 | 3.984 | 3.842 | 3.704 | 3.572 | 3.444 | 3.321 | 3.203 | 3.088 |

E_{ABT} percent

W_i/W_{iBT}	70.	71.	72.	73.	74.	75.	76.	77.	78.	79.
1.0	1.000	1.000	1.000	1.000	1.000	1.000	1.000	1.000	1.000	1.000
1.2	1.193	1.193	1.193	1.193	1.193	1.193	1.193	1.194	1.194	1.194
1.4	1.374	1.375	1.375	1.375	1.376	1.376	1.376	1.377	1.377	1.377
1.6	1.547	1.548	1.548	1.549	1.550	1.550	1.551	1.551	1.552	1.552
1.8	1.713	1.714	1.715	1.716	1.717	1.718	1.719	1.720	1.720	1.721
2.0	1.872	1.874	1.875	1.877	1.878	1.880	1.881	1.882	1.884	1.885
2.2	2.027	2.029	2.031	2.033	2.035	2.037	2.039	2.040	2.042	2.044
2.4	2.178	2.180	2.183	2.185	2.188	2.190	2.192	2.195	1.197	
2.6	2.325	2.328	2.331	2.334	2.337	2.340				
2.8	2.469	2.473	2.476	2.480						
3.0	2.610	2.614								

Values of W_i/W_{iBT} at which E_A = 100 percent

| 2.978 | 2.872 | 2.769 | 2.670 | 2.575 | 2.483 | 2.394 | 2.309 | 2.226 | 2.147 |

E_{ABT} percent

W_i/W_{iBT}	80.	81.	82.	83.	84.	85.	86.	87.	88.	89.
1.0	1.000	1.000	1.000	1.000	1.000	1.000	1.000	1.000	1.000	1.000
1.2	1.194	1.194	1.194	1.194	1.194	1.194	1.194	1.194	1.194	1.194
1.4	1.377	1.378	1.378	1.378	1.378	1.379	1.379	1.379	1.379	1.379
1.6	1.553	1.553	1.554	1.555	1.555	1.555	1.556	1.556	1.557	1.557
1.8	1.722	1.723	1.724	1.725	1.725	1.726	1.727	1.728		
2.0	1.886	1.887	1.888	1.890						
2.2	2.045									

Values of W_i/W_{ibt} at which E_A = 100 percent

| 2.070 | 1.996 | 1.925 | 1.856 | 1.790 | 1.726 | 1.664 | 1.605 | 1.547 | 1.492 |

E_{ABT} percent

W_i/W_{iBT}	90.	91.	92.	93.	94.	95.	96.	97.	98.	99.
1.0	1.000	1.000	1.000	1.000	1.000	1.000	1.000	1.000	1.000	1.000
1.2	1.194	1.195	1.195	1.195	1.195	1.195	1.195	1.195	1.195	1.195
1.4	1.380	1.380	1.380	1.380	1.381					
1.6	1.558									

Values of W_i/W_{iBT} at which E_A = 100 percent

| 1.439 | 1.387 | 1.338 | 1.290 | 1.244 | 1.199 | 1.157 | 1.115 | 1.075 | 1.037 |

Willhite (1986) proposed an analytical expression for determining the value of the ratio (Q_i/Q_{iBT}) at any value of (W_{inj}/W_{iBT}) for a given E_{ABT} :

$$\frac{Q_i}{Q_{iBT}} = 1 + a_1\, e^{-a_1} \left[Ei(a_2) - Ei(a_1) \right] \qquad (14\text{-}68)$$

where

$$a_1 = 3.65\, E_{ABT}$$

$$a_2 = a_1 + \ln \frac{W_{inj}}{W_{iBT}}$$

and $Ei(x)$ is the Ei function as approximated by:

$$Ei(x) = 0.57721557 + \ln(x) + \sum_{n=1}^{\infty} \frac{x^n}{n(n!)}$$

To include the areal sweep efficiency in waterflooding calculations, the proposed methodology is divided into the following three phases:

1. Initial calculations
2. Recovery performance calculations to breakthrough
3. Recovery performance calculations after breakthrough

The specific steps of each of the above three phases are summarized below.

Phase 1: Initial Calculations ($S_{gi} = 0$, $E_V = 100\%$)

Step 1. Express the relative permeability data as relative permeability ratios and plot them versus their corresponding water saturations on a semi-log scale. Describe the resulting straight line by the following relationship:

$$\frac{k_{ro}}{k_{rw}} = a\, e^{bS_w}$$

Step 2. Calculate and plot f_w versus S_w.

Step 3. Draw a tangent to the fractional flow curve as originated from S_{wi} and determine:

- Point of tangency (S_{wf}, f_{wf}), i.e., (S_{wBT}, f_{wBT})
- Average water saturation at breakthrough $\overline{S}_{wBT}$
- Slope of the tangent $\left(\dfrac{df_w}{S_w}\right)_{S_{wf}}$

Step 4. Using S_{wi} and $\overline{S}_{wBT}$, determine the corresponding values of k_{ro} and k_{rw}. Designate these values $k_{ro}@S_{wBT}$ and $k_{rw}@\overline{S}_{wBT}$, respectively.

Step 5. Calculate the mobility ratio as defined by Equation 14-61:

$$M = \frac{k_{rw}@\overline{S}_{wBT}}{k_{ro}@S_{wi}}\frac{\mu_o}{\mu_w}$$

Step 6. Select several water saturations S_{w2} between S_{wf} and $(1 - S_{or})$ and numerically or graphically determine the slope $\left(\dfrac{df_w}{dS_w}\right)_{S_{w2}}$ at each saturation.

Step 7. Plot $\left(\dfrac{df_w}{dS_w}\right)_{S_{w2}}$ versus S_{w2} on a Cartesian scale.

Phase 2: Recovery Performance to Breakthrough

Assuming that the vertical sweep efficiency E_V and initial gas saturation S_{gi} are 100 and 0%, respectively, the required steps to complete the calculations of this phase are summarized below:

Step 1. Calculate the areal sweep efficiency at breakthrough E_{ABT} from Figure 14-35 or Equation 14-64.

Step 2. Calculate pore volumes of water injected at breakthrough by applying Equation 14-41:

$$Q_{iBT} = \frac{1}{\left(\dfrac{df_w}{dS_w}\right)_{S_{wf}}} = \left(\overline{S}_{wBT} - S_{wi}\right)$$

Step 3. Calculate cumulative water injected at breakthrough W_{iBT} from Equation 14-43 or 14-44:

$$W_{iBT} = (PV)(\overline{S}_{wBT} - S_{wi})E_{ABT} = (PV)(Q_{iBT})E_{ABT}$$

Step 4. Assuming a constant water-injection rate i_w, calculate time to breakthrough t_{BT}:

$$t_{BT} = \frac{W_{iBT}}{i_w}$$

Step 5. Calculate the displacement efficiency at breakthrough E_{DBT} from Equation 14-50:

$$E_{DBT} = \frac{\overline{S}_{wBT} - S_{wi}}{1 - S_{wi}}$$

Step 6. Compute the cumulative oil production at breakthrough from Equation 14-51:

$$\left(N_p\right)_{BT} = N_s E_{DBT} E_{ABT}$$

Notice that when $S_{gi} = 0$, the cumulative oil produced at breakthrough is equal to cumulative water injected at breakthrough, or:

$$\left(N_p\right)_{BT} = \frac{W_{iBT}}{B_o}$$

Step 7. Divide the interval between 0 and W_{iBT} into any arbitrary number of increments and set the following production data for each increment:

$$Q_o = i_w/B_o$$
$$Q_w = 0$$
$$WOR = 0$$
$$N_p = W_{inj}/B_o$$
$$W_p = 0$$
$$t = W_{inj}/i_w$$

Step 8. Express steps 1 through 7 in the following tabulated form:

W_{inj}	$t = W_{inj}/i_w$	$N_P = W_{inj}/B_o$	$Q_o = i_w/B_o$	WOR_s	$Q_w = Q_oWOR_s$	W_p
0	0	0	0	0	0	0
•	•	•	•	0	0	0
•	•	•	•	0	0	0
•	•	•	•	0	0	0
W_{iBT}	t_{BT}	$(N_P)_{BT}$	•	WOR_s	•	0

Phase 3: Recovery Performance After Breakthrough ($S_{gi} = 0$, $E_V = 100\%$)

Craig et al. (1955) point out that after water breakthrough, the displacing fluid continues to displace more oil from the already swept zone (behind the front) and from newly swept regions in the pattern. Therefore, the producing water–oil ratio WOR is estimated by separating the displaced area into two distinct zones:

1. Previously swept area of the flood pattern
2. Newly swept zone that is defined as the region that was just swept by the displacing fluid

The previously swept area contains all reservoir regions where water saturation is greater than S_{wf} and continues to produce both oil and water. With continuous water injection, the injected water contacts more regions as the area sweep efficiency increases. This newly swept zone is assumed to produce only oil. Craig et al. (1955) developed an approach for determining the producing WOR that is based on estimating the incremental oil produced, $(\Delta N_P)_{newly}$, from the newly swept region for 1 bbl of total production. The authors proposed that the incremental oil produced from the newly swept zone is given by:

$$\left(\Delta N_p\right)_{newly} = E\lambda \tag{14-69}$$

with

$$E = \frac{S_{wf} - S_{wi}}{E_{ABT}\left(\overline{S}_{wBT} - S_{wi}\right)}$$

$$\lambda = 0.2749\left(\frac{W_{iBT}}{W_{inj}}\right)$$

Notice that the parameter E is constant, whereas the parameter λ is decreasing with continuous water injection. Craig et al. (1955) expressed the producing water–oil ratio as:

$$
\text{WOR}_s = \frac{f_{w2}\left(1-\left(\Delta N_p\right)_{newly}\right)}{1-\left[f_{w2}\left(1-\left(\Delta N_p\right)_{newly}\right)\right]}\left(\frac{B_o}{B_w}\right) \tag{14-70}
$$

where WOR_s = surface water–oil ratio, STB/STB
W_{iBT} = cumulative water injected at breakthrough, bbl
W_{inj} = cumulative water injected at any time after breakthrough, bbl
f_{w2} = water cut at the producing well, bbl/bbl

Note that when the areal sweep efficiency EA reaches 100%, the incremental oil produced from the newly swept areal is zero, i.e., $(\Delta N_P)_{newly} = 0$, which reduces the above expression to Equation 14-28:

$$
\text{WOR}_s = \frac{f_{w2}}{1-f_{w2}}\left(\frac{B_o}{B_w}\right) = \frac{B_o}{B_w\left(\dfrac{1}{f_{w2}}-1\right)}
$$

The recommended methodology for predicting the recovery performance after breakthrough is summarized in the following steps:

Step 1. Select several values of $W_{inj} > W_{iBT}$.

Step 2. Assuming constant injection rate i_w, calculate the time t required to inject W_{inj} barrels of water.

Step 3. Calculate the ratio W_{inj}/W_{iBT} for each selected W_{inj}.

Step 4. Calculate the areal sweep efficiency E_A at each selected W_{inj} by applying Equation 14-65 or 14-66:

$$E_A = E_{ABT} + 0.633 \log\left(\frac{W_{inj}}{W_{iBT}}\right) = E_{ABT} + 0.2749 \ln\left(\frac{W_{inj}}{W_{iBT}}\right)$$

Step 5. Calculate the ratio Q_i/Q_{iBT} that corresponds to each W_{inj}/W_{iBT} from Table 14-1. The ratio Q_i/Q_{iBT} is a function of E_{ABT} and W_{inj}/W_{iBT}.

Step 6. Determine the total pore volumes of water injected by multiplying each ratio of Q_i/Q_{iBT} (obtained in step 5) by Q_{iBT}, or:

$$Q_i = \left(\frac{Q_i}{Q_{iBT}}\right) Q_{iBT}$$

Step 7. From the definition of Q_i, as expressed by Equation 14-46, determine the slope $(df_w/dS_w)_{S_{w2}}$ for each value of Q_i by:

$$\left(\frac{df_w}{dS_w}\right)_{S_{w2}} = \frac{1}{Q_i}$$

Step 8. Read the value of S_{w2}, i.e., water saturation at the producing well, that corresponds to each slope from the plot of $(df_w/dS_w)_{S_{w2}}$ vs. S_{w2} (see phase 1, step 7).

Step 9. Calculate the reservoir water cut at the producing well f_{w2} for each S_{w2} from Equation 14-24 or 14-37.

$$f_{w2} = \frac{1}{1 + \dfrac{\mu_w}{\mu_o} \dfrac{k_{ro}}{k_{rw}}}$$

or

$$f_{w2} = \frac{1}{1 + \left(\dfrac{\mu_w}{\mu_o}\right) a\, e^{bS_{w2}}}$$

Step 10. Determine the average water saturation in the swept area $\overline{S}_{w2}$ by applying Equation 14-45:

$$\overline{S}_{w2} = S_{w2} + \frac{1 - f_{w2}}{\left(\dfrac{df_w}{dS_w}\right)_{S_{w2}}}$$

Step 11. Calculate the displacement efficiency E_D for each $\overline{S}_{w2}$:

$$E_D = \frac{\overline{S}_{w2} - S_{wi}}{1 - S_{wi}}$$

Step 12. Calculate cumulative oil production from Equation 14-6:

$$N_p = N_S E_D E_A E_V$$

For 100% vertical sweep efficiency:

$$N_p = N_S E_D E_A$$

Step 13. Calculate cumulative water production from Equation 14-53 or 14-54:

$$W_p = \frac{W_{inj} - N_p B_o}{B_w}$$

$$W_p = \frac{W_{inj} - (\overline{S}_{w2} - S_{wi})(PV)E_A}{B_w}$$

Step 14. Calculate the surface water–oil ratio WOR_s that corresponds to each value of f_{w2} from Equation 14-70:

$$WOR_s = \frac{f_{w2}\left(1-(\Delta N_P)_{newly}\right)}{1-\left[f_{w2}\left(1-(\Delta N_P)_{newly}\right)\right]}\left(\frac{B_o}{B_w}\right)$$

Step 15. Calculate the oil and water flow rates from Equations 14-55 and 14-50, respectively:

$$Q_o = \frac{i_w}{B_o + B_w WOR_s}$$

$$Q_w = Q_o WOR_s$$

Steps 1 through 15 could be conveniently performed in the following worksheet form:

W_{inj}	$t = \dfrac{W_{inj}}{i_w}$	$\dfrac{W_{inj}}{W_{iBT}}$	E_A	$\dfrac{Q_i}{Q_{iBT}}$	Q_i	$\left(\dfrac{df_w}{dS_w}\right)_{Sw2}$	S_{w2}	f_{w2}	$\bar{S}_{w2}$	E_D	N_P	W_P	WOR_S	Q_o	Q_w
W_{iBT}	t_{BT}	1.0	E_{ABT}	1.0	Q_{iBT}	—	S_{wBT}	f_{wBT}		E_{DBT}	—	—	—	—	—
•	•	•	•	•	•	•	•	•	•	•	•	•	•	•	•
•	•	•	•	•	•	•	•	•	•	•	•	•	•	•	•
•	•	•	•	•	•	•	•	•	•	•	•	•	•	•	•

Example 14-11[1]

An oil reservoir is under consideration for waterflooding. The relative permeability data and the corresponding water cut are given below:

S_w	0.100	0.300	0.400	0.450	0.500	0.550	0.600	0.650	0.700
k_{ro}	1.000	0.373	0.210	0.148	0.100	0.061	0.033	0.012	0.000
k_{rw}	0.000	0.070	0.169	0.226	0.300	0.376	0.476	0.600	0.740
f_w	0.000	0.2729	0.6168	0.7533	0.8571	0.9250	0.9665	0.9901	1.0000

[1]From *The Reservoir Engineering Aspects of Waterflooding*, Craig, Dallas: Society of Petroleum Engineers, 1971, p. 116.

Reservoir properties are as follows:

Flood area, acres	= 40
Thickness, ft	= 5
Average permeability, md	= 31.5
Porosity, %	= 20
Initial water saturation, %	= 10
Connate water saturation, %	= 10
Current gas saturation, %	= 0
Water viscosity, cp	= 0.5
Oil viscosity, cp	= 1.0
Reservoir pressure, psi	= 1000
Constant B_o, bbl/STB	= 1.20
Flood pattern	= 5 spot
Wellbore radius, ft	= 1.0

Predict the recovery performance under a constant water injection rate of 269 bbl/day.

Solution

Phase 1. Initial Calculations

Step 1. Calculate pore volume and oil volume at start of flood:

$$(PV) = 7758(40)(5)(0.20) = 310,320 \, bbl$$
$$N_S = 310,320(1-0.1)/1.20 = 232,740 \, STB$$

Step 2. Plot f_w vs. S_w on a Cartesian scale, as shown in Figure 14-37, and determine:

$$S_{wf} = S_{wBT} = 0.469 \qquad Q_{iBT} = \frac{1}{2.16} = 0.463$$
$$f_{wf} = f_{wBT} = 0.798 \qquad \overline{S}_{wBT} = 0.563$$
$$(df_w/dS_w)_{Swf} = 2.16$$

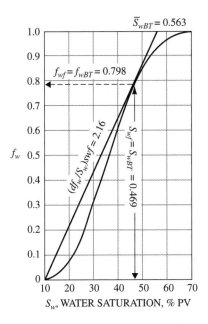

Figure 14-37. The f_w curve for Example 14-11. *(Permission to publish by the Society of Petroleum Engineers.)*

Step 3. Determine k_{ro} and k_{rw} at S_{wi} and $\overline{S}_{wBT}$ from the relative permeability data, to give:

$$k_{ro} @ 0.1 = 1.00$$
$$k_{rw} @ 0.563 = 0.40$$

Step 4. Calculate the mobility ratio M from Equation 14-61:

$$M = \frac{0.4}{1.0} \frac{1.0}{0.5} = 0.8$$

Step 5. Calculate the areal sweep efficiency at breakthrough from Equation 14-64 or Figure 14-35:

$$E_{ABT} = 0.71702$$

Step 6. Select several values of S_{w2} between 0.469 and 0.700 and determine the slope, graphically or numerically, at each selected saturation:

S_{w2}	f_{w2}	df_w/dS_w
0.469	0.798	2.16
0.495	0.848	1.75
0.520	0.888	1.41
0.546	0.920	1.13
0.572	0.946	0.851
0.597	0.965	0.649
0.622	0.980	0.477
0.649	0.990	0.317
0.674	0.996	0.195
0.700	1.000	0.102

Step 7. Plot df_w/dS_w vs. S_{w2} as shown in Figure 14-38.

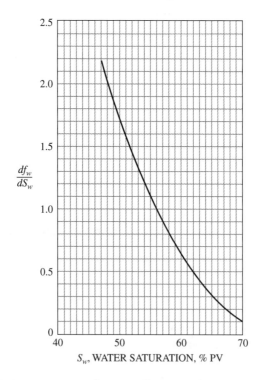

Figure 14-38. Derivative curve for Example 14-11. (*Permission to publish by the Society of Petroleum Engineers.*)

Phase 2. Calculation of Recovery Performance to Breakthrough

Step 1. Calculate Q_{iBT} using Equation 14-41:

$$Q_{iBT} = \left(\overline{S}_{wBT} - S_{wi}\right)$$
$$Q_{iBT} = (0.563 - 0.1) = 0.463$$

Step 2. Calculate cumulative water injected at breakthrough from Equation 14-43 or 14-44:

$$W_{iBT} = (PV)Q_{iBT}E_{ABT}$$
$$W_{iBT} = (310,320)(0.463)(0.71702) = 103,020\,\text{bbl}$$

Step 3. Calculate time to breakthrough:

$$t_{BT} = W_{iBT}/i_w$$
$$t_{BT} = 103,020/269 = 383\,\text{days}$$

Step 4. Calculate the displacement efficiency at breakthrough E_{DAB} from Equation 14-50:

$$E_{DBT} = \frac{\overline{S}_{wBT} - S_{wi}}{1 - S_{wi}}$$
$$E_{DBT} = \frac{0.563 - 0.1}{1 - 0.1} = 0.5144$$

Step 5. Calculate cumulative oil production at breakthrough by using Equation 14-51.

$$\left(N_p\right)_{BT} = N_S E_{DBT} E_{ABT}$$
$$\left(N_p\right)_{BT} = (232,740)(0.5144)(0.717) = 85,850\,\text{STB}$$

Notice that when $S_{gi} = 0$, the cumulative water injected at breakthrough W_{iBT} will displace an equivalent volume of oil, i.e.:

$$\left(N_p\right)_{BT} = \frac{W_{iBT}}{B_o} = \frac{103,020}{1.2} = 85,850\,\text{STB}$$

Step 6. Calculate the surface water cut WOR_s exactly at breakthrough from Equation 14-70:

$$E = \frac{S_{wf} - S_{wi}}{E_{ABT}\left(\overline{S}_{wBT} - S_{wi}\right)} = \frac{0.469 - 0.1}{0.717(0.563 - 0.1)} = 1.1115$$

$$\lambda = 0.2749\left(\frac{W_{iBT}}{W_{inj}}\right) = 0.2749\left(\frac{103,020}{103,020}\right) = 0.2749$$

$$\left(\Delta N_P\right)_{newly} = E\lambda = (1.1115)(0.2749) = 0.30555$$

$$WOR_s = \frac{f_{wf}\left[1 - \left(\Delta N_P\right)_{newly}\right]}{1 - f_{wf}\left[1 - \left(\Delta N_P\right)_{newly}\right]}\left(\frac{B_o}{B_w}\right) = \frac{0.798[1 - 0.30555]}{1 - 0.798[1 - 0.30555]}\left(\frac{1.2}{1}\right)$$

$$= 1.49\,\text{STB}\,/\,\text{STB}$$

Step 7. Set up the following table to describe the oil recovery performance to breakthrough (remember, $S_{gi} = 0$):

W_{inj}	$t = \dfrac{W_{inj}}{i_w}$	$N_P = \dfrac{W_{inj}}{B_o}$	$Q_o = \dfrac{i_w}{B_o}$	WOR_s	$Q_w = Q_o WOR_s$	W_P
bbl	days	STB	STB	STB/STB	STB/day	STB
0	0	0	0	0	0	0
20,000	74.34	16,667	224	0	0	0
40,000	148.7	33,333	224	0	0	0
60,000	223.0	50,000	224	0	0	0
80,000	297.4	66,667	224	0	0	0
103,020	383.0	85,850	224	1.49	334	0

Phase 3. Oil Recovery Calculations After Breakthrough

A step-by-step description of the oil recovery calcuations as well as a convienent worksheet to perform the computations after breakthrough are given below:

Column 1: Select several values of W_{inj}.

Column 2: For a constant injection rate, calculate the time t required to W_{inj} barrels of water.

Column 3: Divide values of W_{inj} in column 1 by W_{iBT}.

(1)	(2)	(3)	(4)	(5)	(6)	(7)	(8)	(9)	(10)	(11)	(12)	(13)	(14)	(15)	(16)
W_{inj}	$t = \dfrac{W_{inj}}{i_w}$	$\dfrac{W_{inj}}{W_{iBT}}$	E_A	$\dfrac{Q_i}{Q_{iBT}}$	$\left(\dfrac{Q_i}{Q_{iBT}}\right) Q_{iBT}$	$\left(\dfrac{df_w}{dS_w}\right) = \dfrac{1}{Q_i}$	S_{W2}	f_{W2}	$\bar{S}_{w2}$	E_D	N_p	W_p	WOR_s	Q_o	Q_w
(Assumed)	(days)		Eq. 14-65	Table 14-1			Fig. 14-38	Fig. 14-37	Eq. 14-45	Eq. 14-10	Eq. 14-6	Eq. 14-53	Eq. 14-70	Eq. 14-55	Eq. 14-56
103,020	383	1.0	0.717	1.000	0.463	2.159	0.470	0.800	0.563	0.514	85,850	0	1.49	99.6	149.4
123,620	460	1.2	0.767	1.193	0.552	1.810	0.492	0.843	0.579	0.532	100,292	3,270	2.03	83.3	169.1
144,230	536	1.4	0.809	1.375	0.636	1.570	0.507	0.870	0.590	0.544	106,986	15,847	2.55	71.73	182.9
164,830	613	1.6	0.846	1.548	0.717	1.394	0.524	0.893	0.601	0.557	113,820	28,246	3.12	62.3	194.4
185,440	689	1.8	0.879	1.715	0.794	1.259	0.534	0.905	0.610	0.567	119,559	41,969	3.63	55.7	202.2
206,040	766	2.0	0.906	1.875	0.869	1.151	0.543	0.920	0.163	0.570	128,417	51,940	4.24	49.4	209.5
257,550	958	2.5	0.969	2.256	1.046	0.956	0.562	0.937	0.628	0.587	136,618	93,608	5.56	39.8	221.3
309,060	1,149	3.0	1.000	2.619	1.214	0.823	0.575	0.944	0.637	0.597	138,946	142,325	22.33	11	255
412,080	1,532	4.0	1.000	3.336	1.545	0.647	0.597	0.963	0.653	0.614	142,902	240,598	31.23	8	259
515,100	1,915	5.0	1.000	4.053	1.877	0.533	0.611	0.973	0.660	0.622	144,764	341,383	43.24	6	262
618,120	2,298	6.0	1.000	4.770	2.208	0.453	0.622	0.980	0.664	0.627	145,928	443,006	58.8	4	264
824,160	3,064	8	1.000	6.204	2.872	0.348	0.637	0.985	0.676	0.640	148,954	645,415	78.8	3	265
1,030,200	3,830	10	1.000	7.638	3.536	0.283	0.650	0.990	0.683	0.648	150,816	849,221	119	2	267
1,545,300	5,745	15	1.000	11.223	5.199	0.192	0.677	0.995	0.697	0.663	154,307	1,360,132	239	1	268

Column 4: Calculate E_A from Equation 14-65 for value of W_{inj}/W_{iBT}.

Column 5: Determine the values of the ratio Q_i/Q_{iBT} from Table 14-1 for each value of W_{inj}/W_{iBT} in column 4.

Column 6: Obtain Q_i by multiplying column 5 by Q_{iBT}.

Column 7: The term $(df_w/dS_w)_{S_{w2}}$ is the reciprocal of column 6, i.e., $1/Q_i$.

Column 8: Determine the value of S_{w2} from the plot of df_w/dS_w vs. S_w as given in Figure 14-38.

Column 9: Calculate the value of f_{w2} that corresponds to each value of S_{w2} in column 8 by using Equation 14-24 or Figure 14-37.

Column 10: Calculate the average water saturation in the swept area $\overline{S}_{w2}$ by applying Equation 14-45.

Column 11: Calculate the displacement efficiency E_D by using Equation 14-10 for each value of $\overline{S}_{w2}$ in column 10.

Column 12: Calculate cumulative oil production N_p by using Equation 14-53.

Column 13: Calculate the cumulative water production W_p from Equation 14-53.

Column 14: Calculate the surface water-oil ratio WOR_s from Equation 14-70.

Column 15: Calculate the oil flow rate Q_o by using Equation 14-55.

Column 16: Determine the water flow rate Q_w by multiplying column 14 by column 15.

Results of the above waterflooding calculations are expressed graphically in Figure 14-39.

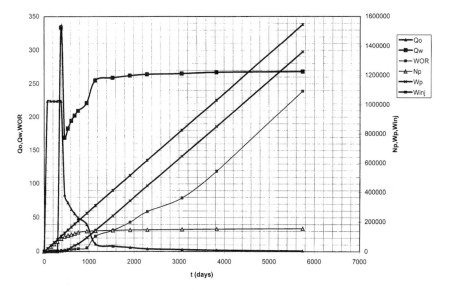

Figure 14-39. Performance curves for Example 14-11.

Note that all the areal sweep efficiency correlations that have been presented thus far are based on idealized cases with severe imposed assumptions on the physical characteristics of the reservoir. These assumptions include:

- Uniform isotropic permeability distribution
- Uniform porosity distribution
- No fractures in reservoir
- Confined patterns
- Uniform saturation distribution
- Off-pattern wells

To understand the effect of eliminating any of the above assumptions on the areal sweep efficiency, it has been customary to employ laboratory models to obtain more generalized numerical expressions. However, it is virtually impossible to develop a generalized solution when eliminating all or some of the above assumptions.

Landrum and Crawford (1960) have studied the effects of directional permeability on waterflood areal sweep efficiency. Figures 14-40 and 14-41 illustrate the impact of directional permeability variations on areal sweep efficiency for a line drive and five-spot pattern flood.

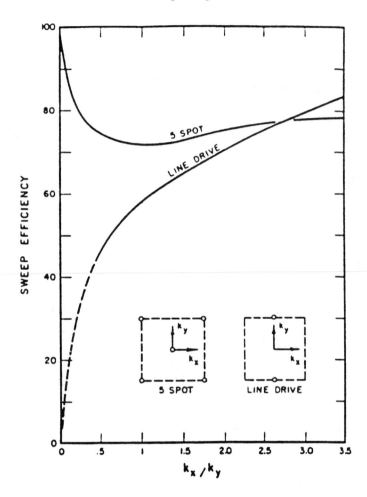

Figure 14-40. Effect of directional permeability on E_A. (*Permission to publish by the Society of Petroleum Engineers.*)

Two key elements affect the performance of waterflooding that must be included in recovery calculations: (1) Water injection rate, i.e., fluid injectivity, and (2) Effect of initial gas saturation on the recovery performance.

These key elements are discussed next.

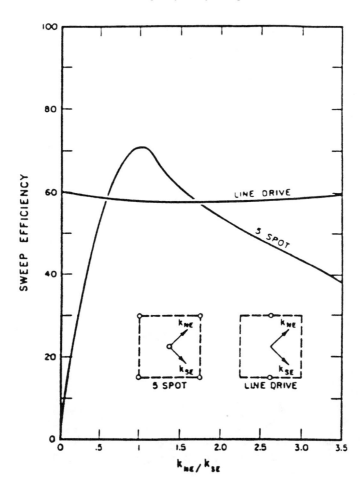

Figure 14-41. Effect of directional permeability on E_A. (*Permission to publish by the Society of Petroleum Engineers.*)

Fluid Injectivity

Injection rate is a key economic variable that must be considered when evaluating a waterflooding project. The waterflood project's life and, consequently, the economic benefits will be directly affected by the rate at which fluid can be injected and produced. Estimating the injection rate is also important for the proper sizing of injection equipment and pumps. Although injectivity can be best determined from small-scale pilot floods, empirical methods for estimating water injectivity for regular pattern floods have been proposed by Muskat (1948) and Deppe (1961). The authors derived their correlations based on the following assumptions:

- Steady-state conditions
- No initial gas saturation
- Mobility ratio of unity

Water injectivity is defined as the ratio of the water injection to the pressure difference between the injector and producer, or:

$$I = \frac{i_w}{\Delta P}$$

where I = injectivity, bbl/day/psi
       i_w = injection rate, bbl/day
      ΔP = difference between injection pressure and producing well bottom hole flowing pressure.

When the injection fluid has the same mobility as the reservoir oil (mobility ratio $M = 1$), the initial injectivity at the start of the flood is referred to as I_{base}, or:

$$I_{base} = \frac{i_{base}}{\Delta P_{base}}$$

where i_{base} = initial (base) water injection rate, bbl/day
      ΔP_{base} = initial (base) pressure difference between injector and producer

For a five-spot pattern that is completely filled with oil, i.e., $S_{gi} = 0$, Muskat (1948) proposed the following injectivity equation:

$$I_{base} = \frac{0.003541\, h\, k\, k_{ro} \Delta P_{base}}{\mu_o \left[\ln \dfrac{d}{r_w} - 0.619 \right]} \tag{14-71}$$

or

$$\left(\frac{i}{\Delta P} \right)_{base} = \frac{0.003541\, h\, k\, k_{ro}}{\mu_o \left[\ln \dfrac{d}{r_w} - 0.619 \right]} \tag{14-72}$$

where i_{base} = base (initial) water injection rate, bbl/day
         h = net thickness, ft
         k = absolute permeability, md
       k_{ro} = oil relative permeability as evaluated at S_{wi}
   ΔP_{base} = base (initial) pressure difference, psi
         d = distance between injector and producer, ft
       r_w = wellbore radius, ft

Several studies have been conducted to determine the fluid injectivity at mobility ratios other than unity. All of the studies concluded the following:

- At favorable mobility ratios, i.e., M < 1, the fluid injectivity declines as the areal sweep efficiency increases.
- At unfavorable mobility ratios, i.e., M > 1, the fluid injectivity increases with increasing areal sweep efficiency.

Caudle and Witte (1959) used the results of their investigation to develop a mathematical expression that correlates the fluid injectivity with the mobility ratio and areal sweep efficiency for five-spot patterns. The correlation may only be used in a *liquid-filled system*, i.e., $S_{gi} = 0$. The authors presented their correlation in terms of the **conductance ratio** γ, which is defined as the ratio of the fluid injectivity at any stage of the flood to the initial (base) injectivity, i.e.:

$$\gamma = \frac{\text{Fluid injectivity at any stage of the flood}}{\text{Base (initial)fluid injectivity}}$$

$$\gamma = \frac{\left(\dfrac{i_w}{\Delta P}\right)}{\left(\dfrac{i}{\Delta P}\right)_{base}} \tag{14-73}$$

Caudle and Witte presented the variation in the conductance ratio with EA and M in graphical form as shown in Figure 14-42. Note again that if an initial gas is present, the Caudle-Witte conductance ratio will not be applicable until the gas is completely dissolved or the system becomes liquid filled (fill-up occurs). The two possible scenarios for the practical use of Equation 14-73 follow:

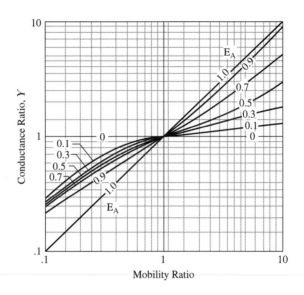

Figure 14-42. Conductance ratio curve. *(Permission to publish by the Society of Petroleum Engineers.)*

Scenario 1: Constant Injection Pressure and Variable Injection Rate

At constant injection pressure, i.e., $\Delta P_{base} = \Delta P$, the conductance ratio as expressed by Equation 14-73 can be written as:

$$\gamma = \frac{i_w}{i_{base}}$$

or

$$i_w = \gamma \, i_{base} \qquad (14\text{-}74)$$

where i_w = Water injection rate, bbl/day
  i_{base} = Base (initial) water injection rate, bbl/day

Scenario 2: Constant Injection Rate and Variable Injection Pressure

When the water injection rate is considered constant, i.e., $i_w = i_{base}$, the conductive ratio is expressed as:

$$\gamma = \frac{\Delta P_{base}}{\Delta P}$$

or

$$\Delta P = \frac{\Delta P_{base}}{\gamma} \qquad (14\text{-}75)$$

where ΔP_{base} = initial (base) pressure difference, psi
ΔP = pressure difference at any stage of flood, psi

The usefulness of the conductance ratio in determining the pressure and injectivity behavior of the five-spot system can be best described by the following example.

Example 14-12

Estimate the water-injection rate for the waterflood in Example 14-11 at 60,000 and 144,230 bbl of water injected. Assume that the pressure between the injector and producer will remain constant at 3000 psi.

Solution

Step 1. Calculate the distance between the injector and producer as shown in Figure 14-43, to give:

$$d = \sqrt{(660)^2 + (660)^2} = 933\,\text{ft}$$

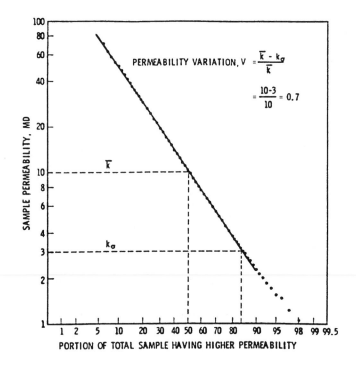

Figure 14-43. Forty-acre, five-spot spacing.

Step 2. Calculate the initial (base) injection rate from Equation 14-71:

$$\dot{i}_w = \frac{0.003541(5)(31.5)(1)(3000)}{(1)\left(\ln\dfrac{933}{1} - 0.619\right)} = 269.1\,\text{bbl}/\text{day}$$

Step 3. Notice that the cumulative water injected of 60,000 bbl is less than the amount of cumulative water injected at breakthrough of 103,020 bbl; therefore, M = 0.8 (remains constant until breakthrough) and E_A from Equation 14-63 is:

$$E_A = \frac{60,000}{(310,320)(0.563 - 0.10)} = 0.418$$

Step 4. Calculate the conductance ratios from Figure 14-42, to give $\gamma = 0.92$.

Step 5. Calculate the water-injection rate when the cumulative water injected reaches 60,000 bbl from Equation 14-74:

$$i_w = (269.1)(0.92) = 247.6 \, bbl/\, day$$

Step 6. After breakthrough when the cumulative water injected reaches 144,230 barrels of water, the average water saturation in the swept area is 59% (see Example 14-11), or

$$\overline{S}_{w2} = 0.59.$$

Step 7. Determine the water relative permeability k_{rw} at 0.59 water saturation (data of Example 14-11), to give $k_{rw} = 0.45$.

Step 8. Calculate the mobility ratio after breakthrough when $W_{inj} = 144,230$ from Equation 14-62:

$$M = \frac{0.45}{1} \frac{1}{0.5} = 0.9$$

Step 9. Calculate the areal sweep efficiency when $W_{inj} = 144,230$ from Equation 14-65: $E_A = 0.845$.

Step 10. Determine the conductance ratio from Figure 14-42: $\gamma = 0.96$.

Step 11. Calculate the water injection rate from Equation 14-76:

$$i_w = (269.1) \, (0.96) = 258.3 \, bbl/day$$

The conductance ratio can be expressed more conveniently in a mathematical form as follows. For an areal sweep efficiency of 100%, i.e., $E_A = 1.0$:

$$\gamma = M \qquad\qquad (14\text{-}76)$$

where γ = conductance ratio
         M = mobility ratio

For $1 < E_A < 100\%$:

$$\gamma = a_1 + (a_2 + a_3 E_A) M^{(a_4 + a_5 E_A)} + a_6 \left(\frac{M}{E_A}\right)^2 + a_7 M \qquad (14\text{-}77)$$

where the coefficients a_1 through a_7 are given below:

Coefficients	M < 1	M > 1
a_1	0.060635530	0.4371235
a_2	−2.039996000	0.5804613
a_3	0.025367490	−0.004392097
a_4	1.636640000	0.01001704
a_5	−0.624070600	1.28997700
a_6	−0.0002522163	0.00002379785
a_7	2.958276000	−0.015038340

Effect of Initial Gas Saturation

When a solution-gas-drive reservoir is under consideration for water-flooding, substantial gas saturation usually exists in the reservoir at the start of the flood. It is necessary to inject a volume of water that approaches the volume of the pore space occupied by the free gas before the oil is produced. This volume of water is called the **fill-up volume**. Because economic considerations dictate that waterflooding should occur at the highest possible injection rates, the associated increase in the reservoir pressure might be sufficient to redissolve all of the trapped gas S_{gt} back in solution. Willhite (1986) points out that relatively small increases in pressure frequently are required to redissolve the trapped gas (see Figure 14-2). Thus in waterflooding calculations, it is usually assumed that the trapped (residual) gas saturation is zero. A description of the displacement mechanism occurring under a five-spot pattern will indicate the nature of other secondary recovery operations. The five-spot pattern uses a producing well and four injection wells. The four injectors drive the crude oil inward to the centrally located producer. If only one five-spot pattern exists, the ratio of injection to producing wells is 4:1; however, on a full-field scale it includes a large number of adjacent five spots. In such a case, the number of injection wells compared to producing wells approaches a 1:1 ratio.

At the start of the waterflood process in a solution-gas-drive reservoir, the selected flood pattern is usually characterized by a high initial gas saturation of S_{gi} and remaining liquid saturations of S_{oi} and S_{wi}. When initial gas saturation exists in the reservoir, Craig, Geffen, and Morse (1955) developed a methodology that is based on dividing the flood performance into four stages. The method, known as the CGM method after the authors, was developed from experimental data in horizontal laboratory models representing a quadrant of a five spot. Craig et al. identified the following four stages of the waterflood as:

1. Start—interference
2. Interference—fill-up
3. Fill-up—water breakthrough
4. Water breakthrough—end of the project

A detailed description of each stage of the flood is illustrated schematically in Figures 14-44 through 14-46 and described below:

Stage 1: Start—Interference

At the start of the water-injection process in the selected pattern area of a solution-gas-drive reservoir, high gas saturation usually exists in the flood area as shown schematically in Figure 14-44. The current oil production at the start of the flood is represented by point **A** on the conventional flow rate–time curve of Figure 14-45. After the injection is initiated and a certain amount of water injected, an area of high water saturation called the **water bank** is formed around the injection well at the start of the flood. This stage of the injection is characterized by a **radial flow** system for both the displacing water and displaced oil. With continuous water injection, the water bank grows radially and displaces the oil phase that forms a region of high oil saturation that forms an **oil**

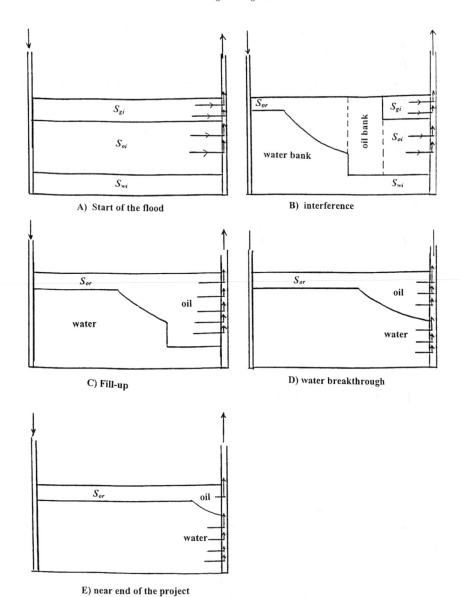

Figure 14-44. Stages of waterflooding.

bank. This radial flow continues until the oil banks, formed around adjacent injectors, meet. The place where adjacent oil banks meet is termed **Interference**, as shown schematically in Figure 14-46. During this stage of the flood, the condition around the producer is similar to that of the beginning of the flood, i.e., no changes are seen in the well flow rate Q_o as indicated in Figure 14-45 by point **B**. Craig, Geffen, and Morse (1955) summarized the computational steps during this stage of the flood, where radial flow prevails, in the following manner:

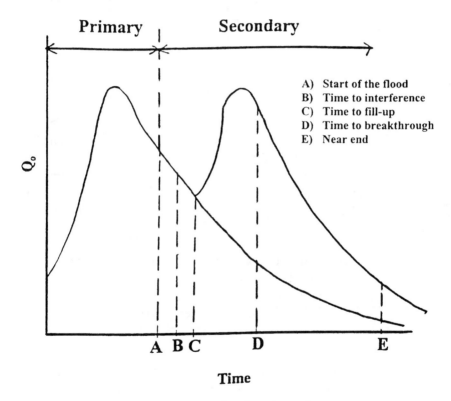

Figure 14-45. Predicted production history.

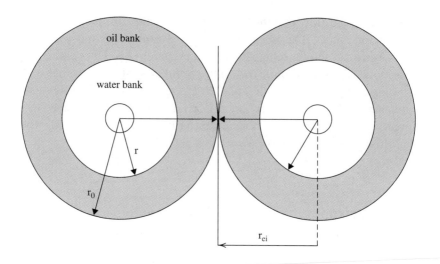

Figure 14-46. Interference of oil banks.

Step 1. Calculate the cumulative water injected to interference W_{ii} from the following expression:

$$W_{ii} = \frac{\pi h \phi S_{gi} r_{ei}^2}{5.615}$$

(14-78)

where W_{ii} = cumulative water injected to interference, bbl
S_{gi} = initial gas saturation
ϕ = porosity
r_{ei} = half the distance between adjacent injectors, ft

Step 2. Assume several successive values of cumulative water injected W_{inj}, ranging between 0 and W_{ii}, and calculate the water-injection rate at each assumed value of W_{inj} from:

$$i_w = \frac{0.00707\, h\, k\, \Delta P}{\left(\dfrac{\mu_w}{k_{rw}} \ln \dfrac{r}{r_w} + \dfrac{\mu_o}{k_{ro}} \ln \dfrac{r_o}{r} \right)}$$

(14-79)

where i_w = water injection, bbl/day

ΔP = pressure difference between injector and producer, psi

k = absolute permeability, md

k_{ro} = relative permeability of oil at S_{wi}

k_{rw} = relative permeability of water at $\overline{S}_{wBT}$

r_o = outer radius of the oil bank, ft

r = outer radius of the water bank, ft

r_w = wellbore radius, ft

The outer radii of the oil and water banks are calculated from:

$$r_o = \sqrt{\frac{5.615 W_{inj}}{\pi h \phi S_{gi}}} \qquad (14\text{-}80)$$

$$r = r_o \sqrt{\frac{S_{gi}}{\overline{S}_{wBT} - S_{wi}}} \qquad (14\text{-}81)$$

The flood performance from the start to interference, i.e., stage 1, is further discussed in the following example.

Example 14-13

Use the data given in Example 14-11 and determine the performance of the flood from the start to interference. The following additional data are available to reflect the assumption that a free gas exists at the start of the flood:

Initial oil saturation $S_{oi} = 0.75$

Initial gas saturation $S_{gi} = 0.15$

Initial water saturation $S_{wi} = 0.10$

Constant pressure difference $\Delta P = 3,000$ psi

Half distance between injectors $r_{ei} = 660$ ft

Distance between injector and producer $d = 932$ ft

Mobility ratio $M = 0.8$

$E_{ABT} = 0.717$

$Q_{iBT} = 0.463$

Pore volume = 310,320 bbl

Solution

Step 1. Calculate stock-tank oil in place at start of flood, NS:

$$N_S = \frac{(PV)S_{oi}}{B_{oi}} = \frac{(310,320)(1-0.15-0.10)}{1.20} = 193,950\,STB$$

Step 2. Calculate injected water at interference W_{ii} from Equation 14-78:

$$W_{ii} = \frac{\pi h \phi S_{gi} r_{ei}^2}{5.615} = \frac{\pi(5)(0.20)(0.15)(660)^2}{5.615} = 36,572\,bbl$$

Step 3. Simplify the calculations by expressing outer radii of the oil and water banks (Equations 14-80 and 14-81) as follows:

$$r_o = \sqrt{\frac{5.615 W_{inj}}{\pi h \phi S_{gi}}} = \sqrt{\frac{5.615 W_{inj}}{\pi(5)(0.2)(0.15)}} = 3.452\sqrt{W_{inj}}$$

$$r = r_o\sqrt{\frac{S_{gi}}{S_{wBT} - S_{wi}}} = r_o\sqrt{\frac{0.15}{0.563 - 0.10}} = 0.562 r_o$$

Step 4. Express the injectivity equation as represented by Equation 14-79 by:

$$i_w = \frac{0.00707 hk\Delta P}{\left(\dfrac{\mu_w}{k_{rw}}\ln\dfrac{r}{r_w} + \dfrac{\mu_o}{k_{ro}}\ln\dfrac{r_o}{r}\right)} = \frac{0.00707(5)(31.5)(3000)}{\dfrac{0.5}{0.4}\ln\left(\dfrac{r}{1}\right) + \dfrac{1.0}{1.0}\ln\left(\dfrac{r_o}{r}\right)}$$

$$i_w = \frac{3,340}{1.25\ln(r) + \ln\left(\dfrac{r_o}{r}\right)}$$

Step 5. Perform the required calculation for "stage one" in the following tabulated form:

W_{inj} (Assume)	r_o	r	i_w	$(i_w)_{avg}$	$\Delta_t = \Delta W_{inj} / (i_w)_{avg}$	$t = \Sigma(\Delta t)$
500	77.2	43.9	631.1		0.79	0.79
5,000	244.1	138	496.2	563.8	7.98	8.77
10,000	345.2	196.5	466.2	481.2	10.39	19.16
15,000	422.8	240.7	450.3	458.3	10.91	30.07
20,000	488.2	277.9	439.6	445.0	11.24	41.31
25,000	545.8	310.7	431.7	435.7	11.48	52.79
30,000	597.9	340.3	425.5	428.6	11.67	64.61
35,000	645.8	367.6	420.3	422.9	11.82	76.43
36,572	660	375.7	418.9	419.6	3.75	80.18

The above calculations indicate that time to interference t_{ii} will occur at 80.18 days after the start of the flood with a water-injection rate at interference i_{wi} of 418.9 bbl/day. Prior to oil bank interference, the injection rate i_w (or injectivity $i_w/\Delta P$) decreases because the radii of the oil and water banks, i.e., r_o and r, are continuously increasing. Notice that the **reservoir will not respond to the waterflood during this stage**. This delay in the reservoir response is mainly due to the fact that the injected water and the displaced oil are essentially moved to fill up part of the gas pore space. As described previously in Example 14-11, an immediate reservoir response to the waterflood can **only** occur when no gas exists at the start of the flood, i.e., $S_{gi} = 0$.

Stage 2: Interference—Fill-Up

This stage describes the period from interference until the fill-up of the preexisting gas space. **Fill-up** is the start of oil production response as illustrated in Figure 14-44 and by point **C** on Figure 14-45. The flow during this time is not strictly radial and is generally complex to quantify mathematically. Therefore, the flood performance can only be determined at the time of fill-up.

The required performance calculations at the fill-up are summarized in the following steps:

Step 1. Calculate the cumulative water injected at fill-up W_{if} by applying the following expression:

$$W_{if} = (PV)S_{gi} \qquad (14\text{-}82)$$

where W_{if} = cumulative water injected at fill-up, bbl
(PV) = total flood pattern pore volume, bbl
S_{gi} = initial gas saturation

The above equation suggests that while fill-up is occurring, the oil production rate is either zero or negligible, compared with the water injection rate. If the oil production rate Q_o prior to fill-up is significant, the cumulative water injected at the fill-up W_{if} must be increased by the total volume of oil produced from the start of injection to fill-up, i.e.:

$$W_{if} = (PV)S_{gi} + \frac{N_P}{B_o} \qquad (14\text{-}83)$$

where N_p = cumulative oil production from start of flood to fill-up, STB
B_o = oil formation volume factor, bbl/bbl

Equation 14-83 indicates that the fill-up time will also increase; in addition, it causes the fill-up time calculation to be iterative.

Step 2. Calculate the areal sweep efficiency at fill-up by using Equation 14-63, or:

$$E_A = \frac{W_{inj}}{(PV)\left(\overline{S}_{wBT} - S_{wi}\right)}$$

at fill-up:

$$E_A = \frac{W_{if}}{(PV)\left(\overline{S}_{wBT} - S_{wi}\right)}$$

Step 3. Using the mobility ratio and the areal sweep efficiency at fill-up, determine the conductance ratio γ from Figure 14-42 or Equation 14-77. Note that the conductance ratio can only be determined when the flood pattern is completely filled with liquids, which occurs at the fill-up stage.

Step 4. For a constant pressure difference, the initial (base) water injection rate i_{base} from Equation 14-71 is:

$$i_{base} = \frac{0.003541 \, h \, k \, k_{ro} \Delta P}{\mu_o \left[\ln \dfrac{d}{r_w} - 0.619 \right]}$$

Step 5. Calculate the water injection at fill-up i_{wf} and thereafter from Equation 14-74:

$$i_{wf} = \gamma \, i_{base}$$

The above expression is only valid when the system is filled with liquid, i.e., from the fill-up point and thereafter.

Step 6. Calculate the incremental time occurring from interference to fill-up from:

$$\Delta t = \frac{W_{if} - W_{ii}}{\dfrac{i_{wi} + i_{wf}}{2}}$$

The above expression suggests that the **fill-up will occur after interference.**

Example 14-14

Using the data given from Example 14-13, calculate the flood performance at fill-up. Results of Example 14-13 show:

- Time to interference $t_{ii} = 80.1$ days
- Cumulative water injected to interference $W_{ii} = 36,572$ bbl
- Water injection rate at interference $i_{wi} = 418.9$ bbl/day

Solution

Step 1. Calculate the cumulative water injected at fill-up from Equation 14-81:

$$W_{if} = (PV) S_{gi} = 310,320(0.15) = 46,550 \, \text{bbl}$$

Step 2. Calculate the areal sweep efficiency at fill-up from Equation 14-63:

$$E_A = \frac{W_{inj}}{(PV)(\overline{S}_{wBT} - S_{wi})} = \frac{W_{inj}}{310,320(0.563 - 0.10)} = \frac{W_{inj}}{143,678}$$

at fill-up:

$$E_A = \frac{W_{if}}{143,678} = \frac{46,550}{143,678} 0.324$$

Step 3. Given a mobility ratio M of 0.8 (Example 14-11) and E_A of 0.324, calculate the conductance ratio at the fill-up from Figure 14-42: $\gamma = 0.96$.

Step 4. Calculate the initial (base) injection rate from Equation 14-71:

$$i_{base} = \frac{0.003541(5)(31.5)(1)(3,000)}{(1.0)\left(\ln \dfrac{932}{1} - 0.619 \right)} = 269.1\, bbl/day$$

Step 5. Calculate the water injection rate at fill-up i_{wf} from Equation 14-74:

$$i_{wf} = \gamma i_{base} = (0.96)(269.1) = 258.2\, bbl/day$$

Step 6. Calculate the average water injection rate from interference to fill-up:

$$(i_w)_{avg} = \frac{i_{wi} + i_{wf}}{2} = \frac{418.9 + 258.2}{2} = 338.55\, bbl/day$$

Step 7. Calculate the incremented time occurring from interference to fill-up:

$$\Delta t = \frac{W_{if} - W_{ii}}{(i_w)_{avg}} = \frac{46,550 - 36,572}{338.55} = 29.5\, days$$

Thus the time to fill-up t_f is:

$$t_f = 80.2 + 29.5 = 109.7\, days$$

Stage 3: Fill-up—Water Breakthrough

The time to fill-up, as represented by point **C** on Figures 14-44 and 14-15, marks the following four events:

1. No free gas remaining in the flood pattern
2. Arrival of the oil-bank front to the production well
3. Flood pattern response to the waterflooding
4. Oil flow rate Q_o equal to the water injection rate i_w

During this stage, the oil production rate is essentially equal to the injection due to the fact that no free gas exists in the swept flood area. With continuous water injection, the leading edge of the water bank eventually reaches the production well, as shown in Figure 14-44, and marks the time to breakthrough. At breakthrough the water production rises rapidly.

The waterflood performance calculations are given by the following steps:

Step 1. Calculate cumulative water injected at breakthrough by using Equation 14-43 or 14-44 :

$$W_{iBT} = (PV)(\overline{S}_{wBT} - S_{wi})E_{ABT} = (PV)(Q_{iBT})E_{ABT}$$

Step 2. Assume several values of cumulative water injected W_{inj} between W_{if} and W_{iBT} and calculate the areal sweep efficiency at each Winj from Equation 14-63:

$$E_A = \frac{W_{inj}}{(PV)(\overline{S}_{wBT} - S_{wi})}$$

Step 3. Determine the conductance ratio γ for each assumed value of W_{inj} from Figure 14-42.

Step 4. Calculate the water injection rate at each W_{inj} by applying Equation 14-74:

$$i_w = \gamma i_{base}$$

Step 5. Calculate the oil flow rate Q_o during this stage from:

$$Q_o = \frac{i_w}{B_o} \qquad (14\text{-}84)$$

Step 6. Calculate cumulating oil production N_P from the following expression:

$$N_P = \frac{W_{inj} - W_{if}}{B_o} \qquad (14\text{-}85)$$

Example 14-15

Using the data given in Example 14-14, calculate the flood performance from the fill-up to breakthrough. Results of Example 14-14 show:

- Cumulative water injected to fill-up W_{if} = 46,550 bbl
- Water injection rate at fill-up i_{wf} = 358.2 bbl/day
- Time to fill-up t_f = 109.7 days

Solution

Step 1. Calculate cumulative water injected at breakthrough from Equation 14-43:

$$W_{iBT} = 310{,}320(0.563 - 0.1)(0.717) = 103{,}020 \, \text{bbl}$$

Step 2. Perform the required computations in the following tabulated form:

(1) W_{inj} (Assume)	(2) E_A	(3) γ	(4) i_w	(5) $(i_w)_{avg}$	(6) $\Delta t = \dfrac{\Delta W_{inj}}{(i_w)_{avg}}$	(7) $t = \Sigma \Delta t$	(8) $Q_o = i_w/B_o$	(9) $N_p = (W_{inj} - W_{if})/B_o$
46,550	0.324	0.96	258.6			109.7	215.5	0
50,000	0.348	0.95	255.6	257.1	13.27	123.0	213.0	2,844
60,000	0.418	0.94	253.0	254.3	39.32	162.3	210.8	11,177
70,000	0.487	0.94	253.0	253.0	39.53	201.8	210.8	19,511
80,000	0.557	0.93	251.7	251.7	39.73	241.6	208.6	27,844
90,000	0.626	0.92	247.6	249.0	40.16	281.7	206.3	36,177
100,000	0.696	0.92	247.6	247.6	40.39	322.1	206.3	44,511
103,020	0.717	0.91	244.9	246.3	12.26	334.4	204.1	47,027

The above calculations indicate that the time to breakthrough will occur
after 334.4 days from the start of flood with cumulative oil produced of
47,027 STB.

Stage 4: Water Breakthrough—End of the Project

After breakthrough, the water–oil ratio increases rapidly with a notice-
able decline in the oil flow rate as shown in Figure 14-45 by point **D**. The
swept area will continue to increase as additional water is injected. The
incrementally swept area will contribute additional oil production, while
the previously swept area will continue to produce both oil and water.

As represented by Equation 14-70, the WOR is calculated on the basis of the amounts of oil and water flowing from the swept region and the oil displaced from the newly swept portion of the pattern. It is assumed the oil from the newly swept area is displaced by the water saturation just behind the stabilized zone, i.e., S_{wf}.

The calculations during the fourth stage of the waterflooding process are given below:

Step 1. Assume several values for the ratio W_{inj}/W_{iBT} that correspond to the values given in Table 14-1, i.e., 1, 1.2, 1.4, etc.

Step 2. Calculate the cumulative water injected for each assumed ratio of (W_{inj}/W_{iBT}) from:

$$W_{inj} = \left(\frac{W_{inj}}{W_{iBT}} \right) W_{iBT}$$

Step 3. Calculate the areal sweep efficiency at each assumed (W_{inj}/W_{iBT}) from Equation 14-65:

$$E_A = E_{ABT} + 0.633 \log\left(\frac{W_{inj}}{W_{iBT}} \right)$$

Step 4. Calculate the ratio (Q_i/Q_{iBT}) that corresponds to each value of (W_{inj}/W_{iBT}) from Table 14-1 or Equation 14-69.

Step 5. Determine the total pore volumes of water injected by multiplying each ratio of Q_i/Q_{iBT} by Q_{iBT}, or:

$$Q_i = \left(\frac{Q_i}{Q_{iBT}} \right) Q_{iBT}$$

Step 6. From the definition of Q_i, as expressed by Equation 14-46, determine the slope $(df_w/dS_w)_{Sw2}$ for each value of Q_i by:

$$\left(\frac{df_w}{dS_w} \right)_{Sw2} = \frac{1}{Q_i}$$

Step 7. Read the value of S_{w2}, i.e., water saturation at the producing well, that corresponds to each slope from the plot of $(df_w/dS_w)S_{w2}$ vs. S_{w2} (see Example 14-11).

Step 8. Calculate the reservoir water cut at the producing well f_{w2} for each S_{w2} from Equation 14-24 or 14-37:

$$f_{w2} = \frac{1}{1 + \dfrac{\mu_w}{\mu_o} \dfrac{k_{ro}}{k_{rw}}}$$

or

$$f_{w2} = \frac{1}{1 + \left(\dfrac{\mu_w}{\mu_o}\right) a e^{b S_{w2}}}$$

Step 9. Determine the average water saturation in the swept area $\overline{S}_{w2}$ by applying Equation 14-45:

$$\overline{S}_{w2} = S_{w2} + \frac{1 - f_{w2}}{\left(\dfrac{df_w}{dS_w}\right)_{S_{w2}}}$$

Step 10. Calculate the surface water–oil ratio WOR_s that corresponds to each value of f_{w2} by applying Equation 14-70:

$$WOR_s = \frac{f_{w2}\left[1 - (\Delta N_P)_{newly}\right]}{1 - f_{w2}\left[1 - (\Delta N_P)_{newly}\right]} \left(\frac{B_o}{B_w}\right)$$

Step 11. Craig, Geffen, and Morse (1955) point out when calculating cumulative oil production during this stage that one must account for the oil lost to the unswept area of the flood pattern. To account for the lost oil, the authors proposed the following expression:

$$N_P = N_s E_D E_A - \frac{(PV)(1 - E_A)S_{gi}}{B_o} \qquad (14\text{-}86)$$

where E_D is the displacement efficiency and is given by Equation 14-9 as:

$$E_D = \frac{\overline{S}_w - S_{wi} - S_{gi}}{1 - S_{wi} - S_{gi}}$$

Step 12. Calculate cumulative water from the expression:

Water produced = Water injected – Oil produced
– Fill-up volume

or

$$W_P = \frac{w_{inj} - N_P B_o - (PV) S_{gi}}{B_w}$$

Step 13. Calculate $k_{rw} \, \overline{S}_{w2}$ at and determine the mobility ratio M **after breakthrough** from Equation 14-62:

$$M = \frac{k_{rw} @ \overline{S}_{w2}}{k_{ro} @ S_{wi}} \left(\frac{\mu_o}{\mu_w} \right)$$

Step 14. Calculate the conductance ratio γ from Figure 14-42.

Step 15. Determine the water injection rate from Equation 14-74
$i_w = \gamma \, i_{base}.$

Step 16. Calculate the oil and water production rates from Equations 14-55 and 14-56, respectively:

$$Q_o = \frac{i_w}{B_o + B_w WOR_s}$$

$$Q_w = Q_o WOR_s$$

Example 14-16

Complete the waterflooding performance calculation for Example 14-11 by predicting the performance of a producing WOR of 50 STB/STB, given:

W_{iBT}	= 103,020 bbl
$(N_p)_{BT}$	= 47,027 bbl
t_{BT}	= 334.4 days
E_{ABT}	= 0.717
$\underline{S}_{wf}$	= 0.469
$\overline{S}_{wBT}$	= 0.563
S_{wi}	= 0.10

Solution

The required calculations are conveniently performed in the following worksheet:

(1) W_{inj}/W_{iBT} (assume)	(2) $W_{inj}=$ (1) $\times W_{iBT}$	(3) E_A Eq. 14-65	(4) Q_i/Q_{iBT} Table 14-1	(5) $Q_i=$ (4) $\times Q_{iBT}$	(6) df_w/dS_w $=1/Q_i$	(7) S_{w2} Fig. 14-38	(8) f_{w2} Fig. 14-37	(9) $\bar{S}_{w2}$ Eq. 14-45
1.0	103,020	0.717	1.000	0.463	2.159	0.470	0.800	0.563
1.2	123,620	0.767	1.193	0.552	1.810	0.492	0.843	0.579
1.4	144,230	0.809	1.375	0.636	1.570	0.507	0.870	0.590
1.6	164,830	0.46	1.548	0.717	1.394	0.524	0.893	0.601
1.8	185,440	0.879	1.715	0.794	1.259	0.534	0.905	0.610
2.0	206,040	0.906	1.875	0.869	1.151	0.543	0.920	0.613
2.5	257,550	0.969	2.256	1.046	0.956	0.562	0.937	0.628
3.0	309,060	1.000	2.619	1.214	0.823	0.575	0.949	0.637
4.0	412,080	1.000	3.336	1.545	0.647	0.597	0.963	0.653
5.0	515,100	1.000	4.053	1.877	0.533	0.611	0.973	0.660
6.0	618,120	1.000	4.770	2.208	0.453	0.622	0.980	0.664

(10) $(\Delta N_p)_{newly}$ Eq. 14-69	(11) WOR_S Eq. 14-70	(12) E_D Eq. 14-9	(13) N_p Eq. 14-86	(14) W_p Eq. 14-87	(15) $k_{rw}@\bar{S}_{w2}$	(16) M Eq. 14-62	(17) γ Fig. 14-42	(18) i_w Eq. 14-76
0.3056	1.5*	0.4173	47,027	0	0.400	0.800	0.91	244.9
0.2545	2.03	0.4387	56,223	9,604	0.430	0.860	0.94	252.9
0.2182	2.55	0.4533	63,716	21,223	0.450	0.900	0.96	258.3
0.1910	3.12	0.4680	70,816	33,303	0.480	0.960	0.98	263.7
0.1697	3.63	0.480	77,138	46,326	0.500	1.000	1.0	269.1
0.1528	4.24	0.484	81,400	61,812	0.510	1.020	1.02	274.5
0.1223	5.56	0.504	93,518	98,780	0.542	1.084	1.08	287.9
0.000	22.3†	0.516	100,078	142,418	0.560	1.120	1.12	301.4
0.0000	31.2	0.5373	104,209	240,481	0.600	1.200	1.20	322.9
0.0000	43.2	0.5467	106,032	341,314	0.625	1.250	1.25	336.4
0.0000	58.8	0.5520	107,060	443,100	0.635	1.270	1.27	341.8

(19)	(20)	(21) $\Delta t =$ (20)÷(19)	(22) $t = \Sigma(\Delta t)$ days	(23) Q_o Eq. 14-55	(24) $Q_w =$ (11)×(22)
$(i_w)_{avg}$	ΔW_{inj}				
			334.4	90.7	136.1
248.9	20,600	82.7	417.1	78.6	159.6
255.6	20,610	80.6	497.7	69.6	177.5
261.0	20,600	79.0	576.5	61.5	191.9
266.4	20,610	77.3	663.8	56.5	205.1
271.8	20,600	75.9	729.7	51.5	218.4
282.2	51,510	183.0	912.7	43.1	239.6
295.6	103,020	349.0	1261.7	12.8	285.4
312.7	103,020	330.0	1591.5	9.9	308.9
330.6	103,020	312.0	1903.5	7.6	328.3
339.1	103,020	305.0	2208.3	5.7	335.2

*Equation 14-70.
†Equation 14-20.

To illustrate the use of Equation 14-70 in calculating the WOR_s values of column 11, the value of the surface water–oil ratio when W_{inj}/W_{iBT} reaches 2 bbl/bbl is calculated below:

Step 1. Calculate the coefficient E, which remains constant for all the values of W_{inj}/W_{iBT}:

$$E = \frac{S_{wf} - S_{wi}}{E_{ABT}(\overline{S}_{wBT} - S_{wi})} = \frac{0.469 - 0.1}{0.717(0.563 - 0.1)} = 1.1115$$

Step 2. Calculate the parameter λ:

$$\lambda = 0.2749\left(\frac{W_{iBT}}{W_{inj}}\right) = 0.2749\left(\frac{1}{2}\right) = 0.13745$$

Step 3. Calculate the incremental oil produced from the newly swept area when $(W_{inj}/W_{iBT}) = 2$ from Equation 14-69:

$$\left(\Delta N_p\right)_{newly} = E\,\lambda = (1.1115)(0.13745) = 0.1528\,\text{bbl/bbl}$$

Step 4. Calculate WOR_s from Equation 14-70:

$$WOR_s = \frac{0.920(1 - 0.1528)}{1 - 0.920(1 - 0.1528)}\left(\frac{1.20}{1.00}\right) = 4.24\,\text{STB/STB}$$

Figure 14-47 documents results of Examples 14-15 and 14-16 graphically.

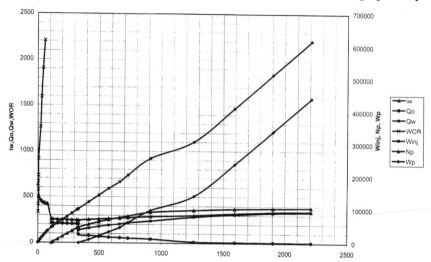

Figure 14-47. Performance curves for Example 14-16.

Water Fingering and Tonguing

In thick, dipping formations containing heavy viscous oil, water tends to advance as a "tongue" at the bottom of the pay zone. Similarly, displacement of oil with a gas will result in the gas attempting to overrun the oil due to gravity differences unless stopped by a shale barrier within the formation or by a low overall effective vertical permeability. In linear laboratory experiments, it was observed that the fluid interface remains horizontal and independent of fluid velocity when the viscosities of the two phases are equal. If the oil and water have different viscosities, the original horizontal interface will become tilted.

In a dipping reservoir, Dake (1978) developed a gravity segregation model that allows the calculation of the critical injection rate I_{crit} that is required to propagate a stable displacement. The condition for stable displacement is that the angle between the fluid interface and the direction of flow should remain constant throughout the displacement as shown in Figure 14-48. Dake introduced the two parameters, the Dimensionless Gravity Number "G" and the End-point Mobility Ratio M*, that can be used to define the stability of displacement. These two parameters are defined by the following relationships:

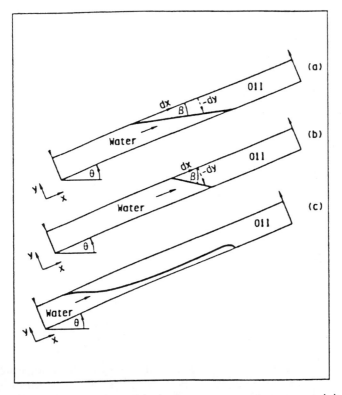

Figure 14-48. Stable and unstable displacement in gravity segregated displacement: (a) stable: $G > m - 1$, $M > 1$, and $\beta < \theta$; (b) stable: $G > M - 1$, $M < 1$, $\beta > \theta$; and (c) unstable: $G < M - 1$. (Courtesy of Elsevier.)

1. **Dimensionless gravity number.** The dimensionless gravity number G is given by:

$$G = \frac{7.853 \times 10^{-6} k \, k_{rw} A (\rho_w - \rho_o) \sin(\theta)}{i_w \mu_w} \qquad (14\text{-}87)$$

where
k = absolute permeability, md
k_{rw} = relative permeability to water as evaluated at S_{or}
A = cross-sectional area
ρ_w = water density, lb/ft^3
θ = dip angle

2. **End-point mobility ratio.** The end-point mobility ratio M^* is defined by:

$$M^* = \frac{k_{rw} \, @ \, S_{or}}{k_{ro} \, @ \, S_{wi}} \frac{\mu_o}{\mu_o} \qquad (14\text{-}88)$$

Dake used the above two parameters to define the following stability criteria:

- **If M* > 1.** The displacement is stable if G > (M* − 1), in which case the fluid interface angle β < θ. The displacement is unstable if G < (M* − 1).

- **If M* = 1.** This is a very favorable condition, because there is no tendency for the water to bypass the oil. The displacement is considered unconditionally stable and is characterized by the fact that the interface rises horizontally in the reservoir, i.e., β = θ.

- **If M* < 1.** When the end-point mobility ratio M* is less than unity, the displacement is characterized as unconditionally stable displacement with B > θ (Figure 14-48b).

The author also defined the critical flow rate, i_{crit} by:

$$i_{crit} = \frac{7.853 \times 10^{-6} k\, k_{rw} A (\rho_w - \rho_o) \sin(\alpha)}{\mu_w (M * -1)} \qquad (14\text{-}89)$$

where i_{crit} = critical water injection rate, bbl/day
$\quad\quad k_{rw}$ = relative permeability to water @ S_{or}
$\quad\quad \mu_w$ = water viscosity, cp
$\quad\quad k$ = absolute permeability, md
$\quad\quad \theta$ = dip angle

Example 14-17

A tilted linear reservoir is under consideration for waterflooding. The rock and fluid properties are given below:

Cross-sectional area A = 31,250 ft^2
Absolute permeability k = 70 md
Dip angle θ = 20°
Water density ρ_w = 63 lb/ft^3
Oil density ρ_o = 35 lb/ft^3
Water viscosity μ_w = 0.5 cp
Oil viscosity μ_o = 3.0 cp
k_{rw} @ S_{or} = 0.35
k_{ro} @ S_{wi} = 1.00
Water-injection rate = 800 bbl/day

Calculate the critical water injection rate for water displacing oil updip.

Solution

Step 1. Calculate the end-point mobility ratio from Equation 14-88:

$$M* = \frac{0.35}{1.00} \frac{3.0}{0.5} = 2.0$$

Step 2. Calculate the critical injection rate by using Equation 14-89:

$$i_{crit} = \frac{7.853 \times 10^{-6}(70)(0.35)(31,250)(63-35)\sin(20)}{0.5(2.1-1)} = 106\,bbl/day$$

The above example indicates that the water injection rate must be 106 bbl/day to ensure a stable displacement, which, when compared with the proposed injection rate of 800 bbl/day, is perhaps not economically feasible to maintain.

Dake (1978) and Willhite (1986) presented a comprehensive treatment of water displacement under segregated flow conditions.

III. VERTICAL SWEEP EFFICIENCY

The vertical sweep efficiency, E_V, is defined as the fraction of the vertical section of the pay zone that is the injection fluid. This particular sweep efficiency depends primarily on (1) the mobility ratio and (2) total volume injected. As a consequence of the nonuniform permeabilities, any injected fluid will tend to move through the reservoir with an irregular front. In the more permeable portions, the injected water will travel more rapidly than in the less permeable zone.

Perhaps the area of the greatest uncertainty in designing a waterflood is the quantitative knowledge of the permeability variation within the reservoir. The degree of permeability variation is considered by far the most significant parameter influencing the vertical sweep efficiency.

To calculate the vertical sweep efficiency, the engineer must be able to address the following three problems:

1. How to describe and define the permeability variation in mathematical terms
2. How to determine the minimum number of layers that are sufficient to model the performance of the fluid
3. How to assign the proper average rock properties for each layer (called the **zonation problem**)

A complete discussion of the above three problems is given below.

Reservoir Vertical Heterogeneity

As pointed out in Chapter 4, one of the first problems encountered by the reservoir engineer is that of organizing and utilizing the large amount of data available from core and well logging analyses. Although porosity and connate water saturation may vary aerially and vertically within a reservoir, the most important rock property variation to influence water-flood performance is permeability. Permeabilities pose particular problems because they usually vary by more than an order of magnitude between different strata.

Dykstra and Parsons (1950) introduced the concept of the permeability variation V, which is designed to describe the **degree of heterogeneity** within the reservoir. The value of this uniformity coefficient ranges between zero for a completely homogeneous system and one for a completely heterogeneous system. Example 4-18 of Chapter 4 illustrates the required computational steps for determining the coefficient V that is given by Equation 4-70, as:

$$V = \frac{k_{50} - k_{84.1}}{k_{50}}$$

To further illustrate the use of the Dykstra and Parsons permeability variation, Craig (1971) proposed a hypothetical reservoir that consists of 10 wells (wells A through J) with detailed permeability data given for each well, as shown in Table 14-2. Each well is characterized by 10 values of permeability with each value representing 1 ft of pay.

Arranging all of these permeability values, i.e., the entire 100 permeability values, from maximum to minimum, Craig (1971) obtained the permeability distribution as shown in the log-probability scale of Figure 14-49. The resulting permeability distribution indicates that this hypothetical reservoir is characterized by a permeability variation of 70%, or:

$$V = \frac{k_{50} - k_{84.1}}{k_{50}} = \frac{10 - 3}{10} = 0.7$$

Table 14-2
Ten-Layer Hypothetical Reservoir
(Permission to publish by the Society of Petroleum Engineers)

CORE ANALYSIS FOR HYPOTHETICAL RESERVOIR
Cores from 10 Wells, A Through J; Each Permeability Value (md) Represents 1 ft of Pay

Depth (ft)	A	B	C	D	E	F	G	H	I	J
6791	2.9	7.4	30.4	3.8	8.6	14.5	39.9	2.3	12.0	29.0
6792	11.3	1.7	17.6	24.6	5.5	5.3	4.8	3.0	0.6	99.0
6793	2.1	21.2	4.4	2.4	5.0	1.0	3.9	8.4	8.9	7.6
6794	167.0	1.2	2.6	22.0	11.7	6.7	74.0	25.5	1.5	5.9
6795	3.6	920.0	37.0	10.4	16.5	11.0	120.0	4.1	3.5	33.5
6796	19.5	26.6	7.8	32.0	10.7	10.0	19.0	12.4	3.3	6.5
6797	6.9	3.2	13.1	41.8	9.4	12.9	55.2	2.0	5.2	2.7
6798	50.4	35.2	0.8	18.4	20.1	27.8	22.7	47.4	4.3	66.0
6799	16.0	71.5	1.8	14.0	84.0	15.0	6.0	6.3	44.5	5.7
6800	23.5	13.5	1.5	17.0	9.8	8.1	15.4	4.6	9.1	60.0

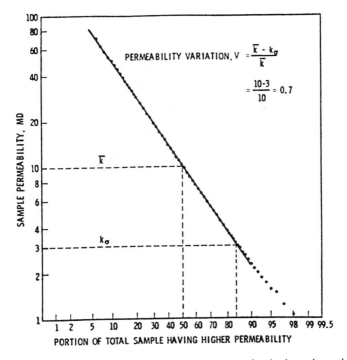

Figure 14-49. Determination of permeability variation for the hypothetical reservoir. *(Permission to publish by the Society of Petroleum Engineers.)*

Minimum Number of Layers

Based on a computer study, Craig (1971) outlined some guidelines for selecting the minimum number of layers needed to predict the performance of a reservoir under waterflooding operation. The author simulated the performance of a waterflood five-spot pattern that is composed of 100 layers with permeability variations ranging from 0.4 to 0.8. The minimum number of layers required to match results of the 100-layer model was determined as a function of mobility ratio M and permeability variation V. Tables 14-3 through 14-5 summarize results of these simulations and provide a guide to selection of the number of layers for five-spot patterns.

Table 14-3
Minimum Number of Layers for WOR > 2.5
(Permission to publish by the Society of Petroleum Engineers)

Mobility Ratio	Permeability Variation							
	0.1	0.2	0.3	0.4	0.5	0.6	0.7	0.8
0.05	1	1	2	4	10	20	20	20
0.1	1	1	2	4	10	20	100	100
0.2	1	1	2	4	10	20	100	100
0.5	1	2	2	4	10	20	100	100
1.0	1	3	3	4	10	20	100	100
2.0	2	4	4	10	20	50	100	100
5.0	2	5	10	20	50	100	100	100

Table 14-4
Minimum Number of Layers for WOR > 5
(Permission to publish by the Society of Petroleum Engineers)

Mobility Ratio	Permeability Variation							
	0.1	0.2	0.3	0.4	0.5	0.6	0.7	0.8
0.05	1	1	2	4	5	10	10	20
0.1	1	1	2	4	10	10	10	100
0.2	1	1	2	4	10	10	20	100
0.5	1	2	2	4	10	10	20	100
1.0	1	2	3	4	10	10	20	100
2.0	2	3	4	5	10	10	50	100
5.0	2	4	5	10	20	100	100	100

Table 14-5
Minimum Number of Layers for WOR > 10
(Permission to publish by the Society of Petroleum Engineers)

Mobility Ratio	Permeability Variation							
	0.1	0.2	0.3	0.4	0.5	0.6	0.7	0.8
0.05	1	1	1	2	4	5	10	20
0.1	1	1	1	2	5	5	10	20
0.2	1	1	2	3	5	5	10	20
0.5	1	1	2	3	5	5	10	20
1.0	1	1	2	3	5	10	10	50
2.0	1	2	3	4	10	10	20	100
5.0	1	3	4	5	10	100	100	100

Example 14-18

A reservoir is under consideration for waterflooding. The heterogeneity of the reservoir is described by a permeability variation V of 40%. The mobility ratio is determined as 2.0. Determine the minimum number of layers required to perform waterflooding calculations.

Solution

Table 14-4 shows that the minimum number of layers required to match the performance of the 100-layer computer model with a producing WOR above 10 STB/STB is 4 layers.

The Zonation Problem

In waterflooding calculations, it is frequently desirable to divide the reservoir into a number of layers that have equal thickness but different permeabilities and porosities. Traditionally, two methods are used in the industry to assign the proper average permeability for each layer: (1) the positional method or (2) the permeability ordering method.

Positional Method

The positional method describes layers according to their relative location within the vertical rock column. This method assumes that the injected fluid remains in the same elevation (layer) as it moves from the injector to the producer. Miller and Lents (1966) successfully demonstrated this concept in predicting the performance of the Bodcaw Reservoir Cycling Project. The authors proposed that the average permeability in a selected layer (elevation) should be calculated by applying the geometric-average permeability as given by Equation 4-54 or 4-55:

$$k_{avg} = \exp\left[\frac{\sum_{i=1}^{n} h_i \ln(k_i)}{\sum_{i=1}^{n} h_i}\right]$$

If all the thicknesses are equal, then:

$$k_{avg} = \left(k_1 k_2 k_3 \ldots k_n\right)^{1/n}$$

Example 14-19

Using the core analysis data given in Table 14-2 for the 10-well system, assign the proper average permeability for each layer if the reservoir is divided into:

a. 10 equal-thickness layers, each with a 1-ft thickness
b. 5 equal-thickness layers, each with a 2-ft thickness

Solution

a. Using the positional method approach and applying Equation 4-55, calculate the permeability for each 1-ft layer:

Layer $1 = [(2.9)(7.4)(30.4)(3.8)(8.6)(14.5)(39.9)(2.3)(12.0)(29.0)]^{1/10} = 10$ md

A similar approach for calculating the permeability for the remaining layers yields:

Layer #	Permeability, md
1	10.0
2	6.8
3	4.7
4	10.4
5	20.5
6	12.1
7	8.6
8	18.4
9	14.3
10	10.9

b. Five equal-thickness layers:

Step 1. Calculate the arithmetic-average permeability for each layer per location:

Depth	A	B	C	D	E	F	G	H	I	J
6791–92	7.10	4.55	24.00	14.20	7.05	9.90	22.35	2.65	6.30	64.00
93–94	84.55	11.20	3.50	12.20	8.35	3.85	38.95	16.95	5.20	6.75
95–96	11.55	473.30	22.40	21.20	13.60	10.50	69.50	8.25	3.40	20.00
97–98	28.65	19.20	6.95	30.10	14.75	20.35	38.95	24.70	4.75	34.35
99–00	19.75	42.50	1.65	15.50	46.90	13.05	10.70	5.45	26.80	32.85

Step 2. Use the geometric-average method to calculate the permeability in each layer:

$$\text{Layer } 1 = \left[(7.1)(4.5)(24.0)(14.2)(7.05)(9.9)(22.35)(2.65)(6.3)(64.0) \right]^{1/10} = 10.63$$

Remaining layers are treated in the same fashion to give:

Layer #	Permeability, md
1	10.63
2	11.16
3	20.70
4	18.77
5	15.26

Permeability Ordering Method

The permeability ordering method is essentially based on the Dykstra and Parsons (1950) permeability sequencing technique. The core analysis permeabilities are arranged in a decreasing permeability order and a plot like that shown in Figure 14-49 is made. The probability scale is divided into equal-percent increments with each increment representing a layer. The permeability for each layer is assigned to the permeability value that **corresponds to the midpoint of each interval.**

Example 14-20

For the 10-layer system of Example 14-19, determine the permeability for each layer by using the permeability ordering approach.

Solution

From Figure 14-49, determine the permeability for each of the 10 layers by reading the permeability at the following midpoints: 5, 15, 25, 35, 45, 55, 65, 75, 85, and 95%:

Layer #1	Permeability Ordering	Positional Approach
1	84.0	10.0
2	37.0	6.8
3	23.5	4.7
4	16.5	10.4
5	12.0	20.5
6	8.9	12.1
7	6.5	8.6
8	4.6	18.4
9	3.0	14.3
10	1.5	10.9

Porosity assignments for the selected reservoir layers may also be treated in a similar manner to that of the permeability ordering approach. All porosity measurements are arranged in decreasing order and a plot of the porosity versus percentage of thickness with greater porosity is made on a Cartesian-probability scale (rather than a log-probability scale). The porosity of each layer can then be obtained for each interval of thickness selected.

The permeability ordering technique is perhaps the most widely used approach in the petroleum industry when determining the vertical sweep efficiency.

Calculation of Vertical Sweep Efficiency

Basically two methods are traditionally used in calculating the vertical sweep efficiency EV: (1) Stiles' method and (2) the Dykstra–Parsons method. These two methods assume that the reservoir is composed of an idealized layered system, as shown schematically in Figure 14-50. The layered system is selected based on the permeability ordering approach with layers arranged in order of descending permeability. The common assumptions of both methods are:

- No cross-flow between layers
- Immiscible displacement
- Linear flow
- The distance water has traveled through each layer is proportional to the permeability of the layer
- Piston-like displacement

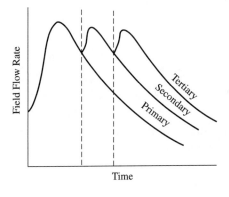

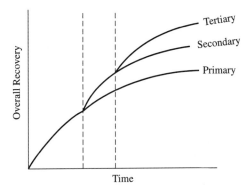

Figure 14-50. Idealized layered system.

The basic idea used in Stiles' method and the Dykstra–Parsons method is to determine the frontal position in each layer at the time water breakthrough occurs in successive layers. If the flow capacity of each layer is defined by the product of permeability and thickness, i.e., kh, then the water and oil flow rates from all layers can be calculated to yield the producing water–oil ratio.

Stiles' Method

Stiles (1949) proposed an approach that takes into account the effect of permeability variations in predicting the performance of waterfloods. Stiles assumes that in a layered system, the water breakthrough occurs in a sequence that starts in the layer with the highest permeability. Assuming that the reservoir is divided into n layers that are arranged in a descending permeability order with breakthrough occurring in a layer i, all layers from 1 to i have already been swept by water. The remaining layers obviously have not reached breakthrough.

Based on the above concept, Stiles proposed that the vertical sweep efficiency can be calculated from the following expression:

$$E_V = \frac{k_i \sum_{j=1}^{i} h_j + \sum_{j=i+1}^{n} (kh)_j}{k_i h_t} \qquad (14\text{-}90)$$

where   i = breakthrough layer, i.e., i = 1,2,3, . . . n
n = total number of layers
E_V = vertical sweep efficiency
h_t = total thickness, ft
h_i = layer thickness, ft

If the values of the porosity vary between layers, Equation 14-90 can be written:

$$E_V = \frac{\left(\dfrac{k}{\phi}\right)_i \sum_{j=1}^{i} (\phi h)_j + \sum_{j=i+1}^{n} (kh)_j}{\left(\dfrac{k}{\phi}\right)_i \sum_{j=1}^{n} (\phi h)_j} \qquad (14\text{-}91)$$

Stiles also developed the following expression for determining the surface water–oil ratio as breakthrough occurs in any layer:

$$WOR_s = A \left[\frac{\sum_{j=1}^{i} (kh)_j}{\sum_{j=i+1}^{n} (kh)_j} \right] \tag{14-92}$$

with

$$A = \frac{k_{rw}}{k_{ro}} \frac{\mu_o B_o}{\mu_w B_w} \tag{14-93}$$

where WOR_s = surface water–oil ratio, STB/STB
  k_{rw} = relative permeability to water at S_{or}
  k_{ro} = relative permeability to oil at S_{wi}

Both the vertical sweep efficiency and surface WOR equations are used simultaneously to describe the sequential breakthrough as it occurs in layer 1 through layer n. It is usually convenient to represent the results of these calculations graphically in terms of $\log(WOR_s)$ as a function of E_V.

Example 14-21

The Dykstra and Parsons (1950) permeability ordering approach is used to describe a reservoir by the following five-layer system:

Layer	k, md	h, ft
1	120	15
2	90	15
3	70	10
4	55	10
5	30	10

The reservoir is under consideration for further development by water injection. The following additional information is available:

$k_{rw} @ S_{or} = 0.3$
$k_{ro} @ S_{wi} = 0.9$
  $\mu_o = 2.0$ cp
  $\mu_w = 0.5$ cp
  $B_o = 1.20$ bbl/STB
  $B_w = 1.01$ bbl/STB
  $h_t = 60$ ft

Calculate the vertical sweep efficiency and surface water-oil ratio using Stiles' method:

Solution

Step 1. Calculate parameter A using Equation 14-93:

$$A = \frac{0.3}{0.9} \frac{(2.0)(1.20)}{(0.5)(1.01)} = 1.584$$

Step 2. Calculate E_V and WOR_s when breakthrough occurs in the first layer, i.e., $i = 1$, by applying Equations 14-90 and 14-92:

$$E_V = \frac{k_i \sum\limits_{j=1}^{i} h_j \sum\limits_{j=2}^{5} (kh)_j}{k_i h_t}$$

$$E_V = \frac{k_1 h_1 + [k_2 h_2 + k_3 h_3 + k_4 h_4 + k_5 h_5]}{k_i h_t}$$

$$E_V = \frac{(120)(15) + [(90)(15) + (70)(10) + (55)(10) + (30)(10)]}{(120)(60)} = 0.653$$

$$WOR_s = (1.584) \frac{\sum\limits_{j=1}^{1} (kh)_j}{\sum\limits_{j=2}^{5} (kh)_j}$$

$$WOR_s = (1.584) \frac{(kh)_1}{(kh)_2 + (kh)_3 + (kh)_4 + (kh)_5}$$

$$WOR_s = (1.584) \frac{(120)(15)}{(90)(15) + (70)(10) + (55)(10) + (30)(10)}$$

$$= 0.983 \; STB / STB$$

Step 3. Calculate E_V and WOR_s when water breakthrough occurs in the second layer, i.e., $i = 2$:

$$E_V = \frac{k_2 \sum_{j=1}^{2} h_j + \sum_{j=3}^{5} (kh)_j}{k_2 h_t}$$

$$E_V = \frac{k_2 (h_1 + h_2) + \left[(kh)_3 + (kh)_4 + (kh)_5 \right]}{k_2 h_t}$$

$$E_V = \frac{90(15+15) + \left[(70)(10) + (55)(10) + (30)(10) \right]}{(90)(60)} = 0.787$$

$$WOR_s = (1.584) \frac{\sum_{j=1}^{2} (kh)_j}{\sum_{j=3}^{5} (kh)_j}$$

$$WOR_s = (1.584) \frac{(kh)_1 + (kh)_2}{(kh)_3 + (kh)_4 + (kh)_5}$$

$$WOR_s = (1.584) \frac{(120)(15) + (90)(15)}{(70)(10) + (55)(10) + (30)(10)} = 3.22 \text{ STB/STB}$$

Step 4. The required calculations can be performed more conveniently in the following worksheet:

(1)	(2)	(3)	(4)	(5)	(6)	(7)	(8)	(9)	(10)
Layer	k_i	h_i	Σh_i	$k_i \Sigma h_i$	$k_i h_i$	$\Sigma k_i h_i$	$h_t k_i$	$E_V = \dfrac{(5) + [\text{sum} + (7)]}{8}$	$WOR_s = 1.584 \left[\dfrac{(7)}{\text{sm} - (7)} \right]$
1	120	15	15	1,800	1,800	1,800	7,200	0.653	0.983
2	90	15	30	2,700	1,350	3,150	5,400	0.787	3.22
3	70	10	40	2,800	700	3,850	4,200	0.869	7.17
4	55	10	50	2,750	550	4,400	3,300	0.924	23.23
5	30	10	60	1,800	300	4,700	1,800	1.000	—
				sum = 4700					

Figure 14-51 shows the resulting relationship between the vertical sweep efficiency and producing WOR. The curve can be extended to WOR = 0 to give the vertical sweep efficiency at breakthrough E_V.

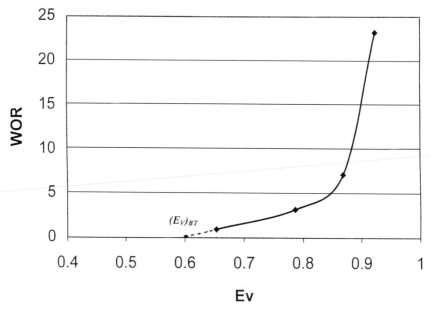

Figure 14-51. WOR vs. E_V.

The Dykstra–Parsons Method

Dykstra and Parsons (1950) correlated the vertical sweep efficiency with the following parameters:

- Permeability variation V
- Mobility ratio M
- Water–oil ratio WOR_r as expressed in bbl/bbl

The authors presented their correlation in a graphical form for water–oil ratios of 0.1, 0.2, 0.5, 1, 2, 5, 10, 25, 50, and 100 bbl/bbl. Figure 14-52 shows Dykstra and Parsons' graphical correlation for a WOR of 50 bbl/bbl. Using a regression analysis model, de Souza and Brigham (1981) grouped the vertical sweep efficiency curves for $0 \le M \le 10$ and $0.3 \le V \le 0.8$ into one curve as shown in Figure 14-53. The authors used a combination of WOR, V, and M to define the correlation parameter Y of Figure 14-53:

$$Y = \frac{(WOR + 0.4)(18.948 - 2.499V)}{(M - 0.8094V + 1.137)10^x} \qquad (14\text{-}94)$$

with

$$x = 1.6453V^2 + 0.935V - 0.6891 \qquad (14\text{-}95)$$

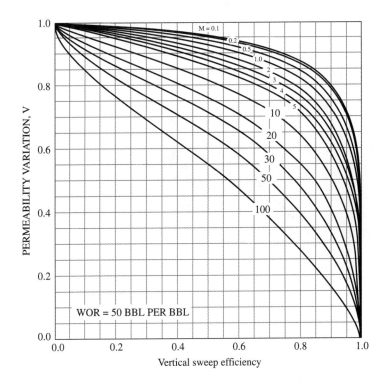

Figure 14-52. Vertical sweep efficiency curves for WOR = 50. (*Permission to publish by the Society of Petroleum Engineers.*)

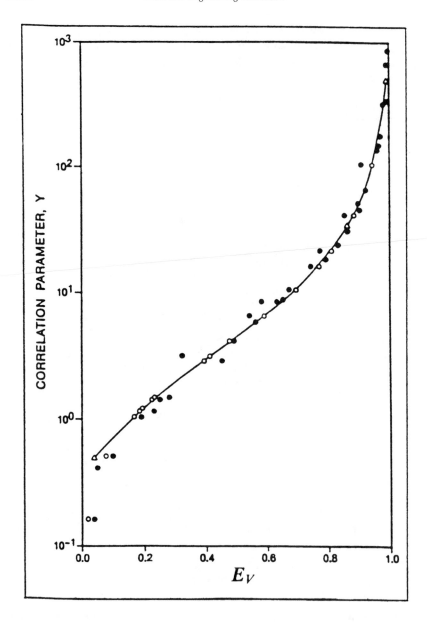

Figure 14-53. E_V versus the correlating parameter Y. (*Permission to publish by the Society of Petroleum Engineers.*)

The specific steps involved in determining the vertical sweep efficiency as a function of water–oil ratios are summarized below:

1. Calculate the mobility ratio M and permeability variation V.
2. Select several values for the WOR, e.g., 1, 2, 5, 10, and calculate the correlating parameter Y at each selected WOR.
3. Enter Figure 14-53 with each value of Y and determine the corresponding values of the vertical sweep efficiency E_V.
4. Plot WOR versus E_V.

To further simplify the calculations for determining E_V, Fassihi (1986) curve-fitted the graph of Figure 14-53 and proposed the following non-linear function, which can be solved iteratively for the vertical sweep efficiency E_V:

$$a_1 E_V^{a_2} \left(1 - E_V\right)^{a_3} - Y = 0 \qquad (14\text{-}96)$$

where $a_1 = 3.334088568$
$\quad\quad a_2 = 0.7737348199$
$\quad\quad a_3 = 1.225859406$

The Newton–Raphson method is perhaps the appropriate technique for solving Equation 14-96. To avoid the iterative process, the following expression could be used to estimate the vertical sweep efficiency using the correlating parameter Y:

$$E_V = a_1 + a_2 \ln(Y) + a_3 \left[\ln(Y)\right]^2 + a_4 \left[\ln(Y)\right]^3 + a_5 / \ln(Y) + a_6 Y$$

With the coefficients a_1 through a_6 as given by:

$a_1 = 0.19862608$ $\qquad\qquad a_2 = 0.18147754$
$a_3 = 0.01609715$ $\qquad\qquad a_4 = -4.6226385 \times 10^{-3}$
$a_5 = -4.2968246 \times 10^{-4}$ $\qquad a_6 = 2.7688363 \times 10^{-4}$

Example 14-22

A layered reservoir is characterized by a permeability variation V of 0.8. Calculate the vertical sweep efficiency E_V when the producing water–oil ratio reaches 50 bbl/bbl assuming a mobility ratio of 10.0.

Solution

Step 1. Calculate the parameter x by applying Equation 14-95:

$$x = 1.6453(0.8)^2 + 0.9735(0.8) - 0.6891 = 1.1427$$

Step 2. Calculate the correlation parameter Y from Equation 14-96:

$$Y = \frac{(50+0.4)[18.948 - 2.499(0.8)]}{[10 - 0.8094(0.8) + 1.137]10^{1.1427}} = 5.863$$

Step 3. From Figure 14-53, determine E_V to give:

$$E_V = 0.56$$

METHODS OF PREDICTING RECOVERY PERFORMANCE FOR LAYERED RESERVOIRS

To account for the reservoir vertical heterogeneity when predicting reservoir performance, the reservoir is represented by a series of layers with no vertical communication, i.e., no cross-flow between layers. Each layer is characterized by a thickness h, permeability k, and porosity ϕ. The heterogeneity of the entire reservoir is usually described by the permeability variation parameter V. Three of the methods that are designed to predict the performance of layered reservoirs are discussed below.

Simplified Dykstra–Parsons Method

Dykstra and Parsons (1950) proposed a correlation for predicting waterflood oil recovery that uses the mobility ratio, permeability variation, and producing water–oil ratio as correlating parameters. Johnson (1956) developed a simplified graphical approach for the Dykstra and Parsons method that is based on predicting the overall oil recovery R at water–oil ratios of 1, 5, 25, and 100 bbl/bbl. Figure 14-54 shows the proposed graphical charts for the four selected WOR_s. The correlating parameters shown in Figure 14-54, are:

R = overall oil recovery factor
S_{wi} = initial water saturation
M = mobility ratio
V = permeability variation

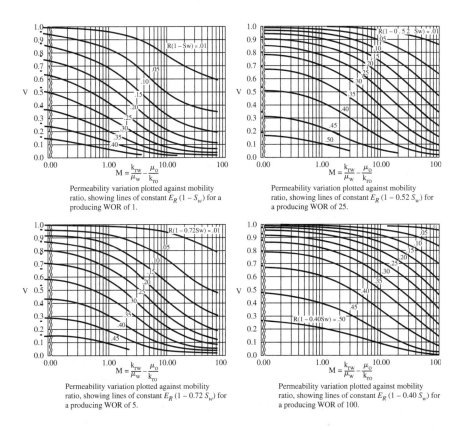

Figure 14-54. Simplified Dykstra and Parsons curves. (*Permission to publish by the Society of Petroleum Engineers.*)

The practical application of the simplified Dykstra and Parsons method is outlined below:

1. Calculate the permeability variation V and mobility ratio M.
2. Using the permeability ratio and mobility ratio, calculate the overall oil recovery factor R from the four charts at WOR of 1, 5, 25, 100 bbl/bbl. For example, to determine the oil recovery factor when the WOR reaches 5 bbl/bbl for a flood pattern that is characterized by a V and M of 0.5 and 2, respectively:
   - Enter the appropriate graph with these values, i.e., 0.5 and 2.
   - The point of intersection shows that $R(1-0.72 S_{wi}) = 0.25$.
   - If the initial water saturation S_{wi} is 0.21, solve for the recovery factor to give R = 0.29.

3. Calculate the cumulative oil production N_P at each of the four water–oil ratios, i.e., 1, 5, 25, and 100 bbl/bbl, from:

$$N_P = N_S R$$

4. If the water–oil ratio is plotted against the oil recovery on semi-log paper and a Cartesian scale, the oil recovery at breakthrough can be found by extrapolating the line to a very low value of WOR, as shown schematically in Figure 14-55.

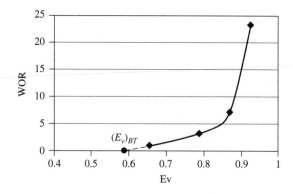

Figure 14-55. WOR versus E_V relationship.

5. For a constant injection rate, adding the fill-up volume W_{if} to the cumulative oil produced at breakthrough and dividing by the injection rate can estimate the time to breakthrough.
6. The cumulative water produced at any given value of WOR is obtained by finding the area under the curve of WOR versus N_P, as shown schematically in Figure 14-56.
7. The cumulative water injected at any given value of WOR is calculated by adding cumulative oil produced to the produced water and fill-up volume, or:

$$W_{inj} = N_P B_o + W_P B_w + W_{if}$$

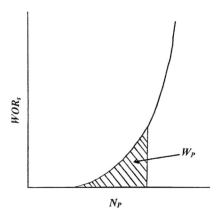

Figure 14-56. Cumulative water production from WOR vs. N_P curve.

Example 14-23

A reservoir is characterized by the following parameters:

Initial oil-in-place, N_S = 12 MMSTB
Permeability variation, V = 0.8
Mobility ratio, M = 2.0
Initial water saturation, S_{wi} = 0.25

Predict the cumulative oil production as a function of the producing water–oil ratio.

Solution

Using Johnson's graphical approach, perform the required calculations in the following worksheet:

WOR	Figure 14-54	R	$N_p = N_S R$
1	$R (1 - S_{wi}) = 0.049$	0.065	0.78 MMSTB
5	$R (1 - 0.72 S_{wi}) = 0.100$	0.122	1.464 MMSTB
25	$R (1 - 0.32 S_{wi}) = 0.200$	0.217	2.604 MMSTB
100	$R (1 - 0.40 S_{wi}) = 0.377$	0.419	5.028 MMSTB

Modified Dykstra–Parsons Method

Felsenthal, Cobb, and Heuer (1962) extend the work of Dykstra and Parsons to account for the presence of initial gas saturation at the start of flood. Assuming a constant water injection rate i_w, the method is summarized in the following steps:

Step 1. Perform the following preliminary calculations to determine:
- Pore volume PV and oil in place at start of flood N_S
- Water cut f_w as a function of S_w
- Slope (df_w/dS_w) as a function of S_w
- Average water saturation at breakthrough $\overline{S}_{wBT}$
- Mobility ratio M from Equation 14-61
- Vertical sweep efficiency at breakthrough E_{DAB} from Equation 14-9
- Areal sweep efficiency at breakthrough E_{ABT} from Equation 14-64
- Permeability variation V from Equation 14-70
- Fill-up volume W_{if} from Equation 14-82

Step 2. Using Equations 14-94 and 14-96, calculate the vertical sweep efficiency at assumed water–oil ratios of 1, 2, 5, 10, 15, 20, 25, 50, and 100 bbl/bbl.

Step 3. Plot WOR versus E_V on a Cartesian scale, as shown schematically in Figure 14-55, and determine the vertical sweep efficiency at breakthrough E_{VBT} by extrapolating the WOR versus E_V curve to WOR = 0.

Step 4. Calculate cumulative water injected at breakthrough by using Equation 14-43:

$$W_{iBT} = (PV)(\overline{S}_{wBT} - S_{wi})E_{ABT}E_{VBT}$$

where W_{iBT} = cumulative water injected at breakthrough, bbl
PV = pattern pore volume
E_{VBT} = vertical sweep efficiency at breakthrough
E_{ABT} = areal sweep efficiency at breakthrough

Step 5. Calculate cumulative oil produced at breakthrough from the following expression:

$$\left(N_P\right)_{BT} = \frac{W_{iBT} - W_{if}E_{VBT}}{B_o} \qquad (14\text{-}97)$$

Step 6. Calculate the time to breakthrough t_{BT} from:

$$t_{BT} = \frac{W_{iBT}}{i_w}$$

Step 7. Assume several values for water–oil ratios WOR_r, e.g., 1, 2, 5, 10, 15, 20, 25, 50, and 100 bbl/bbl.

Step 8. Determine E_V for each assumed value of WOR (see step 3).

Step 9. Convert the assumed values of WOR_r to water cut f_{w2} and surface WOR from Equations 14-25 and 14-29, respectively:

$$f_{w2} = \frac{WOR_r}{WOR_r + 1}$$

$$WOR_s = WOR_r\left(\frac{B_o}{B_w}\right)$$

where f_{w2} = water cut at the sand face of producer, bbl/bbl
    WOR_s = surface water–oil ratio, STB/STB
    WOR_r = reservoir water–oil ratio, bbl/bbl

Step 10. Determine the water saturation S_{w2} for each value of f_{w2} from the water cut curve.

Step 11. Using Equation 14-67 or Figure 14-36, determine the areal sweep efficiency E_A for each value of f_{w2}.

Step 12. Using Equation 14-67 or Figure 14-36, determine the areal sweep efficiency E_A for each value of f_{w2}.

Step 13. Determine the average water saturation $\overline{S}_{w2}$ for each value of f_{w2} from Equation 14-45.

Step 14. Calculate the displacement efficiency E_D for each $\overline{S}_{w2}$ in step 13 by applying Equation 14-9.

Step 15. Calculate cumulative oil production for each WOR from:

$$N_P = N_S E_D E_A E_V - \frac{(PV)S_{gi}(1 - E_A E_V)}{B_o} \qquad (14\text{-}98)$$

Step 16. Plot the cumulative oil production N_p versus WOR_s on Cartesian coordinate paper, as shown schematically in Figure 14-56, and calculate the area under the curve at several values of WOR_s. The area under the curve represents the cumulative water production W_p at any specified WOR_s, i.e., $(W_p)_{WOR}$.

Step 17. Calculate the cumulative water injected W_{inj} at each selected WOR from:

$$W_{inj} = (N_P)_{WOR} B_o + (W_P)_{WOR} B_w + (PV)S_{gi}(E_v)_{WOR} \qquad (14\text{-}99)$$

where W_{inj} = cumulative water injected, bbl
S_{gi} = initial gas saturation
$(N_P)_{WOR}$ = cumulative oil production when the water–oil ratio reaches WOR, STB
$(E_V)_{WOR}$ = vertical sweep efficiency when the water–oil ratio reaches WOR

Step 18. Calculate the time to inject W_{inj}:

$$t = \frac{W_{inj}}{i_w}$$

Step 19. Calculate the oil and water flow rates from Equations 14-55 and 14-56, respectively:

$$Q_o = \frac{i_w}{B_o + B_w WOR_S}$$
$$Q_w = Q_o WOR_S$$

Craig–Geffen–Morse Method

With the obvious difficulty of incorporating the vertical sweep efficiency in oil recovery calculations, Craig et al. (1955) proposed performing the calculations for only one selected layer in the multilayered system. The selected layer, identified as the **base layer,** is considered to have a 100% vertical sweep efficiency. The performance of each of the remaining layers can be obtained by "sliding the timescale" as summarized in the following steps:

Step 1. Divide the reservoir into the appropriate number of layers.

Step 2. Calculate the performance of a single layer, i.e., the base layer, for example, layer n.

Step 3. Plot cumulative liquid volumes (N_P, W_P, W_{inj}) and liquid rates (Q_o, Q_w, i_w) as a function of time t for the base layer, i.e., layer n.

Step 4. For each layer (including the base layer n) obtain:
- (k / ϕ)
- (ϕh)
- $(k h)$

Step 5. To obtain the performance of layer i, select a succession of times t and obtain plotted values N_p^*, W_p^*, W_{inj}^*, Q_o^*, Q_w^*, and i_w^* by reading the graph of step 3 at time t*:

$$t_i^* = t \frac{\left(\dfrac{k}{\phi}\right)_i}{\left(\dfrac{k}{\phi}\right)_n} \qquad (14\text{-}100)$$

Then calculate the performance of layer i at any time t from:

$$N_P = N_p^* \frac{(\phi h)_i}{(\phi h)_n} \qquad (14\text{-}101)$$

$$W_P = W_p^* \frac{(\phi h)_i}{(\phi h)_n} \qquad (14\text{-}102)$$

$$W_{inj} = W_{inj}^* \frac{(\phi h)_i}{(\phi h)_n} \tag{14-103}$$

$$Q_o = Q_o^* \frac{(k/\phi)_i}{(k/\phi)_n} \tag{14-104}$$

$$Q_w = Q_w^* \frac{(k/\phi)_i}{(k/\phi)_n} \tag{14-105}$$

$$i_w = i_w^* \frac{(k/\phi)_i}{(k/\phi)_n} \tag{14-106}$$

where
$$n = \text{base layer}$$
$$i = \text{layer } i$$
$$N_p^*, W_p^*, W_{inj}^* = \text{volumes at } t^*$$
$$Q_o^*, Q_w^*, \text{ and } i_w^* = \text{rates at } t^*$$

Step 6. The composite performance of the flood pattern at time t is obtained by summation of individual layer values.

Example 14-24

Results of Example 14-16 are shown graphically in Figure 14-47. Assume that the reservoir has an *additional* four layers that are characterized by the following properties:

Layer	k, md	h, ft	ϕ
"Original (base)" 1	31.5	5	0.20
2	20.5	5	0.18
3	16.0	4	0.15
4	13.0	3	0.14
5	10.9	2	0.10

Calculate N_P, W_P, W_{inj}, Q_o, Q_w, and i_w for the remaining four layers at t = 730 days.

Solution

Step 1. Calculate k/ϕ, ϕh, and kh for each layer.

Layer	k/ϕ	ϕh	$k h$
1	157.5	1.00	157.5
2	113.9	0.90	102.5
3	106.7	0.60	64.0
4	92.8	0.42	39.3
5	109.0	0.20	21.8

Step 2. At $t = 730$, calculate t^*:

Layer	$\dfrac{(k/\phi)_i}{(k/\phi)_n}$	$t^* = 730\left[\dfrac{(k/\phi)_i}{(k/\phi)_n}\right]$
1	1.000	730
2	0.723	528
3	0.677	495
4	0.589	430
5	0.692	505

Step 3. Read the values of N_p^*, W_p^*, W_{inj}^*, Q_o^*, Q_w^*, and i_w^* at each t^* from Figure 14-47:

Layer	t^*	N_P^*	W_P^*	W_{inj}^*	Q_o^*	Q_w^*	i_w^*
1	730	81,400	61,812	206,040	51.5	218.4	274.5
2	528	68,479	19,954	153,870	65	191	261
3	495	63,710	21,200	144,200	68	175	258
4	430	57,000	9,620	124,000	75	179	254
5	505	74,763	41,433	177,696	58	200	267

Step 4. Calculate N_P, W_P, W_{inj}, Q_o, Q_w, and i_w for each layer after 730 days by applying Equations 14-101 through 14-106:

Layer	$(\phi h)_i/(\phi h)_1$	N_P	W_P	W_{inj}
1	1.00	81,400	61,812	206,040
2	0.90	61,631	11,972	138,483
3	0.60	38,226	12,720	86,520
4	0.42	23,940	4,040	52,080
5	0.20	14,953	8,287	35,539
Total		220,150	98,831	466,582

Layer	$(\phi h)_i/(\phi h)_1$	Q_o	Q_w	i_w
1	1.000	51.5	218.4	274.5
2	0.651	42.3	124.3	169.9
3	0.406	27.6	71.1	104.7
4	0.250	18.8	44.8	63.5
5	0.138	8.0	27.6	36.8
Total		148.2	486.2	649.4

$$\text{Producing WOR} = \frac{486.2}{148.2} = 3.28 \text{ STB / STB}.$$

Step 5. Steps 2 and 3 are repeated for a succession of times t and the composite reservoir performance curves are generated to describe the entire reservoir performance.

PROBLEMS

1. A saturated oil reservoir exists at its bubble-point pressure of 2840 psi. The following pressure-production data are given:

P	N_P, STB	G_P, Mscf	B_t, bbl/STB	B_g, ft^3/scf	R_S, scf/STB
2840	0	0	1.528	—	827
2660	36,933	37,851	1.563	0.00618	772
2364	65,465	74,137	1.636	0.00680	680
2338	75,629	91,910	1.648	0.00691	675
2375	85,544	115,256	1.634	0.00674	685
2305	96,100	148,200	1.655	0.00702	665

The following information is also available:

$S_{wi} = 25\%$ and $S_{or} = 35\%$.

a. Calculate the reduction in the residual gas saturation if a waterflooding project were to start at 2364 psi.

b. Calculate the injection that is required to dissolve the trapped gas.

2. The relative permeability data for a core sample taken from the Vu-Villa Field are given below:

S_w	k_{rw}	k_{ro}
0.16	0	1.00
0.20	0.0008	0.862
0.26	0.0030	0.670
0.32	0.0090	0.510
0.40	0.024	0.330
0.50	0.064	0.150
0.60	0.140	0.040
0.66	0.211	0.010
0.72	0.30	0.00
1.00	1.00	0.00

$S_{wi} = 0.16$ $S_{or} = 0.28$ $\mu_w = 0.75$ cp
$\mu_o = 2.00$ cp $\rho_w = 1.0$ g/cm^3 $\rho_o = 0.83$ g/cm^3

a. Neglecting gravity and capillary pressure terms, develop the fractional flow curve.
b. Assuming a water-injection rate of 0.08 bbl/day/ft^2 and a dip angle of 15°, develop the fractional flow curve.
c. Determine from both curves:
   - f_w at the front
   - S_w at the front, i.e., S_{wf}
   - Average water saturation behind the front

3. A linear reservoir system is characterized by the following data:

$L = 500$ ft $A = 10,000$ ft^2 $S_{wi} = 24\%$ $B_w = 1.01$ bbl/STB
$S_{or} = 20\%$ $i_w = 100$ ft^3/hr $B_o = 1.25$ bbl/STB $\phi = 15\%$

S_w	f_w
0.30	0.0181
0.40	0.082
0.50	0.247
0.60	0.612
0.70	0.885
0.75	0.952
0.80	1.000

Determine:

a. Water saturation profile after 10, 100, and 300 hr
b. Oil recovery at breakthrough
c. Time to breakthrough
d. Cumulative water injected at breakthrough

4. The following relative permeability data[2] are available on a rock sample taken from the R-Field:

S_w	k_{rw}	k_{ro}
0.25	0.00	0.495
0.35	0.015	0.327
0.40	0.030	0.260
0.45	0.068	0.200
0.50	0.110	0.148
0.55	0.149	0.102
0.60	0.213	0.064
0.65	0.277	0.032
0.70	0.350	0.000

Additional data:

$S_{wi} = 25\%$	$S_{oi} = 75\%$	$S_{gi} = 0\%$
$B_o = 1.2$ bbl/STB	$B_w = 1.0$ bbl/STB	$\mu_w = 0.90$ cp
$\mu_o = 10.0$ cp	$A = 28{,}934$ ft^2	$i_w = 100$ bbl/day
$E_A = 100\%$	$E_V = 100\%$	Area = 10 acres
$h = 31$ ft	$k = 50$ md	

Distance between injector and producer = 467 ft.

a. Calculate the water saturation profile after 100, 200, 500, 1000, 2000, and 5000 days. Plot your results.
b. Calculate the mobility ratio.
c. Calculate the time to breakthrough
d. Calculate and plot N_p, Q_o, Q_w, and W_{inj} as a function of time.

---

[2]Data from *Reservoir Engineering Manual,* Cole, Houston, TX: Gulf Publishing Company, 1969.

5. An oil reservoir is under consideration for waterflooding. The relative permeability data are given by the following expression:

$$k_{rw} = 0.4(1 - S_{WD})^2$$

$$k_{ro} = 0.3(S_{WD})^{2.5}$$

where

$$S_{WD} = \frac{S_w - S_{wi}}{1 - S_{or} - S_{wi}}$$

Other reservoir data are given below:

Flood pattern	= Five-spot
Flood area	= 40 acres
Oil viscosity	= 2 cp
Water viscosity	= 0.5 cp
B_o	= 1.3 bbl/STB
B_w	= 1.05 bbl/STB
S_{oi}	= 0.75
S_{wi}	= 0.25
S_{or}	= 0.35
S_{wi}	= 0.25
ϕ	= 15%
k	= 50 md
r_w	= 0.3 ft
h	= 20 ft
P_i	= 1000 psi
E_V	= 100%

Assuming a constant water injection rate of 800 bbl/day, predict the recovery performance and express results in a graphical form.

6. An oil reservoir is under consideration for further development by water injection. The relative permeability data are given below:

S_w	0.10	0.20	0.30	0.40	0.70	0.85
k_{rw}	0.00	0.00	0.02	0.05	0.35	0.60
k_{ro}	1.00	0.93	0.60	0.35	0.05	0.00

Additional data are given below:

Flood pattern	= Five-spot
Absolute permeability	= 70 md
Thickness	= 20 ft
Porosity	= 15%
S_{gi}	= 15%
S_{wi}	= 20%
μ_o	= 3.1 cp
μ_w	= 1.0 cp
B_o	= 1.25 bbl/STB
B_w	= 1.01 bbl/STB
Pattern area	= 40 acres
r_w	= 1.0 ft
$\Delta p = (P_{inj} - P_{wf})$	= 1000 psi (constant)

a. Calculate and plot the reservoir performance during the following stages:
- Start—interference
- Interference—fill-up
- Fill-up—breakthrough
- After breakthrough

b. Show on your graph time to: interference, fill-up, and breakthrough.

c. Plot water injectivity and areal sweep efficiency as a function of time.

7. The following core analysis is available on a reservoir that is being considered for a waterflooding project:

Sample	h, ft	k, md
1	2	14
2	2	39
3	1	108
4	2	77
5	2	28
6	1	212
7	1	151
8	3	10
9	2	20
10	3	55

Other data:

$$
\begin{aligned}
i_w &= 1000 \text{ bbl/day} \\
\mu_o &= 9.0 \text{ cp} \\
\mu_w &= 0.95 \text{ cp} \\
M &= 4.73 \\
N_s &= 6 \text{ MMSTB} \\
B_o &= 1.02 \text{ bbl/STB} \\
B_w &= 1.00 \text{ bbl/STB} \\
S_{wi} &= 0.2 \\
S_{oi} &= 0.8
\end{aligned}
$$

Using the simplified Dykstra–Parsons method, determine the following recovery parameters as a function of time:

- Q_o
- Q_w
- WOR
- N_p
- W_p

Show your results graphically.

8. The following core and relative permeability data are given:

Depth, ft	k, md
4100–4101	295
2	762
3	88
4	87
5	148
6	71
7	62
8	187
9	349
10	77
11	127
12	161
13	50
14	58
15	109
16	228

Depth, ft	k, md
17	282
18	776
19	87
20	47
21	16
22	35
23	47
24	54
25	273
26	454
27	308
28	159
29	178

S_w	k_{rw}	k_{ro}
0.10	0.00	0.96
0.20	0.00	0.89
0.24	0.00	0.80
0.30	0.01	0.60
0.40	0.03	0.30
0.50	0.06	0.12
0.60	0.12	0.06
0.70	0.19	0.015
0.76	0.25	0.00
0.80	0.32	0.00
0.90	0.65	0.00

$S_{oi} = 0.59$ $\mu_o = 40.34$ cp

$S_{wi} = 0.24$ $\mu_w = 0.82$ cp

$S_{gi} = 0.17$ $B_o = 1.073$ bbl/STB

$h = 29$ ft $B_w = 1.00$ bbl/STB

$A = 40$ acres $\phi = 19\%$

Using the modified Dykstra–Parsons method, generate the performance curves for this reservoir under a constant water-injection rate of 700 bbl/day.

9. Using Stiles' and the Dykstra-Parsons methods, calculate the vertical sweep efficiency as a function of producing water–oil ratio, given:

Layer	h, ft	k, md
1	2	5.0
2	2	7.0
3	2	11.0
4	2	4.0
5	2	14.0
6	2	21.0
7	2	68.0
8	2	13.0

μ_o = 8.0 cp
μ_w = 0.9 cp
M = 1.58
k_{ro} = 0.45
k_{rw} = 0.08
ϕ = 15%
B_o = 1.2 bbl/STB
B_w = bbl/STB

Show your results graphically.

10. An oil reservoir is characterized by the following six-layer system:

Layer	h, ft	k, md	ϕ
1	10	50	0.20
2	10	40	0.16
3	5	35	0.12
4	5	30	0.12
5	5	25	0.10
6	5	20	0.10

The first layer is identified as the *base layer* with the following relative permeability data:

S_w	0.0	0.1	0.2	0.3	0.4	0.5	0.6	0.7	0.75	0.8	0.9	0.10
k_{rw}	0	0	0	0	0.04	0.11	0.20	0.30	0.36	0.44	0.68	1.00
k_{ro}	1.00	1.00	1.00	1.00	0.94	0.80	0.44	0.16	0.045	0	0	0

The other rock and fluid properties are given below:

S_{oi}	= 0.65
S_{wi}	= 0.30
S_{gi}	= 0.05
μ_o	= 1.5 cp
μ_w	= 0.8 cp
B_o	= 1.2 bbl/STB
B_w	= 1.01 bbl/STB
N_S	= 12 MMSTB
Constant $(P_{inj} - P_{wf})$	= 950 psi
Wellbore radius	= 0.3 ft

a. Generate the performance curves for the base layer.

b. Generate the composite (overall) performance curves for the reservoir.

REFERENCES

1. Buckley, S., and Leverett, M., "Mechanism of Fluid Displacement in Sands," *Trans. AIME,* 1962, Vol. 146, p. 107.

2. Caudle, B., and Witte, M., "Production Potential Changes During Sweep-out in a Five-Spot System," *Trans. AIME,* 1959, Vol. 216, pp. 446–448.

3. Cole, F., *Reservoir Engineering Manual.* Houston, TX: Gulf Publishing Company, 1969.

4. Craft, B., and Hawkins, M., *Applied Petroleum Reservoir Engineering.* Englewood Cliffs, NJ: Prentice Hall, 1959.

5. Craig, Jr., F. F., *The Reservoir Engineering Aspects of Waterflooding.* Dallas: Society of Petroleum Engineers of AIME, 1971.

6. Craig, F., Geffen, T., and Morse, R., "Oil Recovery Performance of Pattern Gas or Water Injection Operations from Model Tests," *JPT,* Jan. 1955, pp. 7–15, *Trans. AIME,* p. 204.

7. Dake, L., *Fundamentals of Reservoir Engineering.* Amsterdam: Elsevier Scientific Publishing Company, 1978.

8. Deppe, J., "Injection Rates—The Effect of Mobility Ratio, Areal Sweep, and Pattern," *SPEJ,* June 1961, pp. 81–91.

9. de Souza, A., and Brigham, W., "A Study on Dykstra–Parsons Curves," TR29. Palo Alto, CA: Stanford University Petroleum Research Institute, 1981.

10. Dyes, A., Caudle, B., and Erickson, R., "Oil Production After Breakthrough as Influenced by Mobility Ratio," *JPT,* April 1954, pp. 27–32; *Trans. AIME,* p. 201.

11. Dykstra, H., and Parsons, R., "The Prediction of Oil Recovery by Water Flood," in *Secondary Recovery of Oil in the United States,* 2nd ed. Washington, DC: American Petroleum Institute, 1950, pp. 160–174.

12. Fassihi, M., "New Correlations for Calculation of Vertical Coverage and Areal Sweep Efficiency," *SPERE,* Nov. 1986, pp. 604–606.

13. Felsenthal, M., Cobb, T., and Heur, G, "A Comparison of Waterflooding Evaluation Methods," SPE Paper 332 presented at the SPE 5th Biennial Secondary Recovery Symposium, Wichita Falls, TX, May 7, 1962.

14. Johnson, C., "Prediction of Oil Recovery by Waterflood—A Simplified Graphical Treatment of the Dykstra-Parsons Method," *Trans. AIME,* 1956, Vol. 207, pp. 345–346.

15. Khelil, C., "A Correlation of Optimum Free Gas Saturation," SPE Paper 01983, 1983.

16. Landrum, B., and Crawford, P., "Effect of Directional Permeability on Sweep Efficiency and Production Capacity," *JPT,* Nov. 1960, pp. 67–71.

17. Miller, M., and Lents, M., "Performance of Bodcaw Reservoir, Cotton Valley Field Cycling Project, New Methods of Predicting Gas-Condensate Reservoirs," *SPEJ,* Sep. 1966, p. 239.

18. Muskat, M., *Flow of Homogeneous Fluids Through Porous Systems.* Ann Arbor, MI: J. W. Edwards, 1946.

19. Muskat, M., "The Theory of Nine-Spot Flooding Networks," *Prod. Monthly,* March 1948, Vol. 13, No. 3, p. 14.

20. Stiles, W., "Use of Permeability Distribution in Waterflood Calculations," *Trans. AIME,* 1949, Vol. 186, p. 9.

21. Terwilliger, P., Wilsey, L., Hall, H., Bridges, P., and Morse, R., "An Experimental and Theoretical Investigation of Gravity Drainage Performance," *Trans. AIME,* 1951, Vol. 192, pp. 285–296.

22. Thomas, C. E., Mahoney, C. F., and Winter, G. W., *Petroleum Engineering Handbook.* Dallas: Society of Petroleum Engineers, 1989.

23. Welge, H., "A Simplified Method for Computing Oil Recovery By Gas or Water Drive," *Trans. AIME,* 1952, pp. 91–98.

24. Willhite, G. P., *Waterflooding.* Dallas: Society of Petroleum Engineers, 1986.

C H A P T E R   1 5

VAPOR–LIQUID PHASE EQUILIBRIA

A phase is defined as that part of a system that is uniform in physical and chemical properties, homogeneous in composition, and separated from other coexisting phases by definite boundary surfaces. The most important phases occurring in petroleum production are the hydrocarbon liquid phase and the gas phase. Water is also commonly present as an additional liquid phase. These can coexist in equilibrium when the variables describing change in the entire system remain constant with time and position. The chief variables that determine the state of equilibrium are system temperature, system pressure, and composition.

The conditions under which these different phases can exist are a matter of considerable practical importance in designing surface separation facilities and developing compositional models. These types of calculations are based on the concept of equilibrium ratios.

VAPOR PRESSURE

A system that contains only one component is considered the simplest type of hydrocarbon system. The word *component* refers to the number of molecular or atomic species present in the substance. A single-component system is composed entirely of one kind of atom or molecule. We often use the word *pure* to describe a single-component system. The qualitative understanding of the relationship that exists between temperature T, pressure p, and volume V of pure components can provide an excellent basis for understanding the phase behavior of complex hydrocarbon mixtures.

Consider a closed evacuated container that has been partially filled with a pure component in the liquid state. The molecules of the liquid are in constant motion with different velocities. When one of these molecules reaches the liquid surface, it may possess sufficient kinetic energy to overcome the attractive forces in the liquid and pass into the vapor spaces above. As the number of molecules in the vapor phase increases, the rate of return to the liquid phase also increases. A state of equilibrium is eventually reached when the number of molecules leaving and returning is equal. The molecules in the vapor phase obviously exert a pressure on the wall of the container and this pressure is defined as the vapor pressure, p_v. As the temperature of the liquid increases, the average molecular velocity increases with a larger number of molecules possessing sufficient energy to enter the vapor phase. As a result, the vapor pressure of a pure component in the liquid state increases with increasing temperature.

A method that is particularly convenient for expressing the vapor pressure of pure substances as a function of temperature is shown in Figure 15-1. The chart, known as the *Cox chart,* uses a logarithmic scale for the vapor pressure and an entirely arbitrary scale for the temperature in °F. The vapor pressure curve for any particular component, as shown in Figure 15-1, can be defined as the dividing line between the area where vapor and liquid exists. If the system pressure exists at its vapor pressure, two phases can coexist in equilibrium. Systems represented by points located below that vapor pressure curve are composed only of the vapor phase. Similarly, points above the curve represent systems that exist in the liquid phase. These statements can be conveniently summarized by the following expressions:

- $p < p_v \rightarrow$ system is entirely in the vapor phase
- $p > p_v \rightarrow$ system is entirely in the liquid phase
- $p = p_v \rightarrow$ vapor and liquid coexist in equilibrium

where p is the pressure exerted on the pure component. Note that the above expressions are valid only if the system temperature T is below the critical temperature T_c of the substance.

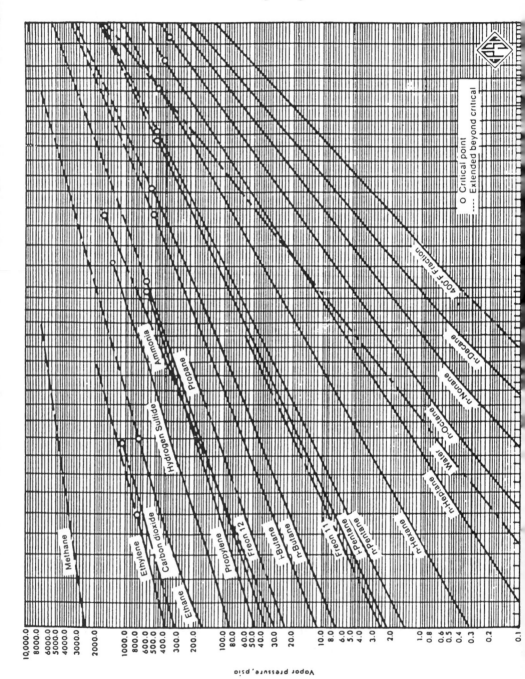

Figure 15-1. Vapor pressures for hydrocarbon components. (*Courtesy of the Gas Processors Suppliers Association, Engineering Book, 10th Ed., 1987.*)

The vapor pressure chart allows a quick determination of p_v of a pure component at a specific temperature. For computer and spreadsheet applications, however, an equation is more convenient. Lee and Kesler (1975) proposed the following generalized vapor pressure equation:

$$p_v = p_c \exp(A + \omega B)$$

with

$$A = 5.92714 - \frac{6.09648}{T_r} - 1.2886 \ln(T_r) + 0.16934(T_r)^6$$

$$B = 15.2518 - \frac{15.6875}{T_r} - 13.4721 \ln(T_r) + 0.4357(T_r)^6$$

where p_v = vapor pressure, psi
  p_c = critical pressure, psi
  T_r = reduced temperature (T / T_c)
  T = system temperature, °R
  T_c = critical temperature, °R
  ω = acentric factor

EQUILIBRIUM RATIOS

In a multicomponent system, the equilibrium ratio K_i of a given component is defined as the ratio of the mole fraction of the component in the gas phase y_i to the mole fraction of the component in the liquid phase x_i. Mathematically, the relationship is expressed as:

$$K_i = \frac{y_i}{x_i} \tag{15-1}$$

where K_i = equilibrium ratio of component i
  y_i = mole fraction of component i in the gas phase
  x_i = mole fraction of component i in the liquid phase

At pressures below 100 psia, Raoult's and Dalton's laws for ideal solutions provide a simplified means of predicting equilibrium ratios. Raoult's law states that the partial pressure p_i of a component in a multicomponent system is the product of its mole fraction in the liquid phase x_i and the vapor pressure of the component p_{vi}, or:

$$p_i = x_i \, p_{vi} \qquad\qquad (15\text{-}2)$$

where p_i = partial pressure of a component i, psia
              p_{vi} = vapor pressure of component i, psia
              x_i = mole fraction of component i in the liquid phase

Dalton's law states that the partial pressure of a component is the product of its mole fraction in the gas phase y_i and the total pressure of the system p, or:

$$p_i = y_i \, p \qquad\qquad (15\text{-}3)$$

where   p = total system pressure, psia.

At equilibrium and in accordance with the above stated laws, the partial pressure exerted by a component in the gas phase must be equal to the partial pressure exerted by the same component in the liquid phase. Therefore, equating the equations describing the two laws yields:

$$x_i p_{vi} = y_i p$$

Rearranging the above relationship and introducing the concept of the equilibrium ratio gives:

$$\frac{y_i}{x_i} = \frac{p_{vi}}{p} = K_i \qquad\qquad (15\text{-}4)$$

Equation 15-4 shows that for ideal solutions and regardless of the overall composition of the hydrocarbon mixture, the equilibrium ratio is only a function of the system pressure p and the temperature T since the vapor pressure of a component is only a function of temperature (see Figure 15-1).

It is appropriate at this stage to introduce and define the following nomenclatures:

         z_i = mole fraction of component in the entire hydrocarbon mixture
         n = total number of moles of the hydrocarbon mixture, lb-mol
         n_L = total number of moles in the liquid phase
         n_v = total number of moles in the vapor (gas) phase

By definition:

$$n = n_L + n_v \qquad\qquad (15\text{-}5)$$

Equation 15-5 indicates that the total number of moles in the system is equal to the total number of moles in the liquid phase plus the total number of moles in the vapor phase. A material balance on the i'th component results in:

$$z_i n = x_i n_L + y_i n_v \qquad (15\text{-}6)$$

where $\quad z_i n$ = total number of moles of component i in the system
$\qquad\quad x_i n_L$ = total number of moles of component i in the liquid phase
$\qquad\quad y_i n_v$ = total number of moles of component i in the vapor phase

Also by the definition of mole fraction, we may write:

$$\sum_i z_i = 1 \qquad (15\text{-}7)$$

$$\sum_i x_i = 1 \qquad (15\text{-}8)$$

$$\sum_i y_i = 1 \qquad (15\text{-}9)$$

It is convenient to perform all phase-equilibria calculations on the basis of 1 mol of the hydrocarbon mixture, i.e., n = 1. That assumption reduces Equations 15-5 and 15-6 to:

$$n_L + n_v = 1 \qquad (15\text{-}10)$$

$$x_i n_L + y_i n_v = z_i \qquad (15\text{-}11)$$

Combining Equations 15-4 and 15-11 to eliminate y_i from Equation 15-11 gives:

$$x_i n_L + (x_i K_i) n_v = z_i$$

Solving for x_i yields:

$$x_i = \frac{z_i}{n_L + n_v K_i} \qquad (15\text{-}12)$$

Equation 15-11 can also be solved for y_i by combining it with Equation 15-4 to eliminate x_i:

$$y_i = \frac{z_i K_i}{n_L + n_v K_i} = x_i K_i \qquad (15\text{-}13)$$

Combining Equation 15-12 with 15-8 and Equation 15-13 with 15-19 results in:

$$\sum_i x_i = \sum_i \frac{z_i}{n_L + n_v K_i} = 1 \qquad (15\text{-}14)$$

$$\sum_i y_i = \sum_i \frac{z_i K_i}{n_L + n_v K_i} = 1 \qquad (15\text{-}15)$$

Since

$$\sum_i y_i - \sum_i x_i = 0$$

Therefore,

$$\sum_i \frac{z_i K_i}{n_L + n_v K_i} - \sum_i \frac{z_i}{n_L + n_v K_i} = 0$$

or

$$\sum_i \frac{z_i (K_i - 1)}{n_L + n_v K_i} = 0$$

Replacing n_L with $(1 - n_v)$ yields:

$$f(n_v) = \sum_i \frac{z_i (K_i - 1)}{n_v (K_i - 1) + 1} = 0 \qquad (15\text{-}16)$$

The above set of equations provides the necessary phase relationships to perform volumetric and compositional calculations on a hydrocarbon system. These calculations are referred to as *flash calculations* and are discussed next.

FLASH CALCULATIONS

Flash calculations are an integral part of all reservoir and process engineering calculations. They are required whenever it is desirable to know the amounts (in moles) of hydrocarbon liquid and gas coexisting in a reservoir or a vessel at a given pressure and temperature. These calculations are also performed to determine the composition of the existing hydrocarbon phases.

Given the overall composition of a hydrocarbon system at a specified pressure and temperature, flash calculations are performed to determine:

- Moles of the gas phase n_v
- Moles of the liquid phase n_L
- Composition of the liquid phase x_i
- Composition of the gas phase y_i

The computational steps for determining n_L, n_v, y_i, and x_i of a hydrocarbon mixture with a known overall composition of z_i and characterized by a set of equilibrium ratios K_i are summarized in the following steps:

Step 1. **Calculation of n_v:** Equation 15-16 can be solved for n_v by using the Newton–Raphson iteration techniques. In applying this iterative technique:

- Assume any arbitrary value of n_v between 0 and 1, e.g., $n_v = 0.5$. A good assumed value may be calculated from the following relationship, providing that the values of the equilibrium ratios are accurate:

$$n_v = A/(A - B)$$

where $\quad A = \sum_i \left[z_i (K_i - 1) \right]$

$$B = \sum_i \left[z_i (K_i - 1)/K_i \right]$$

- Evaluate the function $f(n_v)$ as given by Equation 15-16 using the assumed value of n_v.

- If the absolute value of the function $f(n_v)$ is smaller than a preset tolerance, e.g., 10^{-15}, then the assumed value of n_v is the desired solution.
- If the absolute value of $f(n_v)$ is greater than the preset tolerance, then a new value of n_v is calculated from the following expression:

$$(n_v)_n = n_v - f(n_v)/f'(n_v)$$

with

$$f' = -\sum_i \left[\frac{z_i(K_i - 1)^2}{[n_v(K_i - 1) + 1]^2} \right]$$

where $(n_v)_n$ is the new value of n_v to be used for the next iteration.
- The above procedure is repeated with the new values of n_v until convergence is achieved.

Step 2. **Calculation of n_L:** Calculate the number of moles of the liquid phase from Equation 15-10, to give:

$$n_L = 1 - n_v$$

Step 3. **Calculation of x_i:** Calculate the composition of the liquid phase by applying Equation 15-12:

$$x_i = \frac{z_i}{n_L + n_v K_i}$$

Step 4. **Calculation of y_i:** Determine the composition of the gas phase from Equation 15-13:

$$y_i = \frac{z_i K_i}{n_L + n_v K_i} = x_i K_i$$

Example 15-1

A hydrocarbon mixture with the following overall composition is flashed in a separator at 50 psia and 100°F.

Component	z_i
C_3	0.20
$i - C_4$	0.10
$n - C_4$	0.10
$i - C_5$	0.20
$n - C_5$	0.20
C_6	0.20

Assuming an ideal solution behavior, perform flash calculations.

Solution

Step 1. Determine the vapor pressure for the Cox chart (Figure 15-1) and calculate the equilibrium ratios from Equation 15-4.

Component	z_i	p_{vi} at 100°F	$K_i = p_{vi}/50$
C_3	0.20	190	3.80
$i - C_4$	0.10	72.2	1.444
$n - C_4$	0.10	51.6	1.032
$i - C_5$	0.20	20.44	0.4088
$n - C_5$	0.20	15.57	0.3114
C_6	0.20	4.956	0.09912

Step 2. Solve Equation 15-16 for n_v by using the Newton–Raphson method, to give:

Iteration	n_v	$f(n_v)$
0	0.08196579	3.073 E-02
1	0.1079687	8.894 E-04
2	0.1086363	7.60 E-07
3	0.1086368	1.49 E-08
4	0.1086368	0.0

Step 3. Solve for n_L:

$$n_L = 1 - n_v$$

$$n_L = 1 - 0.1086368 = 0.8913631$$

Step 4. Solve for x_i and y_i to yield:

Component	z_i	K_i	$x_i = z_i/(0.8914 + 0.1086K_i)$	$y_i = x_iK_i$
C_3	0.20	3.80	0.1534	0.5829
$i - C_4$	0.10	1.444	0.0954	0.1378
$n - C_4$	0.10	1.032	0.0997	0.1029
$i - C_5$	0.20	0.4088	0.2137	0.0874
$n - C_5$	0.20	0.3114	0.2162	0.0673
C_6	0.20	0.09912	0.2216	0.0220

Notice that for a binary system, i.e., two-component system, flash calculations can be performed without restoring to the above iterative technique by applying the following steps:

Step 1. **Solve for the composition of the liquid phase x_i.** From equations 15-8 and 15-9:

$$\sum_i x_i = x_1 + x_2 = 1$$

$$\sum_i y_i = y_1 + y_2 = K_1x_1 + K_2x_2 = 1$$

Solving the above two expressions for the liquid compositions x_1 and x_2 gives:

$$x_1 = \frac{1 - K_2}{K_1 - K_2}$$

and

$$x_2 = 1 - x_1$$

where x_1 = mole fraction of the first component in the liquid phase
x_2 = mole fraction of the second component in the liquid phase
K_1 = equilibrium ratio of the first component
K_2 = equilibrium ratio of the second component

Step 2. **Solve for the composition of the gas phase y_i.** From the definition of the equilibrium ratio, calculate the composition of the gas as follows:

$$y_1 = x_1 K_1$$

$$y_2 = x_2 K_2 = 1 - y_1$$

Step 3. **Solve for the number of moles of the vapor phase n_v.** Arrange Equation 15-12 to solve for n_v, to give:

$$n_v = \frac{z_1 - x_1}{x_1 (K_1 - 1)}$$

and

$$n_1 = 1 - n_v$$

where z_1 = mole fraction of the first component in the entire system
x_1 = mole fraction of the first component in the liquid phase
K_1 = equilibrium ratio of the first component
K_2 = equilibrium ratio of the second component

EQUILIBRIUM RATIOS FOR REAL SOLUTIONS

The equilibrium ratios, which indicate the partitioning of each component between the liquid phase and gas phase, as calculated by Equation 15-4 in terms of vapor pressure and system pressure, proved to be inadequate. The basic assumptions behind Equation 15-4 are that:

- The vapor phase is an ideal gas as described by Dalton's law
- The liquid phase is an ideal solution as described by Raoult's law

The above combination of assumptions is unrealistic and results in inaccurate predictions of equilibrium ratios at high pressures.

For a real solution, the equilibrium ratios are no longer a function of the pressure and temperature alone, but also a function of the composition of the hydrocarbon mixture. This observation can be stated mathematically as:

$$K_i = K(p, T, z_i)$$

Numerous methods have been proposed for predicting the equilibrium ratios of hydrocarbon mixtures. These correlations range from a simple mathematical expression to a complicated expression containing several composition-dependent variables. The following methods are presented:

- Wilson's correlation
- Standing's correlation
- Convergence pressure method
- Whitson and Torp correlation

Wilson's Correlation

Wilson (1968) proposed a simplified thermodynamic expression for estimating K values. The proposed expression has the following form:

$$K_i = \frac{p_{ci}}{p} \exp\left[5.37(1 + \omega_i)\left(1 - \frac{T_{ci}}{T}\right)\right] \qquad (15\text{-}17)$$

where　p_{ci} = critical pressure of component i, psia
　　　　p = system pressure, psia
　　　　T_{ci} = critical temperature of component i, °R
　　　　T = system temperature, °R
　　　　ω_i = acentric factor of component i

The above relationship generates reasonable values for the equilibrium ratio when applied at low pressures.

Standing's Correlation

Hoffmann et al. (1953), Brinkman and Sicking (1960), Kehn (1964), and Dykstra and Mueller (1965) suggested that any pure hydrocarbon or nonhydrocarbon component could be uniquely characterized by combining its boiling-point temperature, critical temperature, and critical pressure into a characterization parameter that is defined by the following expression:

$$F_i = b_i \left[1/T_{bi} - 1/T \right] \tag{15-18}$$

with

$$b_i = \frac{\log(p_{ci}/14.7)}{\left[1/T_{bi} - 1/T_{ci} \right]} \tag{15-19}$$

where F_i = component characterization factor
T_{bi} = normal boiling point of component i, °R

Standing (1979) derived a set of equations that fit the equilibrium ratio data of Katz and Hachmuth (1937) at pressures of less than 1000 psia and temperatures below 200°F. The proposed form of the correlation is based on an observation that plots of $\log(K_i p)$ vs. F_i at a given pressure often form straight lines. The basic equation of the straight-line relationship is given by:

$$\log(K_i p) = a + cF_i$$

Solving for the equilibrium ratio K_i gives:

$$K_i = \frac{1}{p} 10^{(a+cF_i)} \tag{15-20}$$

where the coefficients a and c are the intercept and the slope of the line, respectively.

From a total of six isobar plots of $\log(K_i p)$ vs. F_i for 18 sets of equilibrium ratio values, Standing correlated the coefficients a and c with the pressure, to give:

$$a = 1.2 + 0.00045p + 15\left(10^{-8}\right)p^2 \tag{15-21}$$

$$c = 0.89 - 0.00017p - 3.5\left(10^{-8}\right)p^2 \tag{15-22}$$

Standing pointed out that the predicted values of the equilibrium ratios of N_2, CO_2, H_2S, and C_1 through C_6 can be improved considerably by changing the correlating parameter b_i and the boiling point of these components. The author proposed the following modified values:

Component	b_i	$T_{bi} \, °R$
N_2	470	109
CO_2	652	194
H_2S	1136	331
C_1	300	94
C_2	1145	303
C_3	1799	416
$i - C_4$	2037	471
$n - C_4$	2153	491
$i - C_5$	2368	542
$n - C_5$	2480	557
C_6*	2738	610
$n - C_6$	2780	616
$n - C_7$	3068	669
$n - C_8$	3335	718
$n - C_9$	3590	763
$n - C_{10}$	3828	805

Lumped Hexanes-fraction.

When making flash calculations, the question of the equilibrium ratio to use for the lumped heptanes-plus fraction always arises. One rule of thumb proposed by Katz and Hachmuth (1937) is that the K value for C_{7+} can be taken as 15% of the K of C_7, or:

$$K_{C_{7+}} = 0.15 K_{C_{7+}}$$

Standing (1979) offered an alternative approach for determining the K value of the heptanes and heavier fractions. By imposing experimental equilibrium ratio values for C_{7+} on Equation 15-20, Standing calculated the corresponding characterization factors F_i for the plus fraction. The calculated F_i values were used to specify the pure normal paraffin hydrocarbon having the K value of the C_{7+} fraction.

Standing suggested the following computational steps for determining the parameters b and T_b of the heptanes-plus fraction.

Step 1. Determine, from the following relationship, the number of carbon atoms n of the normal paraffin hydrocarbon having the K value of the C_{7+} fraction,

$$n = 7.30 + 0.0075(T - 460) + 0.0016p \qquad (15\text{-}23)$$

Step 2. Calculate the correlating parameter b and the boiling point T_b from the following expression:

$$b = 1{,}013 + 324n - 4.256n^2 \tag{15-24}$$

$$T_b = 301 + 59.85n - 0.971n^2 \tag{15-25}$$

The above calculated values can then be used in Equation 15-18 to evaluate F_i for the heptanes-plus fraction, i.e., F_{C7+}. It is also interesting to note that experimental phase equilibria data suggest that the equilibrium ratio for carbon dioxide can be closely approximated by the following relationship:

$$K_{CO_2} = \sqrt{K_{C_1} K_{C_2}}$$

where K_{CO_2} = equilibrium ratio of CO_2
K_{C_1} = equilibrium ratio of methane
K_{C_2} = equilibrium ratio of ethane

Example 15-2

A hydrocarbon mixture with the following composition is flashed at 1000 psia and 150°F.

Component	z_i
CO_2	0.009
N_2	0.003
C_1	0.535
C_2	0.115
C_3	0.088
$i - C_4$	0.023
$n - C_4$	0.023
$i - C_5$	0.015
$n - C_5$	0.015
C_6	0.015
C_{7+}	0.159

If the molecular weight and specific gravity of C_{7+} are 150.0 and 0.78, respectively, calculate the equilibrium ratios by using:

a. Wilson's correlation
b. Standing's correlation

Solution

Step 1. Calculate the critical pressure, critical temperature, and acentric factor of C_{7+} by using the characterization method of Riazi and Daubert discussed in Chapter 1. Example 1-1, page 27, gives:

$$T_c = 1139.4\,°R, \quad p_c = 320.3\,psia, \quad \omega = 0.5067$$

Step 2. Apply Equation 15-17 to give:

Component	P_c, psia	T_c, °R	ω	$K_i = \dfrac{P_{ci}}{1000} \exp\left[5.37(1+\omega_i)\left(1 - \dfrac{T_{ci}}{610}\right)\right]$
CO_2	1,071	547.9	0.225	2.0923
N_2	493	227.6	0.040	16.343
C_1	667.8	343.37	0.0104	7.155
C_2	707.8	550.09	0.0986	1.236
C_3	616.3	666.01	0.1542	0.349
$i-C_4$	529.1	734.98	0.1848	0.144
$n-C_4$	550.7	765.65	0.2010	0.106
$i-C_5$	490.4	829.1	0.2223	0.046
$n-C_5$	488.6	845.7	0.2539	0.036
C_6	436.9	913.7	0.3007	0.013
C_{7+}	320.3	1139.4	0.5069	0.00029

b.

Step 1. Calculate coefficients a and c from Equations 15-21 and 15-22 to give:

$$a = 1.2 + 0.00045(1000) + 15\left(10^{-8}\right)(1000)^2 = 1.80$$

$$c = 0.89 - 0.00017(1000) - 3.5\left(10^{-8}\right)(1000)^2 = 0.685$$

Step 2. Calculate the number of carbon atoms n from Equation 15-23 to give:

$$n = 7.3 + 0.0075(150) + 0.0016(1000) = 10.025$$

Step 3. Determine the parameter b and the boiling point T_b for the hydrocarbon component with n carbon atoms by using Equations 15-24 and 15-25 to yield:

$$b = 1013 + 324(10.025) - 4.256(10.025)^2 = 3833.369$$

$$T_b = 301 + 59.85(10.025) - 0.971(10.025)^2 = 803.41\ °R$$

Step 4. Apply Equation 15-20, to give:

Component	b_i	T_{bi}	F_i Eq. 15-18	K_i Eq. 15-20
CO_2	652	194	2.292	2.344
N_2	470	109	3.541	16.811
C_1	300	94	2.700	4.462
C_2	1145	303	1.902	1.267
C_3	1799	416	1.375	0.552
$i-C_4$	2037	471	0.985	0.298
$n-C_4$	2153	491	0.855	0.243
$i-C_5$	2368	542	0.487	0.136
$n-C_5$	2480	557	0.387	0.116
C_6	2738	610	0	0.063
C_{7+}	3833.369	803.41	− 1.513	0.0058

Convergence Pressure Method

Early high-pressure phase-equilibria studies have revealed that when a hydrocarbon mixture of a fixed overall composition is held at a constant temperature as the pressure increases, the equilibrium values of all components converge toward a common value of unity at certain pressure. This pressure is termed the convergence pressure P_k of the hydrocarbon mixture. The convergence pressure is essentially used to correlate the effect of the composition on equilibrium ratios.

The concept of the convergence pressure can be better appreciated by examining Figure 15-2. The figure shows a schematic diagram of a typical set of equilibrium ratios plotted versus pressure on log-log paper for a hydrocarbon mixture held at a constant temperature. The illustration shows a tendency of the equilibrium ratios to converge isothermally to a value of $K_i = 1$ for all components at a specific pressure, i.e., convergence pressure. A different hydrocarbon mixture may exhibit a different convergence pressure.

The Natural Gas Processors Suppliers Association (NGPSA) correlated a considerable quantity of K-factor data as a function of temperature, pressure, component identity, and convergence pressure. These correlation charts were made available through the NGPSA's *Engineering Data Book* and are considered to be the most extensive set of published equilibrium ratios for hydrocarbons. They include the K values for a number of convergence pressures, specifically 800, 1000, 1500, 2000, 3000, 5000, and 10,000 psia. Equilibrium ratios for methane through decane and for a convergence pressure of 5000 psia are given in Appendix A.

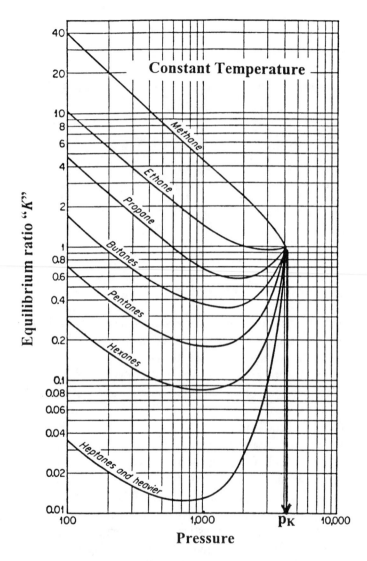

Figure 15-2. Equilibrium ratios for a hydrocarbon system.

Several investigators observed that for hydrocarbon mixtures with convergence pressures of 4000 psia or greater, the values of the equilibrium ratio are essentially the same for hydrocarbon mixtures with system pressures of less than 1000 psia. This observation led to the conclusion that the overall composition of the hydrocarbon mixture has little effect on equilibrium ratios when the system pressure is less than 1000 psia.

The problem with using the NGPSA equilibrium ratio graphical correlations is that the convergence pressure must be known before selecting the appropriate charts. Three of the methods of determining the convergence pressure are discussed next.

Hadden's Method

Hadden (1953) developed an iterative procedure for calculating the convergence pressure of the hydrocarbon mixture. The procedure is based on forming a "binary system" that describes the entire hydrocarbon mixture. One of the components in the binary system is selected as the lightest fraction in the hydrocarbon system and the other is treated as a "pseudo-component" that lumps all the remaining fractions. The binary system concept uses the binary system convergence pressure chart, as shown in Figure 15-3, to determine the p_k of the mixture at the specified temperature.

The equivalent binary system concept employs the following steps for determining the convergence pressure:

Step 1. Estimate a value for the convergence pressure.

Step 2. From the appropriate equilibrium ratio charts, read the K values of each component present in the mixture by entering the charts with the system pressure and temperature.

Step 3. Perform flash calculations using the calculated K values and system composition.

Step 4. Identify the lightest hydrocarbon component that comprises at least 0.1 mol % in the liquid phase.

Step 5. Convert the liquid mole fraction to a weight fraction.

Step 6. Exclude the lightest hydrocarbon component, as identified in step 4, and normalize the weight fractions of the remaining components.

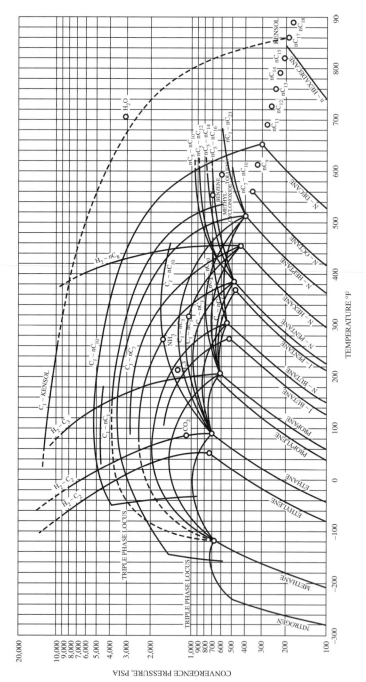

Figure 15-3. Convergence pressures for binary systems. *(Courtesy of the Gas Processors Suppliers Association, Engineering Data Book, 10th Ed., 1987.)*

Step 7. Calculate the weight average critical temperature and pressure of the lumped components (pseudo-component) from the following expressions:

$$T_{pc} = \sum_{i=2} w_i^* T_{ci}$$

$$p_{pc} = \sum_{i=2} w_i^* p_{ci}$$

where w_i^* = normalized weight fraction of component i
T_{pc} = pseudo-critical temperature, °R
p_{pc} = pseudo-critical pressure, psi

Step 8. Enter Figure 15-3 with the critical properties of the pseudo-component and trace the critical locus of the binary consisting of the light component and the pseudo-component.

Step 9. Read the new convergence pressure (ordinate) from the point at which the locus crosses the temperature of interest.

Step 10. If the calculated new convergence pressure is not reasonably close to the assumed value, repeat steps 2 through 9.

Note that when the calculated new convergence pressure is between values for which charts are provided, interpolation between charts might be necessary. If the K values do not change rapidly with the convergence pressure, i.e., $p_k \gg p$, then the set of charts nearest to the calculated p_k may be used.

Standing's Method

Standing (1977) suggested that the convergence pressure can be roughly correlated linearly with the molecular weight of the heptanes-plus fraction. Whitson and Torp (1981) expressed this relationship by the following equation:

$$p_k = 60M_{C_{7+}} - 4200 \tag{15-26}$$

where $M_{C_{7+}}$ is the molecular weight of the heptanes-plus fraction.

Rzasa's Method

Rzasa, Glass, and Opfell (1952) presented a simplified graphical correlation for predicting the convergence pressure of light hydrocarbon mixtures. They used the temperature and the product of the molecular weight and specific gravity of the heptanes-plus fraction as correlating parameters. The graphical illustration of the proposed correlation is shown in Figure 15-4.

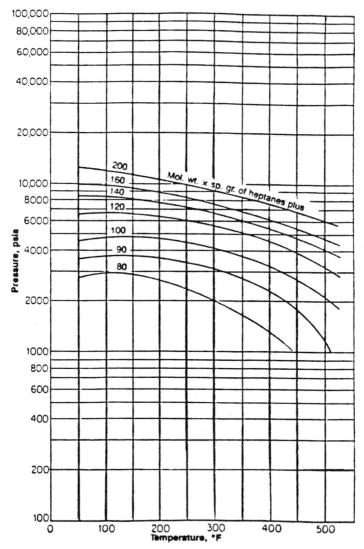

Figure 15-4. Rzasa's convergence pressure correlation. *(Courtesy of the American Institute of Chemical Engineers.)*

The graphical correlation is expressed mathematically by the following equation:

$$p_k = -2{,}381.8542 + 46.341487[M\gamma]_{C_{7+}} + \sum_{i=1}^{3} a_i \left[\frac{(M\gamma)_{C_{7+}}}{T - 460} \right]^i \qquad (15\text{-}27)$$

where $(M)_{C_{7+}}$ = molecular weight of C_{7+}
$(\gamma)_{C_{7+}}$ = specific gravity of C_{7+}
T = temperature, °R
$a_1 - a_3$ = coefficients of the correlation with the following values:
$a_1 = 6{,}124.3049$
$a_2 = -2{,}753.2538$
$a_3 = 415.42049$

The above mathematical expression can be used for determining the convergence pressure of hydrocarbon mixtures at temperatures in the range of 50 to 300°F.

Whitson and Torp Correlation

Whitson and Torp (1981) reformulated Wilson's equation (Equation 15-17) to yield accurate results at higher pressures. Wilson's equation was modified by incorporating the convergence pressure into the correlation, to give:

$$K_i = \left(\frac{p_{ci}}{p_k} \right)^{A-1} \left(\frac{p_{ci}}{p} \right) \exp\left[5.37A(1+\omega_i)\left(1 - \frac{T_{ci}}{T} \right) \right] \qquad (15\text{-}28)$$

with

$$A = 1 - \left(\frac{p}{p_k} \right)^{0.7}$$

where p = system pressure, psig
p_k = convergence pressure, psig
T = system temperature, °R
ω_i = acentric factor of component i

Example 15-3

Rework Example 15-2 and calculate the equilibrium ratios using the Whitson and Torp method.

Solution

Step 1. Determine the convergence pressure from Equation 15-27 to give $P_k = 9,473.89$.

Step 2. Calculate the coefficient A:

$$A = 1 - \left(\frac{1000}{9474}\right)^{0.7} = 0.793$$

Step 3. Calculate the equilibrium ratios from Equation 15-28 to give:

Component	p_c, psia	T_c, °R	ω	$K_i = \left(\frac{P_{ci}}{9474}\right)^{0.793^{-1}} \frac{P_{ci}}{1000} \exp\left[5.37A(1+\omega_i)\left(1-\frac{T_{ci}}{610}\right)\right]$
CO_2	1071	547.9	0.225	2.9
N_2	493	227.6	0.040	14.6
C_1	667.8	343.37	0.0104	7.6
C_2	707.8	550.09	0.0968	2.1
C_3	616.3	666.01	0.1524	0.7
$i-C_4$	529.1	734.98	0.1848	0.42
$n-C_4$	550.7	765.65	0.2010	0.332
$i-C_5$	490.4	829.1	0.2223	0.1749
$n-C_5$	488.6	845.7	0.2539	0.150
C_6	436.9	913.7	0.3007	0.0719
C_{7+}	320.3	1139.4	0.5069	$0.683(10^{-3})$

EQUILIBRIUM RATIOS FOR THE PLUS FRACTION

The equilibrium ratios of the plus fraction often behave in a manner different from the other components of a system. This is because the plus fraction in itself is a mixture of components. Several techniques have been proposed for estimating the K value of the plus fractions. Some of these techniques are presented here.

Campbell's Method

Campbell (1976) proposed that the plot of the log of K_i versus T_{ci}^2 for each component is a linear relationship for any hydrocarbon system. Campbell suggested that by drawing the best straight line through the points for propane through hexane components, the resulting line can be extrapolated to obtain the K value of the plus fraction. He pointed out that the plot of log K_i versus $1/T_{bi}$ of each heavy fraction in the mixture is also a straight-line relationship. The line can be extrapolated to obtain the equilibrium ratio of the plus fraction from the reciprocal of its average boiling point.

Winn's Method

Winn (1954) proposed the following expression for determining the equilibrium ratio of heavy fractions with a boiling point above 210°F.

$$K_{C_+} = \frac{K_{C_7}}{\left(K_{C_2}/K_{C_7}\right)^b} \tag{15-29}$$

where K_{C_+} = value of the plus fraction

K_{C_7} = K value of n-heptane at system pressure, temperature, and convergence pressure

K_{C_7} = K value of ethane

b = volatility exponent

Winn correlated, graphically, the volatility component b of the heavy fraction, with the atmosphere boiling point, as shown in Figure 15-5.

This graphical correlation can be expressed mathematically by the following equation:

$$b = a_1 + a_2\left(T_b - 460\right) + a_3\left(T - 460\right)^2 + a_4\left(T_b - 460\right)^3$$
$$+ a_5/(T - 460) \tag{15-30}$$

where T_b = boiling point, °R

$a_1 - a_5$ = coefficients with the following values:

$a_1 = 1.6744337$

$a_2 = -3.4563079 \times 10^{-3}$

$a_3 = 6.1764103 \times 10^{-6}$

$a_4 = 2.4406839 \times 10^{-6}$

$a_5 = 2.9289623 \times 10^2$

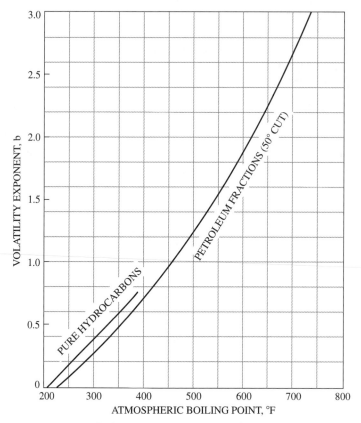

Figure 15-5. Volatility exponent. *(Courtesy of the Petroleum Refiner.)*

Katz's Method

Katz et al. (1957) suggested that a factor of 0.15 times the equilibrium ratio for the heptane component will give a reasonably close approximation to the equilibrium ratio for heptanes and heavier. This suggestion is expressed mathematically by the following equation:

$$K_{C_{7+}} = 0.15 K_{C_7} \tag{15-31}$$

APPLICATIONS OF THE EQUILIBRIUM RATIO IN RESERVOIR ENGINEERING

The vast amount of experimental and theoretical work that has been performed on equilibrium ratio studies indicates their importance in solving phase equilibrium problems in reservoir and process engineering. Some of their practical applications are discussed next.

Dew-Point Pressure

The dew-point pressure p_d of a hydrocarbon system is defined as the pressure at which an infinitesimal quantity of liquid is in equilibrium with a large quantity of gas. For a total of 1 lb-mol of a hydrocarbon mixture, i.e., $n = 1$, the following conditions are applied at the dew-point pressure:

$$n_L = 0$$
$$n_v = 1$$

Under these conditions, the composition of the vapor phase y_i is equal to the overall composition z_i. Applying the above constraints to Equation 15-14 yields:

$$\sum_i \frac{z_i}{K_i} \qquad (15\text{-}32)$$

where z_i = total composition of the system under consideration.

The solution of Equation 15-32 for the dew-point pressure p_d involves a trial-and-error process. The process is summarized in the following steps:

Step 1. Assume a trial value of p_d. A good starting value can be obtained by applying Wilson's equation (Equation 15-17) for calculating K_i to Equation 15-32 to give:

$$\sum_i \left[\frac{z_i}{\dfrac{p_{ci}}{p_d} \exp\left[5.37(1+\omega_i)\left(1 - \dfrac{T_{ci}}{T}\right)\right]} \right] = 1$$

Solving for p_d yields:

$$\text{initial } p_d = \frac{1}{\displaystyle\sum_i \left[\frac{z_i}{p_{ci} \exp\left[5.37(1+\omega_i)\left(1 - \dfrac{T_{ci}}{T}\right)\right]} \right]} \qquad (15\text{-}33)$$

Another simplified approach for estimating the dew-point pressure is to treat the hydrocarbon mixture as an ideal system with the equilibrium ratio K_i as given by Equation (15-4):

$$K_i = \frac{p_{vi}}{p}$$

Substituting the above expression into Equation (15-29) gives:

$$\sum_i \left[z_i \left(\frac{p_d}{p_{vi}} \right) \right] = 1.0$$

Solving for p_d yields:

$$\text{initial } p_d = \frac{1}{\sum\limits_{i=1} \left(\frac{z_i}{p_{vi}} \right)}$$

Step 2. Using the assumed dew-point pressure, calculate the equilibrium ratio, K_i, for each component at the system temperature.

Step 3. Compute the summation of Equation 15-33.

Step 4. If the sum is less than 1, steps 2 and 3 are repeated at a higher initial value of pressure; conversely, if the sum is greater than 1, repeat the calculations with a lower initial value of p_d. The correct value of the dew-point pressure is obtained when the sum is equal to 1.

Example 15-4

A natural gas reservoir at 250°F has the following composition:

Component	z_i
C_1	0.80
C_2	0.05
C_3	0.04
$i - C_4$	0.03
$n - C_4$	0.02
$i - C_5$	0.03
$n - C_5$	0.02
C_6	0.005
C_{7+}	0.005

If the molecular weight and specific gravity of C_{7+} are 140 and 0.8, calculate the dew-point pressure.

Solution

Step 1. Calculate the convergence pressure of the mixture from Rzasa's correlation, i.e., Equation 15-27, to give:

$$p_k = 5000 \text{ psia}$$

Step 2. Determine an initial value for the dew-point pressure from Equation 15-33 to give:

$$p_d = 207 \text{ psia}$$

Step 3. Using the K-value curves in Appendix A, solve for the dew-point pressure by applying the iterative procedure outlined previously, and by using Equation 15-32, to give:

Component	z_i	K_i at 207 psia	z_i/K_i	K_i at 300 psia	z_i/K_i	K_i at 222.3 psia	z_i/K_i
C_1	0.78	19	0.0411	13	0.06	18	0.0433
C_2	0.05	6	0.0083	4.4	0.0114	5.79	0.0086
C_3	0.04	3	0.0133	2.2	0.0182	2.85	0.0140
$i-C_4$	0.03	1.8	0.0167	1.35	0.0222	1.75	0.0171
$n-C_4$	0.02	1.45	0.0138	1.14	0.0175	1.4	0.0143
$i-C_5$	0.03	0.8	0.0375	0.64	0.0469	0.79	0.0380
$n-C_5$	0.02	0.72	0.0278	.55	0.0364	0.69	0.029
C_6	0.005	0.35	0.0143	0.275	0.0182	0.335	0.0149
C_{7+}	0.02	0.255*	0.7843	0.02025*	0.9877	0.0243*	0.8230
			0.9571		1.2185		1.0022

*Equation 15-29

The dew-point pressure is therefore 222 psia at 250°F.

Bubble-Point Pressure

At the bubble point p_b the hydrocarbon system is essentially liquid, except for an infinitesimal amount of vapor. For a total of 1 lb-mol of the hydrocarbon mixture, the following conditions are applied at the bubble-point pressure:

$$n_L = 1$$
$$n_v = 0$$

Obviously, under the above conditions, $x_i = z_i$. Applying the above constraints to Equation 15-15 yields:

$$\sum_i (z_i K_i) = 1 \qquad (15\text{-}34)$$

Following the procedure outlined in the dew-point pressure determination, Equation 15-34 is solved for the bubble-point pressure p_b by assuming various pressures and determining the pressure that will produce K values that satisfy Equation 15-34.

During the iterative process, if:

$$\sum_i (z_i K_i) < 1 \quad \rightarrow \quad \text{the assumed pressure is high}$$

$$\sum_i (z_i K_i) > 1 \quad \rightarrow \quad \text{the assumed pressure is low}$$

Wilson's equation can be used to give a good starting value for the iterative process:

$$\sum_i \left[z_i \frac{p_{ci}}{p_b} \exp\left[5.37(1+\omega)\left(1 - \frac{T_{ci}}{T}\right)\right]\right] = 1$$

Solving for the bubble-point pressure gives:

$$p_b = \sum_i \left[z_i p_{ci} \exp\left[5.37(1+\omega)\left(1 - \frac{T_{ci}}{T}\right)\right]\right] \qquad (15\text{-}35)$$

Assuming an ideal solution behavior, an initial guess for the bubble-point pressure can also be calculated by replacing the K_i in Equation 15-34 with that of Equation 15-4 to give:

$$\sum_i \left[z_i \left(\frac{p_{vi}}{p_b}\right)\right] = 1$$

or

$$p_b = \sum_i (z_i p_{vi}) \qquad (15\text{-}36)$$

Example 15-5

A crude oil reservoir has a temperature of 200°F and a composition as given below. Calculate the bubble-point pressure of the oil.

Component	x_i
C_1	0.42
C_2	0.05
C_3	0.05
$i - C_4$	0.03
$n - C_4$	0.02
$i - C_5$	0.01
$n - C_5$	0.01
C_6	0.01
C_{7+}	0.40*

*$(M)_{C_{7+}} = 216.0$
$(\gamma)_{C_{7+}} = 0.8605$
$(T_b)_{C_{7+}} = 977°R$

Solution

Step 1. Calculate the convergence pressure of the system by using Standing's correlation (Equation 15-26):

$$p_k = (60)(216) - 4200 = 8760 \text{ psia}$$

Step 2. Calculate the critical pressure and temperature by the Riazi and Daubert equation (Equation 1-2), to give:

$$p_c = 230.4 \text{ psia}$$
$$T_c = 1,279.8°R$$

Step 3. Calculate the acentric factor by employing the Edmister correlation (Equation 1-3) to yield:

$$\omega = 0.653$$

Step 4. Estimate the bubble-point pressure from Equation 15-35 to give:

$$p_b = 3,924 \text{ psia}$$

Step 5. Employing the iterative procedure outlined previously and using the Whitson and Torp equilibrium ratio correlation gives:

Component	z_i	K_i at 3924 psia	z_iK_i	K_i at 3950 psia	z_iK_i	K_i at 4,329 psia	z_iK_i
C_1	0.42	2.257	0.9479	2.242	0.9416	2.0430	0.8581
C_2	0.05	1.241	0.06205	2.137	0.0619	1.1910	0.0596
C_3	0.05	0.790	0.0395	0.7903	0.0395	0.793	0.0397
$i-C_4$	0.03	0.5774	0.0173	0.5786	0.0174	0.5977	0.0179
$n-C_4$	0.02	0.521	0.0104	0.5221	0.0104	0.5445	0.0109
$i-C_5$	0.01	0.3884	0.0039	0.3902	0.0039	0.418	0.0042
$n-C_5$	0.01	0.3575	0.0036	0.3593	0.0036	0.3878	0.0039
C_6	0.01	0.2530	0.0025	0.2549	0.0025	0.2840	0.0028
C_{7+}	0.40	0.227	0.0091	0.0232	0.00928	0.032	0.0138
Σ			1.09625		1.09008		1.0099

The calculated bubble-point pressure is 4330 psia.

Separator Calculations

Produced reservoir fluids are complex mixtures of different physical characteristics. As a well stream flows from the high-temperature, high-pressure petroleum reservoir, it experiences pressure and temperature reductions. Gases evolve from the liquids and the well stream changes in character. The physical separation of these phases is by far the most common of all field-processing operations and one of the most critical. The manner in which the hydrocarbon phases are separated at the surface influences the stock-tank oil recovery. The principal means of surface separation of gas and oil is the conventional stage separation.

Stage separation is a process in which gaseous and liquid hydrocarbons are flashed (separated) into vapor and liquid phases by two or more separators. These separators are usually operated in series at consecutively lower pressures. Each condition of pressure and temperature at which hydrocarbon phases are flashed is called a *stage of separation*. Examples of one- and two-stage separation processes are shown in Figure 15-6. Traditionally, the stock-tank is normally considered a separate stage of separation. Mechanically, there are two types of gas-oil separation: (1) differential separation and (2) flash or equilibrium separation.

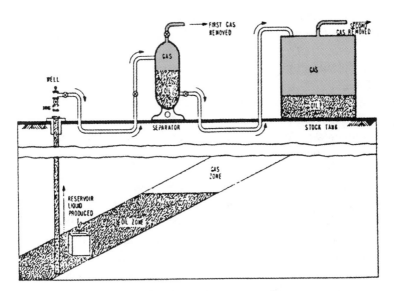

One-stage separation

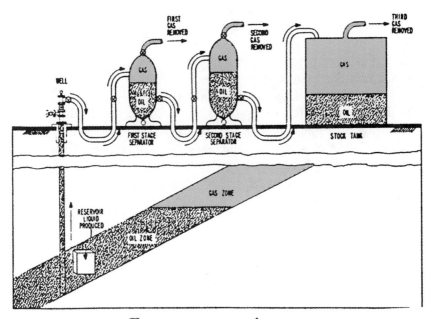

Two-stage separation

Figure 15-6. Schematic drawing of one- and two-stage separation processes. (*After Clark, 1951*)

To explain the various separation processes, it is convenient to define the composition of a hydrocarbon mixture by three groups of components:

1. The very volatile components ("lights"), such as nitrogen, methane, and ethane
2. The components of intermediate volatility ("intermediates"), such as propane through hexane
3. The components of less volatility, or the "heavies," such as heptane and heavier components

In the differential separation, the liberated gas (which is composed mainly of lighter components) is removed from contact with the oil as the pressure on the oil is reduced. As pointed out by Clark (1960), when the gas is separated in this manner, the maximum amount of heavy and intermediate components will remain in the liquid, minimum shrinkage of the oil will occur, and, therefore, greater stock-tank oil recovery will occur. This is due to the fact that the gas liberated earlier at higher pressures is not present at lower pressures to attract the intermediate and heavy components and pull them into the gas phase.

In the flash (equilibrium) separation, the liberated gas remains in contact with oil until its instantaneous removal at the final separation pressure. A maximum proportion of intermediate and heavy components is attracted into the gas phase by this process, and this results in a maximum oil shrinkage and, thus, a lower oil recovery.

In practice, the differential process is introduced first in field separation when gas or liquid is removed from the primary separator. In each subsequent stage of separation, the liquid initially undergoes a flash liberation followed by a differential process as actual separation occurs. As the number of stages increases, the differential aspect of the overall separation becomes greater.

The purpose of stage separation then is to reduce the pressure on the produced oil in steps so that more stock-tank oil recovery will result. Separator calculations are basically performed to determine:

- Optimum separation conditions: separator pressure and temperature
- Compositions of the separated gas and oil phases
- Oil formation volume factor
- Producing gas–oil ratio
- API gravity of the stock-tank oil

Note that if the separator pressure is high, large amounts of light components will remain in the liquid phase at the separator and be lost along with other valuable components to the gas phase at the stock tank. On the other hand, if the pressure is too low, large amounts of light components will be separated from the liquid and they will attract substantial quantities of intermediate and heavier components. An intermediate pressure, called *optimum separator pressure,* should be selected to maximize the oil volume accumulation in the stock tank. This optimum pressure will also yield:

- A maximum stock-tank API gravity
- A minimum oil formation volume factor (i.e., less oil shrinkage)
- A minimum producing gas–oil ratio (gas solubility)

The concept of determining the optimum separator pressure by calculating the API gravity, B_o, and R_s is shown graphically in Figure 15-7. The computational steps of the separator calculations are described below in conjunction with Figure 15-8, which schematically shows a bubble-point reservoir flowing into a surface separation unit consisting of n stages operating at successively lower pressures.

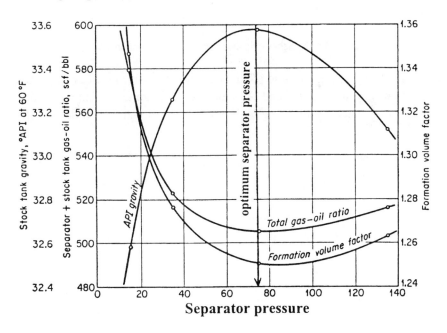

Figure 15-7. Effect of separator pressure on API, B_o, and GOR. (*After Amyx, Bass, and Whiting, 1960.*)

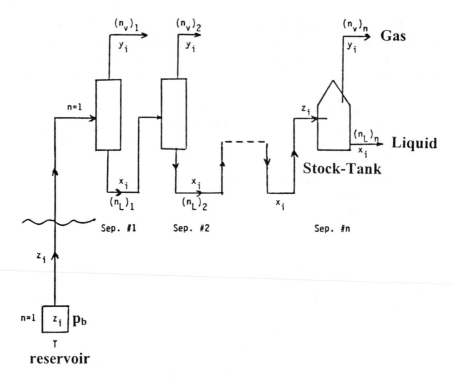

Figure 15-8. Schematic illustration of n separation stages.

Step 1. Calculate the volume of oil occupied by 1 lb-mol of crude at the reservoir pressure and temperature. This volume, denoted V_o, is calculated by recalling and applying the equation that defines the number of moles to give:

$$n = \frac{m}{M_a} = \frac{\rho_o V_o}{M_a} = 1$$

Solving for the oil volume gives:

$$V_o = \frac{M_a}{\rho_o} \tag{15-37}$$

where   m = total weight of 1 lb-mol of crude oil, lb/mol
        V_o = volume of 1 lb-mol of crude oil at reservoir conditions, ft³/mol
        M_a = apparent molecular weight
        ρ_o = density of the reservoir oil, lb/ft³

Step 2. Given the composition of the feed stream z_i to the first separator and the operating conditions of the separator, i.e., separator pressure and temperature, calculate the equilibrium ratios of the hydrocarbon mixture.

Step 3. Assuming a total of 1 mol of the feed entering the first separator and using the above calculated equilibrium ratios, perform flash calculations to obtain the compositions and quantities, in moles, of the gas and the liquid leaving the first separator. Designating these moles as $(n_L)_1$ and $(n_v)_1$, the actual number of moles of the gas and the liquid leaving the first separation stage are:

$$\left[n_{v1}\right]_a = (n)(n_v)_1 = (1)(n_v)_1$$

$$\left[n_{L1}\right]_a = (n)(n_L)_1 = (1)(n_L)_1$$

where $[n_{v1}]_a$ = actual number of moles of vapor leaving the first separator

                 $[n_{L1}]_a$ = actual number of moles of liquid leaving the first separator

Step 4. Using the composition of the liquid leaving the first separator as the feed for the second separator, i.e., $z_i = x_i$, calculate the equilibrium ratios of the hydrocarbon mixture at the prevailing pressure and temperature of the separator.

Step 5. Based on 1 mol of the feed, perform flash calculations to determine the compositions and quantities of the gas and liquid leaving the second separation stage. The actual number of moles of the two phases are then calculated from:

$$\left[n_{v2}\right]_a = \left[n_{L1}\right]_a (n_v)_2 = (1)(n_L)_1 (n_v)_2$$

$$\left[n_{L2}\right]_a = \left[n_{L1}\right]_a (n_L)_2 = (1)(n_L)_1 (n_L)_2$$

where $[n_{v2}]_a$, $[n_{L2}]_a$ = actual moles of gas and liquid leaving separator 2

                $(n_v)_2$, $(n_L)_2$ = moles of gas and liquid as determined from flash calculations

Step 6. The previously outlined procedure is repeated for each separation stage, including the stock-tank storage, and the calculated moles and compositions are recorded. The total number of moles of gas off all stages are then calculated as:

$$(n_v)_t = \sum_{i=1}^{n}(n_{va})_i = (n_v)_1 + (n_L)_1(n_v)_2 + (n_L)_1(n_L)_2(n_v)_3$$
$$+ \dots + (n_L)_1 \dots (n_L)_{n-1}(n_v)_n$$

In a more compacted form, the above expression can be written:

$$(n_v)_t = (n_v)_1 + \sum_{i=2}^{n}\left[(n_v)_i \prod_{j=1}^{i-1}(n_L)_j\right] \tag{15-38}$$

where $(n_v)_t$ = total moles of gas off all stages, lb-mol/mol of feed

n = number of separation stages

Total moles of liquid remaining in the stock tank can also be calculated as:

$$(n_L)_{st} = n_{L1}n_{L2}\dots n_{Ln}$$

or

$$(n_L)_{st} = \prod_{i=1}^{n}(n_L)_i \tag{15-39}$$

where

$(n_L)_{st}$ = number of moles of liquid remaining in the stock tank.
$(n_L)_i$ = moles of liquid off ith stage.

Step 7. Calculate the volume, in scf, of all the liberated solution gas from:

$$V_g = 379.4(n_v)_t \tag{15-40}$$

where V_g = total volume of the liberated solution gas scf/mol of feed.

Step 8. Determine the volume of stock-tank oil occupied by $(n_L)_{st}$ moles of liquid from:

$$(V_o)_{st} = \frac{(n_L)_{st}(M_a)_{st}}{(\rho_o)_{st}} \qquad (15-41)$$

where $(V_o)_{st}$ = volume of stock-tank oil, ft³/mol of feed
$(M_a)_{st}$ = apparent molecular weight of the stock-tank oil
$(\rho_o)_{st}$ = density of the stock-tank oil, lb/ft³

Step 9. Calculate the specific gravity and the API gravity of the stock-tank oil by applying these expressions:

$$\gamma_o = \frac{(\rho_o)_{st}}{62.4}$$

$$°API = \frac{141.5}{\gamma_o} - 131.5$$

Step 10. Calculate the total gas–oil ratio (or gas solubility R_s) :

$$GOR = \frac{V_g}{(V_o)_{st}/5.615} = \frac{(5.615)(379.4)(n_v)_t}{(n_L)_{st}(M)_{st}/(\rho_o)_{st}}$$

$$GOR = \frac{2,130.331(n_v)_t(\rho_o)_{st}}{(n_L)_{st}(M)_{st}} \qquad (15-42)$$

where GOR = gas–oil ratio, scf/STB.

Step 11. Calculate the oil formation volume factor from the relationship:

$$B_o = \frac{V_o}{(V_o)_{st}}$$

Combining Equations 15-37 and 15-41 with the above expression gives:

$$B_o = \frac{M_a(\rho_o)_{st}}{\rho_o(n_L)_{st}(M_a)_{st}}$$

where B_o = oil formation volume factor, bbl/STB
M_a = apparent molecular weight of the feed
$(M_a)_{st}$ = apparent molecular weight of the stock-tank oil
ρ_o = density of crude oil at reservoir conditions, lb/ft³

The separator pressure can be optimized by calculating the API gravity, GOR, and B_o in the manner outlined above at different assumed pressures. The optimum pressure corresponds to a maximum in the API gravity and a minimum in gas–oil ratio and oil formation volume factor.

Example 15-6

A crude oil, with the composition given below, exists at its bubble-point pressure of 1708.7 psia and at a temperature of 131°F. The crude oil is flashed through two-stage and stock-tank separation facilities. The operating conditions of the three separators are:

Separator	Pressure, psia	Temperature, °F
1	400	72
2	350	72
Stock tank	14.7	60

The composition of the crude oil is given below:

Component	z_i
CO_2	0.0008
N_2	0.0164
C_1	0.2840
C_2	0.0716
C_3	0.1048
$i - C_4$	0.0420
$n - C_4$	0.0420
$i - C_5$	0.0191
$n - C_5$	0.0191
C_6	0.0405
C_{7+}	0.3597

The molecular weight and specific gravity of C_{7+} are 252 and 0.8429. Calculate B_o, R_S, stock-tank density, and the API gravity of the hydrocarbon system.

Solution

Step 1. Calculate the apparent molecular weight of the crude oil to give $M_a = 113.5102$.

Step 2. Calculate the density of the bubble-point crude oil by using the Standing and Katz correlation to yield $\rho_o = 44.794$ lb/ft^3.

Step 3. Flash the original composition through the first separator by generating the equilibrium ratios by using the Standing correlation (Equation 15-20) to give:

Component	z_i	K_i	x_i	y_i
CO_2	0.0008	3.509	0.0005	0.0018
N_2	0.0164	39.90	0.0014	0.0552
C_1	0.2840	8.850	0.089	0.7877
C_2	0.0716	1.349	0.0652	0.0880
C_3	0.1048	0.373	0.1270	0.0474
$i-C_4$	0.0420	0.161	0.0548	0.0088
$n-C_4$	0.0420	0.120	0.0557	0.0067
$i-C_5$	0.0191	0.054	0.0259	0.0014
$n-C_5$	0.0191	0.043	0.0261	0.0011
C_6	0.0405	0.018	0.0558	0.0010
C_{7+}	0.3597	0.0021	0.4986	0.0009

With $n_L = 0.7209$ and $n_v = 0.29791$.

Step 4. Use the calculated liquid composition as the feed for the second separator and flash the composition at the operating condition of the separator.

Component	z_i	K_i	x_i	y_i
CO_2	0.0005	3.944	0.0005	0.0018
N_2	0.0014	46.18	0.0008	0.0382
C_1	0.089	10.06	0.0786	0.7877
C_2	0.0652	1.499	0.0648	0.0971
C_3	0.1270	0.4082	0.1282	0.0523
$i-C_4$	0.0548	0.1744	0.0555	0.0097
$n-C_4$	0.0557	0.1291	0.0564	0.0072
$i-C_5$	0.0259	0.0581	0.0263	0.0015
$n-C_5$	0.0261	0.0456	0.0264	0.0012
C_6	0.0558	0.0194	0.0566	0.0011
C_{7+}	0.4986	0.00228	0.5061	0.0012

With $n_L = 0.9851$ and $n_v = 0.0149$.

Step 5. Repeat the above calculation for the stock-tank stage to give:

Component	z_i	K_i	x_i	y_i
CO_2	0.0005	81.14	0000	0.0014
N_2	0.0008	1,159	0000	0.026
C_1	0.0784	229	0.0011	0.2455
C_2	0.0648	27.47	0.0069	0.1898
C_3	0.1282	6.411	0.0473	0.3030
$i-C_4$	0.0555	2.518	0.0375	0.0945
$n-C_4$	0.0564	1.805	0.0450	0.0812
$i-C_5$	0.0263	0.7504	0.0286	0.0214
$n-C_5$	0.0264	0.573	0.02306	0.0175
C_6	0.0566	0.2238	0.0750	0.0168
C_{7+}	0.5061	0.03613	0.7281	0.0263

With $n_L = 0.6837$ and $n_v = 0.3163$.

Step 6. Calculate the actual number of moles of the liquid phase at the stock-tank conditions from Equation 15-39:

$$\left(n_L\right)_{st} = (1)(0.7209)(0.9851)(0.6837) = 0.48554$$

Step 7. Calculate the total number of moles of the liberated gas from the entire surface separation system:

$$n_v = 1 - \left(n_L\right)_{st} = 1 - 0.48554 = 0.51446$$

Step 8. Calculate apparent molecular weight of the stock-tank oil from its composition to give $(M_a)_{st} = 200.6$.

Step 9. Calculate the density of the stock-tank oil by using the Standing correlation to give:

$$\left(\rho_o\right)_{st} = 50.920$$
$$\gamma = 50.920/62.4 = 0.816 \ \ 60°/60°$$

Step 10. Calculate the API gravity of the stock-tank oil:

$$API = \left(141.5/0.816\right) - 131.5 = 41.9$$

Step 11. Calculate the gas solubility from Equation 15-42 to give:

$$R_s = \frac{2130.331(0.51446)(50.92)}{0.48554(200.6)} = 573.0 \text{ scf/STB}$$

Step 12. Calculate B_o from Equation 15-43 to give:

$$B_o = \frac{(113.5102)(50.92)}{(44.794)(0.48554)(200.6)} = 1.325 \text{ bbl/STB}$$

To optimize the operating pressure of the separator, the above steps should be repeated several times under different assumed pressures and the results, in terms of API, B_o, and R_s, should be expressed graphically and used to determine the optimum pressure.

Note that at **low pressures**, e.g., $p < 1000$, equilibrium ratios are nearly independent of the overall composition z_i or the convergence pressure and can be considered only a function pressure and temperature. Under this condition, i.e, $p < 1000$, the equilibrium ratio for any component i can be expressed as:

$$K_i = \frac{A_i}{p}$$

The temperature-dependent coefficient A_i is a characterization parameter of component i that accounts for the physical properties of the component. The above expression suggests that the K_i varies linearly at a constant temperature with $1/p$. For example, suppose that a hydrocarbon mixture exists at 300 psi and 100°F. Assume that the mixture contains methane and we want to estimate the equilibrium ratio of methane (or any other components) when the mixture is flashed at 100 psi and at the same temperature of 100°F. The recommended procedure is summarized in the following steps:

Step 1. Because at low pressure the equilibrium ratio is considered independent of the overall composition of the mixture, use the equilibrium ratio charts of Appendix A to determine the K_i value of methane at 300 psi and 100°F:

$$K_{C_1} = 10.5$$

Step 2. Calculate the characterization parameter A_i of methane from the above proposed relationship:

$$10.5 = \frac{A_i}{500}$$
$$A_i = (10.5)(300) = 3,150$$

Step 3. Calculate the K_i of methane at 100 psi and 100°F from:

$$K_{C_1} = \frac{3,150}{100} = 31.5$$

In many low-pressure applications of flash calculations at constant temperature, it might be possible to characterize the entire hydrocarbon mixture as a binary system, i.e., two-component system. Because methane exhibits a linear relationship with pressure of a wide range of pressure values, one of the components that forms the binary system should be methane. The main advantage of such a binary system is the simplicity of performing flash calculations because it does not require an iterative technique.

Reconsider Example 15-6 where flash calculations were performed on the entire system at 400 psia and 72°F. To perform flash calculations on the feed for the second separator at 350 psi and 72°F, follow these steps:

Step 1. Select methane as one of the binary systems with the other component defined as ethane-plus, i.e., C_{2+}, which lumps the remaining components. Results of Example 15-6 show:

- $K_{C_1} = 8.85$
- $y_{C_1} = 0.7877$
- $x_{C_2} = 0.089$
- $y_{C_{2+}} = 1.0 - 0.7877 = 0.2123$
- $x_{C_{2+}} = 1.0 - 0.089 = 0.911$

Step 2. From the definition of the equilibrium ratio, calculate the K value of C_{2+}:

$$K_{C_{2+}} = \frac{y_{C_{2+}}}{x_{C_{2+}}} = \frac{0.2123}{0.9110} = 0.2330$$

Step 3. Calculate the characterization parameter A_i for methane and C_{2+}:

$$A_{C_1} = K_{C_1} p = (8.85)(400) = 3,540$$
$$A_{C_{2+}} = K_{C_{2+}} p = (0.233)(400) = 93.2$$

The equilibrium ratio for each of the two components (at a constant temperature) can then be described by:

$$K_{C_1} = \frac{3,540}{p}$$

$$K_{C_{2+}} = \frac{93.2}{p}$$

Step 4. Calculate the K_i value for each component at the second separator pressure of 350 psi:

$$K_{C_1} = \frac{3,540}{350} = 10.11$$

$$K_{C_{2+}} = \frac{93.2}{350} = 0.266$$

Step 5. Using the flash calculations procedure as outlined previously for a binary system, calculate the composition and number of moles of the gas and liquid phase at 350 psi:

- Solve for x_{C1} and x_{C2+}:

$$x_{C_1} = \frac{1-K_2}{K_1-K_2} = \frac{1.0-0.266}{10.11-0.266} = 0.0746$$

$$x_{C_{2+}} = 1-x_{C_1} = 1.0-0.0746 = 0.9254$$

- Solve for y_{C_1} and $y_{C_{2+}}$:

$$y_{C_1} = x_{C_1}K_1 = (0.0746)(10.11) = 0.754$$

$$y_{C_{2+}} = 1-y_{C_1} = 1.0-0.754 = 0.246$$

- Solve for number of moles of the vapor and liquid phase:

$$n_v = \frac{z_1-x_1}{x_1(K_1-1)} = \frac{0.089-0.0746}{0.0746(10.11-1)} = 0.212$$

$$n_L = 1-n_v = 1.0-0.212 = 0.788$$

The above calculations are considered meaningless without converting moles of liquid n_l into volume, which requires the calculation of the liquid density at separator pressure and temperature. Notice:

$$V = \frac{n_L M_a}{\rho_o}$$

where M_a is the apparent molecular weight of the separated liquid and is given by (for a binary system):

$$M_a = x_{C_1} M_{C_1} + x_{C_{2+}} M_{C_{2+}}$$

Density Calculations

The calculation of crude oil density from its composition is an important and integral part of performing flash calculations. The best known and most widely used calculation methods are those of Standing-Katz (1942) and Alani-Kennedy (1960). These two methods are presented below:

The Standing-Katz Method

Standing and Katz (1942) proposed a graphical correlation for determining the density of hydrocarbon liquid mixtures. The authors developed the correlation from evaluating experimental, compositional, and density data on 15 crude oil samples containing up to 60 mol% methane. The proposed method yielded an average error of 1.2% and maximum error of 4% for the data on these crude oils. The original correlation did not have a procedure for handling significant amounts of nonhydrocarbons.

The authors expressed the density of hydrocarbon liquid mixtures as a function of pressure and temperature by the following relationship:

$$\rho_o = \rho_{sc} + \Delta\rho_p - \Delta\rho_T$$

where ρ_o = crude oil density at p and T, lb/ft^3
           ρ_{sc} = crude oil density (with all the dissolved solution gas) at standard conditions, i.e., 14.7 psia and 60°F, lb/ft^3
           $\Delta\rho_p$ = density correction for compressibility of oils, lb/ft^3
           $\Delta\rho_T$ = density correction for thermal expansion of oils, lb/ft^3

Standing and Katz correlated graphically the liquid density at standard conditions with:

• The density of the propane-plus fraction $\rho_{C_{3+}}$
• The weight percent of methane in the entire system $(m_{C_1})_{C_{1+}}$
• The weight percent of ethane in the ethane-plus $(m_{C_2})_{C_{2+}}$

This graphical correlation is shown in Figure 15-9. The following are the specific steps in the Standing and Katz procedure of calculating the liquid density at a specified pressure and temperature.

Step 1. Calculate the total weight and the weight of each component in 1 lb-mol of the hydrocarbon mixture by applying the following relationships:

$$m_i = x_i M_i$$
$$m_t = \Sigma x_i M_i$$

where m_i = weight of component i in the mixture, lb/lb-mol
$\qquad x_i$ = mole fraction of component i in the mixture
$\qquad M_i$ = molecular weight of component i
$\qquad m_t$ = total weight of 1 lb-mol of the mixture, lb/lb-mol

Step 2. Calculate the weight percent of methane in the entire system and the weight percent of ethane in the ethane-plus from the following expressions:

$$\left(m_{C_1}\right)_{C_{1+}} = \left[\frac{x_{C_1} M_{C1}}{\sum\limits_{i=1}^{n} x_i M_i} \right] 100 = \left[\frac{m_{C_1}}{m_t} \right] 100$$

and

$$\left(m_{C_2}\right)_{C_{2+}} = \left[\frac{m_{C_2}}{m_{C_{2+}}} \right] 100 = \left[\frac{m_{C_2}}{m_t - m_{C_1}} \right] 100$$

where $(m_{C_1})_{C_{1+}}$ = weight percent of methane in the entire system
$\qquad m_{C_1}$ = weight of methane in 1 lb-mol of the mixture, i.e., $x_{C_1} M_{C_1}$
$\qquad (m_{C_2})_{C_{2+}}$ = weight percent of ethane in ethane-plus
$\qquad m_{C_2}$ = weight of ethane in 1 lb-mol of the mixture, i.e., $x_{C_2} M_{C_2}$
$\qquad M_{C_1}$ = molecular weight of methane
$\qquad M_{C_2}$ = molecular weight of ethane

Step 3. Calculate the density of the propane-plus fraction at standard conditions by using the following equations:

$$\rho_{C_{3+}} = \frac{m_{C_3}}{V_{C_{3+}}} = \frac{\displaystyle\sum_{i=C_3}^{n} x_i M_i}{\displaystyle\sum_{i=C_3}^{n} \frac{x_i M_i}{\rho_{oi}}}$$

with

$$m_{C_{3+}} = \sum_{i=C_3} x_i M_i$$

$$V_{C_{3+}} = \sum_{i=C_3} V_i = \sum_{i=C_3} \frac{m_i}{\rho_{oi}}$$

where $\rho_{C_{3+}}$ = density of the propane and heavier components, lb/ft^3
$\quad\quad m_{C_{3+}}$ = weight of the propane and heavier fractions, lb/lb-mol
$\quad\quad V_{C_{3+}}$ = volume of the propane-plus fraction, ft^3/lb-mol
$\quad\quad V_i$ = volume of component i in 1 lb-mol of the mixture
$\quad\quad m_i$ = weight of component i, i.e., $x_i M_i$, lb/lb-mole
$\quad\quad \rho_{oi}$ = density of component i at standard conditions, lb/ft^3

Density values for pure components are tabulated in Table 1-2 in Chapter 1, but the density of the plus fraction must be measured.

Step 4. Using Figure 15-9, enter the $\rho_{C_{3+}}$ value into the left ordinate of the chart and move horizontally to the line representing $(m_{C_2})_{C_{2+}}$; then drop vertically to the line representing $(m_{C_1})_{C_{1+}}$. The density of the oil at standard condition is read on the right side of the chart. Standing (1977) expressed the graphical correlation in the following mathematical form:

$$\rho_{sc} = \rho_{C_{2+}} \left[1 - 0.012 \left(m_{c_1} \right)_{C_{1+}} - 0.000158 \left(m_{c_1} \right)^2_{C_{1+}} \right]$$
$$+ 0.0133 \left(m_{c_1} \right)_{C_{1+}} + 0.00058 \left(m_{c_1} \right)^2_{C_{2+}}$$

with

$$\rho_{C_{2+}} = \rho_{C_{3+}} \left[1 - 0.01386 \left(m_{c_2} \right)_{C_{2+}} - 0.000082 \left(m_{c_2} \right)^2_{C_{2+}} \right]$$
$$+ 0.379 \left(m_{c_2} \right)_{C_{2+}} + 0.0042 \left(m_{c_2} \right)^2_{C_{2+}}$$

where $\rho_{C_{2+}}$ = density of ethane-plus fraction.

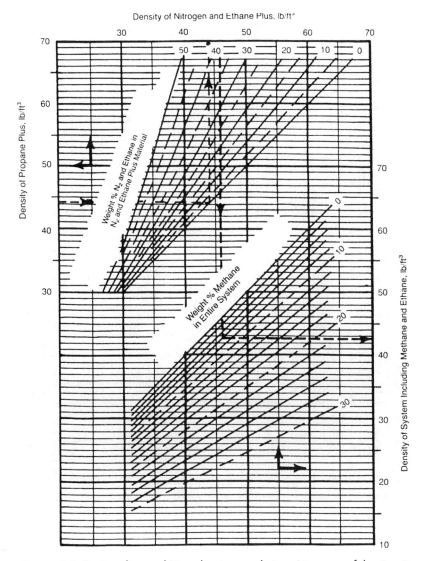

Density of Nitrogen and Ethane Plus, lb/ft³

Figure 15-9. Standing and Katz density correlation. *(Courtesy of the Gas Processors Suppliers Association, Engineering Data Book, 10th ed., 1987.)*

Step 5. Correct the density at standard conditions to the actual pressure by reading the additive pressure correction factor, $\Delta\rho_p$, from Figure 15-10, or using the following expression:

$$\Delta\rho_p = \left[0.000167 + (0.016181)10^{-0.0425\rho_{sc}}\right]p - \left(10^{-8}\right)\left[0.299 + (263)10^{-0.0603\rho_{sc}}\right]p^2$$

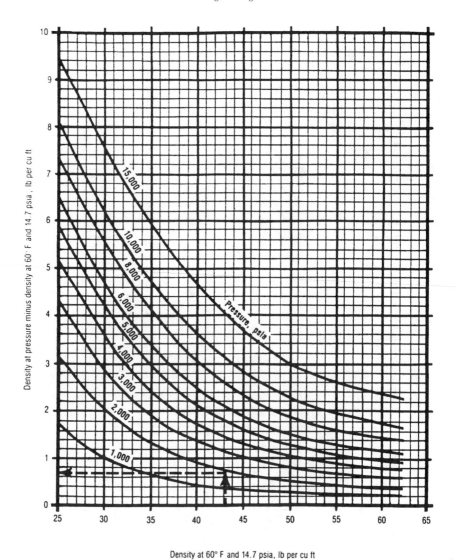

Density at 60° F and 14.7 psia, lb per cu ft

Figure 15-10. Density correction for compressibility of crude oils. *(Courtesy of the Gas Processors Suppliers Association, Engineering Data Book, 10th ed., 1987.)*

Step 6. Correct the density at 60°F and pressure to the actual temperature by reading the thermal expansion correction term, $\Delta\rho_T$, from Figure 15-11, or from:

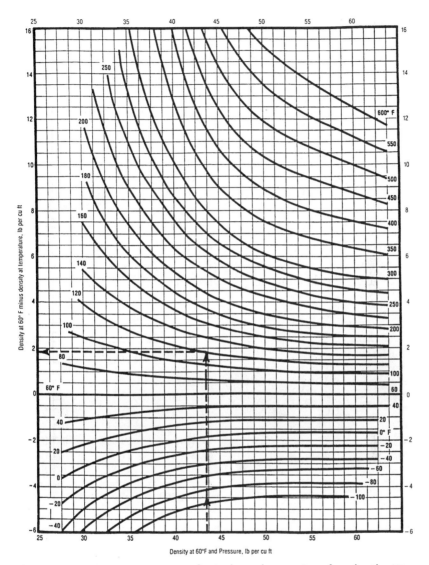

Figure 15-11. Density correction for isothermal expansion of crude oils. *(Courtesy of the Gas Processors Suppliers Association, Engineering Data Book, 10th ed., 1987.)*

$$\Delta\rho_T = (T - 520)\left[0.0133 + 152.4\left(\rho_{sc} + \Delta\rho_p\right)^{-2.45}\right] -$$

$$(T - 520)^2\left[8.1\left(10^{-6}\right) - (0.0622)10^{-0.0764\left(\rho_{sc} + \Delta\rho_p\right)}\right]$$

where T is the system temperature in °R.

Example 15-7

A crude oil system has the following composition.

Component	x_i
C_1	0.45
C_2	0.05
C_3	0.05
C_4	0.03
C_5	0.01
C_6	0.01
C_{7+}	0.40

If the molecular weight and specific gravity of C_{7+} fractions are 215 and 0.87, respectively, calculate the density of the crude oil at 4000 psia and 160°F by using the Standing and Katz method.

Solution

Component	x_i	M_i	$m_i = x_i M_i$	ρ_{oi}, lb/ft³*	$V_i = m_i/\rho_{oi}$
C_1	0.45	16.04	7.218	—	—
C_2	0.05	30.07	1.5035	—	—
C_3	0.05	44.09	2.2045	31.64	0.0697
C_4	0.03	58.12	1.7436	35.71	0.0488
C_5	0.01	72.15	0.7215	39.08	0.0185
C_6	0.01	86.17	0.8617	41.36	0.0208
C_{7+}	0.40	215.0	86.00	54.288†	1.586
			$m_t = 100.253$		$V_{C_{3+}} = 1.7418$

From Table 1-2.
†$\rho_{C_{7+}} = (0.87)(62.4) = 54.288.$

Step 1. Calculate the weight percent of C_1 in the entire system and the weight percent of C_2 in the ethane-plus fraction:

$$\left(m_{C_1}\right)_{C_{1+}} = \left[\frac{7.218}{100.253}\right]100 = 7.2\%$$

$$\left(m_{C_2}\right)_{C_{2+}} = \left[\frac{1.5035}{100.253 - 7.218}\right]100 = 1.616\%$$

Step 2. Calculate the density of the propane-plus fraction:

$$\rho_{C_{3+}} = \frac{100.253 - 7.218 - 1.5035}{1.7418} = 52.55 \text{ lb/ft}^3$$

Step 3. Determine the density of the oil at standard conditions from Figure 15-9:

$$\rho_{sc} = 47.5 \text{ lb/ft}^3$$

Step 4. Correct for the pressure by using Figure 15-10:

$$\Delta\rho_p = 1.18 \text{ lb/ft}^3$$

Density of the oil at 4000 psia and 60°F is then calculated by the expression:

$$\rho_{p,60} = \rho_{sc} + \Delta\rho_p = 47.5 + 1.18 = 48.68 \text{ lb/ft}^3$$

Step 5. From Figure 15-11, determine the thermal expansion correction factor:

$$\Delta\rho_T = 2.45 \text{ lb/ft}^3$$

Step 6. The required density at 4000 psia and 160°F is:

$$\rho_0 = 48.68 - 2.45 = 46.23 \text{ lb/ft}^3$$

The Alani–Kennedy Method

Alani and Kennedy (1960) developed an equation to determine the molar liquid volume V_m of pure hydrocarbons over a wide range of temperature and pressure. The equation was then adopted to apply to crude oils with the heavy hydrocarbons expressed as a heptanes-plus fraction, i.e., C_{7+}.

The Alani–Kennedy equation is similar in form to the Van der Waals equation, which takes the following form:

$$V_m^3 - \left[\frac{RT}{p} + b\right]V_m^2 + \frac{aV_m}{p} - \frac{ab}{p} = 0 \tag{15-43}$$

where R = gas constant, 10.73 psia ft^3/lb-mol °R

  T = temperature, °R

  p = pressure, psia

  V_m = molecular volume, ft^3/lb-mol

  a, b = constants for pure substances

Alani and Kennedy considered the constants a and b to be functions of temperature and proposed these expressions for calculating the two parameters:

$$a = Ke^{n/T}$$

$$b = mT + c$$

where K, n, m, and c are constants for each pure component. Values of these constants are tabulated in Table 15-1. Table 15-1 contains no constants from which the values of the parameters a and b for heptanes-plus can be calculated. Therefore, Alani and Kennedy proposed the following equations for determining a and b of C_{7+}.

Table 15-1
Alani and Kennedy Coefficients

Components	K	n	m × 10^4	c
C_1 70° – 300°F	9,160.6413	61.893223	3.3162472	0.50874303
C_1 301° – 460°F	147.47333	3,247.4533	–14.072637	1.8326659
C_2 100° – 249°F	46,709.573	–404.48844	5.1520981	0.52239654
C_2 250° – 460°F	17,495.343	34.163551	2.8201736	0.62309877
C_3	20,247.757	190.24420	2.1586448	0.90832519
$i - C_4$	32,204.420	131.63171	3.3862284	1.1013834
$n - C_4$	33,016.212	146.15445	2.902157	1.1168144
$i - C_5$	37,046.234	299.62630	2.1954785	1.4364289
$n - C_5$	37,046.234	299.62630	2.1954785	1.4364289
$n - C_6$	52,093.006	254.56097	3.6961858	1.5929406
H_2S*	13,200.00	0	17.900	0.3945
N_2*	4,300.00	2.293	4.490	0.3853
CO_2*	8,166.00	126.00	1.8180	0.3872

Values for non-hydrocarbon components as proposed by Lohrenz et al. (1964).

$$\ln\left(a_{C_{7+}}\right) = 3.8405985\left(10^{-3}\right)(M)_{C_{7+}} - 9.5638281\left(10^{-4}\right)\left(\frac{M}{\gamma}\right)_{C_{7+}}$$

$$+ \frac{261.80818}{T} + 7.3104464\left(10^{-6}\right)(M)_{C_{7+}}^2 + 10.753517$$

$$b_{C_{7+}} = 0.03499274(M)_{C_{7+}} - 7.275403(\gamma)_{C_{7+}} + 2.232395\left(10^{-4}\right)T$$

$$- 0.016322572\left(\frac{M}{\gamma}\right)_{C_{7+}} + 6.2256545$$

where $M_{C_{7+}}$ = molecular weight of C_{7+}
$\gamma_{C_{7+}}$ = specific gravity of C_{7+}
$a_{C_{7+}}, b_{C_{7+}}$ = constants of the heptanes-plus fraction
T = temperature in °R

For hydrocarbon mixtures, the values of a and b of the mixture are calculated using the following mixing rules:

$$a_m = \sum_{i=1}^{C_{7+}} a_i x_i$$

$$b_m = \sum_{i=1}^{C_{7+}} b_i x_i$$

where the coefficients a_i and b_i refer to pure hydrocarbons at existing temperature, and x_i is the mole fraction in the mixture. The values of a_m and b_m are then used in Equation 15-43 to solve for the molar volume V_m. The density of the mixture at pressure and temperature of interest is determined from the following relationship:

$$\rho_o = \frac{M_a}{V_m}$$

where ρ_o = density of the crude oil, lb/ft³
M_a = apparent molecular weight, i.e., $M_a = \sum x_i M_i$
V_m = molar volume, ft³/lb-mol

The Alani and Kennedy method for calculating the density of liquids is summarized in the following steps:

Step 1. Calculate the constants a and b for each pure component from:

$$a = Ke^{n/T}$$
$$b = mT + c$$

Step 2. Determine $a_{C_{7+}}$ and $b_{C_{7+}}$.

Step 3. Calculate the values of coefficients a_m and b_m.

Step 4. Calculate molar volume V_m by solving Equation 15-43 for the smallest real root:

$$V_m^3 - \left[\frac{RT}{p} + b_m \right] V_m^2 + \frac{a_m V_m}{p} - \frac{a_m b_m}{p} = 0$$

Step 5. Compute the apparent molecular weight, M_a.

Step 6. Determine the density of the crude oil from:

$$\rho_0 = \frac{M_a}{V_m}$$

Example 15-8

A crude oil system has the composition:

Component	x_i
CO_2	0.0008
N_2	0.0164
C_1	0.2840
C_2	0.0716
C_3	0.1048
$i - C_4$	0.0420
$n - C_4$	0.0420
$i - C_5$	0.0191
$n - C_5$	0.0191
C_6	0.0405
C_{7+}	0.3597

The following additional data are given:

$$M_{C_{7+}} = 252$$
$$\gamma_{C_{7+}} = 0.8424$$
$$\text{Pressure} = 1708.7 \text{ psia}$$
$$\text{Temperature} = 591°R$$

Calculate the density of the crude oil.

Solution

Step 1. Calculate the parameters $a_{C_{7+}}$ and $b_{C_{7+}}$:

$$a_{C_{7+}} = 229269.9$$
$$b_{C_{7+}} = 4.165811$$

Step 2. Calculate the mixture parameters a_m and b_m:

$$a_m = \sum_{i=1}^{C_{7+}} a_i x_i$$
$$a_m = 99111.71$$
$$b_m = \sum_{i=1}^{C_{7+}} b_i x_i$$
$$b_m = 2.119383$$

Step 3. Solve Equation 15-43 for the molar volume:

$$V_m^3 - \left[\frac{RT}{p} + b_m \right] V_m^2 + \frac{a_m V_m}{p} - \frac{a_m b_m}{p} = 0$$
$$V_m = 2.528417$$

Step 4. Determine the apparent molecular weight of this mixture:

$$M_a = \Sigma x_i M_i$$
$$M_a = 113.5102$$

Step 5. Compute the density of the oil system:

$$\rho_0 = \frac{M_a}{V_m}$$
$$\rho_0 = \frac{113.5102}{2.528417} = 44.896 \, lb/ft^3$$

EQUATIONS OF STATE

An equation of state (EOS) is an analytical expression relating the pressure p to the temperature T and the volume V. A proper description of this PVT relationship for real hydrocarbon fluids is essential in determining the volumetric and phase behavior of petroleum reservoir fluids and in predicting the performance of surface separation facilities.

The best known and the simplest example of an equation of state is the ideal gas equation, expressed mathematically by the expression:

$$p = \frac{RT}{V} \tag{15-44}$$

where V = gas volume in cubic feet per 1 mol of gas. This PVT relationship is only used to describe the volumetric behavior of real hydrocarbon gases at pressures close to the atmospheric pressure for which it was experimentally derived.

The extreme limitations of the applicability of Equation 15-44 prompted numerous attempts to develop an equation of state (EOS) suitable for describing the behavior of real fluids at extended ranges of pressures and temperatures.

The main objective of this chapter is to review developments and advances in the field of empirical cubic equations of state and demonstrate their applications in petroleum engineering.

The Van der Waals Equation of State

In developing the ideal gas EOS (Equation 15-44), two assumptions were made:

- **First assumption**: The volume of the gas molecules is insignificant compared to the volume of the container and distance between the molecules.
- **Second assumption:** There are no attractive or repulsive forces between the molecules or the walls of the container.

Van der Waals (1873) attempted to eliminate these two assumptions by developing an empirical equation of state for real gases. In his attempt to eliminate the first assumption, van der Waals pointed out that the gas molecules occupy a significant fraction of the volume at higher pressures and proposed that the volume of the molecules, as denoted by the parameter b, be subtracted from the actual molar volume V in Equation 15-44, to give:

$$p = \frac{RT}{V-b}$$

where the parameter b is known as the co-volume and is considered to reflect the volume of molecules. The variable V represents the actual volume in cubic feet per 1 mol of gas.

To eliminate the second assumption, van der Waals subtracted a corrective term, as denoted by a/V^2, from the above equation to account for the attractive forces between molecules. In a mathematical form, van der Waals proposed the following expression:

$$p = \frac{RT}{V-b} - \frac{a}{V^2} \tag{15-45}$$

where p = system pressure, psia
 T = system temperature, °R
 R = gas constant, 10.73 psi-ft^3/lb-mol = °R
 V = volume, ft^3/mol

The two parameters a and b are constants characterizing the molecular properties of the individual components. The symbol a is considered a measure of the intermolecular attractive forces between the molecules. Equation 15-45 shows the following important characteristics:

1. At low pressures, the volume of the gas phase is large in comparison with the volume of the molecules. The parameter b becomes negligible in comparison with V and the attractive forces term a/V^2 becomes insignificant; therefore, the van der Waals equation reduces to the ideal gas equation (Equation 15-44).
2. At high pressure, i.e., $p \rightarrow \infty$, volume V becomes very small and approaches the value b, which is the actual molecular volume.

The van der Waals or any other equation of state can be expressed in a more generalized form as follows:

$$p = p_{repulsive} - p_{attractive}$$

where the repulsive pressure term $p_{repulsive}$ is represented by the term $RT/(V-b)$ and the attractive pressure term $p_{attractive}$ is described by a/V^2.

In determining the values of the two constants a and b for any pure substance, van der Waals observed that the critical isotherm has a horizontal slope and an inflection point at the critical point, as shown in Figure 15-12. This observation can be expressed mathematically as follows:

$$\left[\frac{\partial p}{\partial V}\right]_{T_C,p_C} = 0, \quad \left[\frac{\partial^2 p}{\partial V^2}\right]_{T_C,p_C} = 0 \tag{15-46}$$

Differentiating Equation 15-45 with respect to the volume at the critical point results in:

$$\left[\frac{\partial p}{\partial V}\right]_{T_C,p_C} = \frac{-RT_C}{\left(V_C - b\right)^3} + \frac{2a}{V_C^3} = 0 \tag{15-47}$$

$$\left[\frac{\partial^2 p}{\partial V^2}\right]_{T_C,p_C} = \frac{2RT_C}{\left(V_C - b\right)^3} + \frac{6a}{V_C^4} = 0 \tag{15-48}$$

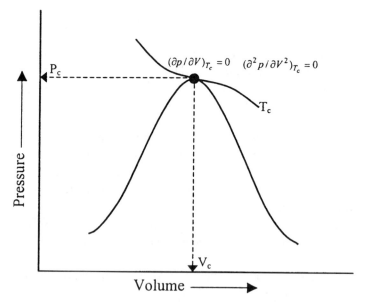

Figure 15-12. An idealized pressure–volume relationship for a pure compound.

Solving Equations 15-47 and 15-48 simultaneously for the parameters a and b gives:

$$b = \left(\frac{1}{3}\right)V_C \tag{15-49}$$

$$a = \left(\frac{8}{9}\right)RT_C V_C \tag{15-50}$$

Equation 15-49 suggests that the volume of the molecules b is approximately 0.333 of the critical volume V_C of the substance. Experimental studies reveal that the co-volume b is in the range of 0.24 to 0.28 of the critical volume and pure component.

By applying Equation 15-45 to the critical point (i.e., by setting $T = T_c$, $p = p_c$, and $V = V_c$) and combining with Equations 15-49 and 15-50, we get:

$$p_C V_C = (0.375)RT_C \tag{15-51}$$

Equation 15-51 shows that regardless of the type of substance, the van der Waals EOS produces a universal critical gas compressibility factor Z_c of 0.375. Experimental studies show that Z_c values for substances range between 0.23 and 0.31.

Equation 15-51 can be combined with Equations 15-49 and 15-50 to give a more convenient and traditional expression for calculating the parameters a and b to yield:

$$a = \Omega_a \frac{R^2 T_c^2}{p_c} \tag{15-52}$$

$$b = \Omega_b \frac{RT_c}{p_c} \tag{15-53}$$

where R = gas constant, 10.73 psia-ft³/lb-mol-°R
          p_c = critical pressure, psia
          T_c = critical temperature, °R
          $\Omega_a = 0.421875$
          $\Omega_b = 0.125$

Equation 15-45 can also be expressed in a cubic form in terms of the volume V as follows:

$$V^3 - \left(b + \frac{RT}{p}\right)V^2 + \left(\frac{a}{p}\right)V - \left(\frac{ab}{p}\right) = 0 \qquad (15\text{-}54)$$

Equation 15-54 is usually referred to as the van der Waals two-parameter cubic equation of state. The term *two-parameter* refers to the parameters a and b. The term *cubic equation of state* implies an equation that, if expanded, would contain volume terms to the first, second, and third power.

Perhaps the most significant feature of Equation 15-54 is its ability to describe the liquid-condensation phenomenon and the passage from the gas to the liquid phase as the gas is compressed. This important feature of the van der Waals EOS is discussed below in conjunction with Figure 15-13.

Consider a pure substance with a p-V behavior as shown in Figure 15-13. Assume that the substance is kept at a constant temperature T below its critical temperature. At this temperature, Equation 15-54 has three real roots (volumes) for each specified pressure p. A typical solution of Equation 15-54 at constant temperature T is shown graphically by the dashed isotherm: the constant temperature curve DWEZB in Figure 15-13. The three values of V are the intersections B, E, and D on the horizontal line, corresponding to a fixed value of the pressure. This dashed calculated line (DWEZB) then appears to give a continuous

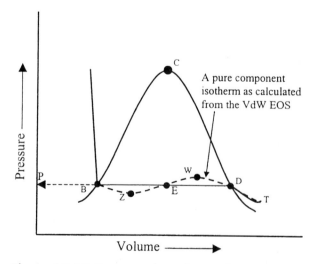

Figure 15-13. Pressure–volume diagram for a pure component.

transition from the gaseous phase to the liquid phase, but in reality, the transition is abrupt and discontinuous, with both liquid and vapor existing along the straight horizontal line DB. Examining the graphical solution of Equation 15-54 shows that the largest root (volume), as indicated by point D, corresponds to the volume of the saturated vapor, while the smallest positive volume, as indicated by point B, corresponds to the volume of the saturated liquid. The third root, point E, has no physical meaning. Note that these values become identical as the temperature approaches the critical temperature T_c of the substance.

Equation 15-54 can be expressed in a more practical form in terms of the compressibility factor Z. Replacing the molar volume V in Equation 15-54 with ZRT/p gives:

$$Z^3 - (1+B)Z^2 + AZ - AB = 0 \qquad (15\text{-}55)$$

where

$$A = \frac{ap}{R^2 T^2} \qquad (15\text{-}56)$$

$$B = \frac{bp}{RT} \qquad (15\text{-}57)$$

Z = compressibility factor
p = system pressure, psia
T = system temperature, °R

Equation 15-55 yields one real root[1] in the one-phase region and three real roots in the two-phase region (where system pressure equals the vapor pressure of the substance). In the latter case, the largest root corresponds to the compressibility factor of the vapor phase Z^V, while the smallest positive root corresponds to that of the liquid Z^L.

An important practical application of Equation 15-55 is for calculating density calculations, as illustrated in the following example.

Example 15-9

A pure propane is held in a closed container at 100°F. Both gas and liquid are present. Calculate, by using the van der Waals EOS, the density of the gas and liquid phases.

---

[1] In some supercritical regions, Equation 15-55 can yield three real roots for Z. From the three real roots, the largest root is the value of the compressibility with physical meaning.

Solution

Step 1. Determine the vapor pressure p_v of the propane from the Cox chart. This is the only pressure at which two phases can exist at the specified temperature:

$$p_v = 185 \text{ psi}$$

Step 2. Calculate parameters a and b from Equations 15-52 and 15-53, respectively.

$$a = \Omega_a \frac{R^2 T_c^2}{p_c}$$

$$a = 0.421875 \frac{(10.73)^2 (666)^2}{616.3} = 34,957.4$$

and

$$b = \Omega_b \frac{RT_c}{p_c}$$

$$b = 0.125 \frac{10.73(666)}{616.3} = 1.4494$$

Step 3. Compute coefficients A and B by applying Equations 15-56 and 15-57, respectively.

$$A = \frac{ap}{R^2 T^2}$$

$$A = \frac{(34,957.4)(185)}{(10.73)^2 (560)^2} = 0.179122$$

$$B = \frac{bp}{RT}$$

$$B = \frac{(1.4494)(185)}{(10.73)(560)} = 0.044625$$

Step 4. Substitute the values of A and B into Equation 15-55 to give:

$$Z^3 - (1+B)Z^2 + AZ - AB = 0$$

$$Z^3 - 1.044625Z^2 + 0.179122Z - 0.007993 = 0$$

Step 5. Solve the above third-degree polynomial by extracting the largest and smallest roots of the polynomial by using the appropriate direct or iterative method to give:

$$Z^v = 0.72365$$

$$Z^L = 0.07534$$

Step 6. Solve for the density of the gas and liquid phases by using Equation 2-17:

$$\rho_g = \frac{pM}{Z^v RT}$$

$$\rho_g = \frac{(185)(44.0)}{(0.72365)(10.73)(560)} = 1.87 \, lb/ft^3$$

and

$$\rho_L = \frac{pM}{Z^L RT}$$

$$\rho_L = \frac{(185)(44)}{(0.7534)(10.73)(560)} = 17.98 \, lb/ft^3$$

The van der Waals equation of state, despite its simplicity, provides a correct description, at least qualitatively, of the PVT behavior of substances in the liquid and gaseous states. Yet it is not accurate enough to be suitable for design purposes.

With the rapid development of computers, the EOS approach for the calculation of physical properties and phase equilibria proved to be a powerful tool, and much energy was devoted to the development of new and accurate equations of state. These equations, many of them a modification of the van der Waals equation of state, range in complexity from simple expressions containing 2 or 3 parameters to complicated forms containing more than 50 parameters. Although the complexity of any equation of state presents no computational problem, most authors prefer to retain the simplicity found in the van der Waals cubic equation while improving its accuracy through modifications.

All equations of state are generally developed for pure fluids first, and then extended to mixtures through the use of mixing rules. These mixing rules are simply means of calculating mixture parameters equivalent to those of pure substances.

Redlich-Kwong Equation of State

Redlich and Kwong (1949) demonstrated that by a simple adjustment, the van der Waals attractive pressure term a/V^2 could considerably improve the prediction of the volumetric and physical properties of the vapor phase. The authors replaced the attractive pressure term with a generalized temperature dependence term. Their equation has the following form:

$$p = \frac{RT}{V-b} - \frac{a}{V(V+b)\sqrt{T}} \tag{15-58}$$

where T is the system temperature in °R.

Redlich and Kwong (1949), in their development of the equation, noted that as the system pressure becomes very large, i.e., $p \to \infty$, the molar volume V of the substance shrinks to about 26% of its critical volume regardless of the system temperature. Accordingly, they constructed Equation 15-58 to satisfy the following condition:

$$b = 0.26V_c \tag{15-59}$$

Imposing the critical point conditions (as expressed by Equation 15-46) on Equation 15-58 and solving the resulting equations simultaneously gives:

$$a = \Omega_a \frac{R^2 T_c^{2.5}}{p_c} \tag{15-60}$$

$$b = \Omega_b \frac{RT_c}{p_c} \tag{15-61}$$

where Ω_a = 0.42747 and Ω_b = 0.08664. Equating Equation 15-61 with 15-59 gives:

$$p_c V_c = 0.333RT_c \tag{15-62}$$

Equation 15-62 shows that the Redlich-Kwong EOS produces a universal critical compressibility factor (Z_c) of 0.333 for all substances. As indicated earlier, the critical gas compressibility ranges from 0.23 to 0.31 for most of the substances.

Replacing the molar volume V in Equation 15-58 with ZRT/p gives:

$$Z^3 - Z^2 + (A - B - B^2)Z - AB = 0 \qquad (15\text{-}63)$$

where

$$A = \frac{ap}{R^2 T^{2.5}} \qquad (15\text{-}64)$$

$$B = \frac{bp}{RT} \qquad (15\text{-}65)$$

As in the van der Waals EOS, Equation 15-63 yields one real root in the one-phase region (gas-phase region or liquid-phase region), and three real roots in the two-phase region. In the latter case, the **largest root** corresponds to the compressibility factor of the gas phase Z^v while the **smallest positive root** corresponding to that of the liquid Z^L.

Example 15-10

Rework Example 15-9 by using the Redlich-Kwong equation of state.

Solution

Step 1. Calculate the parameters a, b, A, and B:

$$a = 0.42747 \frac{(10.73)^2 (666)^{2.5}}{616.3} = 914,110.1$$

$$b = 0.08664 \frac{(10.73)(666)}{616.3} = 1.0046$$

$$A = \frac{(914,110.1)(185)}{(10.73)^2 (560)^{2.5}} = 0.197925$$

$$B = \frac{(1.0046)(185)}{(10.73)(560)} = 0.03093$$

Step 2. Substitute parameters A and B into Equation 15-63, and extract the largest and the smallest root, to give:

$$Z^3 - Z^2 + 0.1660384Z - 0.0061218 = 0$$

Largest Root $Z^v = 0.802641$

Smallest Root $Z^L = 0.0527377$

Step 3. Solve for the density of the liquid phase and gas phase:

$$\rho = \frac{pM}{ZRT}$$

$$\rho^L = \frac{(185)(44)}{(0.0527377)(10.73)(560)} = 25.7\,lb/ft^3$$

$$\rho^v = \frac{(185)(44)}{(0.802641)(10.73)(560)} = 1.688\,lb/ft^3$$

Redlich and Kwong extended the application of their equation to hydrocarbon liquid or gas mixtures by employing the following mixing rules:

$$a_m = \left[\sum_{i=1}^{n} x_i \sqrt{a_i}\right]^2 \tag{15-66}$$

$$b_m = \sum_{i=1}^{n} \left[x_i b_i\right] \tag{15-67}$$

where   n = number of components in mixture

   a_i = Redlich-Kwong a parameter for the i'th component as given by Equation 15–60

   b_i = Redlich-Kwong b parameter for the i'th component as given by Equation 15–61

   a_m = parameter a for mixture

   b_m = parameter b for mixture

   x_i = mole fraction of component i in the liquid phase

To calculate a_m and b_m for a hydrocarbon gas mixture with a composition of y_i, use Equations 15-66 and 15-67 and replace x_i with y_i:

$$a_m = \left[\sum_{i=1}^{n} y_i \sqrt{a_i}\right]^2$$

$$b_m = \sum_{i=1}^{n} \left[y_i b_i\right]$$

Equation 15-63 gives the compressibility factor of the gas phase or the liquid with the coefficients A and B as defined by Equations 15-64 and 15-65.

The application of the Redlich-Kwong equation of state for hydrocarbon mixtures can be best illustrated through the following two examples.

Example 15-11

Calculate the density of a crude oil with the following composition at 4000 psia and 160°F. Use the Redlich-Kwong EOS.

Component	x_i	M	p_c	T_c
C_1	0.45	16.043	666.4	343.33
C_2	0.05	30.070	706.5	549.92
C_3	0.05	44.097	616.0	666.06
$n - C_4$	0.03	58.123	527.9	765.62
$n - C_5$	0.01	72.150	488.6	845.8
C_6	0.01	84.00	453	923
C_{7+}	0.40	215	285	1287

Solution

Step 1. Determine the parameters a_i and b_i for each component by using Equations 15-60 and 15-61.

Component	a_i	b_i
C_1	161,044.3	0.4780514
C_2	493,582.7	0.7225732
C_3	914,314.8	1.004725
C_4	1,449,929	1.292629
C_5	2,095,431	1.609242
C_6	2,845,191	1.945712
C_{7+}	1.022348E7	4.191958

Step 2. Calculate the mixture parameters a_m and b_m from Equations 15-66 and 15-67 to give:

$$a_m = \left[\sum_{i=1}^{n} x_i \sqrt{a_i} \right]^2 = 2,591,967$$

and

$$b_m = \sum_{i=1}^{n} \left[x_i b_i \right] = 2.0526$$

Step 3. Compute the coefficients A and B by using Equations 15-64 and 15-65 to produce:

$$A = \frac{a_m p}{R^2 T^{2.5}} = \frac{2,591,967(4000)}{10.73^2 (620)^{2.5}} = 9.406539$$

$$B = \frac{b_m p}{RT} = \frac{2.0526(4000)}{10.73(620)} = 1.234049$$

Step 4. Solve Equation 15-63 for the largest positive root to yield:

$$Z^3 - Z^2 + 6.93845Z - 11.60813 = 0$$

$$Z^L = 1.548126$$

Step 5. Calculate the apparent molecular weight of the crude oil:

$$M_a = \Sigma x_i M_i$$

$$M_a = 100.2547$$

Step 6. Solve for the density of the crude oil:

$$\rho^L = \frac{p M_a}{Z^L RT}$$

$$\rho^L = \frac{(4000)(100.2547)}{(10.73)(620)(1.548120)} = 38.93 \text{ lb/ft}^3$$

Notice that liquid density, as calculated by Standing's correlation, gives a value of 46.23 lb/ft^3.

Example 15-12

Calculate the density of a gas phase with the following composition at 4000 psia and 160°F. Use the Redlich-Kwong EOS.

Component	y_i	M	p_c	T_c
C_1	0.86	16.043	666.4	343.33
C_2	0.05	30.070	706.5	549.92
C_3	0.05	44.097	616.0	666.06
C_4	0.02	58.123	527.9	765.62
C_5	0.01	72.150	488.6	845.8
C_6	0.005	84.00	453	923
C_{7+}	0.005	215	285	1287

Solution

Step 1. Calculate a_m and b_m by using Equations 15-66 and 15-67 to give:

$$a_m = \left[\sum_{i=1}^{n} y_i \sqrt{a_i}\right]^2$$

$$a_m = 241,118$$

$$b_m = \Sigma b_i x_i$$

$$b_m = 0.5701225$$

Step 2. Calculate the coefficients A and B by applying Equations 15-64 and 15-65 to yield:

$$A = \frac{a_m p}{R^2 T^{2.5}} = \frac{241,118(4000)}{10.73^2(620)^{2.5}} = 0.8750$$

$$B = \frac{b_m p}{RT} = \frac{0.5701225(4000)}{10.73(620)} = 0.3428$$

Step 3. Solve Equation 15-63 for Z^V to give:

$$Z^3 - Z^2 + 0.414688Z - 0.29995 = 0$$

$$Z^V = 0.907$$

Step 4. Calculate the apparent density of the gas mixture:

$$M_a = \Sigma y_i M_i = 20.89$$

$$\rho^v = \frac{p M_a}{Z^v RT}$$

$$\rho^v = \frac{(4000)(20.89)}{(10.73)(620)(0.907)} = 13.85 \text{ lb/ft}^3$$

Soave-Redlich-Kwong Equation of State and Its Modifications

One of the most significant milestones in the development of cubic equations of state was the publication by Soave (1972) of a modification to the evaluation of parameter a in the attractive pressure term of the Redlich-Kwong equation of state (Equation 15-68). Soave replaced the term $a/T^{0.5}$ in Equation 15-58 with a more generalized **temperature-dependent term,** as denoted by ($a\alpha$), to give:

$$p = \frac{RT}{V - b} - \frac{a\alpha}{V(V + b)} \tag{15-68}$$

where α is a dimensionless factor that becomes unity at $T = T_c$. At temperatures other than critical temperature, the parameter α is defined by the following expression:

$$\alpha = \left[1 + m\left(1 - \sqrt{T_r}\right)\right]^2 \tag{15-69}$$

The parameter m is correlated with the acentric factor to give:

$$m = 0.480 + 1.574\omega - 0.176\omega^2 \tag{15-70}$$

where T_r = reduced temperature T/Tc
          ω = acentric factor of the substance
          T = system temperature, °R

For any pure component, the constants a and b in Equation 15-68 are found by imposing the classical van der Waals critical point constraints (Equation 15-46) on Equation 15-68, and solving the resulting equations, to give:

$$a = \Omega_a \frac{R^2 T_c^2}{P_c} \tag{15-71}$$

$$b = \Omega_b \frac{RT_c}{P_c} \tag{15-72}$$

where Ω_a and Ω_b are the Soave-Redlich-Kwong (SRK) dimensionless pure component parameters and have the following values:

$$\Omega_a = 0.42747 \quad \text{and} \quad \Omega_b = 0.08664$$

Edmister and Lee (1986) showed that the two parameters a and b can be determined more conveniently by considering the critical isotherm:

$$\left(V - V_c\right)^3 = V^3 - \left[3V_c\right]V^2 + \left[3V_c^2\right]V - V_c^3 = 0 \qquad (15\text{-}73)$$

Equation 15-27 can also be put into a cubic form to give:

$$V^3 - \left[\frac{RT}{p}\right]V^2 + \left[\frac{a\alpha}{p} - \frac{bRT}{p} - b^2\right]V - \left[\frac{(a\alpha)b}{p}\right] = 0 \qquad (15\text{-}74)$$

At the critical point, the coefficient $\alpha = 1$ and the above two expressions are essentially identical. Equating the like terms gives:

$$3V_c = \frac{RT_c}{p_c} \qquad (15\text{-}75)$$

$$3V_c^2 = \frac{a}{p_c} - \frac{bRT_c}{p_c} - b^2 \qquad (15\text{-}76)$$

and

$$V_c^3 = \frac{ab}{p_c} \qquad (15\text{-}77)$$

Solving the above equations for parameters a and b yields expressions for the parameters as given by Equations 15-71 and 15-72.

Equation 15-75 indicates that the SRK equation of state gives a universal critical gas compressibility factor of 0.333. Combining Equation 15-34 with 15-72 gives:

$$b = 0.26V_c$$

Introducing the compressibility factor Z into Equation 15-33 by replacing the molar volume V in the equation with (ZRT/p) and rearranging, gives:

$$Z^3 - Z^2 + \left(A - B - B^2\right)Z - AB = 0 \qquad (15\text{-}78)$$

with

$$A = \frac{(a\alpha)p}{(RT)^2} \qquad (15\text{-}79)$$

$$B = \frac{bp}{RT} \qquad (15\text{-}80)$$

where p = system pressure, psia

T = system temperature, °R

R = 10.730 psia ft³/lb-mol-°R

Example 15-13

Rework Example 15-9 and solve for the density of the two phases by using the SRK EOS.

Solution

Step 1. Determine the critical pressure, critical temperature, and acentric factor from Table 1-2 of Chapter 1 to give:

$$T_c = 666.01°R$$
$$p_c = 616.3 \text{ psia}$$
$$\omega = 0.1524$$

Step 2. Calculate the reduced temperature.

$$T_r = 560/666.01 = 0.8408$$

Step 3. Calculate the parameter m by applying Equation 15-70 to yield:

$$m = 0.480 + 1.574\omega - 0.176\omega^2$$
$$m = 0.480 + 1.574(0.1524) - 0.176(1.524)^2 = 0.7051$$

Step 4. Solve for the parameter a by using Equation 15-69 to give:

$$\alpha = \left[m + \left(1 - \sqrt{T_r} \right) \right]^2 = 1.120518$$

Step 5. Compute the coefficients a and b by applying Equations 15-71 and 15-72 to yield:

$$a = 0.42747 \frac{10.73^2 (666.01)^2}{616.3} = 35,427.6$$

$$b = 0.08664 \frac{10.73(666.01)}{616.3} 1.00471$$

Step 6. Calculate the coefficients A and B from Equations 15-79 and 15-80, to produce:

$$A = \frac{(a\alpha)p}{R^2T^2}$$

$$A = \frac{(35,427.6)(1.120518)185}{10.73^2(560)^2} = 0.203365$$

$$B = \frac{bp}{RT}$$

$$B = \frac{(1.00471)(185)}{(10.73)(560)} = 0.034658$$

Step 7. Solve Equation 15-78 for Z^L and Z^V:

$$Z^3 - Z^2 + \left(A - B - B^2\right)Z + AB = 0$$

$$Z^3 - Z^2 + \left(0.203365 - 0.034658 - 0.034658^2\right)Z + (0.203365)(0.034658) = 0$$

Solving the above third-degree polynomial gives:

$$Z^L = 0.06729$$

$$Z^V = 0.80212$$

Step 8. Calculate the gas and liquid density to give:

$$\rho = \frac{pM}{ZRT}$$

$$\rho^v = \frac{(185)(44.0)}{(0.802121)(10.73)(560)} = 1.6887 \text{ lb/ft}^3$$

$$\rho^L = \frac{(185)(44.0)}{(0.06729)(10.73)(560)} = 20.13 \text{ lb/ft}^3$$

To use Equation 15-78 with mixtures, mixing rules are required to determine the terms $(a\alpha)$ and b for the mixtures. Soave adopted the following mixing rules:

$$(a\alpha)_m = \sum_i \sum_j \left[x_i x_j \sqrt{a_i a_j \alpha_i \alpha_j}\left(1 - k_{ij}\right)\right] \tag{15-81}$$

$$b_m = \sum_i \left[x_i b_i\right] \tag{15-82}$$

with

$$A = \frac{(a\alpha)_m p}{(RT)^2} \tag{15-83}$$

and

$$B = \frac{b_m p}{RT} \qquad (15\text{-}84)$$

The parameter k_{ij} is an empirically determined correction factor (called the binary interaction coefficient) that is designed to characterize any binary system formed by component i and component j in the hydrocarbon mixture.

These binary interaction coefficients are used to model the intermolecular interaction through empirical adjustment of the $(a\alpha)_m$ term as represented mathematically by Equation 15-81. They are dependent on the difference in molecular size of components in a binary system and they are characterized by the following properties:

- The interaction between hydrocarbon components increases as the relative difference between their molecular weights increases:

$$k_{i, j+1} > k_{i, j}$$

- Hydrocarbon components with the same molecular weight have a binary interaction coefficient of zero:

$$k_{i, i} = 0$$

- The binary interaction coefficient matrix is symmetric:

$$k_{j, i} = k_{i, j}$$

Slot-Petersen (1987) and Vidal and Daubert (1978) presented a theoretical background to the meaning of the interaction coefficient and techniques for determining their values. Graboski and Daubert (1978) and Soave (1972) suggested that no binary interaction coefficients are required for hydrocarbon systems. However, with nonhydrocarbons present, binary interaction parameters can greatly improve the volumetric and phase behavior predictions of the mixture by the SRK EOS.

In solving Equation 15-73 for the compressibility factor of the liquid phase, the composition of the liquid x_i is used to calculate the coefficients A and B of Equations 15-83 and 15-84 through the use of the mixing rules as described by Equations 15-81 and 15-82. For determining the compressibility factor of the gas phase Z^v, the above outlined procedure is used with composition of the gas phase y_i replacing x_i.

Example 15-14

A two-phase hydrocarbon system exists in equilibrium at 4000 psia and 160°F. The system has the following composition:

Component	x_i	y_i
C_1	0.45	0.86
C_2	0.05	0.05
C_3	0.05	0.05
C_4	0.03	0.02
C_5	0.01	0.01
C_6	0.01	0.005
C_{7+}	0.40	0.005

The heptanes-plus fraction has the following properties:

$M = 215$
$p_c = 285$ psia
$T_c = 700°F$
$\omega = 0.52$

Assuming $k_{ij} = 0$, calculate the density of each phase by using the SRK EOS.

Solution

Step 1. Calculate the parameters α, a, and b by applying Equations 15-64, 15-71, and 15-72.

Component	α_i	a_i	b_i
C_1	0.6869	8,689.3	0.4780
C_2	0.9248	21,040.8	0.7725
C_3	1.0502	35,422.1	1.0046
C_4	1.1616	52,390.3	1.2925
C_5	1.2639	72,041.7	1.6091
C_6	1.3547	94,108.4	1.9455
C_{7+}	1.7859	232,367.9	3.7838

Step 2. Calculate the mixture parameters $(a\alpha)_m$ and b_m for the gas phase and liquid phase by applying Equations 15-81 and 15-82 to give:

- **For the gas phase using y_i:**

$$(a\alpha)_m = \sum_i \sum_j \left[y_i y_j \sqrt{a_i a_j \alpha_i \alpha_j} \left(1 - k_{ij}\right) \right] = 9219.3$$

$$b_m = \sum_i \left[y_i b_i \right] = 0.5680$$

- **For the liquid phase using x_i:**

$$(a\alpha)_m = \sum_i \sum_j \left[x_i x_j \sqrt{a_i a_j \alpha_i \alpha_j} \left(1 - k_{ij}\right) \right] = 104,362.9$$

$$b_m = \sum_i \left[x_i b_i \right] = 0.1.8893$$

Step 3. Calculate the coefficients A and B for each phase by applying Equations 15-83 and 15-84 to yield:

- **For the gas phase:**

$$A = \frac{(a\alpha)_m P}{R^2 T^2} = \frac{(9219.3)(4000)}{(10.73)^2 (620)^2} = 0.8332$$

$$B = \frac{b_m P}{RT} = \frac{(0.5680)(4000)}{(10.73)(620)} = 0.3415$$

- **For the liquid phase:**

$$A = \frac{(a\alpha)_m P}{R^2 T^2} = \frac{(104,362.9)(4000)}{(10.73)^2 (620)^2} = 9.4324$$

$$B = \frac{b_m P}{RT} = \frac{(1.8893)(4000)}{(10.73)(620)} = 1.136$$

Step 4. Solve Equation 15-78 for the compressibility factor of the gas phase to produce:

$$Z^3 - Z^2 + \left(A - B - B^2\right)Z + AB = 0$$

$$Z^3 - Z^2 + \left(0.8332 - 0.3415 - 0.3415^2\right)Z + (0.8332)(0.3415) = 0$$

Solving the above polynomial for the largest root gives:

$$Z^v = 0.9267$$

Step 5. Solve Equation 15-78 for the compressibility factor of the liquid phase to produce:

$$Z^3 - Z^2 + (A - B - B^2)Z + AB = 0$$

$$Z^3 - Z^2 + (9.4324 - 1.136 - 1.136^2)Z + (9.4324)(1.136) = 0$$

Solving the above polynomial for the smallest root gives:
$$Z^L = 1.4121$$

Step 6. Calculate the apparent molecular weight of the gas phase and liquid phase from their composition, to yield:

- **For the gas phase:**
$$M_a = \sum y_i M_i = 20.89$$

- **For the liquid phase:**
$$M_a = \sum x_i M_i = 100.25$$

Step 7. Calculate the density of each phase:

$$\rho = \frac{pM_a}{RTZ}$$

- **For the gas phase:**
$$\rho^v = \frac{(4000)(20.89)}{(10.73)(620)(0.9267)} = 13.556 \text{ lb/ft}^3$$

- **For the liquid phase:**
$$\rho^L = \frac{(4000)(100.25)}{(10.73)(620)(1.4121)} = 42.68 \text{ lb/ft}^3$$

It is appropriate at this time to introduce and define the concept of the fugacity and the fugacity coefficient of the component. The **fugacity f** is a measure of the molar Gibbs energy of a real gas. It is evident from the definition that the fugacity has the units of pressure; in fact, the fugacity may be looked on as a vapor pressure modified to correctly represent the escaping tendency of the molecules from one phase into the other. In a mathematical form, the fugacity of a pure component is defined by the following expression:

$$f = p \exp\left[\int_o^p \left(\frac{Z-1}{p}\right) dp\right] \qquad (15\text{-}85)$$

where   f = fugacity, psia
        p = pressure, psia
        Z = compressibility factor

The ratio of the fugacity to the pressure, i.e., f/p, is called the **fugacity coefficient** Φ and is calculated from Equation 15-85 as:

$$\frac{f}{p} = \Phi = \exp\left[\int_o^p \left(\frac{Z-1}{p}\right) dp\right]$$

Soave applied the above-generalized thermodynamic relationship to Equation 15-68 to determine the fugacity coefficient of a pure component:

$$\ln\left(\frac{f}{p}\right) = \ln(\Phi) = Z - 1 - \ln(Z - B) - \frac{A}{B}\ln\left[\frac{Z+B}{Z}\right] \qquad (15\text{-}86)$$

In practical petroleum engineering applications we are concerned with the phase behavior of the hydrocarbon liquid mixture which, at a specified pressure and temperature, is in equilibrium with a hydrocarbon gas mixture at the same pressure and temperature.

The component fugacity in each phase is introduced to develop a criterion for thermodynamic equilibrium. Physically, the fugacity of a component i in one phase with respect to the fugacity of the component in a second phase is a measure of the potential for transfer of the component between phases. The phase with the lower component fugacity accepts the component from the phase with a higher component fugacity. Equal fugacities of a component in the two phases results in a zero net transfer. A zero transfer for all components implies a hydrocarbon system that is in thermodynamic equilibrium. Therefore, the condition of the thermodynamic equilibrium can be expressed mathematically by:

$$f_i^v = f_i^L \quad 1 \le i \le n \qquad (15\text{-}87)$$

where f_i^v = fugacity of component i in the gas phase, psi
        f_i^L = fugacity of component i in the liquid phase, psi
        n = number of components in the system

The fugacity coefficient of component i in a hydrocarbon liquid mixture or hydrocarbon gas mixture is a function of:

- System pressure
- Mole fraction of the component
- Fugacity of the component

For a component i in the gas phase, the fugacity coefficient is defined as:

$$\Phi_i^v = \frac{f_i^v}{y_i p} \tag{15-88}$$

For a component i in the liquid phase, the fugacity coefficient is:

$$\Phi_i^L = \frac{f_i^L}{x_i p} \tag{15-89}$$

where Φ_i^v = fugacity coefficient of component i in the vapor phase
Φ_i^L = fugacity coefficient of component i in the liquid phase

It is clear that at equilibrium $f_i^L = f_i^v$, the equilibrium ratio K_i as previously defined by Equation 15-1, i.e., $K_i = y_i/x_i$, can be redefined in terms of the fugacity of components as:

$$K_i = \frac{\left[f_i^L/(x_i p) \right]}{\left[f_i^v/(y_i p) \right]} = \frac{\Phi_i^L}{\Phi_i^v} \tag{15-90}$$

Reid, Prausnitz, and Sherwood (1977) defined the fugacity coefficient of component i in a hydrocarbon mixture by the following generalized thermodynamic relationship:

$$\ln(\Phi_i) = \left(\frac{1}{RT} \right) \left[\int_v^\infty \left(\frac{\partial p}{\partial n_i} - \frac{RT}{V} \right) dV \right] - \ln(Z) \tag{15-91}$$

where V = total volume of n moles of the mixture
n_i = number of moles of component i
Z = compressibility factor of the hydrocarbon mixture

By combining the above thermodynamic definition of the fugacity with the SRK EOS (Equation 15-68), Soave proposed the following expression for the fugacity coefficient of component i in the liquid phase:

$$\ln\left(\Phi_i^L\right) = \frac{b_i\left(Z^L - 1\right)}{b_m} - \ln\left(Z^L - B\right) - \left(\frac{A}{B}\right)\left[\frac{2\Psi_i}{(a\alpha)_m} - \frac{b_i}{b_m}\right]\ln\left[1 + \frac{B}{Z^L}\right] \quad (15\text{-}92)$$

where

$$\Psi_i = \sum_j \left[x_j \sqrt{a_i a_j \alpha_i \alpha_j}\left(1 - k_{ij}\right)\right] \quad (15\text{-}93)$$

$$(a\alpha)_m = \sum_i \sum_j \left[x_i x_j \sqrt{a_i a_j \alpha_i \alpha_j}\left(1 - k_{ij}\right)\right] \quad (15\text{-}94)$$

Equation 15-92 is also used to determine the fugacity coefficient of component in the gas phase Φ_i^v by using the composition of the gas phase y_i in calculating A, B, Z^v, and other composition-dependent terms, or:

$$\ln\left(\Phi_i^v\right) = \frac{b_i\left(Z^v - 1\right)}{b_m} - \ln\left(Z^v - B\right) - \left(\frac{A}{B}\right)\left[\frac{2\Psi_i}{(a\alpha)_m} - \frac{b_i}{b_m}\right]\ln\left[1 + \frac{B}{Z^v}\right]$$

where

$$\Psi_i = \sum_j \left[y_j \sqrt{a_i a_j \alpha_i \alpha_j}\left(1 - k_{ij}\right)\right]$$

$$(a\alpha)_m = \sum_i \sum_j \left[y_i y_j \sqrt{a_i a_j \alpha_i \alpha_j}\left(1 - k_{ij}\right)\right]$$

Modifications of the SRK EOS

To improve the pure component vapor pressure predictions by the SRK equation of state, Graboski and Daubert (1978) proposed a new expression for calculating parameter m of Equation 15-70. The proposed relationship originated from analyses of extensive experimental data for pure hydrocarbons. The relationship has the following form:

$$m = 0.48508 + 1.55171\omega - 0.15613\omega^2 \quad (15\text{-}95)$$

Sim and Daubert (1980) pointed out that because the coefficients of Equation 15-95 were determined by analyzing vapor pressure data of low-molecular-weight hydrocarbons it is unlikely that Equation 15-95 will suffice for high-molecular-weight petroleum fractions. Realizing that the acentric factors for the heavy petroleum fractions are calculated from an equation such as the Edmister correlation or the Lee and Kessler (1975) correlation, the authors proposed the following expressions for determining the parameter m:

- If the acentric factor is determined by using the Edmister correlation, then:

$$m = 0.431 + 1.57\omega_i - 0.161\omega_i^2 \qquad (15\text{-}96)$$

- If the acentric factor is determined by using the Lee and Kessler correction, then:

$$m = 0.315 + 1.60\omega_i - 0.166\omega_i^2 \qquad (15\text{-}97)$$

Elliot and Daubert (1985) stated that the optimal binary interaction coefficient k_{ij} would minimize the error in the representation of all thermodynamic properties of a mixture. Properties of particular interest in phase equilibrium calculations include bubble-point pressure, dew-point pressure, and equilibrium ratios. The authors proposed a set of relationships for determining interaction coefficients for asymmetric mixtures[2] that contain methane, N_2, CO_2, and H_2S. Referring to the principal component as i and the other fraction as j, Elliot and Daubert proposed the following expressions:

- For N_2 systems:

$$k_{ij} = 0.107089 + 2.9776 k_{ij}^\infty \qquad (15\text{-}98)$$

- For CO_2 systems:

$$k_{ij} = 0.08058 - 0.77215 k_{ij}^\infty - 1.8404 \left(k_{ij}^\infty \right)^2 \qquad (15\text{-}99)$$

- For H_2S systems:

$$k_{ij} = 0.07654 + 0.017921 k_{ij}^\infty \qquad (15\text{-}100)$$

- For methane systems with compounds of 10 carbons or more:

$$k_{ij} = 0.17985 - 2.6958 k_{ij}^\infty - 10.853 \left(k_{ij}^\infty \right)^2 \qquad (15\text{-}101)$$

where

$$k_{ij}^\infty = \frac{-\left(\varepsilon_i - \varepsilon_j \right)^2}{2\varepsilon_i \varepsilon_j} \qquad (15\text{-}102)$$

---

[2] An asymmetric mixture is defined as one in which two of the components are considerably different in their chemical behavior. Mixtures of methane with hydrocarbons of 10 or more carbon atoms can be considered asymmetric. Mixtures containing gases such as nitrogen or hydrogen are asymmetric.

and

$$\varepsilon_i = \frac{0.480453\sqrt{a_i}}{b_i} \qquad (15\text{-}103)$$

The two parameters a_i and b_i in Equation 15-103 were previously defined by Equations 15-71 and 15-72.

The major drawback in the SRK EOS is that the critical compressibility factor takes on the unrealistic universal critical compressibility of 0.333 for all substances. Consequently, the molar volumes are typically overestimated and, hence, densities are underestimated.

Peneloux et al. (1982) developed a procedure for improving the volumetric predictions of the SRK EOS by introducing a volume correction parameter c_i into the equation. This third parameter does not change the vapor-liquid equilibrium conditions determined by the unmodified SRK equation, i.e., the equilibrium ratio K_i, but it modifies the liquid and gas volumes. The proposed methodology, known as the **volume translation method,** uses the following expressions:

$$V_{corr}^{L} = V^{L} - \sum_{i}(x_i c_i) \qquad (15\text{-}104)$$

$$V_{corr}^{v} = V^{v} - \sum_{i}(y_i c_i) \qquad (15\text{-}105)$$

where V^L = uncorrected liquid molar volume, i.e., $V^L = Z^L RT/p$, ft^3/mol
V^v = uncorrected gas molar volume $V^v = Z^v RT/p$, ft^3/mol
V_{corr}^{L} = corrected liquid molar volume, ft^3/mol
V_{corr}^{v} = corrected gas molar volume, ft^3/mol
x_i = mole fraction of component i in the liquid phase
y_i = mole fraction of component i in the gas phase

The authors proposed six different schemes for calculating the correction factor c_i for each component. For petroleum fluids and heavy hydrocarbons, Peneloux and coworkers suggested that the best correlating parameter for the correction factor c_i is the Rackett compressibility factor Z_{RA}. The correction factor is then defined mathematically by the following relationship:

$$c_i = 4.43797878(0.29441 - Z_{RA})T_{ci}/p_{ci} \qquad (15\text{-}106)$$

where c_i = correction factor for component i, ft^3/lb-mol

$\quad\quad T_{ci}$ = critical temperature of component i, °R

$\quad\quad p_{ci}$ = critical pressure of component i, psia

The parameter Z_{RA} is a unique constant for each compound. The values of Z_{RA} are in general not much different from those of the critical compressibility factors Z_c. If their values are not available, Peneloux et al. (1982) proposed the following correlation for calculating c_i:

$$c_i = (0.0115831168 + 0.411844152\omega)\left(\frac{T_{ci}}{p_{ci}}\right) \quad\quad (15\text{-}107)$$

where ω_i = acentric factor of component i.

Example 15-15

Rework Example 15-14 by incorporating the Peneloux volume correction approach in the solution. Key information from Example 15-14 includes:

- For gas: $Z^v = 0.9267$, Ma = 20.89
- For liquid: $Z^L = 1.4121$, Ma = 100.25
- T = 160°F, p = 4000 psi

Solution

Step 1. Calculate the correction factor c_i using Equation 15-107:

Component	c_i	x_i	$c_i x_i$	y_i	$c_i y_i$
C_1	0.00839	0.45	0.003776	0.86	0.00722
C_2	0.03807	0.05	0.001903	0.05	0.00190
C_3	0.07729	0.05	0.003861	0.05	0.00386
C_4	0.1265	0.03	0.00379	0.02	0.00253
C_5	0.19897	0.01	0.001989	0.01	0.00198
C_6	0.2791	0.01	0.00279	0.005	0.00139
C_{7+}	0.91881	0.40	0.36752	0.005	0.00459
sum			0.38564		0.02349

Step 2. Calculate the uncorrected volume of the gas and liquid phase by using the compressibility factors as calculated in Example 15-14:

$$V^v = \frac{(10.73)(620)(0.9267)}{4000} = 1.54119 \text{ ft}^3/\text{mol}$$

$$V^L = \frac{(10.73)(620)(1.4121)}{4000} = 2.3485 \text{ ft}^3/\text{mol}$$

Step 3. Calculate the corrected gas and liquid volumes by applying Equations 15-104 and 15-105:

$$V^L_{corr} = V^L - \sum_i (x_i c_i) = 2.3485 - 0.38564 = 1.962927 \text{ ft}^3/\text{mol}$$

$$V^v_{corr} = V^v - \sum_i (y_i c_i) = 1.54119 - 0.02349 = 1.5177 \text{ ft}^3/\text{mol}$$

Step 4. Calculate the corrected compressibility factors:

$$Z^v_{corr} = \frac{(4000)(1.5177)}{(10.73)(620)} = 0.91254$$

$$Z^L_{corr} = \frac{(4000)(1.962927)}{(10.73)(620)} = 1.18025$$

Step 5. Determine the corrected densities of both phases:

$$\rho = \frac{pM_a}{RTZ}$$

$$\rho^v = \frac{(4000)(20.89)}{(10.73)(620)(0.91254)} = 13.767 \text{ lb/ft}^3$$

$$\rho^L = \frac{(4000)(100.25)}{(10.73)(620)(1.18025)} = 51.07 \text{ lb/ft}^3$$

Peng–Robinson Equation of State and Its Modifications

Peng and Robinson (1976a) conducted a comprehensive study to evaluate the use of the SRK equation of state for predicting the behavior of naturally occurring hydrocarbon systems. They illustrated the need for an improvement in the ability of the equation of state to predict liquid densities and other fluid properties particularly in the vicinity of the critical region. As a basis for creating an improved model, Peng and Robinson proposed the following expression:

$$p = \frac{RT}{V-b} - \frac{a\alpha}{(V+b)^2 - cb^2}$$

where a, b, and α have the same significance as they have in the SRK model, and the parameter c is a whole number optimized by analyzing the values of the two terms Z_c and b/V_c as obtained from the equation. It is generally accepted that Z_c should be close to 0.28 and that b/V_c should be approximately 0.26. An optimized value of c = 2 gave $Z_c = 0.307$ and $(b/V_c) = 0.253$. Based on this value of c, Peng and Robinson proposed the following equation of state:

$$p = \frac{RT}{V-b} - \frac{a\alpha}{V(V+b) + b(V-b)} \tag{15-108}$$

Imposing the classical critical point conditions (Equation 15-46) on Equation 15-108 and solving for parameters a and b yields:

$$a = \Omega_a \frac{R^2 T_c^2}{p_c} \tag{15-109}$$

$$b = \Omega_b \frac{RT_c}{p_c} \tag{15-110}$$

where $\Omega_a = 0.45724$ and $\Omega_b = 0.07780$. This equation predicts a universal critical gas compressibility factor Z_c of 0.307 compared to 0.333 for the SRK model. Peng and Robinson also adopted Soave's approach for calculating the temperature-dependent parameter α:

$$\alpha = \left[1 + m\left(1 - \sqrt{T_r}\right) \right]^2 \tag{15-111}$$

where

$$m = 0.3796 + 1.54226\omega - 0.2699\omega^2$$

Peng and Robinson (1978) proposed the following modified expression for m that is recommended for heavier components with acentric values $\omega > 0.49$:

$$m = 0.379642 + 1.48503\omega - 0.1644\omega^2 + 0.016667\omega^3 \tag{15-112}$$

Rearranging Equation 15-108 into the compressibility factor form gives:

$$Z^3 + (B-1)Z^2 + \left(A - 3B^2 - 2B\right)Z - \left(AB - B^2 - B^3\right) = 0 \tag{15-113}$$

where A and B are given by Equations 15-79 and 15-80 for pure components and by Equations 15-83 and 15-84 for mixtures.

Example 15-16

Using the composition given in Example 15-14, calculate the density of the gas phase and liquid phase by using the Peng–Robinson EOS. Assume $k_{ij} = 0$.

Solution

Step 1. Calculate the mixture parameters $(a\alpha)_m$ and b_m for the gas and liquid phase, to give:

- **For the gas phase:**

$$(a\alpha)_m = \sum_i \sum_j \left[y_i y_j \sqrt{a_i a_j \alpha_i \alpha_j} \left(1 - k_{ij}\right) \right] = 10,423.54$$

$$b_m = \sum_i (y_i b_i) = 0.862528$$

- **For the liquid phase:**

$$(a\alpha)_m = \sum_i \sum_j \left[x_i x_j \sqrt{a_i a_j \alpha_i \alpha_j} \left(1 - k_{ij}\right) \right] = 107,325.4$$

$$b_m = \sum_i (y_i b_i) = 1.69543$$

Step 2. Calculate the coefficients A and B, to give:

- **For the gas phase:**

$$A = \frac{(a\alpha)_m p}{R^2 T^2} = \frac{(10,423.54)(4000)}{(10.73)^2 (620)^2} = 0.94209$$

$$B = \frac{b_m p}{RT} = \frac{(0.862528)(4000)}{(10.73)(620)} = 0.30669$$

- **For the liquid phase:**

$$A = \frac{(a\alpha)_m p}{R^2 T^2} = \frac{(107,325.4)(4000)}{(10.73)^2 (620)^2} = 9.700183$$

$$B = \frac{b_m p}{RT} = \frac{(1.636543)(4000)}{(10.73)(620)} = 1.020078$$

Step 3. Solve Equation 15-113 for the compressibility factor of the gas phase and the liquid phase to give:

$$Z^3 + (B-1)Z^2 + \left(A - 3B^2 - 2B\right)Z - \left(AB - B^2 - B^3\right) = 0$$

- **For the gas phase:** Substituting for A = 0.94209 and B = 0.30669 in the above equation gives:

$$Z^v = 0.8625$$

- **For the liquid phase:** Substituting for A = 9.700183 and B = 1.020078 in the above equation gives:

$$Z^L = 1.2645$$

Step 4. Calculate the density of both phases:

$$\rho^v = \frac{(4,000)(20.89)}{(10.73)(620)(0.8625)} = 14.566 \text{ lb/ft}^3$$

$$\rho^L = \frac{(4,000)(100.25)}{(10.73)(620)(1.2645)} = 47.67 \text{ lb/ft}^3$$

Applying the thermodynamic relationship, as given by Equation 15-86, to Equation 15-109 yields the following expression for the fugacity of a pure component:

$$\ln\left(\frac{f}{p}\right) = \ln(\Phi) = Z - 1 - \ln(Z - B) - \left[\frac{A}{2\sqrt{2}B}\right] \ln\left[\frac{Z + \left(1 + \sqrt{2}\right)B}{Z + \left(1 - \sqrt{2}\right)B}\right] \quad (15\text{-}114)$$

The fugacity coefficient of component i in a hydrocarbon liquid mixture is calculated from the following expression:

$$\ln\left(\frac{f^L}{x_i p}\right) = \ln\left(\Phi_i^L\right) = \frac{b_i\left(Z^L - 1\right)}{b_m} - \ln\left(Z^L - B\right)$$

$$-\left[\frac{A}{2\sqrt{2}B}\right]\left[\frac{2\Psi_i}{(a\alpha)_m} - \frac{b_i}{b_m}\right]\ln\left[\frac{Z^L + \left(1 + \sqrt{2}\right)B}{Z^L - \left(1 - \sqrt{2}\right)B}\right] \quad (15\text{-}115)$$

where the mixture parameters b_m, B, A, Ψ_i, and $(a\alpha)_m$ are as defined previously.

Equation 15-115 is also used to determine the fugacity coefficient of any component in the gas phase by replacing the composition of the liquid phase x_i with the composition of the gas phase y_i in calculating the composition-dependent terms of the equation, or:

$$\ln\left(\frac{f^v}{y_i p}\right) = \ln\left(\Phi_i^v\right) = \frac{b_i\left(Z^v - 1\right)}{b_m} - \ln\left(Z^v - B\right)$$

$$-\left[\frac{A}{2\sqrt{2}B}\right]\left[\frac{2\Psi_i}{(a\alpha)_m} - \frac{b_i}{b_m}\right]\ln\left[\frac{Z^v + \left(1+\sqrt{2}\right)B}{Z^v - \left(1-\sqrt{2}\right)B}\right]$$

The set of binary interaction coefficients k_{ij} on page 1117 is traditionally used when predicting the volumetric behavior of a hydrocarbon mixture with the Peng and Robinson (PR) equation of state.

To improve the predictive capability of the PR EOS when describing mixtures containing N_2, CO_2, and CH_4, Nikos et al. (1986) proposed a generalized correlation for generating the binary interaction coefficient k_{ij}. The authors correlated these coefficients with system pressure, temperature, and the acentric factor. These generalized correlations were originated with all the binary experimental data available in the literature. The authors proposed the following generalized form for k_{ij}:

$$k_{ij} = \delta_2 T_{rj}^2 + \delta_1 T_{rj} + \delta_0 \qquad (15\text{-}116)$$

where i refers to the principal components N_2, CO_2, or CH_4; and j refers to the other hydrocarbon component of the binary. The acentric factor-dependent coefficients δ_0, δ_1, and δ_2 are determined for each set of binaries by applying the following expressions:

- **For nitrogen-hydrocarbons:**

$$\delta_0 = 0.1751787 - 0.7043\log\left(\omega_j\right) - 0.862066\left[\log\left(\omega_i\right)\right]^2 \qquad (15\text{-}117)$$

$$\delta_1 = -0.584474 + 1.328\log\left(\omega_j\right) + 2.035767\left[\log\left(\omega_i\right)\right]^2 \qquad (15\text{-}118)$$

and

$$\delta_2 = 2.257079 + 7.869765\log\left(\omega_j\right) + 13.50466\left[\log\left(\omega_i\right)\right]^2$$
$$+ 8.3864\left[\log\left(\omega\right)\right]^3 \qquad (15\text{-}119)$$

They also suggested the following pressure correction:

$$k_{ij}' = k_{ij}\left(1.04 - 4.2 \times 10^{-5}p\right) \qquad (15\text{-}120)$$

where p is the pressure in pounds per square inch.

- **For methane-hydrocarbons:**

$$\delta_0 = -0.01664 - 0.37283\log\left(\omega_j\right) + 1.31757\left[\log\left(\omega_i\right)\right]^2 \qquad (15\text{-}121)$$

$$\delta_1 = 0.48147 + 3.35342\log\left(\omega_j\right) - 1.0783\left[\log\left(\omega_i\right)\right]^2 \qquad (15\text{-}122)$$

Binary Interaction Coefficients* k_{ij} for the Peng and Robinson EOS

	CO$_2$	N$_2$	H$_2$S	C$_1$	C$_2$	C$_3$	i-C$_4$	n-C$_4$	i-C$_5$	n-C$_5$	C$_6$	C$_7$	C$_8$	C$_9$	C$_{10}$
CO$_2$	0	0	0.135	0.105	0.130	0.125	0.120	0.115	0.115	0.115	0.115	0.115	0.115	0.115	0.115
N$_2$		0	0.130	0.025	0.010	0.090	0.095	0.095	0.100	0.100	0.110	0.115	0.120	0.120	0.125
H$_2$S			0	0.070	0.085	0.080	0.075	0.075	0.070	0.070	0.070	0.060	0.060	0.060	0.055
C$_1$				0	0.005	0.010	0.035	0.025	0.050	0.030	0.030	0.035	0.040	0.040	0.045
C$_2$					0	0.005	0.005	0.010	0.020	0.020	0.020	0.020	0.020	0.020	0.020
C$_3$						0	0.005	0.005	0.005	0.005	0.010	0.005	0.005	0.005	0.005
i-C$_4$							0	0.000	0.005	0.005	0.005	0.005	0.005	0.005	0.005
n-C$_4$								0	0.005	0.005	0.005	0.005	0.005	0.005	0.005
i-C$_5$									0	0.000	0.000	0.000	0.000	0.000	0.005
n-C$_5$										0	0.000	0.000	0.000	0.000	0.005
C$_6$											0	0.000	0.000	0.000	0.000
C$_7$												0	0.000	0.000	0.000
C$_8$													0	0.000	0.000
C$_9$														0	0.000
C$_{10}$															0

* Notice that $k_{ij} = k_{ji}$.

and

$$\delta_2 = -0.4114 - 3.5072 \log(\omega_j) - 1.0783 [\log(\omega_i)]^2 \qquad (15\text{-}123)$$

- **For CO_2-hydrocarbons:**

$$\delta_0 = 0.4025636 + 0.1748927 \ \log(\omega_j) \qquad (15\text{-}124)$$

$$\delta_1 = -0.94812 - 0.6009864 \ \log(\omega_j) \qquad (15\text{-}125)$$

and

$$\delta_2 = 0.741843368 + 0.441775 \ \log(\omega_j) \qquad (15\text{-}126)$$

For the CO_2 interaction parameters, the following pressure correction is suggested:

$$k'_{ij} = k_{ij}(1.044269 - 4.375 \times 10^{-5} p) \qquad (15\text{-}127)$$

Stryjek and Vera (1986) proposed an improvement in the reproduction of vapor pressures of pure components by the PR EOS in the reduced temperature range from 0.7 to 1.0 by replacing the m term in Equation 15-111 with the following expression:

$$m_0 = 0.378893 + 1.4897153 - 0.17131848\omega^2 + 0.0196554\omega^3 \quad (15\text{-}128)$$

To reproduce vapor pressures at reduced temperatures below 0.7, Stryjek and Vera further modified the m parameter in the PR equation by introducing an adjustable parameter m_1 characteristic of each compound to Equation 15-111. They proposed the following generalized relationship for the parameter m:

$$m = m_0 + \left[m_1 \left(1 + \sqrt{T_r}\right)(0.7 - T_r) \right] \qquad (15\text{-}129)$$

where T_r = reduced temperature of the pure component
m_0 = defined by Equation 15-128
m_1 = adjustable parameter

For all components with a reduced temperature above 0.7, Stryjek and Vera recommended setting $m_1 = 0$. For components with a reduced temperature greater than 0.7, the optimum values of m_1 for compounds of industrial interest are tabulated below:

Parameter m₁ of Pure Compounds

Compound	m_1	Compound	m_1
Nitrogen	0.01996	Nonane	0.04104
Carbon dioxide	0.04285	Decane	0.04510
Water	–0.06635	Undecane	0.02919
Methane	–0.00159	Dodecane	0.05426
Ethane	0.02669	Tridecane	0.04157
Propane	0.03136	Tetradecane	0.02686
Butane	0.03443	Pentadecane	0.01892
Pentane	0.03946	Hexadecane	0.02665
Hexane	0.05104	Heptadecane	0.04048
Heptane	0.04648	Octadecane	0.08291
Octane	0.04464		

Due to the totally empirical nature of the parameter m_1, Stryjek and Vera (1986) could not find a generalized correlation for m_1 in terms of pure component parameters. They pointed out that the values of m_1 given above should be used without changes.

Jhaveri and Youngren (1984) pointed out that when applying the Peng–Robinson equation of state to reservoir fluids, the error associated with the equation in the prediction of gas-phase Z factors ranged from 3 to 5%, and the error in the liquid density predictions ranged from 6 to 12%. Following the procedure proposed by Peneloux and coworkers (see the SRK EOS), Jhaveri and Youngren introduced the volume correction parameter c_i to the PR EOS. This third parameter has the same units as the second parameter b_i of the unmodified PR equation and is defined by the following relationship:

$$c_i = S_i b_i \qquad (15\text{-}130)$$

where S_i = dimensionless parameter and is called the shift parameter
b_i = Peng–Robinson co-volume as given by Equation 15-110

The volume correction parameter c_i does not change the vapor-liquid equilibrium conditions, i.e., equilibrium ratio K_i. The corrected hydrocarbon phase volumes are given by the following expressions:

$$V_{corr}^L = V^L - \sum_{i=1} (x_i c_i)$$

$$V_{corr}^v = V^v - \sum_{i=1} (y_i c_i)$$

where V^L, V^v = volumes of the liquid phase and gas phase as calculated by unmodified PR EOS, ft³/mol

V^L_{corr}, V^v_{corr} = corrected volumes of the liquid and gas phase

Whitson and Brule (2000) point out that the volume translation (correction) concept can be applied to any two-constant cubic equation, thereby eliminating the volumetric deficiency associated with application of EOS. Whitson and Brule extended the work of Jhaveri and Youngren (1984) and proposed the following shift parameters for selected pure components:

Shift Parameters for the PR EOS and SRK EOS

Compound	PR EOS	SRK EOS
N_2	−0.1927	−0.0079
CO_2	−0.0817	0.0833
H_2S	−0.1288	0.0466
C_1	−0.1595	0.0234
C_2	−0.1134	0.0605
C_3	−0.0863	0.0825
$i - C_4$	−0.0844	0.0830
$n - C_4$	−0.0675	0.0975
$i - C_5$	−0.0608	0.1022
$n - C_5$	−0.0390	0.1209
$n - C_6$	−0.0080	0.1467
$n - C_7$	0.0033	0.1554
$n - C_8$	0.0314	0.1794
$n - C_9$	0.0408	0.1868
$n - C_{10}$	0.0655	0.2080

Jhaveri and Youngren (1984) proposed the following expression for calculating the shift parameter for the C_{7+}:

$$S = 1 - \frac{d}{(M)^e}$$

where M = molecular weight of the heptanes-plus fraction

d, e = positive correlation coefficients

The authors proposed that in the absence of the experimental information needed for calculating e and d, the power coefficient e can be set equal to 0.2051 and the coefficient d adjusted to match the C_{7+} density with the values of d ranging from 2.2 to 3.2. In general, the following values may be used for C_{7+} fractions:

Hydrocarbon Family	d	e
Paraffins	2.258	0.1823
Naphthenes	3.044	0.2324
Aromatics	2.516	0.2008

To use the Peng and Robinson equation of state to predict the phase and volumetric behavior of mixtures, one must be able to provide the critical pressure, the critical temperature, and the acentric factor for each component in the mixture. For pure compounds, the required properties are well defined and known. Nearly all naturally occurring petroleum fluids contain a quantity of heavy fractions that are not well defined. These heavy fractions are often lumped together as the heptanes-plus fraction. The problem of how to adequately characterize the C_{7+} fractions in terms of their critical properties and acentric factors has been long recognized in the petroleum industry. Changing the characterization of C_{7+} fractions present in even small amounts can have a profound effect on the PVT properties and the phase equilibria of a hydrocarbon system as predicted by the Peng and Robinson equation of state.

The usual approach for such situations is to "tune" the parameters in the EOS in an attempt to improve the accuracy of prediction. During the tuning process, the critical properties of the heptanes-plus fraction and the binary interaction coefficients are adjusted to obtain a reasonable match with experimental data available on the hydrocarbon mixture.

Recognizing that the inadequacy of the predictive capability of the PR EOS lies with the improper procedure for calculating the parameters a, b, and α of the equation for the C_{7+} fraction, Ahmed (1991) devised an approach for determining these parameters from the following two readily measured physical properties of C_{7+}: molecular weight, M_{7+}, and specific gravity, γ_{7+}.

The approach is based on generating 49 density values for the C_{7+} by applying the Riazi and Daubert correlation. These values were subsequently subjected to 10 temperature and 10 pressure values in the range of 60 to 300°F and 14.7 to 7000 psia, respectively. The Peng and Robinson EOS was then applied to match the 4900 generated density values by optimizing the parameters a, b, and α using a nonlinear regression model. The optimized parameters for the heptanes-plus fraction are given by the following expressions.

For the parameter a of C_{7+}:

$$\alpha = \left[1 + m[1 - \sqrt{\frac{520}{T}})\right]^2 \tag{15-131}$$

with m defined by:

$$m = \frac{D}{A_0 + A_1 D} + A_2 M_{7+} + A_3 M_{7+}^2 + \frac{A_4}{M_{7+}} + A_5 \gamma_{7+}$$
$$+ A_6 \gamma_{7+}^2 + \frac{A_7}{\gamma_{7+}} \tag{15-132}$$

with the parameter D defined by the ratio of the molecular weight to the specific gravity of the heptanes-plus fraction, or:

$$D = \frac{M_{7+}}{\gamma_{7+}}$$

where M_{7+} = molecular weight of C_{7+}

        γ_{7+} = specific gravity of C_{7+}

     $A_0 - A_7$ = coefficients as given in Table 15-2

For the parameters a and b of C_{7+}, the following generalized correlation is proposed:

$$a \text{ or } b = \left[\sum_{i=0}^{3} \left(A_i D^i \right) \right] + \frac{A_4}{D} \left[\sum_{i=5}^{6} \left(A_i \gamma_{7+}^{i-4} \right) \right] + \frac{A_7}{\gamma_{7+}}, \tag{15-133}$$

The coefficients A_0 through A_7 are included in Table 15-2.

To further improve the predictive capability of the Peng–Robinson EOS, the author optimized coefficients a, b, and m for nitrogen, CO_2, and methane by matching 100 Z-factor values for each of these components. Using a nonlinear regression model, the following optimized values are recommended:

Table 15-2 Coefficients for Equations 15-132 and 15-133

Coefficient	a	b	m
A_0	-2.433525×10^7	-6.8453198	-36.91776
A_1	8.3201587×10^3	1.730243×10^{-2}	$-5.2393763 \times 10^{-2}$
A_2	-0.18444102×10^2	$-6.2055064 \times 10^{-6}$	1.7316235×10^{-2}
A_3	3.6003101×10^{-2}	9.0910383×10^{-9}	$-1.3743308 \times 10^{-5}$
A_4	3.4992796×10^7	13.378898	12.718844
A_5	2.838756×10^7	7.9492922	10.246122
A_6	-1.1325365×10^7	-3.1779077	-7.6697942
A_7	6.418828×10^6	1.7190311	-2.6078099

Component	a	b	m in Eq. 15-131
CO_2	1.499914×10^4	0.41503575	−0.73605717
N_2	4.5693589×10^3	0.4682582	−0.97962859
C_1	7.709708×10^3	0.46749727	−0.549765

To provide the modified PR EOS with a consistent procedure for determining the binary interaction coefficient k_{ij}, the following computational steps are proposed:

Step 1. Calculate the binary interaction coefficient between methane and the heptanes-plus fraction from:

$$k_{C_1-C_{7+}} = 0.00189T - 1.167059$$

where the temperature T is in °R.

Step 2. Set:

$$k_{CO_2-N_2} = 0.12$$
$$k_{CO_2\text{-hydrocarbon}} = 0.10$$
$$k_{N_2\text{- hydrocarbon}} = 0.10$$

Step 3. Adopting the procedure recommended by Petersen (1989), calculate the binary interaction coefficients between components heavier than methane (e.g., C_2, C_3) and the heptanes-plus fraction from:

$$k_{C_n-C_{7+}} = 0.8k_{C_{(n-1)}-C_{7+}}$$

where n is the number of carbon atoms of component C_n; e.g.:

Binary interaction coefficient between C_2 and C_{7+} is
$$k_{C_2-C_{7+}} = 0.8\, k_{C_1-C_{7+}}$$

Binary interaction coefficient between C_3 and C_{7+} is
$$k_{C_3-C_{7+}} = 0.8\, k_{C_2-C_{7+}}$$

Step 4. Determine the remaining k_{ij} from:

$$k_{ij} = k_{i-C_{7+}} \left[\frac{(M_j)^5 - (M_i)^5}{(M_{C_{7+}})^5 - (M_i)^5} \right]$$

where M is the molecular weight of any specified component. For example, the binary interaction coefficient between propane C_3 and butane C_4 is:

$$k_{C_3-C_4} = k_{C_3-C_{7+}} \left[\frac{\left(M_{C_4}\right)^5 - \left(M_{C_3}\right)^5}{\left(M_{C_{7+}}\right)^5 - \left(M_{C_3}\right)^5} \right]$$

APPLICATIONS OF THE EQUATION OF STATE IN PETROLEUM ENGINEERING

Determination of the Equilibrium Ratios

A flow diagram is presented in Figure 15-14 to illustrate the procedure of determining equilibrium ratios of a hydrocarbon mixture. For this type of calculation, the system temperature T, the system pressure p, and the overall composition of the mixture z_i must be known. The procedure is summarized in the following steps in conjunction with Figure 15-14.

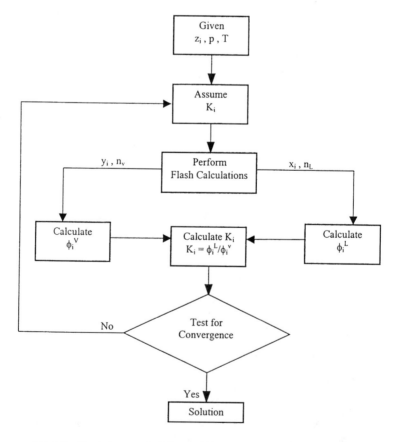

Figure 15-14. Flow diagram of the equilibrium ratio determination by an equation of state.

Step 1. Assume an initial value of the equilibrium ratio for each component in the mixture at the specified system pressure and temperature. Wilson's equation can provide the starting K_i values.

$$K_i^A = \frac{p_{ci}}{p} \exp\left[5.37(1+\omega_i)(1-T_{ci}/T)\right]$$

where K_i^A = assumed equilibrium ratio of component i.

Step 2. Using the overall composition and the assumed K values, perform flash calculations to determine x_i, y_i, n_L, and n_V.

Step 3. Using the calculated composition of the liquid phase x_i, determine the fugacity coefficient Φ_i^L for each component in the liquid phase.

Step 4. Repeat Step 3 using the calculated composition of the gas phase y_i to determine Φ_i^V.

Step 5. Calculate the new set of equilibrium ratios from:

$$K_i = \frac{\Phi_i^L}{\Phi_i^V}$$

Step 6. Check for the solution by applying the following constraint:

$$\sum_{i=1}^{n}\left[K_i/K_i^A - 1\right]^2 \le \varepsilon$$

where ε = preset error tolerance, e.g., 0.0001

n = number of components in the system

If the above conditions are satisfied, then the solution has been reached. If not, steps 1 through 6 are repeated by using the calculated equilibrium ratios as initial values.

Determination of the Dew-Point Pressure

A saturated vapor exists for a given temperature at the pressure at which an infinitesimal amount of liquid first appears. This pressure is referred to as the dew-point pressure p_d. The dew-point pressure of a mixture is described mathematically by the following two conditions:

$$y_i = z_i \quad 1 \le i \le n \tag{15-134}$$

$$n_v = 1$$

and:

$$\sum_{i-1}^{n} \left[\frac{z_i}{K_i} \right] = 1 \tag{15-135}$$

Applying the definition of K_i in terms of the fugacity coefficient to Equation 15-135 gives:

$$\sum_{i=1}^{n} \left[\frac{z_i}{k_i} \right] = \sum_{i=1}^{n} \left[\frac{z_i}{\left(\Phi_i^L / \Phi_i^v \right)} \right] = \sum \left[\left(\frac{z_i}{\Phi_i^L} \right) \frac{f_i^v}{z_i p_d} \right] = 1$$

or

$$p_d = \sum_{i=1}^{n} \left[\frac{f_i^v}{\Phi_i^L} \right]$$

The above equation is arranged to give:

$$f(p_d) = \sum_{i=1}^{n} \left[\frac{f_i^v}{\Phi_i^L} \right] - p_d = 0 \tag{15-136}$$

where p_d = dew-point pressure, psia

f_i^v = fugacity of component i in the vapor phase, psia

Φ_i^L = fugacity coefficient of component i in the liquid phase

Equation 15-136 can be solved for the dew-point pressure by using the Newton–Raphson iterative method. To use the iterative method, the derivative of Equation 15-136 with respect to the dew-point pressure p_d is required. This derivative is given by the following expression:

$$\frac{\partial f}{\partial p_d} = \sum_{i=1}^{n} \left[\frac{\Phi_i^L \left(\partial f_i^v / \partial p_d \right) - f_i^v \left(\partial \Phi_i^L / \partial p_d \right)}{\left(\Phi_i^L \right)^2} \right] - 1 \tag{15-137}$$

The two derivatives in the above equation can be approximated numerically as follows:

$$\frac{\partial f^v}{\partial p_d} = \left[\frac{f_i^v \left(p_d + \Delta p_d \right) - f_i^v \left(p_d - \Delta p_d \right)}{2 \Delta p_d} \right] \tag{15-138}$$

and

$$\frac{\partial f_i^L}{\partial p_d} = \left[\frac{\Phi_i^L(p_d + \Delta p_d) - \Phi_i^L(p_d - \Delta p_d)}{2\Delta p_d} \right] \tag{15-139}$$

where Δp_d = pressure increment, 5 psia, for example
$f_i^V (p_d + \Delta p_d)$ = fugacity of component i at $(p_d + \Delta p_d)$
$f_i^V (p_d - \Delta p_d)$ = fugacity of component i at $(p_d - \Delta p_d)$
$\Phi_i^L (p_d + \Delta p_d)$ = fugacity coefficient of component i at $(p_d + \Delta p_d)$
$\Phi_i^L (p_d - \Delta p_d)$ = fugacity coefficient of component i at $(p_d - \Delta p_d)$
Φ_i^L = fugacity coefficient of component i at p_d

The computational procedure of determining p_d is summarized in the following steps:

Step 1. Assume an initial value for the dew-point pressure p_d^A.

Step 2. Using the assumed value of p_d^A, calculate a set of equilibrium ratios for the mixture by using any of the previous correlations, e.g., Wilson's correlation.

Step 3. Calculate the composition of the liquid phase, i.e., composition of the droplets of liquid, by applying the mathematical definition of K_i, to give:

$$x_i = \frac{z_i}{K_i}$$

Note that $y_i = z_i$.

Step 4. Calculate f_i^V using the composition of the gas phase z_i and Φ_i^L using the composition of liquid phase x_i at the following three pressures:

- p_d^A
- $p_d^A + \Delta p_d$,
- $p_d^A - \Delta p_d$

where p_d^A is the assumed dew-point pressure and Δp_d is a selected pressure increment of 5 to 10 psi.

Step 5. Evaluate the function $f(p_d)$, i.e., Equation 15-136, and its derivative by using Equations 15-137 through 15-139.

Step 6. Using the values of the function $f(p_d)$ and the derivative $\partial f / \partial p_d$ as determined in step 5, calculate a new dew-point pressure by applying the Newton–Raphson formula:

$$p_d = p_d^A - f(p_d) / [\partial f / \partial p_d] \qquad (15\text{-}140)$$

Step 7. The calculated value of p_d is checked numerically against the assumed value by applying the following condition:

$$|p_d - p_d^A| \le 5$$

If the above condition is met, then the correct dew-point pressure p_d has been found. Otherwise, steps 3 through 6 are repeated by using the calculated p_d as the new value for the next iteration. A set of equilibrium ratios must be calculated at the new assumed dew-point pressure from:

$$k_i = \frac{\Phi_i^L}{\Phi_i^V}$$

Determination of the Bubble-Point Pressure

The bubble-point pressure p_b is defined as the pressure at which the first bubble of gas is formed. Accordingly, the bubble-point pressure is defined mathematically by the following equations:

$$x_i = z_i \quad 1 \le i \le n \qquad (15\text{-}141)$$
$$n_L = 1.0$$

and

$$\sum_{i=1}^{n} [z_i K_i] = 1 \qquad (15\text{-}142)$$

Introducing the concept of the fugacity coefficient into Equation 15-142 gives:

$$\sum_{i=1}^{n} \left[z_i \frac{\Phi_i^L}{\Phi_i^V} \right] = \sum_{i=1}^{n} \left[z_i \frac{\left(\dfrac{f_i^L}{z_i p_b} \right)}{\Phi_i^V} \right] = 1$$

Rearranging,

$$p_b = \sum_{i=1}^{n} \left[\frac{f_i^L}{\Phi_i^v} \right]$$

or

$$f(p_b) = \sum_{i=1}^{n} \left[\frac{f_i^L}{\Phi_i^v} \right] - p_b = 0 \tag{15-143}$$

The iteration sequence for calculation of p_b from the above function is similar to that of the dew-point pressure, which requires differentiating the above function with respect to the bubble-point pressure, or:

$$\frac{\partial f}{\partial p_b} = \sum_{i=1}^{n} \left[\frac{\Phi_i^v \left(\partial f_i^L / \partial p_b \right) - f_i^L \left(\partial \Phi_i^v / \partial p_b \right)}{\left(\Phi_i^v \right)^2} \right] - 1 \tag{15-144}$$

Three-Phase Equilibrium Calculations

Two- and three-phase equilibria occur frequently during the processing of hydrocarbon and related systems. Peng and Robinson (1976b) proposed a three-phase equilibrium calculation scheme of systems that exhibit a water-rich liquid phase, a hydrocarbon-rich liquid phase, and a vapor phase.

Applying the principle of mass conservation to 1 mol of a water-hydrocarbon in a three-phase state of thermodynamic equilibrium at a fixed temperature T and pressure p gives:

$$n_L + n_w + n_v = 1 \tag{15-145}$$

$$n_L x_i + n_w x_{wi} + n_v y_i = z_i \tag{15-146}$$

$$\sum_{i}^{n} x_i = \sum_{i=1}^{n} x_{wi} = \sum_{i=1}^{n} y_i = \sum_{i=1}^{n} z_i = 1 \tag{15-147}$$

where n_L, n_w, n_v = number of moles of the hydrocarbon-rich liquid, the water-rich liquid, and the vapor, respectively

x_i, x_{wi}, y_i = mole fraction of component i in the hydrocarbon-rich liquid, the water-rich liquid, and the vapor, respectively.

The equilibrium relations between the compositions of each phase are defined by the following expressions:

$$K_i = \frac{y_i}{x_i} = \frac{\Phi_i^L}{\Phi_i^v} \tag{15-148}$$

and

$$K_{wi} = \frac{y_i}{x_{wi}} = \frac{\Phi_i^w}{\Phi_i^v} \tag{15-149}$$

where K_i = equilibrium ratio of component i between vapor and
hydrocarbon-rich liquid

K_{wi} = equilibrium ratio of component i between the vapor and
water-rich liquid

Φ_i^L = fugacity coefficient of component i in the hydrocarbon-
rich liquid

Φ_i^v = fugacity coefficient of component i in the vapor phase

Φ_i^w = fugacity coefficient of component i in the water-rich liquid

Combining Equations 15-145 through 15-149 gives the following conventional nonlinear equations:

$$\sum_{i=1} x_i = \sum_{i=1} \left[\frac{z_i}{n_L(1-K_i) + n_w\left(\dfrac{K_i}{K_{wi}} - K_i\right) + K_i} \right] = 1 \tag{15-150}$$

$$\sum_{i=1} x_{wi} = \sum_{i=1} \left[\frac{z_i\left(K_i/K_{wi}\right)}{n_L(1-K_i) + n_w\left(\dfrac{K_i}{K_{wi}} - K_i\right) + K_i} \right] = 1 \tag{15-151}$$

$$\sum_{i=1} y_i = \sum_{i=1} \left[\frac{z_i K_i}{n_L(1-K_i) + n_w\left(\dfrac{K_i}{K_{wi}} - K_i\right) + K_i} \right] = 1 \tag{15-152}$$

Assuming that the equilibrium ratios between phases can be calculated, the above equations are combined to solve for the two unknowns n_L and n_v, and hence x_i, x_{wi}, and y_i. It is the nature of the specific equilibrium calculation that determines the appropriate combination of Equations

15-150 through 15-152. The combination of the above three expressions can then be used to determine the phase and volumetric properties of the three-phase system.

There are essentially three types of phase behavior calculations for the three-phase system:

1. Bubble-point prediction
2. Dew-point prediction
3. Flash calculation

Peng and Robinson (1980) proposed the following combination schemes of Equations 15-150 through 15-152.

- **For the bubble-point pressure determination:**

$$\sum_i x_i - \sum_i x_{wi} = 0 \qquad \left[\sum_i y_i\right] - 1 = 0$$

Substituting Equations 15-150 through 15-152 in the above relationships gives:

$$f(n_L, n_w) = \sum_i \left[\frac{z_i\left(1 - K_i/K_{wi}\right)}{n_L\left(1 - K_i\right) + n_w\left(K_i/K_{wi} - K_i\right) + K_i}\right] = 0 \qquad (15\text{-}153)$$

and

$$g(n_L, n_w) = \sum_i \left[\frac{z_i K_i}{n_L\left(1 - K_i\right) + n_w\left(K_i/K_{wi} - K_i\right) + K_i}\right] - 1 = 0 \quad (15\text{-}154)$$

- **For the dew-point pressure:**

$$\sum_i x_{wi} - \sum_i y_i = 0 \qquad \left[\sum_i x_i\right] - 1 = 0$$

Combining with Equations 15-150 through 15-152 yields:

$$f(n_L, n_w) = \sum_i \left[\frac{z_i K_i\left(1/K_{wi} - 1\right)}{n_L\left(1 - K_i\right) + n_w\left(K_i/K_{wi} - K_i\right) + K_i}\right] = 0 \qquad (15\text{-}155)$$

and

$$g(n_L, n_w) = \sum_i \left[\frac{z_i}{n_L\left(1 - K_i\right) + n_w\left(K_i/K_{wi} - K_i\right) + K_i}\right] - 1 = 0 \quad (15\text{-}156)$$

- **For flash calculations:**

$$\sum_i x_i - \sum_i y_i = 0 \qquad \left[\sum_i x_{wi}\right] - 1 = 0$$

or

$$f(n_L, n_w) = \sum_i \left[\frac{z_i(1 - K_i)}{n_L(1 - K_i) + n_w(K_i/K_{wi} - K_i) + K_i}\right] = 0 \qquad (15\text{-}157)$$

and

$$g(n_L, n_w) = \sum_i \left[\frac{z_i K_i/K_{wi}}{n_L(1 - K_i) + n_w(K_i/K_{wi} - K_i) + K_i}\right] - 1.0 = 0 \quad (15\text{-}158)$$

Note that in performing any of the above property predictions, we always have two unknown variables, n_L and n_w, and between them, two equations. Providing that the equilibrium ratios and the overall composition are known, the equations can be solved simultaneously by using the appropriate iterative technique, e.g., the Newton–Raphson method. The application of this iterative technique for solving Equations 15-157 and 15-158 is summarized in the following steps:

Step 1. Assume initial values for the unknown variables n_L and n_w.

Step 2. Calculate new values of n_L and n_w by solving the following two linear equations:

$$\begin{bmatrix} n_L \\ n_w \end{bmatrix}^{new} = \begin{bmatrix} n_L \\ n_w \end{bmatrix} - \begin{bmatrix} \partial f/\partial n_L & \partial f/\partial n_w \\ \partial g/\partial n_L & \partial g/\partial n_w \end{bmatrix}^{-1} \begin{bmatrix} f(n_L, n_w) \\ g(n_L, n_w) \end{bmatrix}$$

where $f(n_L, n_w)$ = value of the function $f(n_L, n_w)$ as expressed
                  by Equation 15-157

         $g(n_L, n_w)$ = value of the function $g(n_L, n_w)$ as expressed
                  by Equation 15-158

The first derivative of the above functions with respect to n_L and n_w are given by the following expressions:

$$\left(\partial f / \partial n_L\right) = \sum_{i=1} \left[\frac{-z_i\left(1-K_i\right)^2}{\left[n_L\left(1-K_i\right)+n_w\left(K_i/K_{wi}-K_i\right)+K_i\right]^2} \right]$$

$$\left(\partial f / \partial n_w\right) = \sum_{i=1} \left[\frac{-z_i\left(1-K_i\right)\left(K_i/K_{wi}-K_i\right)}{\left[n_L\left(1-K_i\right)+n_w\left(K_i/K_{wi}-K_i\right)+K_i\right]^2} \right]$$

$$\left(\partial g / \partial n_L\right) = \sum_{i=1} \left[\frac{-z_i\left(K_i/K_{wi}\right)\left(1-K_i\right)}{\left[n_L\left(1-K_i\right)+n_w\left(K_i/K_{wi}-K_i\right)+K_i\right]^2} \right]$$

$$\left(\partial g / \partial n_w\right) = \sum_{i=1} \left[\frac{-z_i\left(K_i K_{wi}\right)\left(K_i/K_{wi}-K_i\right)}{\left[n_L\left(1-K_i\right)+n_w\left(K_i/K_{wi}-K_i\right)+K_i\right]^2} \right]$$

Step 3. The new calculated values of n_L and n_w are then compared with the initial values. If no changes in the values are observed, then the correct values of n_L and n_w have been obtained. Otherwise, the above steps are repeated with the new calculated values used as initial values.

Peng and Robinson (1980) proposed two modifications when using their equation of state for three-phase equilibrium calculations. The first modification concerns the use of the parameter α as expressed by Equation 15-111 for the water compound. Peng and Robinson suggested that when the reduced temperature of this compound is less than 0.85, the following equation is applied:

$$\alpha = \left[1.0085677 + 0.82154\left(1 - T_r^{0.5}\right)\right]^2 \tag{15-159}$$

where T_r is the reduced temperature $(T/T_c)_{H_2O}$ of the water component.

The second modification concerns the application of Equation 15-81 for the water-rich liquid phase. A temperature-dependent binary interaction coefficient was introduced into the equation to give:

$$\left(a\alpha\right)_m = \sum_i \sum_j \left[x_{wi}x_{wj}\left(a_i a_j \alpha_i \alpha_j\right)^{0.5}\left(1 - \tau_{ij}\right)\right] \tag{15-160}$$

where τ_{ij} is a temperature-dependent binary interaction coefficient. Peng and Robinson proposed graphical correlations for determining this parameter for each aqueous binary pair. Lim et al. (1984) expressed these graphical correlations mathematically by the following generalized equation:

$$\tau_{ij} = a_1 \left[\frac{T}{T_{ci}}\right]^2 \left[\frac{p_{ci}}{p_{cj}}\right]^2 + a_2 \left[\frac{T}{T_{ci}}\right]\left[\frac{p_{ci}}{p_{cj}}\right] + a_3 \qquad (15\text{-}161)$$

where   T = system temperature, °R
T_{ci} = critical temperature of the component of interest, °R
p_{ci} = critical pressure of the component of interest, psia
p_{cj} = critical pressure of the water compound, psia

Values of the coefficients a_1, a_2, and a_3 of the above polynomial are given below for selected binaries:

Component i	a_1	a_2	a_3
C_1	0	1.659	−0.761
C_2	0	2.109	−0.607
C_3	−18.032	9.441	−1.208
$n - C_4$	0	2.800	−0.488
$n - C_6$	49.472	−5.783	−0.152

For selected nonhydrocarbon components, values of interaction parameters are given by the following expressions:

• **For N_2-H_2O binary:**

$$\tau_{ij} = 0.402\left(T/T_{ci}\right) - 1.586 \qquad (15\text{-}162)$$

where τ_{ij} = binary parameter between nitrogen and the water compound
T_{ci} = critical temperature of nitrogen, °R

• **For CO_2-H_2O binary:**

$$\tau_{ij} = -0.074\left[\frac{T}{T_{ci}}\right]^2 + 0.478\left[\frac{T}{T_{ci}}\right] - 0.503 \qquad (15\text{-}163)$$

where T_{ci} is the critical temperature of CO_2.

In the course of making phase equilibrium calculations, it is always desirable to provide initial values for the equilibrium ratios so the iterative procedure can proceed as reliably and rapidly as possible. Peng and Robinson (1980) adopted Wilson's equilibrium ratio correlation to provide initial K values for the hydrocarbon-vapor phase.

$$K_i = p_{ci}/p \quad exp\left[5.3727(1+\omega_i)(1-T_{ci}/T)\right]$$

while for the water-vapor phase, Peng and Robinson proposed the following expression:

$$K_{wi} = 10^6\left[p_{ci}T/(T_{ci}p)\right]$$

Vapor Pressure from Equation of State

The calculation of the vapor pressure of a pure component through an EOS is usually made by the same trial-and-error algorithms used to calculate vapor–liquid equilibria of mixtures. Soave (1972) suggests that the van der Waals (vdW), Soave–Redlich–Kwong (SRK), and the Peng–Robinson (PR) equations of state can be written in the following generalized form:

$$p = \frac{RT}{v-b} - \frac{a\alpha}{v^2 = \mu vb + wb^2} \tag{15-164}$$

with

$$a = \Omega_a \frac{R^2 T_c^2}{p_c}$$

$$b = \Omega_b \frac{RT_c}{p_c}$$

where the values of u, w, Ω_a, and Ω_b for three different equations of state are given below:

EOS	u	w	Ω_a	Ω_b
vdW	0	0	0.421875	0.125
SRK	1	0	0.42748	0.08664
PR	2	−1	0.45724	0.07780

Soave (1972) introduced the reduced pressure p_r and reduced temperature T_r to the above equations to give:

$$A = \frac{a\alpha p}{R^2 T^2} = \Omega_a \frac{\alpha p_r}{T_r} \qquad (15\text{-}165)$$

$$B = \frac{bp}{RT} = \Omega_b \frac{p_r}{T_r} \qquad (15\text{-}166)$$

and

$$\frac{A}{B} = \frac{\Omega_a}{\Omega_b}\left(\frac{\alpha}{T_r}\right) \qquad (15\text{-}167)$$

where:

$$p_r = p / p_c$$
$$T_r = T / T_c$$

In the cubic form and in terms of the Z factor, the above three equations of state can be written:

$$\text{vdW:} \ \ Z^3 - Z^2(1+B) + ZA - AB = 0$$

$$\text{SRK:} \ \ Z^3 - Z^2 + Z(A - B - B^2) - AB = 0 \qquad (15\text{-}168)$$

$$\text{PR:} \ \ Z^3 - Z^2(1-B) + Z(A - 3B^2 - 2B) - (AB - B^2 - B^3) = 0$$

and the pure component fugacity coefficient is given by:

$$\text{VdW:} \ \ \ln(f/p) = Z - 1 - \ln(Z - B) - \frac{A}{Z}$$

$$\text{SRK:} \ \ \ln(f/p) = Z - 1 - \ln(Z - B) - \left(\frac{A}{B}\right)\ln\left(1 + \frac{B}{Z}\right)$$

$$\text{PR:} \ \ \ln(f/p) = Z - 1 - \ln(Z - B) - \left(\frac{A}{2\sqrt{2}B}\right)\ln\left(\frac{Z + (1+\sqrt{2})B}{Z - (1-\sqrt{2})B}\right)$$

A typical iterative procedure for the calculation of pure component vapor pressure at any temperature T through one of the above EOS is summarized below:

Step 1. Calculate the reduced temperature, i.e., $T_r = T/T_c$.

Step 2. Calculate the ratio A/B from Equation 15-167.

Step 3. Assume a value for B.

Step 4. Solve Equation (15-168) and obtain Z^L and Z^v, i.e., smallest and largest roots, for both phases.

Step 5. Substitute Z^L and Z^v into the pure component fugacity coefficient and obtain $\ln(f/p)$ for both phases.

Step 6. Compare the two values of f/p. If the isofugacity condition is not satisfied, assume a new value of B and repeat steps 3 through 6.

Step 7. From the final value of B, obtain the vapor pressure from Equation 15-166, or:

$$B = \Omega_b \frac{(p_v/p_c)}{T_r}$$

Solving for p_v gives

$$p_v = \frac{BT_rP_c}{\Omega_b}$$

SPLITTING AND LUMPING SCHEMES OF THE PLUS-FRACTION

The hydrocarbon plus fractions that comprise a significant portion of naturally occurring hydrocarbon fluids create major problems when predicting the thermodynamic properties and the volumetric behavior of these fluids by equations of state. These problems arise due to the difficulty of properly characterizing the plus fractions (heavy ends) in terms of their critical properties and acentric factors.

Whitson (1980) and Maddox and Erbar (1982, 1984), among others, have shown the distinct effect of the heavy fractions characterization procedure on PVT relationship prediction by equations of state. Usually, these undefined plus fractions, commonly known as the C_{7+} fractions, contain an undefined number of components with a carbon number higher than 6. Molecular weight and specific gravity of the C_{7+} fraction may be the only measured data available.

In the absence of detailed analytical data for the plus fraction in a hydrocarbon mixture, erroneous predictions and conclusions can result if the plus fraction is used directly as a single component in the mixture phase behavior calculations. Numerous authors have indicated that these errors can be substantially reduced by "splitting" or "breaking down" the plus fraction into a manageable number of fractions (pseudo-components) for equation of state calculations.

The problem, then, is how to adequately split a C_{7+} fraction into a number of psuedo-components characterized by:

- Mole fractions
- Molecular weights
- Specific gravities

These characterization properties, when properly M_{7+} combined, should match the measured plus fraction properties, i.e., $(M)_{7+}$ and $(\gamma)_{7+}$.

Splitting Schemes

Splitting schemes refer to the procedures of dividing the heptanes-plus fraction into hydrocarbon groups with a single carbon number (C_7, C_8, C_9, etc.) and are described by the same physical properties used for pure components.

Several authors have proposed different schemes for extending the molar distribution behavior of C_{7+}, i.e., the molecular weight and specific gravity. In general, the proposed schemes are based on the observation that lighter systems such as condensates usually exhibit exponential molar distribution, while heavier systems often show left-skewed distributions. This behavior is shown schematically in Figure 15-15.

Three important requirements should be satisfied when applying any of the proposed splitting models:

1. The sum of the mole fractions of the individual pseudo-components is equal to the mole fraction of C_{7+}.
2. The sum of the products of the mole fraction and the molecular weight of the individual pseudo-components is equal to the product of the mole fraction and molecular weight of C_{7+}.
3. The sum of the product of the mole fraction and molecular weight divided by the specific gravity of each individual component is equal to that of C_{7+}.

The above requirements can be expressed mathematically by the following relationship:

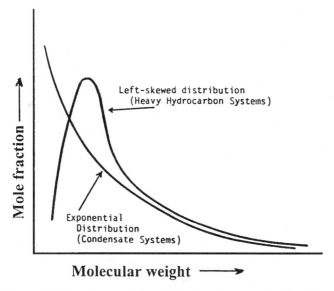

Figure 15-15. Exponential and left-skewed distribution functions.

$$\sum_{n=7}^{N+} z_n = z_{7+} \qquad (15\text{-}169)$$

$$\sum_{n=7}^{N+} \left[z_n M_n \right] = z_{7+} M_{7+} \qquad (15\text{-}170)$$

$$\sum_{n=7}^{N+} \frac{z_n M_n}{\gamma_n} = \frac{z_{7+} M_{7+}}{\gamma_{7+}} \qquad (15\text{-}171)$$

where z_{7+} = mole fraction of C_{7+}

         n = number of carbon atoms

         N_+ = last hydrocarbon group in C_{7+} with n carbon atoms, e.g., 20+

         z_n = mole fraction of psuedo-component with n carbon atoms

   M_{7+}, γ_{7+} = measure of molecular weight and specific gravity of C_{7+}

     M_n, γ_n = Molecular weight and specific gravity of the psuedo-component with n carbon atoms

Several splitting schemes have been proposed recently. These schemes, as discussed below, are used to predict the compositional distribution of the heavy plus fraction.

Katz's Method

Katz (1983) presented an easy-to-use graphical correlation for break-ing down into pseudo-components the C_{7+} fraction present in condensate systems. The method was originated by studying the compositional behavior of six condensate systems using detailed extended analysis. On a semi-log scale, the mole percent of each constituent of the C_{7+} fraction versus the carbon number in the fraction was plotted. The resulting rela-tionship can be conveniently expressed mathematically by the following expression:

$$z_n = 1.38205 z_{7+} e^{-0.25903n} \qquad (15\text{-}172)$$

where z_{7+} = mole fracture of C_{7+} in the condensate system

n = number of carbon atoms of the psuedo-component

z_n = mole fraction of the pseudo-component with number of carbon atoms of n

Equation 15-172 is repeatedly applied until Equation 15-169 is satisfied. The molecular weight and specific gravity of the last pseudo-component can be calculated from Equations 15-170 and 15-171, respectively.

The computational procedure of Katz's method is best explained through the following example.

Example 15-17

A naturally occurring condensate gas system has the following compo-sition:

Component	z_i
C_1	0.9135
C_2	0.0403
C_3	0.0153
$i - C_4$	0.0039
$n - C_4$	0.0043
$i - C_5$	0.0015
$n - C_5$	0.0019
C_6	0.0039
C_{7+}	0.0154

The molecular weight and specific gravity of C_{7+} are 141.25 and 0.797, respectively.

a. Using Katz's splitting scheme, extend the compositional distribution of C_{7+} to the pseudo-fraction C_{16+}.
b. Calculate M, γ, T_b, p_c, T_c, and ω of C_{16+}.

Solution

a. Applying Equation 15-172 with $z_{7+} = 0.0154$ gives

n	Experimental z_n	Equation 15-172 z_n
7	0.00361	0.00347
8	0.00285	0.00268
9	0.00222	0.00207
10	0.00158	0.001596
11	0.00121	0.00123
12	0.00097	0.00095
13	0.00083	0.00073
14	0.00069	0.000566
15	0.00050	0.000437
16+	0.00094	0.001671*

*This value is obtained by applying Equations 15-169, i.e., $0.0154 - \sum_{n=7}^{15} z_n = 0.001671$.

b.

Step 1. Calculate the molecular weight and specific gravity of C_{16+} by solving Equations 15-170 and 15-171 for these properties:

$$M_{16+} = z_{7+}M_{7+} - \left[\left(\frac{1}{z_{16+}} \right) \sum_{n=7}^{15} (z_n \cdot M_n) \right]$$

and

$$\gamma_{16+} = \frac{z_{16+}M_{16+}}{\left(z_{7+}M_{7+}/\gamma_{7+} \right)} - \sum_{n=7}^{15} \left(\frac{z_n M_n}{\gamma_n} \right)$$

where M_n, γ_n = molecular weight and specific gravity of the hydrocarbon group with n carbon atoms. The calculations are performed in the following tabulated form:

n	z_n	M_n (Table 1-1)	$z_n M_n$	γ_n (Table 1-1)	$z_n \cdot M/\gamma_n$
7	0.00347	96	0.33312	0.727	0.4582
8	0.00268	107	0.28676	0.749	0.3829
9	0.00207	121	0.25047	0.768	0.3261
10	0.001596	134	0.213864	0.782	0.27348
11	0.00123	147	0.18081	0.793	0.22801
12	0.00095	161	0.15295	0.804	0.19024
13	0.00073	175	0.12775	0.815	0.15675
14	0.000566	190	0.10754	0.826	0.13019
15	0.000437	206	0.09002	0.836	0.10768
16+	0.001671	—	—	—	—
			1.743284		2.25355

$$M_{16+} = \frac{(0.0154)(141.25) - 1.743284}{0.001671} = 258.5$$

$$\gamma_{16+} = \frac{(0.001671)(258.5)}{\dfrac{(0.0154)(141.25)}{(0.797)}} - 2.25355 = 0.908$$

Step 2. Calculate the boiling points, critical pressure, and critical temperature of C_{16+} by using the Riazi–Daubert correlation to give:

$$T_b = 1{,}136°R$$
$$p_c = 215 \text{ psia}$$
$$T_c = 1{,}473°R$$

Step 3. Calculate the acentric factor of C_{16+} by applying the Edmister correlation to give $\omega = 0.684$.

Lohrenz's Method

Lohrenz et al. (1964) proposed that the heptanes-plus fraction could be divided into pseudo-components with carbon numbers ranging from 7 to 40. They mathematically stated that the mole fraction z_n is related to its number of carbon atoms n and the mole fraction of the hexane fraction z_6 by the expression:

$$z_n = z_6 e^{A(n-6)^2 + B(n-6)} \tag{15-173}$$

The constants A and B are determined such that the constraints given by Equations 15-169 through 15-171 are satisfied.

The use of Equation 15-173 assumes that the individual C_{7+} components are distributed through the hexane mole fraction and tail off to an extremely small quantity of heavy hydrocarbons.

Example 15-18

Rework Example 15-17 by using the Lohrenz splitting scheme and assuming that a partial molar distribution of C_{7+} is available. The composition is given below:

Component	z_i
C_1	0.9135
C_2	0.0403
C_3	0.0153
$i - C_4$	0.0039
$n - C_4$	0.0043
$i - C_5$	0.0015
$n - C_5$	0.0019
C_6	0.0039
C_7	0.00361
C_8	0.00285
C_9	0.00222
C_{10}	0.00158
C_{11+}	0.00514

Solution

Step 1. Determine the coefficients A and B of Equation 15-173 by the least-squares fit to the mole fractions C_6 through C_{10} to give A = 0.03453 and B = 0.08777.

Step 2. Solve for the mole fraction of C_{10} through C_{15} by applying Equation 15-173 and setting $z_6 = 0.0039$:

Component	Experimental z_n	Equation 15-173 z_n
C_7	0.00361	0.00361
C_8	0.00285	0.00285
C_9	0.00222	0.00222
C_{10}	0.00158	0.00158
C_{11}	0.00121	0.00106
C_{12}	0.00097	0.00066
C_{13}	0.00083	0.00039
C_{14}	0.00069	0.00021
C_{15}	0.00050	0.00011
C_{16+}	0.00094	0.00271*

*Obtained by applying Equation 15-169.

Step 3. Calculate the molecular weight and specific gravity of C_{16+} by applying Equations 15-170 and 15-171 to give $(M)_{16+} = 233.3$ and $(\gamma)_{16+} = 0.943$.

Step 4. Solve for T_b, p_c, T_c, and ω by applying the Riazi–Daubert and Edmister correlations, to give:

$T_b = 1{,}103°R$
$p_c = 251 \text{ psia}$
$T_c = 1{,}467°R$
$\omega = 0.600$

Pedersen's Method

Pedersen et al. (1982) proposed that, for naturally occurring hydrocarbon mixtures, an exponential relationship exists between the mole fraction of a component and the corresponding carbon number. They expressed this relationship mathematically in the following form:

$$z_n = e^{(n-A)/B} \tag{15-174}$$

where A and B are constants.

For condensates and volatile oils, Pedersen and coworkers suggested that A and B can be determined by a least-squares fit to the molar distribution of the lighter fractions. Equation 15-174 can then be used to calculate the molar content of each of the heavier fractions by extrapolation. The classical constraints as given by Equations 15-169 through 15-171 are also imposed.

Example 15-19

Rework Example 15-18 using the Pedersen splitting correlation.

Solution

Step 1. Calculate coefficients A and B by the least-squares fit to the molar distribution of C_6 through C_{10} to give A = –14.404639 and B = –3.8125739.

Step 2. Solve for the mole fraction of C_{10} through C_{15} by applying Equation 15-176.

Component	Experimental z_n	Calculated z_n
C_7	0.000361	0.00361
C_8	0.00285	0.00285
C_9	0.00222	0.00222
C_{10}	0.00158	0.00166
C_{11}	0.00121	0.00128
C_{12}	0.00097	0.00098
C_{13}	0.00083	0.00076
C_{14}	0.00069	0.00058
C_{15}	0.00050	0.00045
C_{16+}	0.00094	0.00101*

*From Equation 15-169.

Ahmed's Method

Ahmed et al. (1985) devised a simplified method for splitting the C_{7+} fraction into pseudo-components. The method originated from studying the molar behavior of 34 condensate and crude oil systems through detailed laboratory compositional analysis of the heavy fractions. The only required data for the proposed method are the molecular weight and the total mole fraction of the heptanes-plus fraction.

The splitting scheme is based on calculating the mole fraction z_n at a progressively higher number of carbon atoms. The extraction process continues until the sum of the mole fraction of the pseudo-components equals the total mole fraction of the heptanes-plus (z_{7+}).

$$z_n = z_{n+} \left[\frac{M_{(n+1)+} - M_{n+}}{M_{(n+1)+} - M_n} \right] \tag{15-175}$$

where z_n = mole fraction of the pseudo-component with a number of carbon atoms of n (z_7, z_8, z_9, etc.)

  M_n = molecular weight of the hydrocarbon group with n carbon atoms as given in Table 1-1 in Chapter 1

  M_{n+} = molecular weight of the n+ fraction as calculated by the following expression:

$$M_{(n+1)+} = M_{7+} + S(n-6) \tag{15-176}$$

where n is the number of carbon atoms and S is the coefficient of Equation 15-178 with these values:

Number of Carbon Atoms	Condensate Systems	Crude Oil Systems
$n \leq 8$	15.5	16.5
$n > 8$	17.0	20.1

The stepwise calculation sequences of the proposed correlation are summarized in the following steps:

Step 1. According to the type of hydrocarbon system under investigation (condensate or crude oil), select appropriate values for the coefficients.

Step 2. Knowing the molecular weight of C_{7+} fraction (M_{7+}), calculate the molecular weight of the octanes-plus fraction (M_{8+}) by applying Equation 15-176.

Step 3. Calculate the mole fraction of the heptane fraction (z_7) using Equation 15-175.

Step 4. Apply steps 2 and 3 repeatedly for each component in the system (C_8, C_9, etc.) until the sum of the calculated mole fractions is equal to the mole fraction of C_{7+} of the system.

The splitting scheme is best explained through the following example.

Example 15-20

Rework Example 15-19 using Ahmed's splitting method.

Solution

Step 1. Calculate the molecular weight of C_{8+} by applying Equation 15-176:

$$M_{8+} = 141.25 + 15.5(7-6) = 156.75$$

Step 2. Solve for the mole fraction of heptane (z_7) by applying Equation 15-175:

$$z_7 = z_{7+} \left[\frac{M_{8+} - M_{7+}}{M_{8+} - M_7} \right] = 0.0154 \left[\frac{156.75 - 141.25}{156.75 - 96} \right] = 0.00393$$

Step 3. Calculate the molecular weight of C_{9+} from Equation 15-178:

$$M_{9+} = 141.25 + 15.5(8-6) = 172.25$$

Step 4. Determine the mole fraction of C_8 from Equation 15-177:

$$z_8 = z_{8+} \left[(M_{9+} - M_{8+})/(M_{9+} - M_8) \right]$$
$$z_8 = (0.0154 - 0.00393) \left[(172.5 - 156.75)/(172.5 - 107) \right]$$
$$= 0.00276$$

Step 5. This extracting method is repeated as outlined in the above steps to give:

Component	n	M_{n+} Equation 15-176	M_n (Table 1-1)	z_n Equation 15-175
C_7	7	141.25	96	0.000393
C_8	8	156.25	107	0.00276
C_9	9	175.25	121	0.00200
C_{10}	10	192.25	134	0.00144
C_{11}	11	209.25	147	0.00106
C_{12}	12	226.25	161	0.0008
C_{13}	13	243.25	175	0.00061
C_{14}	14	260.25	190	0.00048
C_{15}	15	277.25	206	0.00038
C_{16+}	16+	294.25	222	0.00159*

*Calculated from Equation 15-169.

Step 6. The boiling point, critical properties, and the acentric factor of C_{16+} are then determined by using the appropriate methods, to

$$M = 222$$
$$\gamma = 0.856$$
$$T_b = 1174.6°R$$
$$p_c = 175.9 \text{ psia}$$
$$T_c = 1449.3°R$$
$$\omega = 0.742$$

Lumping Schemes

The large number of components necessary to describe the hydrocarbon mixture for accurate phase behavior modeling frequently burdens EOS calculations. Often, the problem is either lumping together the many experimentally determined fractions, or modeling the hydrocarbon system when the only experimental data available for the C_{7+} fraction are the molecular weight and specific gravity.

Generally, with a sufficiently large number of pseudo-components used in characterizing the heavy fraction of a hydrocarbon mixture, a satisfactory prediction of the PVT behavior by the equation of state can be obtained. However, in compositional models, the cost and computing time can increase significantly with the increased number of components in the system. Therefore, strict limitations are placed on the maximum number of components that can be used in compositional models and the original components have to be lumped into a smaller number of pseudo-components.

The term *lumping* or *pseudoization* then denotes the reduction in the number of components used in EOS calculations for reservoir fluids. This reduction is accomplished by employing the concept of the pseudo-component. The pseudo-component denotes a group of pure components lumped together and represented by a single component.

Several problems are associated with "regrouping" the original components into a smaller number without losing the predicting power of the equation of state. These problems include:

- How to select the groups of pure components to be represented by one pseudo-component each
- What mixing rules should be used for determining the EOS constants (p_c, T_c, and ω) for the new lumped pseudo-components

Several unique techniques have been published that can be used to address the above lumping problems; notably the methods proposed by:

- Lee et al. (1979)
- Whitson (1980)
- Mehra et al. (1983)
- Montel and Gouel (1984)
- Schlijper (1984)
- Behrens and Sandler (1986)
- Gonzalez, Colonomos, and Rusinek (1986)

Several of these techniques are presented in the following discussion.

Whitson's Lumping Scheme

Whitson (1980) proposed a regrouping scheme whereby the compositional distribution of the C_{7+} fraction is reduced to only a few multiple-carbon-number (MCN) groups. Whitson suggested that the number of MCN groups necessary to describe the plus fraction is given by the following empirical rule:

$$N_g = Int[1 + 3.3\log(N - n)]$$ (15-177)

where N_g = number of MCN groups
  Int = integer
  N = number of carbon atoms of the last component in the hydrocarbon system
  n = number of carbon atoms of the first component in the plus fraction, i.e., n = 7 for C_{7+}.

The integer function requires that the real expression evaluated inside the brackets be rounded to the nearest integer. Whitson pointed out that for black-oil systems, one could reduce the calculated value of N_g.

The molecular weights separating each MCN group are calculated from the following expression:

$$M_I = M_{C7}\left(\frac{M_{N+}}{M_{C7}}\right)^{I/N_g}$$ (15-178)

where $(M)_{N+}$ = molecular weight of the last reported component in the extended analysis of the hydrocarbon system

$$M_{C7} = \text{molecular weight of } C_7$$
$$I = 1, 2, \ldots, N_g$$

Components with molecular weight falling within the boundaries of M_{I-1} to M_I are included in the I'th MCN group. Example 15-21 illustrates the use of Equations 15-177 and 15-178.

Example 15-21

Given the following compositional analysis of the C_{7+} fraction in a condensate system, determine the appropriate number of pseudo-components forming in the C_{7+}.

Component	z_i
C_7	0.00347
C_8	0.00268
C_9	0.00207
C_{10}	0.001596
C_{11}	0.00123
C_{12}	0.00095
C_{13}	0.00073
C_{14}	0.000566
C_{15}	0.000437
C_{16+}	0.001671

$M_{16+} = 259$.

Solution

Step 1. Determine the molecular weight of each component in the system:

Component	z_i	M_i
C_7	0.00347	96
C_8	0.00268	107
C_9	0.00207	121
C_{10}	0.001596	134
C_{11}	0.00123	147
C_{12}	0.00095	161
C_{13}	0.00073	175
C_{14}	0.000566	190
C_{15}	0.000437	206
C_{16+}	0.001671	259

Step 2. Calculate the number of pseudo-components from Equation 15-178:

$$N_g = \text{Int}\left[1 + 3.3\log(16 - 7)\right]$$
$$N_g = \text{Int}\left[4.15\right]$$
$$N_g = 4$$

Step 3. Determine the molecular weights separating the hydrocarbon groups by applying Equation 15-179:

$$M_I = 96\left[\frac{259}{96}\right]^{I/4}$$
$$M_I = 96[2.698]^{I/4}$$

I	(M)$_I$
1	123
2	158
3	202
4	259

- **First pseudo-component:** The first pseudo-component includes all components with molecular weight in the range of 96 to 123. This group then includes C_7, C_8, and C_9.
- **Second pseudo-component:** The second pseudo-component contains all components with a molecular weight higher than 123 to a molecular weight of 158. This group includes C_{10} and C_{11}.
- **Third pseudo-component:** The third pseudo-component includes components with a molecular weight higher than 158 to a molecular weight of 202. Therefore, this group includes C_{12}, C_{13}, and C_{14}.
- **Fourth pseudo-component:** This pseudo-component includes all the remaining components, i.e., C_{15} and C_{16+}.

Group I	Component	z_i	z_l
1	C_7	0.00347	0.00822
	C_8	0.00268	
	C_9	0.00207	
2	C_{10}	0.001596	0.002826
	C_{11}	0.00123	
3	C_{12}	0.00095	0.002246
	C_{13}	0.00073	
	C_{14}	0.000566	
4	C_{15}	0.000437	0.002108
	C_{16+}	0.001671	

It is convenient at this stage to present the mixing rules that can be employed to characterize the pseudo-component in terms of its pseudo-physical and pseudo-critical properties. Because there are numerous ways to mix the properties of the individual components, all giving different properties for the pseudo-components, the choice of a correct mixing rule is as important as the lumping scheme. Some of these mixing rules are given next.

Hong's Mixing Rules

Hong (1982) concluded that the weight fraction average w_i is the best mixing parameter in characterizing the C_{7+} fractions by the following mixing rules:

- Pseudo-critical pressure $p_{cL} = \sum_{}^{L} w_i p_{ci}$

- Pseudo-critical temperature $T_{cL} = \sum_{}^{L} w_i T_{ci}$

- Pseudo-critical volume $V_{cL} = \sum_{}^{L} w_i V_{ci}$

- Pseudo-acentric factor $\omega_L = \sum_{}^{L} w_i \omega_i$

- Pseudo-molecular weight $M_L = \sum_{}^{L} w_i M_i$

- Binary interaction coefficient $K_{kL} = 1 - \sum_{i}^{L} \sum_{j}^{L} w_i w_j \left(1 - k_{ij}\right)$

with:

$$w_i = \frac{z_i M_i}{\sum\limits^{L} z_i M_i}$$

where: w_i = average weight fraction

K_{kL} = binary interaction coefficient between the k'th component and the lumped fraction

The subscript L in the above relationship denotes the lumped fraction.

Lee's Mixing Rules

Lee et al. (1979), in their proposed regrouping model, employed Kay's mixing rules as the characterizing approach for determining the properties of the lumped fractions. Defining the normalized mole fraction of the component i in the lumped fraction as:

$$\phi_i = z_i / \sum\limits^{L} z_i$$

the following rules are proposed:

$$M_L = \sum\limits^{L} \phi_i M_i \qquad\qquad (15\text{-}179)$$

$$\gamma_L = M_L / \sum\limits^{L} [\phi_i M_i / \gamma_i] \qquad\qquad (15\text{-}180)$$

$$V_{cL} = \sum\limits^{L} [\phi_i M_i V_{ci} / M_L] \qquad\qquad (15\text{-}181)$$

$$P_{cL} = \sum\limits^{L} [\phi_i P_{ci}] \qquad\qquad (15\text{-}182)$$

$$T_{cL} = \sum\limits^{L} [\phi_i T_{ci}] \qquad\qquad (15\text{-}183)$$

$$\omega_L = \sum\limits^{L} [\phi_i \omega_i] \qquad\qquad (15\text{-}184)$$

Example 15-22

Using Lee's mixing rules, determine the physical and critical properties of the four pseudo-components in Example 15-21.

Solution

Step 1. Assign the appropriate physical and critical properties to each component:

Group	Comp.	z_i	z_I	M_i	γ_i	V_{ci}	p_{ci}	T_{ci}	ω_i
	C_7	0.00347		96^*	0.272^*	0.06289^*	453^*	985^*	0.280^*
1	C_8	0.00268	0.00822	107	0.748	0.06264	419	1036	0.312
	C_9	0.00207		121	0.768	0.06258	383	1058	0.348
2	C_{10}	0.001596	0.002826	134	0.782	0.06273	351	1128	0.385
	C_{11}	0.00123		147	0.793	0.06291	325	1166	0.419
	C_{12}	0.00095		161	0.804	0.06306	302	1203	0.454
3	C_{13}	0.00073	0.002246	175	0.815	0.06311	286	1236	0.484
	C_{14}	0.000566		190	0.826	0.06316	270	1270	0.516
4	C_{15}	0.000437	0.002108	206	0.826	0.06325	255	1304	0.550
	C_{16+}	0.001671		259	0.908	$0.0638^\dagger$	$215^\dagger$	1467	$0.68^\dagger$

*From Table 1-1.
†Calculated.

Step 2. Calculate the physical and critical properties of each group by applying Equations 15-179 through 15-184 to give:

Group	Z_I	M_L	γ_L	V_{cL}	p_{cL}	T_{cL}	ω_L
1	0.00822	105.9	0.746	0.0627	424	1020	0.3076
2	0.002826	139.7	0.787	0.0628	339.7	1144.5	0.4000
3	0.002246	172.9	0.814	0.0631	288	1230.6	0.4794
4	0.002108	248	0.892	0.0637	223.3	1433	0.6531

PROBLEMS

1. A hydrocarbon system has the following composition:

Component	z_i
C_1	0.30
C_2	0.10
C_3	0.05
$i-C_4$	0.03
$n-C_4$	0.03
$i-C_5$	0.02
$n-C_5$	0.02
C_6	0.05
C_{7+}	0.40

Given the following additional data:

$$\begin{aligned}
\text{System pressure} &= 2{,}100 \text{ psia} \\
\text{System temperature} &= 150°F \\
\text{Specific gravity of } C_{7+} &= 0.80 \\
\text{Molecular weight of } C_{7+} &= 140
\end{aligned}$$

Calculate the equilibrium ratios of the above system.

2. A well is producing oil and gas with the compositions given below at a gas–oil ratio of 500 scf/STB:

Component	x_i	y_i
C_1	0.35	0.60
C_2	0.08	0.10
C_3	0.07	0.10
$n-C_4$	0.06	0.07
$n-C_5$	0.05	0.05
C_6	0.05	0.05
C_{7+}	0.34	0.05

Given the following additional data:

$$\begin{aligned}
\text{Current reservoir pressure} &= 3000 \text{ psia} \\
\text{Bubble-point pressure} &= 2800 \text{ psia} \\
\text{Reservoir temperature} &= 120°F \\
M \text{ of } C_{7+} &= 125 \\
\text{Specific gravity of } C_{7+} &= 0.823
\end{aligned}$$

Calculate the composition of the reservoir fluid.

3. A saturated hydrocarbon mixture with the composition given below exists in a reservoir at 234°F:

Component	z_i
C_1	0.3805
C_2	0.0933
C_3	0.0885
C_4	0.0600
C_5	0.0378
C_6	0.0356
C_{7+}	0.3043

Calculate:

a. The bubble-point pressure of the mixture.
b. The compositions of the two phases if the mixture is flashed at 500 psia and 150°F.
c. The density of the liquid phase.
d. The compositions of the two phases if the liquid from the first separator is further flashed at 14.7 psia and 60°F.
e. The oil formation volume factor at the bubble-point pressure.
f. The original gas solubility.
g. The oil viscosity at the bubble-point pressure.

4. A crude oil exists in a reservoir at its bubble-point pressure of 2520 psig and a temperature of 180°F. The oil has the following composition:

Component	x_i
CO_2	0.0044
N_2	0.0045
C_1	0.3505
C_2	0.0464
C_3	0.0246
$i - C_4$	0.0683
$n - C_4$	0.0083
$i - C_5$	0.0080
$n - C_5$	0.0080
C_6	0.0546
C_{7+}	0.4824

The molecular weight and specific gravity of C_{7+} are 225 and 0.8364. The reservoir contains initially 12 MMbbl of oil. The surface facilities consist of two separation stages connecting in series. The first separation stage operates at 500 psig and 100°F. The second stage operates under standard conditions.

a. Characterize C_{7+} in terms of its critical properties, boiling point, and acentric factor.
b. Calculate the initial oil in place in STB.
c. Calculate the standard cubic feet of gas initially in solution.
d. Calculate the composition of the free gas and the composition of the remaining oil at 2495 psig, assuming the overall composition of the system will remain constant.

5. A pure n-butane exists in the two-phase region at 120°F. Calculate the density of the coexisting phase by using the following equations of state:

a. Van der Waals
b. Redlich–Kwong
c. Soave–Redlich–Kwong
d. Peng–Robinson

6. A crude oil system with the following composition exists at its bubble-point pressure of 3,250 psia and 155°F:

Component	x_i
C_1	0.42
C_2	0.08
C_3	0.06
C_4	0.02
C_5	0.01
C_6	0.04
C_{7+}	0.37

If the molecular weight and specific gravity of the heptanes-plus fraction are 225 and 0.823, respectively, calculate the density of the crude oil by using:

a. Van der Waals EOS
b. Redlich–Kwong EOS

c. SRR EOS

d. PR EOS

7. Calculate the vapor pressure of propane at 100°F by using:

a. Van der Waals EOS

b. SRK EOS

c. PR EOS

Compare the results with that obtained from the Cox chart.

8. A natural gas exists at 2000 psi and 150°F. The gas has the following composition:

Component	y_i
C_1	0.80
C_2	0.10
C_3	0.07
$i - C_4$	0.02
$n - C_4$	0.01

Calculate the density of the gas using the following equations of state:

a. VdW

b. RK

c. SRK

d. PR

9. The heptanes-plus fraction in a condensate gas system is characterized by a molecular weight and specific gravity of 190 and 0.8, respectively. The mole fraction of the C_{7+} is 0.12. Extend the molar distribution of the plus fraction to C_{20+} by using:

a. Katz's method

b. Ahmed's method

Determine the critical properties of C_{20+}.

10. A naturally occurring crude oil system has a heptanes-plus fraction with the following properties:

$M_{7+} = 213$
$\gamma_{7+} = 0.8405$
$x_{7+} = 0.3497$

Extend the molar distribution of the plus fraction to C_{25+} and determine the critical properties and acentric factor of the last component.

11. A crude oil system has the following composition:

Component	x_i
C_1	0.3100
C_2	0.1042
C_3	0.1187
C_4	0.0732
C_5	0.0441
C_6	0.0255
C_{7+}	0.3243

The molecular weight and specific gravity of C_{7+} are 215 and 0.84, respectively.

a. Extend the molar distribution of C_{7+} to C_{20+}.
b. Calculate the appropriate number of pseudo-components necessary to adequately represent the composition from C_7 to C_{20+} and characterize the resulting pseudo-components in terms of:

- Molecular weight
- Specific gravity
- Critical properties
- Acentric factor

REFERENCES

1. Ahmed, T., "A Practical Equation of State," *SPERE,* Feb. 1991, Vol. 291, pp. 136–137.

2. Ahmed, T., Cady, G., and Story, A., "A Generalized Correlation for Characterizing the Hydrocarbon Heavy Fractions," SPE Paper 14266 presented at

the SPE 60th Annual Technical Conference, Las Vegas, NV, Sep. 22–25, 1985.

3. Alani, G. H., and Kennedy, H. T., "Volume of Liquid Hydrocarbons at High Temperatures and Pressures," *Trans. AIME,* 1960, Vol. 219, pp. 288–292.

4. Amyx, J., Bass, D., and Whitney, R., *Petroleum Reservoir Engineering.* New York: McGraw-Hill Book Company, 1960.

5. Behrens, R., and Sandler, S., "The Use of Semi-Continuous Description to Model the C_{7+} Fraction in Equation of State Calculation," SPE/DOE Paper 14925 presented at the 5th Annual Symposium on EOR, Tulsa, OK, April 20–23, 1986.

6. Brinkman, F. H., and Sicking, J. N., "Equilibrium Ratios for Reservoir Studies," *Trans. AIME,* 1960, Vol. 219, pp. 313–319.

7. Campbell, J. M., *Gas Conditioning and Processing,* Vol. 1. Norman, OK: Campbell Petroleum Series, 1976.

8. Chueh, P. and Prausnitz, J., "Vapor-Liquid Equilibria at High Pressures: Calculation of Critical Temperatures, Volumes, and Pressures of Nonpolar Mixtures," *AIChE Journal,* 1967, Vol. 13, No. 6, pp. 1107–1112.

9. Clark, N., *Elements of Petroleum Reservoirs.* Dallas: Society of Petroleum Engineers, 1960.

10. Clark, N., "A Review of Reservoir Engineering," World Oil, June 1951.

11. Dykstra, H., and Mueller, T. D., "Calculation of Phase Composition and Properties for Lean- or Enriched-Gas Drive," *SPEJ,* Sep. 1965, pp. 239–246.

12. Edmister, W., and Lee, B., *Applied Hydrocarbon Thermodynamics,* Vol. 1, 2nd ed. Houston: Gulf Publishing Company, 1986, p. 52.

13. Elliot, J., and Daubert, T., "Revised Procedure for Phase Equilibrium Calculations with Soave Equation of State," *Ind. Eng. Chem. Process Des. Dev.,* 1985, Vol. 23, pp. 743–748.

14. Gibbons, R., and Laughton, A., "An Equation of State for Polar and Non-Polar Substances and Mixtures," *J. Chem. Soc.,* 1984, Vol. 80, pp. 1019–1038.

15. Gonzalez, E., Colonomos, P., and Rusinek, I., "A New Approach for Characterizing Oil Fractions and for Selecting Pseudo-Components of Hydrocarbons," *Canadian JPT,* March-April 1986, pp. 78–84.

16. Graboski, M. S., and Daubert, T. E., "A Modified Soave Equation of State for Phase Equilibrium Calculations 1. Hydrocarbon System," *Ind. Eng. Chem. Process Des. Dev.,* 1978, Vol. 17, pp. 443–448.

17. Hadden, J. T., "Convergence Pressure in Hydrocarbon Vapor-Liquid Equilibria," *Chem. Eng. Progr. Symposium Ser.,* 1953, Vol. 49, No. 7, p. 53.

18. Hariu, O., and Sage, R., "Crude Split Figured by Computer," *Hydrocarbon Process.*, April 1969, pp. 143–148.

19. Heyen, G., "A Cubic Equation of State with Extended Range of Application," paper presented at 2nd World Congress Chemical Engineering, Montreal, Oct. 4–9, 1983.

20. Hoffmann, A. E., Crump, J. S., and Hocott, R. C., "Equilibrium Constants for a Gas-Condensate System," *Trans. AIME,* 1953, Vol. 198, pp. 1–10.

21. Hong, K. C., "Lumped-Component Characterization of Crude Oils for Compositional Simulation," SPE/DOE Paper 10691 presented at the 3rd Joint Symposium on EOR, Tulsa, OK, April 4–7, 1982.

22. Jhaveri, B. S., and Youngren, G. K., "Three-Parameter Modification of the Peng-Robinson Equation of State to Improve Volumetric Predictions," SPE Paper 13118 presented at the 1984 SPE Annual Technical Conference, Houston, Sep. 16–19.

23. Katz, D. L., and Hachmuth, K. H., "Vaporization Equilibrium Constants in a Crude Oil-Natural Gas System," *Ind. Eng. Chem.,* 1937, Vol. 29, p. 1072.

24. Katz, D., et al., "Overview of Phase Behavior of Oil and Gas Production," *JPT,* June 1983, pp. 1205–1214.

25. Katz, D., et al., *Handbook of Natural Gas Engineering.* New York: McGraw-Hill Book Company, 1959.

26. Kehn, D. M., "Rapid Analysis of Condensate Systems by Chromatography," *JPT,* April 1964, pp. 435–440.

27. Kubic, W. L., J., "A Modification of the Martin Equation of State for Calculating Vapor-Liquid Equilibria," *Fluid Phase Equilibria,* 1982, Vol. 9, pp. 79–97.

28. Lee B., and Kesler, G., "A Generalized Thermodynamic Correlation Based on Three-Parameter Corresponding States," *AIChE Journal,* May 1975, Vol 21, No. 3, pp. 510–527.

29. Lee, S., et al., "Experimental and Theoretical Studies on the Fluid Properties Required for Simulation of Thermal Processes," SPE Paper 8393 presented at the SPE 54th Annual Technical Conference, Las Vegas, NV, Sep. 23–26, 1979.

30. Lim, D., et al., "Calculation of Liquid Dropout for Systems Containing Water," SPE Paper 13094 presented at the SPE 59th Annual Technical Conference, Houston, Sep. 16–19, 1984.

31. Lohrenz, J., Bray, B. G., and Clark, C. R., "Calculating Viscosities of Reservoir Fluids from their Compositions," *JPT,* Oct. 1964, p. 1171, Trans. AIME, Vol. 231.

32. Maddox, R. N., and Erbar, J. H., "Improve P-V-T Predictions," *Hydrocarbon Process.*, January 1984, pp. 119–121.

33. Maddox, R. N., and Erbar, J. H., *Gas Conditioning and Processing, Vol. 3— Advanced Techniques and Applications.* Norman, OK: Campbell Petroleum Series, 1982.

34. Mehra, R., et al., "A Statistical Approach for Combining Reservoir Fluids into Pseudo Components for Compositional Model Studies," SPE Paper 11201 presented at the SPE 57th Annual Meeting, New Orleans, Sep. 26–29, 1983.

35. Montel, F., and Gouel, P., "A New Lumping Scheme of Analytical Data for Composition Studies," SPE Paper 13119 presented at the SPE 59th Annual Technical Conference, Houston, Sep. 16–19, 1984.

36. Nikos, V., et al., "Phase Behavior of Systems Comprising North Sea Reservoir Fluids and Injection Gases," *JPT,* Nov. 1986, pp. 1221–1233.

37. Patel, N., and Teja, A., "A New Equation of State for Fluids and Fluid Mixtures," *Chem. Eng. Sci.,* 1982, Vol. 37, No. 3, pp. 463–473.

38. Pedersen, K., Thomassen, P., and Fredenslund, A., "Phase Equilibria and Separation Processes," Report SEP 8207. Denmark: Institute for Kemiteknik, Denmark Tekniske Hojskole (July 1982).

39. Peneloux, A., Rauzy, E., and Freze, R., "A Consistent Correlation for Redlich–Kwong–Soave Volumes," *Fluid Phase Equilibria,* 1982, Vol. 8, pp. 7–23.

40. Peng, D., and Robinson, D., "Two and Three Phase Equilibrium Calculations for Coal Gasification and Related Processes," ACS Symposium Series, No. 133, *Thermodynamics of Aqueous Systems with Industrial Applications.* Washington, DC: American Chemical Society, 1980.

41. Peng, D., and Robinson, D., "A New Two Constant Equation of State," *Ind. Eng. Chem. Fund.,* 1976a, Vol. 15, No. 1, pp. 59–64.

42. Peng, D., and Robinson, D., "Two and Three Phase Equilibrium Calculations for Systems Containing Water," *Canadian J. Chem. Eng.,* 1976b, Vol. 54, pp. 595–598.

43. Peterson, C. S., "A Systematic and Consistent Approach to Determine Binary Interaction Coefficients for the Peng-Robinson Equations of State," *SPERE,* Nov. 1989, pp. 488–496.

44. Redlich, O., and Kwong, J., "On the Thermodynamics of Solutions. An Equation of State. Fugacities of Gaseous Solutions," *Chem. Rev.,* Vol. 44, 1949, pp. 233–247.

45. Reid, R., Prausnitz, J. M., and Sherwood, T., *The Properties of Gases and Liquids, Third Edition.* New York: McGraw-Hill Book Company, 1977, p. 21.

46. Robinson, D. B., and Peng, D. Y., "The Characterization of the Heptanes and Heavier Fractions," *Research Report 28,* Tulsa: GPA, 1978.

47. Rzasa, M. J., Glass, E. D., and Opfell, J. B., "Prediction of Critical Properties and Equilibrium Vaporization Constants for Complex Hydrocarbon Systems," *Chem. Eng. Progr. Symposium Ser.,* 1952, Vol. 48, No. 2, p. 28.

48. Schlijper, A. G., "Simulation of Compositional Process: The Use of Pseudo-Components in Equation of State Calculations," SPE/DOE Paper 12633 presented at the SPE/DOE 4th Symposium on EOR, Tulsa, OK, April 15–18, 1984.

49. Schmidt, G., and Wenzel, H., "A Modified Van der Waals Type Equation of State," *Chem. Eng. Sci.,* 1980, Vol. 135, pp. 1503–1512.

50. Sim, W. J., and Daubert, T. E., "Prediction of Vapor-Liquid Equilibria of Undefined Mixtures," *Ind. Eng. Chem. Process Des. Dev.,* 1980, Vol. 19, No. 3, pp. 380–393.

51. Slot-Petersen, C., "A Systematic and Consistent Approach to Determine Binary Interaction Coefficients for the Peng–Robinson Equation of State," SPE Paper 16941 presented at the SPE 62 Annual Technical Conference, Dallas, Sep. 27–30, 1987.

52. Soave, G., "Equilibrium Constants from a Modified Redlich–Kwong Equation of State," *Chem. Eng. Sci.,* 1972, Vol. 27, pp. 1197–1203.

53. Spencer, C., Daubert, T., and Danner, R., "A Critical Review of Correlations for the Critical Properties of Defined Mixtures," *AIChE Journal,* 1973, Vol. 19, No. 3, pp. 522–527.

54. Standing, M. B., "A Set of Equations for Computing Equilibrium Ratios of a Crude Oil/Natural Gas System at Pressures Below 1,000 psia," *JPT,* Sep. 1979, pp. 1193–1195.

55. Standing, M. B., *Volumetric and Phase Behavior of Oil Field Hydrocarbon Systems.* Dallas: Society of Petroleum Engineers of AIME, 1977.

56. Standing, M. B., and Katz, D. L., "Density of Crude Oils Saturated with Natural Gas," *Trans. AIME,* 1942, Vol. 146, pp. 159–165.

57. Stryjek, R., and Vera, J. H., "PRSV: An Improvement to the Peng-Robinson Equation of State for Pure Compounds and Mixtures," *Canadian J. Chem. Eng.,* April 1986, Vol. 64, pp. 323–333.

58. Valderrama, J., and Cisternas, L., "A Cubic Equation of State for Polar and Other Complex Mixtures," *Fluid Phase Equilibria,* 1986, Vol. 29, pp. 431–438.

59. Van der Waals, J. D., "On the Continuity of the Liquid and Gaseous State," Ph.D. Dissertation, Sigthoff, Leiden, 1873.

60. Vidal, J., and Daubert, T., "Equations of State—Reworking the Old Forms," *Chem. Eng. Sci.,* 1978, Vol. 33, pp. 787–791.

61. Whitson, C., "Characterizing Hydrocarbon Plus Fractions," EUR Paper 183 presented at the European Offshore Petroleum Conference, London, Oct. 21–24, 1980.

62. Whitson, C. H., and Torp, S. B., "Evaluating Constant Volume Depletion Data," SPE Paper 10067 presented at the SPE 56th Annual Fall Technical Conference, San Antonio, TX, Oct. 5-7, 1981.

63. Willman, B. T., and Teja, B. T., "Continuous Thermodynamics of Phase Equilibria Using a Multivariable Distribution Function and an Equation of State," *AIChE Journal,* Dec. 1986, Vol. 32, No. 12, pp. 2067–2078.

64. Wilson, G., "A Modified Redlich–Kwong EOS, Application to General Physical Data Calculations," Paper 15C presented at the Annual AIChE National Meeting, Cleveland, OH, May 4–7, 1968.

65. Winn, F. W., "Simplified Nomographic Presentation, Hydrocarbon Vapor–Liquid Equilibria," *Chem. Eng. Progr. Symposium Ser.,* 1954, Vol. 33, No. 6, pp. 131–135.

66. Whitson, C., and Brule, M., *Phase Behavior.* Richardson, TX: Society of Petroleum Engineers, Inc., 2000.

APPENDIX

PRESSURE, PSIA ⟶

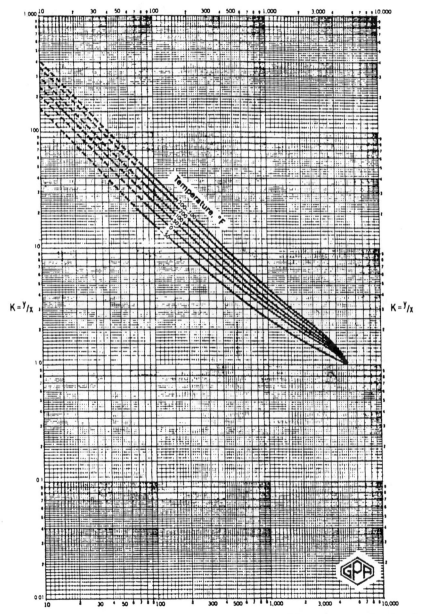

Figure 1. Methane. Conv. press. 5,000 psia. Courtesy of the Gas Processors Suppliers Association. Published in the GPSA Engineering Data Book, Tenth Edition, 1987.

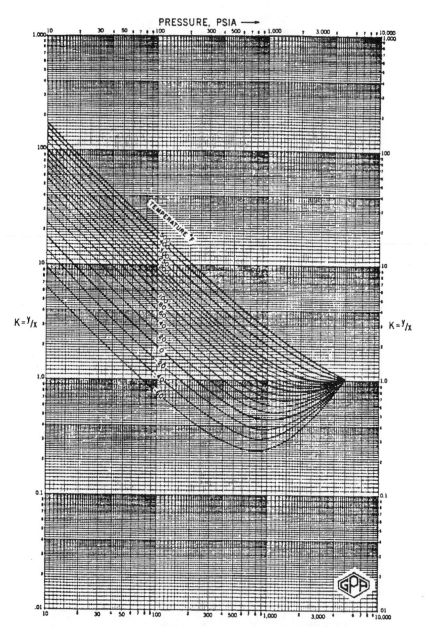

Figure 2. Ethane. Conv. press. 5,000 psia. Courtesy of the Gas Processors Suppliers Association. Published in the GPSA Engineering Data Book, Tenth Edition, 1987.

PRESSURE, PSIA ⟶

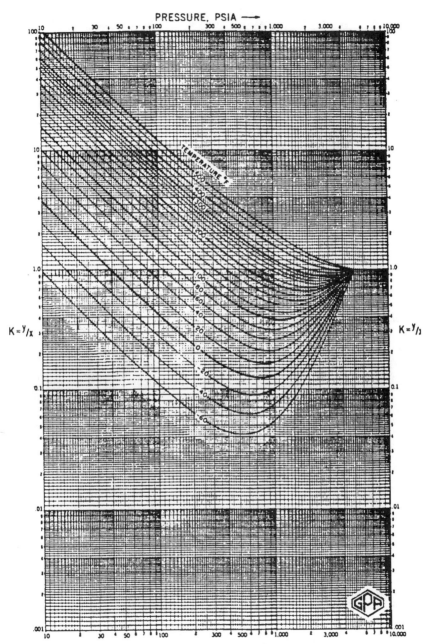

$K = \frac{y}{x}$

$K = \frac{y}{x}$

Figure 3. Propane. Conv. press. 5,000 psia. Courtesy of the Gas Processors Suppliers Association. Published in the GPSA Engineering Data Book, Tenth Edition, 1987.

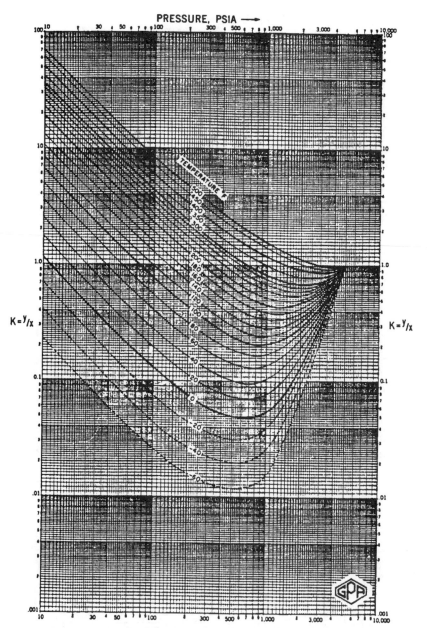

Figure 4. i-Butane. Conv. press. 5,000 psia. Courtesy of the Gas Processors Suppliers Association. Published in the GPSA Engineering Data Book, Tenth Edition, 1987.

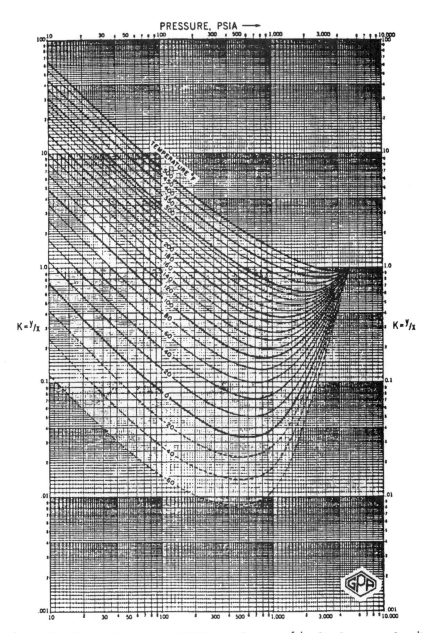

Figure 5. n-Butane. Conv. press. 5,000 psia. Courtesy of the Gas Processors Suppliers Association. Published in the GPSA Engineering Data Book, Tenth Edition, 1987.

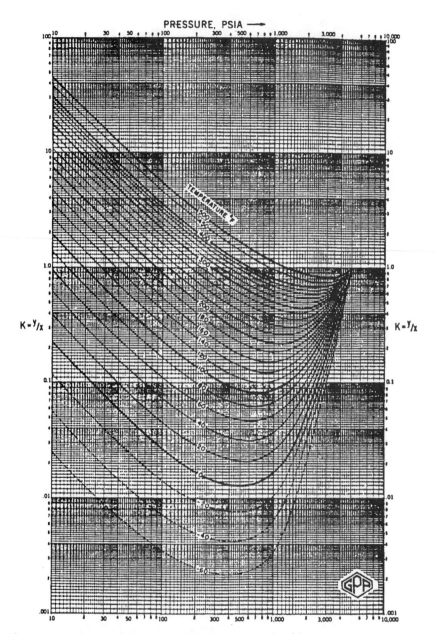

Figure 6. i-Pentane. Conv. press. 5,000 psia. Courtesy of the Gas Processors Suppliers Association. Published in the GPSA Engineering Data Book, Tenth Edition, 1987.

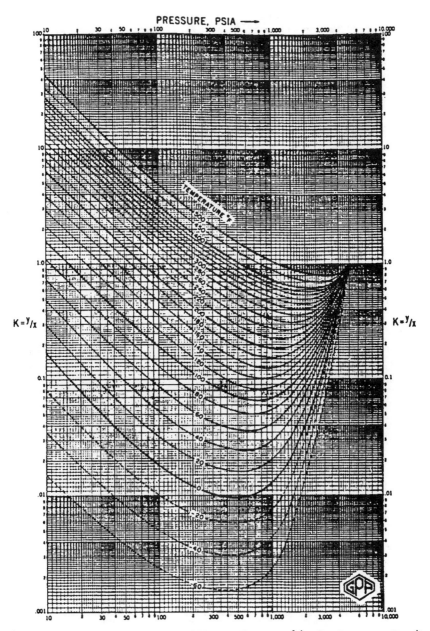

Figure 7. n-Pentane. Conv. press. 5,000 psia. Courtesy of the Gas Processors Suppliers Association. Published in the GPSA Engineering Data Book, Tenth Edition, 1987.

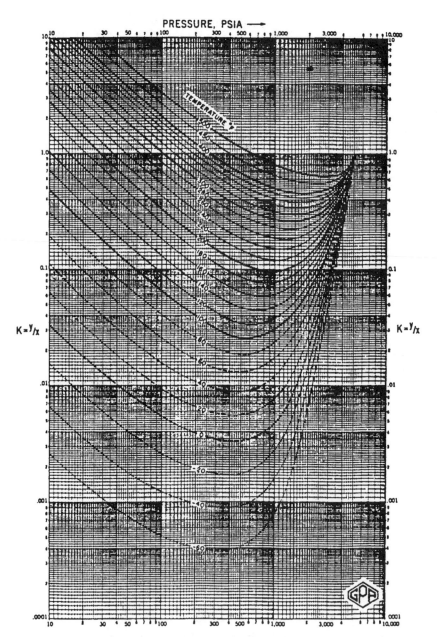

Figure 8. Hexane. Conv. press. 5,000 psia. Courtesy of the Gas Processors Suppliers Association. Published in the GPSA Engineering Data Book, Tenth Edition, 1987.

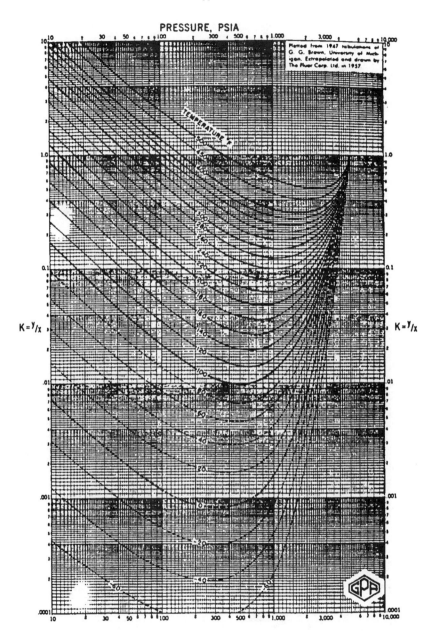

PRESSURE, PSIA

$K = {}^Y/_X$

Figure 9. Heptane. Conv. press. 5,000 psia. Courtesy of the Gas Processors Suppliers Association. Published in the GPSA Engineering Data Book, Tenth Edition, 1987.

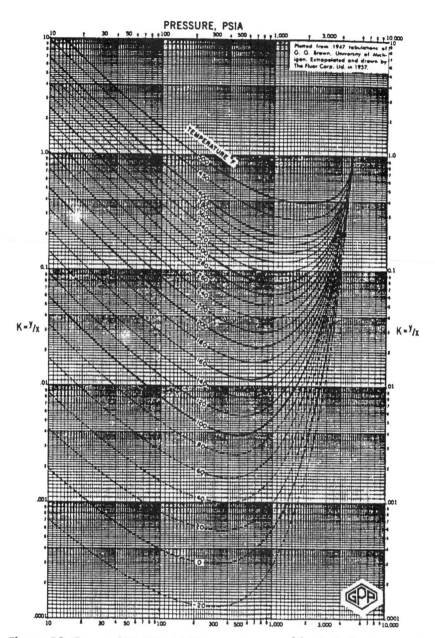

PRESSURE, PSIA

K = ᵞ/ₓ

Figure 10. Octane. Conv. press. 5,000 psia. Courtesy of the Gas Processors Suppliers Association. Published in the GPSA Engineering Data Book, Tenth Edition, 1987.

PRESSURE, PSIA

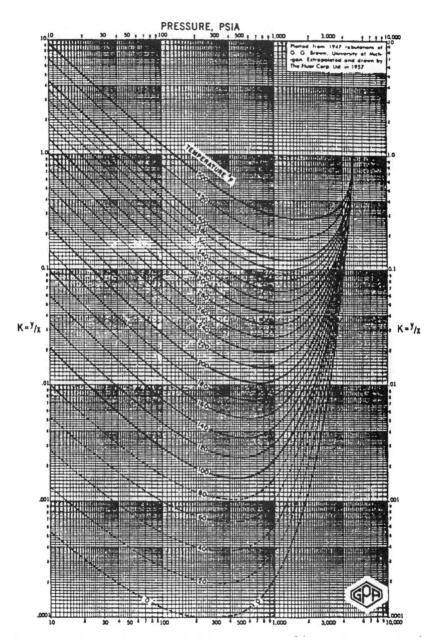

Figure 11. Nonane. Conv. press. 5,000 psia. Courtesy of the Gas Processors Suppliers Association. Published in the GPSA Engineering Data Book, Tenth Edition, 1987.

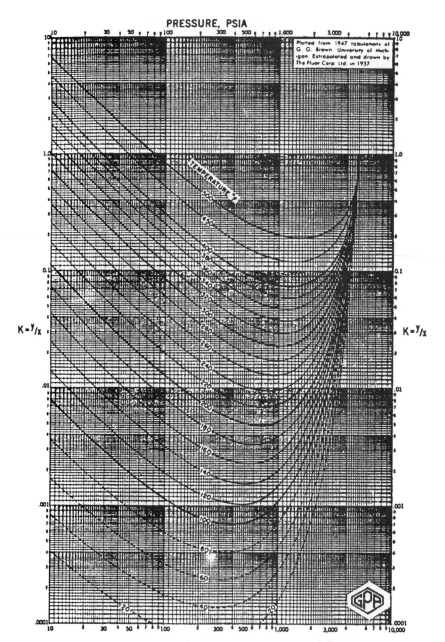

PRESSURE, PSIA

Figure 12. Decane. Conv. press. 5,000 psia. Courtesy of the Gas Processors Suppliers Association. Published in the GPSA Engineering Data Book, Tenth Edition, 1987.

INDEX

ABOUT THE AUTHOR

Tarek Ahmed, Ph.D., P.E., is a Professor and the Chairman of the Petroleum Engineering Department at Montana Tech of the University of Montana. Dr. Ahmed also holds the Union Pacific Resources Endowed Chair. He has a Ph.D. from the University of Oklahoma, an M.S. from the University of Missouri-Rolla, and a B.S. from the Faculty of Petroleum (Egypt)—all degrees in petroleum engineering. Dr. Ahmed is also the author of 29 technical papers, in addition to a book entitled *Hydrocarbon Phase Behavior* (Gulf Publishing Company, 1989).